Plant Genetic Conservation

Plant diversity sustains all animal life, and the genetic diversity within plants underpins global food security. This text provides a practical and theoretical introduction to the strategies and actions to adopt for conserving plant genetic variation, as well as explaining how humans can exploit this diversity for sustainable development. Notably readable, it initially offers current knowledge on the policy context of plant genetic resources. The authors then discuss strategies from *in situ* and *ex situ* conservation to crop breeding, exploring how plant genetic resources can be used to improve food security in the face of increasing agro-biodiversity loss, human population growth and climate change. Each chapter draws on examples from the literature or the authors' research and includes further reading references. Containing other useful features such as a glossary, it is invaluable for professionals and undergraduate and graduate students in plant sciences, ecology, conservation, genetics and natural resource management.

Nigel Maxted is Professor of Plant Genetic Conservation in the School of Biosciences, University of Birmingham, UK. He works on conservation planning and implementation, and has published more than 350 scientific papers and book chapters and 24 books. He is chair of the IUCN SSC Crop Wild Relative Specialist Group, UK Plant Genetic Resources Committee, and European Cooperative Programme for PGR *In Situ* Working Group. He is also the International Scientific Advisor for Bioversity International.

Danny Hunter is Senior Scientist at Bioversity International, Rome, Italy. He works on conservation and sustainable use of crop and tree genetic resources and their role in linking sustainable agriculture, environment, health and nutrition. He has 25 years' experience working with partners and family farmers in more than 30 countries in sub-Saharan Africa, Latin America, South and South East Asia, Central Asia and the Pacific.

Rodomiro Ortiz Ríos is Professor of Genetics and Plant Breeding at the Swedish University of Agricultural Sciences (SLU), Alnarp, Sweden. He has written more than 800 publications and was the leader of a multidisciplinary team working on the plantain and banana improvement programme for which the Consultative Group for International Agricultural Research (CGIAR) awarded the 1994 King Baudouin Award to the International Institute of Tropical Agriculture (IITA). He was also Co-Principal Investigator for a SLU/ICARDA-led project that won the 2017 Olam Prize for Innovation in Food Security.

Plant Genetic Conservation

NIGEL MAXTED
University of Birmingham

DANNY HUNTER
Bioversity International

RODOMIRO ORTIZ RÍOS
Swedish University of Agricultural Sciences

CAMBRIDGE
UNIVERSITY PRESS

CAMBRIDGE
UNIVERSITY PRESS

University Printing House, Cambridge CB2 8BS, United Kingdom

One Liberty Plaza, 20th Floor, New York, NY 10006, USA

477 Williamstown Road, Port Melbourne, VIC 3207, Australia

314–321, 3rd Floor, Plot 3, Splendor Forum, Jasola District Centre, New Delhi – 110025, India

79 Anson Road, #06-04/06, Singapore 079906

Cambridge University Press is part of the University of Cambridge.

It furthers the University's mission by disseminating knowledge in the pursuit of education, learning, and research at the highest international levels of excellence.

www.cambridge.org
Information on this title: www.cambridge.org/9780521806565
DOI: 10.1017/9781139024297

First published 2020

Printed in the United Kingdom by TJ International Ltd, Padstow Cornwall

A catalogue record for this publication is available from the British Library.

ISBN 978-0-521-80656-5 Hardback
ISBN 978-0-521-00130-4 Paperback

Additional resources for this publication at www.cambridge.org/9780521806565

FRUITS OF THE WORLD
IN
DANGER

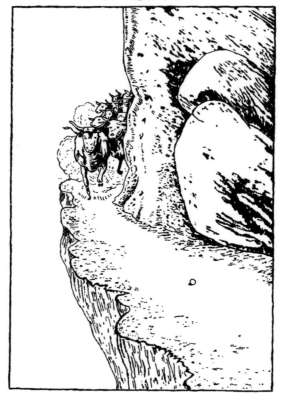

Number 10   The Apricot

CONTENTS

The plate section can be found between pages 274 and 275

FOREWORD

The year 2020 will hopefully be seen as the year the world awoke from its slumber and started to realize the extent of the global crisis facing biodiversity. Both *The State of the World's Biodiversity for Food and Agriculture* report, the first global assessment of biodiversity for food and agriculture worldwide, and the Intergovernmental Science-Policy Platform on Biodiversity and Ecosystem Services (IPBES)'s *Global Assessment Report on Biodiversity and Ecosystem Services* present alarming pictures of unprecedented biodiversity loss and future scenarios of accelerated rates of loss if we fail to act at this point in time. We have clearly reached a point where business as usual is not a viable option if we are to have a long-term future as a species.

The State of the World report warns that, despite the growing evidence of biodiversity's key role in food security and nutrition, the diversity of production systems worldwide is in decline. Of the thousands of plant species cultivated for food, fewer than 200 contribute substantially to global food output, and only 9 account for 66% of total crop production.

The IPBES assessment highlights that local varieties of domesticated plants, as well as wild relatives of our crops, are disappearing globally and that this loss of genetic diversity poses a serious risk to global food security and undermines the resilience of many agricultural systems to threats such as pests, pathogens and climate variability. Fewer and fewer crop species and varieties are being cultivated, raised, traded and maintained around the world, despite many efforts, mostly local and uncoordinated. The IPBES global assessment reports that the drastic loss of agro-biodiversity is the result of land use change, traditional knowledge loss, market preferences and globalized trade, and that too many hotspots of plant genetic resources have been lost, remain under threat or do not receive the formal protection they urgently require.

What can the global community do in the face of such an onslaught? The Sustainable Development Goals (SDGs) and the UN Convention on Biological Diversity (CBD) post-2020 biodiversity framework can provide us with a renewed opportunity to commit and mobilize, provide real political will and leadership, and coalesce action around the conservation and use of plant genetic resources for sustainable agriculture, food systems and sustainable and healthy diets. Mainstreaming plant genetic resources to this end is central if we are to achieve the SDGs by 2030.

We also need to find more creative ways in which to use plant genetic resources in agriculture, food systems and supply chains as well as to explore innovative approaches that incentivize a greater demand and desirability for a more diverse use of plant genetic resources, driven by consumers. Only when we are actively using plant genetic resources for sustainable farming and food systems that deliver diverse, nutritious foods, can we hope to effectively safeguard this green gold for future generations and reverse current alarming rates of biodiversity loss.

Plant Genetic Conservation is a long awaited, much-needed textbook, which I welcome very warmly. It provides up-to-date, state-of-the-art information and knowledge essential to safeguard our global heritage and wealth of plant genetic resources, including a historical perspective, the scientific underpinnings of applied taxonomy, genetics, diversity assessment and measurement needed for effective conservation planning, strategies and actions. Further the text presents a comprehensive overview of germplasm collection, conservation and evaluation, and its importance for plant breeding and general utility.

This book will make an enormous contribution to ensuring plant genetic resources are available for future food production and climate scenarios and the global nutrient needs. I am convinced it will be widely used by universities to build vital capacity and skills for the task ahead, as well as by decision-makers and practitioners working in the field to more effectively conserve the breadth of plant genetic diversity. It is a clear contribution to the IPBES report, which recommends '*ensuring the adaptive capacity of food production incorporates measures that conserve the* *diversity of genes, varieties, cultivars, landraces and species which also contribute to diversified, healthy and culturally-relevant nutrition*'.

Juan Lucas Restrepo
Director General – Bioversity International
CEO-Designate – Alliance of Bioversity
International and the International Center for
Tropical Agriculture (CIAT)

PREFACE

Purpose of the Book

We live in critical times for the world's biological diversity. Plants, as the foundation of the food chain, are essential for sustaining all animal life and the habitats in which animals can thrive, and underpin many ecosystem services. They also supply humans directly with the bulk of our food, construction materials, fibres, medicines, dyes and many other products. Plants are an essential and precious resource for humankind. The International Union for Conservation of Nature (IUCN) estimates that out of the total 16 000 plant species found in India, people use at least 5 000. Yet it is universally agreed that, largely as a result of recent human activity, there is a catastrophic loss of biological diversity occurring at present, and we are living through the sixth global biodiversity extinction with species and, equally importantly, genes and alleles, being lost in perpetuity. It has been estimated that approximately 20% of plant species will be lost in the near future, and although more difficult to estimate, up to 35% of plant genetic diversity is likely to be lost over the same time period. The signing of the Convention on Biological Diversity at the 'Earth Summit' in Rio de Janeiro in 1992, together with the establishment of the International Treaty for Plant Genetic Resources for Food and Agriculture in 2001, the Food and Agriculture Organization (FAO) Global Plan of Action for Conservation and Use of PGR in 1996, the Strategic Plan for Biodiversity 2011–2020, including Aichi Biodiversity Targets in 2010, the CBD Global Strategy for Plant Conservation 2011 – 2020 in 2010, the UN Sustainable Development Goals in 2015, and more recently the first State of the World Report on Biodiversity for Food and Agriculture and IPBES report each drew attention to the need to conserve the world's natural resources, the need to link biodiversity conservation to sustainable exploitation and human development, and the requirement to ensure equitable sharing of the benefits arising from the exploitation of biological diversity. The exploitation of plant genetic resources is of fundamental importance to the survival of humankind; therefore, the need to focus on conserving plant diversity and sustainable exploitation is critical to all our futures.

Yet thus far plans made, or targets established, are consistently failing. At the time of writing, the UN estimates there are 7.7 billion humans on Earth, 78% living in developing countries, and it is predicted there will be 9.6 billion by 2050, 86% of them in developing countries (primarily Africa). FAO estimates that to feed the human population in 2050 will require food supplies to increase by 70% globally and at least 100% in developing countries, while the Intergovernmental Panel on Climate Change (IPCC) estimates that climate change may reduce agricultural production by 2% each decade by 2050. It can logically be argued that the planet is beyond its carrying capacity for its human population. In this state, we are over-exploiting the planet, and our unsustainable actions are directly causing an exponential loss of plant genetic diversity throughout the world. This in turn is having direct negative economic, political and social consequences for humanity.

This is not our only option; we could alter our path and explore fully the positive benefits that are likely to result from systematically conserving and sustainably exploiting the world's plant genetic resources. There have been strong movements to halt this loss of plant diversity and enhance its utilization for the benefit of all humanity since the 1960s, largely led by the FAO and Consultative Group on International Agricultural Research (CGIAR) institutes. This text provides a theoretical and practical introduction to plant genetic conservation, the diversity included, the various strategies and techniques available for conservation, and how humankind can exploit plant genetic diversity sustainably. We have previously published several books on various aspects of this topic, *in situ*, *ex situ*, crop wild relative conservation, crop landrace conservation and breeding, but here we attempt to draw these multiple strands of Plant Genetic Resources for Food and Agriculture (PGRFA) conservation and use together into one textbook. A book we hope will enthuse the next generation to take the actions we ourselves have unfortunately been unable to see through to fruition. With our collective experience, we realize there are no easy answers, but we hope to promote an informed debate of the scientific principles underlying plant genetic conservation and create a more sustainable impact. It looks increasingly as though the outcome for humankind itself may depend on the success of this debate and its implementation.

Precis of Contents

We have approached the subject of systematic conservation and sustainable utilization of plant genetic diversity from theoretical and practical viewpoints, drawing together ideas from published and unpublished sources, and from our own extensive practical field experience, conserving plant genetic diversity and using that diversity to sustain food security around the world. The book is divided into 18 chapters that fall into four parts: *Part I Introduction* (Chapters 1 and 2); *Part II Scientific Background* (Chapters 3–5); *Part III Conservation Practice* (Chapters 6–13); and *Part IV Plant Exploitation* (Chapters 14–18). The aim of each chapter is as follows:

- **Chapter 1: Introduction** – provides an introduction and overview to the subject of biological diversity (biodiversity), and more specifically the systematic conservation and sustainable utilization of plant genetic diversity.
- **Chapter 2:** Establishing the Social, Political and Ethical Context – reviews the social, ethical and policy context of PGRFA conservation and use. This is achieved by providing the historical background and introducing the major conventions, treaties and agreements, as well as the key stakeholders from the formal and informal

sectors that are concerned with plant conservation and use and the equitable sharing of benefits.
- **Chapter 3: Plant Taxonomy** – provides an elucidation of relationships between taxa (families, genera, species or subspecific taxa), the production of classifications that reflect their evolutionary relationships and how other taxonomic products are derived, including descriptions, synonyms, distribution maps, identification aids and nomenclature.
- **Chapter 4: Plant Population Genetics** – introduces the factors determining, maintaining and changing variation in populations, including the factors affecting changes in allele frequency and how population genetic knowledge can be used to improve *ex situ* and *in situ* conservation and use of plant genetic diversity.
- **Chapter 5: Genetic Diversity Measurement** – provides an overview of genetic diversity and variation, and how to measure this when studying plant genetic resources. It also describes the use of DNA markers for assessing polymorphism, studying diversity and revealing trait associations relevant for searching variation of target characters in plant breeding.
- **Chapter 6: Planning Plant Conservation** – focuses on taxon prioritization, establishing the geographic and taxonomic breadth and most important conservation actions using ecogeographic and gap analysis techniques and the production of strategies or action plans to aid conservation implementation.
- **Chapter 7: Conservation Strategies and Techniques** – introduces the two fundamentally distinct conservation strategies, *in situ* and *ex situ*, and within each, the range of techniques available to maximize the range of plant genetic diversity maintained, the relative advantages of each technique and the fact that *in situ* and *ex situ* should be applied in a complementary manner to maximize impact and conserve resource backup.
- **Chapter 8: *In Situ* Conservation** – reviews the conservation of plant species where they naturally occur and involves the planning, design, establishment, management and monitoring of the viable plant populations to be conserved.

- Chapter 9: On-Farm Conservation – provides an overview of the concept of on-farm conservation of plant genetic resources, the maintenance of farmer landraces or traditional varieties in agro-ecosystems. The chapter outlines some of the key steps that need to be considered when implementing an on-farm conservation project.
- Chapter 10: Community-Based Conservation – provides a historical perspective on the development of community-based conservation and how approaches are grounded in changes that not only occurred in conservation practice but also in agriculture and rural development.
- Chapter 11: Germplasm Collecting – provides an introduction and overview of how to plan and undertake the collection of plant genetic samples in the field, and once samples are collected, how the samples should be processed, stored and made available for utilization.
- Chapter 12: Seed Gene Bank Conservation – deals with how to store seeds of crops and wild plant species for medium and long-term conservation, because seed conservation in gene banks remains the most efficient method for conserving plant germplasm of the cultigen and its wild relatives for the majority of plant species.
- Chapter 13: Whole Plant, Plantlet and DNA Conservation – reviews the role of both field gene banks and living collections in botanical gardens and arboreta in *ex situ* conservation of plant diversity, and how *in vitro* conservation and cryopreservation techniques are increasingly used for the conservation of vegetatively propagated species and species with recalcitrant seeds.
- Chapter 14: Plant Uses – reviews past, present and future use of plant genetic resources, including their use by traditional and indigenous cultures, as well as their overexploitation and barriers preventing sustainable exploitation.
- Chapter 15: Germplasm Evaluation – outlines how use of conserved resources are often based on diversity and variability analysis relying on

characterization or evaluation data, DNA markers or both, as well as the use of core subsets of the cultigen pool as representatives of the entire collection that ensures the differences among them are truly genetic and reflect overall diversity.
- Chapter 16: Plant Breeding – provides an introduction and overview of how plant diversity is used in crop improvement. It describes how collecting, conserving and using plant genetic diversity is linked with the basic principles of plant breeding and thereafter how that diversity is used for crossing and selecting segregating offspring with desired characters to generate new cultivars.
- Chapter 17: Participatory Plant Breeding – reviews alternative models of breeding that unite farmers, scientists, extension officers and other actors to generate new cultivars using participatory plant breeding and a participatory varietal selection approach, as well as discussing the advantages and disadvantages of the participatory approach compared to the conventional approach to plant breeding.
- Chapter 18: Conservation Data Management – reviews the kinds of data associated with plant genetic conservation, definition of the types of data, how data are recorded, the advantage of data standardization and how efficient conservation data management helps improve conservation outcomes. The importance of access to and ownership of the various kinds of data is emphasized as fundamental to good data management practice.

The book concludes with a list of Acronyms and Abbreviations, a Glossary of scientific terms, a list of References and an Index. Additional information including a General Bibliographies for each Chapter and Useful Websites lists are provided under the Resources tab on the webpage for the book (see www.cambridge.org/9780521806565).

ACKNOWLEDGEMENTS

We acknowledge with gratitude many colleagues who have provided help and encouragement during the preparation of this book.

Nigel would like to acknowledge that the original book was to be a University of Birmingham authored text, and as such we thank Antonia Eastwood, Brian Ford-Lloyd, Shelagh Kell, Mike Lawrence and Harpel Pooni for their initial involvement. We also thank several colleagues for reviewing drafts of the text (in whole or in part): Ehsan Dulloo (Bioversity International), Nur Fatihah (University of Putra Malaysia), Jose Iriondo (University of Rey Juan Carlos), Shelagh Kell (University of Birmingham), Joana Magos Brehm (University of Birmingham), Valeria Negri (University of Perugia), Sara Oldfield (BGCI), Jade Phillips (University of Birmingham), Lulu Rico Acre (Royal Botanic Gardens, Kew) and Suzanne Sharrock (Botanic Gardens Conservation International). We thank Glen Baxter for allowing us to include one of his Fruits of the World in Danger drawings; Nina Lauridsen (Bioversity International) for assistance in obtaining the permissions for inclusion of figures and images in our text; and Bioversity International itself for use of photographs from its image database.

Danny would like to acknowledge the friendship and mentoring of many colleagues including the late Joe Lennard and Bhuwon Sthapit, and also Danny Bradley, Jimmie Rogers, Mary Taylor, Luigi Guarino, Charles Delp, Prem Mathur, Pablo Eyzaguirre Susan Bragdon, Jessica Fanzo, Braulio Dias, Lidio Coradin and Toby Hodgkin. He dedicates this book to Callum and Imogen, the future guardians and stewards, tasked with making the planet a better, more equitable and compassionate home for the diversity of life that finds itself here.

All three authors have spent a life-time working to improve the conservation and use of plant genetic diversity, primarily working in the centres of plant and crop diversity in developing countries. As such, we acknowledge the unerring support of our work provided by local communities, and therefore we dedicate this text to those communities who provided hospitality and help with our efforts, in the hope this text will be used to improve their lives and livelihoods.

Furthermore, it is obvious that the next generation of environmental campaigners are frustrated with the limited progress we have made in biodiversity conservation thus far, as are the three of us. In the words of Greta Thunberg at the recent United Nations General Assembly (UNGA) Climate Action Summit:

> You have stolen my dreams and my childhood with your empty words. And yet I'm one of the lucky ones. People are suffering. People are dying. Entire ecosystems are collapsing. We are in the beginning of a mass extinction, and all you can talk about is money and fairy tales of eternal economic growth. How dare you!

We applaud her passion and dedicate this text to the next generation of environmental campaigners in the hope they will use our text to succeed where we have failed to make the necessary difference for all humanity.

Part

I

Introduction

Introduction

1.1 Context

The conservation and sustainable use of plant genetic diversity is the basis of human well-being and food security. Today we face a stark challenge – either we learn to conserve biological diversity and practice sustainable use of its components or we ourselves are likely to face extinction. Thus, as biologists our specific challenge is to classify existing biological diversity and halt ecosystem, habitat, species and genetic diversity loss, while feeding the ever-increasing human population. Further as scientists we would be failing if we did not also warn society about the excessive consumption rates of a relatively small proportion of humankind, and the resulting gross inequality and poverty. World population is projected to grow from 6.1 billion in 2000 to 9.8 billion in 2050, an increase of 38% (Figure 1.1). Future population growth is highly dependent on the path that future fertility takes. The average annual population growth rate over this half-century will be 0.77%, substantially lower than the 1.76% average growth rate from 1950 to 2000. Future population growth is highly dependent on the path that future fertility takes. If fertility levels continue to decline, the world population is expected to reach 10.1 billion in 2100, increasing by about 35 million persons each year, according to the medium variant (United Nations, 2011). Even if human population levels do begin to level off, it can be argued that the planet is already beyond its human carrying capacity as evidenced by the current over-exploitation of our natural resources and the dominance of unsustainable environmental management practices.

The exponential loss of plant diversity that is currently occurring has been well documented: habitats, species, gene combinations and alleles are being lost. The State of the World's Plants 2016 report (RBG Kew, 2016) estimates that 21% of global plant species fall into the threatened IUCN Red list criteria, and they conclude in their 2017 report (RBG Kew, 2017) that 'Despite ongoing efforts to increase the rate at which plants are evaluated for their extinction risk, there is widespread recognition that many plants may become extinct before they have been recognized as being at risk, and perhaps even before they have been discovered'. It is perhaps easiest to undertake threat assessment at the plant species level because species are relatively discrete, and, in many cases, the necessary data sets are available. Conversely, loss of genetic diversity may be characterized as a 'silent risk', because unlike habitats and species the loss of genetic diversity is difficult to observe and quantify and often passes unnoticed. Yet loss of genetic diversity will always be greater than habitat and species loss because genetic diversity will be entirely lost from extinct habitat and species but there will also be genetic diversity loss from the habitats and species that remain extant (Maxted et al., 1997a). However, the conservation of plant genetic diversity is of critical importance to the survival of humanity itself due to the pivotal role plants play in the functioning of all natural ecosystems and the direct benefits to humanity that can arise from their sustainable exploitation of plant diversity (Frankel et al., 1995). Humankind has since the earliest times exploited plant diversity in numerous ways, such as the development of new agricultural and horticultural crops, and medicinal drugs, as well as the numerous other ways humans use plants (Lewington, 1990). In contrast to the economic, political and social benefits of active plant conservation linked to sustainable exploitation, the consequences of our careless disregard for loss of diversity or unsustainable exploitation, combined with population growth, will be catastrophic for the planet, our fellow creatures and humanity itself.

The importance of biological diversity conservation, its sustainable utilization and the link to human development were central to the United

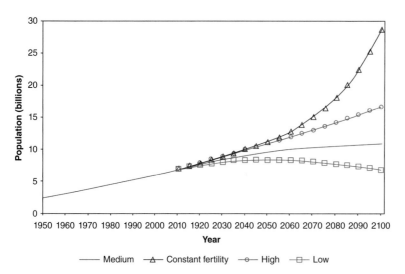

Figure 1.1 Human population 1950–2100.
(United Nations, 2011)

Nations Conference on the Environment and Development (UNCED) held in Rio de Janeiro, Brazil, in 1992. The Conference saw the adoption of the Convention on Biological Diversity (CBD, 1992), whose three key objectives, stated in Article 1, remain a cornerstone of plant genetic conservation today:

The objectives of this convention . . . are the conservation of biological diversity, the sustainable use of its components and the fair and equitable sharing of the benefits arising out of the utilization of genetic resources. . .

Subsequent to signing and ratification of the Convention, steps were taken toward conserving microbial, animal and plant species and genetic diversity, as well as the habitats and ecosystems in which they live. In April 2002, the CBD Conference of the Parties (COP) made a commitment to achieve by 2010 a 'significant reduction of the current rate of biodiversity loss at the global, regional and national level as a contribution to poverty alleviation and to the benefit of all life on Earth' (CBD, 2002). However, it must be admitted that this target was not or even nearly met. In response to this failure, in October 2010, the CBD COP adopted a revised and updated Strategic Plan for Biodiversity, including the Aichi Biodiversity Targets, for the 2011–2020 period (CBD, 2010b). The vision was that humankind should be

'Living in Harmony with Nature' and 'By 2050, biodiversity is valued, conserved, restored and wisely used, maintaining ecosystem services, sustaining a healthy planet and delivering benefits essential for all people'. The rationale for the new plan was that biological diversity underpins ecosystem functioning and these ecosystem services are essential for human well-being. Furthermore, it provides for food security, human health, and the provision of clean air and water, and is essential for the achievement of the Sustainable Development Goals, including poverty reduction. Target 13 of the Aichi Biodiversity Targets specifically addresses genetic conservation:

Target 13: By 2020, the genetic diversity of cultivated plants and farmed and domesticated animals and of wild relatives, including other socio-economically as well as culturally valuable species, is maintained, and strategies have been developed and implemented for minimizing genetic erosion and safeguarding their genetic diversity.

Intermediate progress was assessed in the Global Biodiversity Outlook 4 (CBD, 2014) (Table 1.1).

In parallel to the recent development of the new Strategic Plan, the CBD has also developed the Global Strategy for Plant Conservation 2011–2020 (CBD, 2010a), which aims to achieve the three objectives of the Convention particularly for plant diversity. It should be implemented within the broader framework

Table 1.1. Target 'dashboard'—a summary of progress towards the Aichi Biodiversity Targets, broken down into their components (CBD, 2014). Note The assessment uses a five-point scale and the assessment of level of confidence is indicated by stars (★★).

Legend	Description
5	On track to exceed target (we expect to achieve Target by deadline)
4	On track to achieve target (if we continue on our current trajectory, (we expect to achieve target by 2020)
3	Progress towards target but at an insufficient rate (unless we increase efforts target will not be met by deadline)
2	No significant overall progress (overall, we are neither moving towards nor away from target)
1	Moving away from target (things are getting worse rather than better).

TARGET ELEMENT	STATUS	COMMENT
Target 1		
People are aware of the values of biodiversity	3 ★	Limited geographical coverage of indicators. Strong regional differences
People are aware of the steps they can take to conserve and sustainably use biodiversity	3 ★	Evidence suggests a growing knowledge of actions available, but limited understanding of which will have positive impacts
Target 2		
Biodiversity values integrated into national and local development and poverty reduction strategies	3 ★★	Differences between regions. Evidence largely based on poverty reduction strategies
Biodiversity values integrated into national and local planning processes	3 ★★	The evidence shows regional variation and it is not clear if biodiversity is actually taken into consideration
Biodiversity values incorporated into national accounting, as appropriate	3 ★★★	Initiatives such as WAVES show growing trend towards such incorporation
Biodiversity values incorporated into reporting systems	3 ★★	Improved accounting implies improvement in reporting

Table 1.1. (*cont.*)

Target	Element	Rating	Status
Target 3	Incentives, including subsidies, harmful to biodiversity, eliminated, phased out or reformed in order to minimize or avoid negative impacts	★★★ (T) 2	No significant overall progress, some advances but some backward movement. Increasing recognition of harmful subsidies but little action
	Positive incentives for conservation and sustainable use of biodiversity developed and applied	★★★ (T) 3	Good progress but better targeting needed. Too small and still outweighed by perverse incentives
Target 4	Governments, business and stakeholders at all levels have taken steps to achieve, or have implemented, plans for sustainable production and consumption ...	★★★ (T) 3	Many plans for sustainable production and consumption are in place, but they are still limited in scale
	... and have kept the impacts of use of natural resources well within safe ecological limits	★★★ (T) 2	All measures show an increase in natural resource use
Target 5	The rate of loss of forests is at least halved and where feasible brought close to zero	★★ (T) 3	Deforestation significantly slowed in some tropical areas, although still great regional variation
	The loss of all habitats is at least halved and where feasible brought close to zero	★★ (T) 2	Varies among habitat types, data scarce for some biomes
	Degradation and fragmentation are significantly reduced	★★ (T) 1	Habitats of all types, including forests, grasslands, wetlands and river systems, continue to be fragmented and degraded.

Target	Statement	Progress	Comment
Target 6	All fish and invertebrate stocks and aquatic plants are managed and harvested sustainably, legally and applying ecosystem based approaches	(3) ★★★	Great regional variation, positive for some countries but data limited for many developing countries
	Recovery plans and measures are in place for all depleted species	(3) ★★★	Variable, progress in some regions
	Fisheries have no significant adverse impacts on threatened species and vulnerable ecosystems	(2) ★★★	Some progress e.g. on long-lining used in tuna fisheries, but practices still impacting vulnerable ecosystems
	The impacts of fisheries on stocks, species and ecosystems are within safe ecological limits, i.e. overfishing avoided	(2) ★★	Overexploitation remains an issue globally, but with regional variation
Target 7	Areas under agriculture are managed sustainably, ensuring conservation of biodiversity	(3) ★★	Increasing area under sustainable management, based on organic certification and conservation agriculture. Nutrient use flattening globally. No-till techniques expanding
	Areas under aquaculture are managed sustainably, ensuring conservation of biodiversity	(3) ★★	Progress with sustainability standards being introduced, but in the context of very rapid expansion. Questions about sustainability of expansion of freshwater aquaculture
	Areas under forestry are managed sustainably, ensuring conservation of biodiversity	(3) ★★	Increasing forest certification and criterion indicators. Certified forestry mostly in northern countries, much slower in tropical countries

Table 1.1. (*cont.*)

Target 8	Pollutants (of all types) have been brought to levels that are not detrimental to ecosystem function and biodiversity	*No clear evaluation*	Highly variable between pollutants
	Pollution from excess nutrients has been brought to levels that are not detrimental to ecosystem function and biodiversity	★★ (1→T)	Nutrient use levelling off in some regions, e.g. Europe and North America, but at levels that are still detrimental to biodiversity. Still rising in other regions. Very high regional variation
Target 9	Invasive alien species identified and prioritized	★★★ (3→T)	Measures taken in many countries to develop lists of invasive alien species
	Pathways identified and prioritized	★★★ (3→T)	Major pathways are identified, but not efficiently controlled at a global scale
	Priority species controlled or eradicated	★ (3→T)	Some control and eradication, but data limited
	Introduction and establishment of IAS prevented	★★ (2→T)	Some measures in place, but not sufficient to prevent continuing large increase in IAS
Target 10	Multiple anthropogenic pressures on coral reefs are minimized, so as to maintain their integrity and functioning	★★★ (1→T)	Pressures such as land-based pollution, uncontrolled tourism still increasing, although new marine protected areas may ease overfishing in some reef regions
	Multiple anthropogenic pressures on other vulnerable ecosystems impacted by climate change or ocean acidification are minimized, to maintain their integrity and functioning	*Not evaluated*	Insufficient information was available to evaluate the target for other vulnerable ecosystems including seagrass habitats, mangroves and mountains

Target 11

At least 17 per cent of terrestrial and inland water areas are conserved

★★★

Extrapolations show good progress and the target will be achieved if existing commitments on designating protected areas are implemented. Inland water protection has distinct issues.

At least 10 per cent of coastal and marine areas are conserved

★★★

Marine protected areas are accelerating but extrapolations suggest we are not on track to meet the target. With existing commitments, the target would be met for territorial waters but not for exclusive economic zones or high seas

---

Areas of particular importance for biodiversity and ecosystem services conserved

biodiversity ★★★
ecosystem services ★

Progress for protected Key Biodiversity Areas, but still important gaps. No separate measure for ecosystem services

---

Conserved areas are ecologically representative

terrestrial and marine ★★★
inland waters ★

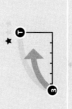

Progress, and possible to meet this target for terrestrial ecosystems if additional protected areas are representative. Progress with marine and freshwater areas, but much further to go

---

Conserved areas are effectively and equitably managed

★

Reasonable evidence of improved effectiveness, but small sample size. Increasing trend towards community involvement in protection. Very dependent on region and location

---

Conserved areas are well connected and integrated into the wider landscape and seascape

★

Initiatives exist to develop corridors and transboundary parks, but there is still not sufficient connection. Freshwater protected areas remain very disconnected

Table 1.1. (*cont.*)

Target 12	Extinction of known threatened species has been prevented	★	Further extinctions likely by 2020, e.g. for amphibians and fish. For bird and mammal species some evidence measures have prevented extinctions
	The conservation status of those species most in decline has been improved and sustained	★★★	Red List Index still declining, no sign overall of reduced risk of extinction across groups of species. Very large regional differences
Target 13	The genetic diversity of cultivated plants is maintained	★★★	*Ex situ* collections of plant genetic resources continue to improve, albeit with some gaps. There is limited support to ensure long term conservation of local varieties of crops in the face of changes in agricultural practices and market preferences
	The genetic diversity of farmed and domesticated animals is maintained	★★★	There are increasing activities to conserve breeds in their production environment and in gene banks, including through *in-vitro* conservation, but to date, these are insufficient
	The genetic diversity of wild relatives is maintained	★★	Gradual increase in the conservation of wild relatives of crop plants in *er situ* facilities but their conservation in the wild remains largely insecure, with few protected area management plans addressing wild relatives
	The genetic diversity of socio-economically as well as culturally valuable species is maintained	*Not evaluated*	Insufficient data to evaluate this element of the target
	Strategies have been developed and implemented for minimizing genetic erosion and safeguarding genetic diversity	★★★	The FAO Global Plans of Action for plant and animal genetic resources provide frameworks for the development of national and international strategies and action plans

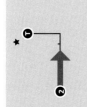

Target 14

Ecosystems that provide essential services, including services related to water, and contribute to health, livelihoods and well–being, are restored and safeguarded . . .

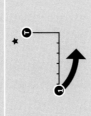

High variation across ecosystems and services. Ecosystems particularly important for services, e.g. wetlands and coral reefs, still in decline

. . . taking into account the needs of women, indigenous and local communities, and the poor and vulnerable

Poor communities and women especially impacted by continuing loss of ecosystem services

Target 15

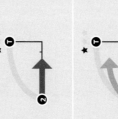

Ecosystem resilience and the contribution of biodiversity to carbon stocks have been enhanced through conservation and restoration

Despite restoration and conservation efforts, there is still a net loss of forests, a major global carbon stock

At least 15 per cent of degraded ecosystems are restored, contributing to climate change mitigation and adaptation, and to combating desertification

Many restoration activities under way, but hard to assess whether they will restore 15% of degraded areas

Target 16

The Nagoya Protocol is in force

The Nagoya Protocol will enter into force on 12 October 2014, ahead of the deadline set.

The Nagoya Protocol is operational, consistent with national legislation

Given progress that has been made, it is likely that the Nagoya Protocol will be operational by 2015 in those countries that have ratified it

Table 1.1. (*cont.*)

Target 17	Submission of NBSAPs to Secretariat by (end of) 2015		For those Parties for which information is available, about 40% are expected to have completed their NBSAP by October 2014 and about 90% by the end of 2015
	NBSAPs adopted as effective policy instrument		The adequacy of available updated NBSAPs in terms of following COP guidance is variable
	NBSAPs are being implemented		The degree of implementation of updated NBSAPs is variable
Target 18	Traditional knowledge, innovations and practices of indigenous and local communities are respected		Processes are under way internationally and in a number of countries to strengthen respect for, recognition and promotion of, traditional knowledge and customary sustainable use
	Traditional knowledge, innovations and practices are fully integrated and reflected in implementation of the Convention …		Traditional knowledge and customary sustainable use need to be further integrated across all relevant actions under the Convention
	… with the full and effective participation of indigenous and local communities		Efforts continue to enhance the capacities of indigenous and local communities to participate meaningfully in relevant processes locally, nationally and internationally but limited funding and capacity remain obstacles

Target 19

Knowledge, the science base and technologies relating to biodiversity, its values, functioning, status and trends, and the consequences of its loss, are improved

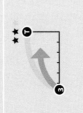

Significant effort on delivery of information and knowledge relevant to decision makers is being made, and relevant processes and institutions are in place

Biodiversity knowledge, the science base and technologies are widely shared and transferred and applied

Improvements in analysis and interpretation of data gathered from disparate collecting and monitoring systems. However, coordination to guarantee models and technologies that can integrate this knowledge into functional applied systems needs to be improved

Target 20

Mobilization of financial resources implementing the Strategic Plan for Biodiversity 2011–2020 from all sources has increased substantially from 2010 levels

Limited information on many funding sources, including domestic funding, innovative financial mechanisms, and the private sector. General increase in bilateral ODA against 2006–2010 baseline.

From CBD (2014).

of the Strategic Plan for Biodiversity 2011–2020 and establishes 16 plant-related Targets to be achieved by 2020:

Objective I: Plant diversity is well understood, documented and recognized
    Target 1: An online Flora of all known plants.
    Target 2: An assessment of the conservation status of all known plant species, as far as possible, to guide conservation action.
    Target 3: Information, research and associated outputs, and methods necessary to implement the Strategy developed and shared.
Objective II: Plant diversity is urgently and effectively conserved
    Target 4: At least 15% of each ecological region or vegetation type secured through effective management and/or restoration.
    Target 5: At least 75% of the most important areas for plant diversity of each ecological region protected with effective management in place for conserving plants and their genetic diversity.
    Target 6: At least 75% of production lands in each sector managed sustainably, consistent with the conservation of plant diversity.
    Target 7: At least 75% of known threatened plant species conserved *in situ.*
    Target 8: At least 75% of threatened plant species in *ex situ* collections, preferably in the country of origin, and at least 20% available for recovery and restoration programmes.
    Target 9: 70% of the genetic diversity of crops including their wild relatives and other socio-economically valuable plant species conserved, while respecting, preserving and maintaining associated indigenous and local knowledge.
    Target 10: Effective management plans in place to prevent new biological invasions and to manage important areas for plant diversity that are invaded.
Objective III: Plant diversity is used in a sustainable and equitable manner
    Target 11: No species of wild flora endangered by international trade.

    Target 12: All wild-harvested plant-based products sourced sustainably.
    Target 13: Indigenous and local knowledge innovations and practices associated with plant resources maintained or increased, as appropriate, to support customary use, sustainable livelihoods, local food security and health care.
Objective IV: Education and awareness about plant diversity, its role in sustainable livelihoods and importance to all life on Earth is promoted
    Target 14: The importance of plant diversity and the need for its conservation incorporated into communication, education and public awareness programmes.
Objective V: The capacities and public engagement necessary to implement the Strategy have been developed
    Target 15: The number of trained people working with appropriate facilities is enough according to national needs to achieve the targets of this Strategy.
    Target 16: Institutions, networks and partnerships for plant conservation established or strengthened at national, regional and international levels to achieve the targets of this Strategy.

Target 9 addresses the genetic conservation of crop-related diversity and firmly places the conservation of the genetic diversity associated with socio-economically important species within the broader plant conservation agenda.

Allied to the development of biodiversity conservation policy has been initiatives within the Food and Agriculture Organization (FAO) of the United Nations (UN) to promote parallel policies that specifically relate to plant genetic resource conservation. The Global Plan of Action (GPA) for the Conservation and Sustainable Utilization of PGRFA was formally adopted in 1996 by representatives of 150 countries during the Fourth International Technical Conference on Plant Genetic Resources in Leipzig, Germany (FAO, 1996) and was revised in 2011 (FAO, 2011e). It provides a strategic framework for the conservation and sustainable use of the plant genetic diversity on which food and agriculture depends, provides a means of identifying priority actions, to ensure the conservation of plant genetic resources for food and agriculture (PGRFA) as a basis for food security, sustainable agriculture and poverty reduction, and promotes sustainable use and exchange of PGRFA and the fair and equitable sharing of the benefits arising from their use. Further it provides a basis for international collaboration, the strengthening of national PGRFA programmes and information sharing. The Second GPA has 18 priority activities organized into four key subjects: *In Situ* Conservation and Management, *Ex Situ* Conservation, Sustainable Use, and Building Sustainable Institutional and Human Capacities.

The GPA is complemented by the International Treaty on Plant Genetic Resources for Food and Agriculture (FAO, 2001), which aims to promote the conservation and sustainable use of PGRFA and the fair and equitable sharing of the benefits arising out of their use, in harmony with the Convention on Biological Diversity (CBD), for sustainable agriculture and food security. Generally, it has a similar structure to the CBD but in relation to PGRFA. Specifically, in Article 9, it recognizes the enormous contribution that the local and indigenous communities and farmers make to the conservation and development of PGRFA and requests governments to implement Farmers' Rights that ensure the protection of their traditional knowledge, their right to equitably sharing benefits arising from PGRFA utilization and their right to participate in national PGRFA-related issues. Articles 10-14 establish the Multilateral System of Access and Benefit Sharing (MLS). This provides scientific institutions and private sector plant breeder's with access and opportunity to exploit materials stored in gene banks or fields by providing a framework for research, innovation and exchange of information, while at the same time safeguarding the rights of genetic resource providers. The coverage of the International Treaty is not universal but applies to 35 food staples, 15 forage legumes, 12 forage grasses and 2 other forage complexes selected on a basis of food security and interdependence and listed in Annex I. As a means of assessing the current condition of PGRFA and monitoring the impact of the

Global Plan and International Treaty, FAO periodically produces a report that summarizes the current status of PGRFA conservation and use globally based on country reports, information gathering, regional syntheses, thematic background studies and the literature. The first State of the World's PGRFA (SoW) report was published in 1998 (FAO, 1998), the second in 2010 (FAO, 2010a) and the first State of the World Report on Biodiversity for Food and Agriculture in 2019 (FAO, 2019).

1.2 Plant Biodiversity

'Biological diversity' or 'biodiversity' is the result of 3000 million years of biotic evolution on Earth; the term being initially associated with the American conservationist Edward Wilson (Wilson, 1992). His definition of biodiversity is:

The variety of organisms considered at all levels, from genetic variants belonging to the same species through arrays of species to arrays of genera, families, and still higher taxonomic levels; including the variety of ecosystems, which comprise both communities of organisms within particular habitats and the physical conditions under which they live.

While the CBD uses the following definition in Article 2:

The variability among living organisms from all sources including, *inter alia*, terrestrial, marine and other aquatic ecosystems and the ecological complexes of which they are part; this includes diversity within species, between species and of ecosystems.

Such biodiversity includes ecosystems, which encompass both living organisms and their physical environment, species and the genetic diversity within species (Figure 1.2). Diversity at the community level may be referred to as ecogeographic diversity, at the species level as taxonomic diversity and at the gene level as genetic diversity. However, at whatever level biological diversity is considered, it is vast; a fact that may be illustrated by the numbers of described and estimated plant species (Table 1.2). Plants can be defined as multicellular eukaryotes (organisms with

Table 1.2 **Kingdom Plantae**

Sub-kingdom	Division	Common name	Number of species	Proportion of group known	Size	Mode of life
BRYATA (non-vascular plants)	Bryophyta	Mosses	ca. 10 000	Moderate/high	Low growing	Terrestrial (moist habitats)
	Hepatophyta	Liverworts	ca. 6000	Moderate	Low growing	Terrestrial (moist and dry habitats)
	Anthocerophyta	Horned liverworts	ca. 100	Moderate	Low growing	Terrestrial (moist habitat)
TRACHEATA (vascular plants) Pteridophytes (fern allies)	Lycophyta	Club mosses	ca. 1300	Moderate	Mainly low growing	Terrestrial, (moist and dry habitats)
	Psilophyta	Whisk fern	10	Moderate/high	Small herbs	Terrestrial
	Sphenophyta	Horsetails	15	High	Herbs	Terrestrial
	Filicinophyta	Ferns	ca. 12 000	High	Few centimetres to 25 m	Terrestrial (few freshwater taxa)
GYMNOSPERMS (naked seed plants)	Cycadophyta	Cycads	145	High	Shrubs to small trees ≤18 m tall	Terrestrial
	Ginkgophyta	Ginkgo	1	High	Tree ≤30 m tall	Terrestrial
	Coniferophyta	Conifers	630	High	Shrub or tree ≤100 m tall	Terrestrial
	Gnetophyta	Gnetophytes	ca. 70	High	Shrubs, vines and small trees	Terrestrial
ANGIOSPERMS (enclosed seed plants)	Anthophyta	Flowering plants, Angiosperms	ca. 369 000	High	<1 mm to >100 m tall	Most habitats

Adapted from Groombridge and Jenkins (2000) and RBG Kew (2017).

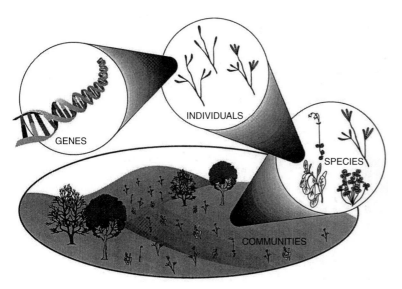

Figure 1.2 Diversity – from genes to communities.
(Reproduced from Frankel *et al.*, 1995)

nucleated cells and membrane-bound organelles), where the fertilized egg develops into a diploid multicellular embryo, there are alternating spore-producing and haploid egg- or sperm-producing generations, and virtually all are terrestrial, photosynthetic autotrophs.

Biological diversity is not only apparent in the numbers of different species and the types of different ecosystems in which they exist but can also be observed between individual of a species. The measurement of biodiversity is related to the various levels of biological organization – genetic, taxonomic and ecological diversity (Table 1.3). Genetic diversity is the heritable variation that is observed within and between populations. The basic genetic component is the gene, and they are found in the nuclei of all cells of all organisms: plants, fungi, bacteria, viruses and animals. Genes are made up of DNA and are situated along chromosomes. Ultimately, it is the variation in the sequence of four nucleotide base pairs (A, T, C and G), which, as components of nucleic acid, constitute the genetic code, which is passed from generation to generation. New genetic variation arises in individuals from gene and chromosome mutations, and in organisms that reproduce sexually, by recombination. There can be various distinct forms of the same gene, referred to as alleles.

Individuals in a plant population or species vary genetically for a range of characteristics or traits. Such genetically significant traits might include: height, fecundity, pathogen or pest resistance, or tolerance/adaptation to extreme environmental conditions such as drought. This variation, which may for instance be expressed morphologically, behaviourally or physiologically, is referred to as the phenotype. The phenotype results from a combination of the individual's genotype (its genetic composition), interacting with the environment in which it is found. The genetic pool of variation present within an inter-breeding population is acted upon by selection. Genetic diversity is not constant for all species; individuals and species vary in the amount and geographic pattern of their genetic variation. Interestingly generally the more highly 'bred' the individual, the less genetic (allelic) diversity is encountered, because bred organisms have passed through the domestication bottleneck where only a limited number of individuals are domesticated from the original ancestral stock. Perhaps the record for the number of alleles per gene locus goes to red clover (*Trifolium pratense*), where self-incompatibility is controlled by a single, multi-allelic gene expressed in the pollen, and it has been estimated that more than 200 alleles exist for this one gene (Lawrence, 1996). It

Table 1.3 **The composition and levels of plant biodiversity**

Genetic diversity	Taxonomic diversity	Ecological diversity
	Kingdom	
	Division	
	Class	Biomes
	Order	Bioregions
	Families	Landscapes
	Genera	Ecosystems
	Species	Habitats
	Subspecies/ varieties	Niches
Populations	Populations	Populations
Individuals	Individuals	
Chromosomes		
Genes		
Alleles		
Nucleotides		

From Heywood and Watson (1995).

is the genetic variation within and between individuals and populations of the same species that ensures the species can adapt and change in response to natural (e.g. changing environment) and artificial (e.g. breeder's selection criteria) selection pressures. Therefore, if a virulent form of a pathogen evolves, such as Ug99, a race of wheat stem rust (*Puccinia graminis* Pers. f. sp. *tritici* Eriks. & Henn.) to which 80–90% of global wheat cultivars are susceptible, it can cause catastrophic loss of wheat grain yield of 70% or more (FAO, 2013b). However, natural genetic diversity within wheat populations means some individuals will be resistant. Notably some resistant individuals have been found to help maintain wheat production (Endresen *et al.*, 2012; Tadesse *et al.*, 2012). Hence, genetic diversity enables natural

evolution and adaptation of species within a changing environment and provides a source of traits for breeders to overcome new virulent strains of pathogens. It is essential for the long-term survival of any species in the wild and for providing food security for humankind.

Taxonomic diversity is diversity at the taxonomic level where organisms are grouped into classes, families, tribes, genera and species using the taxonomic hierarchy. Central to the concept of taxonomic diversity is the species, and for practical purposes, species are the most common targets for biodiversity research and management. The species is however not a standard unit of measurement, since there are several different concepts of what constitutes a species and the level of distinction that constitutes a species in one plant group may be different to that accepted in another group. While genes provide the blueprint for the construction of organisms, they are only expressed through the form and function of species. Similarly, ecosystems are essentially manifestations of the interactions between organisms. It follows that neither genes nor ecosystems can be manipulated or managed without attention to the requirements of species; they are the entities in nature that adapt and evolve, occupy space and become extinct.

Ecological diversity describes biodiversity in terms of the relationship between organisms from population level and upwards through niches, habitats, ecosystems, landscapes and bioregions, to biomes (Table 1.3). At the largest scale, a 'biome' describes any of a group of major regional terrestrial communities with its own type of climate, vegetation and animal life. They are not sharply separated but merge gradually into one another. Examples include tundra, temperate deciduous forest and desert. At the smallest scale, ecological diversity can describe 'populations', which are local communities of potentially inter-breeding organisms. Conservationists refer to habitat, ecosystem or landscape conservation, although these terms are unfortunately often used interchangeably; the ecosystem is the most widely accepted unit of ecological conservation. An ecosystem is defined by the CBD as:

a dynamic complex of plant, animal and micro-organism communities and their non-living environment interacting as a functional unit.

Each level of diversity has its own specific measure: communities, such as savannah grassland, mangrove swamp or steppe, are measured in terms of ecogeographic characteristics, species diversity, and biotic and abiotic interaction; species are measured in terms of representative population density, frequency and cover; individual populations or organisms are measured in terms of their intrinsic genetic diversity; and genes are measured in allelic diversity. Although when people consider plant diversity, they often think in terms of number of higher plant species, but it is important to consider all levels of biodiversity. However, species are commonly considered to have special intrinsic validity because they can be more objectively defined, as a potentially inter-breeding group of individuals, therefore numbers of species are often used to compare diversity with communities or higher taxonomic groups.

Kingdom Plantae is composed of 10 major divisions with approximately 300 000 species, the vast majority of which are flowering plants (or angiosperms), and these are the species most widely exploited by humankind. The diversity within angiosperms has been classified by numerous eminent botanists over the centuries into orders, classes, families and tribes; each classification has its advantages and disadvantages. However, as our knowledge of plant diversity progresses, so successive classification will it is hoped provide a better approximation of the underlying natural classification that exists in nature. Currently a collaborative group of taxonomists, the Angiosperm Phylogeny Group, are producing and iteratively refining a classification of the angiosperms utilizing DNA sequence data. The classification (Angiosperm Phylogeny Group, 2016) recognizes 64 orders and 416 families (Figure 1.3). More information can be obtained from the Angiosperm Phylogeny Group (APG) website (www.mobot.org/mobot/research/apweb/).

1.3 Plant Genetic Resources for Food and Agriculture

Traditionally, plant genetic conservation has focused almost explicitly on crops and, lately, their wild related species. The diversity within these species has been recognized as a tangible, economic resource, thus they were referred to initially as 'plant genetic resources', which may be defined as:

Plant genetic resources are the taxonomic and genetic diversity of plants that is of value as a resource for the present and future generations of people.

(IPGRI, 1993)

Crops may here be broadly defined as any cultivated species, so including those used for food, food additives, feed (animal food), fibre, fuel, feedstocks, bio-based materials, fun (ornamentals and turf-grass), medicine, environmental uses, poisons and gene sources (Cook, 1995). For millennia, humans have exploited the variation within these species. Subsistence farmers would, for example, annually save plants that had larger heads or pest resistance to sow in the following year. This process is as important today as it was for the earliest farmers. However, more recently, to narrow the focus of conservation and exploitation PGRFA have been distinguished from the broader plant genetic resources (PGR) for those species most directly associated with feeding humankind. PGR may also be defined by the nature of the resource utilized that form a continuum between the most advanced cultivars and wild species (Figure 1.4). These include:

- **Modern cultivars:** Genetically uniform or clonal crop varieties bred by plant breeders and currently sown by farmers that become a genetic resource once commercially obsolete.
- **Obsolete cultivars:** Former cultivars that are no longer commercially grown and do not appear on national variety lists, but which may possess genes useful to plant breeders.
- **Breeding lines, clones, populations and genetic stocks:** Material used by plant breeders to develop modern cultivars by means of crossbreeding or use of biotechnology tools.

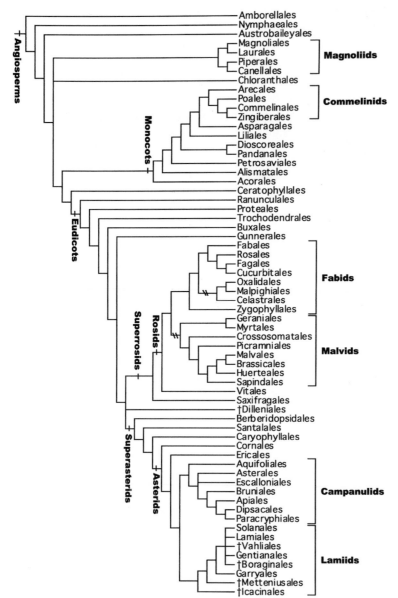

Figure 1.3 Classification of flowering plants.
(Reproduced from Angiosperm Phylogeny Group, 2016)

- **Crop landraces**: Genetically diverse crop varieties that are the product of traditional seed saving systems rather than modern plant breeding, commonly associated with local adaptation, and traditional agricultural practices in more marginal agricultural environments.
- **Weedy races**: Wild species that occur as part of crop–weed complexes as result of hybridization between the crop and wild species, the crop and wild species being evolved from the same ancestor or as the crop's progenitor, often found in gene centres but which hybridize freely with the crop and may introgress useful genes from wild species.
- **Related wild species**: Wild species that are relatively closely related to a crop and may be crossed with the crop either using conventional or genetic engineering techniques to introduce desirable traits from the wild species to the crop.

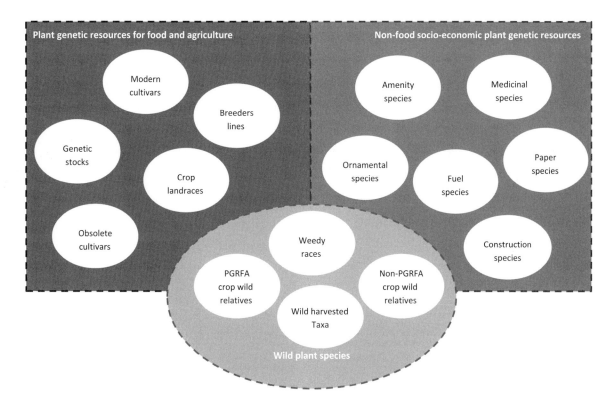

Figure 1.4 The diversity of plant genetic resources.

- **Non-food socio-economic species**: Species whose value is associated with non-agricultural exploitation, such as species with medical, forestry, recreational or ornamental value.
- **Other wild species**: Species of less immediate utilization potential in terms of trait acquisition but which form the basis of natural communities.

Although domesticated plant species represent only a small proportion of the Earth's total biodiversity, they are of fundamental importance to humankind. We have been selecting plants from the wild, domesticating them and adapting them to our needs for around 10 000 years. This process of domestication has led to the existence of an enormous number of different cultivars (product of plant breeding) and landraces (product of farmer-based selection and breeding). Many landraces have been grown in specific localities for extended periods and so grow to withstand local conditions, e.g. altitude, rainfall, drought; they are genetically adapted to the local environment and are referred to as an ecotype.

Increasingly the techniques developed initially for plant genetic resource characterization and genetic conservation are being applied to the broader conservation of wild plant species that are only remotely related to any form of human utilization. The application of biotechnology and systematic bioprospecting has also meant that any plant species has the potential to be of use to humankind in the future. Thus, the boundary between what may and may not be regarded as a plant genetic resource is breaking down and becoming of limited semantic importance. In the future a more appropriate definition might be the total genetic diversity found both between and within all plant species.

1.4 Where Are Plants Found?

The initial response to this question might be throughout the world, but plant diversity and more specifically plant genetic resources are not distributed evenly across the surface of the world, let alone across the terrestrial regions. Their distribution tends to be shaped by four main criteria:

- Historical: organisms that have evolved in isolation from other groups due to historic geographic changes, such as shifts in tectonic plates leading to changes in land mass formation, the rise of mountain ranges or sinking of an isthmus.
- Causal: organisms respond and adapt to specific environmental factors where they are located, e.g. temperature, precipitation and soil conditions.
- Casual: when organisms originally arrive in a location by accident or human intervention, once in that location they evolve to fill all available ecological niches, such as Darwin's finches on the Galápagos Islands or alien introductions.
- Functional: when organisms live together in certain areas, they interact and form communities, each organism within the community playing a specific role, e.g. primary producers, which produce organic material from nutrients in the soil and in the atmosphere with the help of light.

Complicated patterns of biodiversity can originate from the operation and interaction between these four criteria. But if plant diversity is measured in terms of species richness then they tend to vary geographically according to a series of rules applicable for terrestrial environments (Table 1.4). Therefore, plant diversity tends to be concentrated in particular regions near the equator and decreases towards the poles (Figures 1.5). Table 1.5 indicates the distribution of higher plants on a continental basis, and it can be clearly seen that plant diversity increases as you move away from the poles towards the equator.

The concentration of biodiversity in certain regions was noted by Norman Myers, who proposed the concept of biodiversity hotspots (Myers, 1988), defined for plants as an area with a high level of plant species and endemism (0.5% of all vascular plants or 1500 endemic species) and threat (25% or less of original vegetation left intact). He argued that plants should be the baseline for hotspot selection because all other life depends on them. Originally Myers designated 10 hotspots, but following further study Myers (1990) added eight additional hotspots. Thereafter these have been further expanded to the 34 hotspots currently recognized by Mittermeier

Table 1.4 **Biogeographic factors influencing plant concentration**

Biogeographic factor	More biodiversity	Less biodiversity
Latitude	Near equator	Away from equator
Temperature	Warmer	Colder
Rainfall	Wetter	Drier
Seasonality	Less seasonal	More seasonal
Topography	More varied	More uniform
Altitude	Lower	Higher
Area size	Larger	Smaller
Geographic isolation	More endemics	Fewer endemics
Geologic isolation	More endemics	Fewer endemics

et al. (1999) (Table 1.6). These biodiversity hotspots hold 50% of the world's plant species and especially high numbers of endemic species yet covers only 2.3% of the Earth's land surface. Each hotspot is facing extreme threat from habitat mismanagement and has already lost ≥70% of its original natural vegetation. Myers et al.'s (2000) definition of diversity within hotspots is based on species assessment, but even if we consider habitat diversity, these same regions are also rich in habitat diversity, containing tropical rain forest, tropical montane forest, tropical moist forest, warm temperate and Mediterranean vegetation. Further, if plant hotspots (Figure 1.5) are compared to biodiversity hotspots (Figure 1.6) they are correlated with highest plant species concentrations being primarily found in the montane regions of the tropics. But are these areas also rich in genetic diversity?

Generally, we do not yet have enough genetic diversity data to answer this question, except perhaps for PGRFA. The Russian geneticist N.I. Vavilov was one of the first scientists to make the connection between the conservation of genetic diversity and its use in underpinning food security. He also noted that the genetic diversity of PGRFA was

Table 1.5 **Regional distribution of higher plants**

Continent	Sub-Region	Number of Species Continent	Number of Species Sub-Region	Endemics Number	Endemics %
Europe		12 500		3 500	28
Americas		133–138 000			
	North America		20 000	4,198	21
	Middle America		30–35 000	14–19 000	46–54
	South America		70 000	55 000	78.5
	Caribbean Is.		13 000	6,555	50
Africa		40–45 000		35 000	77–87.5
	North Africa		10 000		
	Tropical Africa		21 000		
	Southern Africa		21 000		
Asia					
	Southwest Asia & Middle East		23 000	7,100	31
	Central & North		17 500	2,500	14
	Indian Subcontinent		25 000	12 000	48
	Southeast Asia (Malesia)		42–50 000	29–40 000	70–80
	China & East Asia		45 000	18 650	41.5
Australasia	Australia & New Zealand	17 580		16 202	90
Oceania	Pacific islands	11–12 000		7,000	58–63

Adapted from Davis *et al.* (1995).

concentrated in certain regions (Vavilov, 1917), but his formulated the fundamental concept of the 'Centres of Origin' of crop diversity was published a few years later following extensive field observation and plant collecting across five continents (Vavilov, 1926, 1951). He recognized eight centres (Table 1.7 and Figure 1.7) based on crop presence, within crop landrace and crop wild relative diversity. Vavilov (1965) noted:

The regions of maximum variation, usually including a number of endemic forms and characteristics, can usually also be considered as the centre of type-formation . . . The presence in northern Africa and south-western Asia of large groups of endemic plants, both

species and varieties of cultivated plants, based on which independent agricultural civilizations arose.

He used the term 'Centres of Origin' himself extending Willis's age and area hypothesis (1922), stating that the greater the number of related species occurring in an area, the greater their genetic diversity in that location and the more likely that this was the crop's centre of ancient origin. As Hawkes (1983) notes this is an oversimplification in terms of crop origin, but it proved a very useful concept in terms of identifying that crop species and genetic diversity are distributed unevenly across the terrestrial surface of the Earth but in eight relatively restricted locations. It is also interesting to note that biodiversity in general and

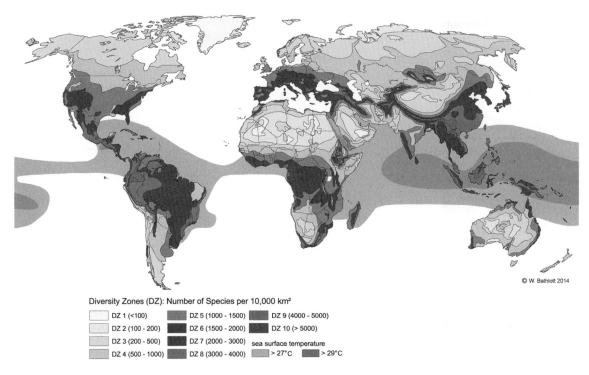

Diversity Zones (DZ): Number of Species per 10,000 km²

DZ 1 (<100)   DZ 5 (1000 - 1500)   DZ 9 (4000 - 5000)
DZ 2 (100 - 200)   DZ 6 (1500 - 2000)   DZ 10 (> 5000)
DZ 3 (200 - 500)   DZ 7 (2000 - 3000)   sea surface temperature
DZ 4 (500 - 1000)   DZ 8 (3000 - 4000)   > 27°C   > 29°C

Figure 1.5 Species diversity globally of vascular plants. (A black and white version of this figure will appear in some formats. For the colour version, refer to the plate section.)
(Reproduced from Barthlott *et al.*, 2014)

plant species diversity are not congruent with the 'Centres of Crop Diversity'; the Vavilov centres do not include the Polynesian and Micronesian Islands, Brazilian Atlantic Forest, Caribbean Antilles, South African Cape region, Madagascar, Southwestern Australia and New Caledonia, but they do include Chiloe in Chile, Ethiopia, Central Asia and China all of which are not biodiversity or plant diversity hotspots.

Whether conserving plant diversity or more specifically plant genetic resources, hotspots are a natural foci of conservation activity, because they contain:

- relatively high species numbers
- high endemicity, whether of common or unusual lineages
- unusual combinations of community ecological characteristics
- super speciose taxa (e.g. the high number of fruit flies in Hawaii)

It might be expected that genetic diversity within a species is spread evenly throughout its range, but it seems that the pattern of genetic diversity is at least

partially independent of geographic patterns. Therefore, it is necessary to have genetic and geography knowledge for efficient conservation.

1.5 Why Does Plant Biodiversity Need Conservation?

The answer to this fundamental question is: plant biodiversity has economic, social and ethical value for humankind, it is a finite natural resource and it is currently being eroded or lost by careless, unsustainable human practices (FAO, 2019; IPBES, 2019). This loss of plant biodiversity can occur at each biodiversity level: genes, species and communities. If we use species extinction to illustrate the point, estimates of the precise number of species and precise rates of species extinction vary, but Lugo (1988) produced a consensus view based on a multiple estimate that 15–20% of all species would become extinct between 1988 and the turn of the century.

Table 1.6 **The 34 hotspots, their characteristics and plant diversity**

Hot spots	Biome(s)	Original extent (km²)	Remaining intact (km²)	(%)	Area protected (km²)	(%)	Diversity	Endemism	% Endemic
		Geographic Area					**Indicative biodiversity**		
		Original extent	Remaining intact		Area protected		Vascular plant species		
Tropical Andes	Tropical and subtropical moist broadleaf forests; montane grassland and shrubland	1 542 644	385 661	25.0	246 871	16.0	30 000	15 000	50
Tumbes-Chocó-Magdalena	Tropical and subtropical moist broadleaf forests	274 597	65 903	24.0	34 338	12.5	11 000	2750	25
Atlantic forest	Tropical and subtropical moist broadleaf forests	1 233 875	99 944	8.0	50 370	4.1	20 000	8000	40
Cerrado	Tropical dry forest, woodland savannah, open savannah	2 031 990	432 514	22.0	111 051	5.5	10 000	4400	44
Chilean Winter Rainfall-Valdivian Forest	Mediterranean forests; woodlands and shrubs; temperate broadleaf and mixed forests	397 142	119 143	30.0	50 745	12.8	3892	1957	50
Mesoamerica	Tropical and subtropical moist broadleaf forests	1 130 019	226 004	20.0	142 103	12.6	17 000	2941	17
Madrean Pine-Oak Woodlands	Tropical and subtropical coniferous forests	461 265	92 253	20.0	27 361	5.9	5300	3975	75
Caribbean islands	Tropical and subtropical dry broadleaf forests	229 549	22 955	10.0	29 605	12.9	13 000	6550	50
California Floristic Province	Mediterranean forests; woodlands and shrubs; temperate coniferous forests	293 804	73 451	25.0	108 715	37.0	3488	2124	61
Guinean forest of West Africa	Tropical and subtropical moist broadleaf forests	620 314	93 047	15.0	108 104	17.4	9000	1800	20
Cape Floristic Province	Mediterranean forests; woodlands and shrubs	78 555	15 711	20.0	10 859	13.8	9000	6210	69
Succulent Karoo	Deserts and xeric shrubland	102 691	29 780	29.0	2567	2.5	6356	2439	38
Maputaland-Pondoland-Albany	Tropical and subtropical moist broadleaf forests; montane grassland and shrubland	274 136	67 163	25.0	23 051	8.4	8100	1900	23

Table 1.6 (*cont.*)

		Geographic Area					Indicative biodiversity		
		Original extent	Remaining intact		Area protected		Vascular plant species		
Hot spots	Biome(s)	(km^2)	(km^2)	(%)	(km^2)	(%)	Diversity	Endemism	% Endemic
Coastal Forests of Eastern Africa	Tropical and subtropical moist broadleaf forests	291 250	29 125	10.0	50 889	17.5	4000	1750	44
Eastern Afromontane	Tropical and subtropical moist broadleaf forests; montane grassland and shrubland	1 017 806	106 870	11.0	154 132	15.1	7598	2356	31
Horn of Africa	Tropical and subtropical grassland; savannas, and shrublands	1 659 363	82 068	5.0	145 322	8.8	5000	2750	55
Madagascar and the other Indian Ocean Islands	Tropical and subtropical moist broadleaf forests	600 461	60 046	10.0	18 482	3.1	13 000	11 600	89
Mediterranean Basin	Mediterranean forests, woodlands and shrubs	2 085 292	98 009	5.0	90 242	4.3	22 500	11 700	52
Caucasus	Temperate broadleaf and mixed forest	532 658	143 818	27.0	42 721	8.0	6400	1600	25
Irano-Anatolian	Temperate broadleaf and mixed forest	899 773	134 966	15.0	56 193	6.2	6000	2500	42
Mountains of Central Asia	Temperate grassland; savannas, and shrublands; montane grassland and shrubland	863 362	172 672	20.0	59 563	6.9	5500	1500	27
Western Ghats and Sri Lanka	Tropical and subtropical moist broadleaf forests	189 611	43 611	23.0	26 130	13.8	5916	3049	52
Himalaya	Tropical and subtropical coniferous forests; montane grassland and shrubland	741 706	185 427	25.0	112 578	15.2	10 000	3160	32
Mountains of Southwest China	Temperate coniferous forests	262 446	20 996	8.0	14 034	5.3	12 000	3500	29
Indo-Burma	Tropical and subtropical moist broadleaf forests	2 373 057	118 653	5.0	235 758	9.9	13 500	7000	52
Sundaland	Tropical and subtropical moist broadleaf forests	1 501 063	100 571	7.0	179 723	12.0	25 000	15 000	60

Table 1.6 (*cont.*)

| | | Geographic Area | | | | | Indicative biodiversity | | |
| | | Original extent | Remaining intact | | Area protected | | Vascular plant species | | |
Hot spots	Biome(s)	(km²)	(km²)	(%)	(km²)	(%)	Diversity	Endemism	% Endemic
Wallacea	Tropical and subtropical moist broadleaf forests	338 494	50 774	15.0	24 387	7.2	10 000	1500	15
Philippines	Tropical and subtropical moist broadleaf forests	297 179	20 803	7.0	32 404	10.9	9253	6091	66
Japan	Temperate broadleaf and mixed forest	373 490	74 698	20.0	62 025	16.6	5000	1950	35
Southwest Australia	Mediterranean forests, woodlands and shrubs	356 717	107 015	30.0	38 379	10.8	5.571	2948	53
East Melanesian Islands	Tropical and subtropical moist broadleaf forests	99 384	29 815	30.0	5677	5.7	8000	3000	38
New Zealand	Temperate broadleaf and mixed forest	270 197	59 443	22.0	74 260	27.5	2300	1865	81
New Caledonia	Tropical and subtropical moist broadleaf forests	18 972	5122	5.0	4192	22.1	3270	2432	74
Polynesia/ Micronesia	Tropical and subtropical moist broadleaf forests	47 239	10 015	21.0	2436	5.2	5330	3074	58
Galápagos	Xerophytic shrubland	7882	4931	62.6	7278	92.3	541	224	41
Juan Fernandez Islands	Temperate forest	100	–	–	91	91.0	209	126	60
Totals		23 498 083	3 384 177	–	2 382 636	–			
Total endemics								131 399	
% Global diversity								43.8	

From Mittermeier *et al.* (1999, 2004).

More recent estimates suggest 20% of species will become extinct within the 30-year period between 1998 and 2028 (American Museum of Natural History, 1998). However, proving a species is extinct is very difficult, particularly for a plant with the possibility of a long-lived soil seed bank, and the IUCN prefer to highlight the fact that species extinction is occurring at unprecedented levels – currently it is estimated that extinction rates are up to 1000 and 10 000 times the 'background' or natural rate (Chivian and Bernstein, 2008).

It is even more difficult, if not impossible, to estimate the precise rates of the loss of genetic diversity from within species. It must, however, always be faster than the loss of species, because there will be some genetic erosion (loss of genetic diversity) from the species that remain extant and complete loss of genetic diversity from species that become extinct.

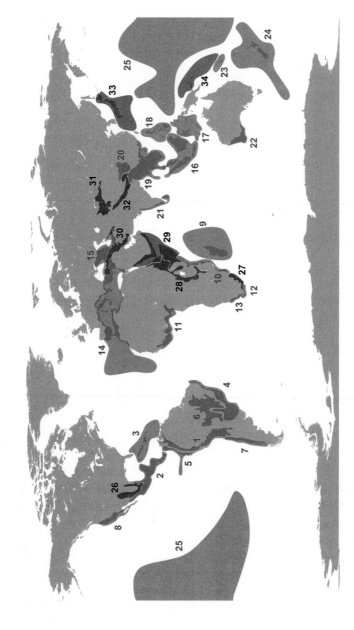

Figure 1.6 The location of areas of exceptionally high biodiversity richness – biodiversity hotspots. (A black and white version of this figure will appear in some formats. For the colour version, refer to the plate section.)
(Reproduced from Mittermeier *et al.*, 1999)

Table 1.7 **World centres of cultivated plant diversity**

Centre	Countries	Crop diversity
I China	Central and West China, Korea, Japan and Taiwan	*Panicum miliaceum* – Chinese millet; *Sesamum indicum* – sesame *Avena nuda* – naked oat; *Fagopygrum esculentum* - buckwheat; *Glycine hispida* - soybean; *Phaseolus angularis* – adzuki bean; *Raphanus sativus* - radish; *Brassica* species; *Colocasia esculenta* – taro yam; *Allium* species; *Cucurbita moschata* – butternut squash; *Phyllostachys* spp. – small bamboos; many temperate (*Pyrus, Malus, Prunus, Chaenomeles* spp.) and tropical fruit trees (*Citrus* spp.); *Camellia sinensis* – Chinese tea; as well as plants producing oils, spices, medicines and fibres
II India	India	*Oryza sativa* – rice; *Eleusine corocana* – finger millet; *Cicer arientinum* – chickpea; *Cajanus cajan* – pigeon pea; *Phaseolus acontifolius* – moth bean, *P. calcaratus* – rice bean; *Vigna sinensis* – asparagus bean; *Dolichos biflorus* – horse gram; *Trigonella foenum-graecum* – fenugreek; *Solanum melongena* – eggplant; several *Amaranthus* species; *Colocasia esculenta* – taro yam; *Dioscorea alata* – yam; tropical fruits (*Citrus* spp., *Musa* spp., *Mangifera* spp.); oil-producing species, fibres (*Corchorus olitorius* – jute), spices (*Piper nigrum* – pepper), stimulants and dye plants; sugar plants such as *Saccharum officinarum* – sugarcane.
IIa Indo-Malaya	South China, South-East Asia	*Dioscorea* spp. – yams; *Citrus maxima* – pomelo; *Musa* spp. – banana; *Cocos nucifera* – coconut
III Inner Asia	Afghanistan, Central Asia and Northwest India	*Triticum vulgare* – wheat; *Pisum sativum* – garden pea; *Lens culinaris* – lentil; *Brassica, Eruca* and *Lepidium* species; *Linum, Sesamum* and *Coriandrum* (one of their centres); *Carthamus tinctorius; Cannabis indica; Gossypium herbaceum*; various vegetables and melon species, spice crops, etc.; fruit and nut trees in the genera *Malus, Pyrus, Prunus, Pistacia, Amygdalus, Juglans, Corylus*, etc.
IV Asia Minor	Turkey, Transcaucasia, Turkmenistan and Iran	*Triticum monococcum, T. durum, T. turgidum* and *T. aestivum* – wheats; *Secale cereale* – rye; *Avena byzantina* – red oat, *A. sativa* – oat; *Cicer arietinum* - chickpea; *Lens culinaris* – lentil; *Vicia ervilia* – bitter vetch; *Pisum sativum* – garden pea; forages (*Medicago sativa* – lucerne, *Trifolium resupinatum* – strawberry clover, *Trigonella foenum-graecum* – fenugreek, *Onobrychis viciifolia* – sainfoin, *Lathyrus cicera* – chickling vetch and *Vicia sativa* – common vetch); oil-producing plants (*Sesamum, Linum, Brassica, Camelina, Eruca* spp.); melons (*Cucumis* and *Cucurbita* spp.); vegetables (*Lepidium, Brassica, Daucus, Eruca, Allium, Petroselinum, Lactuca* and *Portulaca* spp.); fruit crops (*Malus, Pyrus, Punica, Ficus, Cydonia, Cerasus, Amygdalus, Vitis, Pistacia* spp.); dye plants (*Crocus sativus* and *Rubia tinctorum*)

Table 1.7 (*cont.*)

Centre	Countries	Crop diversity
V Mediterranean	Countries bordering the Mediterranean Sea	*Vicia faba* – fababean, *Lathyrus ochrus* – Cyprus vetch, *Vicia sativa* – common vetch, large-seeded *Cicer arientinum* – chickpea, *Hedysarum coronarium* – Italian sainfoin, *Ornithopus vicifoliia* – sainfoin; various oil-producing plants and spices; *Olea europaea* – olive and *Ceratonia siliqua* – carob; *Beta vulgaris* – beets, *Brassica oleracea* – cabbages, *Portulaca oleracea* – purslane, *Allium* spp. – onions, *Asparagus* – asparagus, *Lactuca* – lettuce, *Pastinaca* – parsnip, *Tragopogon* – salsify; ethereal oil species and spices
VI Abyssinia	Ethiopia, Eritrea	*Triticum aestivum* – wheats, *Hordeum vulgare* – barley, *Sorghum bicolor* – sorghum, *Cicer arietinum* – chickpea; *Lens culinaris* – lentil; *Vicia ervilia* – bitter vetch; *Pisum sativum* – garden pea, *Trigonella foenum-graecum* – fenugreek, *Brassica oleracea* – cabbages, *Allium* spp. – onions, *Lepidium latifolium* – peppergrass, *Vigna unguiculata* – cowpea, *Lupinus* spp. – lupins, *Linum usitatissimum* – flax; plus indigenous cereal *Eragrostis tef* – teff and *Eleusine coracana* – African millet; oil-bearing *Guizotia abyssinica* – Niger; *Coffea arabica* – coffee, *Catha edulis* – khat and *Musa ensete* – Abyssinian banana
VII Mesoamerica	South Mexico and Central America	*Zea mays* – corn/maize; *Phaseolus vulgaris* – common bean, *P. coccineus* – runner bean, *P. acutifolius* – tepary bean; *Chenopodium berlandieri* – hauzontle and *Amaranthus cruentus* – purple amaranth; *Cucurbita*, *Sechium* and *Capsicum* spp. (*C. annum*) bell and mostly mild hot pepper; *Pachyrhizus tuberosa* – yam bean, *Ipomaea batatas* – sweet potato and *Maranta arundiacea* – arrowroot; *Gossypium hirsutum* – cotton; many tropical and temperate fruits; *Nicotiana tabacum* – tobacco, *Bixa orellana* – annatto and *Theobroma cacao* – cocoa
VIII South America	Peru, Ecuador and Bolivia	*Solanum tuberosum* – potato, *Oxalis tuberosa* – oca, *Tropaeolum tuberosum* – anu and *Ullucus tuberosus* – ulluco; *Solanum lycopersicum* – tomato, *Solanum muricatum* – Peruvian pepino, *Cyclanthera pedata* – achocha or caigua, *Physalis peruviana* – Cape gooseberry and *Cucurbita maxima* – pumpkin; Phaseolus *vulgaris* – common bean, *P. lunatus* – Lima bean, *Lupinus mutabilis* – pearl lupin, *Capsicum* spp. (*C. baccatum*, *C. chinense* and *C. pubescens*) – hot peppers, *Gossypium barbadense* – cotton *Chenopodium quinoa* – quinoa, *Chenopodium pallidicaudale* – kañiwa, *Amaranthus caudatus* – foxtail amaranth, *Erythroxylum coca* – coca and *Lepidium meyeii* – maca
VIIIa Chiloe	Chile	*Solanum tuberosum* – potato; *Madia sativa* – Chilean oilplant, *Bromus mango* and *Fragaria chiloensis* – beach strawberry
VIIIb Brazil and Paraguay	Brazil and Paraguay	*Manihot utilissima* – manioc, *Arachis hypogaea* – peanut, *Theobroma cacao* – cocoa, *Hevea brasiliensis* – rubber plant and *Ilex paraguayense* – mate

From Vavilov (1926).

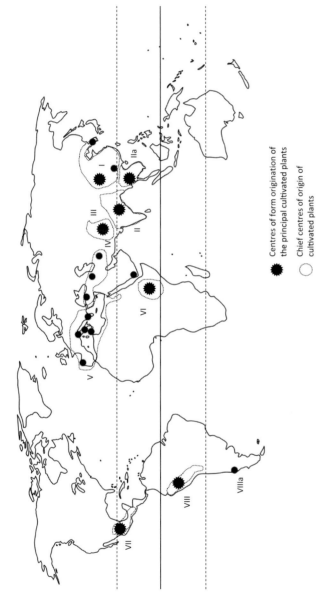

Figure 1.7 The centres of crop diversity.
(From Vavilov, 1951: amended by Hawkes, 1983)

Loss of any genetic diversity means that plants may not be able to adapt to changing conditions quite so readily in the future. Although, as already stressed, rates of genetic erosion cannot be quantified accurately, it seems likely that virtually all species are currently suffering loss of genetic variation to varying degrees. If the figures are correct for species extinction, where 100% of genetic diversity will be lost, then approximately a further 10–15% of plant and animal genetic diversity could be lost over the same period due to genetic erosion (Maxted *et al.*, 1997a).

Obviously, plant diversity has many forms of value and the aesthetic and ethical values cannot be underestimated, particularly as these values are easily identifiable by the public who ultimately fund most conservation action and research. But it is difficult to quantify the economic value of aesthetic and ethical reasons to conserve plant diversity. One estimate of the value of the introduction of new genes from wild relatives to crops is $115 billion per year worldwide (Pimentel *et al.*, 1997), more recently PWC (2013) estimated a similar value for use of CWR in breeding of the 26 top global crops alone. This estimate of the value of wild species for one form of direct use highlights why we need plant diversity. It also underlines the continued need plant breeders have for access to plant genetic diversity if they are to keep their breeding options open. Plant diversity that is lost through genetic erosion or extinction means that diversity is unavailable for exploitation. It also remains the case that 90% of the world's calorie consumption is still based on 30 crops and that all these species originated in the Vavilov centres of diversity primarily located in developing countries (FAO, 1998). Many of these countries are both the home of such critically important plant diversity and are also at risk of food insecurity (Figure 1.8). The Food Security Risk Index is an indication of relative risk of famine based on an evaluation of the accessibility and availability of food and the stability of food supplies across 197 countries. It also takes into consideration the nutritional and health elements of populations (Maplecroft, 2013).

There is a continual requirement for plant breeders to produce novel cultivars that combat evolving pests and pathogens, and less overt demands such as climate change and changing consumer requirements. If the plant breeder is to retain the upper hand, he or she must maintain access to as wide a genetic gene pool as possible. Hence it is important to note that no single country is sufficiently wealthy in native genetic diversity to make it independent of this requirement for the genetic resources of other countries. The reason is that the most species cultivated in any country are rarely native to that country; they were imported historically from the diverse centres of crop diversity worldwide. Take, for example, the botanically rich country of Brazil; more than three-quarters of its calorie consumption is based on crops originating in another continent (Table 1.8). Therefore, there is a need for plant conservation and continued access to the conserved diversity for each country no matter how botanically rich that country may be.

1.6 Threats to Plant Biodiversity

Species extinction and genetic erosion are natural events, just as species and genetic evolution are natural; nature is, and it seems has always been, dynamic in this respect. However, the contemporary situation concerning species extinction and genetic erosion is quite different from that which existed in the past. Humankind now can drastically alter the world environment in ways not previously possible, and it is these anthropogenic changes that have increased the speed of species extinctions and genetic loss. Many species are unable to naturally evolve sufficiently quickly to adapt to the new changing environments created by humans.

The kind of anthropogenic changes that may lead to extinction of taxa (taxonomic erosion) or genetic diversity (genetic erosion) may be broadly grouped under the following general headings:

- Habitat destruction, degradation and fragmentation of natural habitats – leading to direct eradication of taxa as a result of road and reservoir building, changes in land usage, etc.

Verisk Maplecroft Food Security Index 2019-Q1

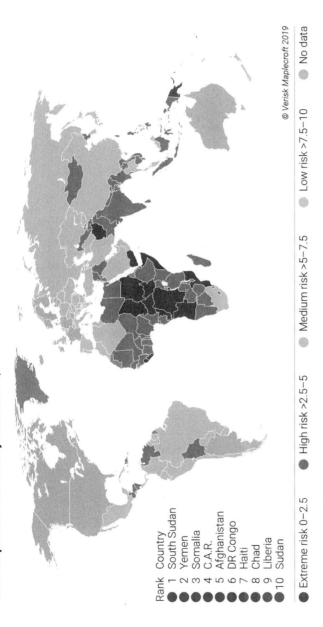

Rank Country
1 South Sudan
2 Yemen
3 Somalia
4 C.A.R.
5 Afghanistan
6 DR Congo
7 Haiti
8 Chad
9 Liberia
10 Sudan

© Verisk Maplecroft 2019

● Extreme risk 0–2.5 ● High risk >2.5–5 ● Medium risk >5–7.5 ● Low risk >7.5–10 ● No data

Figure 1.8 Food security risk index 2013. (A black and white version of this figure will appear in some formats. For the colour version, refer to the plate section.) (From Maplecroft, 2013)

Table 1.8 **Source of plant-derived calories consumed in Brazil**

Crop	Share of plant-derived calories (%)	Centre of origin
Sugar	15.76	Indochina
Wheat	15.76	West and Central Asia
Rice (paddy)	14.45	Asia
Soybean	13.79	China–Japan
Maize	8.58	Central America
Beans	6.20	Andes
Cassava	4.06	Brazil–Paraguay
Coconut	2.09	Indomalaya
Bananas	2.05	Indochina
Oil palm	1.77	West Africa

From FAO (2016).

- Over-exploitation – plant extraction from the wild for food, material, medicines and fuel-wood or overgrazing of plants *in situ*.
- Invasive alien species – human-mediated introduction of exotic species to areas outside of their native range where they compete with, prey on or hybridize with native species. The human-mediated introduction of exotic diseases to areas where the taxa have not previously been subject to the disease can also have devastating effects on susceptible populations.
- Human socio-economic changes and upheaval – resulting in extinction of tribal cultures, urban sprawl, land clearances, wars and human food shortages all of which can negatively impact on local taxa.
- Unsustainable changes in agricultural practices and land use – which can lead to the displacement of landraces by modern cultivars, a shift to monoculture or cash cropping where previously weeds were tolerated is now unacceptable. Other changes may cause incidental extinction, for

example where land drainage leads to the unintentional loss of marsh-loving taxa from drained habitats.
- Calamities – anthropogenic changes that are often not a direct consequence of human action but are an unforeseen by-product resulting in a dramatic effect on biodiversity such as droughts, floods, landslides and of course climate change, which is already having a dramatic global impact.

Finally, climate change is a significant driver of threats to all forms of biodiversity and results from anthropogenic mismanagement of the environment, but its impact is seen through each of the changes listed above rather than having a discrete impact of its own.

It should be noted that the threat to botanical diversity as a result of anthropogenic changes is not universal for all species. Some species are under greater threat of genetic erosion or even of complete extinction than others. Rare and geographically or ecologically restricted species are more highly threatened and likely to become extinct or eroded; that is why the flora of oceanic islands are so vulnerable. Threat is also dynamic, meaning that levels of threat often change rapidly and unexpectedly. Thus, an endemic species or habitat may, for example, suddenly come under the threat of industrial development, road building or logging.

The IUCN has developed a means of assessing relative threat to a taxon and categories of perceived threat, the so-called IUCN Red Data List Categories (Figure 1.9). The assessment is based on the numbers of mature individuals, population size trends, population fluctuations and distributions of populations, demographic patterns, and extinction probabilities in the wild. The IUCN Red List Categories for individual plant species at global (IUCN, 2001) and regional levels (IUCN, 2003) are held in a web-enabled database, the IUCN Red List of Threatened Species (www.iucnredlist.org/). These categories can be applied at local regional as well as global scales to assess comparative threat and so help in prioritizing where conservation effort should be focused.

IUCN Red List Assessment is a widely accepted means of assessing relative threat to species; the Red

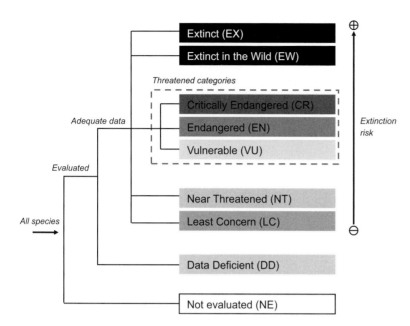

Figure 1.9 Structure of IUCN Red List
Categories. (A black and white version
of this figure will appear in some
formats. For the colour version, refer
to the plate section.)
(Reproduced from IUCN, 2001)

List Criteria have been applied to 17 604 plant species,
and of these, 9829 are listed as threatened (Table 1.9).
Hence, 56% of the species assessed are regarded as
threatened, and 95% of all plant species have yet to be
assessed. However, in terms of the number of species
threatened the percentage threatened figure should
not be extrapolated to all plant species as the species
selected were not selected randomly and they are
likely to have been selected because they were known
a priori to be threatened. A more reliable statistic is
the Sampled Red List Index for Plants (Brummitt and
Bachman, 2010). This index was generated by
selecting approximately 1500 species at random for
each of four major plant groups, monocotyledons,
legumes, pteridophytes and conifers/cycads
(bryophytes are to be added subsequently), and
assessing each species against the IUCN Red List
Categories and Criteria. The result was that 21.5% of
the index plants are currently threatened with
extinction, with 4% of them being Critically
Endangered, 7% Endangered, 10.5% Vulnerable, 10%
Near Threatened and 64% Least Concern, with 4.5%
Data Deficient meaning there was insufficient data
available to undertake an assessment. In the broader
biodiversity context these figures indicate plants are

more threatened than birds, experience a similar level
of threat to mammals but are less threatened than
amphibians.

IUCN Red List Assessment could not be used for
domesticated species as their occurrence is dependent
on humankind but there has recently been a
comprehensive assessment of 572 European crop wild
relative species (Bilz *et al.*, 2011; Kell *et al.*, 2012).
Results of this study show that at least 11.5% (66) of
the species are threatened, with 3.3% (19) of them
being Critically Endangered, 4.4% (22) Endangered
and 3.8% (25) Vulnerable. A further 4.5% (26) of the
species are classified as Near Threatened and one
species (*Allium jubatum* J.F. Macbr.) is Regionally
Extinct. The remaining species were regionally
assessed as Data Deficient (DD) (29%) or Least
Concern (LC) (54.7%). As a group, the most threatened
crop complex was the brassica complex, but multiple
wild relatives of beet, lettuce, wheat and alliums were
also threatened. Kell *et al.* (2012) analyzed the factors
threatening CWR diversity and reported 31 distinct
threats, the most frequent being 'livestock farming
and ranching', 'tourism and recreation areas' and
'housing and urban areas'. However, the authors note
that we should not conclude that farming *per se* is

Table 1.9 **Summary of IUCN Red List Category for plant divisions**

Class[a]	EX[b]	EW	Subtotal (EX+EW)	CR	EN	VU	Subtotal (threatened spp.)	NT	LR/ cd	DD	LC	Total
Anthocerotopsida	0	0	0	0	2	0	2	0	0	0	0	2
Bryopsida	2	0	2	12	13	7	32	1	0	3	3	41
Charophyaceae	0	0	0	0	0	0	0	0	0	3	8	11
Chlorophyceae	0	0	0	0	0	0	0	0	0	1	0	1
Cycadopsida	0	4	4	53	65	74	192	63	0	3	45	307
Florideophyceae	1	0	1	6	0	3	9	0	0	44	4	58
Ginkgoopsida	0	0	0	0	1	0	1	0	0	0	0	1
Gnetopsida	0	0	0	0	1	3	4	7	0	10	76	97
Jungermanniopsida	1	0	1	10	11	12	33	1	0	0	10	45
Liliopsida	9	4	13	501	812	717	2030	352	10	669	2585	5659
Lycopodiopsida	0	0	0	13	11	16	40	9	0	8	29	86
Magnoliopsida	107	26	133	2192	3453	4889	10534	1372	174	1456	6443	20112
Marchantiopsida	0	0	0	1	3	2	6	0	0	4	1	11
Pinopsida	0	0	0	29	96	79	204	98	0	7	298	607
Polypodiopsida	2	1	3	62	69	78	209	26	0	54	180	472
Sphagnopsida	0	0	0	0	0	2	2	0	0	0	0	2
Takakiopsida	0	0	0	0	0	1	1	0	0	0	0	1
Ulvophyceae	0	0	0	0	0	0	0	0	0	1	0	1
Total	122	35	157	2879	4537	5883	13299	1929	184	2263	9682	27514

[a] Anthocerotopsida (hornworts); Bryopsida, Sphagnopsida and Takakiopsida (true mosses); Charophyaceae, Chlorophyceae and Ulvophyceae (green algae); Cycadopsida (cycads); Florideophyceae (red algae); Ginkgoopsida (ginkgo); Gnetopsida (gnetums); Jungermanniopsida and Marchantiopsida (liverworts); Liliopsida (monocotyledons); Lycopodiopsida (club mosses and spike mosses); Magnoliopsida (dicotyledons); Polypodiopsida (ferns, horsetails and quillworts); Pinopsida (conifers).
[b] IUCN Red List Categories: EX – Extinct, EW – Extinct in the Wild, CR – Critically Endangered, EN – Endangered, VU – Vulnerable, LR/cd – Lower Risk/conservation dependent, NT – Near Threatened (includes LR/nt – Lower Risk/ near threatened), DD – Data Deficient, LC – Least Concern (includes LR/lc – Lower Risk, Least Concern). From IUCN (2018).

threatening CWR diversity; in fact, farmed areas (including arable land and pasture) are one of the primary habitats of CWR species. Rather it is unsustainable farming practices, such as severe overgrazing, conversion of land to monocultures, and the heavy application of fertilizers and herbicides, that are the major threats to CWR that grow in agricultural areas (Kell *et al.*, 2012). IUCN Red List assessments do not directly assess threats posed by climate change as the impacts are often less direct and so cannot be unequivocally attributed to climate change. What is recorded is overgrazing, increased fires or competition from alien species, each of which may have at its foundation changes in the biotic or abiotic environment themselves attributable to climate change.

Using the IUCN Red List Categories and Criteria does have a limitation: it only applies to threat assessment at the taxonomic (primarily species) level, and it cannot be used for assessment at the ecosystem or genetic levels. Ecosystem threat assessment is gradually being developed both at the global and national scales, but thus far there is no widely agreed methodology. The most comprehensive global assessment was the Millennium Ecosystem Assessment (MEA, 2005), which found human mismanagement of the world's ecosystems are already causing significant harm to humankind and diminishing the potential long-term benefits we obtain from ecosystems. It noted that approximately 60% (15 out of 24) of the ecosystem services examined during the Millennium Ecosystem Assessment are being degraded or used unsustainably, including fresh water, capture fisheries, air and water purification, and the regulation of regional and local climate, natural hazards and pests. There is also some evidence that changes being made in ecosystems are increasing the likelihood of nonlinear changes in ecosystems (including accelerating, abrupt and potentially irreversible changes) that have important consequences for human well-being, such as changes in pest and disease emergence dates. Further harmful effects of the degradation of ecosystem services (the persistent decrease in the capacity of an ecosystem to deliver services) are being borne disproportionately by the poor, resulting in growing inequities and

disparities between global regions. Perhaps not surprisingly the ecosystems that are being most rapidly eroded are those of highest exploitation value to humankind, such that virtually 100% of natural grasslands in the USA have been lost since 1942 and more than 90% of natural wetlands in New Zealand have been lost since European settlement (Spellerberg, 1996). Furthermore, between 1900 and 2005 the annual rate of forest loss was 14.5M hectares or $145\,000\,km^2$ per year (FAO, 2011b).

While in terms of loss of genetic diversity there are very few examples that quantify the loss of genetic diversity, normally the population numbers or size is taken as a proxy for genetic diversity but there is unlikely to be a direct relationship between loss of populations and loss of genetic diversity. The evidence that is available is largely drawn from loss of diversity in agro-biodiversity where within a crop the disappearance of landraces is likely to be strongly correlated with loss of diversity. In the State of the World's PGRFA report (FAO, 1998) it is noted that the proportion of the wheat grown in Greece contributed by old, indigenous landraces declined from 80% in 1930 to less than 10% in 1970. Furthermore, in Kampuchea rice landraces were lost in the 1970s when war disrupted agricultural production, but in this case a partial duplicate had been preserved in the International Rice Research Institute gene bank in the Philippines and so could be repatriated. While in Mexico and Guatemala, urbanization has displaced about 50% of the populations of teosinte (*Zea mexicana*), the closest relative of maize (Wilkes, 2007), and in the USA there has been loss of vegetable and fruit landraces (Figure 1.10).

Much of this domesticated diversity is being lost due to the replacement of older inherently diverse varieties with modern, high-yielding cultivars (either inbred lines or F_1 hybrids), but the high-yielding cultivars are genetically uniform, thus gradually, locally adapted variation is lost and the domesticate gene pool is shrinking. The aim of domesticated crop conservation is to conserve these landraces and their wild relatives, which retain the essential genetic diversity that is required for breeding new cultivars. We cannot hope to counter or mitigate the effect of climatic change without this breadth of diversity.

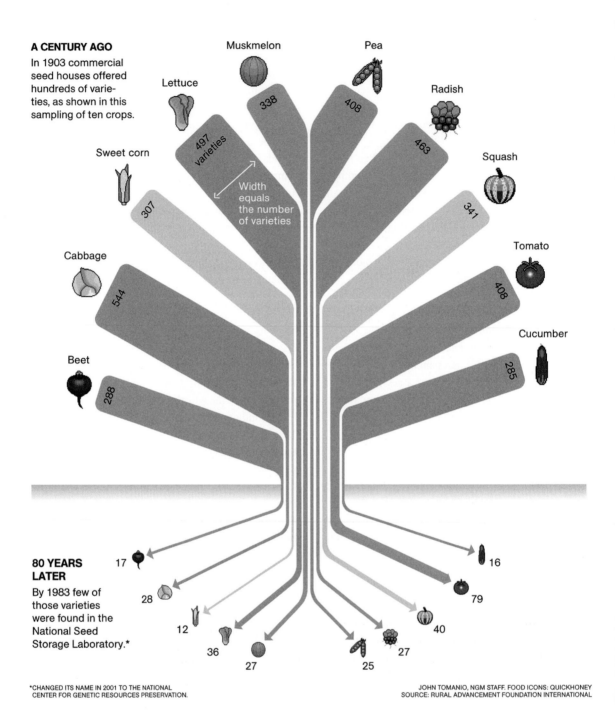

A CENTURY AGO
In 1903 commercial seed houses offered hundreds of varieties, as shown in this sampling of ten crops.

Muskmelon

Pea

Lettuce

Radish

Sweet corn

497 varieties

Width equals the number of varieties

338

408

463

Squash

307

341

Cabbage

544

Tomato

408

Beet

288

Cucumber

285

80 YEARS LATER

By 1983 few of those varieties were found in the National Seed Storage Laboratory.*

17

16

28

79

12

40

36

27

27

25

27

*CHANGED ITS NAME IN 2001 TO THE NATIONAL CENTER FOR GENETIC RESOURCES PRESERVATION.

JOHN TOMANIO, NGM STAFF. FOOD ICONS: QUICKHONEY
SOURCE: RURAL ADVANCEMENT FOUNDATION INTERNATIONAL

Figure 1.10 Loss of vegetable varieties in the United States.
(From RAFI, 1983; National Geographic, 2013)

1.7 Why Do We Need PGRFA?

Genetic vulnerability is a term used to describe the adverse effects of a lack of genetic diversity, and this deficit means that a species or population is unable to respond to change in its biotic or abiotic environment. For example, in an area where soil salinity is increasing gradually those individuals that are less able to survive will be lost from the population and the allele frequency will change to reflect this selection pressure; it is only possible for this to occur if there was genetic variation in the original population and the needed salinity resistance was present. Genetic vulnerability is often a problem for agricultural crops, which are deliberately bred for uniformity to ensure yield and performance stability. These genetically uniform varieties may not have the inherent genetic diversity necessary to withstand adverse pest and pathogen attack or environmental hazards, and they are therefore uniformly susceptible. Genetic diversity is necessary to decrease vulnerability to new races of pest or pathogen, or environmental or cultural changes. The problems resulting from genetic uniformity are highlighted in the following examples:

- The history of potato cultivation in Europe illustrates the necessity for diversity. Potato breeders in the 19th century were worried about the narrow genetic base of the potato in Europe; they used phrases expressing the need for 'new blood' and lamenting the potato's 'degeneration'. It is believed that this resulted from the fact that all European potatoes existing in the 19th century resulted from selection over two centuries earlier of two initial introductions. It is not surprising therefore that the potato crop in Ireland was devastated by epidemics of late blight in the 1840s.
- A more recent, but less publicized, example of genetic vulnerability was that of the Soviet wheat cultivar 'Bezostaja', which was grown on about 15 million hectares in 1972. The cultivar originated in the Ukraine during a period of relatively mild winters. Then in 1972 a very severe winter occurred causing losses of millions of tonnes of winter wheat throughout the Soviet Union. The genetic

uniformity of the cultivar meant it was universally unable to cope with cold conditions.
- The value of local landraces or long-established cultivars and the diversity of genes they may hold often remain unappreciated until they compete against new foreign cultivars. The genetically uniform semi-dwarf wheat cultivars of the Green Revolution when first grown in Mexico were overcome by the fungal diseases black stem rust and stripe rust, while the tried and tested local varieties with their intrinsic genetic diversity were able to resist the attack.
- Genetically uniform upland cotton introduced from the USA to western Tanzania in the early 20th century was deemed unproductive as a result of the insect pest cotton jassid and bacterial blight. Cotton breeders were only able to solve these problems when resistance to jassid attack was found to be related to the length and density of hairs on the underside of the leaves. In Tanzania, genetic variation for hairiness was found to be present in the locally adapted cottons and was rapidly exploited to give jassid-resistant varieties. Genetic variation in local landraces for resistance to bacterial blight was also exploited to produce new and highly successful cultivars.

1.8 How Do We Conserve Plant Genetic Diversity?

Plant genetic conservation aims to maintain the taxonomic and genetic diversity of plants, the habitats or ecosystems in which they live, and the interrelationships between plants, other organisms and their environment. It aims to enhance or maintain diversity and halt habitat, species and genetic erosion by establishing and implementing conservation programmes. To achieve this goal the conservationist must clearly define and understand the processes involved, and then develop practical techniques to achieve taxonomic and genetic stability, maximizing the likelihood of allelic diversity maintenance. Conservationists, when undertaking conservation, use their knowledge of genetics, ecology, geography, taxonomy and many other disciplines to understand

and manage the biodiversity they wish to conserve. It is important to stress that genetic conservation is not just about maintaining alleles or individual plant populations but includes all levels of biodiversity from ecosystems (a community of organisms and its abiotic environment), through communities (collection of species found in a common environment or habitat), species and populations to genetic diversity within populations. To conserve maximum diversity in a species, populations of the species are likely to require protection in diverse locations, and in each of these the habitat must be maintained that contains the target population.

The practice of conservation tends to diverge between those that take an ecosystem and those that take a genetic approach, though these approaches are viewed as extremes in a continuum of overlapping techniques. Ecological conservation focuses on the conservation of whole communities. Individual survival and extinction are a major concern but are seen in the context of overall community health. This form of whole community conservation was basic to the International Biological Programme in the 1960s and early 1970s and was later exemplified by the 'Man and the Biosphere' programme of UNESCO. The latter established a network of biosphere reserves, representing distinct biomes and ecosystems throughout the world. The clear emphasis is on conservation of overall ecosystems and particularly keystone species that dominate that ecosystem. Other individual species are conserved as part of the entire ecosystem, but it is possible that individual species may be lost within a conserved ecosystem. Genetic conservation focuses more explicitly on individual taxa (most commonly species) and attempts to conserve the full range of genetic (allelic) variation within those taxa. The realization of the importance of conserving genetic diversity arose from the work of early geneticists, such as W. Bateson and N.I. Vavilov, who travelled the world in the 1920s and 1930s collecting the wide genetic variation available of crops and their wild relatives. International genetic conservation of crops and crop relatives gained momentum in the 1960s, spearheaded by the FAO of the UN and a series of technical meeting, which they hosted. In 1974 the International Board for Plant

Genetic Resources was established to help develop and promote national and international PGR activities. The aim of genetic as opposed to ecological conservation is often explicitly utilitarian, and the conservation of genetic diversity is often linked directly to human utilization, as occurred with Vavilov's original work.

The aim of genetic conservation is to maximize the maintenance of genetic diversity, but further it is explicitly utilitarian: there is an intimate link between plant genetic diversity, conservation and utilization (Figure 1.11). The model includes a series of steps starting with the full range of genetic diversity for all plant species, through the prioritization of target taxa, the planning of conservation action and the implementation of the conservation action, and leading through characterization and evaluation to utilization. The application of this model is at the core of food security, poverty alleviation and the well-being of humankind.

Central to the model of plant genetic conservation are two general strategies for conservation, each composed of a range of specific techniques. The two strategies are *ex situ* and *in situ* conservation defined by the Convention on Biological Diversity (CBD, 1992) thus:

Ex situ conservation means the conservation of components of biological diversity outside their natural habitats.

In situ conservation means the conservation of ecosystems and natural habitats and the maintenance and recovery of viable populations of species in their natural surroundings and, in the case of domesticates or cultivated species, in the surroundings where they have developed their distinctive properties.

There is an obvious fundamental difference between these two strategies: *ex situ* conservation involves the location, sampling, transfer and storage of the species away from the original location where they were found, whereas *in situ* conservation involves the location, designation, management and monitoring of species at the location where they grow naturally or are cultivated. The two general strategies may be subdivided into several specific techniques (Table 1.10).

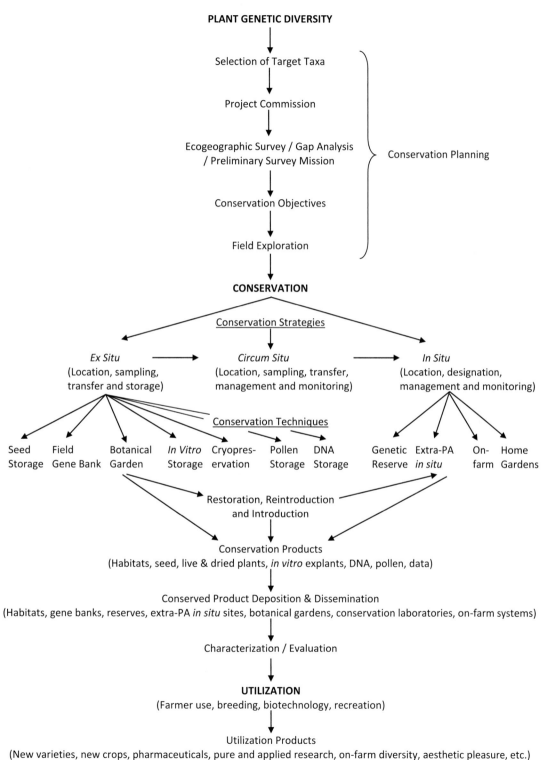

Figure 1.11 Model of plant genetic conservation.
(Adapted from Maxted *et al.*, 1997b)

Table 1.10 **Genetic conservation strategies and techniques**

Strategies	Techniques	Definition
Ex situ conservation	Seed storage	The sampling, transfer and storage of seed samples at a suitably low moisture content ($\approx$5%), and sub-zero temperatures ($\approx-20°$C)
	Cryopreservation	The sampling, transfer and storage of seed samples at ultra-low temperature ($-196°$C)
	In vitro storage	The sampling, transfer and maintenance of explants in a sterile, pathogen-free environment
	DNA/pollen storage	The sampling, transfer and storage of DNA or pollen in sub-zero temperatures ($\approx-20°$C)
	Field gene bank storage	The sampling, transfer and maintenance of living plants under field or plantation conditions
	Botanic garden/ arboretum	The sampling, transfer and maintenance of living plants (tree species for arboreta) in a garden
In situ conservation	Genetic reserve conservation	The location, management and monitoring of genetic diversity of natural wild populations within defined areas designated for active, long-term conservation
	Extra PA *in situ* sites	The location, management and monitoring of genetic diversity of natural wild populations in informal *in situ* conservation sites
	On-farm conservation	The location, management and monitoring of genetic diversity of locally developed traditional crop varieties, with associated wild and weedy species or forms, by farmers within traditional agricultural, horticultural or agri-silvicultural cultivation systems for commercial sale
	Home garden	The location, management and monitoring of genetic diversity of locally developed traditional crop varieties or forms by householder within their individual garden, backyard or orchard cultivation systems for home consumption

Although *in situ* and *ex situ* techniques have been defined and the distinction between the two general strategies emphasized, in practice it may not be possible to make such a clear distinction. Take, for example, the conservation of the legume tree genus, *Leucaena*, where germplasm is often collected from native habitats and then taken *ex situ* to be more easily managed by local communities. The trees are not conserved using field gene bank or arboreta techniques, but within local communities, they are managed using traditional silvi-cultural techniques within an *in situ* on-farm system. This form of 'hybrid' conservation has been termed *circa situm* conservation and is often found in the management of fruit trees in subsistence communities.

The point should be stressed that although nine basic *in situ* and *ex situ* conservation techniques have been outlined, no single technique alone can adequately and completely conserve the genetic diversity found within any single species. A more appropriate methodology is to apply multiple techniques, applied in a complementary fashion to ensure the long-term safety of the entire gene pool of genetic diversity of the target taxon. This is referred to as complementary conservation.

1.9 How Do We Use Plant Genetic Diversity?

As has already been emphasized genetic conservationists often emphasize the link between conservation and use or exploitation. It can be further argued that the ultimate reason for conserving biodiversity, whether 'living' or 'suspended', is to make it available for use by humankind, either now or in the future. However, the ways in which humans use plant diversity are themselves very diverse; plant genetic resources are not just simply used as trait donors by plant breeders. Plants may be used as:

- food, crop species including beverages;
- food additives, including processing agents and other additives used in food and beverage preparation;
- feed (animal foods), the fodder or forage species eaten by vertebrate and invertebrate animals;
- materials, such as wood, fibres, cork, cane, tannins, latex, resins, gums, waxes, oils, etc.;
- fuels, wood, charcoal, etc.;
- poisons;
- medicines, human and veterinary;
- environmental: these will include species that are ornamentals, recreational, hedges, shade plants, windbreaks, soil improvers, plants for regeneration, erosion control, indicator species (e.g. pollution, underground water);
- gene donors: plants that contain desirable traits that can be transferred to other species to improve their use.

As well as these clear examples of human exploitation, there are also more nebulous but equally valid uses, such as those derived from ethical and aesthetic convictions. These may, for example, be as simple as wishing to walk your dog in open countryside, as such diverse habitats provide a pleasant environment – which is a further and valid use of biodiversity. Similarly, you may have been born in a beautiful valley and wish that the valley

retains its basic character, or hopefully you agree that it is wrong for humans to carelessly eradicate species. Defined in these terms all species have a use in some form even if it defined in purely aesthetic or ethical terms. A recent paper in the *British Medical Bulletin* demonstrated the positive effect of biodiversity on human health, as well as the mechanisms and evidence of the positive health effects on humans of diversity in nature and green spaces (Honnay and van Nieuwenhuyse, 2018). These so-called ethical and aesthetic uses are commonly established by the general public and will often be focused on 'flagship' species, for example orchids or cacti for plants, or 'picturesque' environments. This type of value is more difficult to define, but that does not make its worth any less valid, and as much conservation is funded from public sources, it cannot be ignored.

The most fundamental use of plant genetic diversity remains as food crops. In certain cases, the conserved material can be used directly, as is often the case in the selection of new accessions of forage species, where little breeding is undertaken, or in the case of the reintroduction of primitive landrace material following their local extinction. More commonly, however, the first stage of utilization will involve the recording of genetically controlled characteristics (characterization) and the material may be grown out under diverse environmental conditions to evaluate and screen for, say, drought or salt tolerance, or the deliberate infection of the material with diseases or pests to screen for biotic resistance (evaluation).

Having briefly discussed how humankind uses plant diversity, the key point should be stressed here that any form of plant exploitation must be sustainable and non-exploitive. Sustainability, in the sense of continuance, is a fundamental concept for both conservation and utilization within the biosphere, the finite system in which we all live. It is ignorance of this fundamental concept that has resulted in many acute environmental disasters, e.g. desertification of the African Sahel or shrinkage of the Aral Sea in Central Asia. Non-exploitive, in the sense of a subsistence farmer providing a sample of the traditional landrace their family has maintained by cycles planting, cultivation harvesting and seed selection for millennia and finding the landrace had a unique allele for an adaptive trait, then when bred into an elite breeder's line made a new cultigen that had global sales of US$ millions. The signing of a now Standard Material Transfer Agreement by the donating farmer and recipient organization ensures that today benefit would flow back to the farmer, his family and his community. Each of these negative examples is a consequence of policy-makers not thinking sustainably and focusing on selfish short-term benefits, which is not fit practice for the 21st century.

2 Establishing the Social, Political and Ethical Context

2.1 Is There a Political, Social and Ethical Context?

Plant genetic resources for food and agriculture (PGRFA) are critical for human survival and well-being. In addition to their economic importance, and the contribution they make to food security, there are equally important social, ethical and political issues surrounding PGRFA conservation and use, the development of which can be traced from a historical perspective. Such issues range from the long history of biopiracy, or the plundering of genetic resources especially by those countries of the global North, to more recent issues of laws and regulations which prevent farmers from saving seeds or the sharing of benefits with farmers and communities who have long maintained our common heritage of genetic resources. Indeed, ownership and control over plant genetic resources (PGR) are increasingly recognized as critical issues in movements underpinning food sovereignty (Wittman *et al.*, 2010). Many scientists are often reluctant to become involved in or even consider the policy or social consequences of their work and often ignore the historic or ethical foundation. In part this may be due to the way science is often taught within Western education systems, where too frequently science teaching is isolated from what are regarded as social science concerns and scientists are loath to stray outside their field of expertise. Such issues can also be quite contentious and emotive and challenge scientifically accepted research approaches, and they often involve time-consuming lobbying that challenges the status quo. As scientists, we may feel we should focus on pure science and leave social and ethical issues to philosophers, lawyers, activists and others. However, this approach is increasingly untenable in a world where the application of science and technology can be so all pervasive and dramatically alter the way most people live their lives and where systems

thinking approaches are increasingly the norm. Therefore, as scientists, we do have to be aware of our increasing responsibilities in the social and policy context and consider the broader setting of our research. Furthermore, with recent developments in international environmental law, conventions and treaties, such as the United Nations Convention on Biological Diversity (CBD) and the International Treaty of Plant Genetic Resources for Food and Agriculture (ITPGRFA), which clearly lay out many social, ethical and political responsibilities and obligations of governments and other stakeholders, such a position of ignorance or avoidance is no longer possible or acceptable. The history of the conservation and use of plant genetic resources can be used to illustrate the movement toward an increasing social, ethical and political awareness of the impact of technology within society. This process of raising public awareness is encapsulated within the CBD, which has as its core objective: '. . . the conservation of biological diversity, the sustainable use of its components and the fair and equitable sharing of the benefits arising out of the utilization of genetic resources . . .' (Article 1 – CBD: www.cbd.int/convention/articles/?a=cbd-01).

Obviously, the need to conserve biodiversity is central, but here it is explicitly linked to both sustainable use (= sustainable development) and the fair and equitable sharing of the benefits arising from that use. Thus, scientists from the 196 countries which are parties to the CBD (CBD website 20 December 2018, www.cbd.int/information/parties.shtml) for the first time are explicitly obliged to acknowledge and address the social and policy context of their work. Only the United States of America and the Holy See member states of the United Nations are not Parties.

Most recently, the importance of genetic resources to the 2030 Agenda for Sustainable Development has been recognized by the global community in 2015 with their inclusion in the agreed Sustainable

Development Goals (SDGs) as part of Goal 2 'End hunger, achieve food security and improved nutrition and promote sustainable agriculture'. Reference to genetic resources can be found in Target 2.5:

By 2020, maintain the genetic diversity of seeds, cultivated plants and farmed and domesticated animals and their related wild species, including through soundly managed and diversified seed and plant banks at the national, regional and international levels, and ensure access to and fair and equitable sharing of benefits arising from the utilization of genetic resources and associated traditional knowledge, as internationally agreed.

Additionally, Target 2.a draws attention to the need for increasing investment and cooperation for gene banks:

Increase investment, including through enhanced international cooperation, in rural infrastructure, agricultural research and extension services, technology development and plant and livestock gene banks in order to enhance agricultural productive capacity in developing countries, in particular least developed countries and landlocked developing countries, in accordance with their respective programmes of action.

2.2 The Historical Context of PGRFA Conservation and Exploitation

The context is centred on the fundamental link between biological resource conservation and use. Humans use the resources found in the environment that surrounds them, but historically there was a balance or sustainability between conservation and use that allowed natural replenishment of the resources that humankind depended on. Recently, exploitation has entered a new unsustainable phase; our technology has advanced to such a point that we are capable of completely extracting a resource, e.g. fish, wild herbs, timber and fuelwood. Each of these examples illustrates technological exploitation for short-term commercial gain but longer term ecological and financial catastrophe. Within a

botanical context, this unsustainable exploitation of our environment is linked to such anthropogenic factors as: human population growth, poverty, urbanization, exponential development of technology, destruction of natural habitats, universal introduction of exotic species and overexploitation of resources – the general causes of loss of botanical and genetic diversity which is now exacerbated by climate change. There is a paradox at the heart of biological resource conservation and use; we must exploit biological production for the benefit of humankind, but that very enhancement leads to increased loss of botanical and genetic diversity. If this vicious cycle is to be broken, we must link complementary conservation to sustainable exploitation, production and consumption. In other words, we must develop ways of enhancing biological production that do not have a deleterious effect on biological diversity. The understanding of the detrimental link between biological exploitation and loss of biological diversity became apparent for the first time in the early part of the last century and became generally acknowledged in the 1970s, contributing to greater awareness of environmental issues in general and the need for better environmental policy and governance, captured dramatically in the planetary boundaries concept where biodiversity loss is one of the domains which has exceeded safe margins (Steffen *et al.*, 2015) (Figure 2.1). Using this concept, biodiversity loss is one of the elements which has gone beyond the safe operating space to a high-risk zone. Parallel to this, a new geological epoch, the 'Anthropocene', has been proposed as a period in which humans have replaced nature as the defining force influencing environmental change on the planet, though opinion on this differs (Ruddiman *et al.*, 2015).

The key issues around biodiversity loss and genetic erosion of our food crops and wild relatives can also be illustrated by the brief history of PGRFA exploitation and conservation provided by Esquinas-Alcázar and colleagues (2011, box 1.1, pp. 6–7).

2.2.1 Early Exploitation

The beginnings of plant germplasm exploitation are as old as humankind. Hunter-gatherers in

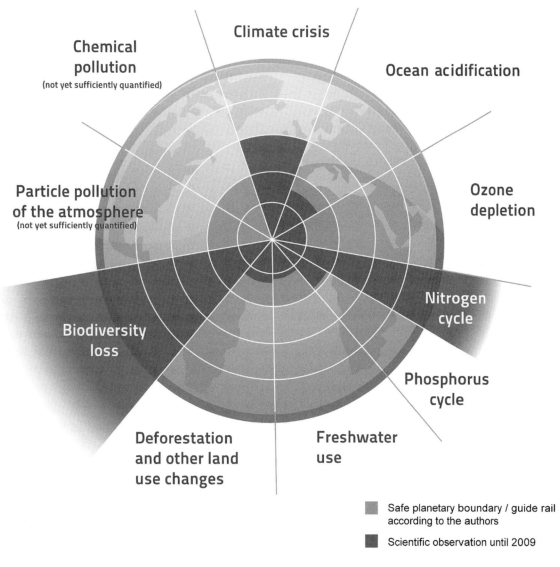

Figure 2.1 The planetary boundaries concept where the red areas represent human activities that have exceeded safe margins. (A black and white version of this figure will appear in some formats. For the colour version, refer to the plate section.) (From Steffen *et al.*, 2015, after Johan Rockström, Stockholm Resilience Centre *et al.* 2009)

pre-agricultural times exploited their environment not only for food, but also for fuel, medicines, building materials, tools, adornment, transport, and recreation. They would have eaten berries, fruits, seeds, flowers, young succulent shoots, underground storage organs and fleshy roots, all found within their local environment (Cunningham, 2001; Ingram *et al.*, 2017).

Evolving in Africa, humans gradually migrated to other parts of the world. During this time, they have demonstrated considerable ecological flexibility and dietary resilience that allowed them to adapt their diets and habitation of ecological zones to a high degree. The Palaeolithic, a period of significant climate change, witnessed major changes in human hunting and dietary patterns. Prey preferences shifted from larger to smaller animals, possibly triggered by population declines of larger prey. Smaller prey proved more difficult to catch and provided a poor return compared to earlier larger prey and combined with increasing population pressure, most likely gave rise to food shortages. It is speculated that such changes prompted major shifts in human behaviour to an increase in foraging and collecting of plants and their increased utilization in the human diet (Murphy, 2007).

These early forms of plant exploitation were *in situ*, in the sense that plants or, for that matter, animals were found and used within the hunter-gatherer's 'home range'. Even with the initial development of seed-based agriculture in the 8th or 7th millennia BC in the Middle East, the early farmers were essentially local plant gatherers and planters – they exploited the plants they found growing in their locality. The early exploitation of crop species is paralleled by that of medicinal plants. It was usually associated with older 'wiser' persons (wise women, witches and later herbalists) or elders. Historical records show that ancient Greeks, Persians, Chinese, Assyrians and Egyptians each used specific plants for their medicinal, benign or malign properties.

These cultures, whether using plants for food, medicines or other forms of use, were predominantly based on local availability of plants. Plant exploration or *ex situ* exploitation is more sophisticated and purposeful – it is the location of plants in one area and their transfer to a second area to be conserved and exploited. This active process involves greater foresight, which may involve scientific, as well as social and economic, considerations. Early examples of *ex situ* exploitation are provided by Sargon in 2500 BC, who crossed the Taurus mountains to the Anatolian Plateau to collect figs, vines and olives. Allied with this is the concept of targeted collecting, where the potential plant genetic resource user searches for a particular 'target taxon' in a particular 'target area'. The earliest recorded

example of a targeted collecting expedition is believed to be that instigated by Queen Hatshepsut of Egypt in 1495 BC, who launched an expedition to Punt (modern-day Somalia) looking for trees that could yield frankincense (*Boswellia carteri*) from their resin. The narrative of the expedition is recorded in wall paintings at her palace in Thebes; they show that living plants were transplanted into pots, transported down the Nile on barges and planted in the palace garden. This demonstrates the early interdependence between countries for PGR.

With the development of routine travel over longer distances, plant germplasm (seeds, bulbs, tubers, corms or any structure that enables the passage of genetic material from one generation to the next) started to be routinely transferred from its area of origin to other areas of the world for exploitation. Other early examples include Alexander the Great in the 4th century BC who took pomegranates from Armenia, and peaches and apples from Central Asia to Macedonia, and the Arabs who took coffee from Ethiopia to Arabia in 900 AD.

2.2.2 Domestication of Plants as Food Crops

The evolution of crop plants began between 5000 and 10 000 years ago, and it is now generally thought that far from a dramatic and sudden 'agricultural revolution', the shift from hunter-gathering to settled agriculture appears to have occurred in a series of transitions over several millennia (Murphy, 2007). That domestication might have taken thousands of years as opposed to a couple of hundred years and was more of an 'agricultural evolution' than the previously believed revolution and also importantly that it was a much more spatially widespread event than previously appreciated (Balter, 2007; Holmes, 2015). It is also likely that certain hunter-gatherer groups would have managed small gardens or plots, which they would periodically tend while continuing to forage and gather plants from their wider environment, as well as other wild foods (Murphy, 2007). The areas where settled agriculture did emerge include the Fertile Crescent of the Near East (wheat, barley, pulses), the Huang He (Yellow

River) region of China (millet) and southern Mexico (maize, pulses, peppers, squashes). This is illustrated by N.I. Vavilov's (1887–1943) hypothesis on the 'centre of origin' of crop plants (Figure 1.7). Vavilov noted that the wild relatives and ancient forms of crop plants are not spread evenly across the land surface of the world, but are restricted to several relatively small and isolated mountainous regions that he referred to as their centres of origin or diversity:

- Tropical Centre (South China, India and Southeast Asia)
- East Asiatic Centre (central and West China, Korea, Japan and Taiwan)
- Central Asia and northwest India (Uzbekistan, Kazakhstan, Kirgizia and India)
- Southwest Asiatic Centre (Turkey, Iran and Afghanistan)
- Mediterranean Centre (countries bordering the Mediterranean Sea)
- Abyssinian Centre (Ethiopia)
- Meso-American Centre (South Mexico and Central America)
- Andean Centre (Peru, Ecuador, Bolivia and Chile)

From these early centres of agriculture, the spread of domesticated plants took place, and following conscious and unconscious selection pressures, individual crops became increasingly diverse:

- In traditional agro-ecosystems newly domesticated plant types and primitive cultivars diverged from their wild ancestors.
- Occasional crosses continued to occur between the early crops and their wild relatives.
- Thus, useful traits continued to be introduced to the crop, such as host plant resistance to pathogens and pests, or tolerance/adaptation to extreme conditions such as drought.

A significant result of the domestication process was the increased reliance of humans on a much-reduced diversity of plant species for their food supply and diet. The number of plants used for food by pre-agricultural human societies is estimated to be around 7000, which is only a tiny fraction of the plant kingdom. Mostly grass and legume species were domesticated and continue to dominate the world's food supply to this day. Modern agriculture, food systems and diets are largely dependent on roughly a dozen plant species originally chosen and domesticated by the early Neolithic farmers (Murphy, 2007). These dozen or so plant species together with 15 mammal and 10 bird species today are estimated to supply 75% of the world's food, while the big three cereal staples, rice, maize and wheat, alone currently provide 60% of the world's food energy intake (Figure 2.2). Today, in population terms, 4 billion people rely on rice, maize or wheat as their staple food, while a further 1 billion people rely on roots and tubers (Millstone and Lang, 2008). Notwithstanding, there remain thousands of plant species with neglected potential utility for humans and which represent one of the most underutilized and underappreciated resources we have (Meldrum and Padulosi, 2017).

With the beginning of human migration and trade, crop plants evolved in new environments and crossed with local wild relatives, leading to even greater genetic diversity. For instance, in West Africa, the indigenous African cultivated rice, *Oryza glaberrima*, has hybridized with the introduced common Asian rice *Oryza sativa*, enriching the rice gene pool in the region. These processes of natural selection in response to new ecological conditions, in conjunction with farmers selecting for characteristics over thousands of years of cultivation, have led to the vast array of crop diversity that we know today. These small farmers, indigenous Peoples and local communities have long been, and continue to be, the custodians of much of the world's crop diversity and its associated ecological knowledge (Sthapit *et al.*, 2017), a fact all too often ignored in our increasingly globalized world (Figure 2.3).

2.2.3 Systematic Exploitation

Christopher Columbus led the waves of European explorers that were sent to search the New World and distant margins of the Old World for exploitable plant resources during the late 15th, 16th and 17th centuries AD. The considerable exchange of plant and

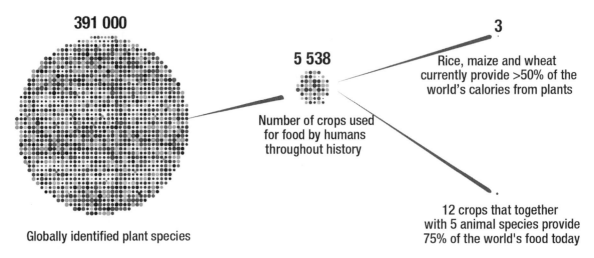

391 000

5 538

3

Rice, maize and wheat currently provide >50% of the world's calories from plants

Number of crops used for food by humans throughout history

12 crops that together with 5 animal species provide 75% of the world's food today

Globally identified plant species

Figure 2.2 The world's reliance on a narrow diversity of food crops. (From FAO, 1995)

Figure 2.3 Farmers and local communities continue to be custodians of most of the world's plant genetic resources. (Credit: Bioversity International/R. Vernoy)

animal species between the Old and the New World that was to follow is often referred to as the Columbian Exchange. The Columbian Exchange was not without its downside, contributing significantly to massive negative impacts on the health of indigenous populations and their social and economic systems including farming systems (Williams, 2017). Columbus himself discovered tobacco (*Nicotiana* spp.) on Cuba in 1492, but he also brought back to Europe corn, allspice and cotton, all of which were collected based on observation of local users. This illustrates another development in the mode of PGRFA exploitation, where the explorer searched systematically for any plants and natural products that could be of immediate or potential use to the colonial powers of the time. The explorer was searching for anything useful rather than a specific species to meet a specific need, and there was certainly no concept of ownership or sharing of any benefits arising from such exploitation with the custodians or communities that had managed these resources for countless generations.

A good example of this form of exploitation is the report of Thomas Heriot in 1588: *A brief and true report of the new found land of Virginia*. He studied the plants of Roanoak (now in North Carolina) and brought potentially useful samples to Britain for further study. Sir Francis Drake, in 1581, similarly brought back numerous samples for the botanist Clusius to study. Clusius had previously been to Spain in 1564 and made detailed descriptions of the plants he saw, noting their local uses. The movement of exploitable plant resources was not only to the home colonial countries, but the colonial powers also took plants from one region of the world to another; for example, the Portuguese took maize and cassava from America to Africa, and the Spanish took maize and sweet potatoes across the Pacific region.

Parallel to the birth of long-distance plant exploration and systematic exploitation came the establishment of the first Physick or botanical gardens in the 16th and 17th centuries to provide the plants required for medicine, as well as botanical curiosities. By 1545 there were gardens in Padua, Florence and Pisa; Leiden was established in 1593, Paris in 1635 and Edinburgh in 1690. The Dutch set up a garden in Cape Town in 1694, the French in Mauritius in 1733 and throughout the 18th century the British established gardens across their empire. However, the golden age of plant exploration was not until the 18th century, when people like Sir Joseph Banks (1743–1820), President of the Royal Society and Keeper of the gardens at Kew, sent out numerous expeditions looking for exotic and useful species. Banks himself sailed with Captain Cook in 1768, but he inspired other men, such as Francis Masson (1741–1806), to collect in South Africa, Madeira, the Canary Islands, the Azores, the West Indies and North America, where he was killed during an expedition.

Famously, David Nelson (died 1789) accompanied Captain Bligh on HMS Bounty's expedition to the Society Islands in the Pacific to collect breadfruit, which were to be used as a cheap food for plantation slaves in the West Indies. When in 1841 Sir William Hooker became the first official Director of the National (later Royal) Botanical Garden based at Kew, a further stimulus was provided for plant exploration, and numerous expeditions were launched to the four corners of the globe. These expeditions largely focused on collecting herbarium specimens, but Kew and some of the other gardens were also specifically interested in collecting living material for exploitation. Kew was instrumental, for example, in the transfer of rubber (*Hevea brasiliensis*) from Brazil via Kew to Malaysia, breadfruit (*Artocarpus altilis*) from the East to the West Indies, oil palm (*Elaeis guineensis*) from West Africa to Southeast Asia, and cocoa (*Theobroma cacao*) from Central America to West Africa.

Although most of the examples cited above are British, it is undoubtedly true that all the colonial powers of the time saw it as their God-given right to scour their territories across the globe looking for potential exploitable resources, while issues of ownership or compensation of local people for use of these resources did not arise. The Spanish crown, for example, made it clear that scientific enquiry, botanic gardens and economic interests were intimately linked. This utilitarian view of the conservation and use of plant genetic resources was driven by the need to improve agriculture at home and in the colonies, as well as providing oddities for inclusion in gardens of exotica.

2.2.4 Genetic Conservation

Commonly among these pre-19th century collectors, the plants collected and transferred *ex situ* for exploitation were only a few per species. The introduced plants had a limited genetic base that we now realize very much reduced their exploitation; for example, until recently the entire African oil palm industry was based on six original tree samples, which seriously limited their disease resistance and yield potential. Likewise, it was the cultivation of a single clonal variety of susceptible Andean potato that contributed to the outbreak in Europe of the pathogen *Phytophthora infestans*. This disease was to have its most devastating impact in Ireland, in the mid-19th century with the Great Famine, where the majority of the rural population depended on the crop as their staple food.

It was Nikolai Ivanovich Vavilov (1887–1943) who first realized the importance of collecting the full range of genetic diversity of crops and their wild relatives rather than collecting a single or few representatives of each species. Although any one accession may be successful or unsuccessful as an introduction to a foreign region, simply because one accession fails it does not follow that all accessions from that species would necessarily fail. The genetic diversity found in any species is a degree of magnitude greater than that found in any one single accession; therefore, the sampling of any one accession cannot hope to reflect the genetic diversity in that species and so the potential for exploitation of the whole species.

Vavilov's studies also led him to propose his 'centre of origin' of crop plants hypothesis. Vavilov led or instigated numerous expeditions to centres of origin or diversity (Figure 1.7) and between his 1916 expedition to the Pamirs and his last expedition to Central Asia in 1939, his travels covered five continents and he collected genetic diversity for most of the major crop species. By 1943, when he died, there were 200 000 collections of crop plants and their wild relatives at his institute in St Petersburg, the largest collection in the world at the time. His work inspired other collectors from the developed world, such as Harry V. Harlan (barley), Jack R. Harlan, Hermann Kuckuck (wheats) and Jack Hawkes (potatoes), to systematically collect crop and crop relative germplasm worldwide. These collectors, like Vavilov, amassed large national collections and their material was used in numerous attempts at crop introductions to foreign lands. In this context, it is interesting to note that agriculture in developed countries remains almost completely based on plant introductions. Less than a third of the crops produced in developing countries in Africa and the American continent, even those with extensive genetic resources of their own, are indigenous.

The predominant ethos of these collectors remained, however, exploitative and utilitarian. Genetic resources were being collected and conserved for their exploitation potential, and the benefit to developing countries was through the stream of superior cultivars released for exploitation in their agriculture or new drugs which they could purchase to fight long-standing diseases in these countries. There was still little discussion of the ownership of the original resource, and the rights and benefits due to the country that supplied the desirable gene or compensation of local people for continuing to grow the diversity of crops that could be effectively exploited by agricultural or pharmaceutical industries of the developed world.

2.2.5 Rise of Agriculture and Environmentalism

The introduction of new crop types, cultivars and animal breeds together with other agricultural innovations laid the foundation for the agricultural revolution and a significant leap in agricultural productivity, particularly in Great Britain, which allowed a growing urban population to be fed. This helped support the industrial revolution and economic growth during the 18th century (Murphy, 2007), a period that also saw many innovations in agricultural technology being increasingly protected by patents, a situation somewhat mirrored in the 20th century by the increased corporate control of plant genetic resources and seed systems. This period culminated in the emergence of the industrial agribusiness and agrifood model that dominates food production today. This model has done much to reduce crop diversity because of demands for uniformity in large-scale monocultural production systems and by global consumers (Murphy, 2007), which today is manifested in reduced diversity in our global food system (Khoury et al., 2014; IPES-Food, 2016). So much so that by the end of the 20th century it has been estimated that only a tenth of the variety of crops developed in the past continue to be cultivated, while many local landraces have been replaced by a small number of improved cultivars (Millstone and Lang, 2008; Jarvis et al., 2016; Massawe et al., 2016). These developments also contributed to substantial loss of large areas of pristine environment and wild biodiversity through agricultural and industrial expansion.

The Green Revolution in the 1960s paved the way for the large increases in the production of cereals that contributed significantly to feeding millions and saving many people from hunger and starvation, though it came at a cost (Murphy, 2007; Massawe *et al.*, 2016; Pingali, 2017). Prominent among such costs was a growing over-reliance on external chemical inputs in large monocrops. Given the reliance on a few improved cultivars, the Green Revolution also contributed to the loss of many traditional cultivars or farmer landraces (Jarvis *et al.*, 2016; Massawe *et al.*, 2016). In fact, there is evidence in India that promotion of improved cultivars as a result of the Green Revolution may have led to the emergence of certain micronutrient deficiencies by contributing to the displacement of traditional pulses, which were previously an important component of the diet. Growing awareness of the negative impacts of the industrial-scale model of agriculture on the environment, biodiversity loss and genetic erosion contributed to the emerging global debate around issues of environmental protection, biodiversity conservation, sustainability and sustainable development that began to gain impetus in the early 1970s and which had been galvanized by the publication of Rachael Carson's *Silent Spring* in 1962. Other important publications and events during this period included: the Club of Rome's *The Limits to Growth* published in 1972; the highly influential Brundtland Report *Our Common Future* published in 1987 and commissioned by the United Nations; and the UN Conference on the Human Environment held in Stockholm in 1972 (out of which the United Nations Environment Programme (UNEP) developed and which importantly drew attention to the need to strengthen global PGRFA conservation activities).

The emergence of international NGOs such as the WWF and the IUCN also occurred around this time. The publication of the *World Conservation Strategy* report by UNEP/IUCN/WWF in 1980 and their follow-up policy statement *Caring for the Earth* in 1991; the pivotal United Nations Conference on the Environment and Development (UNCED, the Rio Conference) in 1992 which gave rise to the Convention on Biological Diversity, and which was followed by Rio+10 in 2002 and most recently Rio+20

in 2012; and finally the *Millennium Ecosystem Assessment* (MEA) report which was completed in 2005 have all been important in shaping the global conservation and environment movement.

Around the same time in the 1970s, intergovernmental negotiations were underway facilitated by the FAO which would eventually lead to the Commission on Plant Genetic Resources for Food and Agriculture (CGRFA) being established in 1983 and the international ratification of the International Treaty on Plant Genetic Resources for Food and Agriculture (ITPGRFA) in 2001. Because of their importance for PGRFA conservation and use, the CBD, CGRFA and ITPGRFA will be revisited in detail in the following section.

What of the future? The debate still rages between the agriculture and biodiversity communities, and between those lobbying for a business as usual technology-driven agriculture, including the application of more biotechnology, as opposed to those promoting a low external input sustainable agriculture based on agroecological principles and practices, and which recognizes crop diversity as one of its fundamental pillars (Rosset and Altieri, 2017). Important questions and challenges remain as to how the world will feed an expected global human population of 9.6 billion by 2050 without the luxury of being able to expand to new lands, and how it might be able to do this in a world that is expected to be much changed as a result of climate change and where poverty persists.

2.3 The Political Context of PGRFA Exploitation and Conservation

2.3.1 International PGRFA Conservation

As mentioned earlier, the Green Revolution led to the development of numerous high-yielding cultivars of the world's major staple food crops, namely rice and wheat. These cultivars often replaced and led to the loss of existing traditional landraces, and inevitably contributed to the diversity of the crop's gene pool. The new improved cultivars were bred to be genetically homogenous whereas the landraces or old cultivars they replaced had evolved or were bred by traditional farmers over millennia to be genetically diverse to suit a range of agroecological and socio-economic conditions. The displacement of landraces with high-yielding cultivars led to some loss of genetic diversity. Although concern over genetic erosion had been initially raised in the 1920s, it was the impacts of the narrowing of the genetic base of crops and the establishment of the Food and Agriculture Organization of the United Nations (FAO) that led to international, concerted action being taken to rectify the situation.

In the 1950s the FAO began to take systematic action to promote conservation, exchange and utilization of PGRFA. FAO began the first international newsletter on crop genetic resources in 1957, which later became the Plant Genetic Resources (PGR) Newsletter, a landmark publication which has recently been re-launched under the aegis of the European Cooperative Programme of Plant Genetic Resources (www.ecpgr.cgiar.org). The FAO also began convening a series of technical meetings and conferences, which acted as a catalyst to promote public awareness of PGRFA issues and the scientific expertise associated with the conservation and use of PGRFA. Following the first technical meeting a panel of Experts on Crop Germplasm Exploration and Introduction was set up in 1965, with Sir Otto Frankel in the chair and Erna Bennett, R.O. Whyte, Jack Harlan, T.T. Chang, Jack Hawkes and others playing an important part. The FAO's Crop Ecology and Genetic Resources Unit started work in 1968 and FAO Technical Conferences on PGRFA were held in 1967, 1973, 1981 and more latterly in 1994. In 1972, the UN Conference on the Human Environment, held in Stockholm, gave the FAO responsibility for the establishment of an International Genetic Resources Programme.

In the 1960s and 1970s growing international concern over the loss of genetic diversity, especially among traditional landraces, focused effort into collecting landrace material and establishing base (to maintain long-term viability) and medium (shorter-term, such as breeders' collections for evaluation and use) storage facilities. Much of this work was spearheaded by the International Board for Plant Genetic Resources (IBPGR, now Bioversity International), which was originally established by the Consultative Group for International Agricultural Research (CGIAR) with a secretariat supplied by FAO in 1974. IBPGR was given the responsibility for and challenge of developing a world plant genetic resources network, with emphasis on food crops. To do this it commissioned PGRFA research, organized or led collection missions throughout the world, and promoted gene bank construction to effectively conserve the world's threatened crop and crop wild relative diversity. This initial work focused primarily on 'emergency missions' to collect crop germplasm in imminent threat of extinction. There was also a need to develop a more comprehensive and coordinated intergovernmental approach to PGR conservation and use. Therefore, the 1983 FAO conference established the intergovernmental FAO Commission on PGRFA as a global forum for plant genetic resource debate.

In 1993, IBPGR became an independent institute within the CGIAR, renamed as the International Plant Genetic Resources Institute (IPGRI), with a mandate to advance the conservation and use of plant genetic resources for the benefit of present and future generations. In 2006, IPGRI became Bioversity International, which continues to work in partnership with other organizations, to undertake research and training, and provide scientific and technical advice and information on PGRFA conservation and use. Though it has broadened of late, Bioversity International's mandate is:

To investigate and promote the use and conservation of agricultural biodiversity in order to achieve better nutrition, improve smallholders' livelihoods and enhance agricultural sustainability.

Many of the other CGIAR centres have active PGR programmes or units. Most recently these centres have come together in the CGIAR Gene bank Platform, led by the Crop Trust, which enables CGIAR gene banks to fulfil their legal obligation to conserve and make available accessions of crops and trees on behalf of the global community under the International Treaty on ITPGRFA.

Prior to this, Bioversity led the System-wide Genetic Resources Programme (SGRP). This programme also included all the CGIAR centres with genetic resource responsibilities and the FAO, and had as its mission:

Through co-ordination among centres of the CGIAR and in collaboration with partner organizations, the SGRP contributes to the global effort to conserve agricultural, forestry and aquatic genetic resources and promotes their use in ways that are consistent with the Convention on Biological Diversity. The SGRP seeks to advance research on policies, strategies and technologies for genetic resources, and to provide information, advice and training to its partners.

(IPGRI, 1997)

SGRP was initiated in 1995 and some of the main outputs were public awareness and research reports, on-going research and the System-wide Information Network for Genetic Resources (SINGER). Through SINGER, a meta-database of CGIAR centres' genetic resource holdings, information was supplied on provenance, characterization and distribution data to potential genetic resource users, so they could better assess which accessions they should request to enhance their utilization. Today, GENESYS, a portal on plant genetic resources, is being developed to improve information exchange about plant genetic resources and acts as a gateway from which germplasm accessions from gene banks around the world can be easily found and ordered. When it was launched in 2011, GENESYS was planned as a one-stop access point to information about one-third of the world's gene bank holdings, including not only information provided by SINGER but also two other genetic resource networks, the European Plant Genetic Resources Search Catalogue and the Genetic Resources Information Network of the United States Department of Agriculture.

One of the major steps forward taken by the establishment of IBPGR and the FAO Commission on PGRFA was a concerted international effort to address the problem of genetic erosion. For the first time, there was an international concerted effort to halt erosion and conserve and make available for use the broad range of plant alleles and genes. Following the success of the early 'emergency missions' to collect crop germplasm, more recently there has been a need to re-

target plant genetic resource conservation, not just on that plant diversity of most immediate use, but also on less obvious, less closely related crop plant diversity, to have sustainable utilization for future generations. The ongoing urgency of this issue has again been highlighted by recent studies on the conservation status of the diversity of crop wild relatives, which demonstrates they are currently poorly represented in the world's gene banks and that systematic efforts are needed to collect, conserve and make available crop wild relative (CWR) germplasm for future crop improvement (Castañeda-Álvarez et al., 2016a).

A second important contribution made by IBPGR and the FAO was to highlight the fact that plant genetic resources are a tangible resource which has a 'real' monetary value to those holding them. This is especially true since the availability of biotechnological techniques and the transfer of genes between accessions and species becomes more routine. As these tangible plant resources were largely concentrated in countries that were financially resource-poor, the 'free' transfer of their plant resources to developed countries for technological exploitation became less appealing to the country of origin and less morally acceptable for all. It is this realization of the 'real' monetary worth of PGRFA that underpins much of the current discussion of the ethical and policy context of biodiversity conservation and use, though there are also social and cultural issues that bear on this.

In the 1990s, as emphasized above, there was increasing international debate over the implications of the transfer of genetic resources across national boundaries. FAO sponsored one 'fair and equitable' solution. In 1989, the FAO Commission requested its Secretariat to prepare an International Code of Conduct for Plant Germplasm Collecting and Transfer (www.fao.org/nr/cgrfa/cgrfa-global/cgrfa-codes/en/). The code was intended to form an important tool in regulating the collecting and transfer of plant genetic resources and their associated information (including indigenous knowledge), with the aim of facilitating access to these resources and promoting their use and development on an equitable basis. The code was formulated after extensive international and regional discussion and was adopted by the Commission in

1993. The code, based on the principle of national sovereignty over plant genetic resources, set out standards and principles to be observed by those countries and institutions that adhered to it. The code was debated at numerous FAO and other international fora, but sufficient international agreement could not be reached to have the document adopted into international legislation. However, it was adopted as a voluntary code by many countries.

Parallel with the development of the Code of Conduct for Plant Germplasm Collecting and Transfer, the FAO Commission also worked on an International Undertaking (IU) on Plant Genetic Resources. The IU is a non-binding intergovernmental agreement to promote the conservation, exchange and utilization of plant genetic resources. The objective of the IU was expressed in Article 1, as follows:

The objective of this undertaking is to ensure that plant genetic resources of economic and/or social interest, particularly for agriculture, will be explored, preserved, evaluated and made available for plant breeding and scientific purposes. This undertaking is based on the universally accepted principle that plant genetic resources are a heritage of mankind and consequently should be available without restriction.

It was adopted by the 1994 Commission meeting and by January 1999, 113 countries had become signatories. However, there were conflicts between the IU and the CBD, most notably the stance over ownership of PGRFA; the IU retained the position of common heritage (ownership) for PGR, whereas the CBD clearly states the primacy of national sovereignty over PGRFA. There were attempts to harmonize IU with the CBD, but these were unresolved because of continued international political wrangling over issues such as scope and access, benefit sharing and farmers' rights.

Following in the footsteps of UNCED and the development of the CBD, but with a more explicitly PGRFA focus, came the Fourth FAO International Technical Conference on Plant Genetic Resources, held in Leipzig in June 1996. At the meeting 150 countries agreed an action plan to conserve PGRFA, and this was incorporated into the Global Plan of Action (GPA) for Conservation and Use of PGR. The Second Global Plan of Action was adopted by the FAO Council at its 143rd Session in 2011 and addresses new challenges, such as climate change and food insecurity, as well as novel opportunities, including information, communication and molecular methodologies. The Second GPA contains 18 priority activities organized in four main groups: *In situ* conservation and management; *Ex situ* conservation; Sustainable use; and Building sustainable institutional and human capacities (FAO, 2011b).

The Aims of the FAO Second Global Plan of Action for Conservation and Use of PGR are:

- to strengthen the implementation of the International Treaty;
- to ensure the conservation of PGRFA as a basis for food security, sustainable agriculture and poverty reduction by providing a foundation for current and future use;
- to promote sustainable use of PGRFA, in order to foster economic development and to reduce hunger and poverty, particularly in developing countries, as well as to provide options for adapting to and mitigating climate change, addressing other global changes and responding to food, feed and other needs;
- to promote the exchange of PGRFA and the fair and equitable sharing of the benefits arising from their use;
- to assist countries, as appropriate and subject to their national legislation, to take measures to protect and promote Farmers' Rights, as provided in Article 9 of the International Treaty;
- to assist countries, regions, the Governing Body of the International Treaty and other institutions responsible for conserving and using PGRFA to identify priorities for action;
- to create and strengthen national programmes, to increase regional and international cooperation, including research, education and training on the conservation and use of PGRFA, and to enhance institutional capacity;
- to promote information sharing on PGRFA among and within regions and countries;
- to set the conceptual bases for the development and adoption of national policies and legislation, as

appropriate, for the conservation and sustainable use of PGRFA;

- to reduce unintended and unnecessary duplication of actions in order to promote cost efficiency and effectiveness in global efforts to conserve and sustainably use PGRFA.

To achieve these aims the following priority activities were proposed:

In Situ Conservation and Management

1. Surveying and inventorying plant genetic resources for food and agriculture
2. Supporting on-farm management and improvement of plant genetic resources for food and agriculture
3. Assisting farmers in disaster situations to restore crop systems
4. Promoting in situ conservation and management of crop wild relatives and wild food plants

Ex Situ Conservation

5. Supporting targeted collecting of plant genetic resources for food and agriculture
6. Sustaining and expanding ex situ conservation of germplasm
7. Regenerating and multiplying ex situ accessions

Sustainable Use

8. Expanding the characterization, evaluation and further development of specific subsets of collections to facilitate use
9. Supporting plant breeding, genetic enhancement and base-broadening efforts
10. Promoting diversification of crop production and broadening crop diversity for sustainable agriculture
11. Promoting development and commercialization of all varieties, primarily farmers' varieties/landraces and underutilized species
12. Supporting seed production and distribution

Building Sustainable Institutional and Human Capacities

13. Building and strengthening national programmes
14. Promoting and strengthening networks for plant genetic resources for food and agriculture

15. Constructing and strengthening comprehensive information systems for plant genetic resources for food and agriculture
16. Developing and strengthening systems for monitoring and safeguarding genetic diversity and minimizing genetic erosion of plant genetic resources for food and agriculture
17. Building and strengthening human resource capacity
18. Promoting and strengthening public awareness of the importance of plant genetic resources for food and agriculture

Preparation for the Fourth Technical Conference also resulted in the First State of the World's Plant Genetic Resources for Food and Agriculture (SoWPGR-1) report in 1998, which discussed the value of agricultural biodiversity, threats to that diversity, what is currently conserved using ex situ and in situ techniques, how conserved accessions are used, institution building requirements, international collaboration, issues of access to conserved plant genetic diversity, and benefit sharing and farmers' rights, along with more technically based annexes. At its Twelfth Session, the now FAO Commission on Genetic Resources for Food and Agriculture (CGRFA) endorsed the second report of the SoWPGR which was published in 2010 (SoWPGR-2). A further State of the World Report was published in 2019 but with a much broader scope to highlight the main findings and gaps that need attention in relation to broader biodiversity for food and agriculture, as opposed to a specific focus on PGRFA.

The adoption, as binding, of the CBD (see next section) was viewed by many as a 'wake-up call' to the agriculture sector and the realization that the IU, because of its voluntary nature, lacked adequate weight to defend the sector's interests (Esquinas-Alcázar et al., 2011). This prompted a rethink of the IU into a more binding agreement which would accommodate equal cooperation with the environment and trade sectors and at the same time guarantee conservation and access to agriculturally important PGRFA for research and breeding through an equitable system for access and benefit sharing (Esquinas-Alcázar et al., 2011). Out of these new negotiations emerged the International Treaty on Plant Genetic Resources for Food and Agriculture (ITPGRFA), which was adopted by the FAO conference in 2001 and entered into force in 2004 and which thus far (2018) has been ratified by 139 contracting parties.

The Treaty provides national authorities with the legal framework to act for the conservation and sustainable use of their crop diversity and provides a mechanism that facilitates multilateral transfers of crop genetic materials and benefit sharing (Esquinas-Alcázar et al., 2011). The first session of the Governing Body of the Treaty resulted in the adoption of a Standard Material Transfer Agreement (SMTA), an important tool which determines the quantity, method and terms of payment related to commercialization of PGR that takes place through the Treaty's Multilateral System of access and benefit sharing (MLS, see Box 2.1). The fourth session of the Governing Body of the Treaty took place in 2011 in Bali, which saw consensus reached on compliance and financial rules and adoption of resolutions on the MLS, Farmers' Rights (Box 2.2), sustainable use, cooperation with other organizations and implementing of the Funding Strategy.

The list of food and forage crops covered under the MLS are listed in Annex 1 of the Treaty and in the year August 2007 to July 2008 almost half a million samples of Annex 1 genetic resources from the CGIAR centres alone were transferred using the SMTA, which represents more than 8500 samples weekly (Esquinas-Alcázar et al., 2011).

The Treaty has also facilitated the establishment of two important 'institutional innovations' that strengthen the world's ability to secure PGRFA long-term as well as providing a 'last resort safety back-up' facility for storage of globally important PGRFA. These are the Global Crop Diversity Trust (GCDT) and the Svalbard Global Seed Vault. An agreement between the GCDT and the Treaty was reached during the first session of the Governing Body of the Treaty (Hawtin and Fowler, 2011). The GCDT (Box 2.3), which is based in Bonn, has established an endowment fund which is used to provide financial support for the maintenance of eligible and priority collections of PGRFA worldwide. The Trust provides support for long-term maintenance for collections

including of aroids, banana, barley, bean, cassava, faba bean, forages, grass pea, pearl millet, rice, sorghum, wheat and yam.

The completion of Svalbard Global Seed Vault in 2008 at a cost of US$9 million provided by the Norwegian Government aims to secure long-term plant genetic resources against major threats to future food security. Built into the permafrost of the island of Svalbard inside the Arctic Circle, the Vault aims to provide insurance against the catastrophic loss of crop diversity held in conventional gene banks around the world. The ultimate aim of the Vault is to safeguard as much of the world's unique crop genetic material as possible, while also avoiding unnecessary duplication. As of the end of 2018, the Vault housed close to one million seed samples, deposited by most countries of the world.

2.3.2 The Convention on Biological Diversity (CBD)

On 22 May 1992 in Nairobi the international community adopted the global Convention on Biological Diversity, and on 5 June 1992 at the UN Conference on Environment and Development, held in Rio de Janeiro, 150 countries signed. Eighteen months later, on 29 December 1993, the Convention entered into force, which was 90 days after the 30th ratification. The Convention was a landmark in many ways, notably because it drew so many Heads of State to the conference, but also because it linked for the first time in the international arena the conservation of the world's biological diversity with sustainable exploitation and the equitable sharing of any profits that might arise from their exploitation in a legally binding manner. It brought to centre stage such issues as access to genetic resources, ownership of biological resources, profit sharing, technology transfer and financial mechanisms, but, most importantly, it acknowledged that conservation of biological diversity must be tied to sustainable development, particularly in areas rich in biodiversity but poor in financial resources.

Although the widespread loss of biodiversity and genetic erosion were first noted in the 1920s and subsequently there were increasing systematic

Box 2.1 | **The Basics of the Multilateral System of the International Treaty**

(ITPGRFA: www.fao.org/plant-treaty/areas-of-work/the-multilateral-system/overview/en/)

The Multilateral System (MLS) is the International Treaty's innovative solution to access and benefit sharing whereby 64 of our most important crops – crops that together account for 80% of all human consumption – will comprise a pool of genetic resources that are accessible to everyone. On ratifying the International Treaty, countries agree to make their genetic diversity and related information about the crops stored in their gene banks available to all through the Multilateral System (MLS). This gives scientific institutions and private sector plant breeders the opportunity to work with, and potentially to improve, the materials stored in gene banks or even crops growing in fields. By facilitating research, innovation and exchange of information without restrictions, this cuts down on the costly and time-consuming need for breeders to negotiate contracts with individual gene banks. The Multilateral System sets up opportunities for developed countries with technical know-how to use their laboratories to build on what the farmers in developing countries have accomplished in their fields. Access to genetic materials is through the collections in the world's gene banks. These can include collections of local seeds kept in small refrigeration units of research labs, national seed collections housed in government ministries or research centre collections that contain all known cultivars or landraces of a crop from around the world. Under the International Treaty and its Multilateral System, collections of local, national and international gene banks that are in the public domain and under the direct control of contracting parties share a set of efficient rules of facilitated access. This includes the vast collections of the CGIAR, a global partnership of 15 international agricultural research centres. In addition, the Global Crop Diversity Trust, a complementary funding mechanism to that established by the Treaty, is committed to raise the funds that will endow the gene banks and ensure their continued viability. Those who access genetic materials through the Multilateral System agree that they will freely share any new developments with others for further research or, if they want to keep the developments to themselves, they agree to pay a percentage of any commercial benefits they derive from their research into a common fund to support conservation and further development of agriculture in the developing world. The fund was established in 2008.

attempts to conserve and sustainably use plant diversity, it was in the early 1980s that biologists more specifically put forward the idea that there was a need for an international convention concerned with the conservation and use of biodiversity. The UN General Assembly recommended in 1984 and 1987 that such a convention should be enacted, and IUCN was given the responsibility of drafting appropriate articles. From the beginning the conservation focus was seen to be at the genetic, species and ecosystem levels and would include both *in situ* and *ex situ* activities.

UNEP established an Ad Hoc Working Group of Experts on Biological Diversity in November 1988 to explore the need for an international convention on biological diversity. Soon after, in May 1989, it

---

Box 2.2 | Farmers' Rights and the International Treaty

(ITPGRFA: www.fao.org/plant-treaty/areas-of-work/farmers-rights/en/)

Farmer' Rights are an attempt to acknowledge and affirm the past, present and future contributions of farmers in all regions of the world, particularly those in centres of origin and diversity, in maintaining, improving and making available plant genetic resources for humanity. Farmers' Rights also seek to support and enable farmers to continue to make this contribution. Farmers' Rights are critical to ensuring the conservation and sustainable use of plant genetic resources for food and agriculture and consequently for food security – today and in the future. In its Article 9, the International Treaty recognizes the enormous contribution that local and indigenous communities and farmers of all regions of the world, particularly those in the centres of origin and crop diversity, have made and will continue to make for the conservation and development of plant genetic resources which constitute the basis of food and agriculture production throughout the world. It gives governments the responsibility for implementing Farmers' Rights, and lists measures that could be taken to protect, promote and realize these rights including:

- the protection of traditional knowledge relevant to plant genetic resources for food and agriculture;
- the right to equitably participate in sharing benefits arising from the utilization of plant genetic resources for food and agriculture;
- the right to participate in making decisions, at the national level, on matters related to the conservation and sustainable use of plant genetic resources for food and agriculture; and
- the right that farmers have to save, use, exchange and sell farm-saved seed/propagating material subject to national law and as appropriate.

---

established the Ad Hoc Working Group of Technical and Legal Experts to prepare an international legal instrument for the conservation and sustainable use of biological diversity. The experts were to take into account 'the need to share costs and benefits between developed and developing countries' as well as 'ways and means to support innovation by local people'. The former working group concluded that existing conventions did not effectively cover all contemporary aspects of conservation and that there was a need for a new consolidating convention. By 1990 IUCN and FAO had already drafted potential articles and UNEP had commissioned several feasibility studies. The first full draft of the convention was originally discussed in February 1991 by the Intergovernmental Negotiating Committee for a Convention on Biological Diversity. The final negotiating meeting of this committee met in Nairobi in May 1992 with a deadline that negotiations must be completed by 22 May, and the final Convention on Biological Diversity (CBD) document was finalized on that day.

The Convention affirms that the conservation and use of the world's biodiversity is a common concern of humanity and, as is stated above, Article 1 of the CBD establishes the three pivotal concerns:

- conservation of biological diversity;
- sustainable use of its components;

Box 2.3 │ **The Global Crop Diversity Trust**

(Hawtin and Fowler, 2011)

The objective of the GCDT is to ensure the long-term conservation and availability of PGRFA, with a view to achieving global food security and sustainable agriculture. More specifically, the Trust aims:

- to safeguard collections of unique and valuable plant genetic resources for food and agriculture held *ex situ*, with priority given to those that are plant genetic resources included in Annex 1 of the Treaty;
- to promote an efficient goal-oriented, economically efficient and sustainable global system of *ex situ* conservation in accordance with the Treaty and the Global Plan of Action;
- to promote the regeneration, characterization, documentation and evaluation of PGRFA and the exchange of related information;
- to promote the availability of PGRFA;
- to promote national and regional capacity building.

- fair and equitable sharing of the benefits arising out of the utilization of genetic resources.

The Convention recognizes the sovereign rights that each country has over the biological resources found within that country, but also underlines the responsibility that each country has to protect and use sustainably its native biological resources. The Convention requires the creation of a system of protected areas, national conservation strategies and appropriate environmental legislation, as well as the identification, regulation and management of activities that are likely to be deleterious to biological diversity.

The CBD operates through a governing body known as the Conference of the Parties (COP). Countries that have ratified the CBD (Parties) meet every 2 years at the COP and take decisions to guide its development and to review implementation. It is a large event, attracting several thousand delegates, and the most recent COP14 was held in Sharm el Sheikh, Egypt in 2018. As well as representatives of all the Parties, it is attended by a wide range of observers including intergovernmental organizations, non-governmental organizations, researchers, the private sector and indigenous Peoples

and their representatives. Decisions of the COP are guided by a second body known as the Subsidiary Body on Scientific, Technical and Technological Advice (SBSTTA). This body is made up of technical experts and scientists from member countries, and also meets every 2 years, between the COP meetings, and makes a series of recommendations for the upcoming COP meetings. An additional body, the Subsidiary Body on Implementation (SBI), reviews progress in implementation, key actions to enhance implementation and operations of the convention and protocols.

The Secretariat of the CBD, based in Montreal, Canada, provides all the administrative support. The COP at any one time has ad hoc working groups to address specific issues, such as access and benefit sharing, and traditional knowledge. The CBD also administers a clearing-house mechanism (CHM) that promotes technical and scientific cooperation between Parties and encourages the exchange of information on biodiversity. Funding for implementation of the CBD is through the Global Environment Facility (GEF), and funds are made available through pledges from donor countries and made available largely to developing countries for implementation of programmes and projects.

Each country is expected to set up a CBD National Focal Point through which information on national CBD implementation can be obtained. To operationalize the CBD at the national level, Governments are required to develop National Biodiversity Strategies and Action Plans (NBSAPs) which should be based on comprehensive surveys of biodiversity and should establish milestones and targets for action including actions and recommendations to target relevant national laws and regulations. For instance, many countries are enacting national or regional legislation to control access to their genetic resources and traditional knowledge, and to ensure that benefits are shared fairly and equitably in return. Although coordinated in most instances by the environment sector, the objective of NBSAPs is to mainstream biodiversity conservation and sustainable use into decision-making across all sectors impacting on the environment including planning, education, forestry, agriculture, fisheries, energy and transport, and more recently health. Each Party is required to report back on progress in implementation of NBSAPs to the COP on their efforts to implement the CBD, as well as submit regular national reports.

To facilitate the implementation of actions to implement its three main objectives, the CBD has established seven thematic programmes of work including: biodiversity of inland waters; biodiversity of dry and sub-humid lands; forest biodiversity; marine and coastal biodiversity; mountain biodiversity; island biodiversity; and, of most importance to genetic resources, on agricultural biodiversity (Box 2.4). These are complimented by a number of cross-cutting issues dealing with traditional knowledge, climate change and biodiversity, biosafety and plant conservation. A 2008 review of the Programme of Work on Agricultural Biodiversity highlighted that the programme is relevant for achieving the objectives of the CBD, and to addressing emerging issues such as climate change, and that major progress had been made in consolidating intergovernmental agendas on agricultural biodiversity.

Working in collaboration with national and international organizations, Parties and NGOs, the CBD developed a Global Strategy for Plant Conservation (GSPC) which was adopted at COP6 in 2002 (see Chapter 1). In 2010 at COP10 the COP adopted the updated Global Strategy for Plant Conservation 2011–2020. The Strategy's vision is to:

halt the continuing loss of plant diversity and to secure a positive, sustainable future where human activities support the diversity of plant life (including the endurance of plant genetic diversity, survival of plant species and communities and their associated habitats and ecological associations), and where in turn the diversity of plants support and improve our livelihoods and well-being.

To achieve this vision, the Strategy has 16 global targets set for 2020 and outlines a framework to facilitate collaboration and cooperation between existing initiatives aimed at plant conservation. Many of these targets are also important for PGRFA. An in-depth review of the GSPC has been carried out to review progress between 2002 and 2008 and has been published as the Plant Conservation Report 2014.

In 2010, the International Year of Biodiversity, the 193 Parties to the CBD and their partners came together for COP10 in Japan and adopted a package of important measures that included the Strategic Plan for Biodiversity 2011–2020 together with its 20 Aichi Biodiversity Targets, as well as the Nagoya Protocol (see Box 2.5) on access and benefit sharing of genetic resources. A long time in the making, the objective of the Nagoya Protocol is:

the fair and equitable sharing of the benefits arising from the utilization of genetic resources, including appropriate access to genetic resources and by appropriate transfer of relevant technologies, taking into account all rights over those resources and to technologies, and by appropriate funding, thereby contributing to the conservation of biological diversity and the sustainable use of its components.

As already mentioned in Chapter 1, Aichi Biodiversity Target 13 is particularly relevant to PGRFA and states:

Box 2.4 | **Programme of Work on Agricultural Biodiversity**

(CBD: www.cbd.int/agro/pow.shtml)

The CBD's Programme of Work (PoW) on Agricultural Biodiversity recognizes the significant contributions of farmers and indigenous communities to the conservation and sustainable use of agricultural biodiversity. The PoW aims to promote:

- the positive effects and mitigate the negative impacts of agricultural practices on biodiversity in agro-ecosystems and their interface with other ecosystems;
- the conservation and sustainable use of genetic resources of value for food and agriculture;
- the fair and equitable sharing of benefits arising out of the utilization of genetic resources.

The PoW is based on four elements:

- assessing the status and trends of the world's agricultural biodiversity, their underlying causes and knowledge of management practices;
- identifying adaptive management techniques, practices and policies;
- building capacity, increasing awareness and promoting responsible action;
- mainstreaming national plans and strategies for the conservation and sustainable use of agricultural biodiversity.

Box 2.5 | **The Nagoya Protocol**

(CBD: www.cbd.int/abs/about/default.shtml/)

The Protocol significantly advances the Convention on Biological Diversity's third objective by providing a strong basis for greater legal certainty and transparency for both providers and users of genetic resources. Specific obligations to support compliance with domestic legislation or regulatory requirements of the Party providing genetic resources and contractual obligations reflected in mutually agreed terms are a significant innovation of the Protocol. These compliance provisions as well as provisions establishing more predictable conditions for access to genetic resources will contribute to ensuring the sharing of benefits when genetic resources leave a Party providing genetic resources. In addition, the Protocol's provisions on access to traditional knowledge held by indigenous and local communities when it is associated with genetic resources will strengthen the ability of these communities to benefit from the use of their knowledge, innovations and practices. By promoting the use of genetic resources and associated traditional knowledge, and by strengthening the opportunities for fair and equitable sharing of benefits from their use, the Protocol will create incentives to conserve biological diversity, sustainably use its components and further enhance the contribution of biological diversity to sustainable development and human well-being.

By 2020, the loss of genetic diversity of cultivated plants and farmed and domesticated animals and of wild relatives, including other socio-economically as well as culturally valuable species is maintained and strategies have been developed and implemented for minimizing genetic erosion and safeguarding their genetic diversity.

Glowka *et al.* (1994) highlight four major issues that are addressed by the Convention, and these are summarized in Box 2.6.

2.3.3 Multilateral and Bilateral PGRFA Collaboration

The 1970s onwards saw the development of numerous multilateral and bilateral PGR collaborations and networks with conservation and use of PGRFA as a focus. The GPA, CBD and ITPGRFA have all given added impetus to greater international cooperation. These partnerships provide links within and between the developing and developed countries. Numerous important projects have been initiated, a selection of which are highlighted here to illustrate the kinds of multilateral and bilateral links that have been successful.

PGR Collaboration between Developing Countries

In the 1980s the multinational Southern African Development Community (SADC), which at that time involved representation from Angola, Botswana, Lesotho, Malawi, Mozambique, Swaziland, Tanzania, Zambia and Zimbabwe (thereafter they have been joined by South Africa, Mauritius and the Seychelles), held a Conference for PGR and noted the urgent need to increase per capita agricultural production and conserve their native PGRFA on which the region's plant breeding industry was based.

IBPGR hosted a consultation meeting in September 1986, and out of this came a 20-year collaborative PGRFA programme, which has been sponsored by the countries in the region and Sida (Swedish International Development Cooperation Agency). The programme established a regional Plant Genetic Resources Centre in Lusaka, Zambia, which holds the base PGRFA collection for the entire region and also assists to coordinate the inventory, collection,

characterization, evaluation, rejuvenation and multiplication of indigenous PGRFA material of the SADC member states. Each country individually established a National Plant Genetic Resources Centre, which is responsible for national inventory, collection, characterization, evaluation, rejuvenation, multiplication and *in situ* conservation activities. The project involved a strong education and training element to ensure there were enough expert personnel in the region to undertake the activities listed above.

The SADC PGR Programme illustrates collaboration between formal sector workers in developing countries. An example of informal sector (non-governmental organization – NGO) collaboration is provided by the Kenya Energy and Environment Organization (KENGO) approach to PGRFA conservation and use. KENGO is a coalition of women's groups, farmers' organizations and other local NGOs involved in environmental conservation, wood-energy use and community development in Kenya. It was founded in 1981 following the UN Conference on New and Renewable Energy held in Nairobi. Through its Seeds and Genetic Resources project, it promotes the conservation and utilization of indigenous plants for their economic usefulness as food, fibres, dyes, fuelwood, fodder and medicines. It attempts to raise awareness among decision-makers and the public about PGRFA issues and the need to halt genetic erosion of traditional landrace material. It is systematically collecting and publishing ethnobotanical data and voucher specimens of indigenous trees in arid and semi-arid regions of Kenya.

The PGRFA awareness campaign and the various KENGO publications led to an explosion of interest in indigenous tree planting and use, which in turn led to research on how best to obtain, plant, harvest and use indigenous trees, as well as the development of curricula, teaching materials and workshops which provide grassroots instruction in PGRFA conservation and use. KENGO have also been active in lobbying government on environmental protection issues, for instance, they were instrumental in the controversial debate on the draining of an important wetland area, the Yala swamp. The project continues to procure and distribute seed from *in situ* conservation stands of

Box 2.6 | **The Four Primary Issues Addressed by the Convention on Biological Diversity**

(Glowka *et al.*, 1994)

- **National sovereignty and the common concern of humanity** – The *status quo* as regards ownership was that biodiversity was the 'common heritage', but this concept is rejected as biodiversity is encountered in areas of national sovereignty. Ownership of biodiversity is firmly placed in the hands of the countries in which that biodiversity is located. It is further recognized that States have the sovereign right to exploit their own resources in the light of their own environmental policies (Articles 3 and 15). However, the CBD emphasizes that conservation of biodiversity is a 'common concern' of all humanity and that the responsibility for conservation is also that of the State possessing the biodiversity (Articles 6, 8 and 10).
- **Conservation and sustainable use** – The CBD places the obligation on States to develop national strategies and plans for the conservation of biodiversity, identify important components of biodiversity and prioritize conservation action, as well as integrating conservation with sustainable use and identifying and monitoring the main threats to efficient conservation and use (Articles 6, 7 and 10). *In situ* conservation is seen as being critical and there is a need to identify, establish, monitor, and maintain viable populations of species in their natural surroundings (Article 8). *Ex situ* conservation is seen primarily as a means of complementing *in situ* (Article 9). Sustainable use is referred to in several articles but primarily in Article 10. The importance of indigenous knowledge held by local people in conserving and using biodiversity is acknowledged and the importance of maintaining their knowledge, but any exploitation of their knowledge should be linked to the equitable sharing of benefits (Articles 8 and 10). Measures to encourage research and training are emphasized in Article 12, and public education and awareness in Article 13.
- **The access issue** – Access may be subdivided into three specific issues: access to genetic resources, access to technology (including biotechnology) and access for donor states to the profits of exploitation. Until the implementation of the CBD, the principle of free access to PGR prevailed and this was recognized in the FAO Undertaking on PGR of 1983. However, as countries realized the financial benefits that were accruing from the exploitation of PGR samples, so access was beginning to be denied and there was a strong movement toward sovereign rights, which is recognized in Article 15. Access to PGR is often closely tied to bilateral negotiation over the fair and equitable sharing of any benefits that might accrue from its exploitation (Article 15.7). The obligations under Articles 15, 16 and 19 for there to be fair and equitable sharing of benefits may, however, prove difficult to administer in practice as it may be difficult to identify the original provenance of materials, especially as in many cases there may be no financial benefit or that it may take decades before real benefits are seen. There is also an obligation in Article 15(3) for technology transfer, but meeting this objective may be in direct conflict with the

Box 2.6 (cont.)

implementation of intellectual property rights in developed countries. It will certainly prove interesting to see how these issues are resolved in practice.

- **Funding** – How is the work involved within the CBD to be funded? The unwritten principle underlying the Convention is that developed countries will provide the bulk of the resources required in the form of some kind of international fund. The funding from the developed countries is to be new and additional (Article 20) and each developing country is to negotiate bilaterally with the international fund for an appropriate level of funding. The detail of the financial mechanism is set out in Article 21, and Article 39 names the Global Environmental Facility as the institution operating the financial mechanism.

native trees and, in 1992, had distributed about 700 kg of seed to 400 users, including schools, local and foreign NGOs, government ministries, individual farmers and research institutes. They have also trained 40 seed collectors and conservationists who now form an informal PGR network throughout Kenya. KENGO has also established a field gene bank on 15 acres of land donated by the Jomo Kenyatta University College of Agriculture and now has 2900 specimens of 86 forest and fruit tree species. Food quality trials have shown that indigenous species, such as *Adansonia digitata* (Baobab) and *Gynandropsis gynandra*, have higher nutritional qualities than exotic vegetables, such as cabbage or kale. The project has helped restore confidence that indigenous knowledge and PGRFA are of value to Kenya and that under appropriate conditions they can out-perform exotic species and germplasm.

The Pacific region provides an additional example of a collaboration between countries on issues related to PGRFA. Faced by constraints related to isolation, high costs, poor infrastructure and limited technical capacity, a highly effective system of regional cooperation has been put in place following the devastating outbreak of taro leaf blight in Samoa in 1993, a disease which has caused significant loss of susceptible taro landraces in the region. To facilitate conservation, utilization and capacity building a regional gene bank, the Centre for Pacific Crops and

Trees (CePaCT), has been established at the Secretariat of the Pacific Community (SPC) in Fiji. This is closely linked to the Pacific Agricultural Plant Genetic Resources Network (PAPGREN) established in 2001, which brings regional countries together in the spirit of cooperation to enhance PGRFA conservation and use. In 2009, SPC placed the yam and taro collections of CePaCT under the scope of the ITPGRFA.

PGR Collaboration between Developed and Developing Countries

It has been estimated that more than 60% of plant species are found in developing countries and of these at least 35 000 have potential medical value. By 2000, it was estimated that the medical value of these species could range between $35 and $47 billion per year (ten Kate and Laird, 1999), so the value of plants in medical and monetary terms is vast. In 1991, Merck, the largest pharmaceutical company in the world, announced they had agreed a 2-year biodiversity exploitation deal with InBio, a not-for-profit NGO in Costa Rica. As part of the deal, Merck would pay US$1.135 million per annum for research funding for InBio and in return receive biodiversity exploitation rights in Costa Rica. InBio in return would supply Merck with 10 000 biodiversity samples from Costa Rica's national parks. Merck would screen these samples for active pharmaceutically exploitable constituents. Any drugs resulting from this screening would be solely owned by Merck, but an undisclosed percentage of the royalties would be repatriated to Costa Rica via InBio. InBio used 10% of their income directly for biodiversity conservation. Subsequently, InBio have established similar deals with several other American, German and Italian pharmaceutical companies. This deal was ground-breaking in the sense that it was the first major example of a large multinational company from a developed country buying rights to the genetic resources of a biodiversity-rich, but resource-poor, developing country.

Views of whether this type of deal is beneficial for Costa Rica differ. Some argued that Costa Rica was selling its birth right cheap. Merck's annual turnover is significantly more than the gross national product of Costa Rica. The average cost of collecting crop germplasm is US$400 per sample. For wild species this would likely be higher because of the added problems of species recognition, species location and transfer of the collected material out of the tropical forest environment to the screening site. Merck were effectively paying US$113 per sample and so making a huge saving on accession acquisition alone, without considering their exploitation rights in perpetuity over the material collected.

The alternative view would be that Costa Rica was far sighted. It realized it had a saleable resource sufficiently early to broker the first large-scale deal with a multinational company. This deal resulted in a significant initial exploration payment, plus free training for national para-taxonomists and other scientists, the establishment of screening laboratories in Costa Rica and the extra income generated if new drugs resulted from the deal. If only 10 new drugs result from the deal, Costa Rica will earn more per year from these royalties than from its entire coffee or banana exports. Whatever way one views the Merck/InBio Agreement, it does illustrate the continuing significance of bilateral deals in an international forum. Those countries not bound by international conventions (such as the USA and their self-exclusion from the CBD) will always be free to strike bilateral resource exploitation deals divergent from, or even in opposition to, international agreement. It is also true

that there are not sufficient large multinational pharmaceutical companies, like Merck, for all the countries that are rich in biodiversity, but poor in financial resources, to establish such biodiversity for cash deals.

PGR Collaboration between Developed Countries

A successful example of a linkage between developed countries is provided by the European Cooperative Programme for Plant Genetic Resources Networks (ECPGR; www.ecpgr.cgiar.org). The programme was originally established as a UNDP/FAO project in 1980 following the recommendation of the European Association for Research on Plant Breeding (EUCARPIA). The objectives of the programme are:

first and foremost, to ensure the long-term conservation and to facilitate and encourage the increased utilisation of plant genetic resources in Europe; to increase the planning of joint activities; strengthen links between east and west European plant genetic resources programmes; to develop joint project proposals to be submitted to funding agencies; to contribute to monitoring the safety of plant genetic resources collections and take appropriate action when required, and to increase public awareness, at all levels, of the importance of plant genetic resources activities.

In 1982 IBPGR was requested by the network to provide a secretariat for the programme. Originally, the project was 50% funded by the European countries and 50% by UN agencies, but since 1986, the project has been entirely funded from member countries and the European Commission. The project has passed through various phases; Phase IX commenced in January 2014. The programme operates through 10 broadly focused networks, with either a crop (cereals, forages, vegetables, grain legumes, fruit, minor crops, industrial crops and potatoes) or general theme (documentation and information, *in situ* CWR conservation, on-farm conservation and technical cooperation) related to PGRFA in Europe. These groups meet regularly and agree unified work plans, which have involved building databases, developing core collections, joint project proposals and research, arranging safety duplication, and collecting, as well as many other activities. However, the principal goal of these activities remains to enhance long-term conservation and use of PGRFA in Europe, while increasing public awareness for policy-makers and the public of the necessity for these activities. Many of the recommendations that are included in the work plan relate to the implementation of the Global Plan of Action for PGR in Europe. The success of ECPGR stimulated European foresters to establish a sister programme, the European Forest Genetic Resources Programme (EUFORGEN), which was established in 1994, wholly funded by contributing countries and with the goal of protecting forests in Europe.

Crop-Specific Collaborations and Thematic Networks

A large number of international crop-specific networks operate globally or regionally and usually have components dealing with conservation of PGRFA, crop improvement, evaluation and characterization of PGRFA, sharing of information, public awareness and capacity building. Some examples include: the International Network for Bamboo and Rattan (INBAR) established in 1997; CacaoNet launched in 2006; the International Network for the Improvement of Banana and Plantain (INIBAP) established in 1985 (and merged in the mid-1990s into today's Bioversity International); the public–private partnership Fondo Latinoamericano para Arroz de Riego (FLAR) established in 1995; the Latin American/Caribbean Consortium on Cassava Research and Development (CLAYUCA) established in 1999; the Taro Genetic Resources: Conservation and Use (TaroGen) project established in 1998; and the International Network for Edible Aroids established in 2011. The role of NGOs in PGR networking has also increased considerably in recent years. A good example is Community Biodiversity Development Conservation (CBDC), a global initiative, which has brought together government and NGO actors from several countries in Africa, Asia and Latin America at local, regional and global levels to focus on strengthening the work of farming communities to

conserve and develop PGR for food security and livelihoods.

International Plant Conservation

Parallel to the establishment of Bioversity International, the FAO CGRFA, the GCDT and events leading up to the development of the ITPGRFA, which specifically focus on the genetic conservation and use of plant diversity, several other organizations were established with a broader more ecologically based plant conservation focus, both within the formal and informal sectors. These are far too numerous to list them all, but some background details for a few key conservation organizations follow.

World Wide Fund for Nature (WWF; www .worldwildlife.org) started in 1961 has become the world's largest and most respected independent conservation organization with offices in around 100 countries, currently running more than 1300 conservation projects. Most of these focus on local issues ranging from school nature gardens in Zambia to initiatives that appear on packaging in local supermarkets. With more than five million supporters distributed throughout five continents, WWF can safely claim to have played a major role in the evolution of the international conservation movement. WWF's aim is to conserve nature and ecological processes by:

- preserving genetic, species, and ecosystem diversity;
- ensuring that the use of renewable natural resources is sustainable both now and in the longer term;
- promoting actions to reduce pollution and the wasteful exploitation and consumption of energy.

WWF has undertaken work on plant genetic resources and conservation including a recent report, *Food Stores: Using Protected Areas to Secure Crop Genetic Diversity* (Dudley *et al.*, 2006), which analyzed the protection status of ecoregions identified by WWF and which are important for the conservation of genetic diversity.

The World Conservation Union (IUCN; www.iucn .org) is one of the world's oldest international conservation organizations. It was originally established in Fontainebleau, France, on 5 October 1948 as the 'International Union for the Protection of Nature'. Today it is a union of governments, government agencies and non-governmental organizations working at the field and policy levels, together with scientists and experts, to protect nature. IUCN is composed of three interconnected strands. One strand is the Union's membership. The second strand is made up of the commissions of expert volunteers that work with IUCN, e.g. Species Survival Commission. The third strand is made up of Secretariat staff that work in IUCN's offices throughout the world. IUCN works with international networks of volunteer experts grouped together in six global commissions:

- The Species Survival Commission
- The World Commission on Protected Areas
- The Commission on Education and Communication
- The Commission on Environmental Law
- The Commission on Ecosystem Management
- The Commission on Environmental, Economic and Social Policy

The Commissions' total membership is composed of more than 8000 technical, scientific and policy experts. IUCN has carried out a considerable body of work in areas related to plant conservation including publications of plant species red lists. The Species Survival Commission operates many Specialist Groups for plants, with the Crop Wild Relative Specialist Group (www.cwrsg.org) being of particular interest for PGRFA.

United Nations Environment Programme (UNEP; www.unenvironment.org/) located in Nairobi, Kenya, was established as the environmental agency of the United Nations following the 1972 Stockholm Conference on the Human Environment. UNEP provides an integrative and interactive mechanism through which it enables efforts by intergovernmental, non-governmental, national and regional bodies in the service of the environment that are reinforced and interrelated. In recent years UNEP has promoted the potential of genetic resources to contribute to overall biodiversity maintenance and ecosystem function as well as to improve nutrition,

increase food security and enhance well-being in rural communities. UNEP over the last 10 years, with support from the GEF, has been assisting 57 countries in mainstreaming genetic diversity conservation and sustainable use in the agriculture production sector. These initiatives are described in the UNEP publication, *Mainstreaming Biodiversity in Production Landscapes* (Mijatovic *et al.*, 2018).

UNEP World Conservation Monitoring Centre (WCMC; www.unep-wcmc.org) is internationally recognized as a centre for the location and management of information on the conservation and sustainable use of the world's living resources. It was created in 1988 by IUCN, WWF and UNEP, and provides information services on conservation and sustainable use of the world's living resources to organizations ranging from UN agencies to multinational corporations. WCMC collates data, maps and statistics on forests, coasts, species, protected areas and national biodiversity with a bibliographic database, and provides access to information on the status, value and management of biological diversity. It also supports directly or indirectly the activities of several international convention and programme secretariats and assists in capacity building and training programmes.

Botanic Gardens Conservation International (BGCI: www.bgci.org) is another leading international conservation organization, which was originally established by IUCN in 1987 as the Botanic Gardens Conservation Secretariat. BGCI exists to ensure the world-wide conservation of threatened plants and represents more than 500 members, largely botanic gardens, in 115 countries. To do this they support the development and implementation of global policy largely through working closely with the Global Strategy for Plant Conservation (GSPC), and their publications include *A Guide to the Global Strategy for Plant Conservation*. BGCI, together with the above three organizations, is part of the Global Partnership for Plant Conservation which brings together international, regional and national organizations to contribute to the implementation of the Global Strategy for Plant Conservation (see Chapter 13). The Global Partnership also includes People and Plants International, Society for Economic Botany, Fauna and Flora International, the Global Biodiversity Informational Facility and Plantlife International.

Indigenous Peoples' Biodiversity Network (IPBN; www.povertyandconservation.info/en/org/o0147) is a global coalition of around 30 Indigenous Peoples groups and comprises a network of indigenous scientists and conservation practitioners, activists, lawyers and community educators. IPBN has provided a forum for negotiations and discussions among Indigenous Peoples concerning relevant issues and opportunities within the Convention on Biological Diversity for promoting, preserving and protecting their rights to manage, control and benefit from their own knowledge and resources. IPBN also plays an important advocacy role in creating awareness among governments, multilateral agencies and non-governmental organizations about the links between cultural and biological diversity and the need to address Indigenous Peoples' rights if the Convention is to succeed in its goals.

The Indigenous Peoples' and Community Conserved Territories and Areas (ICCA; www.iccaconsortium .org) Consortium is a global association dedicated to promoting the appropriate recognition of and support to indigenous territories and areas conserved by Indigenous Peoples and local communities. It comprises Indigenous Peoples' Organizations, community-based organizations and civil society organizations. It is dedicated to conserving critical ecosystems, many of which include priority and threatened crop species and landraces as well as crop wild relatives. The ICCA Consortium also recognizes the cultural and economic livelihoods and rights of millions of Indigenous People. The global coverage of ICCAs has been estimated as being comparable to the formal global state-run protected area network. As an association, the ICCA Consortium collaborates with the CBD Secretariat, GEF, UNEP WCMC, IUCN, research and advocacy organizations, and UN mechanisms promoting human and Indigenous Peoples' rights.

La Via Campesina (viacampesina.org) is an international grassroots movement bringing together millions of peasants, small and medium-size farmers, landless people, women farmers, Indigenous People, migrants and agricultural workers from around the world. It comprises about 164 local and national organizations in 73 countries from Africa, Asia, Europe and the Americas. Altogether, it represents about 200 million farmers. It has a strong focus on biodiversity and genetic resources, and among its actions it defends the interests of its members against corporate control and patents on genetic resources and laws and rules which prevent farmers from saving seeds. It also lobbies at the CBD and the FAO to improve international regimes on access and benefit sharing and farmers' rights in order to make these more farmer-centred. It is a key focus of attention in the grassroots movement for food sovereignty.

Genetic Resources Action International (GRAIN; www.grain.org) is an international non-governmental organization, which was established in 1990 to help further a global movement of popular action against one of the world's most pervasive threats to world food and livelihood security: genetic erosion. GRAIN works to meet its aims to: promote popular control of agricultural biodiversity; stop the destruction of diversity by industrial agriculture; and support grassroots agricultural biodiversity-based programmes globally. GRAIN is registered in Spain as an international, non-profit foundation and has its headquarters in Barcelona.

Rural Advancement Foundation International (RAFI; https://www.rafiusa.org/) is dedicated to the conservation and sustainable improvement of agricultural biodiversity, and to the socially responsible development of technologies useful to rural societies. RAFI is concerned about the loss of genetic diversity, especially in agriculture, and about the impact of intellectual property rights on agriculture and world food security. In its early days it has campaigned successfully on behalf of developing countries with regard to many issues of contemporary importance to PGR, such as farmers' rights, intellectual property rights, biopiracy, food security, terminator technology and what it refers to as bioserfdom (those farmers in developing countries being put in the position of serfs working for large multinational breeding and biotechnology companies), as well as many other issues. RAFI was known for innovative research and aggressive advocacy in both bilateral and multilateral fora, and at attracting media attention for many of these causes. More recently RAFI has been more focused on issues in North America.

2.4 Ethical and Social Context of PGRFA Conservation and Use

As earlier sections of this chapter highlight, small farmers and Indigenous People have been maintaining PGRFA for many millennia with scant acknowledgement or reward for the enormous benefit this has provided for global food production and food security. In developing countries, the majority of small farmers and Indigenous People still act as custodians of the vast majority of PGRFA and where seed saving, seed exchange, reciprocity and equity are elements of an informal system often embedded in cultural and traditional ways of life. However, much of these PGR are threatened, and genetic erosion is believed to be widespread. Not only are we losing genetic resources, especially locally adaptive landraces, but also the knowledge and practices that were associated with these resources. For these reasons, at least in part, international agreements such as the CBD and the ITPGRFA have emerged. This international conservation community has attempted to implement mechanisms which try to ensure that small farmers and indigenous communities can continue to access genetic resources and share in the benefits which arise from their exploitation. However, it is these issues of access and benefit sharing (ABS) that have been most challenging to implement, with relatively few examples of effective and meaningful access and benefit sharing that really benefits small farmers (Farming Matters, 2016).

Some of the criticisms aimed at the formal international regime, which includes both the CBD and the ITPGRFA, and their respective elements in ABS (the Nagoya Protocol and Farmers' Rights), are that they are too bureaucratic, legalistic and theoretical and not in line nor supportive of small holders and Indigenous People. While the CBD recognizes national sovereignty over genetic resources, this is often in contradiction to protecting cultural and human rights of Indigenous People and

small holders, the custodians of genetic resources. Such customary laws and rights can only be recognized under the CBD and the Nagoya Protocol when they are in line with domestic law, which is rarely the case. Furthermore, there are often serious limitations in domestic law in terms of recognition of free and prior informed consent which all too often means that small farmers and Indigenous People are not consulted, or their consent requested when their resources are exploited. Such limitations of domestic law can seriously erode the rights of small farmers and indigenous communities to gain benefits from ABS under the umbrella of the CBD (Pistorius, 2016). However, commercial breeders would argue the opposite case, that the confusion over practical implementation of ABS has halted their access to genetic resources and is now restricting crop improvement (Michiels, 2015).

In contrast to the recognition of national sovereignty over genetic resources in the CBD, the ITPGRFA established the MLS, which was cognizant of the global interdependence on PGR for food security. PGR within the MLS are treated as pooled goods, and individual ownership is not recognized. Benefits arising from the exploitation of PGR goes into a multilateral fund, the Benefit Sharing Fund (BSF), which is shared with other countries. Even so, payments into the BSF have been minimal. Unlike the CBD, the ITPGRFA does recognize Farmers' Rights and the contribution farmers, Indigenous People and their communities have made to the development and management of PGR. However, the responsibility for the realization and implementation of farmers' rights and decisions on whether to embed them in national law or not resides with governments. To date, this has proven to be a challenge especially in developing countries and is due to a combination of factors including limited resources, poor capacity and expertise, and lack of political will and leadership. Even where some progress has been made, as in Nepal and India, these issues remain limiting in terms of practical implementation (Pistorius, 2016).

While the earlier expectations around the implementation of ABS through these formal systems have been slow to materialize, so have been the benefits that were expected to accrue to small farmers and Indigenous People. It has to be appreciated that many of the fundamental values of access, sharing, reciprocity and equity which are culturally embedded in informal systems, such as community seed saving and seed systems, community biodiversity management and seed banks, and participatory plant breeding, still endure. It is these informal systems, which we will turn to in Chapters 8–10 and 17, that can provide time-tested and effective approaches to improving access and benefit sharing to small farmers and Indigenous People.

2.5 Some Contemporary PGRFA Issues

While the previous section highlights that there is still a long way to go to overcome many political, social and ethical issues around PGR such as addressing farmers' rights and other matters of access and benefit sharing, there also remain other pressing contemporary issues for PGR. Some of these issues are described briefly in this section.

2.5.1 *Ex situ* versus *In situ* PGRFA Conservation

The discussion between those championing either the application of *ex situ* or *in situ* conservation techniques is only partially based on scientific principles, although there was and still is a scientific justification for expanding CWR *in situ* and landrace on-farm genetic conservation, both because of the urgent need to protect ecosystems threatened with imminent change and the need to preserve the natural continued evolution of crop diversity within traditional farming systems. Yet it is a fact that today formally recognized CWR *in situ* and landrace on-farm genetic conservation is restricted to a handful of sites globally, and therefore by default about 99% of annual PGR expenditure goes on *ex situ* conservation alone (Maxted *et al.*, 2016a). However, the CBD stresses the need for complementary conservation (Article 9), although giving priority to *in situ* conservation and ecosystem approaches, it clearly states the need for complementary backup *ex situ* application as well. That is, the conservationist should

be using a combination of conservation strategies and techniques to conserve any gene pool, focusing explicitly on neither *ex situ* nor *in situ* techniques but always including both. Therefore, the two strategies are not in opposition to each other but should be necessarily applied together. Further, the current balance of strategies is out of kilter and we need additional investment in CWR *in situ* and landrace on-farm genetic conservation, though with some gene banks fighting to sustain their current budget it is important that funding for *in situ* conservation is not taken from the current *ex situ* budget.

2.5.2 Hotspots of Botanical Diversity and Centres of *Ex situ* Conservation

In recent years there has been extensive policy and ethical discussion concerning the export of germplasm from developing countries to international or regional gene banks primarily located in northern developed countries. The bulk of plant diversity is located in developing countries, but the locations where that diversity is largely conserved *ex situ* are in developed countries. The second FAO State-of-the-World's PGR Report (FAO, 2010a) states there are now more than 1750 gene banks worldwide, about 130 of which hold more than 10 000 accessions each, and that there are relatively fewer in Africa compared with the rest of the world. Table 2.1 indicates the regional distribution of gene banks and their accessions, and that North America and Europe have a disproportionate number of conserved accessions and gene banks compared to the regions that contain the bulk of plant diversity. However, the figures may simply reflect the status of the native plant utilization industry within the regions; Europe does have a more sophisticated plant breeding industry than Africa. To establish whether there is a link between the movements of germplasm from developing to developed countries we would need to establish if the accessions housed in developed countries are predominantly non-indigenous taxa.

Unfortunately, the State-of-the-World Report (FAO, 1998) does not supply provenance information on a regional basis because few countries provided the proportion of indigenous accessions in their national

Table 2.1 **Regional distribution of *ex situ* conserved accessions and gene banks**

Region	Accessions (1998)		Accessions (2010)		Gene bank (1998)	
	Number	%	Number	%	Number	%
Africa	353 523	6	354 196	5	124	10
Latin America and Caribbean	642 405	12	1 023 148	16	227	17
North America	762 061	14	708 107	11	101	8
Asia	1 533 979	28	2 294 060	35	293	22
Europe	1 934 574	35	1 725 315	26	493	38
Near East	327 963	6	460 794	7	67	5
Total regional	5 554 505	100	6 565 620	100	1308	100
CGIAR	593 191		834 380		12	
Total	6 147 696		7 400 000		1 320	

From FAO (1998; 2010).

collections; however, Table 2.2 shows the data available. From these data it can be seen that those countries with a high level of biodiversity conserve primarily indigenous species, while countries with lower native biodiversity but more resources available for exploitation have a higher proportion of exotic accessions conserved. A breakdown of world *ex situ* collections by major crop groups, based on information held in the FAO WIEWS database, is given in Figure 2.4. The fact that the majority of accessions held *ex situ* are of crop species indicates the prime purpose of establishing these gene banks was to facilitate agricultural development and breeding programmes. The figures do tend to suggest that germplasm has been/is being exported from biodiversity-rich developing countries in the South to gene banks primarily located in biodiversity-poor northern developed countries for exploitation.

This debate leads us back to the previous discussion point. In view of the interpretation that can be placed on the data presented, it is not surprising that the discussion concerning the application of *ex situ* and *in situ* conservation techniques has become particularly heated. The point has become especially poignant

since the wider application of biotechnological techniques in the North and bioprospecting in the South, both of which has resulted in increased economic value being placed on wild species as gene sources.

2.5.3 Ownership/Plant Breeders' Rights/ Farmers' Rights

Closely associated with the transfer of germplasm away from their country of origin is the transfer of political and economic control over that material. The historical, social and ethical context, discussed above, indicates that external exploitation of biodiversity has rarely resulted in direct economic benefit to the original country of collection of the initially acquired genetic diversity. The debate over sovereignty and patenting of biological diversity is currently therefore a matter of extensive international debate. One of the major issues implicitly addressed within the CBD is the relationship and potential conflict between the rights of plant breeders and the rights of farmers.

The initial development of Plant Breeders' Rights arose out of patent laws. The principle that plant

Table 2.2 **Percentage of indigenous accessions in national gene banks**

Europe	%	Near East	%
Bulgaria	12	Iran I.R. of	>95
Czech Republic	16	Cyprus	100
Moldovia	40	Iraq	22
Romania	71		
Slovak Republic	8	**Americas**	
Belgium	75	Brazil	24
		Colombia	55
Africa	75	Ecuador	52
Cameroon (roots and tubers)	25	USA	19
Cameroon (fruits)	100		
Ethiopia	100	**Asia**	
Mauritius	100	China	85
Angola	100	Korea, DPR	20
Malawi	100	Korea, Republic of	18
Namibia	10	Sri Lanka	67
Senegal			

From FAO (1998).

breeders were due a royalty payment for the cultivars they developed was established in the USA with the Plant Patent Act of 1930. These rights were gradually extended both within the USA and other developed countries until the 1960s, when the Union for the Protection of New Varieties of Plant (UPOV) was formed in 1961. UPOV currently has 74 member countries and effectively provides a system that recognizes and protects the legal rights of plant breeders in member countries. In 1991 the UPOV convention was strengthened, among other things, to stop farmers from replanting protected cultivars, and in 1992 the US patent office granted the first 'species' patent on genetically modified cotton.

UPOV explains the need to retain Plant Breeders' Rights because:

- experience has shown that it is difficult for a breeder to recover their financial investment if, once a cultivar is released, a competitor can secure supplies of propagating material and in a short time compete with the initial breeder; it may take 10–20 years for the initial breeder to reclaim their financial investment,
- they require a reasonable return on past investments,
- they need an incentive for continued or increased investment in the future, and
- there is a moral right of the innovator to be recognized as such and his economic right to remuneration for his or her efforts.

Accordingly, exclusive rights of exploitation are granted to the breeders of new cultivars to ensure the maintenance of the plant breeding industry and, through them, the stream of new cultivars adapted to a wide range of circumstances for a varied diet and for a wide choice of ornamental and amenity plants.

The Agreement on Trade-Related Aspects of Intellectual Property Rights (TRIPS) is an additional legally binding agreement which sets minimum standards for different forms of intellectual property under WTO member states. It requires member states to provide a system of intellectual property protection for plant varieties which must be based upon patents, an effective *sui generis* system (e.g. Plant Breeder's Rights under UPOV) or a combination of both (FAO, 2011b).

As we have seen, the issue of Farmers' Rights is now embedded in the ITPGRFA and such rights are recognized in the CBD through the Nagoya Protocol. Some commentators contend that it is immoral to allow Plant Breeders' Rights over commercial cultivars without acknowledging the contribution made by farmers in the form of Farmers' Rights. There is general international acceptance for the concept of benefit sharing as outlined in the CBD (Articles 1, 16–20), but there remains no clear mechanism for implementing either bilateral or multilateral benefit sharing.

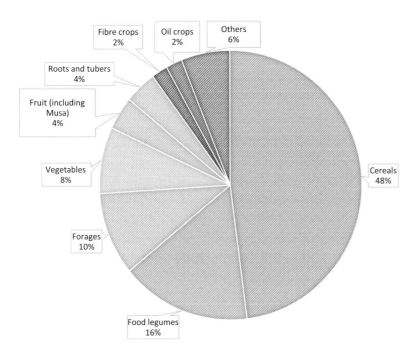

Figure 2.4 Contribution of major crop groups to total *ex situ* collections. (A black and white version of this figure will appear in some formats. For the colour version, refer to the plate section.)
(Reproduced from FAO, 1998)

2.5.4 The Impact of Genetically Modified Organisms

Genetic engineering has made the transfer of genes between species relatively simple, and gene editing is opening a further extension to cultivar 'design'. The development of these bio-techniques has revolutionized the production of new crops with unique characteristics, and it is anticipated that hundreds of these so-called genetically modified organisms (GMOs) will increasingly be released for commercial cultivation. The new technology possesses enormous potential to benefit mankind because it will allow unlimited opportunities for producing plants with specific traits; such as tolerance to herbicides (e.g. cotton, maize), host plant resistance to insects and other pests (e.g. maize, cotton, potatoes, among others), improved water-use efficiency (maize), enhanced food quality and prolonged shelf life (e.g. tomatoes), modified fruit and seed products (e.g. low cholesterol cooking oils in soybean), and even the transfer of novel characteristics from other kingdoms (e.g. animal and human gene products such as haemoglobin).

Associated with these potential applications of the new technology are, however, many potential issues that stem from the introduction of alien genes into cropping systems and their dispersal into the natural flora. The genes that are incorporated into the GMOs could migrate to the crops' nearest relatives through inter-crossing or even transfer to some distant species via the food chain. Excessive selection pressure on the pests can also accelerate their resistance to the toxins produced by the GMOs, and this could lead to the evolution of 'super weeds' and 'super bugs', which may then prove difficult to control. The production and release of GMOs, therefore, pose some biosafety and ethical issues that continue to be passionately debated by well and less well-informed commentators at this very moment. In summary, the proponents of GMOs argue that the problems of genetic pollution and the misuse of biotechnology are very much exaggerated, and the risks are marginal compared to the gains that flow from the commercial exploitation of GMOs. The opponents of GMOs, on the other hand, foresee that unfettered production and release of GMOs will have catastrophic consequences for biological life on this planet and that their production should either be stopped altogether or restricted drastically. They also see it as an extension of corporate control within agriculture. These issues remain a lively and often acrimonious focus of current debate, which should be evidence-based and reliant on knowledge ensuing from science.

2.5.5 Sustainable Conservation and Use of PGRFA

In recent years, particularly post-CBD, the fundamental importance of sustainability in the conservation and use of PGR has finally been recognized. In terms of conservation, sustainability is often associated with the use of a range of conservation techniques (both *in situ* and *ex situ*), and it is becoming apparent that the most efficient conservation activities involve both formal (e.g. state, international) and informal (e.g. NGO, community) sector participation. One area where this formal/informal collaboration is particularly vital is in on-farm conservation. Almost by definition, on-farm conservation is only viable if the two sectors work together. Conservationists must ensure the farmers themselves are aware of the conservation importance of the material they have been growing for millennia. Farmers require appropriate support so they can continue the use of traditional farming systems that have maintained that diversity in the past. Previously, the 'formal' world of research institutes, gene banks and plant breeders has occasionally worked against 'informal' farmer-based approaches, thus negatively affecting the efficiency of these approaches to conserve the PGR the formal system recognizes should be maintained.

2.5.6 Food Sovereignty, Agroecology and Genetic Resources

Resistance to many of the ethical and social issues already discussed such as patenting, ownership, laws and regulations governing saving seed and corporate control of plant genetic resources are likely to intensify over the coming years as these aspects increasingly become the focus of many largely grassroots movements for greater food sovereignty and agro-ecology-based approaches. The emerging discourse for greater empowerment and autonomy within food systems was first documented at the 1996 Rome Food Summit (Trauger, 2015). Many of the debates and negotiations outlined in this chapter may become more acrimonious in such highly contested spaces where governments, business and social movements come together. The ongoing development and evolution of such grassroots movements is likely to strengthen calls for more attention to farm-based conservation.

2.5.7 Healthy Food Systems, Diets and Underutilized PGR

One of the world's greatest challenges will be to secure adequate food that is healthy, safe and of high quality for all, and to do so in an environmentally sustainable manner (Godfray *et al.*, 2010). We are still a long way from this even though there is enough food produced globally. Despite the impressive strides achieved in increasing food production, we still fail to feed about half of humanity in a healthy manner (Kennedy *et al.*, 2017). No single country on Earth is free from one form or other of malnutrition. Increasing the amount of food we produce or increasing the yields (especially of the major nutrients such as proteins and calories) (Frison *et al.*,, 2011) may not necessarily be the most pressing problem humanity faces even though we are bombarded on a daily basis with warnings of the need to feed a global population expected to reach 9.6 billion by 2050 (United Nations, 2011). With about one-third of the global population suffering from hidden hunger, or one form or another of micronutrient deficiency, one of the most pressing challenges facing agriculture and food systems is how we can ensure a supply of food that is of much better nutritional quality. For decades agricultural policies, strategies and programmes have focused on increasing the production of a few staple crops, and their success is measured in terms of the food quantity or dietary energy supply. Yet ample quantity does not necessarily ensure appropriate nutritional quality, with staple crops unable to provide the diversity and adequate amounts of nutrients to meet human requirements, especially those much-needed micronutrients people require to reach their full physical and mental potential. At the same time our knowledge and understanding of the nutritional value of the world's considerable neglected and underutilized plants demonstrates significant intra-specific differences in their nutrient values (Hunter *et al.*, 2015; Kennedy *et al.*, 2017). Oftentimes these variety-specific differences can represent the difference between nutrient deficiencies and nutrient adequacy in populations and individuals. One of the major challenges of the 2030 agenda for sustainable development will be how we can incorporate and mainstream more of these nutritionally diverse plant genetic resources into national agriculture and food systems as well as improving the diversity of diets.

2.5.8 Climate Change and PGRFA under Threat

Climate change has brought major global challenges to many areas of modern life, which the global community must urgently address. The conservation and use of PGR cannot escape this growing threat. Jarvis *et al.* (2008a) using models to predict the impact of climate change on the wild relatives of groundnut or peanut (*Arachis*), potato (*Solanum*) and cowpea (*Vigna*) found that climate change strongly affected all taxa, with an estimated 16–22% of these species predicted to go extinct and most species losing more than 50% of their range size by 2055. Neither can we assume that the occurrence of CWR in those areas receiving state-sanctioned protection, e.g. the global network of protected areas, are safeguarded. Active population management is required to underpin long-term maintenance of any species. Researchers have studied the effects of two climate change scenarios in Mexico on the distribution patterns of eight CWR in protected areas and found a marked contraction in distributions for all eight taxa under both, with maintenance of CWR in only 29 out of the 69 natural protected areas where they currently occur (Lira *et al.*, 2009). Unless drastic steps are taken now, we stand to lose many of these genetic resources forever because few of them are safeguarded long-term in gene banks or maintained *in situ* in sites less likely to be impacted by climate change. The situation regarding the impact of climate change on landraces is less clear though understanding and appreciation of their importance for adaptation and resilience is not. Secondly, if global agriculture is to adapt to the many challenges which climate change will throw up, such as increased temperatures and drought, it will need to be able to draw upon the genetic traits found in CWR and landraces to do so and to use these resources they must be adequately conserved. At the same time, we also know from the limited studies to date that climate change can impact the nutritional quality of PGR (Myers *et al.*, 2014).

2.5.9 Under-Represented PGRFA in Gene Banks

To add to the above challenge, a recent global assessment of the representativeness of CWR in the world's gene banks reveals that there are very significant gaps, and this has major implications for agriculture to adapt to future climate change and for food security. The main findings of the study revealed, in terms of species and geographic regions, is that nearly 29% of the total wild relative plant species analyzed are completely missing from the world's gene banks, while a further 24% of species are represented by less than 10 samples having been collected for each. Low numbers of samples per species means a significant amount of potentially important diversity within these species is still underconserved and threatened by climate change or other threats such as deforestation or urbanization. The study stresses the urgent need for measures to address these gaps, highlighting that more than 70% of the total CWR species need collection and conservation. Furthermore, it highlights that representation in gene banks for more than 95% of CWR species is insufficient to represent the full range of geographic and ecological variation in their native distributions. The study therefore called for critical collecting for gap-filling with novel collections from the Mediterranean and Near East, western and southern Europe, Southeast and East Asia, and South America (Castañeda-Álvarez *et al.*, 2016a).

Part II

Scientific Background

3 Plant Taxonomy

3.1 Introduction

Taxonomy can be defined as the scientific study, organization and understanding of biodiversity and how that diversity arose through time. Taxonomy may be thought of as the fundamental 'language' that enables us to understand and utilize living organisms. Biological diversity, in terms of the number of species, is enormous. There are estimated to be 369 000 species of flowering plants, and of these 177 000 remain poorly known to science (Paton *et al.*, 2008; RBG Kew, 2017). Conservationists can only conserve known taxa that are recognizable in the field. Taxonomy attempts to describe systematically and to organize the enormous range of variation present in living organisms to make that biodiversity itself comprehensible to the users of taxonomy. In a hierarchical and successively inclusive manner taxonomy groups more similar organisms together based on morphological, anatomical and genetic characteristics.

The objective of taxonomy is to describe and understand the relationship within and between groups of organisms. For conservationists, this enables us for example to establish the following:

- How many taxa are there within the group we wish to conserve?
- How do we distinguish between these taxa?
- What are their characteristics?
- Where are centres of multiple taxon diversity or individual taxa to be found?
- Which are the closest relatives?

An understanding of the general principles of taxonomy, how taxon relationships are understood and, specifically, the taxonomy of the target group to be conserved is fundamental to the formulation of effective biodiversity conservation programmes. Taxonomy, through the organism naming system it provides, contributes the referencing system for the whole of biology. It provides the backbone on which pieces of biological information, including conservation data, are attached and so communicated.

Conservation activities will always be limited by the financial and human resources available. If the resources only allow the conservation of a limited set of taxa, then taxonomy can help us decide which taxa to choose in order to maximize the overall range of genetic diversity conserved. In this case, rather than selecting taxa which are closely related to each other, according to the classification of the group, the conservationist could deliberately select those taxa most distantly related to each other to ensure the broadest range of genetic diversity is preserved. Further taxonomy can also assist conservationists to select those taxa most likely to be used by breeders to cross adaptive traits into the crop using the generic classification. Those wild species most closely juxtaposed to the crop in the classification of the genus can be more easily crossed with the crop and so are more likely to be used by breeders looking for novel traits to transfer to the crop. The conservation of species will inevitably involve surveying and either collection or conservation nomination in the field. To do either will involve the conservationist using a range of Flora guides and identification keys to identify the species correctly.

Specifically, in the context of plant genetic resources (PGR) conservation the need for taxonomic skills has never been greater. Such skills have become essential since the rise to prominence in recent years of CWR conservation. A major block to CWR use has been lack of conserved CWR diversity available to the user community, and this blockage often exists due to the lack of skilled PGR scientists available to identify CWR in the field (Maxted, 2011). The problem has at the same time been accentuated by the widespread demise of teaching of taxonomy in School and University (House of Lords, 2002). The need at least for a basic understanding of taxonomy, if you are to conserve plant genetic diversity, is obvious. How can you conserve or exploit plant diversity if we cannot distinguish different species or recognize them in the field?

3.2 Producing Novel Classifications

3.2.1 The Revision

In taxonomy, the experimental study of patterns of diversity is commonly referred to as a revision. The term revision is used because following an experimental study our knowledge of the patterns of diversity within a taxon is itself 'revised', and we have a new concept of the taxa included and their inter-relationships. The revision process involves the recording and analysis of features or characteristics both within and between taxa (populations, species, genera). The results of this analysis are then considered in conjunction with information from the literature and yield various taxonomic products. The primary product is the novel, 'revised' classification of the taxon, which places all the specimens, populations and higher taxa in successively inclusive groups. The novel form of the classification is the primary product because once established it will directly affect the production of the secondary products, such as taxon descriptions, identification aids, accepted nomenclature, etc. Once the range of specimens representing a species have been identified, we can use the characteristics of those specimens to define its genetic amplitude, the range of morphology shown, its distribution and its relative isolation from other taxa or a related crop. If, say, following our investigation we know that in our herbarium we have 30 specimens of a species, then the description of those 30 individual specimens can be summed to produce a general description for the species which each specimen represents. Therefore, placing specimens into species, species into genera, genera into tribes, and successively further up the hierarchy, which delimits the classification of a group, is of fundamental importance, because all the secondary products are subsequently derived directly from the group's classification.

To illustrate the elements of taxonomy that are directly used by conservationists the grain legume winged bean (*Psophocarpus tetragonolobus* (L.) DC.) from the genus *Psophocarpus* Neck. ex DC. (Leguminosae; Papilionoideae) will be used. The genus comprises nine species, eight of which are endemic to West, Central and East Africa, while the ninth species is the important pulse crop winged bean, whose distribution is centred in Asia. The species has recently been introduced into other tropical areas as a crop. According to Fatihah *et al.* (2012), the genus contains nine species: *Psophocarpus grandiflorus* Wilczek, *P. lancifolius* Harms, *P. lecomtei* Tisserant, *P. lukafuensis* (De Wild.) Wilczek, *P. monophyllus* Harms, *P. obovalis* Tisserant, *P. palustris* Desv., *P. scandens* (Endl.) Verdc. and *P. tetragonolobus* (L.) DC.

3.2.2 Using Taxonomic Collections

During the revision of a group of organisms, features or characteristics are recorded from representative specimens and the pattern of their distribution analyzed. This allows conclusions to be drawn about the relationship between and within the taxa being investigated. It would be impractical for taxonomists to conduct revisions on plants directly in the wild, and so they often use specimens from taxonomic collections. Taxonomic collections can be either living plants, for example, a living collection in a botanic garden, or a preserved herbarium collection of dried specimens (see Box 3.1).

For *Psophocarpus* there are very few seed accessions available for genetic analysis and not all species have any seed available, so recent taxonomic studies have focused on morphological studies using herbarium specimens. Due to the distribution of the species and the lack of access to herbaria in West, Central and East Africa, the specimens analyzed were largely drawn from the large ex-colonial herbaria: Royal Botanical Gardens, Kew, England; British Museum (Natural History), London, England; Museum National d'Histoire Naturelle, Laboratoire de Phanerogamie, Paris, France; Jardin Botanique National de Belgique – Nationale Plantentuin van

---

Box 3.1 | Herbaria and Taxonomic Collections

A herbarium (plural, herbaria) is simply a collection of preserved plant specimens, usually arranged according to a classification system. The most common form of preservation is when the plant material is pressed, dried and then mounted on a sheet of stiff paper. Each herbarium specimen will have a label attached to it with information such as for a specimen of *Psophocarpus palustris* Desv., the close wild relative of winged bean:

- Scientific or Latin name with author abbreviation (= *Psophocarpus palustris*)
- Collector's name(s) and number (= Oloruferni & Macauley 1213)
- Date of collection (14.04.1992)
- Collection locality (preferably with latitude and longitude) (= about 5 miles east of Loko village. Near Nasarawa Town, Nasarawa State, Nigeria, Nigeria: 8 degrees 09 minutes 49 seconds North; 7 degrees 43 minutes 57 seconds East)
- Habitat/ecological notes (= Habitat: Savannah beside track and field margin. Altitude: 79 metres)
- Uses (= minor forage, also important as gene donor to winged bean crop)

Herbaria are usually associated with botanic gardens and research institutions such as universities. They vary greatly in size and the number of specimens they hold. For example, the Muséum National d'Histoire Naturelle in Paris, France, has 8 877 300 specimens, whereas the University of Benin Herbarium, Cotonou, Benin, has just 23 500 specimens. Herbarium specimens are usually grouped alphabetically, geographically or according to a classification of the group and are stored in cabinets or boxes. These plant collections are used by taxonomists to conduct systematic research and produce revisions. Herbaria also act as reference centres, helping botanists, ecologists and conservationists identify unknown species. They also house a wealth of botanical information, other than that which can be extracted from the specimens themselves, which is available from taxonomic experts, the associated libraries and databases.

België, Meise, Belgium; Conservatoire et Jardin Botaniques de la Ville de Geneve, Chambesy, Switzerland; National Herbarium of the Netherlands, Leiden, Netherlands; and the National Herbarium and Botanical Garden, Causeway Harare, Zimbabwe.

3.2.3 Characters

The features or characteristics that are recorded and analyzed during a revision are referred to as characters. A formal definition of a character is:

Any attribute (or descriptive phase) referring to form, structure or behaviour which the taxonomist separates from the whole organism for a particular purpose such as comparison or interpretation.

(Davis and Heywood, 1973)

Traditionally, taxonomists have recorded obvious features, such as morphological characters associated with flowers because of their conservatism, defined as their ability to remain relatively unchanged over long periods of evolutionary development and the ease of recording compared to say molecular characteristics. However, within the context of trying to relate taxa we need to study the genotype as much as possible; studying the phenotype which may be impacted by the environment where the plant grows may falsely bias results. Therefore, in recent years the emphasis has moved to studying the genome itself. However, there are many types of characters that have historically been used to classify taxa: morphological, physiological, chemical, behavioural, ecological and distributional, as well as genomic. It is important to distinguish between a character and a character state. The character is what the taxonomist is describing, for example, flower colour or leaf shape. The character state is what the taxonomist records, so for the character flower colour the character states could be yellow, blue and brown.

3.2.4 Analysis of Taxonomic Data

Historically there are two fundamentally different schools of thought in taxonomic data analysis – phenetic and phylogenetic:

- phenetic techniques use the occurrence of character combinations as they are now perceived to construct taxonomic groups, and
- phylogenetic techniques add the dimension of time and ancestry to describing organisms, and the characters selected are those that are thought to have been important in evolution.

There remains some debate over which approach is more scientifically valid but in terms of conservation it might be thought that a phenetic approach might prove more useful in the field as taxa that look similar will be most closely related, but, in terms of utilization of CWR by breeders, taxa that are phylogenetically related are more likely to produce successful crosses with the crop and so ultimately more useful. However, for both schools the analysis of taxonomic data uses multivariate statistics to clarify the relationship between Operational Taxonomic Units (OTU). An OTU is the individual unit of study in that investigation, for example, it could be an individual plant specimen, species or genus. Traditionally, both phenetic and phylogenetic approaches to production of the novel

classification was largely intuitive and subjective, but in recent years the use of computers has improved objectivity and repeatability. We will now go through a sequence of a taxonomic revision using the phylogenetic approach for the legume genus *Psophocarpus* published by Fatihah *et al.* (2012) to illustrate the basic process of producing the novel classification.

Analysis Stage 1: Selection of Characters and States

First the study group has been selected and delimited, in our case the nine species of the genus *Psophocarpus*, then a group of 'good' characters is selected. 'Good' characters are usually defined as those that vary between taxa but remain constant for any taxon. When undertaking a phylogenetic analysis, there is also an implicit assumption that the selected characters are those that will help elucidate the evolutionary path within the study group. Taxonomists may use over a hundred characters when they conduct revisions, and in general terms the more characters used the more powerful the analysis. Each representative specimen within the study group has its character states recorded for each character. For a phylogenetic analysis, the character states need to be polarized into the primitive (= plesiomorphic) or advanced (= apomorphic) states for the study group, referred to technically as the ingroup, which in this case is the *Psophocarpus* species. This was achieved using the character state present in the closely related legume genera *Otoptera*, *Dysolobium* and *Vigna*, which acts as the outgroup, and character states present in the outgroup and *Psophocarpus* species are regarded as primitive in *Psophocarpus*. Therefore, in this case the OTUs are the nine species of *Psophocarpus* plus outgroup exemplar species from *Otoptera*, *Dysolobium* and *Vigna*. A list of the characters and character states for the analysis is given in Table 3.1.

In Table 3.1 some characters are binary (two state), e.g. *Growth form*: (0) perennial, (1) annual, while others are continuous (quantitative measure) but for the analysis are converted to binary characters, e.g. *Leaflet width*: (0) less than 5 cm, (1) equal to or more

Table 3.1 *Psophocarpus* species character number, name and states

1, Growth form: (0) perennial, (1) annual;
2, Growth habit: (0) climber, (1) prostrate;
3, Stipule length: (0) less than 0.5 cm, (1) equal to or more than 0.5 cm;
4, Stipule shape: (0) lanceolate, (1) ovate;
5, Stipule base projection: (0) projected, (1) not projected;
6, Stipule indumentum: (0) pubescent, 1 glabrous;
7, Number of leaflets: (0) trifoliate, (1) unifoliate;
8, Terminal leaflet length: (0) less than 10 cm, (1) equal to or more than 10 cm;
9, Leaflet width: (0) less than 5 cm, (1) equal to or more than 5 cm;
10, Terminal leaflet shape: (0) ovate, (1) elliptic;
11, Shape of the terminal leaflet apex: (0) acute, (1) mucronate;
12, Shape of the terminal leaflet base (0) truncate, (1) cordate;
13, Prominent veins on abaxial leaflet surface: (0) not prominent, (1) prominent;
14, Petiole length: (0) less than 5 cm, (1) equal to or more than 5 cm;
15, Petiolule length: (0) less than 2 cm, (1) equal to or more than 2 cm;
16, Inflorescence type: (0) pseudoraceme, (1) raceme;
17, Number of flowers per inflorescence: (0) less than 10, (1) equal to or more than 10;
18, Peduncle length: (0) less than 20 cm, (1) equal to or more than 20 cm;
19, Bracteole apex shape: (0) acute, (1) obtuse;
20, Bracteole to calyx length ratio: (0) shorter than calyx, (1) as long as or longer than calyx;
21, Calyx indumentum: (0) pubescent, (1) glabrous;
22, Standard shape outline: (0) rounded, (1) obovate;
23, Standard shape at apex: (0) rounded, (1) emarginate;
24, Standard apex form: (0) smooth, (1) crinkled;
25, Standard lobe division at the apex: (0) not divided, (1) weakly divided, (2) strongly divided;
26, Standard appendages: (0) absent, (1) present;
27, Wing shape: (0) shape1, (1) shape2, (2) shape3;
28, Wing apex shape: (0) beaked, (1) rounded;
29, Wing with extra tooth: (0) absent, (1) present;
30, Wing claw shape: (0) simple, (1) T-shaped;
31, Keel spiral at apex: (0) present, (1) absent;
32, Keel pouch: (0) absent, (1) present;
33, Keel sculpturing: (0) absent, (1) present;
34, Vexillary filament attach at the middle of the tube: (0) free, (1) joined;
35, Ovary shape: (0) oblong, (1) linear;
36, Ovary indumentum position: (0) lateral, (1) in a whorl;
37, Stigma position: (0) subterminal, (1) terminal;
38, Style indumentum at the base: (0) absent, (1) present;
39, Style apex shape: (0) not bifid, (1) bifid;
40, Pod exocarp: (0) coriaceous, (1) woody, (2) lignified;
41, Pod outline: (0) linear, (1) oblong, (2) ellipsoid;
42, Pod cross-section: (0) nearly flat, (1) rounded or square;
43, Pod wing prominent: (0) absent, (1) present;
44, Pod wing edge: (0) entire, (1) crinkled;
45, Pod indumenta: (0) glabrous, (1) pubescent;
46, Raphe visibility: (0) not visible, (1) visible or not visible;
47, Hilum length: (0) less than 0.1 cm, (1) equal to or more than 0.1 cm;
48, Position of hilum: (0) not central, (1) central;
49, Hilum concealment: (0) fully concealed by funicular remnant, (1) partially concealed by aril;
50, Seed aril: (0) absent, (1) present;
51, Split grooved hilum: (0) absent, (1) present.

From Fatihah *et al.* (2012).

than 5 cm. The language of plant morphology used to describe the species is complex and for most of us unfamiliar but necessary to precisely describe the plant's morphology. But technical terms should not threaten the conservationist, as there are many good glossaries to these technical terms available (see Stearn, 1966; Hickey and King, 2000; Harris and Woolf Harris, 2001; Beentje, 2010) and working on a specific group of plants means you will soon learn the descriptive vocabulary.

Analysis Stage 2: Recording Character States in Coded Format

The recorded set of character states is converted to codes (Table 3.2) and entered into a computer for analysis. The parsimony analysis for the *Psophocarpus* species data was performed using the maximum parsimony and Wagner approaches, and trees were generated using PAUP (version 4.0b10; Swofford, 2002). This package is specifically designed for the analysis of taxonomic data, but there are many other analysis packages currently available.

Analysis Stage 3: Parsimony Analyses

The PAUP program calculates the shortest parsimonious tree (= cladogram) with a minimum length of 127 steps, a consistency index (CI) of 0.4409 and a retention index (RI) of 0.6802 (Figure 3.1). The genus *Psophocarpus* formed a group with a common evolutionary ancestor (= monophyletic) at node 37 (indicated by an arrow in Figure 3.1) group characterized by the important shared derived characters (= synapomorphies) of stipule length equal to or more than 0.5 cm (#3), stipule base projected (#5), wing shape #3 (#27), wing without extra tooth (#29), vexillary filament joined at the middle of staminal tube (#34), linear ovary (#35) and pod wing prominent (#43). The *Psophocarpus* ingroup was divided into four consistently monophyletic subclades; subclade *Bifidstyllus* (containing *P. obovalis*, *P. monophyllus* and *P. lecomtei*), subclade *Vignopsis* (containing *P. lancifolius* and *P. lukafuensis*), subclade *Psophocarpus* (containing *P. palustris*, *P. tetragonolobus* and *P. scandens*) and subclade X (containing *P. grandiflorus*).

Analysis Stage 4: Proposal of Classification

The results of the parsimony analysis indicate four monophyletic subgroups with three subgroups directly mapping onto existing subgeneric taxa. However, the taxon in *Psophocarpus* considered most primitive (= basal), *P. grandiflorus*, which was suggested by both Verdcourt and Halliday (1978) and Maxted (1990) as a remote member of sect. *Psophocarpus*, compared to *P. palustris*, *P. scandens* and *P. tetragonolobus*, is clearly indicated as being distant to these species. The degree of isolation warranted subgeneric distinction, so Fatihah *et al.* (2012) erected the new subgenus *Longipedunculares* for *P. grandiflorus*. Therefore, the final classification based on the parsimony analysis of 51 characters for the nine *Psophocarpus* species is composed of three subgenera, two sections and nine species (Table 3.3).

3.3 Taxonomic Products

3.3.1 The Classification

The primary and most fundamental product of any revision is the classification. This is usually presented in the form of a list of accepted taxon names with their authors and publication details. Based on the parsimony analysis a classification of the nine *Psophocarpus* species is presented in Table 3.3. Once established, the classification will directly affect the production of each of the secondary products, such as taxon descriptions, synonyms, distribution maps, identification aids and accepted nomenclature, because each secondary product is derived from individual specimen data that are synthesized to generate the species level data, each specimen having been assigned to a species according to the classification used.

3.3.2 Artificial and Natural Classifications

Many historical classifications, even those of Linnaeus (1753), were artificial, meaning they were based on a few characters. Contemporary classifications are more commonly based on a larger number of characters from diverse sources (e.g.

Table 3.2 **Data matrix for 51 morphological characters; Plesiomorphic state = 0, apomorphic state = 1 or 2, missing data = ? (Fatihah et al. 2012).**

	1111111111222222222233333333334444444444455
	12345678901234567890123456789012345678901234567890 1
Vigna racemosa	00100000000000001000001010000010000000?0?????10?1?
V. monophylla	0000011000000??0110000101000001100000000?0?????11?0?
V. vexillata	021000000000000010000101100001000100000?0?????11???
V. mungo	0010000000001100?0100101100001000000000?????????????
V. lasiocarpa	001110000000000000??0010100000010010100?????????????
V. speciosa	000010000000000001000101000001000010100?????????????
V. frutescens	000100000000000010001010100000001?010000?0?????11?0?
Otoptera burchellii	0000000001100000001001000100101111100100000010?000
O. madagascariensis	0000000001100000001001000100101111100100000010?000
Dysolobium lucens	000110011000011010100010111110111000100??0?1??????
D. apoides	0000100000000001?000101011??01000000000??0??????????
D. pisolum	000010000010000010000101011101100101100000101?111
D. grande	000010011000011011100010111110100000100111001 01?111
D. dolichoides	000110011010011010000010101110110000100011001 11?111
Psophocarpus grandiflorus	0010000110000110111000101120011111010001111111?111
P. tetragonolobus	101001001000011000101010112101110110100211110???????
P. palustris	001000001000110010000110112100111110100211100???????
P. scandens	001000001000011011000110212110110110100211111?1?1??
P. obovalis	011000011110101010101101120010011110101110 1??????
P. monophyllus	0110001110111 0?0101011111112100111111101011101??????
P. lecomtei	0110001010110 0?0101011101121001111010102110 1??????
P. lancifolius	001100001000001000101101121001111100000111 0001?111
P. lukafuensis	0011010001000001000011101121001111100000111 00???????

From Fatihah *et al.* (2012).

morphology, anatomy, genomic, etc.). They are referred to as natural classifications and have a higher information content than the artificial ones and can be predictive. A natural classification is considered to present a closer approximation of the intrinsic 'natural' classification that underlies the natural partitioning of biodiversity. Natural classifications are predictive, which means that if we know a taxon is,

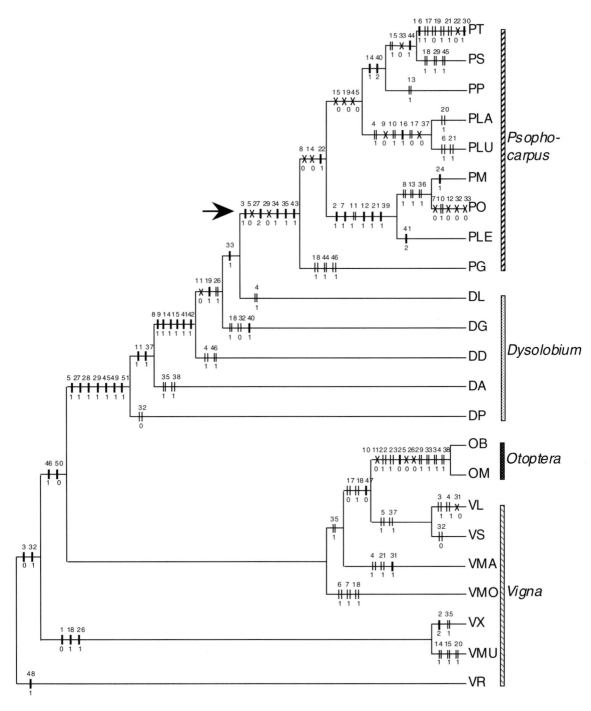

Figure 3.1 Equally parsimonious tree resulting from the cladistic analysis of species' morphological data. Terminal taxa: VR = *V. racemosa*, VX = *V. vexillata* , VMU = *V. mungo*, VMO = *V. monophylla*, VMA = *V. frutescens*, VL = *V. lasiocarpa*, VS = *V. speciosa*, OB = *O. burchellii*, OM = *O. madagascariensis*, DA = *D. apoides*, DP = *D. pilosum*, DD = *D. dolichoides*, DG = *D. grande* , DL = *D. lucens*, PG = *P. grandiflorus*, PLE = *P. lecomtei*, PM = *P. monophyllus*, PO = *P. obovalis*, PLA = *P. lancifolius*, PLU = *P. lukafuensis*, PP = *P. palustris*, PT = *P. tetragonolobus*, PS = *P. scandens*. Bars = synapomorphies; parallel lines = parallelisms; crosses = reversals.
(From Fatihah *et al.*, 2012)

Table 3.3 **Classification of *Psophocarpus* species**

Subgenus *Longipedunculares* H.N. Nur Fatihah, N. Maxted and L. Rico
P. grandiflorus Wilczek
Subgenus *Psophocarpus* (L.) DC.
Section *Vignopsis* (De Wild.) Maxted
P. lancifolius Harms
P. lukafuensis (De Wild.) Wilczek
Section *Psophocarpus* (L.) DC.
P. palustris Desv.
P. tetragonolobus (L.) DC.
P. scandens (Endl.) Verdc.
Subgenus *Bifidstyllus* Maxted
P. obovalis Tisserant
P. monophyllus Harms
P. lecomtei Tisserant

From Fatihah *et al.* (2012).

say, a *Psophocarpus* species, then we can predict many characteristics of that taxon; i.e. that it is a leguminous herb with uni- or tri-folioliate leaves; stipules prolonged below the point of insertion; flowers are borne singly on the flower stalk/peduncle and the ovary has 3–21 ovules; the standard is broadly auriculate with appendages; the keel is beaked at right angles to the axis of the flower; the pods are oblong, with more or less distinct 4 wings along the angles and septate between the seeds (Figure 3.2). A new classification is not automatically accepted by the taxonomic and biological community. However, a classification is more likely to be accepted if it can be shown to be natural and has a high predictability or information content.

3.3.3 The Taxonomic Hierarchy

Within the classification, each individual is placed in a specific taxonomic category; each category is given a

Latin name which distinguishes it from other groups of organisms. Each group of organisms given a name is referred to in general terms as a taxon (plural taxa); this term can be applied to any rank, e.g. species, genus, tribe, family. Those species that share most characters are placed in a group called a genus, genera which share most characters are grouped into a tribe, etc. Therefore, the taxonomic hierarchy is composed of a series of increasingly inclusive groups based on the decreasing similarity or phylogenetic relatedness, and each distinct level is given a name, species, genus, class, etc., as shown in Figure 3.2. The level of the hierarchy used for a taxon is referred to as its taxonomic rank.

An example of the taxonomic hierarchy using the winged bean *Psophocarpus tetragonolobus* is given in Table 3.4. As an aid to identifying to which rank a taxon belongs, the Latin names of the taxa between division and subtribe generally have characteristic endings as indicated, though this is not always the rule as you note from the example of the Family name. The legume specialists conserve in majority the old name Leguminosae, but Fabaceae finds increasing usage by purists who do not condone this exception.

3.3.4 Nomenclature

Having established which species and other taxa exist the taxonomist must provide each taxon with a nomenclaturally correct name. It is important that the conservationist uses the correct, accepted name for a taxon, so that when they communicate the results of their research, others have the same concept of the taxon for which the results apply. The procedure of applying the correct scientific name to an organism is nomenclature (Box 3.2).

There are numerous rules that apply to the naming of plants, and these nomenclature rules are set out in the current edition of the International Code of Nomenclature for algae, fungi and plants (McNeill *et al.*, 2012). Some basic nomenclature rules follow:

- *Binomial* – Species names are a Latin binomial made up of the genus name and specific epithet. For example, the Latin phrase, *Psophocarpus tetragonolobus*: *Psophocarpus* is the genus name and *tetragonolobus* the specific epithet.

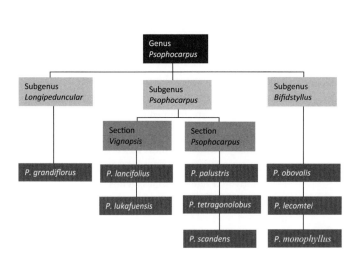

Rank

Genus

Subgenus

Section

Species

Figure 3.2 Organizational view of *Psophocarpus* classification. (A black and white version of this figure will appear in some formats. For the colour version, refer to the plate section.)
(From Fatihah *et al.*, 2012)

- *Priority* – Names included in Linnaeus's *Species Plantarum* 1st Edition have priority (nomenclatural precedence) over earlier or more recently published names. Therefore, if Linnaeus cites a binomial name in the *Species Plantarum* 1st Edition, then that is the valid name for perpetuity, but for names not cited in the *Species Plantarum* 1st Edition the first binomial used for a taxon is valid providing the binomial does not contravene any other nomenclatural rule.
- *Accepted name* – The name used for any taxon should be nomenclaturally correct (follow the rules of Botanical Nomenclature as established in the latest version of the *Code of Botanical Nomenclature*) and be the name accepted by the appropriate international specialists for the group. Generally, taxonomic experts will decide which classification they recognize and then attach the appropriate name to the taxa included. The name used will either be the Linnaean name, if Linnaeus recognized that taxon, or the first name subsequently validly published.

The Latin name used for a taxon does not have to have any biological meaning, but often the author of a name will use a name that does imply some meaning, e.g. *sativa* or *vulgaris* = common, *viridus* = green, *hirsuta* = hairy, *palustris* = marshy, *peruviana* = from Peru. Taxon names for different ranks in the same classification are often drawn from the same stem; for example, the winged bean of the legume family, *Psophocarpus tetragonolobus*, belongs to the legume section *Psophocarpus*, subgenus *Psophocarpus*, of the genus *Psophocarpus*.

Anyone who finds a distinct taxon that has not previously been described can publish it as a taxon new to science. To publish validly a new name, the various nomenclatural codes demand that the author provides certain key details (precise details vary depending on the code), which are shown in Box 3.3.

3.3.5 Synonyms

Another important secondary product of a revision is a list of synonyms for each accepted (valid) name. Synonyms are nomenclaturally illegal names (nomenclaturally incorrect according to the Code of Botanical Nomenclature) which refer to a taxon with a different accepted name. In a study of the temperate forage legume vetches, one species, the field or common vetch, was found to have 149 synonyms referring to the one species. This species was described originally by Linnaeus as *Vicia sativa* in the *Species Plantarum* (1753). Therefore, several or many binomials can refer to a single taxon; however, only one binomial is accepted and correct. Although it may appear arcane, a knowledge

Table 3.4 **A hierarchy of plant taxonomic ranks**

Rank	Common ending	Example
KINGDOM	Plantae	
Subkingdom	- bionta	Embryobionta
DIVISION	- phyta	Tracheophyta
Subdivision	- phytina	Spermatophytina
CLASS	- opsida	Angiospermopsida
Subclass	- idae	Dicotyledonidae
ORDER	- ales	Fabales
Suborder	- ineae	Fabineae
FAMILY	- aceae	Leguminosae (syn. Fabaceae)
Subfamily	- oideae	Papilionoideae
TRIBE	- eae	Phaseoleae
Subtribe	- inae	Phaseolinae
GENUS	-ia, us, ium	*Psophocarpus*
Subgenus	-	*Psophocarpus*
Section	-	*Psophocarpus*
Subsection	-	
Series	-	
Subseries	-	
SPECIES	-	*tetragonolobus*
Subspecies (ssp.)	-	
Variety (var.)	-	
Form (f.)	-	

of at least the commonly used synonymy is important for the conservationist; therefore they can still obtain access to the biological information associated with the taxon via the accepted name, e.g. germplasm accessions held in an *ex situ* collection under any synonymous name.

3.3.6 Citation

Within botany there are standard means of citing a name of a taxon. A full citation includes the binomial, authors and publication details as follows:

Vicia kalakhensis Khattab, Maxted & Bisby, Kew Bull. 43(3): 535 (1988).

This means the name *Vicia kalakhensis* was validly published by three authors, A. Khattab, N. Maxted & F.A. Bisby, in *Kew Bulletin* volume 43(3) page 535 in 1988. While the citation:

Viburnum ternatum Rehder in Sargent

means that Rehder validly published the name *Viburnum ternatum* in a work edited by Sargent. Further:

Gossypium tomentosum Nutt. ex Seem.

means Seeman validly published the name *Gossypium tomentosum*, which had originally been used by Nuttall, but not validly published by him. While:

Lens culinaris subsp. *odemensis* (Ladiz.) M.E. Ferguson, N. Maxted, M. van Slageren & L.D. Robertson

means that Gideon Ladizinsky described and validly published the new species *Lens odemensis* in 1986. In 2000 M.E. Ferguson, N. Maxted, M. van Slageren and L.D. Robertson reassessed the taxonomic rank of the taxon and concluded it deserved subspecific rank and formally lowered Ladizinsky's specific to subspecific rank as *Lens culinaris* subsp. *odemensis*. If, as in this example, two publications are cited, it is referred to as double citation. An even more complex citation is shown by:

Vicia mollis Boiss. & Hausskn. ex Boiss. 1872 non Bentham 1876

This means Boissier validly published the name *Vicia mollis*, which had originally been used by Boissier and Hausknecht, but not validly published by them. Following the valid publication of this name, another author Bentham published the same binomial combination (homonym) for a different taxon, but as Boissier used the name first his name

Box 3.2 | Carl Linnaeus and Taxonomic Nomenclature

Carl Linnaeus (1707–1778) is known as the founder of taxonomic nomenclature. His life's work was to name and describe biological diversity, and he attempted to produce a catalogue of all animals and plants. His work was presented in numerous publications (*Systema Naturae* (Linnaeus, 1759) for animals, plants and minerals and *Species Plantarum* (Linnaeus, 1753) for plants). The first edition of the *Species Plantarum* (Linnaeus, 1753) has subsequently been taken as the starting points for nomenclatural priority. Figure 3.3 shows a facsimile of a double page spread from Linnaeus's *Species Plantarum*. Perhaps unsurprisingly, Linnaeus did not have as good a knowledge of tropical as he did of temperate species, and he did not include any currently recognized *Psophocarpus* species in *Species Plantarum*, but he did include one, *P. tetragonolobus*, under the synonym *Dolichos tetragonolobus*, in *Systema Naturae*.

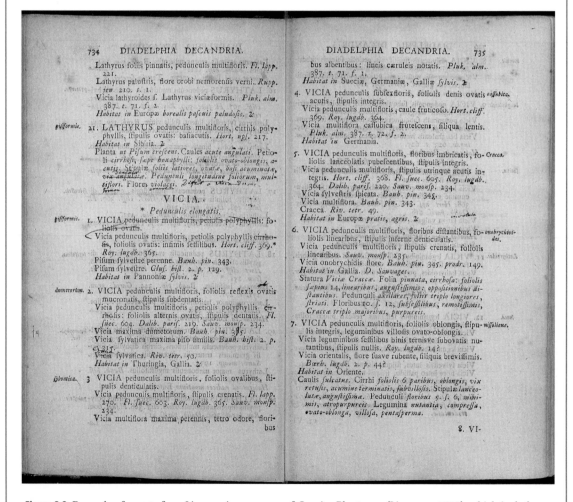

Figure 3.3 Example of a page from Linnaeus's own copy of *Species Plantarum* (Linnaeus, 1753), which includes Linnaeus's hand-written annotations of the text.

Box 3.2 | (cont.)

The pages in Figure 3.3 are written in Latin and show the account for the legume genus *Vicia* (temperate legume vetches). Linnaeus divides the genus in two based on inflorescence length, a distinction still maintained today as the two subgenera of the genus. He then gives an account for each species he recognized with the species name (specific epithet) written in the outer margin. He also provides a brief morphological description, cites specimens and gives its habitat and distribution.

Box 3.3 | How to Publish Validly a New Name

Provide the following:

- a new name combination (binomial); the combination of genus and specific epithet cannot have been used previously,
- the taxonomic rank must be specified,
- a diagnostic (short and distinguishing) description in Latin (or English, post 2012), preferably with an illustration,
- a type (typical representative) specimen(s) or drawing must be designated, and
- this information must then be published in an accepted journal.

stands, and the taxon named by Bentham had to be re-named.

3.3.7 Description and Diagnosis

Having observed a range of specimens during the revision and formed a concept of species, genera, tribes, etc., each specimen can be assigned to a taxon. The range of characters and character states for the specimens that represent each taxon can then be generalized to describe the taxon itself. A logically laid-out statement of the characteristics of a taxon is referred to as a description. An example description for *Psophocarpus grandiflorus* follows:

P. grandiflorus Wilczek – *Habit*: Perennial climbing herb with pubescent stem. *Leaves*: Trifoliolate, leaflets ovate,

5–15 × 4–12 cm, apex acute or acuminate, base rounded or truncate, sparsely pubescent along the veins on both sides but mainly beneath, veins not prominent on lower surface. Petiole 2.5–16 cm long. Petiolule 0.3–0.7 cm. Stipule ovate, obtuse at the apex, base projected, spurred, pubescent on the outside. Stipels 0.45–1.1 cm. *Inflorescence*: 2–3 flowers per node; peduncle 4–23 cm long, pubescent, ridged; rachis 1–3.5 cm long; bract ovate to lanceolate, obtuse at the apex, glabrescent. Bracteole deciduous, ovate, obtuse at the apex, always shorter than calyx. *Flower*: Calyx pubescent on the outside, 4-toothed, lower and lateral teeth either acute or acuminate but upper teeth always emarginate or form a bifid lip. Standard oblong, apex smooth, emarginate with weakly divided lobe, minutely papillate seen on the outside, internally with 2 elongate appendages. Wing rounded at the apex, simply clawed. Keel beaked at the apex, pouched, sculpturing present. Ovary linear, laterally pubescent. Stigma terminal, penicillate. Style apex not bifid, glabrous at the base. *Pods*: Oblong to ellipsoid, square in cross-section, 4–9 × 0.7–1.5 cm, glabrescent or pubescent, prominently winged, the wings 0.3–0.5 cm wide, crinkled edge. *Seeds*: unknown.

Again, it is worth noting that the language used is complex and unfamiliar to many but good glossaries to these technical terms are available (see Stearn, 1966; Hickey and King, 2000; Beentje, 2010), and when working with a specific group of plants, you will soon learn the descriptive vocabulary.

A diagnosis is much shorter than a description, covering only those characters (diagnostic characters) which are necessary to distinguish a particular taxon from closely related taxa. Diagnostic characters are those used in identification aids to distinguish related taxa. Exemplar diagnoses are provided below for the three closely related species of *Psophocarpus* section *Psophocarpus* (L.) DC.

- *P. palustris* Desv. – Plant pubescent, bracteoles shorter than calyx, standard obovate to oblong; wing claw linear, keel apex rounded, pod 2.3–5.5 cm long, with entire wing margin.
- *P. tetragonolobus* (L.) DC. – Plant mostly glabrous, bracteoles shorter than calyx, standard rounded to oblong; wing claw T-shaped, keel apex beaked, pod (6–)8.6–26(–40) cm long, with crinkled wing margin.
- *P. scandens* (Endl.) Verdc. – Plant pubescent, bracteoles as long as or longer than calyx, standard obovate to oblong; wing claw linear, keel apex rounded, pod 3.5–8 cm long, with crinkled wing margin.

3.3.8 Distributions

Just as the range of character states for a group of specimens that represent each taxon can be generalized to produce a description of the taxon, so ecological and geographic location (ecogeographic) details of representative specimens can be synthesized to produce generalized data concerning the taxon's

ecology and geographic distribution. The identification of a species' ecological niches and geographic localities is of vital importance when planning conservation. Geographic data are commonly displayed in the form of distribution maps (Figure 3.4).

3.3.9 Identification

Following the fundamental importance of the classification, probably the second most important product of taxonomy to conservation are tools for plant identification, because without a precise identification of a taxon it is impossible to enact conservation. The correct identification provides access to the wealth of information available for the species and enables communication of novel information to others without confusion. If a specimen is either unidentified or wrongly identified, any data that may be associated with that specimen is likely to be misleading and utilization of the specimen or population is likely to be limited. The traditional tools used to identify specimens are some form of key, usually dichotomous keys found in a Flora of the region or country where the specimen is found. Keys are commonly based on gross morphological features: the characteristics that are readily seen and recorded in the field, laboratory, museum or herbarium. The process of specimen identification or determination involves two steps:

- The decision as to which taxon (e.g. genus, species or subspecies) the specimen represents. This is normally achieved by use of some form of key or identification aid (discussed in detail below).
- The decision as to what is the 'accepted' name to use for that taxon, if more than one name has been used for that taxon. This involves finding out from the latest taxonomic treatment of the group or by consulting a taxonomic expert, what is the accepted name for that taxon.

There are basically two methods of identification, matching and elimination:

- Matching – involves the comparison of the unidentified specimen to a taxon description or

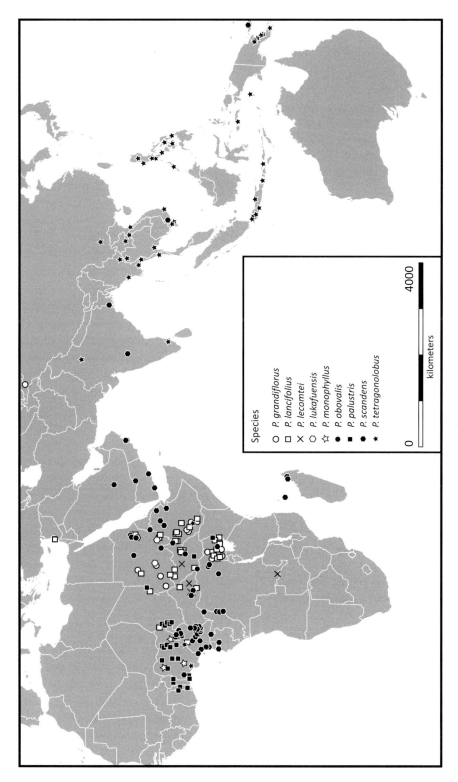

Figure 3.4 Distribution of *Psophocarpus* species.
(From Fatihah *et al.*, 2012)

some form of exemplar, such as a named herbarium sheet. For example, when trying to identify a *Psophocarpus* specimen, the characteristics of the specimen would be compared to each of the nine species descriptions, and the description that provided the best match would indicate the identification. This method is relatively easy for the nine *Psophocarpus* species, but to match a specimen to one of many possible taxa could prove impossibly time-consuming.

- Elimination – is where the user compares a specimen to a set of mutually exclusive short descriptions and decides as to which fits the specimen better. This is repeated for another set of descriptions until only one taxon remains, giving the identification. Returning to the *Psophocarpus* example, one might start by taking the character, e.g. leaflet number; if the specimen you are attempting to identify is unifoliolate, then the seven trifoliolate species would be eliminated leaving the two unifoliolate species *P. lecomtei* and *P. monophyllus,* and further characters could be used to distinguish which of these two species the specimen represents. Often, in practice, identification will begin by elimination and proceed by matching when the range of possible taxa has been narrowed down to manageable proportions.

3.3.10 Single-Access Keys

The most common keys used for identification are single-access keys which share the following features:

- single-access – meaning there is only one point of entry, via the choice between the first pair of brief, alternative descriptions,
- dichotomous – the user is presented with two brief, diagnostic descriptions at each stage in the key, only one of which matches the specimen being identified (rarely keys are written that offer a choice of three or four alternative diagnostic descriptions),
- sequential – the user is faced with a sequence of choices of alternative diagnostic descriptions; once the first choice is made the user is faced with a second and third choice of alternative diagnostic

information until by making the final choice the name for the specimen is obtained, and
- diagnostic – the few characters used in the key to distinguish the taxa included do not constitute a full description of the taxon.

The nature of the single-access key design means that the process of identification follows a specific and structured sequence until the identification is finally made. At each step in the key two alternative descriptions or leads (together referred to as a couplet) are provided and these descriptions should be constant for the species and mutually exclusive. There are two basic styles of single-access key: the parallel (or bracketed) and the yoked (or indented); neither has an obvious advantage over the other. A parallel key to *Psophocarpus* species is provided in Figure 3.5. The same data set is shown in the yoked style in Figure 3.6. Once an identification is achieved using any form of key, the specimen should always be compared to a detailed description of the taxon to confirm the identification and find if mistakes have been made during the keying-out process.

3.3.11 Multi-access Keys

Multi-access keys were developed to overcome some of the problems associated with single-access keys. A multi-access key does not force the user to go through the character set in a specific pre-ordained sequence. It allows the user to ignore a single character and still obtain an identification. For example, if the specimen lacks seed, the identification can be based on vegetative and flower characters alone. Early forms of printed tabular multi-access keys are available; an example of a tabular key to *Psophocarpus* species is provided in Table 3.5, with species in the columns and characters in the rows. The table is filled with coded character states for each taxon. To identify a new specimen, the user records its characteristics using the same codes and row structure as the tabular key and then compares the specimen data with each column of the table. A match between the user's score card and a taxon provides the identification.

1. Leaflets unifoliolate; petiole 0.15–0.4 cm long 2
   Leaflets trifoliolate; petiole 0.4–0.6 cm long 3
2. Prostrate herbs with leaves flat on the ground; leaflets elliptic or ovate, 1.8–8 × 1.5–6 cm, apex mucronate, base cordate with distinct narrow sinus; veins are not prominent on abaxial surface .... *P. lecomtei*
   Plant trailing; leaflets ovate, 6.5–15 × 3.5–9.5 cm, apex rounded to obtuse or slightly emarginate, usually rounded or truncate at the base; veins are not prominent on abaxial surface .... *P. monophyllus*
3. Standard shallowly emarginate at the apex, stigma subterminal; ovary covered with hairs; leaflet narrow 4
   Standard deeply emarginate at the apex to form bilobed lip; stigma terminal; ovary hairs lateral only; leaflet broad 5
4. Stem glabrous; calyx glabrous save for ciliolate teeth, leaflets 1.7–8 × 0.4–1.4 cm .... *P. lukafuensis*
   Stem adpressed-pubescent; calyx usually pubescent; leaflets 2.2–11 × 0.7–3.5(–6) cm .... *P. lancifolius*
5. Leaflets narrowly elliptic or obovate, narrowed to the base .... *P. obovalis*
   Leaflets broadly ovate, broader at the base 6
6. Standard rounded to oblong; keel beaked at the apex 7
   Standard obovate to oblong; keel rounded at the apex 8
7. Plant mostly glabrous; wing has T-shaped claw; pod 6–40 cm long, oblong to linear-oblong; often grown in cultivation in Southeast Asia .... *P. tetragonolobus*
   Plant pubescent; wing has simple claw; pod 4–9 cm, oblong; wild in Ethiopia, E Zaire and Uganda .... *P. grandiflorus*
8. Leaflets much more rhomboid than ovate, pubescent mainly beneath; bract pubescent; bracteoles always shorter than the mature calyx; pod 2.3–5.5 cm long and wing margin often entire; distributed from Senegal to Nigeria, eastwards to Sudan .... *P. palustris*
   Leaflets rhomboid or ovate, glabrous to glabrescent; bract glabrous; bracteoles as long as or longer than calyx; pod often longer, 3.5–8 cm long and wing margin crinkled; distributed from Cameroon to Angola, Zaire, E Africa, Malawi, Madagascar .... *P. scandens*

Figure 3.5 Parallel key to *Psophocarpus* species.

1. Leaflets unifoliolate; petiole 0.15–0.4 cm long 2
   2. Prostrate herbs with leaves flat on the ground; leaflets elliptic or ovate, 1.8–8 × 1.5–6 cm, apex mucronate, base cordate with distinct narrow sinus; veins are not prominent on abaxial surface .... *P. lecomtei*
   2. Plant trailing; leaflets ovate, 6.5–15 × 3.5–9.5 cm, apex rounded to obtuse or slightly emarginate, usually rounded or truncate at the base; veins are not prominent on abaxial surface .... *P. monophyllus*
1. Leaflets trifoliolate; petiole 0.4–0.6 cm long 3
   3. Standard simply emarginate at the apex, stigma subterminal; ovary covered with hairs; leaflet narrow. 4
      4. Stem glabrous; calyx glabrous save for ciliolate tooth, leaflets 1.7–8 × 0.4–1.4 cm .... *P. lukafuensis*
      4. Stem adpressed-pubescent; calyx usually pubescent; leaflets 2.2–11 × 0.7–3.5(–6) cm .... *P. lancifolius*
   3. Standard emarginate at the apex to form bilobed lip; stigma terminal; ovary hairs lateral only; leaflet broad 5
      5. Leaflets narrowly elliptic or obovate, narrowed to the base .... *P. obovalis*
      5. Leaflets broadly ovate, broader at the base 6
         6. Standard rounded to oblong; keel beaked at the apex 7
            7. Plant mostly glabrous; wing has T-shaped claw; pod 6–40 cm long, oblong to linear-oblong; often grown in cultivation in Southeast Asia .... *P. tetragonolobus*
            7. Plant pubescent; wing has simple claw; pod 4–9 cm, oblong; wild in Ethiopia, E Zaire and Uganda ... *P. grandiflorus*
         6. Standard obovate to oblong; keel rounded at the apex 8
            8. Leaflets much more rhomboid than ovate, pubescent mainly beneath; bract pubescent; bracteoles always shorter than the mature calyx; pod 2.3–5.5 cm long and wing margin often entire; distributed Senegal to Nigeria, eastwards to Sudan .... *P. palustris*
            8. Leaflets rhomboid or ovate, glabrous to glabrescent; bract glabrous; bracteoles as long as or longer than calyx; pod often longer 3.5–8 cm long and wing margin crinkled; distributed Cameroon to Angola, Zaire, E Africa, Malawi, Madagascar .... *P. scandens*

Figure 3.6 Yoked key to *Psophocarpus* species.

Table 3.5 **Tabular key to *Psophocarpus* species**

Character	PG	PLE	PM	PO	PLA	PLU	PP	PT	PS
Plant stature	C	P	P	P	C	C	C	C	C
Plant pubescence	Pu	Pu	Pu	Pu	Pu	G	Pu	G	Pu
Stipule shape	La	La	La	La	Ov	Ov	La	La	La
Number of leaflets per leaf	3	1	1	3	3	3	3	3	3
Terminal leaflet length (cm)	≥ 10	<10	≥ 10	≥ 10	<10	<10	<10	<10	<10
Terminal leaflet shape	O	O	O	E	E	E	O	O	O
Terminal leaflet base shape	Tr	Co	Co	Tr	Tr	Tr	Tr	Tr	Tr
Terminal leaflet abaxial vein prominence	A	A	Pr	Pr	A	A	Pr	A	A
Petiole length (cm)	≥ 5	<5	<5	<5	<5	<5	≥ 5	≥ 5	≥ 5
Inflorescence type	Ps	Ps	Ps	Ps	Ra	Ra	Ps	Ps	Ps
Number flowers per inflorescence	≥ 10	≥ 10	≥ 10	≥ 10	<10	<10	≥ 10	<10	≥ 10
Bracteole to calyx length	B$<$C	B$<$C	B$<$C	B$<$C	B$\geq$C	B$<$C	B$<$C	B$<$C	B$<$C
Wing claw T-shaped	A	A	A	A	A	A	A	Pr	A
Keel side pouch	Pr	Pr	Pr	A	Pr	Pr	Pr	Pr	Pr
Keel sculpturing	Pr	Pr	Pr	A	Pr	Pr	Pr	A	A
Stigma position	T	T	T	T	St	St	T	T	T
Style apex shape	Tr	B	B	B	Tr	Tr	Tr	Tr	TR
Pod exocarp	C	C	C	C	C	C	L	L	L
Pod wing edge	Cr	En	En	En	En	En	En	Cr	Cr

Abbreviations: PG = *P. grandiflorus*, PLE = *P. lecomtei*, PM = *P. monophyllus*, PO = *P. obovalis*, PLA = *P. lancifolius*, PLU = *P. lukafuensis*, PP = *P. palustris*, PT = *P. tetragonolobus*, PS = *P. scandens*. C climbing, P prostrate, Pu pubescent, G glabrous, O ovate, E elliptic, Tr truncate, Co cordate, Pr present, A absent, Ps pseudoraceme, Ra raceme, B bract, C calyx, T terminal, St subterminal, C coriaceous, L lignified, En entire, Cr crinkled.

Commonly now computers are used for multi-access identification, so-called interactive identification, meaning the user interacts with the identification program and data set to obtain the identification. The interactive identification program holds a matrix of taxa against characters (essentially like Table 3.5), possibly including both text and images, and the user enters attributes (character state values) of the specimen to be identified. The program eliminates taxa whose attributes do not match those of the specimen. This process is continued until only one taxon remains, giving a provisional identification. As with all identifications, the specimen may then be checked against a description, images or already named specimens to confirm the identification.

There are several interactive identification programs now available. For example, Linnaeus II (www.eti.uva.nl/products/linnaeus.php) and Lucid (www.lucidcentral.com), which supports the creation of taxonomic databases, optimizes the construction of easy-to-use identification keys, expedites the display and comparison of distribution patterns, and promotes the use of taxonomic data for biodiversity studies. Examples of interactive keys for CWR identification for African *Vigna* (Maxted *et al.*, 2004) and *Psophocarpus* species (Fatihah *et al.*, 2012) can be downloaded from the Lucid website (http://keys .lucidcentral.org/keys/African_Vigna/default.htm). The availability of conservation field guides that provide a user-friendly combination of ecogeographic and identification text, images, distribution maps and an interactive key, such as the International Center for Agricultural Research in the Dry Areas (ICARDA)'s Conservation Field Guide to Medics of the Mediterranean Basin (Al-Atawneh *et al.*, 2009), is proving helpful, but such field guides are few and unavailable for many key CWR taxa. Key generation is summarized in Box 3.4.

3.3.12 Illustrations

One of the unfortunate problems associated with, for example, a non-specialist learning to use traditional keys is the amount of technical terminology involved. For example, where is the keel claw in a flower or what shape is a capitate style? One means of avoiding technical terms is to use illustrations (line drawings, photographs, paintings, etc.) of the key features of the species, hence avoiding the use of complex terminology. There is a problem associated with using photographs for identification, in that they can only show what is observed at that time in a two-dimensional image, whereas with a drawing or painting the illustrator can enhance the observed two-dimensional image to include features that may be less obvious on an individual specimen or in that plane of view (Figure 3.7). For this reason, it is less likely that a specimen could be accurately identified by comparison with photographs, and generally, illustrations should be used in conjunction with other aids.

3.3.13 Barcode Identification

The increasingly routine use of genomic techniques had led to suggestions that DNA samples might be used for specimen identification (Kress *et al.*, 2005). So-called DNA barcoding uses a short genetic marker in an organism's DNA to identify it as belonging to a species. While this would not have a field application (at least at the present time), it does mean that if a

---

Box 3.4 | **How to Write a Key**

1. The study group is selected and delimited, e.g. the nine species of the genus *Psophocarpus*.
2. Diagnostic characters are selected which distinguish the taxon from related taxa. (Ideally key characters should be easily scored and have a high information content – enable division of the taxa into roughly equal groups), e.g. see Table 3.1 for *Psophocarpus* species.
3. All taxa (via specimens) are scored for the character set producing a dataset (taxon × characters).
4. The dataset can then be:
   a. Entered into a key generation program (e.g. DELTA) and the key generated automatically or,
   b. Used manually to generate the key.

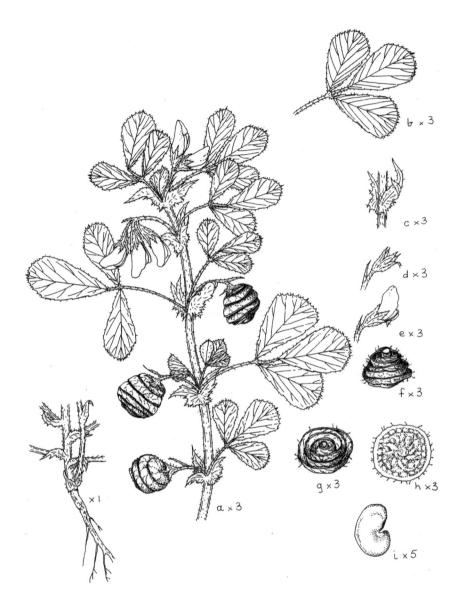

Figure 3.7 Botanical line drawing of *Medicago scutellata* (L.) Miller: a, habit (×3); b, leaflet (×3) c, stipule (×3); d, calyx (×3); e, flower (×3); f, pod three diminution view (×3); g, pod tip view (×3); h, pod venation(×3); i, seed (5×); j, root (1×). (From Al-Atawneh *et al.*, 2009)

specimen cannot be identified using more conventional means, a DNA sample could be taken and compared against a DNA reference collection. Specimens can also be identified from plant parts (e.g. even leaves or roots) when the flowers or fruits normally used for identification are unavailable.

For the system to work, a desirable locus for DNA barcoding needs to be agreed upon, so that large databases of sequences for that locus can be developed for each plant species. The Plant Working Group for the Consortium for the Barcode of Life (CBOL, 2009) have suggested the concatenation of the *rbcL* and *matK* chloroplast genes as the locus for plants. The DNA sequences would be stored in a DNA sequence database, such as GenBank, but there would be a need to link DNA sequences to vouchered specimens to ensure that the sequences are grounded to a named specimen of the taxon (Miller, 2007). Although DNA barcoding has been criticized and there are still technical problems that remain with its practical implementation, it seems likely to be a technique of growing importance in future years.

3.4 Sources of Taxonomic Information

3.4.1 Specialist Publications

The primary sources of taxonomic information are specialist publications (revisions, monographs, checklists, taxonomic journals, Floras and Faunas), educational materials (field guides, lectures, multimedia programmes) and taxonomic experts. Specialist publications are likely to be found in the libraries associated with various kinds of taxonomic collections, e.g. museums, herbaria, botanic gardens, germplasm collections and zoos, while general libraries will provide popular field guides and other media guides.

3.4.2 Taxon Experts

The conservationist may gain access to taxonomic experts by contacting the taxonomic collections or educational establishments with which they are usually associated. Finding the appropriate taxon or geographic expert usually involves a search of the taxonomic literature and identification of who has published literature on that plant or that locality. By contacting them they may be able to provide information on the taxon, identify the currently accepted classification, provide leads into the published and grey literature, and indicate where the major collections are located, as well as other information, such as identifying phytogeographic experts.

3.4.3 Taxonomic Databases

These often contain summary information on the taxonomy of specific groups, e.g. ILDIS (International Legume Database and Information Service) for the legume family Leguminosae, which contains information on legume nomenclature, bibliography, distribution and biodiversity. Other databases, such as Index Kewensis (administered by the Royal Botanic Gardens, Kew), contain nomenclatural information on the world's angiosperms. The databases may also be geographically specific, such as ERIN (Environmental Resources Information Network), which provides a geographically related environmental information system for Australia that can be used to aid environmental impact assessment and monitoring of species, vegetation types and heritage sites.

4 Plant Population Genetics

4.1 The Importance of Population Genetics to Conservation

Population genetics, which is a branch of genetics, provides the genetic basis of evolution (Gillespie, 1998), studies genetic differences within and between populations, and assists our understanding of the variation of a species due to allele frequency changes in a target population at a defined area (Falconer and Mackay, 1996). As noted by Felsenstein (2016), the theory of population genetics – an area of biology depending significantly on mathematics – emerged between the 1920s and 1940s and was further rigorized during the 1970s and 1980s, which explains why most of the basic work was done before the 1980s. Mendelian segregation and random mating are very important in population genetics because they provide its framework, while its focus is the quantification of allelic and genotypic frequencies in successive generations (Holsinger, 2015). Conservation genetics (Box 4.1) uses population genetic theory for measuring diversity in a target population.

4.2 Types of Variation

The plant's genome includes its genetic code, i.e. the deoxyribonucleic acid (DNA) or the individual's blueprint encoded by 'building blocks' or four chemical bases, namely adenine (A), guanine (G), cytosine (C) and thymine (T). This code may vary among individuals of a population within a species, which is its genetic diversity. The sequence (or order) of these four bases defines the available genetic information of a plant and defines the plant itself. The small differences in coding a gene among individuals of the same population lead to distinct genotypes,

---

Box 4.1 | What Is Conservation Genetics?

Conservation genetics measures individuals and populations affected by habitat loss, exploitation or change of environment, and provides knowledge on how such populations may survive further (Frankham *et al.*, 2002). Genetic diversity measurements refer to the allele types of a particular gene(s) in a population, i.e. the variability of genes. It can be estimated by measuring heterozygosity, number of alleles per locus or percentage of polymorphic loci in a population. There will be low genetic diversity if most members of a population share the same alleles, while genetic diversity may be high when a population shows distinct allele types in a particular gene(s). A population whose individuals are the same may be at risk of genetic erosion or become extinct if susceptible to a pathogen or a pest, whereas a population with high genetic diversity will increase its chances of surviving if assuming that some of its individuals bear genes providing host plant resistance against such pathogens or pests, i.e. their 'genetic makeup' allows them to 'fight' and survive. Genetic diversity may change because only surviving variants contribute to diversity in the next generation.

which are reflected in their distinct phenotypes. A population or species with low diversity may result in inbreeding depression, which is the decreased fitness of a population due to its individuals having high homozygosity of deleterious recessive alleles, thus affecting their ability to survive and reproduce.

4.3 Population and Quantitative Genetics

A Mendelian population is a group of individuals of the same species sharing a common set of genes and living in a geographic area sufficiently restricted so that any member has the same potential to mate with another member (of opposite sex) of the same population, i.e. random mating. Population genetics research shows how changes in allele frequency across generations in a population are due to selection, mutation, migration and drift. Hence, as indicated by Hermisson (pers. comm.) it describes heritable changes in biological populations over time.

Gametes give rise to the zygotes of the next generation that may differ from the previous generation due to re-assortments in the gene pool. Let's assume that a gene has two alleles, A and a, and the number of the following genotypes AA, Aa and aa are noted as X, Y and Z; being $X + Y + Z = N$. Their allele frequency (p for A and q for a) are thereafter calculated from the number of individuals of each genotype as follows:

$$p = \frac{\left[X + \frac{1}{2}\, Y \right]}{N}$$

and

$$q = \frac{\left[Z + \frac{1}{2}\, Y \right]}{N}$$

In summary, the frequency of allele A is determined by adding half the frequency of heterozygotes Aa to the frequency of homozygotes AA, while the frequency of allele a is equal to the sum of half the frequency of heterozygotes Aa and the proportion of homozygotes aa. Likewise, as $p + q = 1$, then $q = 1 - p$.

The evolutionary change in a population can therefore be described as modifications in both the allele frequencies and genotype frequencies (Falconer and Mackay, 1996). There will be no evolution at a locus if all individuals of a population are homozygous for the same allele, i.e. $p = 1$. There may be evolution at a locus if there are two alleles, because the frequency of one allele may increase at the expense of that of the other.

In 1908, Godfrey Harold Hardy (English mathematician) and Wilhelm Weinberg (German obstetrician–gynaecologist) demonstrated separately that in a large random mating population whose individuals are equally viable and fertile, inheritance itself does not change the allelic or genotypic frequencies of a given locus (Hardy, 1908; Weinberg, 1908). Random mating between individuals refers to the unsystematic union of gametes in a population of individuals at sexual maturity. However, partial self-fertilization (or selfing) leads to having less frequency of heterozygotes, thus increasing the frequency of homozygotes. The Hardy–Weinberg principle, as known today, assumes that under certain conditions, the description of the system does not change in time once the equilibrium is reached, and that the achievement of the latter can take one or more generations, depending on the physical constraints imposed by the organization of the genome. The certain conditions are:

- a large population size;
- a closed population (there is no migration);
- no mutation;
- normal Mendelian segregation of alleles;
- random union of gametes;
- equal fertilization capacity of gametes;
- equal survival of all genotypes.

This means that any gene located on any chromosome that is not involved with the determination of sex achieves genotypic frequency equilibrium after a generation of random mating when taken separately, but it is not true when considering two or more loci at a time. The tendency towards equilibrium slows down when the linkage is intense. The frequency of heterozygotes is relatively higher as recessive phenotypes becomes rarer in a Hardy–Weinberg equilibrium population, thus showing the difficulty to eliminate recessive deleterious characters because of heterozygosity upon which selection for cannot act.

Most species are subdivided into a few or many Mendelian populations, which have distinct genetic diversity patterns owing to their adaptation to the local environment where they thrive. Hence, it is important to keep such genetic diversity revealing distinct adaptation archetypes distinct, for example having genetic variants suitable to environment(s) brought by the changing climate. The terms heterozygosis observed, expected heterozygosity, number of alleles per locus and polymorphism are often used to describe genetic diversity.

Qualitative variation refers to phenotypes of individuals assigned to a small number of classes because their distinct characters are discontinuous or discrete. The Austro-German Augustinian friar Gregor Mendel, who is the founder of modern genetics, did his inheritance research on such characters in peas (Mendel, 1866). A quantitative morphological, physiological or behavioural character may take different quantifiable values in different individuals and does not follow a pattern of simple Mendelian inheritance. It may show continuous (e.g. plant height or seed weight) or discrete variation (e.g. days to flowering or fruit number). The phenotypes of individuals showing quantitative variation are determined by measuring them, and statistical methods such as mean and variance are used to describe their quantitative variation, which is often determined by many genes whose individual effects are small and affected by the environment. Such characters are of great importance for agriculture, and the study of their inheritance is the subject of quantitative genetics. Sir Ronald Aylmer Fisher – an English statistician and biologist – integrated Mendelian genetics with biometrics by demonstrating that quantitative variation was a natural consequence of Mendelian inheritance (Fisher, 1930). The multi-genes defining quantitative characters show the same features of major genes, i.e. they are on the chromosomes and can display linkage in their inheritance, dominance and epistasis. They only differ from major genes controlling qualitative characters

on the magnitude of their individual effects on the phenotype. In this context please note the following definitions:

- Linkage refers to the association in transmission between genes on the same chromosome instead of displaying independent inheritance. Such genes are closely located on the same chromosome; thus, crossing over does not often occur between their loci that leads to a higher frequency of offspring carrying the parental combination of alleles than those containing non-parental combination of alleles.
- Dominance describes the interaction between alleles of the same gene in a heterozygote. On the allele with this feature the dominant allele masks the expression in the phenotype of the recessive allele at the same locus.
- Epistasis occurs when the phenotypic expression of genes at one locus depends on the genotype at another locus, i.e. a non-allelic interaction.
- Multi-genes are many genes segregating simultaneously and affecting a quantitative character. Multi-genic inheritance arises when one characteristic is controlled by at least two genes.

4.4 Changes of Gene and Genotype Frequencies

As noted before, the allelic and genotype frequencies remain stable in a large random mating population. There are, however, certain conditions that may affect these frequencies (Figure 4.1). Mutation, migration, selection, sampling variation and the mating system cause changes of gene and genotype frequencies in populations (Falconer and Mackay, 1996). They are either directional or dispersive. Directional processes such as migration, mutation and selection change frequencies predictably in quantity and direction, particularly in infinite size populations. In contrast, dispersive processes occur in small populations due to sampling (e.g. genetic drift) or mating systems. Although its frequency changes are predictable in quantity, they are not predictable in direction.

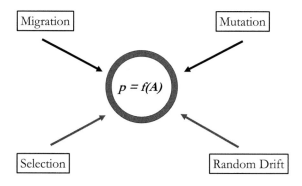

Figure 4.1 Factors affecting gene frequency (*p*) changes in a population.

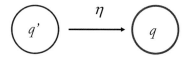

Donor population   Recipient population

Figure 4.2 Allelic frequency changes between a donor (with q' allelic frequency) and recipient population (with q allelic frequency) depend on migration rate (η).

$$q_{t+1} = \eta q' + (1 - \eta)q_t$$

Thus,

$$\Delta q = \eta(q' - q)_t$$

Migration introduces new alleles to a local population and if followed by hybridization and introgression through random recombination, leads to changes in gene frequency and an overall increase of genetic diversity, e.g. after spontaneous crossing of cultigen with crop wild relatives. Plant introduction or assisted migration, i.e. a deliberate migration of plants to a new 'home', will also change gene frequency and increases genetic diversity.

4.4.1 Migration or Gene Flow

Populations often show an amount of gene transfer, which is expected for populations closely related both spatially and genetically. Gene flow among nearby populations of the same species may be large; thus contiguous populations have a more similar genetic composition than those that are more geographically distant. Migration between geographically isolated groups is a key event because adaptive gene complexes are broken, allelic frequencies are altered and genetic differences between populations are reduced. Gene flow effects depend on the structure of the migrant and recipient populations, the amount of migration and the magnitude of the difference in gene frequencies between the two populations.

Allele frequencies in each local population can change independently, and the local population can undergo considerable genetic differentiation in the absence of migration. Such genetic differentiation reveals variation in the frequency of common alleles among local populations and also may lead to having therein rare alleles that are not found in other populations. Migration minimizes genetic differentiation, which may occur in spite of the former if natural selection is large enough.

The changes in allelic frequencies (Δq) are proportional to the frequency differences between the donor (q') and recipient (q) populations and the migration rate (η) (Figure 4.2). For example, the frequency of allele *a* in recipient population at generation t+1 is

4.4.2 Mutation

Mutations and gene edits alter single base units in DNA and transmit the change to subsequent generations, or from deletions, insertions or rearrangements of large sections of genes or chromosomes. They are therefore non-recurrent or occurring only once, or recurrent by arising at a certain rate but being reversible or irreversible. Gene mutations are often recurrent, whereas chromosomal rearrangements are very rare and most likely unique events because breakages occur randomly along the chromosomes, thus it being highly unlikely that such a reordering will be repeated. Non-recurring mutations may remain in the populations depending on the number of descendants that a mutant individual has. Such probability of remaining decreases generation after generation because a single mutation without selective advantage cannot produce a permanent change in the population. There are other mechanisms, in addition to selection, that can help

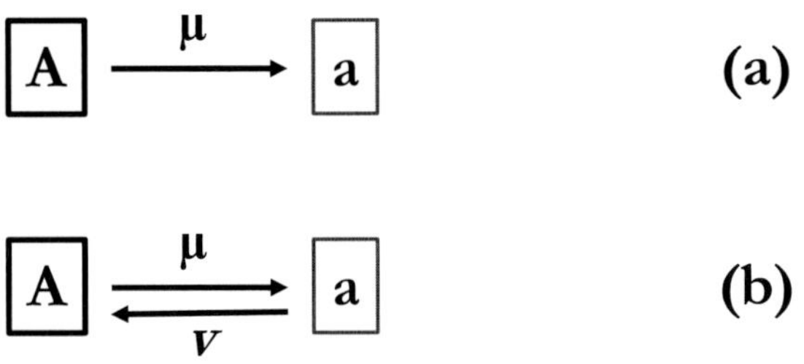

(a)

(b)

Figure 4.3 Mutation (a) at μ rate and reversible mutation at a rate v (b).

non-recurrent mutations to increase in frequency in populations. Meiotic drive favouring the transmission of one allele from a heterozygote over another in the gametes and genetic drift are other mechanisms, beyond selection, helping non-recurrent mutations to increase their frequency in populations (see Figure 4.3).

Let's assume an allele A mutates only forward to a in an irreversible mutation at a μ rate, thus being $1-μ$ the proportion of allele A that does not mutate to a. Hence, the frequency of A in any generation (p_n) is

$$p_n = p_{n-1}(1 - μ)$$

where p_{n-1} is the frequency of A in the previous generation. After successive substitutions

$$p_n = p_0(1 - μ)^n$$

where p_0 is the initial frequency. When the allele A is fixed at time n, then

$$p_n = (1 - μ)^n.$$

When taking into account a reversible mutation (from a to A) occurring at a rate v, at equilibrium

$$pμ = qv$$

or after replacing q by $1-p$,

$$p = \frac{v}{μ + v}$$

Mutation rates are low, and most newly arising mutations are harmful to the organisms and are therefore eliminated from the population in a few generations. Although mutation is a relatively weak force *per se* for allele frequency change, it is still

essential in evolution as the ultimate source of genetic variation.

4.4.3 Selection

Different individuals in a population differ in their viability or fecundity, thus contributing a different number of offspring and genes therein to the next generation. Fitness of an individual is the relative proportion of its offspring that form the next generation. Although fitness seems to be difficult to measure, it may suffice to count the offspring number produced by an individual and compare it with those produced by the rest in the population. The relative survival rates of each genotype allows us to estimate their relative viability. The coefficient of selection (s) is the proportional reduction in fitness of a certain genotype compared to another genotype that is the most favoured by selection. The efficacy of selection depends on the degree of dominance in fitness.

Sir Ronald Fisher (1930), Sewall Wright (1931), a US geneticist, and John Burson Sanderson Haldane (1932), a British-born Indian scientist, provided the basis for the genetical theory of natural selection, which accounts for many adaptive characters of organisms. Their research was key to bringing together genetics and evolution. The occurrence of some genes within a population provides the population with a certain advantage over others to adapt, owing to their high survival rate and reproductive capacity, to environmental changes. Selection is, therefore, the most important factor responsible for the change in gene frequency.

Natural selection is the process in which heritable traits that increase an organism's chances of survival and reproduction are favoured over beneficial traits. It is the driving force of adaptive evolution. The main principles of natural selection relate to variability among members of a population, heritability of the variable characters that help to survive and reproduce, differences in population members' ability to reproduce, and survival of the fittest (when competing among offspring) in terms of variability, heritability and reproduction. Genotypes promoting the survival are therefore present in excess among individuals of reproductive age in every generation, thus

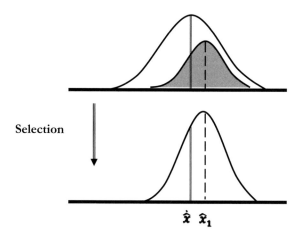

Selection

Figure 4.4 Directional selection (grey shade) changes population mean from $\hat{x}$ to $\hat{x}_1$.

contributing disproportionally to the offspring of the next generation. Hence, those alleles that enhance survival and reproduction increase in frequency from generation to generation, thereby allowing the population to survive and reproduce in the prevailing environment, i.e. evolutionary adaptation.

Directional selection (Figure 4.4) favours the increase of one allele, thus tending to eliminate the alternative allele (or alleles). For example, let's consider the selection against homozygous recessive in a locus with two alleles: *A* and *a*. The fitness will be 1 for both *AA* and *Aa*, and $1-s$ for *aa*. After selection the genotypic frequencies will be $[p^2/\varphi_D]$ for *AA* $[2pq/\varphi_D]$ for *Aa*, and $[q^2(1-s)/\varphi_D]$ for *aa*, where $\varphi_D = 1-sq^2$. Selection only acts, therefore, if there is genetic variability ($pq \neq 0$), while any change in frequency will lead to the disappearance of the unfavourable allele. Furthermore, selection increases the frequency of a new beneficial allele faster when dominant, while the efficacy of selection is lower when the gene is rare.

The selective advantage of heterozygote is noted in balanced selection through the genotypic frequencies as follows: $[p^2(1-s_1)/\varphi_H]$ for *AA*, $[2pq/\varphi_H]$ for Aa, and $[q^2(1-s_2)/\varphi_H]$ for *aa*, where $\varphi_H = 1-s_1p^2-s_2q^2$. Hence, this process (provided that $p \neq q$) depends on the initial *p* and *q*, and the magnitude of the selection coefficient against homozygotes (s_i). Neither allele is eliminated by selection because the heterozygotes produced more offspring than the homozygotes, thus keeping both alleles in the population.

Domesticated plants depend today on humans for their survival because the desired characters which humans select for breeding means they are often maladaptive in nature (e.g. seed retention and non-dispersal). Deliberate selection such as plant domestication and modern plant breeding have changed gene frequency and reduced genetic diversity (Tanksley and McCouch, 1997). The methodical selection for desired character(s) increases systematically the frequency of alleles controlling such character(s) or results in homogenization and standardization of released cultivars due to end-user demands.

Environmental degradation due to pollution, salinization, soil erosion, climate change or weed invasion causes inadvertent selection, thus showing the human impact on gene frequency changes. Furthermore, there may be a strong selection pressure for evolutionary change in weeds after relentless herbicide use. The evolution of crop–weed associations also leads to inadvertent selection due to mimicry (resulting from the similarity of one to another owing to their resemblance) that allows weedy species to evolve characters for their survival in human agricultural systems, thus changing both gene frequency and diversity. Moreover, the deployment of resistance genes in plants changes gene frequency and may increase diversity in the pathogen due to shifts directing towards new virulent races that overcome these host plant resistance genes. Likewise, biological control agents may reduce a pest but may allow other pest forms to extend their spreading and abundance.

Mutation and selection occur simultaneously in nature, and both, as already noted, affect gene frequencies. The population attains a state of equilibrium (eq) if selection and mutation act in opposite directions, i.e.

$$q_{EQ} = \sqrt{\frac{\mu}{s}}$$

This state is known as selection–mutation balance, in which new mutations offset selective elimination, which explains why deleterious or lethal genes associated with low fitness remain present in populations.

4.4.4 Genetic Drift

There are many genes included in the transmission of gametes from the parents to the offspring, which makes allele frequencies fluctuate from generation to generation, particularly in natural populations that have a finite size. Genetic drift leads to changes in allele frequencies in a population over time due to random sampling and chance, which determines whether an individual survives and reproduces. Hence, these changes are not driven by the environment or adaptative pressure and may be beneficial, neutral or detrimental to reproductive success. Random genetic drift may cause variants to disappear completely, thereby reducing overall population genetic diversity (Figure 4.5). This effect is high in small populations, which reveals that genetic drift rate is inversely proportional to the effective population size, i.e. the smaller the effective population size, the faster differences accumulate due to genetic drift. Although selection can oppose genetic drift, it should be very strong in small populations.

Random genetic drift may also account for the narrowing of the genetic base in agricultural systems and for populations subject to habitat destruction and fragmentation or result from weed colonization after accidental anthropogenic-led invasion; each will bring a major decrease of genetic diversity. When establishing a population with very few individuals, the differences in the frequency of many genes with respect to the population of origin can be inflated because the small group of migrants is not genetically representative of the original population. This 'founder effect' is associated with loss of genetic diversity as the new population is a non-random selection from the parent population. Likewise, chance variations in allele frequencies, similar to those of the 'founder effect', occur when populations pass through a 'bottleneck' arising from unfavourable conditions that drastically reduced the population in number. Although such populations may regain their size, genetic drift has already changed significantly their allelic frequencies. A population 'bottleneck' leads to genetic variation loss and frequent mating among closely related individuals. In summary, small populations favour homogeneity due to elimination or fixation of genes, while diversity prevails in large populations.

4.5 Mating Systems and Inbreeding Coefficient

Inbreeding, which is a form of 'consanguinity', is the mating between closely related individuals in a population to which they belong or with itself, thus resulting in a decrease of heterozygotes (*Aa*) and an increase of dominant (*AA*) and recessive (*aa*) homozygotes in the population. Homozygosity may lead to genetic defects associated with recessive genes (*a*), particularly when dominant heterozygous (*Aa*), which are masking them, segregate. Inbreeding can be calculated when knowing the pedigree of any individual. It is estimated as the probability that alleles in any locus of an individual are identical by descent, i.e. an identical replica of an allele at the time of generating the gametes and reunited in an offspring. The inbreeding coefficient (*F*) of an individual measures, therefore, the cumulative probability across all common ancestors in the pedigree of having identical homozygotes at any locus (Wright, 1922). *F*, being a measure of a probability, ranges between 0 and 1.

Inbreeding *per se* does not increase the number of recessive genes in a population, but it unmasks how gene pairs are grouped, i.e. more homozygous and less heterozygous. Inbreeding changes, therefore, genotype frequencies of the population because it reduces the frequency of heterozygotes. Hence, the genotype frequencies under inbreeding are $[p^2 + pqF]$ for *AA*, $[2pq (1-F)]$ for *Aa* and $[q^2 + pqF]$ for *aa*. The increase in *pqF* of each homozygote arises from reducing $2pqF$ in heterozygote frequency, which affects equally both *p* and *q*. Hence, inbreeding only changes genotype frequencies but does not affect alleles.

The likelihood of mating between relatives is very low in a large population, but it may be large in small populations or self-pollinating species. The coefficient of inbreeding is zero in an infinite population under

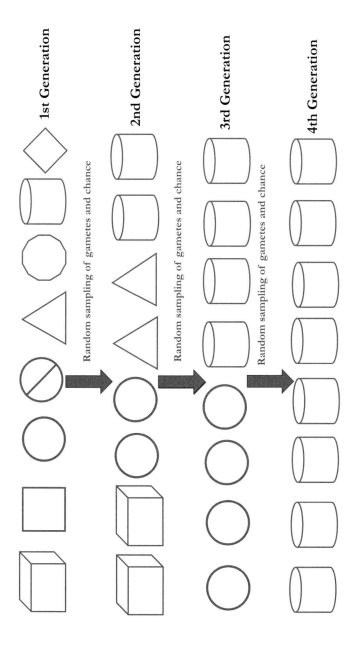

Figure 4.5 Genetic drift due to random sampling of gametes and chance.

random mating, but it may reach a value in a finite population. Hence, F can be estimated in finite populations with unknown kinship relations based on population size. The probability that an individual bears identical alleles in a population of size N is

$$\frac{1}{2N} + \left(1 - \frac{1}{2N}\right)F$$

while F can be estimated as

$$F = 1 - \left(1 - \frac{1}{2N}\right)n$$

where n is the number of generations. F increases progressively as N becomes smaller.

4.6 Effective Population Size

Effective population size or EPS or N_e is a basic parameter in various population genetics models and defined as the number of breeding individuals in an idealized population showing the same amount of dispersion of allele frequencies under random genetic drift or the same amount of inbreeding as the population under consideration. The number of individuals making up the breeding population in an idealized population is known as the breeding effective population size, while the minimum viable population size (MVP) is an estimate of the smallest number of individuals within a population capable of maintaining that population's viability over an extended time period without significant manipulation. Determining an EPS is not an easy task because it is difficult to obtain reliable estimates of the vital statistics of real populations as well as due to the complex way in which factors combine to determine the relationship between N_e and actual population size N. Some useful insights can be obtained, however, by considering the effects of some of these factors separately.

Let's study the effect that variation in family size has on the relationship between N_e and N by assuming that individuals in a population mate randomly, that the number of alleles (equivalent to family size) contributed by the *ith* individual to the next generation is k_i and with a variance V, and differences

in the parental micro-environment or chance cause variation in family size. If N remains constant from one generation to the next, the average number of alleles contributed by each parent to the following generation will be $k = 2$. Thus, the relationship between N_e and N is approximately

$$N_e = \frac{(4N - 2)}{(2 + V)}$$

If reproductive outputs differ randomly, then k_i follows the binomial distribution, thus

$$V = 2\left(1 - \frac{1}{N}\right)$$

and

$$N_e = N$$

In natural populations the distribution of family size is, however, mostly non-random so

$$V > 2\left(1 - \frac{1}{N}\right)$$

thus making

$$N_e < N.$$

Hence, the effective population (N_e) size is often smaller than the absolute population size (N).

When each parent contributes exactly two alleles to the next generation as happens with a 'captive' population in a gene bank, $V = 0$, thus making

$$N_e = 2N - 1.$$

A population may have various N_e according to the features of interest, including different loci. N_e may be measured as within-species genetic diversity divided by four times the mutation rate, because in such an idealized population, the heterozygosity is equal to $4N\mu$.

4.7 Practical Implementation of Genetic Knowledge

4.7.1 Collecting Germplasm

Gene banks collect plant germplasm when gene pools are in danger of genetic erosion or extinction, to address requests from users, after identifying that diversity may not be well represented (i.e. gap-filling), or for doing more research about them. The main collecting sites are often farmers' fields (especially for landraces), kitchen or orchard gardens (particularly for minor crops), markets, and wild habitats for crop-related species (see Chapter 11). Knowledge on, *inter alia*, distribution and variation of species to collect, their breeding system or population structure assists in all conservation planning. For example, there will be homogeneous homozygous (single) genotypes, homogeneous but heterozygous (F_1) genotypes, heterogeneous but only homozygous genotypes (cultivar mixtures), and heterogeneous (homo- and heterozygous) genotypes in an open-pollinated composite, landrace or cultivar.

Following the neutral allele theory (Kimura, 1968), the Japanese biologist Motoo Kimura and US population genetics pioneer James F. Crow (1963) determined that the number of selectively neutral alleles (k) in a sample S of random gametes, from an equilibrium population of size N, at a locus with a mutation rate μ is approximately

$$k \sim \theta \log_e[(S + \theta)/\theta)] + 0.6$$

where $\theta = 4N\mu > 0.1$ and $S > 10$ (Kimura and Crow, 1963). Hence, the number of alleles in a sample depends directly on the $\log_e$ of sample size. The resources required to collect in a single site are, therefore, directly proportional to the sample size. Thus, there will be a diminishing return per unit cost to collect new alleles by increasing the sample size.

A general formula for the sample size (S) considering diploid individuals with an inbreeding coefficient (F) to be 95% certain of obtaining one copy of an allele with frequency p is

$$S = \frac{3}{(F - 2)\log_e(1 - p)}$$

Although this sample may contain a single copy of all the common alleles in a population, it may be inadequate for accurate studies of their population frequency.

As per population genetics theory, the probability (P) that a randomly drawn sample of size N contains at least one of two alleles in a locus (i.e. both alleles are present) is

$$P(A, a) = 1 - p^N - q^N$$

where the probability that all N plants in a sample are *aa* is q^N and the probability that all plants are *AA* is p^N. Hence, a random sample of size 172 plants should be enough for preserving at a very high probability (0.9999) at least one copy of each of the two alleles of a single gene, provided that neither has a frequency below 0.05, irrespective of whether the individuals set all of their seed by selfing, outcrossing or a mixture of both (Lawrence *et al.*, 1995).

In an outcrossing species, such as maize, each field may be further regarded as an experimental unit, thereby estimating the sample size (S_N) for a categorical variable (presence or absence) with a finite population size with the equation below derived by the Scottish statistician William Gemmell Cochran (1977):

$$S_N = \frac{N * Z_{1-\alpha}^2 * p * q}{d^2 * (N - 1) + Z_{1-\alpha}^2 * p * q}$$

where N is the population size or the total number of maize fields, p the estimated proportion of an attribute that is present in the population, $q = 1-p$, d is the precision, α the significance level (0.05), $1-\alpha$ the confidence level and $Z_{1-\alpha}$ a pre-established value. A zig-zag walk may be used for seed collecting and taking at least one seed per plant included in the sampling (see further discussion in Chapter 11).

4.7.2 Gene Bank Seed Regeneration Strategy

The initial propagule viability, the rate of viability loss and the 'rejuvenation' standard (especially for seed-propagates species) determine the frequency for regeneration. For example, a 50% drop in seed germination may cause a significant change in the genetic composition of a heterogeneous accession because of the differential survival rate of genotypes. Although some gene banks set a minimum standard for regeneration, when a drop of 65% could be predicted or actually occurred, Bioversity International set the preferred standard at 85% seed germination rate. Regeneration should also consider, as noted by the Austrian-born Australian geneticist sir Otto Herzberg Frankel, the minimum population size that is likely to yield a required level of variation to afford the flexibility for evolutionary persistence.

Very often, gene banks use up to 30 for keeping polymorphic (p = 0.10) or 300 plants for rare (p = 0.01) alleles, respectively (Crossa et al., 1993). A bi-parental mating procedure should be used for outcrossing species to avoid selfing, to ensure each individual contributes two alleles to the next generation and produced seed by each plant are kept separately. This procedure is also known as paired crossing design, in which a large number of plants (N) are taken at random for further crossing them in pairs to produce 1/2N full-sibs. In each pair one is designated at random as a female parent and the other as the male parent. Female parent's flowers are emasculated – for non-self-incompatible species – and thereafter pollinated by the male parent. Seed are taken separately at harvest from each female parent and put into packets for storing. In the next generation, two plants are raised from the seed of each of the female parents of the previous generation. This procedure is repeated in each round of seed regeneration thereafter. This paired crossing approach ensures that $N_e \approx 2N$, thus minimizing genetic drift for any given population size.

Self-pollination (instead of random mating) allows the production of fresh seed from selfing species, in which inbreeding depression is seldom a problem, and does not need to take into account genetic drift because once a gene is in the sample taken from a source population, it will be preserved in perpetuity if plants used for producing fresh seed are raised from seed taken from each plant of the original sample. The flowers of individuals are usually enclosed in bags to make sure they set seed only after selfing.

Unfortunately, many gene banks prefer bulking when collecting and regenerating seed instead of packing them separately as noted above. Furthermore, plants of outcrossing species are allowed to open pollinate while being isolated from each other, and their ensuing seed bulked rather than kept separately. This lack of control over the contribution of genes that each plant makes to the next generation has a drastic effect on reducing N_e relative to N for the

following reason. Such practices lead to reducing the effective population size, relative to the actual size.

4.7.3 Conserving Crop Wild Relatives *In situ*

The genetic reserve size for *in situ* conservation should be large enough to support target species in a suitable habitat where they are adapted and that allows intra-population gene flow. The steps involved include identifying key species whose loss will affect significantly the reserve's biodiversity, determine the minimum viable population size to ensure a high probability for the long-term survival of these species, and estimate the necessary area to sustain their genetically viable populations. The minimum viable population should be of a size that reduces to a low level the loss of genetic variation and heterozygosity ensuing from inbreeding due to genetic drift (Crow and Denniston, 1988). Heterozygosity declines on average 1/2N per generation, thus the minimum viable population should include 50 individuals at 1% of inbreeding per generation. Furthermore, 500 individuals in the minimum viable population will allow any variation resulting from spontaneous mutations to replace any loss by genetic drift. Both estimates lead to the '50/500 rule' used for defining 5000 as the actual size of minimum viable population for fluctuating natural populations. A population of 5000 for most herbaceous plants will fit in about 500 m^2 (or 1/20th of a hectare).

Another important genetic issue relates to having a single large reserve because populations in small reserves may go extinct. However, risk management against catastrophes suggest otherwise, i.e. to establish more than one reserve. Hence, the best strategy considers having sizeable reserves allowing gene flow among them as well as more than one population per species (see further discussion in Chapter 8).

4.7.4 On-Farm Conservation

In situ conservation on-farm includes preserving crop landraces, often alongside the crop's wild relatives in the centre of diversity (Veteläinen et al., 2009a). Both should be preserved in their 'natural surroundings' to ensure their survival, evolution and adaptation to the changing environment, thus allowing variation to continue being generated in the gene pool. *In situ* conservation keeps, therefore, genetic diversity in its ecological or agricultural context (Jarvis *et al.*, 2000). On-farm conservation refers to maintenance of landraces where they evolved their distinctive characters, along with their pollinators, soil biota and other associated biodiversity (Hunter *et al.*, 2017). The following are some issues related to the genetics of on-farm conservation: (1) differences between landrace populations of the same crop are greater than those between crop wild relative species because of variation in habitat, local preferences and selection practised by farmers; (2) genetic structure varies more widely in landraces than in natural populations; and (3) fluctuations in family size variation and very small number of plants used to harvest seed to further raising a crop. Moreover, it seems that planning and managing on-farm conservation may be a complex task because socio-economics issues should be considered along with those noted above and related to population biology for preserving landraces *in situ* (see further discussion in Chapter 9).

4.8 Conclusion

Population genetic information concerning populations or species should be used when planning *ex situ* and *in situ* conservation to ensure the preservation of the genetic variation of interest. Its usefulness begins when considering planning and collecting plant germplasm, in which theoretically a sample of about 172 plants may be enough to ensure preserving most of the genetic variation of the target species. When sampling populations during collection, ideally seed from each plant should be kept separately and not bulked. Likewise, when refreshing seed of gene bank accessions, each individual of a regeneration population should make the same gene contribution to the next generation. The use of the paired crossing design for regeneration minimizes both genetic drift and inbreeding, thus avoiding genetic variation loss. The minimum viable population size for *in situ* conservation seems to be about 5000 per population, and ideally five populations per species should be conserved.

Genetic Diversity Measurement

5.1 Measuring Diversity and Clustering Germplasm with Morphological Characters

Appropriate classification and setting up relationships between and within taxonomic clusters are important tools for genetic resources management and use. To characterize is to separate or differentiate the genetic and phenotypic variability. With advances in computing technology, numerical taxonomy and multivariate statistical methods for classifying individuals have become powerful tools for clustering and measuring diversity.

Morphological characters are used for characterization and evaluation of plant germplasm. Estimating genetic diversity or relatedness among individuals or populations may therefore rely upon recording morphological characters. Such characters may be affected by the environment, but they are often used for studying similar adaptation patterns and are also useful to define potential divergent heterotic cultigen pools for further production of hybrid cultivars. There are several classification methods that can be used in genetic resources. For example, a distance matrix is a two-dimensional array matrix that includes the distances, taken pairwise, between the elements of a set. A phenotypic distance matrix ensues after calculating the difference between each pair of accessions for each quantitative character. The distance index is thereafter estimated by averaging all the differences in the phenotypic value for each character divided by the respective range. The analysis of variance for the phenotypic distance index based on quantitative morphological characters considers the variation between and within taxonomic clusters. The phenotypic distances are treated as deviations from a taxonomic cluster position mean, and the analysis assumes the squared deviations as variances. This approach allows partitioning the total sum of squares into, between

and within taxonomic clusters. Variance components are thereafter estimated for each of the sources of variation, i.e. between (σ_C^2), and within (σ_W^2) taxonomic clusters. The ratio Φ_{FS}, which measures the degree of population divergence, is calculated as follows

$$\Phi_{FS} = \frac{\sigma_C^2}{\sigma_C^2 + \sigma_W^2}.$$

This Wright's Φ_{FS} is, therefore, the correlation between random genetic accessions within a taxonomic cluster relative to random accessions from the population at large (Wright, 1965). It ranges from 0 to 1, in which a value close to 1 indicates greater partitioning of the population into taxonomic clusters (Box 5.1).

Multivariate data on continuous characters and categorical traits are included in data recording for genetic resources characterization. Individual accessions are located in a multidimensional space, where there is one dimension for each trait. The shape and structure of the groups of accessions in this multidimensional space are unknown, but the association between characters and traits influences the shape of the groups, while the structure is affected by the true composition of the groups. The true shape and structure of the underlying groups may be defined by hierarchical, nonhierarchical and statistical classification methods. The best numerical classification strategy should be that producing the most compact and well-separated groups; i.e. minimum variability within each group and maximum variability among groups.

The average clustering method may also be used after recording qualitative morphological descriptors for grouping gene bank accessions. Data should be standardized to get a mean equal to 0 and variance equal to 1, when the data follows normal distribution. This scaling method is often required because large range differences of descriptors, i.e. normalize each value of a vector to avoid influences from descriptors with wider range of possible values that may bias distance-based classification. Clusters are merged sequentially based on Euclidean distance (E_d) using an algorithm that initially uses each accession as a cluster. The Euclidean distance (E_d) between two accessions P_1 and P_2 is a straight line defined as:

$$E_d(P_1, P_2) = [(Y_{11} - Y_{12})^2 + (Y_{21} - Y_{22})^2 + \ldots + (Y_{p1} - Y_{p2})^2]^{1/2}$$

where Y_{ij} was the frequency of an ith descriptor state in jth individual, which could have the values of 0 (if they match) or 1 (for a mismatch). A dendrogram, or a tree diagram displaying taxonomic relationships, shows the graphic depiction of the Euclidean distance coefficient (Figure 5.2). Accessions with close genetic distances are placed adjacent in this hierarchical analysis.

A classification strategy in two stages uses sequentially geometric and statistical techniques. The initial groups formed by the geometric technique may be based on Ward's minimum variance within groups (Ward, 1963), and thereafter a mixture distribution approach such as the Modified Location Method (MLM) acts upon the previous clusters (Franco *et al.*, 1998), particularly regarding shape, direction and volume of the clouds of points that make up the groups in the *p* dimensional space. MLM assumes that the *m* levels of the *W* variable and the *p*-multinormal variables, for each subpopulation, are independent. The maximization of the likelihood function begins at a point that has been reached using the geometric technique, and it will then reach a peak (which could be local) near the starting point that contains the characteristics of the geometric technique. This two-stage classification approach is called the Ward–MLM method, in which the initial groups are first defined using a hierarchical clustering method based on Ward's minimum variance within groups' principle, and the MLM is subsequently used with those groups because it improves the initial groupings. This approach assumes that the mean vectors of the y_{sj} random vectors do not depend on the multinomial cell in which they appear, but they only depend on the subpopulation. The number of groups is defined using pseudo-*F* and pseudo-t^2 criteria combined with the likelihood profile associated with

---

Box 5.1 | Triploid *Musa* Cultivars

(Ortiz *et al.*, 1998)

A phenotypic diversity index, based on 16 quantitative morphological characters, was used for germplasm clustering and identification of duplicates among 92 triploid plantain and banana accessions (Figure 5.1) available at the gene bank of the International Institute of Tropical Agriculture in Nigeria. There were significant differences for the phenotypic diversity index among 14 taxonomic clusters, defined by fruit type (cooking banana, dessert banana, other banana and plantain) plus inflorescence type and plant size (both only for plantain). Furthermore, σ_C^2 (0.001779) was above σ_W^2 (0.001380), while Wright's Φ_{FS} was 0.563, which suggests little gene flow via pollen among the almost male-sterile, triploid, vegetatively propagated *Musa* taxonomic clusters. Hence, the variation observed within each triploid *Musa* taxonomic cluster most likely arose from mutations accumulated throughout the history of cultivation of banana and plantain. Putative duplicates based on qualitative morphological characters were not regarded as the same accession according to the phenotypic diversity index based on quantitative characters, thus calling on *Musa* gene bank curators to include quantitative morphological characters for the identification of duplicate accessions in banana and plantain.

Box 5.1 | (cont.)

Figure 5.1 Top (left-to-right): 'Gros Michel', 'Cavendish' and 'Prata' dessert bananas. Bottom (left-to-right): cooking banana, plantain and East African highland banana.
Source: Max Ruax, Alliance Bioversity–CIAT, *Musa* Germplasm Information System.

the likelihood-ratio test. The sequential clustering strategy allows modifying the form of the initial groups obtained by the Ward strategy (spherical clusters) to a form (elliptical clusters) that permits the formation of more homogeneous groups.

The Ward–MLM classification strategy has several advantages (Franco and Crossa, 2005). The process responds to the optimization of two related objective functions: in the first stage the sum of squares within groups and in the second stage the likelihood function

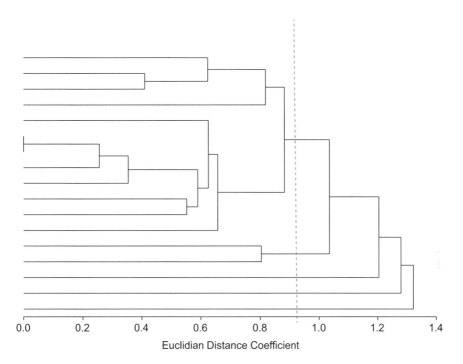

Figure 5.2 Dendrogram resulting after the average linkage clustering method (the perpendicular dashed line indicates the level of cut).

Box 5.2 | **Classifying Domesticated *Capsicum* Peppers**

(Ortiz *et al.*, 2010a)

Multivariate techniques with both qualitative and quantitative characters in the five domesticated species of *Capsicum* (*C. annuum*, *C. baccatum*, *C. chinense*, *C. frutescens* and *C. pubescens*) were used for grouping them after assessing inter- and intra-specific variation. The lack of sound clustering using the average linkage methods and qualitative descriptors suggests that multivariate analysis of qualitative characters that are not key for classifying accessions to their respective species could mislead grouping of *Capsicum* germplasm. Likewise, a limited number of inappropriate descriptors may result in misleadingly simplistic interpretations of dendrograms ensuing from multivariate analysis. Key qualitative characters such as seed colour, corolla colour and spot, calyx constriction, numbers of flowers per node and filament colour served for assigning most accessions to their respective species, whereas intra-specific multivariate diversity was better assessed by quantitative characters such as fruit length/width ratio, numbers of days to flowering, leaf width, and anther, filament and pedicel length. The Ward-modified location model was an adequate method for classifying *Capsicum* accessions using quantitative characters (Figure 5.3). This research confirms previous views that there are significant amounts of genetic diversity within as well as between *Capsicum* species, which can be a useful source for further genetic improvement of hot and sweet peppers.

Box 5.2 | (cont.)

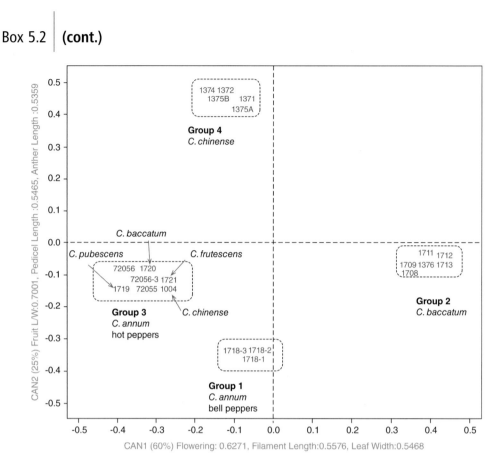

Figure 5.3 Distribution of 22 *Capsicum* accessions based on a bi-plot of the first two canonical variables formed by the Ward–MLM method using 23 quantitative descriptors, which are useful for intra-specific multivariate diversity, while qualitative descriptors assign most accessions to their respective species. Values along CAN axes show the weights for characteristics accounting for most of the variation.

of the observations. Moreover, it is linked to a method for defining the optimum number of groups. Furthermore, this approach allows calculation of a measure of the group's precision because it assigns to each observation a probability of membership to the group. It also uses all the available information, i.e. the continuous variables as well as the categorical variables, thus producing better classification results than the use of only part of the data. Last but not least, the distances between groups are maximized and better differentiated, and more compact groups are obtained.

The Ward–MLM method may also be used for clustering three-way data (cultivar × environment ×

character) in which the vector of each observation for r environments is $1 \times (rp + 1)$. The three-way Ward–MLM method assumes the same trait measured in different environments as different variables (environment–trait combinations), thus clustering individuals with consistent response across all variables in all the environments (Franco *et al.*, 1999). It is therefore reasonable to expect that this three-way cluster strategy should form groups of accessions with negligible group × environment interaction (GEI) for all the characters included therein. The three-way Ward– MLM classification method considers a random $n \times rp$ matrix of n observations where, for each observation, p characters are measured in each

Box 5.3 | Numerical Classification of Peruvian Highland Maize

(Ortiz *et al.*, 2008a, 2008b)

Peruvian highland maize has a high degree of variation owing to its history of cultivation by Andean farmers. Racial classification in maize uses morphological characters that reflect primarily genetic differences among accessions and not differences due to single-gene expression or genotype × environment interaction. Such characters used for maize racial classification may be either qualitative or quantitative. Numerical classification based on six quantitative plant characters and using the three-way Ward–MLM method kept the main structure of the maize races, but reclassified parts of these races into new groups, which were more separated and well defined with a decreasing accession within group × environment interaction.

Numerical classification using internal ear characters (Figure 5.4) maintained the structure of the more differentiated races but identified two distinct accessions in one race and separated them into a homogeneous group. Hence, the three-way Ward–MLM numerical method produced Peruvian highland maize groups with distinct characters in terms of variables.

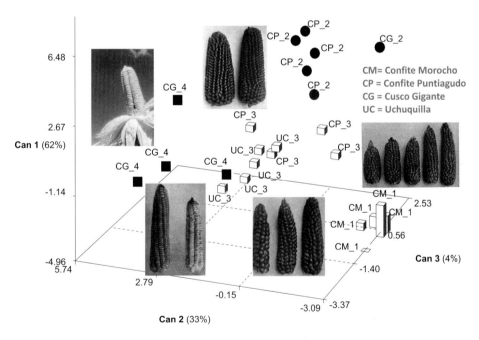

Figure 5.4 Plot of the first three canonical variables (CAN1, CAN2 and CAN3) and distribution of the 24 accessions of Peruvian highland maize.
Photo sources: Grobman et al. (1961) and Luis Narro (CIMMYT, Cali, Colombia).

of the *r* environments, thus forming a matrix with *n* rows (observations) and *rp* columns. These *rp* environment–trait combinations will be named

'variables'. The results of the three-way Ward–MLM classification method can be represented in a graphical display of the canonical analysis, where the accessions comprising the groups are plotted in a two- (or three-) dimensional diagram of the first two (or three) canonical variables (Franco *et al.*, 2003). The biological interpretation is provided by the correlation between the original variables and the canonical variables. The first canonical variable is the one that best separates the accessions; i.e. the best original variables discriminating the accessions are those most correlated with the first canonical variables.

5.2 DNA Markers and Polymorphism

The level of phenotypic polymorphism observed in plant characters is often limited. Hence, the genetic variation in plants can be further characterized using DNA markers, which are a piece of DNA that can be easily detected and whose inheritance is known. Markers based on nuclear DNA, chloroplast DNA (cpDNA) and mitochondrial DNA (mtDNA) reveal sites of polymorphism (i.e. variation in DNA sequence) that are useful for genetic diversity research. These markers add the power to monitor genetic variation at the elemental level of DNA sequences. Furthermore, DNA markers may be tailored to specific plant germplasm to accommodate the differences in breeding systems and relative levels of genetic diversity. They can also be scaled up according to the number of accessions included in the assessment, the genetic loci to be appraised, and what genome sequences will be sampled.

Molecular markers, which are heritable DNA sequence differences (or polymorphisms) among individual organisms, vary according to the nature of the biomolecule being analyzed; i.e. proteins or nucleic acids. Although protein polymorphism based on isoenzyme patterns is the oldest, DNA markers are the most accurate and widely used today. They are broadly used for assessing diversity, identifying duplicate gene bank accessions, probing evolutionary pathways and estimating genetic relationships in germplasm collections.

Various techniques identify DNA markers. For example, hybridization allows finding restriction fragment length polymorphisms (RFLP), which exploits variation among homologous DNA sequences

that differ on their restriction enzyme sites. Herein, the restriction enzyme digests the DNA sample and its resulting restriction fragments are separated as per their length using electrophoresis. Other DNA markers are based on the polymerase chain reaction (PCR), which produced millions of DNA strands in few hours. Random amplified polymorphic DNA (RAPD) and inter-simple sequence repeats (ISSR) are PCR-based markers that use arbitrary or random primers, while amplified fragment length polymorphisms (AFLP) use primers combining restriction enzyme specific sequences and random sequences. Other PCR-based markers require specifically designed primers that are often species-specific, e.g. microsatellites or simple sequence repeats (SSR), cleaved amplified polymorphic sequence (CAPS), sequence characterized amplified regions (SCAR) and single strand conformation polymorphism (SSCP). Single nucleotide polymorphisms (SNPs) are derived after DNA sequencing. SNPs result from variation in single nucleotides in a specific genome site. Table 5.1 lists the advantages and disadvantages of the most commonly used markers in plant research.

The use of DNA markers depends on prior knowledge regarding the genome sequences of the target species, the techniques to be used, the polymorphism level, their inheritance, reproducibility, transferability and available resources to perform the technique. The question seeking an answer, the level of kinship to be analyzed and available resources determine what DNA marker to use. Microsatellites and SNP are evenly and frequently distributed throughout the genome. These DNA markers are also very informative, co-dominant (discriminating homozygous from heterozygous), reproducible, have high transferability among related species, and their respective methods are fast, easy and relatively economical.

5.3 Analysis of DNA Marker Data

Genetic distance is the difference between two accessions due to allelic variation, and is measured by some numerical quantity, be it at the DNA sequence or allele frequency levels (Nei, 1972). Various genetic

Table 5.1 **Main advantages and few disadvantages of most common molecular markers for analyzing plant polymorphisms**

Marker	Main advantages	Few disadvantages
Isozymes	Relatively quick and simple protocol for this co-dominant marker that allows identification of heterozygotes	Small number of isozyme loci with low levels of polymorphism in some plants
Restriction fragment length polymorphisms (RFLP)	Highly reproducible and transferable co-dominant marker with discriminating power using single-locus probes at species or population level, or multi-locus probes at the individual level in any plants	Time-consuming, expensive, not amenable for automation marker requiring technical expertise and often using radioactively labelled probes, which needs expertise in autoradiography
Random amplified polymorphic DNA (RAPD)	Fast, simple and relatively cheap marker for detecting polymorphisms without using radioisotopes	Dominant marker that often lacks reproducibility and has difficulty in interpreting band patterns
Amplified fragment length polymorphisms (AFLP)	Highly sensitive, reproducible and widely applicable marker showing high levels of polymorphisms in plants	Dominant, expensive and technically demanding marker using radioisotopes and requiring high quality and quantity of DNA
Simple sequence repeats (SSR) or microsatellites	Co-dominant, robust and reliable marker with several alleles per locus, whose primer pairs may be used for several species within same genus, and analyzing several loci simultaneously through multiplexing	High costs and time-consuming for primer development
Single nucleotide polymorphisms (SNP)	High-density sequence variation in a genome, and high-throughput method for analysis of variation with co-dominant markers showing high accuracy	Almost always two alleles at locus, thus being less informative than SSR, and multiplexing not possible for all loci. Some essays are costly and labour-oriented

distance measures have been used for diversity analysis using DNA markers. It seems that datasets very often determine the type of genetic distance to be estimated among accessions. Nonetheless, the appropriate choice of a distance measure considering the type of variable and scale of measurement is key for analyzing genetic diversity among a set of accessions.

A diversity index is a mathematical function combining richness and evenness in a single measure, although it may not be often explicitly noted. The most common is the Shannon diversity index (*H*), which is a quantitative measure reflecting how many different types are in a dataset (Shannon, 2001), while simultaneously considering how evenly the basic entities (e.g. individuals) are distributed among those types. It is calculated by multiplying the proportion i relative to the total (p_i), and then multiplied by the natural logarithm of this proportion ($\ln p_i$):

$$H = -\sum_{i=1}^{s} p_i \ln p_i$$

Its minimum value is zero and becomes greater as diversity increases, i.e. *H* increases as richness increases for a given pattern of evenness, and *H* increases as evenness increases for a given richness.

Box 5.4 | **Microsatellite Polymorphism in Sorghum from Botswana**

(Motlhaodi *et al.*, 2014)

Microsatellites were used to characterize genetic diversity in sorghum accessions kept by the national gene bank of Botswana. The number of polymorphic loci ranged from 0 to 8, with an average of 3.3. The average gene diversity over all polymorphic loci for each population ranged from 0 to 0.39. The mean number of alleles per accession across all loci ranged from 1 to 2.4, with an average of 1.6 alleles per accession. The observed heterozygosity for each accession across all loci ranged from 0 to 0.39 with an average of 0.18, which was low due to predominant inbreeding in sorghum. The expected heterozygosity across all loci ranged from 0 to 0.28 with an overall average value of 0.15. Accession-specific rare alleles were detected in 7 out of 30 sorghum accessions. The low allelic richness could be due to the collecting of this sorghum germplasm in a narrow geographic area. Moreover, Botswanan farmers exchange seeds all over the country, thus accounting for the lack of geographic patterns of sorghum distribution. They pursue this activity as a risk management strategy to overcome mainly drought and infestation by the parasitic weed *Striga asiatica*. Furthermore, sorghum uses are very similar between ethnic groups and geographic areas in this country. Nonetheless, microsatellites were able to discriminate between sorghum accessions held by the national gene bank of Botswana, thus indicating the value of such DNA markers for characterizing genetic diversity among closely related individuals.

Genetic diversity measurements include allelic frequencies (p_i), the mean number of alleles per locus (A) or allele richness, the mean number of alleles per polymorphic locus (A_P) in which the frequency of the most common allele is equal to or below 0.99, polymorphism index or the relationship (in %) between number of polymorphic loci versus total, number of unique alleles found only in a given population, and the effective number of alleles in a locus (A_E) that refers to alleles with the ability to move to the next generation and is estimated as

$$A_E = \frac{1}{r} \sum_{j=1}^{r} \frac{1}{1 - D_j}$$

where D_j is the gene diversity of the jth of r loci.

The observed heterozygosity (H_o) for a DNA marker is estimated by the following relationship

$$H_o = \frac{\text{Heterozygous individuals per locus}}{\text{Total number of individuals analyzed per locus}}$$

or the $f(Aa)$, while the probability of null alleles for a given locus is defined as

$$r = \frac{(H_e - H_o)}{(1 + H_e)}$$

in which H_e is the expected heterozygosity (also known as gene diversity) that is calculated for two alleles (A and a) as $2pq$. A negative or very low r suggests that the presence of a single DNA band corresponds to a homozygote instead of a heterozygote.

Nei's coefficient of genetic diversity (H) is the probability that two gametes randomly chosen from a population or sample will differ at a locus (Nei, 1973) and is represented by the equation

$$H = 1 - \sum_{i=1}^{a} p_i^2.$$

H quantifies the evenness amongst the allele frequencies and can be therefore regarded as equivalent to H_e.

The polymorphism information content (PIC) measures how informative a genetic marker is, which depends on the number of alleles for that locus and its relative frequencies. The 'informativeness' of a genetic marker is the probability that a descendant of a pair is informative, i.e. the parental origin of each of the alleles of that locus can be deduced. PIC together with A_E are used to evaluate the utility of a given locus of a molecular marker to distinguish between the basic characterized units (BCUs), which are also known as operative taxonomic units (OTUs) when used taxonomically. A BCU may include a single individual if it has peculiar characters (an accession or a cultivar) or a set of individuals having certain characters in common (plants from a wild population). The BCU is determined by the specific objective of the characterization.

Wright's statistics include the coefficient of inbreeding (F_{IS}) and the fixation index (F_{ST}). F_{IS} measures the reduction in individual heterozygosity

due to deviations from random crosses, while F_{ST} quantifies the reduction of heterozygosity in a subpopulation due to non-random cross-breeding relative to the whole population (Wright, 1951). Hence, fixation means increased homozygosity resulting from inbreeding. Wright's statistics are estimated as follows:

$$F_{IS} = \frac{\overline{H_S} - H_I}{\overline{H_S}}$$

where H_I is the average of the observed heterozygosity in all populations, and H_s is the expected heterozygosity in each sub-population, and

$$F_{ST} = \frac{H_T - \overline{H_S}}{H_T}$$

in which H_T is the expected heterozygosity in the whole population (or the sum across all populations).

Box 5.5 | **Diversity Changes in Nordic Spring Wheat Cultivars Bred in the 20th Century**

(Christiansen *et al.*, 2002)

Microsatellites were assessed in Nordic spring wheat cultivars bred during the 20th century to determine genetic diversity throughout this period. The number of alleles ranged from 1 to 7, with an average of 3.6 alleles per gene. The analysis of molecular variance revealed that only 14% of all variation was due to country (relative geographic isolation), 35% was due to variation among decades within countries, and the 51% remaining variation was within decades within countries. Genetic diversity differed significantly between all seven chromosome groups. All cultivars were discriminated after using a dendrogram resulting from the analysis of the matrix of dissimilarities using the UPGMA, which also revealed clusters of accessions released both from some geographic area in the Nordic Region and the breeding era; i.e. genetic diversity increased from 1900 to 1940 and again from 1960 onwards. The noted loss of diversity (in the 1940s and 1950s) could not be explained by changes in a single genome or in one or a few chromosome sets. There were differences in frequency of some microsatellite alleles within countries, thus revealing the different selection effects therein. Some microsatellite alleles were lost during the first quarter of the century, while several new alleles were introduced in this Nordic spring wheat bred-germplasm during the second half of the 20th century. Genetic diversity in Nordic spring wheat was therefore enhanced by plant breeding in the first quarter of the 20th century, and after its decrease during the second quarter was increased again by plant breeding.

The F_{ST} ranges from 0 (thus indicating lack of differentiation between the whole population and its subpopulations) to a theoretical maximum of 1 (fixation of alternate alleles in different subpopulations). A F_{ST} above 0 but below 0.05 suggests low differentiation, while differentiation will be moderate when F_{ST} ranges from 0.05 to 0.15, and large between 0.15 and 0.25. A F_{ST} above 0.25 and below 1 suggests very large differentiation.

The analysis of molecular variance (AMOVA) is used to study the molecular variation within a species following a hierarchical and nested model (Excoffier et al., 1992, 2005). The AMOVA may contain different evolutionary assumptions without modifying the basic structure of the analysis, and its underlying hypothesis uses permutation methods that do not require the assumption of a normal distribution. Mean squares are computed for groupings at all levels of the hierarchy, which allows hypothesis testing between and within groups' differences at various hierarchical levels. The hierarchical levels of genetic diversity used by the AMOVA are clusters that contain lower hierarchical levels, e.g. geographic regions, areas within a region, populations within a zone or individuals within a population. The AMOVA describes the partitioning of the genetic variation among and within groups and tests user-defined groupings of populations. It examines DNA markers (both the allelic content and frequency of haplotypes, or genotypic data with an unknown gametic phase as is the case for most natural populations based on F_{ST} statistics) and sequence data. The generic AMOVA model for measuring genetic diversity between populations is defined as

$$Y_{ki(j)} = Y + a_k + b_{k(i)} + w_{ki(j)}$$

where a_k is the effect of the kth population with variance σ^2_a, b_{ki} is the effect of the ith individual within the kth population with variance σ^2_b, and $w_{ki(j)}$ is the effect of the jth locus of the ith individual of the kth population, with variance σ^2_w.

Similarity between two accessions within one locus is calculated as the number of common alleles relative to total alleles in the locus observed for the two accessions. If an accession is genetically heterogeneous, i.e. a mixture of several genotypes, it will reveal more than one allele for a co-dominant DNA marker. Similarity for two accessions is the average similarity over all loci, and the coefficient of dissimilarity between them is, therefore, 1 – similarity. The matrix of dissimilarities is used to build the dendrogram or phylogenetic trees using either the unweighted pair-group method with arithmetic average (UPGMA; Sokal and Michener, 1958) or the neighbour-joining method (NJ) (Saitou and Nei, 1987). These trees consist of nodes and branches, in which the former are accessions and the branch lengths between accessions are graphical estimates of the genetic distance between them, thus giving an indication of genetic relationships between accessions. The main difference between NJ and UPGMA is that the former does not assume an equal evolutionary rate for each lineage, while UPGMA trees give an indication of the time of separation (divergence) of accessions: the larger the branch length, the longer the separation period between accessions. UPGMA should be avoided for the analysis of genetic loci under natural selection because the evolutionary rate is not the same for each population. The matrix of dissimilarities may also be subjected to an AMOVA to estimate variance components among hierarchical levels of genetic diversity.

There are other methods for estimating genetic distances or similarities and used within a cluster analysis for assessing relationships between populations, accessions or taxa, e.g. the simple matching coefficient or Dice's coefficient (Dice, 1945), Jaccard's coefficient (Jaccard, 1901) and Sokal–Michener's simple matching coefficient (Sokal and Michener, 1958) are used for developing a distance matrix for phenetic similarity among individuals, while the Mannen's index calculates the distance matrix for genetic similarity between individuals, and Nei's standard genetic distance (Nei, 1972) or Rogers' distance (Rogers, 1972) develop the distance matrix for genetic similarity between populations.

The Dice's coefficient (S_D) expresses the probability that a band of a DNA marker in one individual is also in another. It is the index that is less affected by erratic bands of a DNA marker when the similarity between samples is due more to double presences than to double absences. It is calculated as

$$S_D = \frac{2a}{2a + b + c}$$

where a is the number of bands present in both individuals, while b refers to the number of bands present in the first individual but not in the second, and c to the number of bands present in the second individual but not in the first.

Jaccard's coefficient (S_J) is a statistic used for comparing the similarity and diversity between finite sample sets through this relationship

$$S_J = \frac{a}{a + b + c}$$

Jaccard's distance, which measures dissimilarity between sample sets is obtained by subtracting the Jaccard coefficient from 1.

Sokal–Michener's (S_{SM}) simple matching coefficient considers as doubling factor the double absences, which may lead to errors if there is no identity behind the double absences. Nonetheless, double absences can be considered as identities when intra-specific comparisons are made. It is defined as

$$S_{SM} = \frac{a + d}{a + b + c + d}$$

where d is the number of markers in which the band is simultaneously absent in both. S_J is very similar to S_{SM} when used for binary attributes. The main difference is that the S_{SM} has d in its numerator and denominator, which is absent in S_J, because S_{SM} compares the number of matches with the entire set of the possible attributes.

The Mannen index (GS) calculates the genetic similarity among individuals as follows

$$GS = \frac{2p + q}{2m}$$

where p is the number of loci in which both individuals share the genotype, q is the number of loci in which one individual is homozygous and the other heterozygous, and m is the total number of loci analyzed.

Roger's distance (d_{ij}) estimates the distance matrix for genetic similarity between populations through this geometric relationship

$$d_{ij} = \frac{1}{2m} \sum_{x=1}^{m} \sqrt{\sum_{k=1}^{q} \left(x_{ki} - x_{kj} \right)^2}$$

in which m and q are the number of loci and the number of alleles for locus x, respectively; while x_{ki} and x_{kj} are the frequency of allele k in ith and jth populations, respectively.

Nei's standard genetic distance (D_{ij}) considers changes in allelic frequencies due to both mutations and genetic drift, which makes it very suitable for phylogenetic research. This distance has a biological basis because it expresses the likelihood that two randomly selected alleles from two different populations will be identical with respect to the probability that two alleles will be randomly taken from the same population. D_{ij} is expressed as follows

$$D_{ij} = -\ln \frac{\left(\frac{\sum_{k=1}^{q} |x_{ki} x_{kj}|}{\sqrt{\sum_{k=1}^{q} x_{ki}^2 x_{kj}^2}} \right)}{\sqrt{\frac{2n_i \sum_{k=1}^{q} x_{ki}^2 - m}{2n_i - 1} \frac{2n_j \sum_{k=1}^{q} x_{kj}^2 - m}{2n_j - 1}}}.$$

D_{ij} increases in proportion to divergence time if the rate of genetic change (i.e. amino acid substitution) is constant per year or generation.

Principal coordinate analysis (PCoA) or multidimensional scaling are used to determine relationships among populations based on a distance matrix, thereby supplementing those defined by phylogenetic analysis based on NJ or UPGMA. PCoA should be used with binary distance measures such as Dice's or Jaccard's coefficients. PCoA displays the structure of (complex) distance-like data (i.e. a dissimilarity matrix) in a high dimensional space into a lower dimensional space without losing too much information. PCoA bi-plots sometimes show population clusters more clearly than dendrograms or phylogenetic trees.

Box 5.6 | Genetic Diversity of Common Beans Grown in Kyrgyzstan

(Hegay *et al.*, 2013a)

Common beans were introduced to Kyrgyzstan during the ruling of the Soviet Union in the 20th century and became an important export crop after its fall. Microsatellites were used to estimate genetic diversity in Kyrgyz cultivars and germplasm from Mesoamerica and South America's Andes, where common beans originated. The observed heterozygosity ranged from 0 to 0.11 among accessions, whose average heterozygosity was 0.01; while the average expected heterozygosity was 0.05, thus reflecting the inbreeding of common beans. The analysis of molecular variance revealed that 95% of the total variation was due to differences among cultivars (F_{ST} = 0.947; P < 0.001). There were two groups (Andean and Mesoamerican) after clustering following principal coordinate analysis (Figure 5.5). The Mesoamerican gene pool was further separated, and having Durango and Jalisco races in same group. The Kyrgyz cultivars Lopatka (kidney bean) and Kytayanka (navy bean), which are in the same grouping with Mesoamerican gene pool 1a, were the most divergent accessions, while Mesoamerican accessions were the least diverse. The use of microsatellites along with multidimensional scaling led to assigning modern Kyrgyz cultivars to common bean gene pools or races.

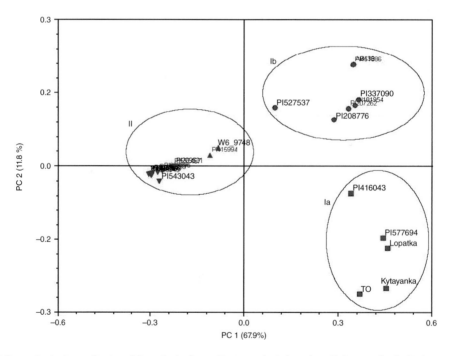

Figure 5.5 Two principal coordinates of the principal coordinate analysis based on Nei's genetic similarity matrix using microsatellites on common bean accessions. Group Ia (upper right) and Ib (bottom right) include the Mesoamerican gene pool, while the Group II (centre left) includes the Andean gene pool. Durango and Jalisco race accessions are in Group Ia, while Mesoamerican accessions belonging to other races are in Group Ib.

5.4 Quantitative Genetic Variation

Genetic variability is a measure of the tendency of the genotypes of a population to differentiate (King *et al.*, 2006). Although individuals of the same species are recognizable as belonging to it, they are not identical because of their differences in form, function and behaviour. A monomorphic population at a locus bears only one allele at such a locus, while a population with two or more alleles at a locus is polymorphic. There will often be variation within the species for each of the characters of any organism. Any observed phenotype (*P*) has two components: genetic (*G*) and the environment (*E*). The relationship between both appears to be complex, but let's assume that the phenotypic value is simply the sum of its components, i.e.

$$P = G + E$$

which may also be

$$P = m + G + E$$

if these effects are related to the mean, so the sums of both the genetic and environmental values are zero. The estimation of these values could be made by exposing genetically identical individuals (i.e. pure lines or clones in plants) to different environments, and solving the least squares model,

$$P_{ijk} = m + G_i + E_j + e_{ijk}$$

where P_{ijk} is the data of kth individual of ith genotype in the jth environment, and e_{ijk} is the residual component accounting for the variation between individuals of the same genotype in the same environment. Random environmental effects, which are caused by several independent factors of small effect each, show a normal distribution. Likewise, multi-genic characters determined by many genes with small effects are distributed normally as well as the phenotypic values according to the Central Limit Theorem, which states that the sampling distribution of the mean of any independent, random variable will be normal or nearly normal, if the sample size is large enough and the effects are small. Such distributions offer many advantages for calculation and interpretation, e.g. if two variables are distributed together normally, the regression of one over the other is linear, and if the correlation is zero, then they are independent.

It is known that the best genotypes in certain environments are not necessarily the best in other environments. For example, genotypes G_1 and G_2 were grown in environments E_1 and E_2, and results for two of their characters (P_1 and P_2) are given in Table 5.2.

Hence, genotype G_1 had a better adaptation than genotype G_2 for both characters P_1 and P_2 across environments E_1 and E_2. Furthermore, the decomposition of the phenotypic value for P_1 should include the genotype–environment interaction (GEI) as noted below

$$P_{ij} = m + G_i + E_j + GEI$$

The phenotypic variability can be split into that arising due to genes and another attributable to the environment. The variance of the phenotypic value (or phenotype variance; σ_P^2) is therefore

$$\sigma_P^2 = \sigma_G^2 + \sigma_E^2 + 2\, cov(GE)$$

Table 5.2 **Results from testing two genotypes across two environments for two characters**

Characters	Genotypes	Environment	
		E_1	E_2
P_1[🍎]	G_1	6🍎	3🍎
	G_2	4🍎	1🍎
P_2[●]	G_1	20●	14●
	G_2	10●	7●

where σ_G^2 and σ_E^2 are the variances due to the genes (or genetic variance) or the environments (or environmental variance), while $cov(GE)$ is the covariance between genotypic and environmental values, and it is non-zero when there is an association between genotypic and environmental values. The proportion of the phenotypic variance due to genes is called the broad-sense heritability (H^2), which is estimated as

$$H^2 = \frac{\sigma_G^2}{\sigma_P^2}$$

Genetic variance may be estimated in plants by using highly inbred lines to separate the variability due to the environment. Further crossings of these inbred lines can be made to have an estimate of the genetic variance assuming that the variability due to the environment is the same for each genotype. For example, in a crossing experiment involving two inbred lines the following data for a multi-genic quantitative character were obtained: the variance of their F_1 hybrid under ith and jth environments (V_{F1}), and the variance of the F_2 when grown in each of the environments V_{F2}. Assuming GEI is negligible (≈ 0), then

$$V_{F1} = \sigma_E^2$$

and

$$V_{F2} = \sigma_G^2 + \sigma_E^2$$

Thus,

$$\sigma_G^2 = V_{F2} - V_{F1}$$

Hence,

$$H^2 = \frac{V_{F2} - V_{F1}}{V_{F2}}.$$

A high broad-sense heritability indicates that the variation observed for a character in the population has genetic causes and the environment has little influence on it. Hence, an individual with high observed values will have, on average, high genetic values. Such an individual will be difficult to detect if broad-sense heritability is low because variation between the different individuals observed in the population is mainly due to environmental causes.

The individual effects of the alleles A and a and their interaction determine the genotypic value of Aa. The analysis of variance (ANOVA) of a factorial design with two effects could estimate the value of these effects and their interaction when the genotypic values could be measured. Let's consider that X_1 is an effect with two levels (A, a), or with more levels if there are more alleles, and X_2 is another effect with two other (or more) levels (A, a). Thus, a two-effect model with interaction within a locus will be defined as

$$G = m + X_1 + X_2 + X_{12}$$

being $X_1 + X_2$ the effects of alleles and X_{12} the interaction between alleles, respectively. The additive value ($\bar{A}$) of the locus is the sum of effects X_1 and X_2, which are the effects of the alleles of the same locus; while the interaction X_{12} between the effects of the alleles of the same locus is the dominant ($\bar{D}$) value of the locus. The model can be expressed as

$$G = m + \bar{A} + \bar{D}.$$

The total genotypic value G may also be divided into components attributable to different causes as follows

$$G = \bar{A} + \bar{D} + \bar{I}$$

where $\bar{I}$ is the deviation due to non-allelic interactions or epistasis (i.e. masking action of a gene by another, thus influencing the phenotype).

Assuming a, d and $-a$ as the values for the genotypes AA, Aa, aa, respectively, the average effect of a gene substitution (α) for a population in equilibrium is the difference caused by changing one allele in an average individual into the other allele as noted below

$$\alpha = a + d(q - p).$$

The average effect for A (α_1) and a (α_2) are the mean deviation from the population mean of individuals that received that allele from one parent, when the other allele is chosen at random from the population, i.e.

$$\alpha_1 = q[a + d(q - p)]$$

and

$$\alpha_2 = -p[a + d(q - p)].$$ The average effects are also derivable from a least squares regression of phenotype

on allele copy number. The breeding value of an individual is measured by the average value of its offspring, which is twice the deviation of the offspring from the population mean because the individual only contributes half of the alleles to its offspring. It is also defined as the sum of the average effects of the individual. The mean breeding value is zero under random mating. The additive (σ_A^2) and dominance variance (σ_D^2) are the variance of the breeding values and of the deviations, respectively. They are defined as

$$\sigma_A^2 = 2\ pq\alpha^2$$

and

$$\sigma_D^2 = (2\ pq\mathrm{d})^2$$

respectively.

Narrow-sense heritability (h^2) is defined as the quotient between the additive and the phenotypic variance, i.e.

$$h^2 = \frac{\sigma_A^2}{\sigma_P^2}$$

Thus, narrow-sense heritability measures part of the variation due to additive values, which is that passing to the offspring because genetic interactions are not inherited. This is therefore a more useful definition of heritability for selection because the breeding value of an individual indicates its ability to transmit 'good' genes to the next generation.

Realized heritability (H_R) is estimated from results of selection. It has the advantage of not being based on the similarity between relatives and therefore of not requiring presuppositions on the effects of dominance, epistasis or maternal effects. H_R is only used after the results of the selection are available, which may require many years in perennial crops. Knowledge regarding both the selection differential ($\bar{S}$) and response ($\bar{R}$) that occur in the population under selection is required to calculate H_R as follows

$$H_R = \frac{\bar{R}}{\bar{S}}$$

where $\bar{R}$ is how much gain to achieve after mating the selected parents, and $\bar{S}$ is the difference of the base population mean and the mean of the selected parents.

5.5 Association Genetics and Genomic Prediction

A quantitative trait locus (QTL) of a gene affects quantitative character variation. A QTL cannot be found through pedigree analysis because their individual effects are masked by the effects of the other genes that influence the trait and by environmental effects. A QTL can be localized if genetically linked to polymorphic DNA markers because the genotype that affects the quantitative character may be related to the genotype of the linked genetic marker(s). The location of the QTL in the genome is important for its use in plant breeding programmes. Polymorphic DNA markers such as microsatellites or SNPs are abundant and distributed throughout the genome. They also possess several co-dominant alleles, thus being ideal for linkage analysis of quantitative characters. The largest possible number of widely distributed DNA markers along with the complex character under study is used in successive generations of a genetically heterogeneous population for QTL research. Biometrics assists on identifying what DNA markers are associated with the complex quantitative character, so that their genotypes are always associated with desired phenotypic effects. These DNA markers identify regions of the genome that contain one QTL or more with significant effects on the quantitative character. The markers are then used to study the segregation of important chromosome regions for plant breeding.

Polymorphic DNA markers associated with quantitative variation for important characters may be found through linkage disequilibrium (LD) between loci (Breseghello and Sorrells, 2006). The physical distance of the loci across chromosomes determines LD. This approach for identifying quantitative trait loci is useful for dissecting complex characters because it offers fine-scale mapping due to historical recombination. It is worthwhile acknowledging that covariance between DNA markers and traits may arise due to population structure, which ensues from admixture, mating system, and genetic drift or by artificial or natural selection during evolution, domestication or plant breeding. False associations may also result from very low frequency alleles in the

---

Box 5.7 | Quantitative Trait Loci for Enhancing Adaptation to Salinity at Reproductive Stage in Rice from a Saudi Arabian Landrace

(Bimpong *et al.*, 2013)

Salinity affects rice productivity in rainfed and irrigated agro-ecosystems. The knowledge and further use of quantitative trait loci controlling adaptation to salinity accelerates breeding rice germplasm for environments prone to this abiotic stress. The salt-tolerant Saudi Arabian landrace 'Hasawi' was the donor parent of three F_2 offspring (each consisting of 500 individuals) with African cultivars that were the recipient parents (RP). The F_2s and $F_{2:3}$s were evaluated for grain yield and other characters under salinity, which caused a 65–73% yield reduction across the F_2-derived offspring. Nonetheless, some offspring had twice the RP's grain yield. The analysis revealed 75 quantitative trait loci for different characters in all genetic backgrounds, 24 of them being common to all, while 31 were noted in two, and 17 in one. 'Hasawi' contributed on average 49% of alleles to these quantitative trait loci. Two grain yield and yield-related quantitative trait loci common in all genetic backgrounds were mapped on the same chromosomal segment suggesting these quantitative trait loci could be stable. Another four quantitative trait loci were strongly associated with salinity tolerance with a peak microsatellite marker representing a potential candidate for marker-aided breeding due to the high LOD score and relatively large effects.

[*Note*: The logarithm of the odds (LOD) – to the base 10 – above 3 is often used to indicate that two loci are near to each other on the chromosome; i.e. a LOD score of 3 means that the odds are 1000 to 1 in favour of genetic linkage.]

---

initial population. Association genetics requires therefore to separate LD due to physical linkage from LD due to population structure, which may be caused by complex pedigrees derived from crosses of parents with different level of relatedness in plant breeding programmes. Marker-based estimation of probability of identity by descent between individuals or coefficient of parentage (COP) that measures the covariance between related individuals in a population can detect relationships between individuals. This association analysis should be based on comprehensive phenotypic data for modelling GEI. Linear mixed model methods for analyzing plant phenotypic data allow accurate prediction of genotypic performance using covariance structures that consider genetic associations between relatives included in the experiment (Crossa *et al.*, 2007). Three datasets from the germplasm set under study are used

for association genetics, namely, phenotyping data from multi-environment trials (*T*), a measure of population structure using neutral markers (Q matrix), coefficient of parentage (*K* matrix) or both (*Q* + *K*), and microsatellite or SNP polymorphisms in candidate gene(s) (*C*) or high-density markers (within LD). Thus,

$$T = C + (Q + K) + E$$

Although many quantitative trait loci were located for various characters in plants, only a limited proportion of the total additive variance of the character is often detected by DNA markers. It was therefore proposed to select all quantitative trait loci affecting character and predicting its breeding value by a large number of DNA markers distributed throughout the genome (Abera Desta and Ortiz, 2014). This genomic selection approach simultaneously

> ## Box 5.8 | Association Analysis of Historical Spring Bread Wheat Germplasm
>
> (Crossa *et al.*, 2007)
>
> Mapped Diversity Array Technology (DArT) markers were used to identify associations with host plant resistance to rusts and powdery mildew, as well as grain yield in five historical wheat international multi-environment trials from the Centro Internacional de Mejoramiento de Maíz y Trigo (CIMMYT). Two linear mixed models were used to assess marker–trait associations incorporating information on population structure and covariance between relatives. There were various linkage disequilibrium clusters bearing multiple host plant resistance genes. Most of the associated markers were in genomic regions previously noted as having genes or quantitative trait loci encoding for the same traits. Nonetheless, association analysis was able to find many new chromosome regions for host plant resistance and grain yield. Phenotyping across environments and years facilitated the modelling of the genotype × environment interaction, thus making possible the identification of markers contributing to both additive and additive × additive interaction effects of traits.
>
> Summary: Number of Significant DNA Markers per Trait
>
> Stem rust resistance: 63 DArT (many already published including 1B.1R with major effect)
> Leaf rust resistance: 87 DArT (>50 already published)
> Yellow rust resistance: 122 DArT (many reported with minor effect)
> Powdery mildew (PM) resistance: 61 DArT (several already known to be associated)
> Grain yield: 213 DArT (7 QTL already found with regions where DArT were significant)

estimates the effect of all DNA markers or positions on chromosomes on the target character(s). Different methods were proposed for estimating the effects of quantitative trait loci, but a Bayesian method, which assumes random effects of quantitative trait loci and assigns a different variance to each of them, appears as the most promising. Its special features are the *a priori* distribution of these variances and assigning a high probability for quantitative trait loci with effects having 0 value. The logic of this distribution is that most DNA markers have a high probability of not being associated with any quantitative trait loci. The accuracy or precision of this type of selection is determined by evaluating the value observed in the offspring and correlating it with the genomic estimated breeding value. This approach may also reveal the masked breeding value of available gene bank accessions (Longin and Reif, 2014), thus

bridging the trait gap between them and elite breeding pools of the cultigen. Accurate genomic prediction models will therefore assist on targeting promising gene bank accessions with high breeding value for further use in plant breeding.

5.6 Conclusion

Genetic markers are very useful and valuable for studying diversity and structure of plant populations. The different measures of genetic diversity and structure are related to each other, and many of them are based on analyzing the differences in allele frequencies. There are other measured based on estimating the proportion of genetic variation within and between populations, usually using the F_{ST} statistic, which provides a quantitative comparison of the existing differentiation between populations. Furthermore, genetic distance measures between pairs of populations allows comparisons at different levels, e.g. between and within populations, among accessions or cultivars. Genetic distances between pairs of populations are also used to reconstruct genealogy of populations, either by clustering methods such as UPGMA or NJ, or through plotting the paired genetic dissimilarity measures as a function of the distance that separates them, which assumes that these genetic differences increase with the distance that separates them, though it may not always be the case as noticed, for example, in lentils (Ferguson *et al.*, 1998b) or white clover (Hargreaves *et al.*, 2010). This knowledge allows to study more accurately the genetic structure and to the molecular characterization among and within populations. DNA markers such as microsatellites are very effective at establishing genetic relationships between and within populations, thus facilitating the global assessment of diversity, which should be considered to implement plans for conserving and enhancing plant genetic resources. Many of the key characters in the evolutionary process or in agriculture are quantitative. The study of their variation is the subject of quantitative genetics, which may tell us if there are significant levels of trait heritability that ensure a response to both natural and artificial selection.

Part

III

Conservation Practice

6 Planning Plant Conservation

6.1 Introduction

There are numerous diverse approaches to planning the conservation of plant genetic diversity, but the fundamental first step is deciding the taxonomic and geographic breadth of the conservation to be planned; deciding which single species or group of species in a single locality, group of localities, country, group of adjoining countries or globally warrants the expenditure of limited conservation resources. The commissioning agency will commonly define the taxonomic and geographic breadth of the conservation and allocate funding to plan the conservation of the taxon or group of taxa within a defined geographic area. In this way, the target taxon or taxa to be conserved within the geographic area are considered of enough interest to warrant conservation priority and expenditure of conservation resources.

6.2 Taxonomic and Geographic Prioritization

Emotionally all of us with an interest in or love of nature believe that each element of biodiversity has intrinsic value, should not be thoughtlessly eroded and should be actively conserved. However, with the limited financial, temporal and technical resources available to conservationists, we are forced to set priorities and select which taxa to focus our scarce conservation resources on. In terms of plant genetic conservation, the species, genera or varieties we focus our conservation activities on are referred to as target taxa, and the choice of target taxa should be as objective as possible, based on logical, scientific and economic principles, in terms of PGR related to the perceived utilitarian value of the species (Ford-Lloyd et al., 2008). Each species or group of species will have a perceived value and the establishment of its 'value' is aided by considering some or all the following

criteria (Maxted et al., 1997a; Hunter and Heywood, 2011):

Current Conservation Status

Before a taxon can be given high priority for conservation action, current conservation activities must be reviewed. If enough genetic diversity is already conserved from a range of agricultural systems, ecological habitats and geographic locations using both *in situ* and/or *ex situ* techniques, then additional conservation efforts may not be necessary. Details of what material is already conserved can be obtained from catalogues, databases and other records of the holdings of gene banks, botanical gardens and *in situ* conservation areas. Much of this information is available via the internet such as: EURISCO (www.ecpgr.cgiar.org/resources/germplasm-databases/eurisco-catalogue/) for European PGR *ex situ* collections; GENESYS (www.genesys-pgr.org) for global PGR *ex situ* collections; GRIN (www.npgsweb.ars-grin.gov/gringlobal) for US PGR *ex situ* collections; Botanic Gardens Conservation International (www.bgci.org/plant_search.php) for global botanic garden *ex situ* collections. Although care should be taken when interpreting current conservation status because: (a) the conserved material may be incorrectly identified, (b) single population samples may be duplicated using several conservation techniques or the same technique in different locations as a safety backup measure, (c) in the case of seed stored *ex situ* the sample may for various reasons be dead, in poor condition or be too small a sample to allow distribution, (d) the sample may lack passport data concerning its provenance or characterization and evaluation information so is unlikely to be available to users, and (e) an effective *ex situ* sample or *in situ* population may be conserved but for administrative reasons may remain unavailable to users. In each case, further conservation would be justified. It should be acknowledged that it is practically much easier to obtain information on existing *ex situ* conservation status than *in situ* due to the widespread lack of inventory data on which plant diversity is conserved *in situ* or on-farm, respectively.

A more objective means of assessing relative conservation status has recently been proposed by Khoury *et al.* (2019). The methodology includes five steps: (1) list taxa to be included in the assessment; (2) collate ecogeographic data for each taxon; (3) model potential distributions for each taxon; (4) comparing model potential distributions for each taxon against existing *ex situ* and *in situ* conservation activities; and (5) calculating the relative indicator score for each taxon based on the modelled versus conserved distributions. Although the *in situ* conservation assessment is based on presence or absence of a taxon in protected areas and it is well established that mere presence in a protected area does not mean the taxon is being actively conserved in that protected area, the use of this indicator does offer a more objective means of future conservation status assessment.

Socio-economic Use

Plant species that have either direct or indirect socio-economic use, e.g. that provide food, fuel, medicines, building materials, tools, adornment, recreation etc., for humankind, are likely to be given priority over species that are not perceived as having these uses when prioritizing taxa for conservation. Crops and their relatives will often be given the highest priority because of their relative socio-economic value in providing food security, and similarly within food crops those with greater commercial value will be prioritized higher (see more detailed discussion in Sections 6.2.1 and 6.2.2).

Perceived Threat

Plant species or crop landraces that are known to have decreased in population numbers or genetic diversity should be prioritized over species or landraces with stable or increasing populations. However, a problem comes from finding enough available time-series data

on population trends or genetic diversity for most species or landraces to make the assessment. Such time-series data is extremely limited for assessing changes in genetic diversity, but there are a few studies giving examples. Akimoto *et al.* (1999) sampled the crop wild relatives (CWR) of Asian rice, *Oryza rufipogon* Griff., populations in Thailand in 1985 and 1994 and found significant loss of genetic diversity and widespread genetic pollution from hybridization with rice cultivars, and in 1996 the *O. rufipogon* populations were extinct! Wilkes (2007) reported a more than 50% reduction in wild teosinte populations (*Zea mays* subsp. *mexicana* (Schrad.) H.H. Iltis) over the last 40 years in Mexico, and Keiša *et al.* (2008) investigated the changes in population sizes of faba bean wild relatives in Syria between 1986 and 2006 and reported a significant reduction in population size of all wild *Vicia* species and the near extinction of *V. kalakhensis* Khattab, Maxted & Bisby at its type and only known location. However, the loss of population size or number may not equate directly to loss of genetic diversity. The most widely used means of assessing threat to plants is using the IUCN Red List Categories and Criteria (IUCN, 2001), which provides a standardized system for the classification of taxa according to their potential extinction risk (see www.iucnredlist.org/). Relatively few CWR have thus far been assessed using these criteria again due to the lack of time series population trend data (see more detailed discussion in Section 6.2.3).

Historically threats to plant diversity have been associated with: habitat destruction, degradation and fragmentation of natural habitats (e.g. road and reservoir building), over-exploitation (where the individual removal rate is greater than replacement e.g. medicinal plant extraction from the wild, fuelwood gathering, over-grazing), introduction of exotic species which compete with, prey on or hybridize with native species, human socio-economic changes and upheaval (e.g. extinction of tribal cultures, urban sprawl, land clearances, food shortages), changes in agricultural practices and land use (e.g. displacement of landraces by modern cultivars or the shift from diverse crop cultivation to monoculture), and calamities, both natural and man-made (e.g. floods, landslides or wars). Although natural calamities are rarely thought to cause extinction, they are more likely to cause local population extinction and lose of population-specific diversity. However, a potentially much more devastating threat to biodiversity at all levels is climate change (IPCC, 2014b). Climate smart conservation (Stein *et al.*, 2014) has been developed as a means of: understanding the implications of climate change, developing and implementing actions that address the additional threats associated with climate change, plan conservation to mitigate its impact, and respond to increasing change and uncertainty. Further, Stein *et al.* (2014) outline the principles of climate smart conservation as: acting with intentionality by linking actions to impacts, accepting change will occur and managing for change, not just persistence, reconsidering conservation goals, not just strategies, and fully integrating adaptation into existing work programmes. In the prioritization context, this means making informed choices of the best options for conservation efficiency using modelling techniques (see more detailed discussion in Section 6.2.4).

Taxonomic and Genetic Distinction

It can be argued that with limited conservation funding available one approach to maximize the breadth of diversity conserved is to use the taxonomic hierarchy to conserve a few examples from each higher-level taxon, assuming conservation of closely related taxa will involve significant allelic duplication. Remotely related taxa will share a lesser proportion of alleles and genes than closely related taxa, so maximizing the conservation of the most remotely related taxa will maximize the overall conservation of genetic diversity. For example, imagine a country which only has three species, two apomictic dandelion species (*Taraxacum officinale* F.H. Wigg. agg.) and *Welwitschia mirabilis* Hook. F. (the monotypic gymnosperm genus comprising only this species). If resources are available to conserve only two of these species, logically it should be one of the dandelions and the *Welwitschia mirabilis* that are chosen. This is because a wider overall range of genetic variation will be conserved by focusing on two distantly related taxa rather than on two closely related ones. In this way, the conservationist can use phylogenetic relationship as a proxy for actual genetic distance and diversity to deliberately select complementary target taxa and maximize conservation value. As such, taxonomic 'outliers' will tend to be of high conservation priority, but the underlying assumption should be acknowledged that phylogenetic distance equates exactly to genetic distance, which is not always the case (Heywood, 1994).

Ecogeographic Distinction

It would appear reasonable to assume that plant species that are widespread, in terms of geographic and ecological range, are under less threat from genetic erosion and extinction than those that are localized or restricted to a distinct habitat (Ford-Lloyd *et al.*, 2008). Conservation priority should therefore be given to species that are restricted in their distribution or habitat requirements. These species are often referred to as endemics, which means their distribution is restricted to a region, usually a country. However, care must be taken if using single country endemism to select target taxa as the size of countries is non-standard. For example, it is impossible to compare a species endemic to the small island of St. Helena in the Atlantic Ocean and one endemic to the Russian Federation. The use of the term 'endemic' ignores the concept of scale in defining the endemism and is therefore only practical when related to some form of standardized scale, such as, for example, surface area grid squares.

Biological Importance

Not all species within an ecosystem play an equally important role in that ecosystem's overall function or processes; these species are usually termed keystone species. They are essential to the overall integrity of the ecosystem and to the survival of other species, and their removal fundamentally and disproportionately changes the ecosystem. Generally, keystone species are critical to the interactions between trophic levels, whether predators, herbivores or mutualists. The

Brazil nut tree (*Bertholletia excelsa*), as well as being economically important, has an important ecological role in that it provides a critical or pivotal food resource (large oil-rich seeds) for many other species, with whose survival it is associated. As such, these biologically important species are often given disproportionate value when considering alternative target taxa because of their pivotal role in maintaining the habitat or ecosystem. So, for example, even when the conservation target is the wild lentil-related species *Lens ervoides* (Brign.) Grande, it would be necessary to actively conserve the *Pinus* species it is often found growing under in Mediterranean woodland, as well the wild lentil itself.

Cultural Importance

Species may also be selected as target taxa because of their symbolic or religious significance in local or national culture. An example of such a species is the Cedar of Lebanon (*Cedrus libani* A. Rich.) in Lebanon, whose area of native forest has declined extensively in the 19th and 20th century. This species is an important national symbol, being represented on the nation's flag, money, stamps, etc. Compared with other coniferous tree species, it has no specific economic value, but because of its symbolic role in national Lebanese culture, it is now being actively conserved. Similar examples are provided by the Banyan tree (*Ficus benghalensis* L.) in India, the Swamp cypress (*Taxodium distichum* (L.) Rich.) in Mexico and (*Bauhinia guianensis* Aubl.) among the Waimiri Indians of Brazil.

Relative Cost of Conservation

The relative cost of either collection and *ex situ* storage or establishing *in situ* conservation for species will affect the selection of target taxa. Faced, as conservationists always are, with a limited conservation budget, and forced to select between two alternative target taxa of otherwise equal 'value', the relative costs of conservation would be a factor affecting the final decision. For example, the UK Biodiversity Steering Group report (Department of the Environment, 1996) costed the effective conservation of 45 of the UK's most threatened or endangered plant species. They estimated that conserving the Killarney fern (*Trichomanes speciosum* Willd.) would cost £33 000 per year, while the starry breck lichen (*Buellia asterelle* L.) would cost by comparison a mere £1000 per year. All other factors being equal, the conservation of the starry breck lichen would therefore be given higher priority over the Killarney fern because the cost of species conservation is 33 times cheaper. Using a similar argument in terms of *in situ* conservation, the number of target taxa it is possible to conserve in a single *in situ* site will also affect the choice of taxa. If it is possible to conserve more than one target taxon in a reserve, simple economics will dictate that species that can be conserved together in one multi-purpose reserve will be given a higher priority than those requiring distinct reserves of their own.

Conservation Sustainability

Conservation, whether *in situ* or *ex situ*, should be, by definition, long-term, and requires a relatively large initial investment of resources, whether sampling and collecting seed samples in a remote location or establishing a new reserve. There would therefore be little value in collecting and storing seed, establishing a genetic reserve or encouraging farmers to practise on-farm conservation unless the conservation activity was likely to succeed and be sustainable in at least the medium to long term. Shaffer (1981) considers the *in situ* conserved population should have a 99% chance of remaining extant for 1000 years to be deemed sustainable or as Traill *et al.* (2007) proposed at least having a >95% probability of persistence for more than 100 years. In the era of climate change and ecosystem instability this time scale might be considered ambitious, but the fact remains that conservation has a significant cost, and if the expenditure is to be justified, it must be as sustainable as possible.

Legislation

Species protected under international, regional or national legislation (which are usually those under

Table 6.1 **Examples of biodiversity conservation legislation**

Geographic scope	Treaty	Place and date of adoption
Global	Ramsar – Convention on Wetlands of International Importance Especially as Waterfowl Habitat	Ramsar, 1971
Global	WHC – Convention Concerning the Protection of the World Cultural and Natural Heritage	Paris, 1972
Global	CITES – Conventional on International Trade in Endangered Species of Wild Fauna and Flora	Washington, 1973
Global	CBD – Convention on Biological Diversity	Rio de Janeiro, 1992
Global	ITTA – International Tropical Timber Agreement	Geneva, 1983
Global	ITPGRFA – International Treaty on Plant Genetic Resources for Food and Agriculture	Rome, 2001
Africa	Protocol Concerning Protected Areas and Wild Fauna and Flora in the Eastern African Region	Nairobi, 1985
	Lusaka Agreement on Cooperative Enforcement Operations Directed at Illegal Trade in Wild Fauna and Flora	Lusaka, 1994
Asia and Pacific	Plant Protection Agreement for Asia and Pacific Region	Rome, 1956
	ASEAN Agreement on the Conservation of Nature and Natural Resources	Kuala Lumpur, 1985
Europe and Central Asia	Convention on the Conservation of European Wildlife and Natural Habitats	Berne, 1979
	Directive on the Conservation of Natural Habitats and of Wild Fauna and Flora	Brussels, 1992
America	Convention on Nature Protection and Wildlife Preservation in the Western Hemisphere	Washington DC, 1940
	Regional Convention for the Management and Conservation of Natural Forest Ecosystems and the Development of Forest Plantations	Guatemala City, 1993
National	UK Wildlife and Countryside Act	London, 1981
	Indian Biological Diversity Act	New Delhi, 2002

greatest threat) will also be given high priority for active conservation, as the signatories of conventions are required to implement their obligation. Some global or regional examples of such biodiversity conservation legislation are provided in Table 6.1, but as well as these, most countries have national biodiversity or even specific PGR-related legislation. The list does not include follow-on actions from Treaties or Conventions, like the CBD Nagoya Protocol or CBD Sustainable Development Goals, which may themselves have significance that equals the initial legislation (see Chapter 2).

Ethical and Aesthetic Considerations

It remains a fact that most conservation is funded from general national taxation funded by tax payers, so the selection of target taxa should not dismiss ethical and aesthetic justifications for biological conservation favoured by the public. The ethical justification for conservation reflects the sympathy, responsibility and concern that most people feel towards fellow species and ecosystems. For many people nature does have an intrinsic value above the utilitarian value of human exploitation; our quality of life is enhanced knowing that biodiversity is 'safe', and this value for humankind should not be overlooked. This may be illustrated by the media attention focused on and relative ease with which funds may be raised to conserve so-called flagship species such as giant pandas, tigers, whales or orchids. It is more difficult to convince the public of the worth of conserving less attractive species, like goat grass, teosinte or Johnson grass, each primary wild relatives of globally important cereals. However, the education of the public in the value of food diversity is having impact as noted by the increase of crop landraces in mainstream supermarkets (Veteläinen *et al.*, 2009a).

Priorities of Conservation Agencies

The conservation priority given to a taxon will be influenced by the mandate and priorities of the conservation agency commissioning the work. The priorities of a country's Ministries of Agriculture, Forestry and the Environment are likely to be quite different. The Ministry of Agriculture will focus on conserving crops and crop relatives, while the Ministry of Forestry will focus on timber trees and the Ministry of Environment may have a more general ecosystem-based, rare or threatened taxon remit. Different levels of priority are also likely to be given to the same taxon by different interest groups, such as ecologists, economists, park managers, plant breeders, sociologists, population geneticists, traditional healers, agriculturalists and taxonomists. There will also be further differences between national, regional and international agencies prioritizing taxa for conservation action. A species may be considered

Least Concern using the IUCN Red List Categories and Criteria of threat internationally, but within a country, perhaps on the edge of its natural distribution, it may be vulnerable and so warrant active conservation in that country.

The various prioritization criteria discussed above provide relative 'value' to plant taxa and thus aid the efficient selection of priority taxa for conservation. It is by weighing each in accordance with the mandate of the commissioning agency that the conservationist will be able to determine relative conservation priorities more objectively. Once the decision has been made over which prioritization criteria are to be applied, they may be coded and the taxa prioritized. The prioritization process may occur in parallel or serially; in parallel meaning all criteria are scored for all taxa first, then the coded scores are summed and the taxa with the higher scores are the prioritized taxa, while serially means the criteria are scored in a sequence using one criteria at a time and only the high-scoring taxa for the first criteria are scored for the second criteria and so on, and following the sequence of applying the various criteria the remaining taxa at the end of the process are the prioritized taxa.

Practically the process of prioritization will involve the selection of the relevant criteria, rarely will all 12 criteria be applied, and the selection of criteria will impact the taxa prioritized. Only three are commonly used for establishing PGR priorities: crop value, crop relatedness and relative threat assessment, although native status is also often used in national PGR conservation planning (Iriondo *et al.*, 2016). Whatever prioritization methodology and criteria are used, the total number of target taxa is likely to be adjusted to a number that can be actively conserved using the available financial and human resources. There is no precise way of estimating the number of target taxa and no set number or proportion that should result from prioritization (Maxted *et al.*, 2013).

6.2.1 Crop Value-Based Prioritization

The assumption here is that not all plant species or even PGR have equal value to humankind; for example, higher value is inferred by CWR or landraces

(LR) related to major crops than that related to minor crops, simply because their value is at least partially dictated by the value of the associated crop to which adaptive traits could be transferred. Hence, there is a need to establish crop value at the country, region or global scale depending on the scale of the conservation being planned. Such statistics exist at all three levels, though the level of detail and crop coverage varies significantly at the country and regional levels. At the global level, which also includes country and regional data, the FAO crop production statistics (FAO, 2016) provide data on crop area harvested (hectares), yield (hg/ha), production quantity, seed produced and gross production value (in local currency or US$) for crops or groups of crops, for countries, regions or globally, per year. Kell *et al.* (2015) used FAOSTAT data on annual production values of human food crops with an average annual value of more than US$500 million over this period to prioritize Chinese CWR for conservation. Khoury *et al.* (2015a) investigated the importance of the geographic provenance of PGR in relation to food security using FAOSTAT data on national food supplies (measured in calories, protein, fat and food weight) and national production systems (measured in production quantity, harvested area and production value) to estimate each country's interdependence in PGR in providing national food security.

6.2.2 Crop Relatedness-Based Prioritization

In terms of prioritizing PGR taxa, not all CWR have equal value. Those taxa that are more likely to be used by breeders or other users will have higher value; therefore wild species from which adaptive traits can be transferred more easily to the crop or wild species that have been previously used for such transfers will be prioritized. Generally, the more closely related the wild taxa are to the crop, the easier the trait transfer. Harlan and De Wet (1971) proposed the crop gene pool concept to help quantify the relationship of the wild taxa to the crop (Figure 6.1). They distinguished three distinct crop gene pools:

- Primary gene pool (GP1): the true biological species including all cultivated (GP1A), wild and weedy forms (GP1B) of a crop species. Hybrids among these taxa are fertile and gene transfer to the crop is simple and direct.
- Secondary gene pool (GP2): the group of species that can be artificially hybridized with the crop, but where gene transfer is difficult. Hybrids may be weak or partially sterile, or chromosomes pair poorly.
- Tertiary gene pool (GP3): including all species that can be crossed with difficulty (e.g. requiring *in vitro* hybrid embryo rescue), and where gene transfer is impossible or requires radical techniques (e.g. radiation-induced chromosome breakage).

Where genetic information is available, taxa can be classified using the gene pool concept and for most major crops the gene pool concept has already been defined. If we apply this concept to barley, as an example, then: *Hordeum vulgare* subsp. *vulgare* and its progenitor *H. vulgare* subsp. *spontaneum* would belong to GP1 as GP1A and GP1B, respectively, *H. bulbosum* to GP2, and all the other *Hordeum* species to GP3. Therefore, the higher value would be ascribed to the GP1A (primarily barley LR), then GP1B within the crop species, then GP2 and finally GP3, giving the relative priorities defined by socio-economic use.

To establish gene pools requires detailed information on the relative success of crossing between the CWR and crop, but this information is absent for the bulk of crops, so it is necessary to find an alternative method of estimating crop relatedness for crops where no gene pool concept is available. Therefore, Maxted *et al.* (2006) proposed the taxon group concept that uses the generic classification of the crop as a proxy for the genetic relation (Figure 6.2):

- TG1a: the crop itself, usually the type subspecies of the species containing the crop (the single black circle in Figure 6.2).
- TG1b: the same species as crop but the non-crop subspecies (also the single black circle in Figure 6.2).
- TG2: other species found in the same series or section as the crop (grey circle in series c, d and e in Figure 6.2).

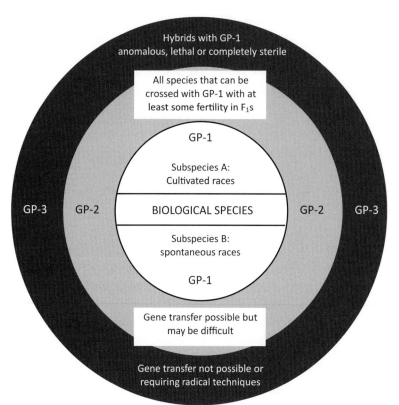

Figure 6.1 Schematic diagram of gene pool concept.
(From Harlan and de Wet, 1971)

- TG3: other species found in the same subgenus as the crop (grey circle in subgenus X, excluding series c, d and e in Figure 6.2).
- TG4: other species found in the other subgenera but the same genus as the crop (grey circle in subgenus Y and Z in Figure 6.2).

In general, the closest wild relatives in GP1B and GP2 or TG1B and TG2 are given priority. However, tertiary wild relatives that are already known as gene donors or have shown promise for crop improvement should also be assigned high priority.

If the genetic conservation goal is not to conserve individual crop gene pools but overall national, continental or global crop gene pool diversity, then the target is not only CWR relatively closely related to the crop but further CWR related to multiple crops. In this context Maxted *et al*. (1997a) developed the concept of the 'gene sea', where each crop is at the centre of its own gene pool that includes related CWR

taxa, and all individual crop gene pools are interrelated in one expanse of genetic diversity – the gene sea; see Figure 6.3. As the goal is to conserve overall CWR genetic diversity, CWR taxa that are present in multiple gene pools within the gene sea would have highest priority for conservation because they would better represent the breadth of the plant genetic diversity in the limited number of populations or accessions likely to be actively conserved.

If neither gene pool nor taxon group concepts can be applied, then the available information on genetic and/or taxonomic distance should be analyzed to make reasoned assumptions about the most closely related taxa. Gene Pool or Taxon Group concepts have been compiled for major food crop gene pools and were collated by Vincent *et al*. (2013) in the Harlan and de Wet Crop Wild Relative Inventory (www .cwrdiversity.org/checklist/). For other crops, a literature survey will be required to ascertain if Gene Pool or Taxon Group concepts have already been

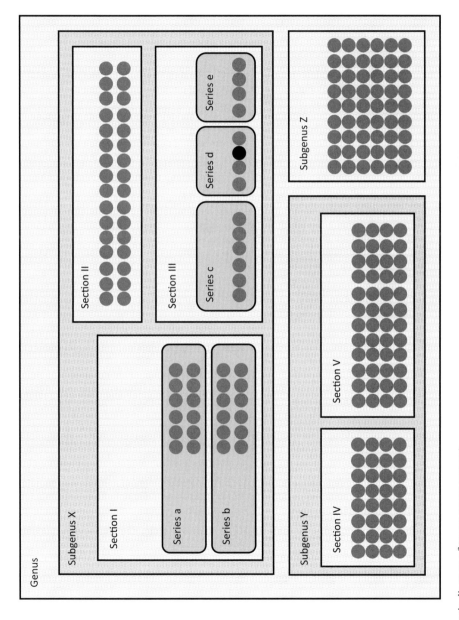

Figure 6.2 Schematic diagram of taxon group concept. (From Maxted *et al.*, 2006)

Figure 6.3 Schematic diagram illustrating the gene sea concept.
(From Maxted *et al.*, 1997a)

established or if taxonomic classification is available to establish new taxon group concepts and so establish the degree of relationship of each wild relative to its associated crop.

6.2.3 Perceived Threat-Based Prioritization

Relative assessment of threat is a very intuitive means of prioritization – the more threatened (i.e. increased likelihood of genetic erosion or actual extinction of the species), the greater the conservation priority. Therefore, a review of existing threat assessments and the level of extinction risk attributed to a taxon will aid prioritization.

The most commonly applied means of assessing the threatened status of wild taxa is the application of the IUCN Red List Categories and Criteria (IUCN, 2001),

which is a standardized system for the classification of taxa according to their extinction risk (or threatened status). As the knowledge about plant taxa has increased, so national Red Lists and Red Books are published. IUCN Red List-based threat assessment can be carried out at different geographic scales (i.e. global and regional, and the latter is usually applied at national scale; IUCN 2001, 2003). Both global and regional assessments should be considered in the prioritization process. The collation of existing threat assessments is a four-stage process: (i) identification of potential sources of information on threat to CWR, (ii) establish if CWR have been Red List assessed, (iii) for the CWR not already assessed gather the necessary data and undertake novel Red List assessment, and (iv) collation of existing threat assessments (at regional or global level). Information on threat assessment of CWR can be obtained from national and regional Red Lists and Red Data Books and the IUCN Red List of Threatened Species (for global Red List assessments, searchable at www.iucnredlist.org/), as well as peer-reviewed papers and reports, and expert knowledge.

The process of Red List assessment of taxa, most often assessed as species, is essentially a two-step process for each taxon being assessed:

- Data of seven types are collated and documented: (i) taxonomic; (ii) distribution; (iii) population; (iv) habitat and ecology; (v) use and trade; (vi) threats; and (vii) conservation actions. These data are gathered from diverse sources, including taxon experts, published and grey literature, databases and websites.
- The taxon is evaluated against the IUCN Red List Criteria (Box 6.1) using the collated data, and based on this evaluation the appropriate Red List Category (Figure 1.9) is selected.

There are five main Red List Criteria: (A) population reduction, (B) geographic range (recorded in terms of extent of occurrence and/or area of occupancy), (C) small population size and decline, (D) very small or restricted population, and (E) quantitative analysis indicating the probability of extinction. Geographic range is based on extent of occurrence (using the area within an enclosed line drawn round the most

Box 6.1 | **Summary of the Five Criteria (A–E) Used to Evaluate Whether a Taxon Belongs in a Threatened Category (Critically Endangered, Endangered or Vulnerable)**

(From IUCN, 2001)

A. **Population size reduction.** Population reduction (measured over the longer of 10 years or 3 generations) based on any of A1–A4

	Critically Endangered	Endangered	Vulnerable
A1	$\geq$ 90%	$\geq$ 70%	$\geq$ 50%
A2, A3 & A4	$\geq$ 80%	$\geq$ 50%	$\geq$ 30%

A1 Population reduction observed, estimated, inferred or suspected in the past where the causes of the reduction are clearly reversible AND understood AND have ceased A2 Population reduction observed, estimated, inferred, or suspected in the past where the causes of reduction may not have ceased OR may not be understood OR may not be reversible A3 Population reduction projected, inferred or suspected to be met in the future (up to a maximum of 100 years) *[(a) cannot be used for A3]*. A4 An observed, estimated, inferred, projected or suspected population reduction where the time period must include both the past and the future (up to a max. of 100 years in future), and where the causes of reduction may not have ceased OR may not be understood OR may not be reversible	*based on any of the following:*	(a) direct observation [except A3] (b) an index of abundance appropriate to the taxon (c) a decline in area of occupancy (AOO), extent of occurrence (EOO) and/or habitat quality (d) actual or potential levels of exploitation (e) effects of introduced taxa, hybridization, pathogens, pollutants, competitors or parasites

B. **Geographic range in the form of either B1 (extent of occurrence) AND/OR B2 (area of occupancy)**

	Critically Endangered	Endangered	Vulnerable
B1. Extent of occurrence (EOO)	< 100 km²	< 5000 km²	< 20 000 km²
B2. Area of occupancy (AOO)	< 10 km²	< 500 km²	< 2000 km²
AND at least two of the following 3 conditions:			
(a) Severely fragmented OR Number of locations	= 1	$\leq$ 5	$\leq$ 10

(b) Continuing decline observed, estimated, inferred or projected in any of: (i) extent of occurrence; (ii) area of occupancy; (iii) area, extent and/or quality of habitat; (iv) number of locations or subpopulations; (v) number of mature individuals

Box 6.1 | (cont.)

(c) Extreme fluctuations in any of: **(i)** extent of occurrence; **(ii)** area of occupancy; **(iii)** number of locations or subpopulations; **(iv)** number of mature individuals

C. Small population size and decline

	Critically Endangered	Endangered	Vulnerable
Number of mature individuals	< 250	< 2500	< 10 000

AND at least one of C1 or C2

	Critically Endangered	Endangered	Vulnerable
C1. An observed, estimated or projected continuing decline of at least (up to a max. of 100 years in future):	25% in 3 years or 1 generation (whichever is longer)	20% in 5 years or 2 generations (whichever is longer)	10% in 10 years or 3 generations (whichever is longer)

C2. An observed, estimated, projected or inferred continuing decline AND at least one of the following 3 conditions:

		Critically Endangered	Endangered	Vulnerable
(a)	(i) Number of mature individuals in each subpopulation	≤ 50	≤ 250	≤ 1000
	(ii) % of mature individuals in one subpopulation =	90–100%	95–100%	100%

(b) Extreme fluctuations in the number of mature individuals

D. Very small or restricted population

	Critically Endangered	Endangered	Vulnerable
D. Number of mature individuals	< 50	< 250	D1. < 1000
D2. *Only applies to the VU category* Restricted area of occupancy or number of locations with a plausible future threat that could drive the taxon to CR or EX in a very short time	–	–	D2. typically: AOO < 20 km² or number of locations ≤ 5

E. Quantitative analysis

	Critically Endangered	Endangered	Vulnerable
Indicating the probability of extinction in the wild to be:	≥ 50% in 10 years or three generations, whichever is longer (100 years max.)	≥ 20% in 20 years or five generations, whichever is longer (100 years max.)	≥ 10% in 100 years

outlying specimens and/or area of occupancy (actual area covered by the taxon), and these areas can be calculated automatically based on georeferenced data for the taxon using the open source program GeoCAT (Bachman *et al.*, 2011). Criterion E is based on modelling or quantitative analysis, and in theory this could be related to an assessment of climate change vulnerability, but the models are still to be developed and Criterion E is still rarely used in Red List assessment. Each main criterion includes sub-criteria against which the species is evaluated (Box 6.1), and an example of its application is given in Box 6.2.

A species should be assessed against all criteria for which there is information available, and if the species meets the criteria in at least one of the main classes, it is assigned one of the threatened categories,

---

Box 6.2 | **Example of IUCN Red Listing of a CWR: Belin Vetchling (*Lathyrus belinensis* Maxted & Goyder)**

Scientific Name: *Lathyrus belinensis*   **Species Authority:** N. Maxted & D. Goyder

Red List Category & Criteria: Critically Endangered A3c+4c; B1ab(i,ii,iii,v)+2ab(i,ii,iii,v)

Year Assessed: 2012   **Assessor/s:** N. Maxted   **Reviewer/s:** M. Bilz

Justification: *Lathyrus belinensis* is a rare CWR of *L. odoratus* (sweet pea). The type population, which is restricted to an area of only 2 km², was discovered in 1987; no further populations have been reported. The species was found growing adjacent to the new main road along the south Turkish coast in an area being developed for tourism. The specific site has been planted with conifers. The population located in 1987 was revisited in 1995, but in 2010 the original type location had been destroyed by earthworks associated with the building of a new police station. Only a few plants remain in nature, but seed samples are held in two seed banks. Eighty per cent of the original population was lost between 1995 and 2010. There is on-going threat from extensive conifer planting and very high levels of sheep and goat grazing in the area. Therefore, it is globally assessed as Critically Endangered.

Range Description: *Lathyrus belinensis* was originally discovered in 1987 while searching near Cavus, Antalya province, Turkey, for food, fodder and forage legume species. The single population was growing alongside a road between Kumluca and Tekirova in Antalya province. Only one location is known with an area of occupancy of 2 km².

Countries: Turkey; Turkey-in-Asia   **Native:** Yes

Population: The single population decreased in size from 5000 in 1995 to 1000 in 2010 due to the building of a new police station or planting of conifers under which the species could not grow. The remnant population exists now on the hillside and road banks; however, the hillside has been planted with conifers and the road may be expanded as it is the main south coast holiday route.

Population Trend: Decreasing   **Habitat and Ecology:** The species grows on a rocky limestone hillside, margins of cultivated land and roadside.

Major Threat(s): The species is a single location. Almost the entire original type location had either been bulldozed to build a new police station or planted with conifers under which the species could not grow. The site is also severely over-grazed by local small-holder livestock farmers.

Conservation Actions: *Lathyrus belinensis* has seed samples conserved *ex situ* in the Aegean Agricultural Research Institute, Menemen, Turkey, and the International Centre for Agricultural Research in the Dry Areas (ICARDA) in Rabat, Morocco, but there is no active *in situ* conservation of the species.

Critically Endangered (CR), Endangered (EN) or Vulnerable (VU). If the species meets the criteria in more than one main class, it is assigned the highest category of threat, but the less threatened category according to the other criterion or criteria is also documented. If the species does not meet any of the criteria A–E needed to evaluate it as threatened, another category is selected; this can be Extinct (EX), Extinct in the Wild (EW), (or Regionally Extinct (RE) if the regional criteria are being used), Near Threatened (NT), Least Concern (LC), Data Deficient (DD) or Not Applicable (NA) (for definitions of the categories, see IUCN, 2001). Therefore, in each assessment taxon or population data is assessed against established criteria and depending on what levels for the criteria are met, an IUCN category is assigned. Hence, using data on the number of mature individuals, population size trends, population fluctuations and distribution of populations, demographic patterns of *ex situ* and *in situ* conserved individuals, along with an analysis of extinction probabilities in the wild, taxa can be assigned a category of threat according to the IUCN Red List Categories. Species that are classified using one of the three 'Threatened' categories will have higher conservation priority than those classified as 'Least Concern' (LC). It should be noted that a taxon assessed as DD may also be threatened, but there is insufficient data to assess the taxon objectively. Also, it should be noted that taxa require regular assessment as their relative threat may change with time. The availability of specific population and threat trend data will limit the possibility of proposing an assessment. In the absence of enough data to achieve a Red List assessment, endemism and relative distribution can be used as an indicator of relative threat (Ford-Lloyd *et al.*, 2008). Inferences from known threats to/loss of habitats/land use types can also be applied, as well as local expert knowledge.

In the context of genetic conservation, it is worth noting that the IUCN Red List Categories focus on threat to species, and the loss of genetic diversity is not specifically addressed. It is possible for a species to remain viable in terms of IUCN Red List Categories, while losing significant genetic diversity. Therefore, as regards genetic conservation, it would be advisable to assess relative threat of genetic erosion when prioritizing taxa for conservation, genetic erosion being: (i) the loss of a crop, variety or allele diversity, (ii) a reduction in richness (in the total number of species, crops, varieties or alleles), and (iii) a reduction in evenness (Maxted *et al.*, 2013b). Ideally this would involve an assessment of significant changes in genetic diversity over time, as demonstrated by Akimoto *et al.* (1999) and Gao *et al.* (2000) for the rice CWR *Oryza rufipogon* Griff., but such studies are rare. Another approach would be to try to identify a proxy or indicator of genetic erosion. Such a methodology was proposed by Guarino (1995) based on the scoring parameters, such as relative taxon distribution, whether the species distribution is declining, increasing or static, degree of farm mechanization, relative use of herbicides and fertilizers, conservation status of the taxon and the extent of its use, etc., but the methodology has not been widely applied, and we await the development of a practical, routine and cheap means of assessing genetic erosion.

Extinction and genetic erosion of LR diversity is equally if not more immediately as detrimental to sustaining food security as loss of CWR, so threat assessment of an early warning system to detect and prevent genetic erosion and extinction is necessary. It can be assessed at two levels: (i) individual LR (i.e. the extinction of individual LR), and (ii) genetic diversity within LR (allelic loss within a LR) (Maxted *et al.*, 2013b). LR threat assessment using the IUCN Red List Categories and Criteria is, unfortunately, not an option as the criteria cannot be applied at the within taxa. Various authors have proposed methods based on alternative categories and criteria (Joshi *et al.*, 2004; Porfiri *et al.*, 2009; Antofie *et al.*, 2010; Padulosi and Dulloo, 2012; Maxted *et al.*, 2013b); however, to date there is no standardized methodology for threat assessment of erosion or extinction for LR even though the need for such a methodology is widely accepted.

6.2.4 Climate Change, Impact and Prioritization

It is well known that climate and plant distribution are strongly correlated (von Humboldt and Bonpland, 1807; Woodward and Williams, 1987).

Climate variability is a natural phenomenon that has existed since earliest geological time. It means that there are periodic changes in the climate attributable to natural causes such as changes in Earth orbit, fluctuation in energy from sun, volcanic eruptions and ocean/atmosphere interactions. But climate change, as defined by the United Nations Framework Convention on Climate Change Article 1, is 'a change of climate which is attributed directly or indirectly to human activity that alters the composition of the global atmosphere and which is in addition to natural climate variability observed over comparable time periods' (UNFCCC, 1992). The International Panel for Climate Change (IPCC, 2007, 2014), using General Circulation Models, predicts that climate change will affect different areas of the world to different degrees but that for the next two decades a warming of 0.3°C is projected to result in further:

- melting of glaciers and polar ice, rising sea level and change to oceanic currents and acidity;
- increased storms, hurricanes, droughts, rain, changes in snow cover and earthquakes;
- desertification or irrigation and extreme salination; and
- agriculture failures, species extinction and ecosystem disruption.

While the IPCC (2013) reported that the years 1983–2012 contained the warmest 30-year period of the last 1400 years, that global temperature increase between 1880 and 2012 was 0.85 (0.65–1.06)°C, but by 2100 it is predicted to be between 1.4 and 5.8°C, global glacial shrinkage was 226 (91–361) Gt per year between 1971 and 2009, but was 275 (140–410) Gt per year between 1993 and 2009, and that global average sea level rose between 1901 and 2010 by 0.19 (0.17–0.21) cm, but since 1971 the figure has been 3.2 (2.8–3.6) mm/year. The future general trend in climate change is for average global temperature to increase as the population rises. This supports the need for climate change assessments of species distributions and the need for increased discussion in how to mitigate negative effects to ensure the minimum loss of species and minimum impact on our lives.

There is already evidence that climate changes are altering plant distributions (Midgley *et al.*, 2003; Kelly and Goulden, 2008), whether species are moving towards the poles (e.g. Norway spruce, Seppä *et al.*, 2009), moving upwards in mountainous regions (Parmesan and Yohe, 2003; Lenoir *et al.*, 2008), or increasing or decreasing in their area of distribution (Gottfried *et al.*, 2012). Populations may be more vulnerable to pests and disease (Newton *et al.*, 2008) as well anthropogenic problems such as pollution and habitat destruction. Furthermore, species will tend to germinate and flower earlier, placing them under increased threat from pests and diseases (Gregory *et al.*, 2009; Chakraborty and Newton, 2011) that arrive earlier and have a longer time to build up their numbers. Increasingly problematic is the threat from invasive species, as their defining characteristics suit rapid population expansion (Bradley *et al.*, 2010). These biotic changes will also be made more critical for plants due to associated abiotic changes, such as increased levels of drought or salinity caused by over-irrigation, which directly makes the environment less suitable for plant survival. For species that do not adapt or expand their distribution it will be incumbent on the farmers and conservationists to minimize the negative impact of such events.

However, Cahill *et al.* (2012) argued that the most crucial factor affecting plant survival under climate change is disrupted interaction between species, rather than a simple, physiological intolerance to increased temperatures or drought. They undertook a meta-analysis of 136 published studies that demonstrated a link between extinction and climate change and found only seven studies that identified specific mechanisms by which climate change caused local extinction of populations, none of which involved a straightforward physiological intolerance to higher temperatures. In fact, most studies suggested that changes in how species interact (e.g. pest or pollinators and host species) is the main cause of extinctions linked to climate change, with declines in food availability being the most common cause. Therefore, it is more likely that changes in plant phenology may perhaps mean that pollinators are not around at the appropriate time for pollination and lead to population decline rather than the plant itself being unable to cope with higher temperature levels or less water availability.

Some species are much more susceptible to climate change impacts than others due to inherent biological traits related to their life history, ecology, behaviour, physiology and genetics (IPCC, 2007). For example, they estimate climate change is likely to impact maize and wheat by reducing their grain yields by as much as 40% at low latitudes and for rice, yields will decrease by up to 30% in China and the Indian Subcontinent. Furthermore, looking at the different continental scenarios, excluding Antarctica, the most adverse impact is likely to be in areas of subsistence or marginal agriculture. In Africa, for instance, by 2020, 75–250 million additional people will have crops exposed to increased water stress, rain-fed agriculture

yield will be reduced by up to 50%, and by 2080 there is predicted to be an increase of 5–8% of arid and semi-arid land (IPCC, 2014a). A summary of the likely predicted impacts of climate change on species is provided in Figure 6.4.

In response to this unprecedented threat, the FAO (2008, 2015a) argued that the existing genetic diversity within PGR can be used to sustain production systems by improving abiotic stress tolerance (e.g. facilitating new timing of sowing or harvesting, and increasing water-use efficiency, heat tolerance and improving use of nutrients) and biotic stress tolerance (e.g. using disease-resistant cultivars or multi-lines to strengthen crop resilience, and using diversified strategies to increase species and genetic diversity farmed) and by promoting community-

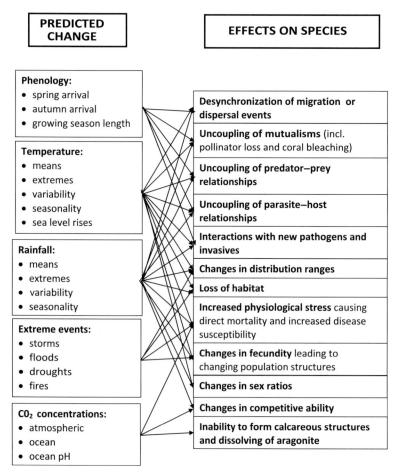

Figure 6.4 Predicted aspects of climate change and examples of their effects. (From Foden *et al.*, 2009)

based management of a wide portfolio of genetic diversity to facilitate adaptive capacity, particularly of local varieties adapted to extreme climatic events. Further, the systematic use of genetic diversity will be required if, with a growing human population, future agricultural production targets are to be achieved (Burke *et al.*, 2009; Zimmerer, 2010) and the necessary resilience to sustain small-scale farming systems is met (Hertel *et al.*, 2010). The FAO (2015b) proposed a set of voluntary guidelines that attempt to integrate genetic resources for food and agriculture (GRFA) into overall national biodiversity adaptation planning by identifying clear national goals for GRFA conservation and priority areas for future investments in conservation and use of GRFA. Changes in abiotic stresses at a location may be mitigated by adapting local varieties to permit changes in the timing of sowing or harvesting as well as to increase water-use efficiency, heat and salinity tolerance, and use of nutrients.

There are two distinct approaches to crop improvement in the context of climate change: (i) use of greater agro-biodiversity (particularly CWR diversity) within formal plant breeding (Tanksley and McCouch, 1997; McCouch *et al.*, 2003); and (ii) farmers' use of diversity through the deployment of pre-bred and therefore already adapted material into their own crop management strategies (FAO, 2008; Padulosi *et al.*, 2015; McDonald *et al.*, 2018). Both promote increased use of agro-biodiversity and encourage community-based management of a wide portfolio of genetic diversity to facilitate greater adaptive capacity. Furthermore, the impacts of extreme climate events might be mitigated by breeding cultivars that delay spring growth to avoid late frosts, the use of genetic diversity of fire-tolerant forest species and community-based conservation of local varieties adapted to extreme climatic events. Changed biotic stresses at a location may also be countered using disease-resistant cultivars or multi-lines to strengthen crop resilience and the promotion of diversification strategies to increase the species and genetic diversity farmed. Ultimately, current crop varieties will need replacement or enhancement to enable them to better suit changing agro-environmental conditions, and the trait basis for

these 'novel' varieties may draw on untapped genetic variation (Deryng *et al.*, 2011; Li *et al.*, 2011; Luck *et al.*, 2011), much of which may be available within a local community but is currently undervalued, poorly conserved and unavailable for sustainable utilization (Esquinas-Alcázar, 1993; Veteläinen *et al.*, 2009b).

The changing climate is likely to increase the importance of the use of CWR in breeding new cultivars to meet the future needs of farmers and so provide increased food production (Maxted *et al.*, 2016a). As wild species tend to contain higher levels of genetic diversity (Oka, 1988; Morden *et al.*, 1990; Murphy and Phillips, 1993; Tanksley and McCouch, 1997; Buckler *et al.*, 2001; Maxted and Kell, 2009) and are found in a much wider range of habitats under variable environmental conditions than crops, they may be critical in maintaining high levels of food production and ecosystem resilience. Overall it appears that climate change is likely to impact both agricultural production directly, by necessitating the production of new cultivars that are climate change resilient, and indirectly, by threatening and reducing the agro-biodiversity available to plant breeders and which contains the traits needed by the breeders to produce the new cultivars.

Plant species may respond to climate change by: (a) adapting to a changed environment (natural evolution), (b) shifting their range (moving with their climatic envelope), (c) being moved or assisted by humans (translocation, assisted migration or population enhancement) or (d) going extinct (IPCC, 2007; Ford-Lloyd *et al.*, 2014). Species vary in their ability to disperse naturally and adapt to new environments, and so if extinction is to be avoided, we need to identify and understand what makes a species vulnerable to climate change and to give conservation priority to these species. For example, we need to know which plant species have a small population size, and where component populations are highly fragmented, have a narrow ecological or geographic range, are poorly adapted to their niche or are poor competitors. Each of these potentially lowers the effective population size for the species below the sustainable minimal viable population number for that species (Foden *et al.*, 2009).

There are at least three approaches to identifying climate change vulnerable taxa: IUCN Red List Assessment, IUCN Climate Change Vulnerability Assessment and Species Distribution Modelling. The IUCN Red List Categories and Criteria (IUCN, 2001, 2003) were developed before the threat of climate change was fully appreciated and a more robust method was required to aid conservation planning (Foden *et al.*, 2009). A plant species-specific life history, biological traits and existing range of genetic diversity will influence its capacity to adapt to novel environmental conditions or to migrate to environments matching the conditions within its current niche (Pearson *et al.*, 2006; Hannah, 2008; Vitt *et al.*, 2010; Loss *et al.*, 2011). Foden *et al.* (2009, 2013) proposed that a species's susceptibility to climate change can be estimated based on specific biological traits (life history, ecology, behaviour, physiology and genetic diversity), and this assessment might guide species prioritization. However, a more direct means of assessing the impact of climate change on plant diversity is through modelling. Climate envelope models use environmental data to identify taxon-specific ecological niches (Phillips *et al.*, 2006) to postulate the locations where a species has been found or are absent, to infer its climatic requirements. These inferred requirements can then be used to compare a taxon's current niche with future climate scenario models to predict the longevity of potential protected area (PA) sites, the likely shift and extent of suitable niche environments available as the climate changes (Hijmans and Graham, 2006) and the probability that suitable conditions will remain within the boundaries of sites managed. Species distribution models (SDM) will identify areas that may be suitable for populations to persist in and prioritize areas for conservation (Guisan and Zimmermann, 2000; Guarino *et al.*, 2002; Guralnick, 2007; Scheldeman and van Zonneveld, 2010; Johnston *et al.*, 2012; Pacifici *et al.*, 2015). To illustrate SDM application in conservation planning an example from Oman (Al Lawati *et al.*, 2016) is provided in Box 6.3.

6.3 Local, National and Global Conservation Planning

While the value of agro-biodiversity for food and livelihood security is widely recognized, there is a lack of knowledge about the diversity that exists, what the best approach to conservation is and precisely how that diversity may be used for crop improvement. Within PGR research CWR and LR inventories are lacking for most countries – without knowledge of how many populations, crops or taxa exist and where they are located, it is difficult to plan for their systematic conservation. At least when planning the conservation of other wild species or habitats, more inventories and associated background data are often available, and there is more certainty over the scope of the conservation units to be conserved.

There are many potential approaches to achieve systematic global or sub-global (regional, national and local) agro-biodiversity conservation. Regardless of the approach, the goal is to maximize the conserved genetic diversity of CWR and LR using a complementary approach that incorporates both *in situ* and *ex situ* techniques (Maxted *et al.*, 2013b). The conservation of genetic diversity usually results from a combination of conservation actions at the macro- and micro-levels (Figure 6.6). Macro-conservation deals with the political, economic and strategic planning issues related to habitat, species or genetic diversity conservation and can be implemented at global, regional, national and local levels. In other words, macro-conservation deals with the development of strategic plans targeting the conservation of specific elements of biodiversity, in this case of CWR and LR, but not its practical implementation. Whereas micro-conservation comprises the distinct, practical conservation actions which are implemented via use of *in situ* and *ex situ* techniques focused on individual habitats, species or intra-specific genetic diversity and that contribute to implementation of the strategies developed at the macro-conservation level. As such, the development and application of conservation strategies for CWR and LR can be thought of as involving macro- and micro-conservation decision-making.

Box 6.3	Exemplar Climate Change Species Distribution Modelling for *Medicago sativa* and *Aerva javanica* in Oman

(Al Lawati *et al.*, 2016)

To illustrate the procedure for climate change modelling a cultivated and a wild species were chosen, *Medicago sativa* and *Aerva javanica*; both species had a relatively large number of presence points available for the analysis (the higher the number of points the greater the accuracy of the prediction). Current climatic data were obtained from WorldClim (Hijmans *et al.*, 2005) and future climate projections from CCAFS (www.ccafs-climate.org). The SRES-A2 emission scenario was selected, as it is considered the most realistic and due to its use in similar studies (Ramírez-Villegas *et al.*, 2014). The General Circulation Model used was the UKMO-HadCM3 (www.metoffice.gov.uk/research/modelling-systems/unified-model/climate-models/hadcm3). The 13 variables used to create the potential species distribution maps were selected using objective methods that included a test for collinearity and principal component analysis (PCA), which resulted in the removal of correlated variables. The resulting variables were checked by an Omani national expert for appropriateness in predicting the distribution of the species. Maxent (Phillips *et al.*, 2006) was used to create the predicted distribution at a resolution of 2.5 arc-minutes (~5 km^2 cell size at the equator). Validation of the model was achieved using a random test percentage of 30%. Modelled distributions were restricted to the 'equal training sensitivity and specificity' threshold. The remaining parameters were kept at default.

The three time projections for *M. sativa* and *A. javanica* are shown in Figures 6.5a to 6.5f. Figure 6.5a shows the distribution of *M. sativa* under current climatic conditions with a distribution of 12% throughout the country and predicted presence in three PAs. Figure 6.5b shows the predicted distribution under climatic conditions in 2020 with 17.4% distribution area and presence in one PA. Figure 6.5c shows the predicted distribution in 2050 with 2.3% distribution area and presence within one PA.

To ensure *in situ* conservation of *M. sativa* it is recommended to set up active on-farm conservation management with local communities within the Al Jabal Al Akhdar Scenic Reserve in the Ad Dakhliyah region. This area appears to be the 'core' of the *M. sativa* distribution as populations are predicted to be present here under all three time projections up to 2050. Figure 6.5d shows the distribution of *A. javanica* under current climatic conditions with a distribution of 10.7% throughout the whole country and predicted presence in three PAs. Figure 6.5e shows the predicted distribution under climatic conditions in 2020 with 35.2% distribution area and presence in one PA. Figure 6.5f shows the predicted distribution in 2050 of 1.6% distribution area and presence in one PA. To ensure *in situ* conservation of *A. javanica* it is recommended to undertake active protected area management within the existing PAs in the Al Jabal Al Akhdar Scenic Reserve in the Ad Dakhliyah regions.

It would also be a priority to increase active management and protection of populations of *M. sativa* found in the south near Salalah and *A. javanica* adjacent to the Yemen border because they both appear to persist over the prediction period and may contain important and different genetic diversity to the populations in the North. It will also be necessary to store *ex situ* samples from the populations that are projected to go extinct due to climate change. However, given a funding constraint, priority would be given to *M. sativa* because of its direct utilization value.

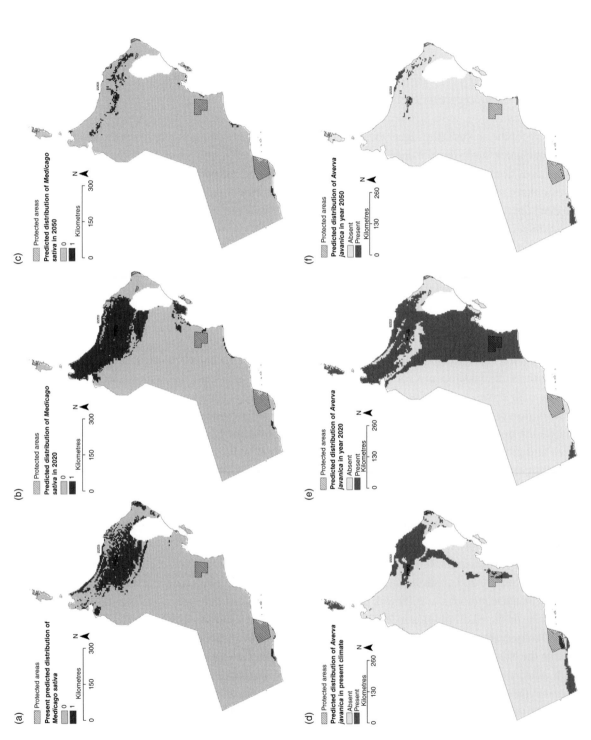

Figure 6.5 Predicted distribution of *Medicago sativa* (a, b and c) and *Aerva javanica* (d, e and f) at present and two future times (red area indicates potential distribution). (A black and white version of this figure will appear in some formats. For the colour version, refer to the plate section.)

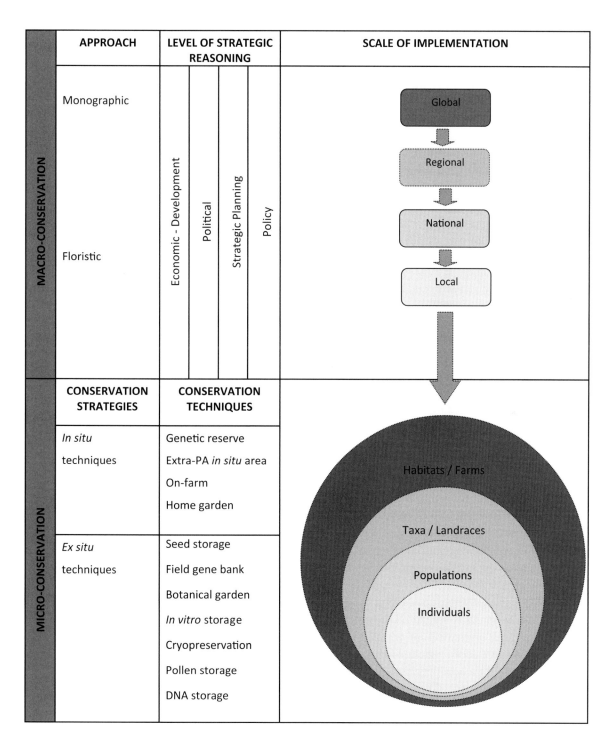

Figure 6.6 Conservation planning overview.
(Adapted from Maxted *et al.*, 2013b)

At the macro-conservation level, a first decision is which of the two possible, distinct approaches to take to develop the conservation plan: whether to adopt a monographic or a floristic approach. The monographic approach focuses on priority crop gene pools and can be applied at different geographic levels (global, regional or national). It is monographic because the methodology is comprehensive for individual target taxa throughout their full geographic range or its full taxonomic diversity within a geographically defined unit such a region or a country. It aims to systematically conserve the selected priority CWR or LR diversity via a network of *in situ* genetic reserves, extra-protected area sites or on-farm sites, with complementary *ex situ* storage and back-up. The floristic approach is taxa/crop comprehensive but for a defined area, because it attempts to encompass all CWR/LR that occur within a geographic unit (i.e. a region, country, subnational unit or sub-national region), regardless of the plant taxa/crops' normal range or extent of occurrence. The full geographic range of an individual taxon may or may not be included, depending on whether it is endemic to the target country or region. The floristic approach is commonly associated with the development of National plans for CWR and LR conservation.

Given the different intrinsic features that characterize CWR (wild species) and LR (crops), the application of the monographic and floristic approaches is similar in concept but may be slightly different in application depending on whether the target is CWR or LR diversity. Regarding the use of the term floristic for LR conservation, it is meant to imply the entire LR diversity found within a defined geographic area (e.g. local, region, country, even continental), just like a botanical Flora encompasses the wild plant diversity found within a defined area. The monographic and floristic approaches, for both CWR and LR, may be strategic in that they are likely to be implemented by national or global conservation agencies or institutions, and should not be an alternative but rather a holistic matrix to maximize the overall CWR or LR diversity conserved.

6.3.1 Global Agro-biodiversity Conservation Planning

A global approach aims at the systematic conservation of CWR and LR diversity as a means of maintaining global food security and meeting continuing consumer choice. At a global level, the monographic approach (targeting specific crops and crop gene pools) must be used since there is currently no global list of CWR taxa or LR diversity. The requirement for a global approach is especially important because CWR and LR diversity, like plant diversity in general, is not evenly spread across the globe, but is concentrated into botanical (Myers *et al.*, 2000; Mittermeier *et al.*, 2004) and crop diversity hotspots (Vavilov, 1926; Hawkes, 1983), and maintaining food security requires a global overview to be successful. Conservation in these highly diverse hotspots is thus necessarily independent of national political borders and needs to be globally coordinated if it is to be effective.

In response to this challenge, the FAO Commission on Genetic Resources for Food and Agriculture (CGRFA) called for the development of a network of *in situ* conservation areas to conserve CWR diversity (Activity 4 of the *Global Plan of Action for the Conservation and Sustainable Utilization of Plant Genetic Resources for Food and Agriculture* – FAO, 1996). Within this context, the CGRFA commissioned a thematic background study on 'the establishment of a global network for the *in situ* conservation of crop wild relatives: status and needs' (Maxted and Kell, 2009). The objective of this study was to provide baseline information for planning the future work of the Commission in the establishment and monitoring of a network of *in situ* conservation areas for CWR, and this was achieved by ecogeographic/gap analysis techniques. The crops included in this background study were those identified as being of major importance for food security in one or more sub-regions of the world (FAO, 1998) and/or that are listed in Annex I of the ITPGRFA (FAO, 2001b): finger millet (*Eleusine coracana*), barley (*Hordeum vulgare*), sweet potato (*Ipomoea batatas*), cassava (*Manihot esculenta*), banana/plantain (*Musa* ssp. AAA, AAB and ABB groups, *M. acuminata* (AA) and *M.*

balbisiana (BB)), rice (*Oryza sativa*), pearl millet (*Pennisetum glaucum*), potato (*Solanum tuberosum* except Group *Phureja*), sorghum (*Sorghum bicolor*), wheat (*Triticum aestivum*), maize (*Zea mays*), cowpea (*Vigna unguiculata*), faba bean (*Vicia faba*) and garden pea (*Pisum sativum*). These priority crops represented different crop groups (cereals, food legumes, roots and tubers), with different breeding systems (cross-pollinating, self-pollinating, clonally propagated), as well as crops of temperate and tropical origin (Maxted and Kell, 2009). Preliminary recommendations for the establishment of a global network of *in situ* conservation areas for the highest priority CWR species from the 14 crop gene pools for which distribution data were available were proposed. Many were in the centres of crop diversity identified by Vavilov (1926), which remain the hotspots of crop and CWR origin/diversity today.

A more comprehensive analysis of ecogeographic/gap analysis techniques is currently being applied by the Global Crop Diversity Trust initiative 'Adapting Agriculture to Climate Change: Collecting, Protecting, and Preparing Crop Wild Relatives' (www.cwrdiversity.org/; Dempewolf *et al.*, 2013). Using published gene pool and taxon group concepts the project developed a prioritized CWR inventory for 173 crops of 1667 CWR taxa, divided between 37 families, 108 genera, 1392 species and 299 sub-specific taxa (Vincent *et al.*, 2013). Ecogeographic data was collated for these taxa and *ex situ* and *in situ* conservation priorities were identified using the Maxted *et al.* (2008a) and Ramírez-Villegas *et al.* (2010) methodologies. The key results of the two analyses of *ex situ* (Castañeda-Álvarez *et al.*, 2016a, 2016b) and *in situ* (Vincent *et al.*, 2019) conservation priorities are shown in Figures 6.7 and 6.8, respectively. For the *ex situ* analysis, the geographic distributions of 1076 unique CWR taxa from 76 genera belonging to 24 plant families was compared against the potential geographic and ecological diversity encompassed in these distributions to those accessions that are currently available in gene banks, and the gaps in conservation were identified. For 313 (29.1% of total) taxa associated with 63 crops, no *ex situ* germplasm accessions exist at all, and a further 257 taxa are represented by fewer than 10 accessions. A total of 765 (71.1%) taxa were ranked as

high priority for further collecting from their natural habitats, 148 (13.8%) as medium priority, 118 (11.0%) as low priority and only 45 (4.2%) as currently sufficiently represented in gene banks. Proposed hotspots for further collecting were identified in the Mediterranean, Near East, and southern and western Europe; Southeast and East Asia; and South America, with a maximum of 43 CWR associated with 23 crops in a single 25-km^2 grid cell. Compared to the identification of *ex situ* priorities, the identification of *in situ* priorities is simpler because there has been no concerted effort to conserve CWR diversity *in situ* thus far (Maxted *et al.*, 2014), and the hotspots identified are the localities where *in situ* CWR should be conserved.

In parallel to the Global Crop Diversity Trust initiative (Dempewolf *et al.*, 2013) to locate and collect for *ex situ* conservation under-collected CWR global priority taxa, the 13th Regular Session of FAO CGRFA (FAO, 2011c) recognized the importance of establishing a global network for *in situ* CWR conservation and LR on-farm management. The FAO initiated a consultation process and held a technical workshop 'Towards the establishment of a global network for *in situ* conservation and on-farm management of PGRFA' in November 2012, to identify options, ways, and means for establishing a global network (FAO, 2014a). The discussion agreed the need to establish a broad, decentralized and participatory global network to support coordination of efforts, help raise resources and create more awareness of the value and necessity of *in situ* conservation of agro-biodiversity; however, further implementation is still being actively discussed.

The systematic *in situ* on-farm conservation of LR is far less advanced at the global, or for that matter at the regional, national or local levels (Brush, 2000; Jarvis *et al.*, 2007, 2016; Veteläinen *et al.*, 2009b; Koohafkhan and Altieri, 2017). Networks of on-farm sites for LR conservation are yet to be systematically implemented. In fact, Veteläinen *et al.* (2009a) highlighted the difficulty of systematically conserving all LR diversity on-farm due to the very high numbers of existing LR but stressed that a coherent global network of on-farm conservation should be established to actively conserve the highest priority LR, either based onz LR concentration or occurrence

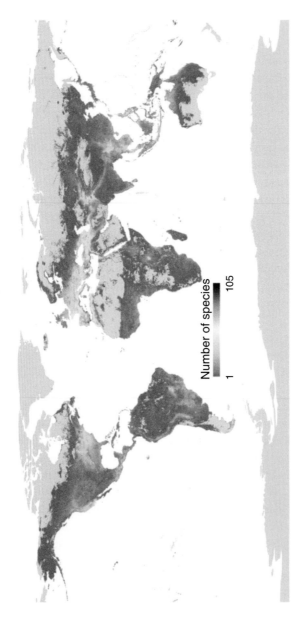

Figure 6.7 Priorities for global *ex situ* CWR conservation. Hotter colours indicate greater CWR concentration. (A black and white version of this figure will appear in some formats. For the colour version, refer to the plate section.)
(From Castañeda-Álvarez *et al.*, 2016a)

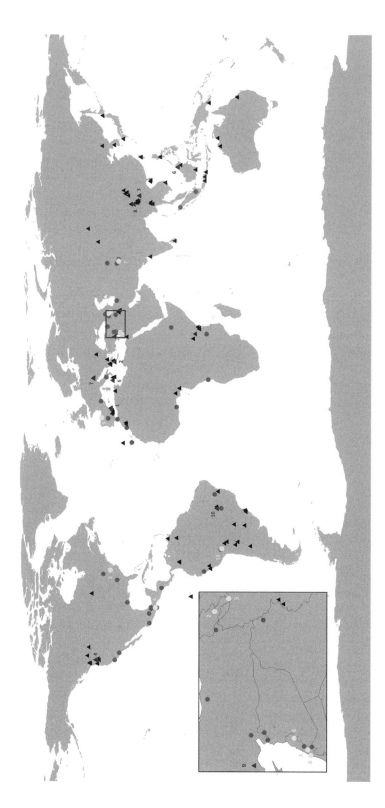

Figure 6.8 Priorities for global *in situ* CWR conservation, with the inset map showing the priority sites in the Fertile Crescent and Caucasus. The top 10 sites within existing protected areas are shown as magenta triangles, and the remaining 90 priority sites within protected areas are represented by blue triangles; the top 10 sites outside of existing protected areas are shown as yellow circles, with the remaining priority 40 sites outside of protected areas represented by turquoise circles. (A black and white version of this figure will appear in some formats. For the colour version, refer to the plate section.) (From Vincent *et al.*, 2019)

of adaptive diversity. A similar point could equally be made nationally for each individual country's priority crops. The importance of establishing such networks was noted and the FAO (2014a) called for the establishment of a global network for *in situ* conservation and on-farm management, recognizing the need for a global initiative focused on LR *in situ* conservation.

There is a strong logic for an intergovernmental institution focusing on biodiversity for food security to cooperate with international partners from environment and agriculture to lead the required research and the establishment of global CWR and LR networks. However, in practice all *in situ* actions are necessarily implemented at specific localities and so national agencies need to take the lead on practical implementation. There is therefore a parallel onus on each country to conserve its CWR and LR diversity *in situ*, and this will require the establishment of national networks of genetic reserves and on-farm sites. This means that governments will maintain their national sovereignty over PGR, but also national sites, as well as being of national value, may also contribute to the regional and in turn contribute to the global *in situ* or on-farm network.

There have also been initiatives to establish regional CWR and LR diversity *in situ* networks in the Middle East (Amri *et al.*, 2007), Europe (Kell *et al.*, 2016), the Nordic countries (Ansebo, 2015; Fitzgerald *et al.*, 2016) and the Southern African Development Community (www.cropwildrelatives.org/sadc-cwr-project/; Bioversity International, 2016). Further in Europe the 13th meeting of the European Cooperative Programme for Plant Genetic Resources Steering Committee (ECPGR, 2012) acknowledged the importance of *in situ* and on-farm conservation and recommended the development of conservation planning concepts for Europe. These were subsequently proposed by Maxted *et al.* (2015) and ECPGR (2017).

6.3.2 National Agro-biodiversity Conservation Planning

Significant progress has been made in recent years in acknowledging both the value of systematic planning of agro-biodiversity conservation and the development of methodologies to generate a national

strategy or action plan. However, as with global initiatives, more progress has been made in *ex situ* as opposed to *in situ* planning and implementation and for CWR rather than LR diversity conservation. The FAO commissioned a 'Resource Book for the Preparation of National Plans for Conservation of Crop Wild Relatives and Landraces' (Maxted *et al.*, 2013b) to address the issue of how to systematically approach national CWR and LR conservation planning and implementation, while emphasizing the need to integrate national actions with local, regional and global levels of implementation. It is important to stress there is no single method for developing strategic plans for CWR and LR conservation because of issues concerning resource and baseline biodiversity data availability, the local community where the plan is to be implemented, as well as the focal area and remit of the agencies which are responsible for formulating and implementing it. Nevertheless, developing national plans for PGR conservation and use can be viewed as a series of decisions and actions that follow the same basic pattern in all countries (see Figure 6.9) and these this will involve:

- Creation of a CWR/LR checklist – the first step is the creation of a CWR or LR checklist from existing botanical or crop data, or from field or farmer surveys.
- Prioritization for conservation action – it is likely that prioritization will be necessary because the number of taxa or LR included in the checklist is too large for immediate conservation action given limited resources, prioritization for CWR is commonly based on crop value, threat and relationship to crop and for LR crop type, crop value and threat.
- Creation of CWR/LR inventory – for each of the prioritized taxa or LR additional provenance, cultivation and characterization data is collated and made available via the inventory.
- Genetic data analysis – for each of the prioritized taxa or LR the pattern of both within and between taxon/LR diversity is established.
- Ecogeographic surveying/ gap/climate change analysis – for each of the prioritized taxa or LR

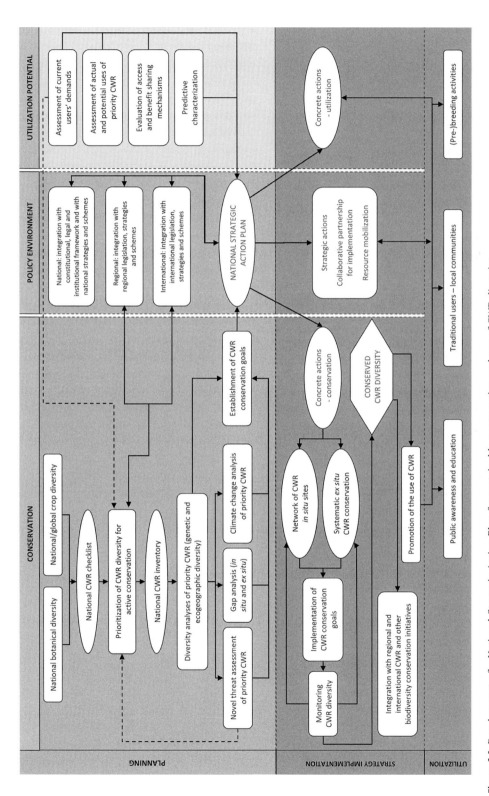

Figure 6.9 Development of a National Strategic Action Plan for sustainable conservation and use of CWR diversity. (Adapted from Magos Brehm et al., 2017a)

where the pattern of genetic diversity is unknown ecogeographic data is interpreted using gap analysis techniques. The initial results are checked against the likely impact of climate change so that the conservation recommendations are climate smart.

- Establishment of *in situ* and *ex situ* conservation goals – the inventory data, genetic, ecogeographic surveying/gap analysis and threat information is used to aid formulation of a national CWR or LR conservation action plan, with clear *in situ* and *ex situ* conservation goals and recommendations for *in situ* and *ex situ* actions.
- Implementation of CWR or LR conservation action plan – the national CWR or LR conservation action plan is enacted, primarily genetic reserves, extra-protected area *in situ* site for CWR and on-farm or home garden sites for LR as well as complementary *ex situ* conservation actions to ensure a safety back-up and user availability of the genetic diversity.
- Monitoring of diversity conserved – the on-going effectiveness of the conservation actions is assessed regularly using specific indicators (e.g. population's levels and trends over time for target taxa), and if necessary, the national CWR or LR conservation action plan is periodically revised.
- Promoting the use of diversity – agro-biodiversity conservation is unique in that the end goal is not just the maintenance of the diversity but also the making of that genetic diversity available to potential users (e.g. farmers and breeders, etc.) and its use in sustaining food production and food security.

The practical implementation of *in situ* genetic reserves within or outside existing PAs should be addressed at the policy level and a strong commitment to conservation site sustainability confirmed. The National management plans for CWR and LR conservation should thus be integrated and linked to the GSPC (through the GSPC national focal point), the ITPGRFA, the National Biodiversity Strategies and Action Plans (NBSAPs) – the principal instruments for implementing the CBD at national level (www.cbd.int/nbsap/) – and to National Plant Conservation Strategies, where they exist. Whether CWR are conserved *in situ* within PAs or outside of them, it is advisable that the sites have some form of legal protection to help prevent sudden threats to conserved populations. On the other hand, local communities living within the target sites where genetic reserves are to be implemented should be actively involved so a holistic and thus efficient approach to conservation of CWR is implemented. Awareness of National management plans for CWR and LR conservation should be raised among the different stakeholders. These can take the form of local community conservation (training) workshops. Agreements with private owners (e.g. tax incentives) could be made, not only to ensure CWR are properly managed but also to recognize the local communities' role in conserving such a valuable resource (Hunter and Heywood, 2011).

Practically, the number of examples of local communities actively using CWR diversity to bolster genetic diversity in their crop maintenance systems are very limited (Hunter and Heywood, 2011). However, the local communities living with CWR diversity, where collections are collected or where active *in situ* CWR conservation is taking place, should be actively involved in CWR maintenance, particularly in less intensive or subsistence agricultural production systems in both developed and developing countries (Veteläinen *et al.*, 2009b; de Boef *et al.*, 2013a; Vernooy *et al.*, 2015). In this case, such activities may be promoted by establishing community seed banks or seed fairs to promote local LR diversity value and its continued use and enhance product market opportunities. Recently an interactive toolkit for CWR conservation planning has been released as a tool to facilitate national CWR conservation planning (Magos Brehm *et al.*, 2017a), and this is described in Box 6.4.

6.4 Ecogeographic Surveys and Gap Analysis

6.4.1 Introduction

Plant species are not randomly distributed across the globe. Each species is found growing within a

Box 6.4 | **Interactive Toolkit for Crop Wild Relative Conservation Planning**

(Magos Brehm *et al.*, 2017a)

The Interactive Toolkit is designed to provide guidance to plan and implement systematic and active *in situ* and *ex situ* conservation of CWR at national level. The conservation recommendations that result from this national CWR conservation planning process can and should be used to develop National Strategic Action Plans (NSAP) (or National Strategies) for the conservation and sustainable use of CWR. It can be used by local institutions (e.g. gene banks, universities, research institutes), individual scientists, NGOs, agencies responsible for planning and implementing NSAP or national strategies, such as national agricultural or environmental agencies, and FAO National Focal Points. There is no single method for planning CWR conservation or for developing an NSAP for the conservation of CWR. This is mainly due to factors concerning financial resources, availability of baseline biodiversity data and the local community where the NSAP is to be implemented, as well as the focal area and remit of the agencies that are responsible for formulating and implementing the NSAP. Nevertheless, CWR conservation planning can be viewed as a series of steps and decisions that follow the same basic pattern in all countries.

The Toolkit thus includes a series of 12 modules (plus an introductory module to 'National systematic CWR conservation planning'), each corresponding to a step in conservation planning. Each module comprises an introduction to that particular step, methodology that explains how to undertake that step, examples and applied use which include case studies showcasing how that particular CWR conservation planning step has been undertaken in different scenarios, the list of references used to compile the texts in that module, and a list of additional materials and resources (bibliographic references, PowerPoint presentations, relevant projects, recordings and video files, software tools, web links and relevant social networks) that help the user visualize and understand how to undertake that particular step. The central element of the Toolkit are the interactive flowcharts present in each methodology section which are generally composed of a series of yes/no questions, helping the user move through the various phases and choose the options that are most appropriate given the user's national context. Detailed explanations of each phase are linked to the interactive flowchart. In addition, there is a static version of the flowchart which can be consulted at any time and a full written text that exactly matches the information presented in the flowchart. The Toolkit can be used either online or offline (by exporting its content) and in order to access the modules of the Toolkit the user needs to register using an email address and providing some additional information used for statistical purposes only.

This Toolkit and its protocols, examples and resources should thus be viewed as a framework and an aid for planning CWR conservation. It can be used for the entire conservation planning process or for individual steps. It should be noted that the Toolkit was designed for CWR conservation planning, but it can be adapted to any plant group.

defined range of localities defined by biotic and abiotic constraints, which is its ecological niche defined in part by its 'ecogeographic envelope' for that species. The ecogeographic envelope, the series of ecological, geographic and environmental constraints that limit a species distribution and maintenance of a viable population, will vary from species to species in its magnitude and range. Some species are generalists and are found in a broad range of habitats and locations, therefore have a broad ecogeographic envelope, while other species are more restricted and found in a more restrictive range or a single habitat or may need other species present to form a viable population. To develop an efficient and effective conservation strategy using both *in situ* and *ex situ* techniques the conservationist must have a clear understanding of the target taxon's ecogeographic envelope, its geographic distribution, its habitat preferences and niche requirements, its genetics and its taxonomy. Therefore, when planning the conservation of a species, the details of the locality and associated environment characteristics of sites where the species has been previously recorded, the so-called passport data, associated with herbarium and germplasm collections or field surveys, is a guide to future conservation activities. For example, if the passport data associated with accessions of *Oryza rufipogon* Griff. (a close wild relative of rice) indicates that it has previously been found in swamps and marshes of South and Southeast Asia (Vaughan, 1994), these areas are likely to contain the species today. Within these areas, localities with a high concentration of individuals would be considered as potential sites for establishing *in situ* genetic reserves and field collection for *ex situ* gene bank storage.

Effectively, the conservationist is identifying the ecogeographic envelope and the ecological, geographic, genetic and taxonomic amplitude of the species and using it to plan the species conservation. This process is referred to as an ecogeographic survey and is defined as:

. . . a process of gathering and synthesizing information on ecological, geographical, taxonomic and genetic

diversity. The results are predictive and can be used to assist in the formulation of complementary *in situ* and *ex situ* conservation priorities.

(Castañeda-Álvarez *et al.*, 2011)

The procedure for undertaking an ecogeographic survey is divided into three phases: project design, data collection and analysis, and production. The process of undertaking an ecogeographic survey is discussed in Maxted *et al.* (1995), Guarino *et al.* (2006) and Castañeda-Álvarez *et al.* (2011).

However, practically when planning conservation as well as having a clear understanding of the target taxon's ecogeographic envelope, there is also a need to take into consideration the current conservation of the taxon. This is done via an associated technique called gap analysis (Burley, 1988; Margules, 1989; Margules and Pressey, 2000), and today the two techniques are often used in parallel. Gap analysis was originally applied to conservation planning of indigenous forests (particularly on small islands rich in endemic species) and has subsequently been extended to agro-biodiversity conservation (Maxted *et al.*, 2008a; Ramírez-Villegas *et al.*, 2010; Parra-Quijano *et al.*, 2011a) encompassing conservation strategies for both *in situ* and *ex situ* genetic diversity. It has been applied to diverse taxa, e.g. African *Vigna* (Maxted *et al.*, 2004), *Hordeum* (Vincent *et al.*, 2012), *Phaseolus* (Ramírez-Villegas *et al.*, 2010), *Glycine* in Australia (González-Orozco *et al.*, 2012), and temperate forage and pulse legume species (Maxted *et al.*, 2012a).

The basis for gap analysis is a comparison of the target taxon's ecogeographic and genetic diversity with the elements of that diversity which are currently actively conserved. The 'gap' is therefore the component of the target taxon's ecogeographic and genetic diversity that is not actively conserved and which then becomes the future conservation priority. As such, the protocol for gap analysis involves four steps: (a) circumscription of taxa, (b) identifying the breadth of ecogeographic and genetic diversity for the target taxa, (c) identifying and matching current *in situ* and *ex situ* conservation actions with the breadth of ecogeographic and genetic diversity to identify the conservation 'gaps', and on that basis (d) formulate a revised *in situ* and *ex situ* conservation strategy (Maxted *et al.*, 2008a). Ramírez-Villegas *et al.* (2010) extended the methodology by using a quantitative approach to determine the requirements for *ex situ* conservation, using three criteria: sampling representativeness, and geographic and environmental coverage. Sampling representativeness consists of a comparison between the total number of populations sampled and those sampled as gene bank accessions. Geographic coverage is estimated by comparing the potential distribution coverage of the taxon with the gene bank geographic coverage using a 'circular area statistic within a 50km radius' (CA50) around each germplasm sample. The environmental coverage representativeness (or environmental representativeness score) is estimated by assessing the environmental range of germplasm accessions in relation to the whole environmental range of the taxon. The final priority score is calculated by averaging the three previous scores, the lower the value the higher priority for conserving that taxon. Parra-Quijano *et al.* (2011a) further extended the approach by developing ecogeographic land characterization (ELC) maps, which characterized the potential adaptive scenarios or habitat preferences for a target taxon or group of taxa based upon abiotic factors (geophysical, edaphic and bioclimatic variables) which are likely to influence distribution. In this context ecogeographic representativeness can be used as a surrogate of genetic representativeness for *in situ* and *ex situ* conservation (Parra-Quijano *et al.*, 2012), and the most genetically divergent sampling will come from a mixture of the most divergent categories. Hence, the populations that occur in ELC categories that are not conserved are gaps and in need of active conservation. The ELCmapas tool of the CAPFITOGEN package (Parra-Quijano *et al.*, 2014, available from www.capfitogen.net/en/) can be used to create ELC maps, and a specific application is demonstrated in Box 6.5. Today conservation planning involves therefore a combination of ecogeographic, gap and climate change analysis, which are summarized in the following sections.

6.4.2 Conservation Project Scope

In practise, all conservation activities are preceded by some form of project commission; an organization (e.g. CGIAR, Global Crop Diversity Trust, Global Environment Facility, national gene bank or agencies) or individual decides that a species or group of species or habitat in a defined area warrants active conservation. The unit to be conserved is currently not or is under-conserved (insufficient genetic diversity is held in gene banks, on-farm or *in situ*), is threatened *in situ*, or there is a specific user requirement for diversity not currently conserved *in situ* or *ex situ*. In each case a group of taxa from a defined geographic area is considered sufficiently of interest to warrant conservation planning and subsequent conservation implementation.

The acquisition of ecogeographic (ecological, geographic, taxonomic and genetic) data required for conservation planning can be enhanced by discussion with appropriate specialists. Taxon or geographic experts may point out relevant genetic diversity studies and other literature, recommend Floras, monographs and taxonomic databases, suggest which herbaria and/or gene banks to visit, and help contact other specialists. Identification of taxon expertise can be achieved by searching for recent publications and internet resources. One critical point it is important to clarify early on is the accepted classification for the target taxa; the classification will prescribe the taxa recognized by the international community and will also often provide significant ancillary information (see Chapter 2). The generally accepted classification can be obtained from target taxon specialists, recent classifications of the group and phylogenetic analyses. The target area of the taxon being studied may be restricted by the commissioning agency (e.g. the genus *Allium* in Uzbekistan or *Helianthus* in the United States), but if it is unspecified, the taxon should be studied throughout its range. Taking as broad a view of the target area as possible should avoid any problems of multiple works on the same taxon not being compatible once completed. Some sources of taxonomic and ecogeographic information and literature are provided in Box 6.6.

Box 6.5 | **Conservation Planning for *Medicago sativa* in Oman**

(Al Lawati *et al.*, 2016)

Ideally when planning plant genetic conservation, it would be desirable to know the relative distribution of genetic diversity within the populations of the target taxon within the target area. Without this knowledge, ecogeographic data is commonly used as a proxy for genetic diversity, based on the assumption that ecogeographic diversity is likely to be correlated with genetic diversity, an assumption that is not always correct but is expedient in the absence of contrary information. An ELC map segregates the regions of the country into geographic zones which are ecogeographically distinct. This can then be overlaid with current *in situ* or *ex situ* conservation actions to help identify gaps in conservation. Figure 6.10 shows the overlay of known *Medicago sativa* localities with the *M. sativa* species ELC map for Oman. *Medicago sativa* specimens have been recorded in seven ELC categories (2, 3, 4, 10, 11, 13 and 14); specimens that occur in four categories (2, 3, 10 and 11) overlap with existing PAs, whereas populations that occur in the remaining three categories (4, 13 and 14) do not overlap with any PA. It may be possible to establish genetic reserves for the conservation of those populations that occur in the ELC categories that overlap with the PAs. *Medicago sativa* as well as being wild in Oman is also cultivated as the forage alfalfa or lucerne, so the process could be repeated with the cultivated accessions and the potential site for on-farm conservation of *M. sativa* populations within the farming system identified. Similarly, populations from ELC categories not already conserved *ex situ* will become priorities for additional collection and *ex situ* storage.

6.4.3 Sources of Taxon Passport Data

In practice, much conservation planning is based on ecogeographic analysis of existing disparate data sets, primarily drawn from herbarium specimen and germplasm accession provenance localities or passport data. Historically this would involve the conservationist visiting herbaria and gene banks to collate the data, but today much data is held on-line. Most notably data is made available via the Global Biodiversity Information Facility (GBIF) (http://data.gbif.org), which provides extensive access to global taxon nomenclature and occurrence data drawn from natural history collections, libraries and databases. GBIF taxon searches can be limited to the country or countries of interest or to a specific species or set of species; data can be downloaded and the necessary records extracted. There are other initiatives that provide access to herbarium specimens and gene bank accessions, as well as national programmes that are digitizing their collections and making the data available via the internet (Table 6.2).

Notable among the data sources for specific CWR-related data is CWR Diversity, the portal associated with the Global Crop Diversity Trust (GCDT) initiative 'Adapting Agriculture to Climate Change: Collecting, Protecting, and Preparing Crop Wild Relatives' (Dempewolf *et al.*, 2013). As part of this project, the gene pool and taxon group concepts were collated for 173 prioritized crop gene pools which included 1667 CWR taxa, divided between 37 families, 108 genera, 1392 species and 299 sub-specific taxa (Vincent *et al.*, 2013). The *Harlan and de Wet CWR Inventory* of priority CWR taxa is available to be queried and downloaded from the portal (www

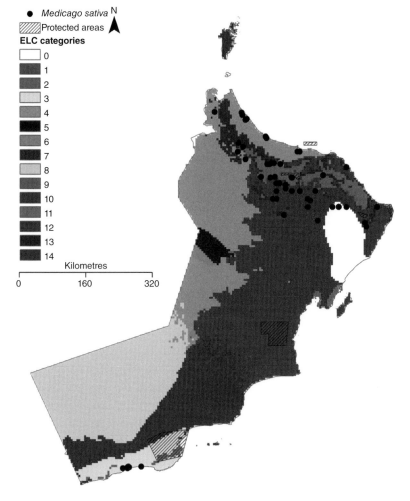

Figure 6.10 Specific ELC map for *M. sativa* locations in Oman. (A black and white version of this figure will appear in some formats. For the colour version, refer to the plate section.)

.cwrdiversity.org/checklist/). Also as part of the gap analysis undertaken by the project, by far the largest digitized ecogeographic data set was collated, which can also be queried and downloaded (www .cwrdiversity.org/checklist/cwr-occurrences.php). Further the portal also provides an atlas (www .cwrdiversity.org/distribution-map/) that plots CWR distributions and *ex situ* collecting priorities using a gap analysis methodology combining sampling, geographic and environmental representativeness scores (Ramírez-Villegas *et al.*, 2010). The data can be displayed by CWR taxon, as a summary of a crop gene pool, or as a summary of all CWR in all assessed crop gene pools. This project has stimulated much CWR conservation research and clear proposals for *ex situ*

(Castañeda-Álvarez *et al.*, 2016a) and *in situ* (Vincent *et al.*, 2016) conservation implementation.

6.4.4 Conservation Data Collection and Analysis

Genetic Diversity Assessment

Central to undertaking ecogeographic surveys and gap analysis is the analysis of georeferenced ecogeographic data, but such ecogeographic data is effectively a proxy for information on the inherent genetic diversity within those taxa and where such data is available, it is a more useful guide for conservation planning. The point may be illustrated with a classic study of the wild lentil gene pool by

Box 6.6 | Some Sources of Taxonomic and Ecogeographic Information and Literature

- *CWR Diversity* – a global quarriable portal providing access to (i) the Harlan and de Wet CWR Inventory of CWR taxa; (ii) the CWR Global Occurrence Database of germplasm accessions, specimens and population occurrence data; and (iii) the CWR Atlas, which provides CWR species distributions and *ex situ* collecting priorities maps (www .cwrdiversity.org/).
- CWR Global Portal – a global quarriable portal providing access to (i) experts working with CWR, (ii) CWR-related projects, (iii) CWR images, (iv) national or regional CWR checklists/ inventories, (v) CWR conservation strategies, (vi) CWR presentations divided by themes (genetic diversity, conservation, climate change), (vii) CWR Publications (papers, books, theses, newsletters), and (viii) CWR online Toolkit which has further CWR resources (www .cropwildrelatives.org/).
- *Global Biodiversity Information Facility (GBIF)* – provides extensive access to global taxon nomenclature and population occurrence data (www.gbif.org/).
- *Tropicos* – a global quarriable database of taxonomic, occurrence and literature data (www .tropicos.org/).
- *International Plant Names Index (IPNI)* – a database of the names and associated basic bibliographical details of all seed plants. Its goal is to eliminate the need for repeated reference to primary sources for basic bibliographic information about plant names. The data are freely available and are gradually being standardized and checked. IPNI is the product of a collaboration between The Royal Botanic Gardens, Kew, The Harvard University Herbaria, and the Australian National Herbarium (www.ipni.org/).
- *National Geospatial-Intelligence Agency (NGA)* – NGA enables *georeferencing* of locality data based on the locality name (geonames.nga.mil/gns/html/).
- *Protected Planet* – a world database of protected areas which can be queried interactively or downloaded (http://protectedplanet.net/).
- *The IUCN Red List* – a database of taxonomic, conservation status and distribution and relative threat information on plants, fungi and animals that have been globally evaluated using the IUCN Red List Threat Categories and Criteria (www.iucnredlist.org/).
- *The Plant List* – a working list of vascular plants (flowering plants, conifers, ferns and their allies, and mosses and liverworts) produced by the Royal Botanic Gardens, Kew and Missouri Botanical Garden. It contains 350 699 accepted names, alongside synonyms and unresolved names (www.theplantlist.org/).
- *Plant Book* – a comprehensive listing of higher plant names with basic botanical information compiled by D.J. Mabberley (2008) and published by Cambridge University Press.
- *Standard World Floras* – Frodin, D.G. (2001). *Guide to the Standard Floras of the World*, 2nd Edition. Cambridge University Press, Cambridge.

Table 6.2 **Data sources for herbarium specimen and germplasm accession passport data**

Name	Description	Website address
JSTOR Plant Science	Images of specimens from 155 institutions	http://plants.jstor.org
Databases on crop gene bank holdings		
GENESYS	A global portal to germplasm accession holdings of plant genetic resources for food and agriculture	www.genesys-pgr.org
CWR Diversity	A global portal providing access to CWR Global Occurrence Database of germplasm accession, herbarium specimens and other population occurrence data	www.cwrdiversity.org/
European Plant Genetic Resources Search Catalogue (EURISCO)	A searchable catalogue of European *ex situ* PGR collections based on European *ex situ* National Inventories (NIs), currently comprises passport data concerning 1.8 million samples. EURISCO is maintained on behalf of the Secretariat of the European Cooperative Programme for Plant Genetic Resources (ECPGR), in collaboration with and on behalf of the National Focal Points for the National Inventories	http://eurisco.ipk-gatersleben.de/
Wild plant species gene bank databases		
Botanic Gardens Conservation International (BGCI) database	A global portal providing access to plant species information associated with botanic gardens, including: research or conservation features and facilities, plant collections, conservation and education programmes	www.bgci.org/plant_search.php
National PGR programmes		
Russia	AgroAtlas	www.agroatlas.ru
Japan	NIAS	www.gene.affrc.go.jp/databases_en.php
Mexico	–	www.biodiversidad.gob.mx/genes/ proyectoMaices.html
USA	GRIN National GR Programme	www.ars-grin.gov
Other (taxonomic groups, herbaria, etc.)		
Millennium Seed Bank, Kew, UK		www.kew.org/science-conservation/ collections/millennium-seed-bank

Table 6.2 (*cont.*)

Name	Description	Website address
Natural History Museum, UK		www.nhm.ac.uk/research-curation/ collections/departmental-collections/ botany-collections/search/index.php
Royal Botanic Gardens, Kew, UK		http://apps.kew.org/herbcat/navigator .do
Royal Botanical Garden, Edinburgh, UK		www.rbge.org.uk/databases
United States Virtual Herbarium, USA		http://uaes.usu.edu/htm/aes-news/us-virtual-herbarium-online/
Virtual Australian Herbarium, Australia		http://plantnet.rbgsyd.nsw.gov.au/ HISCOM/Virtualherb/virtualherbarium .html#Virtual

Adapted from Castañeda-Álvarez *et al.* (2011).

Ferguson *et al.* (1998a). The distribution of *Lens culinaris* subsp. *orientalis* shows that the subspecies is found from Western Turkey to Tajikistan, and if resources were available to establish three genetic reserves for the subspecies, these could be logically located in Turkey, Iran and Tajikistan, assuming geographic distance can be used as a proxy for genetic distance (Figure 6.11). However, genetic analysis found that *L. culinaris* subsp. *orientalis* genetic diversity was partitioned into 10 clusters, but populations representing all 10 clusters are found in the Western Fertile Crescent and outside of this region there is little unique genetic diversity. Therefore, in this case, ecogeographic data would be an inadequate proxy for genetic diversity as it would indicate *ex situ* sampling or *in situ* genetic reserve establishment across the geographic distribution rather than focusing both in the Western Fertile Crescent (Figure 6.11).

Another example of the use of genetic diversity data in conservation planning is provided by van Zonneveld *et al.*'s (2012) use of microsatellite markers to understand the spatial genetic diversity of cherimoya (*Annona cherimola*), an edible fruit-bearing species consumed throughout the Andean region in Peru, Ecuador and Bolivia. The study found the highest genetic diversity in southern Ecuador and

northern Peru and recommended the establishment of *in situ* conservation there and the filling of existing gaps in *ex situ* collections in southern Ecuador.

Such examples are relatively rare, and more often knowledge of inherent patterns of genetic diversity are limited and it may be too resource-intensive to collate de novo. Consequently, genetically based approaches to conservation assessment either in terms of 'richness', the total number of genotypes or alleles present regardless of frequency, or 'evenness' of the frequencies of different alleles or genotypes, can therefore only be applied to the most highly prioritized taxa. Therefore, in the absence of such data, proxy or surrogate measures of genetic diversity are necessary and ecogeography often acts as a proxy for lack of specific genetic diversity data. However, it should be remembered that using any proxy will not be as accurate as using primary data.

Ecogeographic Data

For conservation planning ecogeographic data is collated at two levels, taxon and specimen. Taxon level data is collated via a media survey of geographic, ecological, taxonomic and genetic data related to the target taxa. The sources of information will include monographs, revisions, Floras, scientific

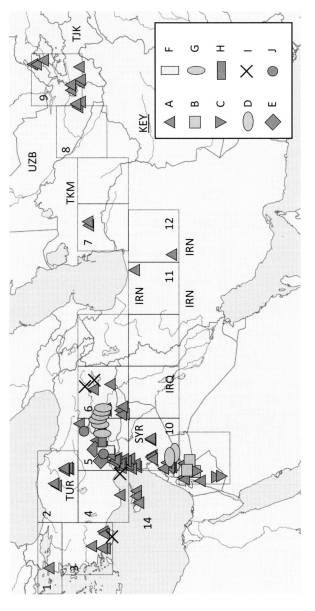

Figure 6.11 Distribution of *Lens culinaris* subsp. *orientalis* divided into 10 genetic diversity clusters (A–J). TUR, Turkey; SYR, Syria; IRQ, Iraq; IRN, Iran; TKM, Turkmenistan; UZB, Uzbekistan, TJK, Tajikistan. (From Ferguson *et al.*, 1998a)

Table 6.3 **Example taxon level ecogeographic data**

Ecogeographic data	Example data
Accepted taxon name	*Psophocarpus grandiflorus* Wilczek
Locally used taxon name	N/A
Distribution	Democratic Republic of the Congo, Ethiopia and Uganda
Phenology	May–June
Habitat preference	Mainly in disturbed areas, grasslands, lake shore, open bush, forest margin, river and road sides
Topographic preference	0–1670 m
Soil preference	Sandy loam
Geological preference	N/A
Climatic and micro-climatic preference	N/A
Breeding system	Allogamous
Genotypic and phenotypic variation	N/A
Biotic interactions (pests, pathogens, herbivores)	N/A
Ethnobotanical information	Use as food and drink
Threat status	Least Concern
Conservation status	Well represented in *ex situ* collections, no active *in situ* conservation; its disjunct distribution in Zaire, Ethiopia and Uganda suggests populations may be found in other countries in the region

papers, soil, vegetation and climatic maps, atlases, etc. As well as conventional media, information can be accessed from websites and on-line databases. The kinds of data that might be obtained from literature and databases sources are listed in Table 6.3 for the winged bean wild relative *Psophocarpus grandiflorus* Wilczek. It should be noted from this table that it is often not possible to obtain data for every taxon for all the possible data fields.

Specimen level data are collated from herbarium specimens, germplasm accessions and field survey data (see Table 6.4 for an example). An extended list of specimen level ecogeographic data is included in Castañeda-Álvarez *et al.* (2011). Note this listing of

potential categories of ecogeographic data is extensive, and it is unlikely that all will be recorded for any single specimen. Care must be taken in accepting scientific names attributed to herbarium specimens or germplasm accessions sheets; the identification should always be checked to ensure the taxonomic identification is correct and misinterpretation of ecogeographic distribution based on wrongly identified specimens is avoided.

Resources and materials availability will always limit the scope of ecogeographic investigations. Herbaria and gene banks contain millions of specimens, and the number of specimens of any target taxon can be vast. A combination of these factors will

Table 6.4 **Example specimen level ecogeographic data**

Ecogeographic data	Example data
Accepted taxon name	*Vicia sativa* L. subspecies *amphicarpa* (L.) Batt.
Herbarium, gene bank or botanic garden where specimen/accession is deposited	AARI – Aegean Agricultural Research Institute
Collector's name and number	Kitiki, Kell and Maxted 4108
Collection date	20.04.96
Phenological data (Does specimen have flowers or fruit?)	Flower – Yes; Fruit – No
Provenance	Province – Hatay; nearest settlement – Saylak; location – 24 km from Kirikhan on road to Tasoluk; latitude 36E 38'; longitude 36E 24
Altitude	295 m
Habitat	Field and field margins
Soil type	Red Mediterranean
Rock type	Limestone
Vegetation type	Low shrubs and grasses
Site slope and aspect	Slope <10%, with an eastern aspect
Land use and/or agricultural practice	Corn field and rough pasture
Plant uses	Pasture species

force the positive selection of representative specimens of the target taxon for inclusion in the analysis. Specimens are likely to be selected on the basis of: (a) presence of detailed ecogeographic passport data, (b) presence of latitude and longitude data or enough location detail that it can be estimated, (c) representation of the geographic and ecological range of the taxon, (d) recently collected (more recently collected specimens usually have more detailed passport data that is typed, so easily read, and the populations from which the specimen was taken is also more likely to be extant), and (e) unusual taxonomic or ecogeographic representation (recording outlying specimens is necessary if the full range of the taxon is to be understood).

One critical question is how many specimens or accessions do you need to include in the analysis for it to be valid? Initially each new specimen will contain a

significant amount of additional ecogeographic information, but when novel ecogeographic combinations no longer occur in the specimens being examined and additional specimens fail to expand the latitudinal/longitudinal extent of the distribution of the species, then the full range of geographic and ecological niches that the taxon inhabits will probably have been included. Castañeda-Álvarez *et al.* (2011) suggest, based on a literature review (Hernandez *et al.*, 2006; Pearson *et al.*, 2007; Wisz *et al.*, 2008) and experience, that at least the passport data of 20 specimens are required to produce a reliable distribution model for a species. An image of a typical herbarium specimen is shown in Figure 6.12.

The ecogeographic data will be collated in a database, but before it can be analyzed it must be searched for errors and corrected. Indexing the database (i.e. re-arranging the records in alphabetical

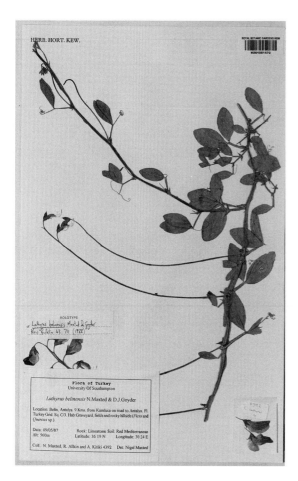

Figure 6.12 An example of a typical herbarium specimen (Royal Botanic Gardens, Kew). (A black and white version of this figure will appear in some formats. For the colour version, refer to the plate section.)

or numerical order using your database management system) on each field in turn may highlight typing errors or invalid entries. Mapping latitude and longitude data may reveal errors if localities are shown up as obvious outliers or in impossible places, e.g. in the sea or a neighbouring country. Herbarium specimen and germplasm collectors often send duplicate sets of their material to different international collections for safety backup so the potential impact of these duplicates on the data analysis should be considered and if necessary, duplicates excluded from the analysis.

Inevitably not all the data required to plan the conservation of the target taxa are available on-line,

so depending on the relative importance of the data it may be necessary to visit key herbaria or gene banks to collate the required data. Although a very large percentage of germplasm records are available on-line, many of the major herbaria are still digitizing their collections, and so those with links with the region where the target taxa are native should ideally be visited. Historically, the conservationist selects specimens (largely based on quality and quantity of ecogeographic data) and then types the selected specimen's passport data directly into a computer, but this often meant an expensive and extensive stay at the herbarium. Such an extensive stay can be avoided by taking a digital photograph of the specimen (Castañeda-Álvarez et al., 2011), and then, back at base, working out the latitude and longitude necessary for the diversity analyses to be carried out during the conservation planning. This approach has the obvious advantage of being relatively quick, therefore reducing the time the conservationist needs to spend at the host herbarium, and it provides a permanent image of the specimen, which can be checked if the identification is thought to be incorrect. It also means that specimen identification in the herbarium is not as critical because the image may later be seen and validated by an expert. Castañeda-Álvarez et al. (2011) also suggest a minimum threshold at 20 specimens per taxon, to produce a reliable distribution model representing the potential geographic areas in which the species might be found.

Often when collating herbarium specimen or gene bank accession passport data, information will be provided on the location where the specimen was collected, nearest inhabitancies, but will lack latitude and longitude data. More recently collected specimens are likely to have precise coordinates taken with a geographic positioning system (GPS) at the sampling site. This data is necessary if geographic information system (GIS) analysis is planned, so the specimens will need to be georeferenced, i.e. its latitude and longitude estimated from the locality (and altitude) data available. Georeferencing is achieved using some form of gazetteer, a book or database of geographic names alongside their latitude and longitude. Manual methods using book gazetteers are still available, but increasingly on-line databases are used, like the

National Geospatial-Intelligence Agency (NGA) (http://earth-info.nga.mil/gns/html).

Ecogeographic Diversity Analysis

The raw ecogeographic data included in the database can be analyzed to help identify the geographic locations and habitats favoured by the target taxa in the following ways.

Frequency Distributions: One of the simplest means of ecogeographic data analysis is to calculate the number of specimens (which can be expressed as percentages) collected from sites characterized by different biotic and abiotic features, e.g. fixed size grid squares, climate zones, soil types, aspect, habitat types, etc. This can be interpreted visually in the form of graphs and pie charts. Data arranged in this fashion will identify the niche occupied by the target taxon, so they can be used to indicate previous collection areas and other areas where the taxon is likely to be found. Frequency distributions can often be compiled directly by querying the database. Correlation of the abundance or frequency of taxa along environmental gradients (such as altitude, latitude and soil pH) will give error terms and can therefore be used predictively. Correlation of morphological characters with environmental conditions will help indicate possible ecotypic adaptation, both in wild and cultivated material.

Mapping of Ecogeographic Data: The data may also be plotted onto some form of map and may be used in conjunction with topographical, climate, geological or soil maps to identify ecological preferences. Plant geographic distribution data can be displayed in two basic ways: (1) shading or enclosing line maps that enclose an area, and (2) using dot distribution maps (Figure 6.13). Dot distribution maps are generally preferred to enclosed line maps as the latter can be ambiguous, as is demonstrated in Figure 6.13, where part of Italy is shaded yet there are no specimens recorded from Italy. Enclosed line maps do not indicate frequency of a taxon within a region and weight the importance of outlying specimens around which the line must be drawn. A dot distribution map

can be developed further by superimposing morphological, ecological or taxonomic information on the symbol indicating the location, to produce an enhanced dot distribution. Note that simply drawing an enclosed line around the sites where *Aegilops triuncialis* was found in North Africa would give a misleading picture of its distribution. The one location in North Libya greatly extends the range of the taxon and could be interpreted as giving a misleading picture of distribution for a taxon located primarily in Western North Africa.

Simple mapping can be enhanced by more sophisticated species distribution modelling (SDM) using GIS. Numerous GIS programmes are currently available (e.g. DIVA-GIS, Hijmans and Graham, 2006; MaxEnt, Phillips *et al.*, 2006; CAPFITOGEN, Parra-Quijano *et al.*, 2014). Each is composed of a relatively sophisticated, computer-based graphics programme for handling digitized cartographic data, interfaced with a database management system for storing and manipulating data associated with map features. Mapping programmes allow the import of georeferenced specimen passport data, latitude and longitude coordinates, from the ecogeographic database and their plotting on to customized maps of the region of the world. Once in the GIS programme, the ecogeographic distribution data for a taxon can be analyzed against known ecological polygons, aerial photographs, field surveys, remote sensing, etc., enabling conservation planning. Such SDM algorithms are tools frequently used in studies of PGR because they allow the prediction of areas that meet the environmental conditions required by a taxon. In particular, GIS has been used in studies of PGR to identify: areas of high diversity (Maxted *et al.*, 2004; Ocampo *et al.*, 2007; Castañeda-Álvarez *et al.*, 2015), suitable locations for in situ conservation (Draper *et al.*, 2003; Maxted and Kell, 2009; Vincent *et al.*, 2012; Phillips *et al.*, 2016; Vincent *et al.*, 2019), potential areas for germplasm collection (Ferguson *et al.*, 1998b; Jarvis *et al.*, 2005; Parra-Quijano *et al.*, 2011b; Khoury *et al.*, 2015b; Castañeda-Álvarez *et al.*, 2016a), ecogeographic representativeness (Parra-Quijano *et al.*, 2012; Parra-Quijano *et al.*, 2014; Rubio Teso *et al.*, 2016) and levels of threat affecting plant species (Jarvis *et al.*, 2008a), as well as creating

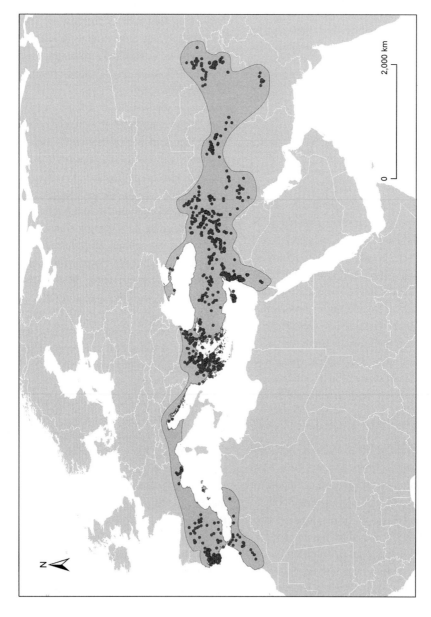

Figure 6.13 Distribution map for *Aegilops triuncialis*; outlined area indicates enclosed line distribution and black circles indicate dot distribution of known population samples.

informative compilations such as atlases (Azurdia *et al.*, 2011; Hijmans *et al.*, 2002). Sources of environmental layers at the global scale include WorldClim (www.worldclim.org), which offers precipitation and temperature layers (Hijmans *et al.*, 2005), SRTM-CIAT (http://srtm.csi.cgiar.org), containing digital elevation data, and Global Land Cover (http://glcf.umiacs.umd.edu).

Rarely when undertaking the analysis outlined above is a 'perfect' dataset available, so the researcher needs to understand the degree of completeness of the database, how fully the inherent pattern of genetic diversity is understood and how far the specimens or accessions sampled reflect the true ecogeographic range of the taxon. If a habitat is under-represented in the database, is it because the taxon is absent from that habitat, or because that type of habitat has not been sampled, or even because the target taxon has not been recognized in that habitat? There the results of analysis must be considered alongside the known biases of the data it is based upon.

Assessing the Impact of Climate Change:
As noted above in Section 6.2.4, climate change is a global concern and agro-biodiversity is and will increasingly suffer genetic erosion and extinction due to climate change. Therefore, whether planning *in situ* or *ex situ* conservation an essential component of ecogeographic diversity analysis will be the identification of climate smart options that maximize long-term population security. Linked to assessing the threat posed by climate change, the value of individual LR and CWR populations as sources of traits that can aid crops adapt to climate change (Farooq and El-Azam, 2004; Hajjar and Hodgkin, 2007) may also be integrated into the conservation planning.

Field Surveys:
SDM is, by definition, modelling so is likely to include a level of error; if the available ecogeographic data for the target taxon is insufficient or limited, the conservationist may not have enough biological knowledge of the target taxon to formulate an effective conservation strategy. In this case, it would be necessary to conduct a field survey to gather the required ecogeographic data. The field survey may

be in the form of 'coarse-grid sampling', which involves travelling throughout the target region and sampling sites at relatively wide intervals over the whole region, possibly followed by 'fine-grid sampling' in the area where the taxon is concentrated to establish more precise ecogeographic associations. The field survey will, by gathering fresh ecogeographic data, determine the distribution of the target taxon, the habitat requirements and biological relationships of the plant, population numbers and sizes, as well as any imminent threats, and so permit conclusions to be drawn about the most appropriate conservation strategy.

6.4.5 Identification of Conservation Gaps

Having used the various techniques described above, a picture of the target taxon's diversity, its richness (total number of taxa present within target area regardless of their frequencies) and evenness (relative abundance of the different taxa making up the richness) should be clarified. But knowing these characteristics alone is insufficient. It is also an important element of gap analysis and conservation planning to understand what diversity is already currently conserved to assess the efficiency of both *in situ* and *ex situ* conservation techniques and so identify the weaknesses – the gaps. If enough genetic material of the target taxon from the target area is already safely conserved either *in situ* and/or *ex situ*, there may be no justification for further conservation activities.

Current *In situ* *Conservation Assessment*
The definition of *in situ* conservation provided by the CBD (1992) includes two distinct conservation techniques: protected area (genetic reserve) conservation for CWR taxa and on-farm conservation for LR. For genetic reserve/protected area assessment this involves a review of existing protected areas and the CWR taxa within them that are being actively managed for conservation. It is important here to stress the distinction between active and passive conservation: active conservation means the CWR population is being actively sustained within the

genetic reserve, while passive conservation means the CWR population exists within the genetic reserve but is not being actively managed or monitored so could decrease or go extinct without the knowledge of the management team. As few centralized databases detail which species are being actively conserved in the world's protected areas. Obtaining detailed knowledge of the protected areas in the target area is likely to involve contacting protected area managers to ascertain if species are present and being actively managed and monitored. By matching the observed distribution or predicted distribution (using SDM) of the CWR taxa with the network of PAs, gaps can be identified, and future genetic reserve designation planned. However, as discussed in Chapter 9, the number of CWR populations already being conserved *in situ* is minimal, so effectively the sites predicted by the SDM to be suitable for *in situ* conservation are likely to be those where *in situ* conservation should be established providing there is support from the biodiversity conservation and local communities. Similarly, the on-farm conservation of LR requires reviewing existing on-farm conservation projects and the crop LR included. The review of on-farm conservation is likely to be simpler than the review of PAs due to the more limited number of on-farm conservation projects and the relative ease of discovering which crops and LR are included.

Current Ex situ *Conservation Assessment*

To assess the completeness of *ex situ* collections a comparison between the coverage of herbarium and gene bank samples needs to be made. Much germplasm holdings information is available from botanic gardens and agricultural gene banks, national and international on-line catalogues, databases and web sites (see Table 6.2). Once the ecogeographic or genetic diversity results are known, they can be compared with the existing conservation actions, and the final element of the gap analysis is to propose the novel priorities and agree them with the scientific and local communities. Such an approach has recently been used to identify the global *ex situ* conservation priorities for 1076 CWR taxa related to 81 crops (Castañeda-Álvarez et al., 2016a). Castañeda-Álvarez

et al. modelled the geographic distributions of the taxa then compared their potential ecogeographic distributions to that known from gene bank accessions only. To aid conservation prioritization they categorized taxa with a final priority score (FPS), calculated using a numeric score of 1–10 based on averaging each taxon's assessed current representation in gene banks regarding overall number of accessions, geographic diversity and ecological diversity. High priority CWR were prioritized for further collecting if they had an FPS $\geq$7 and this group included 72% of the crop gene pools (Figure 6.7). Priority areas for further collecting in order of importance are the Mediterranean and the Near East, Southern Europe, Southeast and East Asia, and South America. As there is effectively so little existing *in situ* conservation for CWR or LR, centres of CWR or LR diversity are, by definition, priority areas for active *in situ* conservation.

6.5 Conservation Strategies and Action Plans

To facilitate agro-biodiversity conservation implementation the process of identifying the appropriate conservation actions and the recommended actions themselves are commonly formalized in a written conservation strategy and action plan. The advantages of an agreed published document are that it gives all stakeholders a chance to formally input into the strategy/action plan, take responsibility and once published it means that progress can be monitored.

6.5.1 Writing Strategies and Action Plans

Conservation strategies and action plans act as a guide for coordination and implementation and are often essential to securing the necessary resources. As most conservation action is planned and implemented at a national level, practically this will often take the form of a National Strategic Action Plan (NSAP). The NSAP may address all agro-biodiversity of the country, all CWR or LR diversity or individual groups of taxa or even a single taxon. The NSAP for various groups of plant taxa should be integrated with each other and with other national initiatives (e.g. on biodiversity conservation, agriculture and development strategies) but also with regional and eventually global programmes into a coordinated holistic approach to ensure that the most important agro-biodiversity resources are conserved and available for crop improvement. There is no widely agreed distinction between a strategy and a plan, but commonly the strategy is more descriptive of the process of conservation planning and defines the vision and associated goal, while a plan is a more succinct set of the actions required to implement the strategy and achieve the vision.

The IUCN/SSC (2008) recognized that simply publishing information on species actions did not engender conservation action and concluded multi-stakeholder agreement of 'prioritized recommendations specifically designed for key players' was necessary. Within the agro-biodiversity context the reasons for establishing a NSAP are:

- to maximize maintenance or enhancement of agro-biodiversity genetic diversity, sustain natural ecological and evolutionary processes, and minimize threats to diversity;
- to provide scientifically based, unbiased recommendations for conservation that minimize intrusive interventions;
- to explain and justify the priorities for conservation actions recommended;
- to provide a baseline to assess the impact of conservation actions and so review implementation;
- to promote access to the conserved resources by the user community;
- to ensure and demonstrate the interests all stakeholders, including the local community, are considered and served;
- to provide a framework for collaboration with other conservation and utilization stakeholders;
- to ensure wider policy goals are achieved as effectively and efficiently as possible;
- to fulfil regional and global convention/treaty agro-biodiversity conservation obligations; and

- to aid fundraising and raise awareness of the value of agro-biodiversity.

To fulfil this role there must be a statement of the conservation actions proposed that can be debated and agreed by the diverse stakeholders involved. The IUCN/SSC (2008) conclude that where conservation planning fails, it is because (a) the target audience is not well defined (is it practical conservationists or policy-makers), (b) the preparation of the plan was under-resourced, so the actions recommended were insufficient, and (c) the actions were insufficiently specific to be successful. To be successful they should be based on sound conservation science and involve all appropriate stakeholders, particularly local communities, as the implementation of the plan is likely to directly impact their environment and livelihood. Such a participatory process will promote broad ownership and increased chances of success.

Strategies/action plans may be written for single species or groups of species, at regional, national or local scales. It is important to stress there is no single method for developing strategic plans for CWR and LR conservation because of issues concerning resource and baseline biodiversity data availability, the local community where the plan is to be implemented, as well as the focal area and remit of the agencies which are responsible for formulating and implementing it. Nevertheless, the process of developing national plans for conservation and use of CWR and LR can be viewed as a series of decisions and actions that follow the same basic pattern in all countries/taxa. The resource book (Maxted *et al.*, 2013b) and CWR Toolkit (Magos Brehm *et al.*, 2017a) should be viewed as a framework and guide for developing such plans, bearing in mind that the suggested steps do not necessarily have to follow the same predefined order, but developed and implemented within the confines of the available data and resources.

Clearly the development of conservation strategies/action plans will involve gap analysis/ecogeographic analysis followed by a formulation of a *Vision* and *Goals* for conserving the agro-biodiversity resource. The vision is a description of the view of the future state of the resource, e.g. systematic complementary (*ex situ* and *in situ*) conservation of the CWR of rice in Indonesia or maize LR in Mexico. Behind this broad vision are very specific scientific goals. For the *ex situ* component, this means active conservation of the number of plants that contain 95% of all the alleles at a random locus occurring in the target population with a frequency greater than 0.05, which was generalized as 50 seeds from each of 50 plants of each target species (Marshall and Brown, 1975). While for the *in situ* component, this means the minimum population size for any given habitat or the smallest population size for a taxon having a 99% chance of remaining extant for 100 years (Shafer, 1990), which is generalized as five populations that represent as far as possible discrete ecogeographic zones (Brown and Briggs, 1991), and each population should contain 5000–10 000 individuals (Hawkes *et al.*, 1997). These *ex situ* and *in situ* goals for the target taxa are ideals, and it must be admitted that they are more achievable for some species (e.g. most crops and wild cereal relatives), than they are for others (e.g. wild fruit tree relatives). The vision is then associated with a set of associated *Goals* which practically enable the vision to be achieved. The *Goals* should be specific at the geographic and temporal scale, focusing the conservation in certain locations and providing a timescale for their achievement. The *Goals* should be SMART – Specific, Measurable, Achievable, Realistic and Time-bound (see Box 6.7). The more 'SMART' the indicator, the more value it will have as a tool for monitoring PGR conservation action and the more precisely further actions can be formulated. The critical point in preparing conservation strategies and action plans or undertaking any form of conservation planning is that they should conclude with clearly formulated conservation priorities, specific and appropriate actions that can be implemented to improve the target taxon's conservation status.

There is an obligation under the CBD (1992) Article 6 under the General Measures for Conservation and Sustainable Use to:

Develop national strategies, plans or programmes for the conservation and sustainable use of biological diversity or adapt for this purpose existing strategies, plans or programmes which shall reflect, inter alia, the measures set out in this Convention relevant to the Contracting Party concerned;

Integrate, as far as possible and as appropriate, the conservation and sustainable use of biological diversity into relevant sectoral or cross-sectoral plans, programmes and policies.

Countries are now periodically reporting information on the measures they have taken to implement the provisions of the CBD and SDG, and their effectiveness in meeting its objectives. The CBD objectives place an onus on CBD signatory countries to integrate the conservation and sustainable use of biological resources into national decision-making. This obligation is particularly pertinent in the context of agro-biodiversity where the linkage of conservation to use is fundamental; we conserve to facilitate utilization, we utilize to ensure food availability and we consume food to sustain our species. Therefore, developing and implementing a NSAP for agro-biodiversity is fundamental to meeting our CBD obligation. The CBD (2011) outlines the suggested content of an NBSAP as shown in Box 6.8.

While specifically in the context of agro-biodiversity there is a requirement under the ITPGRFA (FAO, 2001), *Article 21 - Compliance*, to monitor each country's PGRFA activities acting in accordance with the Treaty, the reporting consists of short questions that are answered every 3 years on a voluntary basis (FAO, 2013b). A more specific requirement is required by the FAO Global Plan of Action (GPA) for Plant Genetic Resources for Food and Agriculture, which provides a strategic framework for the conservation and sustainable use of global PGRFA. The Second GPA (FAO, 2012a) sets out a series of 18 priority activities, 1–4 related to *in situ* conservation and management and 57 related to *ex situ* conservation. Periodically the FAO Commission on Genetic Resources for Food and Agriculture invites countries to report to the FAO on their activities that contribute to the implementation of the Second GPA. More detailed reporting is also required as a contribution to the periodically updated FAO State of the World's Plant Genetic Resources for Food and Agriculture (FAO, 1998, 2010), and State of the World's Biodiversity for Food and Agriculture (FAO, 2019).

Box 6.7 | SMART Management Criteria

Letter	Criterion	Meaning	PGR example
S	Specific	Relates to specific target for improvement	Increased LR diversity maintained on-farm
M	Measurable	Is quantifiable and so comparable and can be used as an indicator of progress towards achieving the specific target	Numbers of LR maintained on-farm
A	Attainable	Specify who is responsible for obtaining the data and doing the assessment and that data collection is feasible	National PGR programme
R	Relevant	State what results can realistically be achieved, given available resources	Part of routine data recording for NPGRP
T	Time-bound	Specify the periodicity of the assessment	Repeat every 5 years

---

Finally, it can be argued that too often conservation is planned and implemented within a short-term project timeframe of 3–5 years rather than as an ongoing commitment to sustaining biodiversity. Such a short-term approach means the project may be completed before it is realized that the goals are too ambitious, the management interventions are inappropriate or there was a lack of support from a key stakeholder. Longer term conservation projects have time to employ adaptive management solutions that may be less of an option within 3- to 5-year projects. Increasingly time limited projects include some form of exit strategy, concerning the post-project management of the conserved resource. This might involve a national agency agreeing to continue to manage the target taxa post-project, facilitate monitoring or include the new reserve within an existing network. Thus, the resources invested in the conservation would not be wasted as the project concludes, and the project has a longer term impact on sustaining biodiversity.

6.5.2 Conservation Indicator Monitoring

Biodiversity and agro-biodiversity conservation have a real and substantial cost. Ineffective conservation is not only a waste of resources but also does nothing to counter the increasing threats that biodiversity faces. In recent years, there has been a movement towards more evidence-based conservation, where the success of alternative conservation actions is monitored to establish their relative effectiveness. Associated with evidence-based conservation is an increasing application of biodiversity indicators. Indicators are key to help assess trends and identify goals for biodiversity conservation, and monitor the effectiveness of conservation (Reid *et al.*, 1993; CBD, 2010c; IUCN, 2012). They help researchers identify changes over time in species, ecosystems or genetic diversity and enable the measurement of progress towards a certain target set by a governing body to conserve biodiversity (Mace and Ballie, 2007). Indicators are also a tool of communication, allowing information to be delivered to a wide audience including the public. Indicators require a coherent relationship between scientists and policy-makers, as their creation relies upon policy-makers' targets and scientists' skills to determine relevant variables of biodiversity (CBD, 2010b).

A biodiversity indicator, being a single measure of achievement or a description of the conditions that

Box 6.8 | **CBD Outline of National Biodiversity Strategy and Action Plan Content**

(Source: CBD, 2011)

I. Introduction

The introduction should present a concise account of the necessary background, set the scene for the updated NBSAP and provide the rationale for the strategy and actions contained in the NBSAP. Where necessary, the NBSAP may be complemented by in-depth studies annexed to it.

1. **Values of biodiversity and ecosystem services in the country and their contribution to human well-being** – Importance of biodiversity for the country. Highlight contribution to human well-being, socio-economic development, including poverty reduction. Include analysis of economic and other values.
2. **Analysis of the causes and consequences of biodiversity loss** – Main threats to biodiversity (and ecosystems) and their underlying causes. Impacts of threats on biodiversity and ecosystems and socio-economic implications of the impacts. Describe the impacts of declining biodiversity and ecosystems on human well-being, livelihoods, poverty reduction, etc. Link the threats (direct drivers) with the underlying causes (indirect drivers) and relate these to the relevant economic sectors.
3. **National constitutional, legal and institutional framework** – Overview of the biodiversity policy and planning framework and relevant broader policy and planning processes (national development plans; poverty reduction strategies; climate change adaptation plans, etc.). Include an outline of any relevant constitutional, legal and institutional elements.
4. **Lessons learned from the earlier NBSAP(s) and the process of developing the updated NBSAP** – A brief account of progress in implementing earlier National Biodiversity Strategy and Action Plan (NBSAP). Summary results of any evaluation of the effectiveness of earlier NBSAPs. What challenges and gaps need to be addressed and main priority areas for the revised NBSAP. Might also develop future scenarios for biodiversity. Might also include brief reflections on the process of developing the previous NBSAP and how it may have influenced its effectiveness. Briefly outline the process of updating the NBSAP, including stakeholder consultations.

II. National Biodiversity Strategy: Principles, Priorities and Targets

The main 'high-level' elements of the Strategy that provide the framework for the NBSAP.

5. **Long-term vision** – Outline the long-term vision for the state of biodiversity in the country. This should be an inspirational statement that reflects the importance of biodiversity for people and is broadly shared across the country. This may be for 2050 (as is the case for the Strategic Plan for Biodiversity 2011–2020) or may be aligned with other long-term national development plans.

Box 6.8   **(cont.)**

6. **Principles governing the strategy** – Core values and beliefs underlying the NBSAP.
7. **Main goals or priority areas** – The most pressing issues that are addressed by the NBSAP. Among these should be goals to ensure the mainstreaming of biodiversity (i.e. integration of biodiversity into broader national policies, strategies and plans).
8. **National Targets (SMART)** – National biodiversity targets in line with the Aichi Biodiversity Targets. These should be strategic, specific, measurable, ambitious but realistic targets that are time-bound (usually for 2020). They may be grouped under the main goals or priority areas.

III. National Action Plan

The details of the Strategy and the Action Plan.

9. **National actions to achieve the strategy, with milestones** – The actions needed to achieve the targets. These should consist largely of strategic actions, such as institutional, legislative, economic or other policy and institutional actions that will provide the enabling conditions and incentives necessary to achieve the goals or priority areas and targets of the NBSAP. More specific actions would be indicative, acknowledging that approaches will need to be adapted in the light of implementation experience. The Plan should determine who does what, where, when and how.
10. **Application of the NBSAP to sub-national entities** – How the NBSAP will be implemented at state/provincial levels (particularly important for federal countries, or quasi-federal countries which devolve territorial management to these entities) and at local or municipal levels (including cities). The national strategy and action plan might be complemented by Local Biodiversity Strategies and Action Plans (LBSAPs) developed separately.
11. **Sectoral action and mainstreaming into development, poverty reduction and climate change plans** – Actions and steps that will be taken to integrate biodiversity into broader national policies, strategies and plans (such as national development plans, poverty reduction strategies, climate change adaptation plans, etc.) and into sectoral policies, strategies and plans, across government, the private sector and civil society.

IV. Implementation Plans

12. **Plan for capacity development for NBSAP implementation, including a technology needs assessment** – The human and technical needs to implement the NBSAP and how they may be mobilized.
13. **Communication and outreach strategy for the NBSAP** – How the NBSAP will be promoted in the country among decision-makers and the public at large. (This is distinct from the Communication, Education and Public Awareness activities of the NBSAP which would be included in the sub-sections on national and sub-national actions.)

Box 6.8 | (cont.)

14. **Plan for resource mobilization for NBSAP implementation** – The financial resources needed to implement the NBSAP and how they will be mobilized through all sources, including the domestic budget, external assistance (where relevant) and innovative financial mechanisms.

V. Institutional, Monitoring and Reporting

15. **National Coordination Structures** – What are the national structures, institutions and partnerships (e.g. national committees, inter-ministerial committees; and secretariat or unit to support these) that will guide, coordinate and clarify the roles and responsibilities of various institutional actors and ensure implementation of the NBSAP? Where relevant, establish coordination mechanisms with local authorities in the development and implementation of local Biodiversity Strategy and Action Plans and/or with regional partners in the case of regional strategies.

16. **Clearing-House Mechanism (CHM)** – Includes the development and/or enhancement of the national CHM and how it is being used to support the development and implementation of the NBSAP; development of a national (and where relevant regional) institutional network for biodiversity.

17. **Monitoring and Evaluation** – How the implementation of the NBSAP will be monitored and evaluated, including provisions for reporting and the identification of indicators to track progress towards national targets.

would show that a conservation action had been implemented successfully, summarizes complex data into simple, standardized and communicable figures that indicate status, trends and pressures on biodiversity and aid identification of appropriate responses. The intended audience for indicators encompasses both international, regional and national communities, from the general public, professional conservationists to policy-makers, but a key feature is that indicators need to be easily understood by all audiences yet specific to a conservation target.

Increasingly all biodiversity is being managed using indicators, and regional and national targets monitored via indicators is routine when implementing NBSAPs or attempting to fulfil commitments set out in the CBD Strategic Plan

(Tittensor *et al.*, 2014). Routine use of indicators in biodiversity is now applied at the global (Biodiversity Indicator Partnership, 2010), regional (e.g. European Commission, 2010; European Environment Agency, 2012) and national (e.g. Joint Nature Conservation Committee, 2014; Brooks and Bubb, 2014) levels.

Internationally the development and implementation of global indicators has been led by the Biodiversity Indicators Partnership (BIP), which aims to prevent biodiversity loss by setting goals and targets and reviewing them on a 10-year cycle (BIP, www.bipindicators.net). BIP is globally responsible for monitoring progress towards achieving the CBD Aichi Biodiversity Targets (CBD, 2010b) and currently has developed indicators for 17 of the 20 Aichi targets. In association with Aichi Target 13 that 'by 2020, the genetic diversity of cultivated plants and farmed and domesticated animals and of wild relatives, including other socio-economically as well as culturally valuable species, is maintained and strategies have been developed and implemented for minimizing genetic erosion and safeguarding their genetic diversity', BIP proposed to measure trends in *ex situ* conserved materials. The indicator is not just based on crude numbers of accessions, but the enrichment index weights originality of accessions compared to existing accessions.

Particularly in the PGR context, use of indicators has lagged behind their use in the broader biodiversity context, but the Second GPA (FAO, 2012a) requests the development of PGRFA indicators for monitoring the implementation of the Plan. Subsequently the FAO (2015c) held a stakeholder workshop and published a set of 63 indicators for monitoring the implementation of the 18 Priority Activities of the Second GPA, and these were subsequently adopted in the CGRFA Fourteenth Regular Session; some examples are included below:

Priority Activity 1: Surveying and inventorying plant genetic resources for food and agriculture.
1. Number of *in situ* (including on-farm) surveys/ inventories of PGRFA carried out.
2. Number of PGRFA surveyed/inventoried.
3. Percentage of PGRFA threatened out of those surveyed/inventoried.

Priority Activity 2: Supporting on-farm management and improvement of plant genetic resources for food and agriculture:
4. Number of farming communities involved in on-farm PGRFA management and improvement activities.
5. Percentage of cultivated land under farmers' varieties/LR in areas of high diversity and/ or risk.
6. Number of farmers' varieties/LR delivered from national or local gene banks to farmers (either directly or through intermediaries).

Priority Activity 4: Promoting *in situ* conservation and management of crop wild relatives and wild food plants.

10. Percentage of national *in situ* conservation sites with management plans addressing crop wild relatives and wild food plants.
11. Number of crop wild relatives and wild food plants *in situ* conservation and management actions with institutional support.
12. Number of crop wild relatives and wild food plant species actively conserved *in situ*.

Priority Activity 5: Supporting targeted collecting of PGRFA:
13. Existence of a strategy for identification of gaps in national gene bank holdings and for targeted collecting missions to fill identified gaps.
15. Number of targeted collecting missions in the country.

Further details on the indicators and how precisely they might be recorded is provided by the FAO (2015c). The FAO has commenced using these indicators for national reporting in the context of the GPA implementation, but it seems likely their use will take some time to bed in, but once routinely recorded, they will prove a valuable tool for global monitoring of PGR activities.

In conclusion herbaria, gene banks, botanic gardens and *in situ* conservation sites are storehouses of botanical data as much as of plants. These data can be used to help plan future plant conservation, and well-planned conservation will be more effective and efficient. Analysis of a taxon's geography, ecology, genetic diversity and taxonomy is a prerequisite for predicting in which areas and habitats a taxon is likely to be found and what is the taxon's current conservation status. The comparison of where diversity exists and what element of that diversity is conserved enables us to locate gaps in our current conservation efforts and so prioritize future actions to ensure systematic genetic conservation. Therefore, *in situ* and *ex situ* conservation actions for the target taxon's genetic diversity can be identified, the diversity secured and made available for use.

7 Conservation Strategies and Techniques

7.1 Introduction

As already discussed in Chapter 1, there are two basic conservation strategies, each comprising a range of techniques that the conservationist can adopt to conserve plant genetic diversity. The two strategies are *in situ* and *ex situ*. To refresh your memory, a definition of these two strategies is given in Box 7.1.

The two conservation strategies are fundamentally distinct in the way they are applied. For plants, *in situ* conservation involves the designation, management and monitoring of target taxa where they thrive or are maintained, whereas *ex situ* conservation involves the sampling, transfer and storage of target taxa from the target area. The two generalized strategies may be further subdivided into several specific techniques (Table 7.1), each one resulting in the active conservation of the genetic diversity of plants.

When considering which conservation strategies and techniques to adopt, the Convention on Biological Diversity (CBD) (1992) stresses that *ex situ* techniques are generally considered as supplementary and supportive to *in situ* techniques, the latter being the 'ideal'. The CBD states that each contracting party shall adopt *ex situ* conservation measures 'predominantly for the purpose of complementing *in situ* measures'. However, in relation to PGR conservation there is a good argument for the case that the two strategies are of equal importance. The distinction between PGR conservation, or for that matter any socio-economic species conservation, is not the conserved resource, but the sustainable use of the conserved resource as clearly indicated in Figure 1.11. In theory use of PGR can come equally from *in situ* and *ex situ* conserved resources but historically and still today, though perhaps not so exclusively in the future, potential germplasm users have obtained accessions from *ex situ* gene banks or direct collection for use. It is also likely that active use will better sustain funding for conservation of PGR. So, in terms of PGR use we prefer to see *in situ* and *ex situ* conservation as being of equal importance. Further, when planning and implementing genetic conservation, neither *in situ* nor *ex situ* conservation techniques should be considered in isolation. To conserve the gene pool adequately, both strategies should be adopted in a complementary manner. The promotion of germplasm use from both *in situ* and *ex situ* conserved resources will be discussed further in Chapter 16.

Box 7.1 | **Definition of *In situ* and *Ex situ* Conservation Strategies Provided by the Convention on Biological Diversity (CBD, 1992)**

In situ conservation means the conservation of ecosystems and natural habitats and the maintenance and recovery of viable populations of species in their natural surroundings and, in the case of domesticated or cultivated species, in the surroundings where they have developed their distinctive properties.

Ex situ conservation means the conservation of components of biological diversity outside their natural habitats.

7.2 *In Situ* Conservation Techniques

7.2.1 Definition of *In Situ* Conservation Techniques

In situ conservation involves the maintenance of genetic diversity in the locality where it is currently found, either where it is naturally located or where it has developed distinctive traits under cultivation (note: this strict definition does not include transportation of the germplasm to a distant location where genetic reserve or on-farm conservation may be easier, which is sometimes referred to as *in situ* conservation in the literature). Although there are four recognized *in situ* conservation techniques, they can be grouped as genetic reserve and extra protected areas (PA) *in situ* conservation which target CWR *in situ* conservation, and on-farm and home garden which target landraces (LR) conservation (see Box 7.2 for definitions).

There are many synonymous terms used in the literature for genetic reserve conservation. In the biodiversity or ecosystem conservation context they would be regarded as protected areas or nature reserves and in the PGR context they have occasionally been referred to as genetic reserve management units, gene management zones, gene or genetic sanctuaries, and crop reservations, as well as other terms. The latter terms are direct synonyms of genetic reserves but the former, protected areas and nature reserves, are only partial synonyms because there is a critical distinction. In each of the four PGR *in situ* conservation techniques (and for that matter PGR *ex situ* conservation techniques), the goal is to maximize the range of genetic diversity conserved, not just the species or taxon *per se*. In protected areas or nature reserves, it is possible for a species to be

Table 7.1 **Plant genetic resources (PGR) conservation techniques**

Strategies	Techniques
In situ	Genetic reserve
	Extra protected areas *in situ*
	On-farm
	Home gardens
Ex situ	Seed storage
	In vitro storage
	DNA storage
	Pollen storage
	Field gene bank
	Botanic garden

Box 7.2 | Definition of the Four *In situ* Conservation Techniques

- **Genetic reserve conservation** – the location, management and monitoring of genetic diversity in natural wild populations within defined areas designated for active, long-term conservation.
- **Extra PA *in situ* conservation** – the location, management and monitoring of genetic diversity of natural wild populations in informally managed *in situ* conservation sites.
- **On-farm conservation** – the sustainable management of genetic diversity of locally developed traditional landraces with associated wild and weedy species or forms by farmers within traditional agricultural, horticultural or agri-silvicultural cultivation systems.
- **Home garden** – the location, management and monitoring of genetic diversity of locally developed traditional landraces and heirloom varieties or forms by householder within their individual garden, backyard or orchard cultivation systems for home consumption.

maintained but genetic diversity to still be lost because of inappropriate management, while in a genetic reserve or extra PA *in situ* site, the active management and monitoring is designed to maintain genetic diversity, and any significant loss of diversity would trigger adaptation of the management to promote genetic diversity maintenance (Maxted *et al.*, 2008b).

7.2.2 Fundamental Difference between CWR and LR *In Situ* Conservation

From the definitions above one can see that the two pairs of techniques have as their focus different groups of target taxa and different roles for the conservationist. On-farm and home garden conservation deal with crop species and the weedy species found growing among a crop, while genetic reserve and extra PA *in situ* conservation focus on wild species exclusively. Another difference between the two pairs of techniques is their mode of operation. Central to on-farm conservation is the role of the farmer and for home garden the maintainers or gardener, and it is they who ultimately conserve the target crop species in their cropping management system. The conservationist's role in on-farm or home garden conservation is to help promote and preserve the conditions in which the farmer or maintainer can maintain genetic diversity of LR within the traditional production systems employed. Further, the role of the conservationist in on-farm or home garden conservation is not to dictate the cropping management system to the farmer/maintainer, that is the prerogative of the farmer/maintainer. In contrast, genetic reserve or extra PA *in situ* site conservation involves a more active role for conservationists, who will often be required to intervene positively to promote the conservation of the target taxon or taxa within the site designated for conservation.

7.2.3 *In Situ* Techniques

Genetic Reserve Conservation
Conservation of wild species in a genetic reserve involves the location, designation, management and monitoring of genetic diversity in a natural or semi-natural location (Maxted *et al.*, 1997b; Iriondo *et al.*,

2008). Location involves identifying, primarily using ecogeographic and gap analysis methods, localities where the target taxon, or more often group of target taxa, is concentrated, and *in situ* conservation is feasible in term of resources availability, access and support from the local community and national authority. Designation means the site is formally recognized by the appropriate national protected area and PGR authority, which is likely to be legally documented. The populations of the target taxon/taxa are being actively managed to promote genetic diversity maintenance, and population and genetic diversity levels are regularly monitored to ensure the active management regime in place is meeting its objective of genetic diversity maintenance. If this is not the case, then adaptive management of the population is triggered that changes the management interventions to better facilitate population maintenance.

Commonly the genetic reserve will be established within an existing protected area for reasons of expediency. Existing protected areas are favoured because: (i) these sites already have an associated long-term conservation ethos and have an associated long-term, stable management regime; (ii) it is relatively easy to amend the existing site management to facilitate genetic conservation of wild plant species and (iii) it circumvents the requirement to create novel conservation sites, thereby avoiding the possibly prohibitive cost of acquiring previously non-conservation-managed land (Maxted *et al.*, 2008b). This technique is the appropriate *in situ* technique for the bulk of wild species, whether closely or distantly related to crop plants and whether they have socio-economic value or not. It also is said to be dynamic in the sense that the conserved resource continues to evolve alongside and in relation to local biotic and abiotic changes. The conservation management focus within the genetic reserve will be on the target taxon, but coincidentally other taxa present at the same locality will also be passively conserved. Further detail is provided in Chapter 8.

Extra PA In Situ Conservation
Historically, protected areas were often designated to conserve charismatic mega-fauna or aesthetically

pleasing habitats, or even simply because the area had limited exploitation value; therefore, perhaps surprisingly there is not always a match between protected areas and high levels of biodiversity. In general, areas with high levels of biodiversity are also areas suitable for agro-silvicultural or some other form of commercial exploitation, so in such biodiverse-rich locations conservation site designation is less common and where sites are protected, they tend be smaller. This, combined with the fact that many CWR are found growing in anthropogenic, disturbed habitats (Jain, 1975; Jarvis *et al.*, 2015), means CWR hotspots do not always match with protected area networks and *in situ* CWR sites may need also to be established outside of existing protected area networks.

The *in situ* conservation of CWR diversity in extra PA *in situ* sites will be located using the same ecogeographic and gap analysis methods as genetic reserves, where CWR hotspots are matched with existing land management regimes that are amenable for long-term target taxon/taxa maintenance. Such sites are often associated with weedy roadsides, field margins, orchards and even fields managed using traditional agro-silvicultural practices. It is necessary to reach agreement with the land owner/manager to retain existing management practices that have led to the sites selection and so retain the target taxon population diversity (Maxted *et al.*, 2008c). In these sites the conservationist would rarely directly manage the site but may influence site management by the provision of some form of incentive to sustain the site management *status quo* and therefore CWR population. This would be reinforced by regularly monitoring to ensure genetic diversity maintenance and potential loss of the incentive for the site owner if target populations' diversity is lost over time.

Perhaps of all conservation techniques, this is the one least well developed, and clearly there is an urgent need for research focused on the governance and management of plants in extra PA *in situ* sites. One recent study by Wainwright *et al.* (2019) investigated payments for agro-biodiversity conservation services in Zambia, where land managers are subsidized to manage their land in a way that would retain CWR populations, by for

example not spraying the crop with herbicide or leaving a set aside strip around the crop. Thirty CWR were identified and in 26 communities and competitive tender bid offers were used to determine the on-farm cost of conserving CWR, specifically in field margins/borders. The conservation costs ranging from US$23 to 91/ha per year in regions of high CWR presence. The study concluded that competitive tendering, coupled with CWR data, can be used to improve the efficiency of extra PA *in situ* CWR conservation.

On-Farm Conservation

Farmer-based conservation involves the maintenance of traditional crop varieties or cropping systems by farmers within traditional agricultural or silvicultural systems (Maxted *et al.*, 1997b). The diverse varieties of traditional crops maintained by farmers have many names, such as heritage, heirloom, primitive, folk and farmer's varieties, but most commonly they are referred to as landraces. It has proven very difficult to precisely define landrace; in fact, Zeven (1998) concluded that providing a clear definition is impossible, a conclusion that does rather preclude conservation of this important resource. Negri *et al.* (2009) review the various definitions and propose:

A landrace of a seed-propagated crop is a variable population, which is identifiable and usually has a local name. It lacks 'formal' crop improvement, is characterized by a specific adaptation to the environmental conditions of the area of cultivation (tolerant to the biotic and abiotic stresses of that area) and is closely associated with the uses, knowledge, habits, dialects, and celebrations of the people who developed and continue to grow it.

However, even this broad definition of LR has not been widely used. Therefore, it may be more useful, rather than attempting to achieve semantic exactitude, to identify a set of potential characteristics that LR often but not always exhibit (Camacho Villa *et al.*, 2005). As such, LR are a dynamic population of a cultivated plant species that has:

- historical origin,
- distinct identity,

- not been formally bred,
- intrinsic genetic diversity,
- locally adapted to a geographic location,
- associations with traditional cultivation systems, and
- cultural associations.

The *in situ* retention of LR diversity is associated with on-farm conservation (Jarvis *et al.*, 2016). Practically, each season the farmer/LR maintainer keeps a proportion of harvested seed for re-sowing in the following season. The LR population is dynamic, unlike *ex situ* conservation techniques which are primarily static, in that genetic diversity will change over time due to evolution in response to changes in local biotic and abiotic interactions and selection/exchange by farmers but will retain its adaptation to the local environment and its distinguishing characteristics. Thus, LR are a rich source of locally adapted alleles or gene complexes (Veteläinen *et al.*, 2009a) both to sustain individual farmers but also for plant breeding. This is perhaps the most recent technique for genetic conservation recognized by conservationists but has been practiced by traditional farmers for millennia, although as a conservation technique it does depend on the ongoing support of farmers and their descendants to sustain cultivation. Further detail is provided in Chapter 9.

Home Garden Conservation

Home (or kitchen) garden conservation is closely allied to on-farm conservation but can be distinguished on included crop diversity and scale. Home garden is practised on a smaller scale, the crops being grown primarily for home consumption, and a greater range of crops are grown and consumed by the household (Eyzaguirre and Linares, 2004). As well as the obvious food, fodder and forage crops, a home garden might also include medicinal, condiment, culinary herb, spice, flavouring and social use species. Orchard gardens may be viewed as even more crop diverse forms of home gardens, including fruit and timber trees, shrubs, pseudo-shrubs such as banana and paw-paw, climbers, and root and tuber crops, as well as food crops. However, like on-farm, practically each season the gardener/LR maintainer keeps a

proportion of harvested seed for re-sowing in the following season. Therefore, again the LR population is dynamic in that it will change over time but will retain its distinguishing characteristics and its adaptation to the local environment.

It is important to stress the fundamental distinction between the genetic reserve and the other three *in situ* conservation techniques described. With the former the conservationist is actively engaged in target taxon population management, whereas with the other three techniques the conservationist can influence population management, but it is the land owner, farmer or householder who actively manages the target populations. This distinction implies that the conservationist, as well as monitoring target populations, may need to supply some form of incentive, Protected Designation of Origin, Protected Geographic Indication, Traditional Speciality Guaranteed schemes or product development schemes to create added value in the case of LR conserved on-farm or even direct or indirect subsidies for the maintenance of conservation varieties or CWR populations. Whatever is provided, population management is at one remove from the conservationist, but the clear benefit is the necessary buy-in from local communities and the obvious public awareness benefit of integrating local communities into conservation and conserved resource utilization. Further detail is provided in Chapters 9 and 10.

7.2.4 Distinction between *In Situ* On-Farm Conservation and Management?

In recent years, some confusion has arisen between those who would regard themselves as focusing on what might be characterized as on-farm conservation and those focusing on on-farm management; increasingly the latter term is being used without a realization that by some at least the two terms are not completely synonymous. The focus of on-farm conservation is the genetic conservation of LR diversity held within on-farm systems. The LR diversity will change gradually over time due to environmental changes and farmer's selection and is used directly by the local farmers, but also has potential for use by external breeders or other users interested in exploiting the full range of adaptive diversity held within LR. Just as it is possible to identify CWR hotspots it is also possible to identify using ecogeographic, gap analysis and genomic techniques LR hotspots (Maxted and Scholten, 2007; Heinonen, 2016; Torricelli *et al.*, 2016) and establish on-farm conservation activities in those key local/ national /regional global locations. In contrast on-farm management (or perhaps more precisely labelled diversity farming) focuses on maximizing the diversity of LR held within any on-farm system; the LR diversity may be indigenous or introduced (allochthonous). It is managed by the farmers to maximize their direct benefit (Suneson, 1956; FAO, 2012a; Ceccarelli, 2014), particularly those in marginal environments (Di Falco and Chavas, 2006; Ceccarelli *et al.*, 2012), and potential use by external breeders or other users is of lesser importance. In such systems allochthonous LR or even cultigen material may be introduced to hybridize with native LR material to help sustain or increase production from the on-farm system. The introduced allochthonous LR or cultigen material introduced may replace and cause loss of native LR material, but overall the local farmer benefits and the on-farm management system is sustained.

On-farm conservation does potentially place a burden on the maintainer; the maintainer is sustaining local LR diversity and sustaining a public good but will potentially be foregoing production development options. Here there is a clear role for the conservationist, and it is likely to involve (a) local participatory LR enhancement that does not threaten local diversity, (b) LR derived product enhancement, (c) value chain development and (d) setting in place some form of stewardship/incentive scheme that sustains LR cultivation. The goal of each option is to ensure the maintainer suffers no dis-benefit from continued cultivation of local LR diversity and operation of existing management interventions. So, while maintaining the on-farm conservation system, the conservationist's role is to help sustain or enhance the farmer's income so she or he is deterred from hybridizing local LR with allochthonous LR diversity or switching entirely to modern crop cultigens.

As described, the distinction is critical and not just of semantic importance; both conservation and management have significant value to LR and on-farm system maintenance and do support food security. However, if the two terms are used synonymously then the specific value of each independently is lost. Breeders or other users increasingly need access to the full range of trait diversity to underpin food security in times of ecosystem instability, so on-farm conservation is essential, but also it has been demonstrated that increasing diversity in on-farms systems is just as beneficial to sustaining local food security and on-farm management is also essential. Just as *in situ* and *ex situ* applications are complementary, so on-farm conservation and on-farm management should be considered as complementary approaches for maintaining and promoting on-farm systems, and both should be supported.

7.3 *Ex Situ* Conservation Techniques

7.3.1 Definition of *Ex Situ* Conservation Techniques

Ex situ conservation involves the maintenance of a sample of the genetic variation present in target taxon populations away from its original location, where it is conserved either as a living whole plant collection in a field gene bank or botanic garden, or as seed, tissue explants, pollen or DNA under special artificial conditions. The techniques are generally appropriate for the conservation of crops, CWR and other wild plant species. See Box 7.3 for definitions.

7.3.2 *Ex Situ* Techniques

Seed Storage Conservation
Ex situ seed collection and storage is the most widely used method of PGR conservation, and it remains the normal route by which many users obtain germplasm. Seeds are the natural dispersal and storage organs for most species and as such are ideally suited to conservation. Therefore, it is the procedure adopted for the bulk of species with orthodox seeds;

Box 7.3 | **Definition of the Six *Ex Situ* Conservation Techniques**

- **Seed storage** – The collection of seed samples in one location and their transfer to a gene bank for storage; orthodox samples are usually dried to a low moisture content of $15 \pm 3\%$, then kept at sub-zero $-18 \pm 3°C$ temperatures, increasingly cryogenically at $-196°C$. Community gene banks may not freeze seed samples, preferring to keep them for shorter periods at $+4-5°C$.
- **DNA storage** – The collection of DNA and storage in appropriate, usually refrigerated conditions.
- **Pollen storage** – The collection of pollen and storage in appropriate, usually refrigerated conditions.
- *In vitro* **storage** – The collection and maintenance of explants in a sterile, pathogen-free environment (suitable for recalcitrant taxa that cannot be conserved in gene banks).
- **Field gene bank storage** – The collecting of a large number of individuals, seeds or living material from one location and its transfer and housing or planting in a second site for conservation (suitable for recalcitrant taxa that cannot be conserved in gene banks).
- **Botanic garden (arboretum)** – The collecting of a relatively small number of individuals, seeds or living material (tree species for arboreta) from one location and its transfer and housing or planting in a second site for public education purposes.

---

definitions of orthodox and recalcitrant species are provided in Box 7.4 (Pritchard and Dickie, 2003). Seed storage involves locating the target taxon crop or wild population, systematic sampling and transfer of the seed samples to the gene bank or community seed bank for processing and storage (Smith, 1995; Hay and Probert, 2011; Pence and Engelmann, 2011). There are about 1750 formal gene banks worldwide and more than 130 with more than 10 000 accessions (FAO, 2010a). The processing usually involves cleaning (removal of any unwanted inflorescence or fruit remnant), drying to low moisture content and storage at sub-zero temperatures (Figure 7.1).

Once stored, the seed requires little maintenance other than continued freezing, periodic germination testing and regeneration when germination levels fall below an acceptable level; a germination level of <75–80% is the common trigger for regeneration (Roberts, 1973; Ellis and Roberts, 1980). Regeneration is costly, particularly if the germplasm is being conserved outside its native range, and each regeneration is associated with a potential loss of original diversity (Frankel *et al.*, 1995), so needs to be kept to a minimum. The relative cost of gene bank conservation has been calculated by Singh *et al.* (2012), who identified three major component costs: capital costs for establishment of the gene bank, costs for acquisition of collections, and costs of processing and storage. The costs varied depending on the plant species, but for average long-term seed storage the annual capital cost was US$0.48–0.72 per plant species accession, the acquisition cost was between US$20.23 and 40.89 per accession plant species, and for processing and storage was US$14.51–37.14 per accession plant species. Once effectively conserved, the germplasm can be made available to potential users.

Within the formal PGR sector, seed is most commonly stored at $-20°C$ in scientifically sophisticated but expensive gene banks that provide

Box 7.4 | Definition of the Three Basic Types of Seed Storage Characteristics

(Pritchard, 2004)

- **Type I or orthodox seeds** can be dried down to 15% relative humidity and stored at low temperatures of $-20°C$ for long periods of time. Most crops have orthodox seeds, a group that includes all major cereals and other grasses from corn, wheat and rice, responsible for 50% of the world's food, to various forage grasses important for raising livestock. Other plants with orthodox seeds are the onion family, carrots, beets, papaya, pepper, chickpea, cucumber, the squashes, soybean, cotton, sunflower, lentil, tomato, various beans, eggplant, spinach and all the brassicas.
- **Type II or intermediate seeds** may lose viability if dried below 80% relative humidity and stored at low temperatures of $-20°C$ for long periods of time, which is shown for monkey puzzle and coffee unless stored for a short period. These species are characterized as being orthodox but with limited desiccation ability.
- **Type III or recalcitrant seeds** cannot be dried without injuring or destroying them. They can only be stored for a few weeks or months – by treating them with a fungicide and keeping them in moist sawdust or charcoal inside a polythene bag with access to oxygen. Therefore, seed banks cannot be used to conserve them. Comparatively few major crops have recalcitrant seeds, but many of these are important, particularly in the tropics: cocoa, coconut, mango, cinnamon, nutmeg, avocado, tea, breadfruit and jackfruit.

For Type II and III crops other *ex situ* techniques (e.g. field gene banks, tissue culture or cryopreservation) can be used, alongside complementary *in situ* conservation techniques.

relatively easy access for all users (farmers and commercial breeders, both local and remote users) (FAO, 2014b). However, formal sector PGR gene banks may seem remote to many local communities, especially when local landraces were being lost due to recurring natural (e.g. floods and droughts) and civil disasters. Therefore, NGOs, such as SEARICE in the Philippines, promoted the establishment of local community-based gene banks that would meet the requirement for a low-tech solution that secured local landrace diversity and to which local communities had easy access (Vernooy, 2013; Vernooy *et al.*, 2017a). There is also rising interest among the NGO seed saver networks, which promote low-tech seed saving (Gough and Moore-Gough, 2011). Such community and NGO-based gene banks are now a

common feature of PGR *ex situ* conservation activities and complement the more formal approach of national gene banks (Shrestha *et al.*, 2013a).

The advantages of seed storage are that it is efficient and reproducible, and feasible for short, medium and long-term secure storage. However, the disadvantages are that there are problems in storing species with intermediate or recalcitrant seeds, and genetic diversity may be lost with each regeneration cycle (but individual cycles can be extended to periods of 20–50 years or more). This technique has also been criticized because of the 'freezing of evolution'. Germplasm held in a gene bank is no longer continuously adapting to changes in the environment, such as new races of pests or disease, or major climatic changes. Species which are clonally

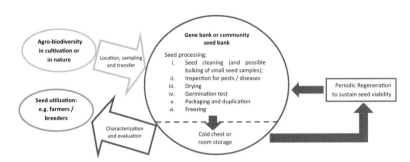

Figure 7.1 Key elements of seed storage conservation.

propagated or do not produce fertile seed cannot be conserved using this technique. Further detail is provided in Chapter 12.

In Vitro *Conservation*

In vitro conservation involves the maintenance of explants in a sterile, pathogen-free environment and is widely used for vegetatively propagated, intermediate and recalcitrant seeded species (Reed *et al.*, 2004). *In vitro* conservation offers a cheaper, less space-consuming alternative to field gene banks. It involves the establishment of tissue cultures of accessions on nutrient agar and their storage under controlled conditions of either slow or suspended growth (Withers, 1995; Pence and Engelmann, 2011). The tissue culture requires periodic sub-culturing onto fresh growth-retarding media. Depending on the protocol being used this could be every 6–36 months (Figure 7.2).

The main advantage is that it offers a solution to the long-term conservation problems of recalcitrant, intermediate, sterile or clonally propagated species. The main disadvantage is the risk of somaclonal variation (genetic and phenotypic variation arising within the cultured tissue over time), the need to develop individual maintenance protocols for most species and risk of culture contamination during necessary sub-culturing. Increasingly the preferred cheap, long-term *in vitro* conservation option that avoids these problems is cryopreservation, which is the storage of frozen tissue cultures at very low temperatures, commonly in liquid nitrogen at –196°C (Figure 7.2). As with ordinary tissue culture, there is a need to develop individual protocols for most species, and even varieties to reduce the damage caused by freezing and thawing, but the

risk of somaclonal variation is minimized and the tissue does not require regular sub-culturing. Cryopreservation offers the possibility of preserving tissue samples indefinitely (Engelmann, 2000; Panis, 2019). Further detail is provided in Chapter 13.pt

Pollen Conservation

The storage of pollen grains is possible in appropriate conditions, allowing their subsequent use through pollination of live plants (Volk, 2011). Pollen may be collected, then desiccated, packaged in a plastic vial and then stored at –20°C, though the precise protocol varies significantly from species to species (Hoekstra, 1995; Towill, 2004; Volk, 2011). It is possible to regenerate haploid plants from pollen samples, but no generalized protocols have yet been developed that could be applied widely for plant species. Pollen conservation has the advantage that it is a relatively low-cost option, but the disadvantage is that only paternal material would be conserved and regenerated. Further detail is provided in Chapter 13.

DNA Storage Conservation

The storage of DNA in appropriate conditions can be achieved given the appropriate level of technology (de Vicente and Andersson, 2006). However, the regeneration of entire plants from DNA cannot be envisaged at present, although single or small numbers of genes can be utilized using biotechnological techniques. Various DNA storage techniques are available for long, medium and short-term storage (de Vicente and Andersson, 2006). Currently the technique is more often used for phylogenetic analysis of taxonomic relationships or,

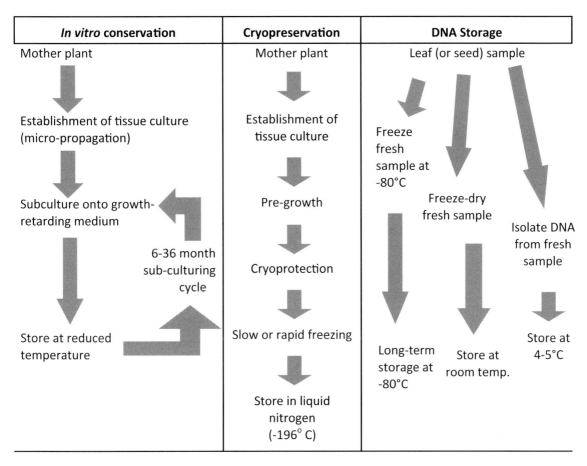

In vitro conservation	Cryopreservation	DNA Storage

Mother plant → Establishment of tissue culture (micro-propagation) → Subculture onto growth-retarding medium → Store at reduced temperature → 6-36 month sub-culturing cycle

Mother plant → Establishment of tissue culture → Pre-growth → Cryoprotection → Slow or rapid freezing → Store in liquid nitrogen (-196° C)

Leaf (or seed) sample → Freeze fresh sample at -80°C → Long-term storage at -80°C; Freeze-dry fresh sample → Store at room temp.; Isolate DNA from fresh sample → Store at 4-5°C

Figure 7.2 Key elements of *in vitro* conservation, cryopreservation and DNA storage.

in association with DNA barcoding, for identification of plant samples held in botanic collections (Chase and Fay, 2009). However, as concluded by Graner *et al.* (2006), the improved availability of DNA samples has the potential to stimulate genome-driven research, improved management of ex situ collections and efficient selection of material for selective breeding programmes. The advantage of this technique is that it is efficient and simple, but the disadvantage lies in problems with subsequent gene isolation, cloning and transfer. Further detail is provided in Chapter 13.

Field Gene Bank Conservation

The conservation of germplasm in field gene banks involves the collecting of material from one location

and the transfer and planting of the material in a second site (Reed *et al.*, 2004). It has traditionally provided the answer for recalcitrant or intermediate species, for sterile seeded species or for those species where it is preferable to store as clonal material. Field gene banks are commonly used for such species as cocoa, rubber, coconut, mango, coffee, banana, cassava, potato, sweet potato and yam. The advantages of field gene banks are that the material is easily accessible for utilization, and that characterization and evaluation can be undertaken while the material is being conserved. The disadvantages are that the material is restricted in terms of genetic diversity, is susceptible to pests, disease and vandalism, severe weather events, and involves large areas of land so involves high maintenance costs. The latter point limits the number of different specimens and therefore the genetic diversity that can be maintained. However, in certain cases there are no viable alternative techniques (Reed *et al.*, 2004; FAO, 2014b). Further detail is provided in Chapter 13.

Botanic Garden Conservation

Recently the role of botanic gardens has broadened from the display of diverse living plant collections and public education to playing an enhanced role in both *in situ* as well as *ex situ* conservation, with specific facilities such as seed banks and tissue culture units alongside herbaria and botanic libraries (Oldfield, 2010; Blackmore and Oldfield, 2017). Globally there are about 2500 botanic gardens, growing 80 000 plant species, about a fifth of all plant species (FAO, 2010a). Historically botanic gardens have cultivated plants of importance to humankind, such as those of medicinal, food and agro-silvicultural and ornamental value. Botanic gardens are rich in CWR taxa. Maxted *et al.* (2010) reviewed European *ex situ* seed collections in botanic gardens via the European ENSCONET portal and found CWR taxa accounted for 61.8% of total germplasm holdings and that the 5756 CWR species included represent 33% of the 17 495 CWR species found in Europe. This is significantly higher than agricultural gene banks, but here they are not included because they represent wild species, and high-priority CWR are still more likely to be found in agricultural gene banks. Individual botanic gardens may also specifically focus on a flora, having species from a specific, exotic region or represent the country's or region's own native flora. Other botanic gardens are associated with universities and so have a strong teaching and research focus. There has been a significant movement recently to establish more botanic gardens in the tropics due to the strong historic disparity between the number of botanic gardens and global biodiversity hotspots (Figure 7.3), but still 36% of botanic gardens are in Europe and conserve only about 4% of the global flora (FAO, 2010a). Yet together, the world's 2500 botanic gardens and arboreta maintain the largest array of living plant diversity outside of natural habitats; however, the focus is on numbers of species, i.e. inter-species diversity not intra-species diversity. Collections of plant species in botanic gardens are often restricted to only a few individual specimens per species so are not representative of the entire genetic diversity present in these species.

With the enhanced role of botanic gardens in conservation their focus has been on conservation of wild species, often building specialist living or seed collections of groups of taxa. However in terms of providing a display that the public appreciate ornamental species, such as orchids, bromeliads, bulbs, cacti or succulents, are often seen as more attractive to the general public as well as providing greater support for the horticultural industry. Another common specialization is rare and endangered species. Species such as the Wood's Cycad (*Encephalartos woodii* Sander), She Cabbage Tree (*Lachanodes arborea* (Roxb.) B.Nord.) and St Helena Redwood (*Trochetiopsis erythroxylon* (G.Forst.) Marais) are Extinct in the Wild, but today exist only as living collections cultivated in a botanic garden (IUCN, 2016). The horticultural expertise present in botanic gardens is particularly important in propagating such species, whose numbers have been reduced to 1 or 2 representatives, or for formulating reintroduction programmes. Many botanic gardens also have associated herbaria, collections of dried voucher specimens often associated with the production of Floras or monographs, which have in recent years proved invaluable as sources of information on distribution of species that can be used in conservation planning (Pearce and Bytebier, 2002). Further detail is provided in Chapter 13.

7.4 *Circa Situ* Conservation

Having stressed the fundamental nature of the two basic conservation strategies, there is a rarely applied third strategy – *circa situ* conservation. Here the target population is located and moved to a host location where it is managed as if an *in situ* population. *Circa situ* conservation is distinguished from *in situ* conservation because the population sample is moved from its natural location to a host location where active conservation will be easier or

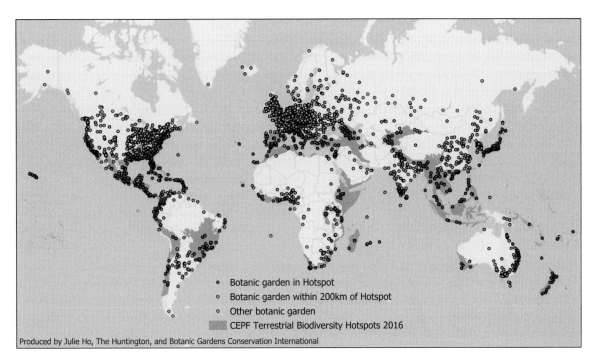

Figure 7.3 Botanic gardens (dots) and biodiversity hotspots (orange areas). (A black and white version of this figure will appear in some formats. For the colour version, refer to the plate section.)
(Image prepared by Julie Ho, The Huntington and BGCI)

more effective (Cooper *et al.*, 1992). Hughes (1998) provides the example of *circa situ* conservation of the legume tree genus, *Leucaena* spp., where population samples are collected from their native habitats and moved *ex situ* to areas more easily managed by local communities in a form of agroforestry conservation-utilization employing traditional silvi-cultural techniques within an *in situ* on-farm system.

More recently Volis and Blecher (2010) proposed *circa situ* (referred to by them as quasi *in situ*) as an alternative to *ex situ* conservation. They argue that current *ex situ* activities are inefficient because (a) space for maintaining the collection is limited, (b) populations are maintained in a usually unsuitable environment, (c) population maintenance and renewal is artificial and (d) cost is high. These disadvantages are overcome by the methodology they propose, which involves five steps: (i) ecogeographic survey and analysis, (ii) *ex situ* sampling of ecogeographically diverse populations, (iii) planting of samples in ecogeographically matching sites and *in situ* maintenance, (iv) record life history traits and

abiotic/biotic effects on population demography and (v) reintroduction of plants or seed to the source location. There are numerous potential criticisms of this proposed approach, not least the fact that it does not take account of the need to link the conserved resource through into use, which *ex situ* does well. However, *circa situ* appears to not appreciate the biological complexity of natural populations. It is not just a case of moving single species samples as in assisted migration methodologies. The target species will have numerous facultative or obligatory relations with other biotic and abiotic factors at the source location, therefore moving a population may also require moving other species alongside the target species to the new host environment. Rarely do we understand these complex inter-species relationships, so recognize the desirable companion species would require significant research. Locating where target populations should be planted may be problematic – existing protected areas may not be ecogeographically suitable and may impact existing included species and buying land for transposition would be prohibitively expensive. Further, as has often been found with habitat restoration, getting the target and companion species established would require significant resources over many years and would not always succeed, particularly compared to *ex situ* gene banking, which is a well-established strategy for orthodox species conservation.

7.5 Strategies and Techniques: Advantages and Disadvantages

Historically there has been a debate between those conservationists who favour the application of either *in situ* or *ex situ* strategies, or even a specific technique. Some criticisms of both strategies are valid, e.g. placing seed samples in a gene bank does 'freeze' the evolution of that germplasm, while populations conserved *in situ* are inherently vulnerable to climate change. However, the consensus is that each plant conservation strategy and technique has its relative advantages and disadvantages, and there is no one perfect technique. The relative advantages and disadvantages of the two main strategies are summarized in Table 7.2. When planning conservation for a specific species or population, the balance of the relative advantages and disadvantages of each technique is considered and the most effective combination of techniques is implemented.

Ex situ conservation involves the conservation of a single target taxon in any one accession, and collectors actively avoid taking mixed taxon collections. Germplasm collecting is a relatively expensive process and for that reason the bulk of *ex situ* conservation activities have been focused on crops or close crop relatives. Although *in situ* conservation may be no less expensive, if the populations are actively managed and monitored, it does allow the possibility of conserving more than a single target taxon in any one genetic reserve or group of traditional farms. For example, the Erebuni State Reserve was established to conserve the wild relatives of cereal grasses (wheat – *Triticum timoheevii* (Zhuk.) Zhuk. subsp. *armeniacum* (Jakubz.) Slageren, *T. urartu* Tumanian ex Gandilyan, *T. monococcum* L. subsp. *aegilopoides* (Link) Thell. and

Aegilops L. spp.; barley – *Hordeum murinum* L. subsp. *glaucum* (Steud.) Tzvelev; and rye – *Secale vavilovii* Grossh.) in Armenia, but because the target taxon is conserved within a community of about 300 other higher flowering plant species, >9% of the Armenian flora and 20 species included in the 'Red Book' of Armenia are also known to be conserved at the same site (Gabrielian and Zohary, 2004). Therefore, it could be argued that the overall expenditure per unit of taxon conservation may be lower in a genetic reserve because many more species can be conserved by the single act of conservation. This lower unit cost of conservation may also mean that wild species not of immediate utilization potential have a greater chance of being conserved, even if passively.

As it can be seen from the discussion of both conservation strategies and techniques, each has its advantages and disadvantages. However, in recent years, particularly post-CBD (1992), precedence has been given to *in situ* techniques and some attention has switched toward the conservation of genetic resources *in situ*. Behind this scientific debate runs a political undercurrent, plant genetic diversity is primarily focused in the Vavilov centres, largely in developing countries, but *ex situ* gene banks are primarily located in developed countries. So implicitly *ex situ* storage involves the transfer of collections and associated resource value from the developing world to OECD member states; i.e. from countries poor in financial resources to wealthy countries that lack plant genetic diversity. A transfer that enabled easy exploitation and wealth generation. Once it was recognized that PGR, as happened in the mid-1980s, were in fact a financial resource with a real rather than a notional value, some began to question this unidirectional transfer of PGR; a point picked-up by the third aim of the CBD, 'the fair and equitable sharing of the benefits arising out of the utilization of genetic resources, including by appropriate access to genetic resources and by appropriate transfer of relevant technologies, taking into account all rights over those resources and to technologies, and by appropriate funding' (CBD, 1992). Therefore, there is often an undeclared benefit to *in situ* conservation, which is that the PGR is maintained in the country of origin and any financial benefit from exploitation will

Table 7.2 **Relative advantages and disadvantages of *in situ* and *ex situ* conservation strategies**

Conservation strategy	Advantages	Disadvantages
In situ	• Dynamic conservation in relation to biotic and abiotic changes • Permits species/pathogen interactions and co-evolution • Provides easy access for evolutionary and genetic studies • Appropriate for recalcitrant or intermediate seeded species • Target taxa conserved alongside associated taxa within a single reserve	• Materials less easily available for utilization • Vulnerable to natural and human-mediated catastrophes, e.g. fire, vandalism, urban and industrial development, climate change • Appropriate management regimes often poorly understood • Require routine active management and monitoring • Limited genetic diversity can be conserved in any single location
Ex situ	• Medium and long-term storage is feasible • Limitless genetic diversity can be conserved in any single location • Easy access for evaluation for resistance to pests and diseases • Easy access to plant breeding, farmers and other forms of utilization • Little routine maintenance once material is conserved (except for field gene banks and botanic garden accessions)	• Impossible to store seeds of recalcitrant or intermediate species, unless via *in vitro* or field gene banks • Practical limitation on storage space for diversity of recalcitrant or intermediate species in field gene banks • Freezes evolutionary development in relation to environmental changes • Genetic diversity is potentially lost with each regeneration cycle • *In vitro* storage may result in loss of diversity and mutation

Adapted from Hawkes *et al.* (2000).

more naturally flow to the people of the country of its origin. Having made this essentially political point, it must be stressed that from a purely scientific standpoint, effective conservation is only possible if a complementary approach is adopted. Some may feel that politics has impinged too far into the science of conservation, but whether the conservationist likes it or not, politics and ethics are and are likely to remain pivotal to contemporary plant conservation.

7.6 Complementary Conservation Strategies

It is worth restressing that *in situ* and *ex situ* conservation strategies should not be viewed as alternatives or in opposition to one another but rather

as being complementary, as is emphasized in Article 9 of the CBD (1992). When formulating an overall conservation strategy for a taxon, the conservationist should apply a combination of *in situ* and *ex situ* techniques in a complementary manner where one strategy or technique backs-up the other (Ford-Lloyd and Maxted, 1993). Each of the specific conservation techniques has as its objective the maintenance of plant genetic diversity, thus the different techniques may be thought to slot together like pieces in a jigsaw puzzle to complete the overall conservation picture. The adoption of this holistic approach requires the conservationist to look at the characteristics and needs of the species or gene pool being conserved and then to assess which combination of techniques offers the most appropriate option to maintain genetic diversity within that species or gene pool.

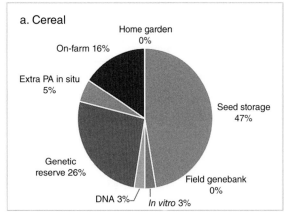
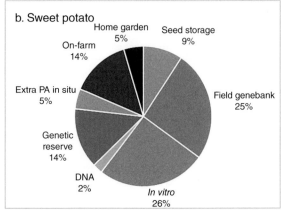
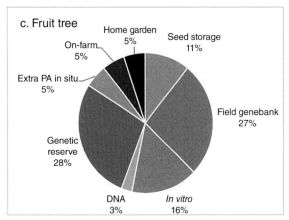
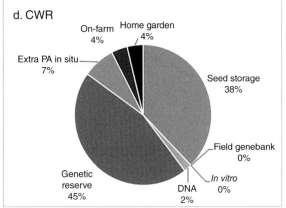

Figure 7.4 Hypothetical representation of the proportions of the gene pool conserved using eight *in situ* and *ex situ* conservation techniques for different plant groups: (a) cereal crop, (b) sweet potato, (c) fruit tree and (d) CWR. (A black and white version of this figure will appear in some formats. For the colour version, refer to the plate section.) (Adapted from Maxted *et al.*, 1997b)

To formulate the plant species or gene pool conservation strategy, the conservationist will need to address not only genetic, but also practical and political questions, such as:

- What are the species storage characteristics?
- What do we know about the species breeding system?
- Do we want to store the germplasm in the short, medium or long term?
- Where is the germplasm located and how accessible is it/does it need to be?
- Are there legal issues relating to access?
- How good is the infrastructure of the gene bank?
- What backup is necessary/desirable?

- How might the resource be best exploited? and even practically,
- What skills and financial resources are available to facilitate implementation?

Given answers to these questions, the appropriate combination of techniques to conserve the gene pool can be applied in a pragmatic and balanced complementary manner (Hawkes *et al.*, 2000). A different balance would be required for different crops, wild species or regions (see Figure 7.4). Figure 7.4a shows the hypothetical situation where the conservation target is a cereal species and the bulk of the gene pool is conserved using seed storage and genetic reserve conservation, lesser proportion of other techniques and there is no field gene bank conservation. In Figure 7.4b the conservation target is sweet potato, a typical root crop – here field gene banks, *in vitro*, genetic reserve and on-farm conservation predominate. Figure 7.4c is for a temperate fruit tree and field gene banks, *in vitro* and genetic reserve predominate. Figure7.4d is a typical CWR species where seed storage, extra PA *in situ* and genetic reserve conservation predominate and there is no field gene bank or *in vitro* storage. The precise combination of techniques is formulated afresh for each species or group of species, demonstrating the flexibility of the holistic approach.

Although this chapter introduces plant conservation strategies and techniques, it is important to establish the context for plant genetic conservation, and that is use. We conserve the maximum range of genetic diversity within species because of its actual or potential value to humankind; we conserve to facilitate utilization, conservation is not seen as an end in itself. Therefore, when planning and implementing conservation strategies and techniques, we need to retain the utilization goal in mind and design actions that facilitate this goal.

8 *In Situ* Conservation

8.1 Introduction

Dudley (2008) defines a protected area as: 'a clearly defined geographic space, recognized, dedicated and managed, through legal or other effective means, to achieve the long-term conservation of nature with associated ecosystem services and cultural values'. Each country has its own network of protected area (PA) sites established to address this long-term goal. Some will be recognized purely at a national level and others will be recognized internationally by the IUCN Commission on National Parks and Protected Areas and form part of the UN List of Protected Areas (UNEP WCMC World Data of PA, available at Protected Planet: www.protectedplanet.net/). Within the internationally recognized system PA are distinguished using the seven IUCN Protected Area Categories based on the PA management objectives, which enable international comparisons to be made (see Box 8.1). As well as these categories of PA, several different governance regimes are also recognized (see Box 8.2).

In addition, there are several other international networks that aim to conserve biodiversity within *in situ* sites, which are considered below.

UNESCO Man and Biosphere Programme (MAB)[*]
The MAB Programme is an Intergovernmental Scientific Programme aiming to set a scientific basis for the improvement of the relationships between people and their environment globally. Launched in the early 1970s, it provides interdisciplinary research and capacity building targeting the ecological, social and economic dimensions of biodiversity loss and the reduction of this loss. The World Network now comprises 669 biosphere reserves in 120 countries, including 16 trans-boundary sites, and they aim to promote sustainable development based on local community efforts and sound science. They seek to reconcile conservation of biological and cultural diversity and economic and social development through partnerships between people and nature. Biosphere reserves are thus globally considered as: sites of excellence where new and optimal practices to manage nature and human activities are tested and demonstrated; and tools to help countries implement conservation convention and treaty policy commitments. Following designation, biosphere reserves remain under national sovereign jurisdiction, yet they share their experience and ideas nationally, regionally and internationally within the World Network of Biosphere Reserves. This network aims to: identify and assess the changes in the biosphere resulting from human and natural activities and the effects of these changes on humans and the environment, in particular in the context of climate change; study and compare the dynamic interrelationships between natural/near-natural ecosystems and socio-economic processes, in particular in the context of accelerated loss of biological and cultural diversity with unexpected consequences that impact the ability of ecosystems to continue to provide services critical for human well-being; ensure basic human welfare and a liveable environment in the context of rapid urbanization and energy consumption as drivers of environmental change; promote the exchange and transfer of knowledge on environmental problems and solutions, and to foster environmental education for sustainable development.

UNESCO World Heritage Sites (WHS)[†]
The Convention concerning the Protection of World Cultural and Natural Heritage, commonly abbreviated to World Heritage Convention, was adopted by UNESCO on

[*] www.unesco.org/new/en/natural-sciences/environment/ ecological-sciences/man-and-biosphere-programme/

[†] http://whc.unesco.org/en/list/

Box 8.1 | IUCN Protected Areas Management Categories

(Dudley, 2008)

Although the goals of most PA meet the general objectives contained in the IUCN PA definition, in practice PA vary significantly in their nature and therefore the way they are managed, and this is reflected in the management categories recognized by the IUCN:

Ia. *Strict Nature Reserve*: Category Ia are strictly protected areas set aside to protect biodiversity and possibly geological/geomorphological features, where human visitation, use and impacts are strictly controlled and limited to ensure protection of the conservation values. Such protected areas can serve as indispensable reference areas for scientific research and monitoring.

Ib. *Wilderness Area*: Category Ib protected areas are usually large unmodified or slightly modified areas, retaining their natural character and influence without permanent or significant human habitation, which are protected and managed to preserve their natural condition.

II. *National Park*: Category II protected areas are large natural or near-natural areas set aside to protect large-scale ecological processes, along with the complement of species and ecosystems characteristic of the area, which also provide a foundation for environmentally and culturally compatible, spiritual, scientific, educational, recreational and visitor opportunities.

III. *Natural Monument or Feature*: Category III protected areas are set aside to protect a specific natural monument, which can be a landform, sea mount, submarine cavern, geological feature such as a cave or even a living feature such as an ancient grove. They are generally quite small protected areas and often have high visitor value.

IV. *Habitat/Species Management Area*: Category IV protected areas aim to protect species or habitats and management reflects this priority. Many Category IV protected areas will need regular, active interventions to address the requirements of species or to maintain habitats, but this is not a requirement of the category.

V. *Protected Landscape/Seascape*: A protected area where the interaction of people and nature over time has produced an area of distinct character with significant ecological, biological, cultural and scenic value: and where safeguarding the integrity of this interaction is vital to protecting and sustaining the area and its associated nature conservation and other values.

VI. *Protected area with sustainable use of natural resources*: Category VI protected areas conserve ecosystems and habitats together with associated cultural values and traditional natural resource management systems. They are generally large, with most of the area in a natural condition, where a proportion is under sustainable natural resource management and where low-level non-industrial use of natural resources compatible with nature conservation is one of the main aims of the area.

Box 8.2 | Different Governance Regimes Associated with *In Situ* Protected Area Management

(Stolton *et al.*, 2006)

- Government-managed protected areas – Protected areas managed by national or local government, occasionally through an officially appointed independent body: i.e. federal or national ministry or agency in charge; local/municipal ministry or agency in charge; or government-delegated management (e.g. to an NGO).
- Co-managed protected areas – Protected areas which involve local communities in the management of government-designated protected areas through active consultation, consensus-seeking, negotiating, sharing responsibility and transferring management responsibility to communities or NGOs, i.e. transboundary management, collaborative management (various forms of pluralist influence) or joint management (pluralist management board).
- Private protected areas – Protected areas managed by private individuals, companies or trusts, i.e. declared and run by an individual landowner, non-profit organization (e.g. NGO), university or cooperative or for-profit organization (e.g. individual or corporate landowners).
- Community conserved areas – Protected areas managed as natural and/or modified ecosystems voluntarily by indigenous, mobile and local communities.

---

16 November 1972 and has subsequently been ratified by 193 states. Significantly the WHC brings together the concepts of nature conservation, the preservation of cultural properties and the balance between the two. It sets out the duties of States Parties in identifying potential sites and their role in protecting and preserving them, as well as defining the kind of natural or cultural sites to be included on the World Heritage List. Sites must be of 'outstanding universal value' and meet at least one of the following 10 selection criteria:

Cultural Criteria:

- Represents a masterpiece of human creative genius
- Exhibits an important interchange of human values, over a span of time, or within a cultural area of the world, on developments in architecture or technology, monumental arts, town-planning or landscape design.
- Bears a unique or exceptional testimony to a cultural tradition or to a civilization which is living, or which has disappeared.

- Is an outstanding example of a type of building, architectural or technological ensemble, or landscape which illustrates a significant stage in human history.
- Is an outstanding example of a traditional human settlement, land use or sea-use which is representative of a culture, or human interaction with the environment especially when it has become vulnerable under the impact of irreversible change.
- Is directly or tangibly associated with events or living traditions, with ideas, or with beliefs, with artistic and literary works of outstanding universal significance.

Natural Criteria:

- Contains superlative natural phenomena or areas of exceptional natural beauty and aesthetic importance.
- Is an outstanding example representing major stages of Earth's history, including the record of life, significant on-going geological processes in the development of landforms, or significant geomorphic or physiographic features.

- Is an outstanding example representing significant on-going ecological and biological processes in the evolution and development of terrestrial, fresh water, coastal and marine ecosystems, and communities of plants and animals.
- Contains the most important and significant natural habitats for *in situ* conservation of biological diversity, including those containing threatened species of outstanding universal value from the point of view of science or conservation.

Sites included have access to the World Heritage Fund to help maintain the unique quality of the site, are a magnet for international cooperation and benefit from the development and implementation of a comprehensive management plan that sets out adequate preservation and monitoring mechanisms. The overall function of the WHC is to preserve our collective heritage from the past, what we live with today and to pass it on to future generations.

IUCN Key Biodiversity Areas (KBA)[‡]

KBA are a relatively new concept being developed by the IUCN that identifies and conserves places of international importance for the conservation of biodiversity through protected areas and other governance mechanisms. Their role is to conserve biodiversity at the habitat, species and genetic level. They are identified nationally using simple, standard criteria, based on their importance in maintaining species populations. As the building blocks for designing the ecosystem approach and maintaining effective ecological networks, key biodiversity areas are the starting point for conservation planning at landscape level. Governments, intergovernmental organizations, NGOs, the private sector and other stakeholders can use key biodiversity areas as a tool for identifying national networks of internationally important sites for conservation. A global KBA structure is not yet in place, but sites will be selected based on one of the following criteria:

A. Threatened biodiversity
B. Geographically restricted biodiversity

C. Biodiversity through outstanding ecological integrity
D. Outstanding biological processes

The *threatened biodiversity criterion* (A) identifies sites contributing significantly to the persistence of:

1. Taxa that are formally assessed as globally threatened or expected to be classified as globally threatened once their risk of extinction is formally assessed; or nationally/regionally endemic taxa that have not been formally globally assessed but have been nationally/regionally assessed as threatened; OR,
2. Ecosystems that are formally assessed as globally threatened or expected to be classified as globally threatened once their risk of collapse is formally assessed.

The *geographically restricted biodiversity criterion* (B) identifies sites contributing significantly to the persistence of:

1. Species with ranges that are permanently or periodically geographically restricted, or highly clumped populations, or which occur at few sites; OR,
2. Assemblages of species with geographically restricted ranges in centres of endemism or genetic distinctness; OR,
3. Ecosystems with geographically restricted distributions or which occur at few sites.

The *ecological integrity criterion* (C) identifies sites contributing significantly to the global persistence of biodiversity because they are exceptional examples of ecological integrity and naturalness, as represented by:

1. Intact species assemblages, comprising the composition and abundance of native species and their interactions, within the bounds of natural ranges of variation; OR,
2. The most outstanding places, within biogeographic regions, of relatively intact:
   a. regionally distinct species assemblages with high contextual species richness; OR,
   b. regionally distinct, contiguous areas of ecosystem and habitat diversity.

The *outstanding biological process criterion* (D) identifies sites contributing significantly to the persistence of:

1. Evolutionary processes of exceptional importance in maintaining biodiversity or driving rapid diversification; OR

[‡] http://www.iucn.org /key-biodiversity-areas/

2. Species at key stages in their life-cycles, such as those which are migratory or congregatory, as indicated by high relative abundance; OR,
3. Ecological processes of exceptional importance in maintaining biodiversity.

All sites should be assessed against all the criteria but meeting any one of the criteria is enough to qualify a site as a KBA.

The KBA identification process involves: Step 1 biological mapping of diversity; and Step 2 identify boundaries to diversity concentration that best support persistence of biodiversity identified incorporating biological/ecological/local knowledge. Delineation of KBAs are considered as iterative, and there is a requirement for adaptive learning.

FAO Globally Important Agricultural Heritage Systems (GIAHS)[§]

GIAHS are the sole international site-based conservation network that have a focus on agro-biodiversity; they are a FAO initiative. GIAHS are defined as 'Remarkable land use systems and landscapes which are rich in globally significant biological diversity evolving from the co-adaption of community with its environment and its needs and aspirations for sustainable development'. GIAHS aim to promote public understanding, awareness, and national and international recognition of agricultural heritage systems. They also aim to safeguard the social, cultural, economic and environmental goods and services associated with agro-biodiversity and support family farmers, smallholders, indigenous peoples and local communities working with this diversity so combining sustainable agriculture and rural development. GIAHS sites are selected based on their provision of (1) food and livelihood security, (2) high levels of agro-biodiversity, (3) extensive local and traditional knowledge systems, (4) significant cultures, value systems and social organizations and (5) special landscapes and seascapes features:

1. *Food and livelihood security*: The proposed agricultural system contributes to food and/or livelihood security of local communities. This includes a wide variety of agricultural types such as self-sufficient and semi-subsistence agriculture where provisioning and exchanges take place among local communities, which contributes to rural economy.
2. *Agro-biodiversity*: Agricultural biodiversity, the variety of animals, plants and micro-organisms that are used directly or indirectly for food and agriculture, including crops, livestock, forestry and fisheries. The system should be endowed with globally significant biodiversity and genetic resources for food and agriculture (e.g. endemic, domesticated, rare, endangered species of crops and animals).
3. *Local and traditional knowledge systems*: The system should maintain local and invaluable traditional knowledge and practices, ingenious adaptive technology and management systems of natural resources, including biota, land, water which have supported agricultural, forestry and/or fishery activities.
4. *Cultures, value systems and social organisations*: Cultural identity and sense of place are embedded in and belong to specific agricultural sites. Social organizations, value systems and cultural practices associated with resource management and food production may ensure conservation of and promote equity in the use and access to natural resources. Such social organizations and practices may take the form of customary laws and practices as well as ceremonial, religious and/or spiritual experiences.
   a. Social organization is defined as individuals, families, groups or communities that play a key role on the agricultural systems' organization and dynamic conservation.
   b. Local social organizations may play a critical role in balancing environmental and socio-economic objectives, creating enhancing resilience and reproducing all elements and processes critical to the functioning of the agricultural systems.
5. *Landscapes and seascapes features*: GIAHS sites should represent landscapes or seascapes that have been developed over time through the interaction between humans and the environment and appear to have stabilized or to evolve very slowly. Their form, shape and interlinkages are characterized by long historical persistence and a strong connection with the local socio-economic systems that produced them.

[§] http://www.fao.org/giahs/en/

Their stability, or slow evolution, is the evidence of integration of food production, the environment and culture in each area or region. They may have the form of complex land use systems, such as land use mosaics, water and coastal management systems.

Thus far, sites have been recognized in Algeria (1), Bangladesh (1), Chile (1), China (15), Egypt (1), India (3), Iran (3), Italy (2), Japan (11), Kenya (1), Mexico (1), Morocco (2), Peru (1), Philippines (1), Portugal (1), South Korea (4), Spain (3), Sri Lanka (1), Tanzania (2), Tunisia (1) and United Arab Emirates (1); in each adaptive management approaches are being developed and implemented, to assist national and local stakeholders in the dynamic conservation of their agricultural heritage systems.

As well as those listed above, the NGO Plantlife is systematically identifying Important Plant Areas (IPA; www.plantlife.org.uk/international/important-plant-areas-international), which areas of landscape that have been identified as being of the highest botanical importance. The concept was first established in the mid-1990s and their goal by designating an IPA is to gain awareness and encourage long-term conservation through an 'ecosystem-based' approach of that site. The identification of IPAs is based on three criteria: presence of threatened plant species, presence of botanical richness and presence of threatened habitats. Initially sites were recognized in the UK and Europe, but more recently the network has been extended to cover nearly 2000 IPA s in 27 countries across Europe, North Africa and the Middle East. Although the CBD has a Programme of Work on Protected Areas (PoWPA; www.cbd.int/protected/), it does not directly manage any protected areas, rather they collaborate with established regional and global networks development PA management tools, organizing capacity building workshops and facilitating PA conservation action. Specifically, in the context of CWR conservation, the IUCN Crop Wild Relative Specialist Group (CWRSG; www.cwrsg.org/) is an international expert group established by IUCN to promote the conservation and sustainable use of CWR diversity. Its mission is: 'to help ensure that crop wild relatives are adequately conserved and utilized, to enhance food security, aid poverty alleviation and

improve the environment worldwide'. It has four primary objectives:

1. Develop effective strategies for gathering, documenting and disseminating baseline information on crop wild relatives.
2. Promote the conservation and use of crop wild relatives.
3. Provide advice, expertise and access to appropriate contacts to enhance the actions of individuals or organizations working on the conservation and use of crop wild relatives.
4. Increase awareness of the importance to agriculture and the environment of crop wild relatives among governments, institutions, decision-makers and the general public.

Therefore, the CWRSG provides a structured and coordinated network of experts in the field of CWR conservation and use and aims to open and maintain the necessary communication channels to share information and experiences and encourage a more strategic approach to CWR conservation (and through conservation, a link to enhanced utilization). Internationally, there has been an increasing consensus of the need to conserve CWR diversity in the era of growing ecosystem instability and climate change. However, despite significant progress in global *ex situ* seed collection of priority CWR (Dempewolf *et al.*, 2013), there is currently no over-arching global network or clearing-house mechanism specifically devoted to the conservation and use of CWR, and progress in implementing *in situ* conservation of CWR diversity is limited.

Most of the world's PA contain CWR populations; however, the PA were established to conserve habitats, and charismatic or threatened species, not CWR. In these PA, CWR are conserved passively, meaning the CWR populations are not actively managed and individual populations may decline or go extinct without the site managers being aware of the loss. Where CWR are actively conserved, this is often because they also happen to be charismatic or threatened and so are prioritized for this reason, not because of their importance as CWR (Maxted *et al.*, 1997c). Even in the few sites where CWR are actively conserved *in situ* (Table 8.1), in most cases, the sites

Table 8.1 **Examples of CWR conserved *in situ* in protected areas**

CWR	Protected area	Country	References
Teosinte (*Zea diploperennis*)	MAB Sierra de Manantlan Biosphere Reserve	Mexico	Sanchez-Velasquez (1991)
Wild emmer wheat (*Triticum turgidum* subsp. *dicoccoides*)	Ammiad, Galilee	Israel	Anikster *et al.* (1997); Safriel *et al.* (1997)
Wild coffee (*Coffea mauritiana, C. macrocarpa, C. myrtifolia*)	Black River Gorges National Park	Mauritius	Dulloo *et al.* (1999)
Wild wheats (*Triticum turgidum* subsp. *dicoccoides, T. monococcum, Ae. tauschii, Ae. speltoides*)	Ceylanpinar	Turkey	Karagöz (1998)
Chestnut (*Castanea sativa*), wild plum (*Prunus cerasifera* var. *divaricata*)	Kazdağ	Turkey	Kűçűk *et al.* (1998)
Wild onions (*Allium columbianum, A. geyeri* and *A. fibrillum*)	Great Basin, Washington State	USA	Hannan and Hellier (1999); Hellier (2000)
Wild grapevine (*Vitis rupestris, V. shuttleworthii, V. monticola*)	Witchita Mountains and Ouachita National Forest, Oklahoma, Clifty Creek, Missouri	USA	Pavek *et al.* (2003)
Wild bean populations (*Phaseolus* spp.)	Central valley	Costa Rica	Baudoin *et al.* (2008)
Medicago spp., *Vicia* spp., *Trifolium* spp., *Lathyrus* spp., *Lens* spp., *Triticum* spp., *Avena* spp., *Hordeum* spp., *Aegilops* spp., *Allium* spp., *Amygdalus* spp., *Prunus* spp., *Pyrus* spp., *Pistacia* spp. and *Olea* spp.	Abu Taha, Sale-Rsheida, Ajloun, Wadi Sair	Lebanon, Syria, Jordan, Palestinian Territories	Al-Atawneh *et al.* (2008)
Wild wheats (*Triticum boeoticum, T. urartu, T. araraticum*)	Erebuni	Armenia	Avagyan (2008)

are not managed in the most appropriate manner to maximize the conservation of the genetic diversity within CWR populations. Commonly, these sites would also not meet the set of quality standards for CWR genetic reserves proposed by Iriondo *et al.* (2012), and their designation has been ad hoc and opportunistic rather than because of any objective scientific programme.

Table 8.1 (*cont.*)

CWR	Protected area	Country	References
Wild beet (*Beta patula*)	Desertas Is.	Portugal	Pinheiro de Carvalho *et al.* (2012)
Wild *Solanum* spp.	Laguna de los Pozuelos Natural Monument and Los Cardones National Park	Argentina	Marfil *et al.* (2015)

From Maxted *et al.* (2016a).

8.2 Genetic Reserve Conservation

8.2.1 The Concept

Genetic reserve conservation is most commonly associated with IUCN Category IV Habitat/Species Management Area (Dudley, 2008). The distinction between genetic reserve conservation and other forms of protected area conservation is the genetic element; the overall goal is to maximize conservation of the gene pool, the full range of genetic diversity of a species or group of species, rather than the species *per se* or the entire ecosystem in which they are found. As such, genetic reserve conservation may be defined as (Maxted *et al.*, 1997b):

The location, management and monitoring of genetic diversity in natural wild populations within defined areas designated for active, long-term conservation.

The process of defining, establishing and managing a genetic reserve involves selecting the most appropriate site for the reserve, designing the reserve, then once established, managing and monitoring the population of the target taxon and ensuring samples are available for use by local or remote stakeholders. It is implied that conservation is active in that the population of the target taxon will be actively (i.e. regularly) managed and monitored, that the management is adaptive, meaning if the CWR population levels decline or other management interventions are more appropriate the interventions implemented will be changed to better promote population maintenance, and that conservation activities at the site are sustainable in the long-term.

The conservation of a plant species, whether *in situ* or *ex situ*, will usually only involve part of the total genetic diversity, as the resources available for conservation are always finite and genetic diversity is usually spread throughout the range of a taxon. The conservationist's objective is to ensure that the maximum possible range of genetic diversity is represented within the minimum number and size of *in situ* protected areas (or for *ex situ* the number and size of *ex situ* collections). This is, however, a complex goal to achieve because it is likely to involve establishing a network of genetic reserves, possibly in multiple countries, and requires detailed information on the amount of genetic variation, population structure, breeding systems, habitat requirements, geographic distribution and the minimum viable population (MVP) size of each taxon. As the core objective of genetic reserve conservation is to maximize the conserved genetic diversity of the various taxa, to achieve this it is necessary to determine the minimum effective population size and hence the MVP size necessary for conservation prior to reserve establishment, but unfortunately the necessary CWR data are not always readily available, even for well-studied species.

The process of locating and establishing genetic reserves, where representative genetic diversity is located and can be conserved, is further complicated because reserves themselves are seldom situated due to purely biological justification and rarely with the aim of conserving a single CWR taxon. Many non-biological, socio-economic and political constraints

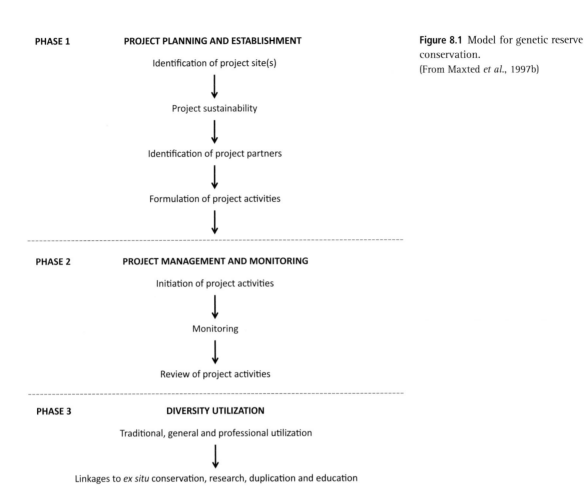

PHASE 1   **PROJECT PLANNING AND ESTABLISHMENT**

Identification of project site(s)

↓

Project sustainability

↓

Identification of project partners

↓

Formulation of project activities

↓

- -

PHASE 2   **PROJECT MANAGEMENT AND MONITORING**

Initiation of project activities

↓

Monitoring

↓

Review of project activities

- -

PHASE 3   **DIVERSITY UTILIZATION**

Traditional, general and professional utilization

↓

Linkages to *ex situ* conservation, research, duplication and education

Figure 8.1 Model for genetic reserve conservation.
(From Maxted *et al.*, 1997b)

are imposed on the location, design and management of the genetic reserve, and the target will often be a group of species, possibly each with different ecogeographic requirements. Also, even though the goal is genetic conservation of a target taxon, there may be a practical requirement to conserve associated species also, such as keystone species, pollinators or seed dispersers, which in turn will help ensure the long-term survival of the target taxon. So, genetic reserve conservation needs to be pragmatically implemented.

The process of establishing and running a genetic reserve is likely to differ for each target species, but there are some basic requirements that will have to be considered for each species. These are summarized in the model for genetic reserve conservation (Maxted *et al.*, 1997d) shown in Figure 8.1 and Box 8.3. It

should be noted that this model for genetic reserve conservation fits within the general model of plant genetic conservation discussed in Chapter 1. Thus, genetic reserve conservation may be summarized as having five integrated, key elements: location, designation, management, monitoring and utilization.

8.2.2 Site Assessment

Maximizing Genetic Diversity

Ideally the site selected will have a substantial and genetically diverse population(s) of the target taxon, substantial so it has maximum chance of long-term survival and genetically diverse so it provides a good representation of the genetic diversity throughout the target taxon's range and the network of reserves

Box 8.3 | **Overview of *In Situ* Genetic Conservation of Wild Plant Species**

(Maxted *et al.*, 1997b)

1. **Selection of target taxa** – Decide which species need active conservation and for which *in situ* genetic reserves is appropriate. If possible, include more than one chosen species in each reserve.
2. **Project commission** – Formulate a clear, concise conservation statement establishing what species, why and in general terms where the species are to be conserved.
3. **Ecogeographic survey/gap analysis/preliminary survey mission** – This facilitates the collation of the basic information for the planning of effective conservation. Survey the distribution of taxonomic and genetic diversity, ecological requirements and the reproductive biology of the chosen species over its entire geographic range. Compare the plant genetic diversity that exists in nature and the sample of that diversity actively conserved *in situ* and *ex situ* to identify gaps/future priorities in the current conservation actions. Where little ecogeographic data are available, a preliminary course grid survey mission to collate the necessary background biological data on the species may be required.
4. **Conservation objectives** – Formulate a clear, concise set of conservation objectives which state the practical steps that must be taken to conserve the species, which is climate smart (the likely impact of climate change has been considered) and specifies the conservation objectives and propose how the conserved diversity is linked to utilization.
5. **Field exploration** – Visit competing potential sites indicated as having high levels of target species and genetic diversity by the ecogeographic survey/gap analysis to 'ground truth' the predictions and identify specific locations where target species and genetic diversity are to be conserved in genetic reserves.
6. **Conservation application for *in situ* genetic reserve** – This involves the designation, management and monitoring of the genetic reserve.
   6.1 Reserve planning and establishment
      6.1.1 **Site assessments** – Within actual locations, establish the sites where genetic reserves will be established; where possible they should cover the range of morphological and genetic diversity, and the ecological amplitude exhibited by the chosen species. Several reserves spread over the geographic range and the ecological environments occupied by the species may be required to cover a sufficiently large fraction of the target CWR species's gene pool. Ensure that each reserve represents the fullest possible ecological range (micro-niches), to help secure maximal genetic variation, and to buffer the protected population against environmental fluctuations, pests and pathogens, and man-made disturbances. As part of this evaluation, prepare a vegetation map of the area, surveying in detail the plant communities (and habitats) in which the target species grows.
      6.1.2 **Assessment of local socio-economic and political factors** – Constraints ranging from economic to scientific and organizational will affect the establishment of the reserve. The simplest way forward in economic and

Box 8.3 | (cont.)

political terms is for countries to establish a network of complementary genetic reserves, as this is likely to be of some benefit to the people and will engender their support.

6.1.3 **Taxon and reserve sustainability** – Establishing and managing an *in situ* genetic reserve is resource expensive and therefore both the taxon and reserve must be deemed sustainable over an extended time or the investment will be forfeit.

6.1.4 **Reserve design** – Sites should be large enough to contain at least (1000–) 5000–10 000 individuals of each target species to prevent natural or anthropogenic catastrophes causing severe genetic drift or population unsustainability. Sites should be selected to maximize environmental heterogeneity. Each reserve site should be surrounded by a buffer zone of the same vegetation type, to facilitate immigration of individuals and gene flow but also where experiments on management regimes might be conducted and visits by the public allowed, under supervision.

6.1.5 **Formulation of the management plan** – The reserve site will have been selected because it contains abundant and hopefully genetically diverse populations of the target taxon/taxa. Therefore, the first step in formulating the management plan is to observe the biotic and abiotic qualities and interactions at the site. Once these ecological dynamics within the reserve are known and understood, a management plan that incorporates these points, at least as they relate to the target taxon/taxa, can be proposed.

6.2 **Reserve management and monitoring**

6.2.1 **Initiation of reserve management plan** – It is unlikely that any management plan will be wholly appropriate when first applied; it will require detailed monitoring of target and associated taxa and experimentation with the site management before a more stable, beneficial plan can evolve. The plan may involve experimentation with several management interventions (a range of grazing practices, tree-felling, burning, etc.) within the reserve to ensure the final plan does meet the conservation objectives, particularly in terms of maintaining the maximum wild plant species and genetic diversity. Genetic reserves conservation is a process-oriented way of maintaining genetic resources. It will maintain the evolutionary potential of a population as well as maintaining the effective population sizes of the species. The actual style of the management plan will vary depending on target taxa, location and implementing agency but is likely to cover: preamble, conservation context, site abiotic description, site biotic description, site anthropogenic description, general taxon description, site-specific taxon description, site management policy, taxon and site population research recommendations and intervention prescription.

Box 8.3 | **(cont.)**

6.2.2 **Reserve monitoring** – Each site should be monitored systematically at a set time interval and the results fed back in an iterative manner to enhance the evolving management regime. The monitoring is likely to take the form of measures of taxon number, diversity and density as measured in permanent transects, quadrats etc.

6.3 **Reserve utilization**

6.3.1 **Traditional, general and professional utilization** – Humans generally conserve because they wish to have actual or potential utilization options; therefore, when designing the reserve, it is necessary to make an explicit link between the material conserved and its current or potential utilization by humankind. There are three basic user communities: traditional or local, the general public and professional users.

6.3.2 **Linkage to *ex situ* conservation, duplication, research and education** – There is a need to form links with *ex situ* conserved material to ensure utilization but also as a form of back-up safety duplication in times of ecosystem instability. The reserve forms a natural platform for ecological and genetic research, as well as providing educational opportunities for the school, higher educational and public level awareness raising.

7. **Conservation products** – These will be populations of live plants held in the reserve, voucher specimens and the passport data associated with the reserve and plant populations.

8. **Conserved product deposition and dissemination** – The main conserved product, the plant populations of the target taxon, are held in the reserve. However, there is a need for safety duplication and a sample of germplasm should also be periodically sampled and deposited in an appropriate *ex situ* collection (gene bank, field gene bank, *in vitro* banks, botanical gardens or conservation laboratory) with the appropriate passport data.

9. **Characterization/evaluation** – The first stage of utilization will involve the recording of genetically controlled characteristics (characterization) and the material may be grown out under diverse environmental conditions to evaluate the populations for biotic and abiotic resistance.

10. **Plant genetic resource utilization** – The conserved material is likely to be used in breeding and biotechnology programmes, and provide food, fuel, medicines and industrial products, as well as a source of recreation and education. Locally the materials held in the reserve may have traditionally been used in construction, craft, adornment, transport or food. This form of traditional utilization of the reserve by local people should be encouraged, providing it is sustainable and not deleterious to the target taxon or taxa, as it is essential to have local support for conservation actions if the reserve is to be sustainable in the medium to long-term. Off-site use will be subject to the ITPGRFA, and therefore there will be a need to establish an access and benefit sharing agreement.

captures the maximum range of genetic diversity of the target species's gene pool overall (Dulloo *et al.*, 2008). Diversity within a population is essential to ensure each taxon can continue to evolve and adapt in the evolving ecosystem where they exist (Frankel and Bennett, 1970). Assessment of a site's suitability to represent the taxon's genetic diversity requires knowledge of the level and patterns of genetic diversity both within and across populations of the target taxon (Neel and Cummings, 2003a), which in turn depends on the target taxon's life history traits, including life form, breeding system and geographic range (Hamrick and Godt, 1996).

However, in practice, it is unlikely for most taxa that information on the amount and distribution of genetic variability will be available (Thomson *et al.*, 2001; Neel and Cummings, 2003b), and studies to generate this information can be too costly and time-consuming to undertake before each genetic reserve is located. So most often a proxy for actual genetic diversity is employed; this application of ecogeographic/gap analysis techniques (Chapter 6) will have predicted areas of potential genetic diversity where genetic reserves might be sited. The SDM techniques used normally identify a 10- or 100-km^2 grid that may contain several potential sites, to establish which is most appropriate. The potential sites will require field surveying to quantify the actual taxonomic and genetic richness/diversity (numbers of and variation) of populations present at those sites and sites worthy of genetic reserve designation determined. If little ecogeographic data are available, a preliminary course grid survey mission (Chapter 11) may be required to collate the necessary ecogeographic data on the species to assist in locating concentrations of diversity. SDM and field work can identify concentrations of taxonomic diversity, or areas where multiple taxa occur. To confirm areas of genetic diversity will require sampling of plants and analysis of their patterns of allelic richness (number), diversity (incorporates richness and abundance) and evenness (distribution of abundances), which in turn will identify ecotypes (locally adapted populations) and their unique diversity (Chapter 4).

Maximizing Taxonomic Diversity

Cost-benefit analysis will dictate that rarely would a genetic reserve be established for a single target taxon; it is more likely that new reserves will be established to conserve multiple target taxa as part of a national, regional or global CWR conservation strategy. To identify the most appropriate sites to establish a multiple target taxon, genetic reserves may be viewed as a four-stage process (Maxted *et al.*, 2007): (a) create a national CWR checklist, (b) prioritize the most important CWR taxa for active conservation and create a CWR inventory, (c) use distributional data for the priority CWR taxa to identify national complementary CWR hotspots and then (d) match CWR hotspots to the existing protected area network to identify existing protected areas where genetic reserves could be established, new reserves established, or novel sites outside of formal PA where CWR *in situ* conservation could be enacted.

Duplication of Sites

Unless a plant species is only present at a single location, no one site can contain all a taxon's genetic diversity; therefore, establishing a network of genetic reserves is required, but each individual site is likely to contain multiple plant taxa and populations, so the total number of sites is not as large as it might at first appear. Duplication of sites for a taxon also facilitates security of conservation into the long-term future, as a single site is vulnerable to changes in site management practice, development projects, climate change or localized stochastic events such as hurricanes or floods, and even political will.

Complementarity of Sites

Genetic variation is strongly correlated with environmental (soil, aspect, drainage, temperature, day length, etc.) and ecological variation (interactions with pathogens, herbivores, pollinators, etc.) as well as factors relating to the population genetics of the species (breeding system, population size, gene flow, etc.) (Hamrick and Godt, 1996). To maximize the conservation of genetic diversity the selected sites should cover as wide a range of habitats and environments as possible. As such, the ecological range and each micro-niche will be included to help buffer the protected population against environmental fluctuations, pests and pathogens, and anthropogenic disturbances. As part of this site evaluation phase, a vegetation map of the area or an ELC map of the taxon (Parra-Quijano *et al.*, 2011a) could be produced to aid complementary site selection. This will help ensure the conservation of adapted genotypes and specific ecotypes within the taxon's gene pool.

Consideration of Ecosystem Instability

One of the criticisms of *in situ* CWR conservation is its vulnerability to climate change, which will increasingly alter plant distributions (Jarvis *et al.*, 2008a; Maxted *et al.*, 2013a; Redden *et al.*, 2015). To ensure its impact is minimized, relative impact on individual taxa (Phillips *et al.*, 2004), on competing potential *in situ* conservation sites (Vincent *et al.*, 2019), or where to build corridors or stepping stones between genetic reserves to facilitate likely population migration (Jarvis *et al.*, 2015) can be modelled. The latter being increasingly important in planning genetic reserves to ensure they are linked to maximize their potential to form a meta-network linked by corridors or stepping stones and facilitate inter-site gene flow.

Location in Existing Protected Areas

Creating new protected areas each time new conservation priorities are established would be prohibitively expensive, especially if that process involved buying the land on which the reserve was to be established. Therefore, where possible, genetic reserves should be in existing protected areas and their management should be incorporated into the overall conservation management programme of the site (Maxted *et al.*, 2008b). This is preferable because: (a) these sites already have an associated long-term conservation ethos and are less prone to hasty management changes associated with private land or roadside where conservation value and sustainability is not normally a consideration, (b) it is relatively easy to amend the existing site management to facilitate genetic conservation of wild plant species and (c) it

means creation of novel conservation sites can be avoided and so the possibly prohibitive cost of acquiring previously non-conservation managed land (Maxted *et al.*, 2008b). However, it does mean that if the conservation of plant genetic resources is to be integrated into an existing protected area, the management plan should be adapted accordingly; to achieve this it will mean persuading the PA manager that the target taxon is worthy of conservation, which has not proved easy for CWR taxa. It should also be noted that not all PA are currently managed effectively. Laurance *et al.* (2012) note that about 50% of PA in the tropics are degraded and so would be less suitable sites to establish genetic reserves. It would be interesting to speculate what proportion of temperate PA would also be considered ineffectively managed, but this consideration does add an additional feature to selecting sites for genetic reserves, as there would be little point in establishing a genetic reserve within a mis- or un-managed PA. Although genetic conservation need not be applied across the entire PA, more effective management could be targeted in selected zone(s) within the protected area.

Cost of Establishment

The relative cost of reserve establishment will also affect the selection of alternative sites. If faced with a choice of equally suitable sites and differing establishment costs, there would be little justification for selecting a site other than the least expensive (Maxted *et al.*, 1997a). The same logic would also apply to the running costs of the reserve once established. This may necessitate the application of some form of cost-benefit analysis of alternative sites prior to actual reserve selection.

8.2.3 Local Socio-economic and Political Factors

The location, design and management of any PA or reserve are rarely decided solely based on biological expedience; social, economic and political factors will all play a part. The planning of a genetic reserve must include an assessment of these factors and will include consultations with local people, landowners and government representatives from such Departments or Ministries as Environment, Agriculture and Forestry (IUCN, 2017).

Social and Cultural Factors

The location of the genetic reserve will almost inevitably be sited on land that is lived on or used by diverse groups of people. The local users may be traditional indigenous people, subsistence farmers or people from a nearby town or city, and their livelihoods may depend on this land (Worboys *et al.*, 2015); for example, it may be traditional hunting ground or pasture land for seasonal grazing of their livestock. Therefore, the needs and aspirations of local people must be considered and incorporated as far as possible into the design of the reserve. It is never as simple as building a fence around a site and keeping people out! The reserve should disrupt as little as possible the lives of traditional indigenous users of the land; in many cases they have been using the resource sustainably for centuries. Activities that may directly threaten habitats and populations of the target taxon may have to be managed or restricted to zones in the reserve. Any restrictions imposed must be introduced in a cooperative manner to minimize conflict between the reserve and local people (Bronkhorst, 2014). A realistic compromise may have to be reached between the conservationist's biological ideal and local people's needs. It is often essential when designating a new PA that it receive public funding to ensure its success and sustainability, and as such, it is essential that local people and the general public support the project. They will only provide such support if they feel they have a stake in the PA, and this is reliant on their direct or indirect involvement in the PA establishment.

Economic and Political Factors

Political and economic factors will put constraints on the design of a protected area and the implementation of the management plan. The location of the reserve may be on privately or publicly owned land, and it is often cheaper and easier to establish reserves on existing state-owned land than attempt to purchase or lease privately owned land for this purpose. However,

whether the land is owned by a traditional farmer, the government or a non-governmental organization, there may be conflicts in resource or land use and the conservation objectives of the reserve that require resolution.

The potential activities that could conflict with the objectives of the reserve may be designated as extrinsic or intrinsic threats depending on whether the threat originates from a remote or local source. Extrinsic threats might include: mining/extraction of oil, coal or minerals, road building, urban development (out of town shopping centres, new housing estates), draining the land for agriculture, dam construction, timber extraction, quarrying for limestone, marble etc.; while intrinsic threats might include: changes in local environmental management (e.g. abandonment of traditional agro- or silvi-cultural practices), destruction of primary habitat (e.g. clearance of forest for agriculture), over-exploitation (e.g. grazing or local resource extraction), introduction and escape of invasive species (e.g. animals, plants, pests and diseases) and increased human population growth. It is worth stressing that intrinsic threats to PA from local communities are rarely on an unsustainable scale and are therefore sufficiently destructive to threaten the sustainability of a protected area; unsustainability is more commonly associated with intensive, external resource exploitation.

Another surprisingly common constraint to genetic reserve implementation is the lack of existing links between the protected area and CWR conservation community. Too often the two communities work independently, but increasingly protected areas are recognizing the advantages of demonstrating provisioning ecosystem services by promoting CWR conservation within the biodiversity context (Hopkins and Maxted, 2010), and at the same time building a cross-sectoral link between the two communities. Some enlightened government policies that recognize the value of CWR conservation in environmental stewardship schemes so directly subsidizing CWR *in situ* conservation are having a significant impact (Maxted *et al.*, 2015). Perhaps one of the key remaining constraints is the lack of adequately trained staff that can plan and implement the conservation and manage genetic materials *in situ*, even if in practice the actual additional expertise required is limited.

Taxon and Reserve Sustainability

Sustainability is fundamental to all conservation implementation (Iriondo *et al.*, 2008), but particularly protected area conservation, where significant resources are invested in a population found in a location. The cost of locating, establishing and running a genetic reserve can be substantial, and as the goal is to conserve the target taxon for an indefinite period, usually defined as 'the population having a 99% chance of remaining extant for 1000 years despite the foreseeable effects of demographic, environmental and genetic stochasticity and natural catastrophes' (Shaffer, 1981) or at least having a >95% probability of persistence for more than 100 years (Traill *et al.*, 2007). The point should be stressed that *in situ* conservation is not an inexpensive option compared to *ex situ* conservation. Even once designated, a reserve will require active and consistent population monitoring, habitat management and site security. This will require a continuing high level of commitment in terms of financial and personnel resources. If the target taxon becomes extinct in the reserve, reintroduction from germplasm conserved *ex situ* is also usually an expensive option requiring extensive research and is not always successful. Therefore, if the target taxon is lost from the reserve, all the resources already expended in establishing and running the reserve will have been wasted. Further, some taxa may not be suitable for conservation in genetic reserves, such as those: (i) species found with very low density, disparate populations, such as many tree species, which would not form a viable population at any spatial scale compatible with a genetic reserve establishment; (ii) strict pioneer species forming metapopulations where the parents and offspring do not necessarily share the same location, e.g. wild lentil (*Lens culinaris* subsp. *orientalis* (Boiss.) Ponert) in Turkey (Karagöz, 1998) or black poplar (*Populus nigra* L.) along river banks in Europe (Lefèvre *et al.*, 2001), or weedy taxa associated with human disturbance,

which is the case for many CWR related to major crops and (iii) highly threatened species with suboptimal population numbers that would not be viable *in situ* (Maxted *et al.*, 2008b). In these cases, *ex situ*, *in situ* restoration or *in situ* conservation in less formally designated reserves, possibly as part of an on-farm conservation project, is a more appropriate solution.

The following factors should be considered to help ensure reserve sustainability:

Development – The reserve should be established in an area where there are no imminent threats from future human development, such as road construction, encroachment from nearby settlements or major hydroelectric projects (IUCN, 2017). For example, the construction of a dam in the same watershed as that of the reserve may affect the water table of the region, resulting in the drying up of wetland habitat or flooding and complete loss of the target populations. An assessment of future government proposals for regional development should be made routinely before locating protected areas.

Climate change – Humans now for the first time in the history of the Earth have the technical facility to radically change ecosystems, even the whole planet (Jackson *et al.*, 2013). The current climatic change due to human-induced pollution is resulting in global warming that is likely to have a dramatic effect on the distribution and ranges of species. Certain species or communities are more likely to be affected by climate change than others, including species found in alpine, coastal, arctic or marginal habitats (Parmesan and Yohe, 2003; Jarvis *et al.*, 2008a; Seppä *et al.*, 2009). An assessment of the differential impact on future climate change on populations proposed for *in situ* conservation should now be routine prior to reserve establishment (see Chapter 6).

Legislation – Once conservation sites are designated, legislation should be enacted to ensure they are maintained and not developed in a manner that would be to the detriment of the target taxon. Although legislation should assist with the security and thus sustainability of the site, experience has shown that it can be circumvented if the political will is sufficiently strong. Evidence of this is provided, for example, by the destruction of very rare CWR in protected areas in Turkey by road building programmes (Maxted, 2012).

Metapopulations and ex situ *conservation* – If multiple reserves are established for each target taxon, they form collectively a metapopulation that enhances their genetic resilience, providing gene flow between populations is possible. It also means the loss or degradation of any one population or reserve will have less overall impact as there will be the possibility to restore the negatively impacted population or add a replacement reserve site into the reserve network. A sample of the target taxon should always be conserved using appropriate *ex situ* technique(s) as a backup to the *in situ* conserved population; this is of particular importance for endangered or restricted distribution taxa.

Flexibility of the management plan – Unforeseen threats and natural catastrophes cannot be planned for and therefore their effects cannot be foreseen. It is fundamental therefore that monitoring is rigorous, and that the management plan is sufficiently adaptable to allow feedback from the monitoring to be acted upon. The management plan may also need to be flexible in terms of the interaction with local stakeholders. To meet the local community's development aspirations the local community should be involved when reviewing the balance between biological expedience and social obligation (IUCN, 2017).

Population exploitation – Further a temptation may be to designate a genetic reserve remote from human habitation to try to avoid conservation/human conflict, but the more remote the site the less likely the germplasm it contains is likely to be actively utilized. Lack of utilization may be correlated with a depreciated value of the population being conserved; therefore, establishment of a utilization plan can be an essential part of a reserve's sustainability.

The factors to consider when locating a genetic reserve are summarized in Box 8.4.

8.2.4 CWR Population Inclusion in Regional or Multi-national Networks

To ensure the scientific validity of any regional or multi-national CWR *in situ* conservation networks and to maximize the CWR diversity conserved *in situ* globally, it is advisable that populations nominated for inclusion in a network meet several criteria. In the context of establishing a European CWR *in situ* conservation network, Maxted *et al.* (2015) proposed a set of criteria, which could be adapted for other regional or global use:

- The CWR population is native at that location, or if introduced, has existed at that location for at least 15 generations.
- The CWR population contains distinct or complementary genetic diversity, ecogeographic diversity as a proxy for genetic diversity or specific traits of interest (e.g. an example of high importance to the CWR user community is beet necrotic yellow vein virus (BNYVV) resistance in *Beta vulgaris* subsp. *maritima* populations from the Kalundborg Fjord area, Denmark; Capistrano *et al.*, 2014) that enhance the overall value of the network.
- The CWR population should not be threatened, so there is a good chance of long-term survival (conventionally thought to mean having a >95% probability of persistence for more than 100 years; Traill *et al.*, 2007), and threats such as development or climate change will need to have been assessed/modelled and found negligible.
- The CWR population is actively and sustainably managed as a long-term *in situ* conservation resource according to the minimum quality standards for genetic reserve conservation (see Box 8.5, and for further discussion see Iriondo *et al.*, 2012).
- The CWR population is sampled and held in a backup *ex situ* facility every 15 generations.
- The CWR population is accessible for research or utilization in accordance with the ITPGRFA from a specified *ex situ* facility as part of the Multilateral System.
- The CWR population is nominated for network inclusion by an appropriate national agency.

Box 8.4 | **Summary of Factors to Consider When Locating a Genetic Reserve**

(Dulloo *et al.*, 2008).

- *Level and pattern of genetic diversity*: The magnitude and geographic pattern of the target taxon's genetic diversity is obtained by genetic characterization and a set of populations/sites selected that maximize within-population as well as between-population genetic diversity. Where genetic information is lacking, ecogeographic or gene-ecological proxies can be applied.
- *Presence in existing protected areas*: Priority is given where possible to establishing genetic reserves in existing PA as such sites can be adapted to conserve genetic diversity through amendment of the management plan and the cost of establishing new PA can be avoided. Where there are no local PA containing the target taxon, either new PA will need to be established or a less formal means of extra PA site established to conserve the target taxon.
- *Size of reserves*: The size of the reserve will depend on the characteristics of the target species and its symbiotic relationships, but size is often dictated by the MVP size for that taxon.
- *Number of populations*: The number of populations representing the target taxon's genetic diversity will vary from taxon to taxon, and the higher the number the more representative genetic diversity to be included in the *in situ* network. Populations known to contain important genetic, chemical or phenotypic variants should be prioritized for inclusion; however in general at least five populations should be selected for inclusion in the genetic reserve network.
- *Number of individuals within the population*: Within each reserve site the target taxon population should be above the taxon's MVP to maintain genetic diversity and population health in the long term. If the MVP is unknown, 5000 individuals is suggested, though practically a figure of 15 000 individual plants per taxon is more desirable and if below this level then reintroduction or assisted should be employed.
- *Site expedience and practicality*: The gap analysis will have identified potential sites suitable to establish as genetic reserves, but it may not have considered current and potential future threat to the site, for example the possible impact of climate change to the population or a local community that have routinely wild harvested the target taxon from the potential site, so sustainability analysis of a competing site should be undertaken. Further it would be inadvisable to attempt to establish a genetic reserve in an area of civil unrest or very remote location, which would hinder access by the conservation team to undertake routine population monitoring.
- *Political and socio-economic factors*: Rarely are genetic reserves located or designed solely based on biological considerations alone. The availability and practicality of potential sites and meeting the needs of local communities must be considered. Cost–benefit analysis of competing sites should be undertaken.
- *Single versus multi-taxa conservation*: The establishment of a network of national sites each selected to contain multiple taxa is likely to be more cost-effective overall for a national conservation programme rather than establishing large numbers of single taxon genetic reserves, unless a high-priority taxon or trait is only located in a single, isolated population.

Box 8.5 | Summary of the Criterion Relates to the Minimum Quality Standards for Genetic Reserve Conservation of Designated CWR Populations Proposed by Iriondo *et al.* (2012)

- Location
  - Located following rigorous scientific process
  - Located in a protected area network.
- Spatial structure
  - Clear boundaries of the genetic reserve should be defined.
  - Sufficient extent to conserve CWR populations and natural processes.
- Target taxa
  - Genetic reserves are designed to capture maximum genetic diversity.
  - Demographic survey of target CWR taxa.
- Populations
  - Population sizes are large enough to sustain long-term populations.
- Management
  - Site recognized by the appropriate national agencies.
  - Management plan formulated.
  - Monitoring plans are designed and implemented.
  - Local community involved in site management.
  - Clearly defined procedure to regulate the use of genetic material.
- Quality standards for the protected areas selected for the establishment of genetic reserves
  - Site has legal foundation.
  - Site management plan acknowledges existence of genetic reserve and genetic conservation within the site.

Once national agencies in the country nominate a site for inclusion in a regional or multi-national network, it would be the responsibility of the authority overseeing the network to review whether the selection criteria for inclusion in the network are met and a decision made as to whether the nominated site is added to the network. Sites included within a network would need to be reviewed periodically to ensure they continue to meet the inclusion criteria, and where this is not the case, recommendations for changes would be made; the sanction of de-selection from the network would be available.

Individual CWR reserves will pragmatically often contain multiple priority species selected using gap analysis techniques, and designated sites may often be found in CWR hotspots. However, the sites included will also need to be complementary to build towards the global network, so may in certain cases contain single CWR populations required to ensure the breadth of CWR diversity coverage. Ideally, designated CWR *in situ* conservation sites would occur within formally designated protected areas for ease of establishment, however, many CWR populations of value occur outside of protected areas. Therefore, a site may be included in a network whether it is within or outside a formally designated protected area. In both cases, active and sustained *in situ* CWR conservation management commitment is pivotal to the long-term success of the *in situ* conservation network.

Experience thus far in working towards regional or multi-national CWR *in situ* conservation networks is that the sites proposed do not meet the set of criteria nominated for inclusion in a network (Maxted *et al.*, 2015) or the quality standards for CWR genetic reserves (Iriondo *et al.*, 2012) because their designation has been ad hoc and opportunistic rather than because of any deliberate scientific decision. Also, they are independent of each other and so do not together constitute an integrated network of genetic reserves. The FAO Commission on Genetic Resources for Food and Agriculture (CGRFA) is leading a discussion of the options for establishing a global network for *in situ* PGRFA conservation (FAO, 2013c). Such a global network would provide the necessary platform to raise awareness of the social and economic value of *in situ* conservation (and on-farm management) in partnerships with national and regional level activities, but it is apparent that such a network requires further debate, a governance structure, an evidence research base and financial security, so is unlikely to be put in place soon.

8.3 Reserve Design

The previous procedures have highlighted how the position of the reserve site is selected. Once selected, the reserve at the site can be designed to maximize genetic conservation of the target taxon or group of taxa and a mutually beneficial relationship established with the local community, farmers and breeders, and other groups of wild plant users.

8.3.1 Reserve Structure

The most widely used model for structuring a genetic reserve is that used in Biosphere Reserve design which divides the reserve into designated zones: core, buffer and transitional zone (Cox, 1993) (see Figure 8.2). The three components may be characterized as follows:

- *Core zone* – The main objective is to protect populations or meta-populations of the target taxon, and it should be sufficiently large to contain at least the MVP of 5000, but ideally 15 000–25 000 potentially inter-breeding individuals, to secure the

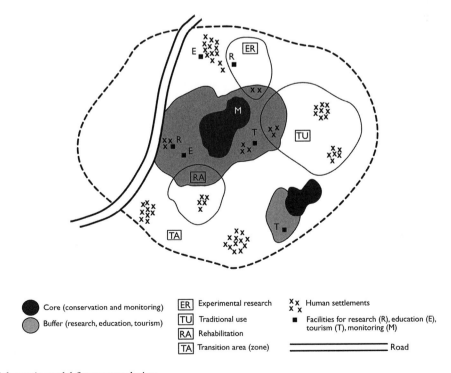

Figure 8.2 Schematic model for reserve design.

long-term health of the target taxon population. This will involve conserving the target population, its habitat and its ecological processes. Only non-destructive scientific monitoring and tested management prescriptions are allowed in the core zone.

- *Buffer zone* – This zone is peripheral to the core zone of the genetic reserve. It has two major advantages:
  - In the biological context, they reduce harmful adverse human impacts, edge effects and help maintain the integrity of the central core by promoting replenishment/enhancement of the target taxon population through immigration from the buffer to the core zone.
  - In the social context, it compensates the local community for any loss of access or harvesting rights to the core zone, so safeguards traditional cultural practices, land rights, recreational and conservation practices, while increasing local conservation-related employment through serving additional educational and ecotourist businesses.
- Activities within the buffer zone must be compatible with the protection of the core zone and do not conflict with the site's/population's conservation objectives. The sort of activities that might be permitted in the buffer zone include: experimental scientific research (e.g. testing of management regimes), education, ecotourism, traditional livestock grazing and crop cultivation, and sustainable harvesting of natural resources. However, these activities will need monitoring and, if necessary, regulated or restricted to ensure they do not conflict with the site's/population's conservation objectives.
- *Transitional zone* – In the transitional zone sustainable development and scientific research are permitted. This area acts as a transition area between the conservation area and external areas; it may include human settlements and agricultural fields.

8.3.2 Reserve Size and Number

In practice reserve size is often dictated by the relative concentration of people and the suitability of the land for human exploitation. However, in the 1970s–1980s there was much theoretical debate on the relative merits of a large single reserve or several small reserves, the so-called SLOSS (Single Large Or Several Small) debate. The debate focused on answering the simple question: is it better for biodiversity conservation to have one large reserve or several small reserves equivalent in area? However, practically large and small reserves both have advantages and disadvantages (see Table 8.2), and the size established will be based on expediency and target taxon characteristics.

The arguments over which was more suitable stemmed from ecological considerations of island biogeography theory and the relationship between species number and area (MacArthur and Wilson, 1967). Much original habitat today exists as remnants or fragments of pre-human climax vegetation, surrounded by anthropogenic modified environment composed of agricultural fields and urban developments. These isolated habitat fragments are analogous to 'oceanic islands' surrounded by the sea. As such, the number of species on an 'island' is an equilibrium between the rate of species extinction and colonization of the island (see Figure 8.3). The rate of immigration is negatively correlated with the distance between the island and the new species source; the more isolated, the lower its immigration rate. In Figure 8.3, Hunter (1990) demonstrates this with the curve for remote islands (far) being lower than the curve for islands that are near the mainland (near). Whereas extinction rates are correlated with island size, populations on large islands tend to be larger and thus less vulnerable to extinction. In Figure 8.3 the extinction curve for large islands is lower than the curve for small islands. Therefore, in practice when designing a genetic reserve, the rates of extinction should equal the rate of colonization per unit area, to maintain the equilibrium and sustain the population over the long-term.

Island biogeography theory has been criticized because in practice studies have demonstrated that several small sites can have as many species or, in some cases, more than a single site of an equivalent area. Traditional island biogeography theory ignores several traits which influence species number,

Table 8.2 **Advantages and disadvantages of relatively large versus small reserves**

Reserve size	Advantages	Disadvantages
Single large	• More ecogeographically diverse • Minimal edge effect • Easier to maintain species and population diversity • Maintains physical integrity of ecosystem (e.g. watersheds, drainage system)	• Impossible to cover all genetic diversity of widely distributed species
Several small	• Site each reserve in a distinct environment • Conservation value of multiple small reserves can be greater than the sum of its individual components • Annuals often naturally found in dense but restricted stands • Usually sited near urban areas so good for public awareness	• Usually sited near urban areas so need more effective buffering • Require more intensive management & monitoring • Impossible to include real habitat diversity • More susceptible to human or natural trauma • Too small or too isolated a population less likely to remain viable

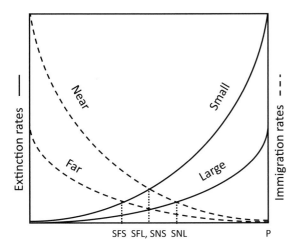

Figure 8.3 Island biogeography theory (Hunter, 1990). In this example, the number of species for four equilibria are represented as follows: SFS, number of species on a far, small island; SFL, far, large island; SNS, near, small island; SNL, near, large island. P is the total number of species that could potentially immigrate to the island from a nearby landmass.

distribution and, most importantly, persistence, notably differences in habitat structure and complexity between sites, and the characteristics of a species or functioning ecosystems (Lomolino *et al.*,

2006; Franzén *et al.*, 2012; Guo, 2014). However, the SLOSS debate was over-shadowed, when discussing the optimum size of a reserve, by the realization that it is more critical to focus on the area which is required to sustain an MVP of the target species (Shaffer, 1981; Traill *et al.*, 2007). As such, whatever the reserve size, if the reserve contains a population of the target species that exceeds the species MVP, the population should be self-sustaining.

The number of reserves designated for species, however, remain an important consideration. If the objective is to conserve overall genetic diversity of the target taxon, then it will almost always be necessary to establish more than one reserve site. Any single population will not contain all the diversity of the whole taxon, unless of course only a single population exists. It is generally assumed that ecogeographic diversity, that is diversity in habitat and geographic location, is correlated with genetic diversity (Maxted *et al.*, 1995), an assumption that appears to hold true for most species, and unless genetic diversity studies indicate otherwise, is assumed to be true for all species. Therefore, by establishing a single reserve it will be impossible to conserve the entire range of genetic diversity.

Multiple reserves will be required to ensure the maximum number of alleles associated with specific habitats or extremes of geographic range are conserved within the *in situ* reserve network.

8.3.3 Minimum Viable Population

However, the key factor in designing a genetic reserve is to ensure the MVP for the target species is at least present and maintained. The MVP defined by Shaffer (1981) is 'the minimum viable population for any given species in any given habitat is the smallest isolated population having a 99% chance of remaining extant for 1000 years despite the foreseeable effects of demographic, environmental and genetic stochasticity and natural catastrophes' or at least having a '>95% probability of persistence for over 100 years' (Traill *et al.*, 2007). Shaffer lists the forms of stochasticity as: (a) *genetic stochasticity* – relates to loss of allelic diversity in populations, particularly a problem of small populations that results in changes in gene frequency, increased inbreeding and accumulated mutations (Van Dyke, 2008); (b) *environmental stochasticity* – refers to fluctuations in external parameters such as herbivore populations, prevalence of disease, short-term climatic changes or changes in the availability of resources; (c) *demographic stochasticity* – relates to the random fluctuations in demographics such as germination and senescence rates, or fecundity, or sex ratio in dioecious species (Van Dyke, 2008) and (d) *natural catastrophes* – natural events like fires, floods or droughts possess the ability to extirpate small populations instantaneously (Shaffer, 1981). It is important to note that these factors are not necessarily functionally independent of each other and can work together to threaten long-term population survival (Burgman and Neet, 1989).

If the population number falls significantly below this number, then allelic diversity will be reduced, which is likely to have negative effects on survival (Shafer, 1990). Maintenance of the MVP will:

- maintain the population's genetic diversity;
- maintain the population's potential for evolutionary adaptation;

- ensure the population is at minimal risk of extinction from demographic fluctuations, environmental variations and potential catastrophe, including over-use.

It is difficult to make precise estimates of the MVP for species in general because each will vary depending on factors such as: ecogeographic breadth, ecotypic differentiation, distribution of genetic diversity across ecogeographic range, breeding success, breeding system, predation, competition, disease, genetic drift, founder effect and even local natural catastrophes (stochastic events), such as fire, drought and flooding, which all may affect population longevity. MVPs may be estimated by: (a) *experimentation* – involving monitoring target taxon populations over time and recording population numbers and persistence, though this may take a long time to provide an answer and even then, the MVP may vary from location to location (Lindenmayer *et al.*, 1993); (b) *ecogeographic zonation* – by analyzing the frequency of extinction and re-colonization in particular ecogeographic zones and determining the length of time the smallest population has persisted an MVP can be estimated (Shaffer, 1981); (c) *modelling* – estimating the probability of persistence based on specific characteristics of the target taxon and its representative population – this is the basis of Populational Viability Analysis discussed below (Lindenmayer *et al.*, 1993) and (d) *genetic principles* – a generalized number of individuals required to maintain genetic fitness and evolutionary potential (Shaffer, 1981).

Few species have an established MVP and even fewer of those are plants, so genetic reserve designers require guidelines on what population sizes they should be aiming for. Therefore, Frankel and Soulé (1981) using genetic principles propose 500–2000 individual plants per population, Hawkes (1980) at least 1000 individuals and Lawrence and Marshall (1997) 5000 individuals to retain the genetic integrity of the target population conserved *in situ*. Further, Franklin (1980) proposed the 50/500 rule suggesting 50 individuals were required to maintain short-term fitness and prevent inbreeding, while 500 were necessary to maintain longer term evolutionary potential. Despite this rule still being widely cited, Frankham *et al.* (2014) recommend these values be at least doubled if they are to be accurate; criticizing the earlier proposal because use of the phrase 'short term' was too vague and that data used to reach an estimate of 50 individuals was derived from captive animals and therefore did not sufficiently acknowledge inbreeding depression in other taxonomic groups. However, whatever generalized MVP figure is proposed, it should be remembered that it is a theoretical minimum population figure. Stochastic catastrophes such as severe drought, hurricanes and fire, or the appearance of a destructive new pest and diseases do occur, so ideally the reserve designer should always aim to site a reserve where there is a population well in excess of the MVP. Having made this point, Dulloo *et al.* (2008) propose as a rule for plant target taxa that the core of the reserve should contain at least 5000 individuals to sustain population viability and genetic integrity.

If population sizes become reduced below this level, allelic diversity will be reduced, and the population will be more prone to the effects of inbreeding depression and genetic drift, both of which reduce biological fitness and survival (see Chapter 4). Small populations, particularly of annual taxa, are particularly vulnerable to stochastic (determined purely randomly) and catastrophic events (Dulloo *et al.*, 2008). There are examples known of populations diminishing to a very small size and yet surviving and expanding later. These 'phoenix' populations that are substantially below 5000 individuals will undoubtedly have suffered extensive genetic drift during the process, and this genetic bottleneck may impact long-term population survival (Lawrence and Marshall, 1997). It is also noticeable that some fruit trees are found in very dispersed populations where regular gene flow between individuals seems unlikely, but here the species are long-lived perennials, and even irregular gene flow is enough to maintain diversity.

Whatever the characteristic of the species, the genetic reserve(s) will logically be established in a site with a healthy target taxon population, and it should aim to maximize genetic diversity at that location and within the network of sites. If the taxon is inbreeding

(reproduces by self-pollination) or annual, as many of the wild relatives of cereal species are, then several small reserves would be more appropriate in maximizing genetic diversity capture. On the other hand, few, larger reserves would be more appropriate for highly outcrossing (reproduces by cross-pollination) or perennial species, such as many temperate forest trees, which tend to have more genetic diversity distributed within populations (Dulloo *et al.*, 2008). Ideally, there should be at least two genetic reserves at two different sites, irrespective of the genetic architecture of the taxon. This will reduce the risk of extinction of the target taxon within the reserves due to a catastrophic event, such as a hurricane.

8.3.4 Populational Viability Analysis

Population viability analysis (PVA) is a modelling tool that uses target taxon population characteristics to estimate the probability that a population will persist for a given amount of time. It is conceptually closely associated with MVP as the latter is often part of the result produced by the former (Lindenmayer *et al.*, 1993). PVA uses computer simulation modelling of demographic, genetic and ecological processes for a specific species/habitat combination; for examples, see VORTEX (www.vortex10.org/Vortex10.aspx; Lacy, 2000). As such, PVA is a valuable tool for investigating current and future risk of endangered plant population decline under specific scenarios of human-mediated activities, locally and globally, which may compromise the population's ability to reproduce successfully and/or survive (Morris and Doak, 2003). Despite PVA's popularity, particularly among mammal conservationists, Lindenmayer *et al.* (1993) criticize its usefulness in being able to estimate the impacts of several factors on a population, identify the most important threats to persistence and guide management decisions. However, Lacy (2000) comments that biodiversity is inherently unpredictable in its detailed behavior and we very rarely fully understand its precise mechanics, and Reed *et al.* (2003) demonstrate that the high uncertainty in many of the inputs means there is significant variance in results, even for the same species. Frankham *et al.* (2014) specifically criticize PVA for bypassing or inadequately considering genetics, leading to systematic underestimations of inbreeding depression and failure to incorporate evolutionary potential. Despite these criticisms, PVA has continued to act as an established tool for many animal conservation programmes, though it has yet to be applied widely for plant *in situ* conservation. Possibly one of the main values of using PVA is to force the conservationist to collate and critically analyze the diverse demographic, genetic and ecological information required, and acknowledge the data gaps when formulating conservation strategies (IUCN, 2017).

8.3.5 Associated Species

Although the prime objective of the genetic reserve is to maximize total genetic diversity of the target taxon in a single site or network of sites, this can only be achieved if we understand the auto-ecology (single species) and syn-ecology (multiple species and their interactions) of the target taxon. Therefore, the core of the reserve should contain not just the MVP of the target taxon but viable populations of the associated species, essential pollinators and dispersers of the target species.

8.3.6 Edge Effects and Reserve Shape

The edge of the reserve will have a different micro-environment from the interior of the reserve. These edge effects include micro-climatic changes in wind, temperature and light. The edge of the reserve is also more vulnerable to fires and invasion by exotic animals and plants and is more liable to human interference. Therefore, the area/edge ratio of the reserve should be maximized to reduce these deleterious edge effects. Round reserves and comparatively large reserves have a larger area/edge ratio than long, elongated or small reserves, so relatively round or larger reserves are preferable unless dealing with islands or narrow valleys with unavoidably defined physical boundaries.

8.3.7 Fragmentation

Fragmentation of the reserve by roads, rail tracks, dams, cultivated fields, etc. will reduce the area of habitat available to the target taxon, lead to habitat isolation and increase edge effects. Shade-tolerant wildflower species of temperate woodland and late successional tree species of tropical forests are threatened by habitat fragmentation. Fragmentation generates physical barriers to species dispersal and colonization thus effectively reducing population size. This is particularly important in plant species which exist as meta-populations and whose survival relies on the colonization of new sites. Associated animal species, which are essential for pollination and dispersal of seed, may not cross open landscape or roads because of the danger of predation. Widespread species that exist in large populations may be subdivided into smaller populations by fragmentation. These smaller populations will be more vulnerable to inbreeding depression, genetic drift and stochastic events.

8.3.8 Habitat Corridors and Stepping Stones

The point is made above that for plant genetic conservation there is a need to establish networks of genetic reserves to maximize genetic diversity representation across the network; however, this means that individual genetic reserves may not all contain 10 000 individuals plus and so may be vulnerable to genetic erosion and loss of allelic diversity. One method of avoiding this fate is to connect fragmented habitats within a reserve or reserves themselves by habitat corridors or stepping stones that facilitate gene flow between subdivided or isolated populations (Dulloo *et al.*, 2008). Habitat corridors are continuous strips of habitat which connect isolated habitat fragments, whereas stepping stones are discontinuous habitat patches that are close enough for pollen or seed and therefore geneflow to pass between otherwise isolated habitat fragments (Saunders and Parfitt, 2005). The relative effectiveness of both corridors and stepping stones will vary depending on the target taxa and their life form characteristics, the distances involved and the relative size of the 'sink' populations (those existing in the corridor or stepping-stone habitat) (Wiens, 1989; Haddad, 2000). The larger the 'sink' population the more likely geneflow will occur along the corridor or between habitat patches reinforcing population and genetic diversity. Linking multiple populations in this manner means they can be effectively managed as one metapopulation level, further maximizing diversity. Habitat corridors or stepping stones, however, have conservation drawbacks, both have a high edge to area ratio, they may facilitate rapid distribution of pests and diseases between reserves, and are not feasible in highly fragmented or disturbed habitats. The concept of corridors and stepping stones can be applied to link isolated populations with a reserve or on a larger scale by connecting individual reserves within a network.

8.4 Reserve Management

Habitat and species management is fundamental to *in situ* genetic conservation. Virtually all habitats, and certainly those rich in plant diversity, are likely to receive some form of anthropogenic management. The site will have been selected because it contains a healthy target population(s), so the management will most often be a case of sustaining that human–site–population(s) interaction. Therefore, any form of protected area will require the continuation of the site's management to ensure a healthy target population(s) or habitat is maintained. The most minimalist management would involve monitoring alone, just to check the taxon or habitat remains stable at the site, and providing this is the case, no further management intervention may be required. However, if the target populations or habitat is suffering deleterious change, then more proactive management of the site is required to reverse the changes. Active management intervention is more likely to be necessary where the protected area:

- is too small or genetically homozygous to support an acceptable population level or range of species;
- is too small to contain enough levels of natural disturbance to maintain evolutionary processes;

- is too fragmented and isolated to permit natural immigration to balance local extinctions;
- is surrounded by hostile anthropogenic environments that produce invasive species (weeds, diseases and generalist predators) and degrading processes (siltation and pollution); or
- is under pressure for development, for release of their natural resources for human use or for use as agricultural lands to feed rapidly increasing human populations.

In each case, active management of the site is required to maintain viable populations, species diversity and habitats, stimulate necessary disturbance, translocate individuals or assist propagation, remove deleterious factors and establish practical compromises with local subsistence or development goals.

The aim of genetic reserve management is to ensure the long-term maintenance or enhancement of the genetic diversity of the target CWR taxa within the reserve (Maxted *et al.*, 2008c). However, it should also be:

- maintaining maximum genetic diversity of key associated species;
- promoting general biodiversity conservation and minimizing threat to all levels of diversity;
- maintaining natural ecological and evolutionary processes that are not deleterious to the target taxon gene pool;
- ensuring that appropriate, but minimally intrusive, management interventions enhance target taxon diversity;
- promoting public awareness of the need for genetic and protected area conservation;
- facilitating the linkage of conservation to sustainable usage by ensuring that diversity is made available for actual or potential utilization.

Maintaining a population's genetic diversity is also central to the IUCN definition of a viable population, which they define as one that: (i) maintains its genetic diversity; (ii) maintains its potential for evolutionary adaptation and (iii) is at minimal risk of extinction from demographic fluctuations, environmental variations and potential catastrophe, including overuse (IUCN, 1994).

8.4.1 The Need for a Management Plan

To help ensure efficient and effective population management, a management plan is required to guide populations, target taxon/taxa and habitat sustainability at the selected site. Active management interventions are likely to be required, particularly if the target taxon/taxa are to be maintained in pre-climax communities, where active intervention is required to halt natural succession to a climax community that might ultimately exclude the target taxon/taxa, succession being the universal, natural process of directional change in vegetation during ecological time, which runs from bare soil through intermediate vegetation types culminating in a climax community when the ecosystem achieves directional stability (Krebs, 2001). As many important plant species are found in pre-climax communities, the only effective way to organize management is via a carefully constructed management plan. Maxted *et al.* (2008b) argue that maintenance of genetic diversity requires more intense management as normal population characteristics (density, frequency and cover) could at least theoretically be maintained while losing genetic diversity.

As genetic reserves are often located pragmatically rather than scientifically, the target population may be too small or fragmented and isolated to support the ideal MVP or permit natural immigration to balance local extinctions. Therefore, initial management will necessarily focus on increasing target populations to viable levels or even the translocation of individuals between management areas. These issues can be reviewed and recommendations made within the management plan. The available reserve site may also be surrounded by hostile anthropogenic environments that result in regular introduction of invasive species (weeds, diseases and generalist predators) and degrading processes (siltation and pollution). Here, again, intervention management outlined in the plan is necessary to minimize or remove such negative influences.

The management plan describes the physical, socio-political-ethnographic and biological context of the reserve in relation to the target taxon and its population within the reserve, so provides a

foundation for site management. It sets out the general conservation objectives and specific goals of the individual reserve and how that reserve sits within institutional, national, regional and possibly global conservation strategies, so helping ensure consistency of implementation across geographic scales. It should anticipate any natural or anthropogenic conflict or problems associated with managing the reserve. It will describe the management objectives and, in as much detail as is possible, the management interventions required for achieving these objectives, as well as the monitoring practices to be implemented. It will assist in organizing human and financial resources, and act as a training guide for new reserve staff (Worboys *et al.*, 2015). It will facilitate communication and collaboration between the individual reserve site and other genetic reserves, protected areas and *ex situ* conservation facilities. It will also act as guidelines for the use of the target taxon and its genetic diversity, and help to raise public awareness of the importance of the specific target taxon and plant conservation as a whole (Maxted *et al.*, 2008b).

8.4.2 Autecology and Synecology

Prior to formulating a management plan for a target taxon, the conservationist must understand the taxon or taxa's biology (autecology) and its ecological relationship with other species and the abiotic environment (synecology). Much of these data should already have been acquired during the ecogeographic survey of the taxon, but missing information can be sought from published literature, grey literature (unpublished reports) and direct observation in the field. The types of information available should be recorded, and the acquisition of knowledge to fill any gaps should be incorporated into the research element of the management plan.

Taxonomy. Published literature on the taxonomy of the target taxon and its close relatives can be collated from revisions, synonyms, descriptions, Floras and identification keys (Castañeda-Álvarez *et al.*, 2011; see Chapter 6). The taxonomic classification and nomenclature used for the target taxon should be that which is internationally accepted and will provide an entry point for broader biological information for the taxa being conserved.

Reproduction Biology. Knowledge of the breeding system of the target taxon is an essential prerequisite to genetic conservation. The nature of the breeding system affects both the distribution of genetic diversity within and between populations. Species that reproduce by self-fertilization (inbreeders or autogamous species) generally contain lower levels of within-population genetic variation than those that reproduce by cross-fertilization (outbreeders or allogamous species). However, autogamous species generally have higher levels of inter-population variation. When plants are dioecious (male and female flowers present on different plants), an imbalance of sex ratios within a population may result in restricted reproduction and reduced effective population size.

Phenology. An understanding of the phenology of the target taxon is important for several reasons. The planning of any research or monitoring will require prior knowledge of when the species flowers, sets seed, germinates, etc. Having plants in flower may be essential for the identification of the taxon when undertaking demographic monitoring of the population. Knowledge of the timing of the collecting window (that period when the seed is ripe but just before dispersal) of the species will also be important when organizing the collection of germplasm for *ex situ* conservation.

Abiotic Features of the Reserve. The basic abiotic features, for example, geology, climate, hydrology of the reserve site and the region should be known.

Habitat. The essential habitat requirement of the target species should be determined by fieldwork as well as from the ecogeographic survey and literature. These requirements may vary according to the stage in the species's life cycle; for example, tree seedlings may require an opening in the forest canopy to establish, whereas mature trees do not. Threats to the habitat, such as grazing from feral animals and livestock or pollution from a chemical plant, should be known and if possible, their effect minimized by appropriate management practices.

Ecosystem Dynamics and Interactions. Knowledge of the associated species that are essential to the survival of the target species, such as specific pollinators and seed dispersers, is critical. The reserve may have to be enhanced or managed to encourage visits by pollinators and dispersers, without which the target taxon could not survive. Strong species interactions also include types of mutualism, such as the specialist pollination system between tropical mistletoes (Loranthaceae) and African sunbirds (*Nectarinia*), and nitrogen-fixing *Rhizobium* species and legumes. Other such interactions might include: do wild or introduced herbivores graze on the target taxon, is a certain level of grazing necessary to sustain the target populations within the habitat, which species compete with the target taxon for nutrients, pollinators or light, or is the target taxon reliant on another species, such as a keystone species? Keystone species play a disproportionately important role in its overall function or process of an ecosystem (Wagner, 2010). They are essential to the overall integrity of the ecosystem and to the survival of the other species found in the ecosystem. Trees are often keystone species because they physically dominate an environment and provide a rich food resource.

Potential Threats to Populations of the Target Species. Populations of the target taxon may be threatened directly within the protected area by various factors, and obviously it is important to have as clear a conception of potential threats as possible. If the threat is unacceptably high, an alternative site should be selected. However, if the risk is deemed low, the reserve manager should be aware of potential threats, any actual threats should be monitored, and the management plan altered accordingly to minimize or eliminate the threat. Potential threats include: incursion of the habitat by invasive plants and animals, unsustainable harvesting (for timber,

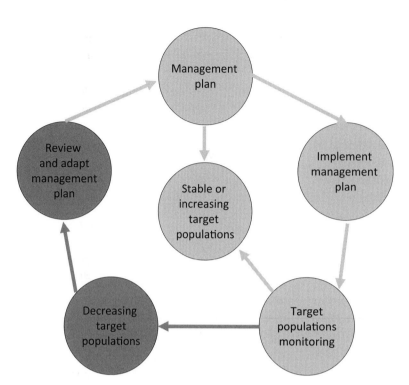

Figure 8.4 The genetic reserve management cycle.

firewood, medicinal herbs), agricultural encroachment from neighbouring villages, fragmentation of populations by road building and other human interventions, lack of pollinators/seed dispersers, changes in habitat, for example, drainage or disturbance, and local stochastic events such as fire, hurricane or flooding

Further Research. Inevitably there will be limited knowledge of certain aspects of the target taxon's life history, ecology, threats and the other types of necessary information listed above. Research to fill these knowledge gaps should be incorporated within the management plan.

8.4.3 Factors to Consider When Preparing a Management Plan

The management plan should be based around a clear conservation goal(s) and a statement of measurable objectives (Iriondo *et al.*, 2008). It should comprise a complete framework for all future work on the

reserve, including any legal or managerial constraints. It should also be 'flexible' and 'adaptive' to allow for natural fluctuations in populations and to incorporate any amendments resulting from monitoring, research or interaction with the local community. The plan should be reviewed on a regular basis so that any amendments can be incorporated (Figure 8.4). The management plan may be broken down into a series of annual plans, whereby the activities for the calendar year are set out.

There are certain points that need to be considered when devising a management plan. The management plan is however broadly focused on maintenance or enhancement of the target taxa within the genetic reserve and more general restoration of the flora and fauna of the site.

Management Goals. Although each reserve is likely to have target taxon as well as site-specific goals, there is one general management goal that is likely to be applicable for all genetic conservation in any protected area, that is 'to maintain maximum genetic diversity of the target taxon along with key associated species within the reserve'. Along with this, secondary goals are likely to include: promotion of general biodiversity conservation, maintenance of local ecological and evolutionary processes, implementing appropriate but minimally intrusive site management, minimizing the external threats to biodiversity, promoting public awareness for the need of genetic and protected area conservation, and ensuring diversity is available for actual or potential utilization.

Management Plan Function. To achieve these goals it will be necessary to write a specific management plan for each protected area. This sets out the objectives and purpose of the reserve, describes the biotic and abiotic environment as well as the target taxon within the reserve, and discusses management and monitoring regimes. Ideally it should also anticipate any conflict or problems associated with managing the reserve. Some management plans are used to organize human and financial resources and act as a training guide for new staff. A formally written management plan will help ensure consistency between the individual reserve and national and regional conservation plans or the policies of parent organizations, and will facilitate communication and collaboration between genetic reserves sites.

Monitoring Activities. This section should include the species, populations and habitats that are to be monitored and their specific locations within the reserve. It should detail when the monitoring is to take place, frequency and type of monitoring, and the sampling strategy to be used.

Management Prescriptions. Establishes any active management interventions at the site, including grazing regimes, assisted propagation protocols, invasive species control or burning, and their frequency should be stated.

Research Activities. As well as general research on the target taxon to assist with writing the

management prescription, there may be a need for specific research associated with the target taxon within the protected area site. The research could be conducted in collaboration with scientific institutes or universities with interests in the same field.

Education and Public Awareness. To promote the conservation value of the reserve and gain public support, the management plan should include details of educational and public awareness initiatives within its remit. This could include open days for schools and colleges, production of leaflets for circulation, erection of notice boards, directed walks through the reserve, construction of a field centre, etc.

Personnel and Resource Management. The allocation of staff and resources to implement the various projects within the areas of research, monitoring, management and education requires careful consideration. Costs for purchasing or hiring equipment, consultancy fees and contracts should all be costed and budgeted for. Collaboration with local schools, voluntary organizations and universities to undertake research or assist in any monitoring or management programme is also desirable.

Utilization. Protected area sustainability is greatly enhanced by utilization, but to ensure conservation is balanced against use it will be necessary to manage and monitor the traditional, professional and local users of the reserve. The infrastructure of the reserve may need to be improved to facilitate user access and prevent trampling of vegetation and soil erosion. Levels of harvesting and cultivation in the buffer and transitional zones of the reserve may have to be monitored and regulated.

Collaboration with Other Reserves and Organizations. The management plan should also incorporate or acknowledge the work and research undertaken by other regional, national or even international institutions and voluntary organizations working in associated projects.

Administration and Funding. The conservation project will have to be administered and a headquarters/base with appropriate facilities should be established. No conservation project will be successful or sustainable if adequate funds are not secured. Fund raising and making grant applications will be an integral part of the project.

As noted above, the actual style of the management plan will vary depending on target taxa, location and implementing agency, but a possible structure is provided in Box 8.6.

8.4.4 Management Prescription

This is possibly the most fundamentally important element of the plan as it sets out the minimum set of interventions required to manage the genetic reserve; minimum because each intervention is likely to require additional resources and any intervention if deleterious could adversely impact the target population. The management prescription for a natural habitat that has evolved with minimum human influence is likely to be relatively simple, because the target taxon is more likely to be suited to the environment in which it grows. The management prescription for semi-natural habitats, those relatively modified by humans over evolutionary time scale, are likely to require more active intervention and be dependent on specific human activities, such as grazing of livestock or burning to avoid natural succession toward the local climax community. Most CWR are predominantly found in semi-natural habitats (Jain, 1975; Maxted *et al.*, 1997b; Jarvis *et al.*, 2015), and therefore active management to prevent succession is nearly always required. Historically, succession may have been halted by periodic grazing of the site, and to maintain the target populations, such interventions need to be

---

Box 8.6 │ Possible Genetic Reserve Management Plan Structure

(Maxted *et al.*, 2008c)

1. *Preamble:* conservation objectives, site ownership and management responsibility, reasons for location of reserve, evaluation of populations of the target taxon, reserve sustainability, factors influencing management (legal, constraints of tenure and access).
2. *Conservation context:* place reserve within broader national conservation strategy for the responsible conservation agency and target taxon, externalities (e.g. climate change, political considerations), obligations to local people (e.g. allowing sustainable harvesting), present conservation activities (*ex situ* and *in situ*), general threat of genetic erosion.
3. *Site abiotic description:* location (latitude, longitude, altitude), map coverage, photographs (including aerial), detailed physical description (geology, geomorphology, climate, hydrology, soils).
4. *Site biotic description:* general biotic description of the vegetation, flora, fauna of the site, focusing on the species that directly interact with the target taxa (keystone species, pollinators, seed dispersers, herbivores, symbionts, predators, diseases, etc.).
5. *Site anthropogenic description*: effects of local human population (both within reserve and around it), land use and land tenure (and history of both), cultural significance, public interest (including educational and recreational potential), bibliography and register of scientific research.

Box 8.6 | (cont.)

6. *General target taxa(on) description*: taxonomy (classification, delimitation, description, iconography, identification aids), wider distribution, habitat preferences, phenology, breeding system, means of reproduction (sexual or vegetative) and regeneration ecology, genotypic and phenotypic variation, local name(s) and uses.

7. *Site-specific target taxon description*: taxa included, distribution, abundance, demography, habitat preference, breeding system, MVP size, and genetic structure and diversity of the target taxon within the site, autecology within the reserve, synecology with associated fauna and flora (particularly pollinators and dispersal agents), specific threats to population(s) (e.g. potential for gene flow between CWR and domesticate).

8. *Site management policy (non-prescriptive)*: site objectives, control of human intervention, allowable sustainable harvesting/hunting by local people and general genetic resource exploitation, educational use, application of material transfer agreements.

9. *Taxon and site population research recommendations*: this may include taxon and reserve description, aut- and synecology, genetic diversity analysis, breeding system, pollination, characterization and evaluation.

10. *Prescription (management interventions)*: details (timing, frequency, duration, etc.) of management interventions for target taxon, population mapping, impact assessment of target taxon prescriptions on other taxa at the site. Staffing requirements and budget, project register.

11. *Monitoring and Feedback (evaluation of interventions)*: demographic, ecological and genetic monitoring plan (including methodology, schedule), monitoring data analysis and trend recognition. Feedback loops resulting from management and monitoring of the site in the context of the site itself and the regional, national and international context.

maintained. Typical management interventions to sustain plant populations include:

Non-intervention – Minimal human disturbance or management is a prescription that may be appropriate for target species of natural habitats such as primary forests maintained as Strict Nature Reserve or Wilderness Area in the sense of the IUCN (Dudley, 2008) categories of protected area.

Fire – Controlled, periodical fires within a reserve can be an effective and essential management tool for certain target species and their habitats. Fires can be used to maintain open habitats at early stages of succession by arresting regeneration of dominating woodland and forest species. Species that may benefit from fire management include grassland, heathland and shrub species. Some species, for example, numerous Australian biota, are adapted to periodic fires (Bowman, 2003), and the production of ethylene during the fire stimulates flowering in several monocotyledonous species (e.g. *Xanthorrhoea* spp.); therefore, many Australian species are dependent on fire for dispersal and regeneration.

Grazing – Grazing is another management tool that can be utilized to manipulate the environment to enhance or create suitable habitats for the target taxon. Grazing management is particularly valuable in semi-natural habitats, which have had a long history of grazing by domesticated animals, such as chalk grasslands and pastures in Europe, prairie in North America or savannah in East Africa. The species found in these habitats are often grazing tolerant or require a certain level of grazing pressure for healthy growth. However, the introduction of livestock into plant communities with no history of mammalian herbivores, for example, many oceanic islands, can have disastrous consequences on the native vegetation.

Tree-felling – The selective felling of trees can be used to create openings in woodland and forest, so assisting natural regeneration. This can create favourable conditions for many open canopy species that are shade intolerant and require light for germination. Exotic invasive tree species, such as the sycamore in ancient woodlands in the UK or *Eucalyptus* species in America and Africa, can be felled as part of routine weed control.

Nutrient control: Many natural ecosystems are nutrient-poor, and pollution may cause eutrophication, significant artificial nutrient enrichment, which favours alien species suited to growing in nutrient-rich habitats. In these cases, the source of the nutrient enrichment needs to be identified and curtailed, to prevent further pollution. However, this is not always easy as the source may be very remote from the reserve site and tracing the source over long distances via wind or water movement is difficult.

Fencing – The construction of enclosure fencing around key populations of the target species could eliminate the detrimental effects of over-grazing by invasive animals like rabbits and goats and limit disturbance by humans.

Control of Invasive species – It is important to distinguish between native, alien and invasive species; native (indigenous) species are found within their natural distributional range, alien species are species found outside of their natural distributional range and their occurrence results directly from and is possibly sustained by human intervention, while invasive species are alien species that become established in natural or semi-natural habitats and act as agents of habitat change

threatening native biodiversity. As such, alien and particularly invasive species can pose a serious threat to target species and ecosystems. There are many examples whereby the introduction of invasive species has resulted in the loss of native species and the complete alteration of community structure. However, not all exotic species within the reserve will be invasive or detrimental to the target species and overall composition of the ecosystem. It is difficult to predict what the effects of an exotic or introduced species will be on an ecosystem; therefore, exotic species within the reserve should be monitored so that those with invasive characteristics can be identified quickly and controlled or eradicated. Invasive plants can be controlled by manual weeding, using biological control methods or with the application of herbicides in selected areas. The introduction or invasion of exotic animals to a protected area can have a deleterious effect by directly altering the grazing regime or indirectly altering the carnivore/herbivore balance within the reserve. Goats, rabbits and sheep are notorious for over-grazing natural vegetation. Invasive animals that are a threat to populations of the target species in the reserve can be eliminated or controlled by culling, selective poisoning or use of fencing. However, control and eradication are rarely easy and often costly, so education concerning the threat to native plant diversity from invasive plant and animal species and prevention of their introduction is a preferable option (Veitch and Clout, 2002). The IUCN has an SSC Invasive Species Specialist Group that aims to reduce threats to natural ecosystems and the native species they contain by increasing awareness of invasions, and of ways to prevent, control or eradicate them. It has a website (www.issg.org/) with guidelines, newsletters and links to databases with information on invasive species.

Habitat restoration: As a genetic reserve may have been established pragmatically in a less than ideal location, an initial intervention may require habitat restoration to improve the likelihood of target population survival. It involves physically putting something back into a habitat or ecosystem or removing negative factors from the habitat or ecosystem to allow the target populations to regain their balance. It may involve, depending on the actual circumstances within the site, the recovery of the target populations to self-sustaining levels with the appropriate range of abiotic and biotic interactions within the habitat, and this may involve reintroduction of individuals of the target to the reserve translocated from other possibly threatened sites. If individuals are going to be introduced to a reserve, then the provenance of the source of material is an important consideration; locally sourced material is best because it is likely to be pre-adapted to the local ecogeographic conditions (Houde *et al.*, 2015). Habitat restoration is a complex and possibly resource-intensive activity, which may require a knowledge of the target taxon's soil, hydrological, vegetation succession, pollinator, seed disperser and other faunal relationships. The process is likely to involve: (a) examining pre-existing, historic and current management practices, (b) development of a restoration plan, (c) obtaining any necessary permits, (d) implementing the plan and (e) monitoring progress and providing feedback for further restoration action (Pullen, 2002). Application of habitat restoration will require reference to specialist literature (Harker *et al.*, 1999; Peel, 2010), particularly in the plant genetic reserve context (Kell *et al.*, 2008). The IUCN also has a Reintroduction Specialist Group, and they have recently revised their guidelines on reintroductions (IUCN/SSC, 2013).

Assisted propagation: Allied to habitat restoration, another intervention to help bulk-up the inherent target population is assisted propagation. This may involve assisted pollination, collection of seed, germination and planting of seedlings, and *in vitro* generation of additional plants and subsequent planting in the reserve. However, even if seed is collected from the site, cultivated and the seedlings planted, the young plants may still fail because of the lack of biotic interactions with associated species or some unknown limiting factor. There may be many reasons why target populations are lower than the MVP at the site, but these will need to be understood and if possible corrected prior to the investment in assisted propagation, which may require extensive resource investment. If, for example, a species is insect-pollinated and the insect species is absent or in decline, assisted propagation of the target taxon alone will be ineffective.

Cultural change: As stressed above, a reserve is likely to fail without the support of local communities; implementing *in situ* conservation is often a compromise between scientific principles and practical constraints, such as retaining the support of local communities in the conservation action and meeting their development aspirations. The compromise should not only be on the conservation side; local people may also be persuaded to change cultural practices to sit better alongside conservation goals (Maxted *et al.*, 2008c). The compromise might involve wild harvesters not halting but reducing their collection of target populations, pastoralists agreeing to graze their herds/flocks at specific times of year after flowering and fruiting, or at specific stocking densities, or communities being persuaded to grow native rather than exotic plant species. It should be emphasized that when attempting to persuade local people to change their cultural practices, careful explanation and sensitivity is critical.

8.4.5 Implementation of the Management Plan

The management of any genetic reserve will involve an element of experimentation, and it is unlikely that the ideal management regime will be known when initially establishing the reserve. For example, how can one accurately estimate the appropriate level of grazing, if grazing is thought a requirement, in the reserve prior to initiating the management plan? Knowledge of current and historic grazing levels will be important, but the precise grazing regime recommended once the reserve is established can only be known through scrupulous experimentation. The initial plan may encompass several different management regimes (a range of grazing practices, tree-felling, burning, etc.) associated with detailed target taxon monitoring within the reserve to provide evidence of which regime is most appropriate at that location. Thus, the implementation of the management plan will require careful introduction, combined with evaluation, revision and refinement in the light of its practical application. Therefore, the initial level of management will be high, because of the intensive and extensive monitoring procedures, which will feed back into the management plan. Such an approach is often referred to as '*adaptive management*'; it involves research into diverse conservation actions and the establishment of an evidence base of successful conservation outcomes that can then be implemented. Specifically, it is the integration of design, management and monitoring to systematically test assumptions and so adapt and learn which, maximizes beneficial outcomes (Salafsky *et al.*, 2001).

This means of improving intervention success is also associated with an '*evidence-based approach*' to conservation, an approach originally derived from medicine (Pullin and Knight, 2001; Woodcock *et al.*, 2014), that has become an increasingly widely applied conservation technique (see www .environmentalevidence.org/). Sutherland (2000) noted that 'practical conservation is not well supported by background knowledge and is largely based on anecdotal evidence. This inhibits the development of scientific management and effective project planning'. The quality of conservation action often reflects the ratio between the information that the conservationist has at hand compared to the sum of relevant information that is potentially available; the more background information (evidence), the better the decision and conservation outcomes. The evidence-based framework aims to inform conservationists about the likely result of applying alternative conservation actions, helping them maximize the likelihood of a successful outcome by choosing those actions that are most appropriate. Pullin and Stewart (2006) provide guidelines on how an evidence-based approach may be applied in the conservation context. The features of such an evidence-based system might include: systematic reviews of conservation application, cost–benefit evaluation, explicit assessment of effectiveness and web delivery to practitioners. The general advantages

of an evidence-based system are that they: (a) are efficient, unbiased, systematic and scientific; (b) provide a formalized method to identify areas where evidence is lacking; (c) give a clear statement of best practice and (d) generate a needs-led research agenda (Pullin and Knight, 2003). Although adopting an evidence-based approach is beneficial, the outcome is still not guaranteed and therefore management interventions need to be monitored to assess success and further interventions revised to further enhance outcomes.

8.5 Reserve Monitoring

Reserve management attempts to maintain a 'healthy' target taxon population, 'healthy' both in terms of maximizing demographic (numbers) and genetic (gene and allelic) diversity over time within the reserve. Demographic monitoring will provide information on population trends, extinction risks, MVP sizes and which factors are most influencing population viability, while genetic monitoring will enable understanding of the ecogeographic partitioning of genetic diversity, the magnitude of natural dynamics and identification of any significant changes in the target populations. Significant changes in either the structure or size of populations of the target taxa/taxon within the reserve or genetic constitution over time will, if negative, undermine the conservation integrity of the reserve. Thus, target populations in the genetic reserve will require regular monitoring to identify any obvious or incipient changes and, if detected, trigger a management review and intervention amendment as required. It may be formally defined as: 'the systematic collection of data over time to detect changes in relevant plant population or habitat attributes, to determine the direction of those changes, and to measure their magnitude' (Iriondo *et al.*, 2008). Therefore, significant changes in the number of individuals, the genetic constitution or reduction in its diversity over time will trigger a management review and amendment. The management plan should include an estimate of the MVP for the target taxa/taxon and an ideal population size for the reserve, as well as a baseline assessment of the inter- and intra-population genetic diversity, and it is against this baseline that the target population is monitored. The monitoring will have a set of specific objectives, such as identification of trends over time in population size, structure, frequency or cover, identification of trends over time in genetic diversity, assessment of threat facing the target population, and assessment of impact of changes to the management regime (Iriondo *et al.*, 2008). Thus, target population monitoring is a routine component of population management (see Figure 8.4).

However, it should be stressed that populations are intrinsically dynamic, and they will naturally vary in demographic and genetic diversity from season to season. Therefore, it is necessary to distinguish between normal population seasonal fluctuations and gradual evolution, sometimes referred to as the '*limit of acceptable change*' (Maxted *et al.*, 2008c), on the one hand and significant deleterious changes that are likely to negatively impact the conservation outcomes on the other. Three basic forms of natural change are recognized:

- Stochastic – resulting from drought, floods, fire, cyclones, hurricanes and epidemics, etc.
- Successional – directional change toward a stable, climax community, often halted by anthropogenic intervention.
- Cyclical – density-dependent interactions between taxa within a community.

Although often superficially dramatic, these natural changes seldom cause species extinctions (Hellawell, 1991), but may result in genetic drift or local extinction (Gillman, 1997). By contrast human activities can have a significantly more devastating impact, e.g. human-mediated introduction of invasive species such as goats or rats had resulted in rapid extinction of many island floras (Sax *et al.*, 2002; Steadman, 2006). However, human activity can also create habitats, e.g. urban wasteland, agriculture land and roadsides are the favoured habitat of ruderal species, and many CWR are found growing in cropped and weedy areas, fertile grassland and lowland woodland, and have a significant association with linear features such as hedgerows, roadsides, field boundaries and field margins (Jarvis *et al.*, 2015).

8.5.1 Practical Demographic Monitoring

It is not practically possible to monitor every target plant in the reserve, so it will be necessary to employ a sampling that is indicative of the entire population. The conservationist will face a series of choices in selecting the sample and the choices made are likely to vary depending on the monitoring objectives, characteristics of the taxa being sampled (e.g. annual, biennial or perennial, in- or outbreeder, clone), reserve location and resources available, so it not possible to give generic advice for all reserves, and each reserve will require a unique monitoring regime. A schematic approach that helps address these questions is outlined in the schematic model shown in Figure 8.5.

Which Taxa to Monitor?

Obviously, the target taxa for which the reserve was established will be central to the time-series collection of monitoring data but monitoring these taxa alone may not provide the whole picture of the taxa's population sustainability. Other species necessary for target taxon population sustainability such as associated pollinators, seed dispersers, grazers and symbionts would need to be monitored, as decline will impact the target taxa. Keystone species, those species that dominate any habitat and removal of which changes the whole ecological balance of the habitat, such as the dominant vegetation that are commonly tree species or large herbivores that maintain a habitat in a pre-climax stage and without which natural succession would change the habitat type, need to have stable population levels. Changes in the keystone species are likely to impact target taxa; therefore, they should also be monitored. In a similar manner species that may threaten the target taxa, such as competition from exotic, invasive plant or animal species that may eat or displace the target taxa should also be monitored. Although the target taxa will be the focus of monitoring activities, the range of ancillary taxa that are monitored will practically be circumscribed by the resources, skills and time available.

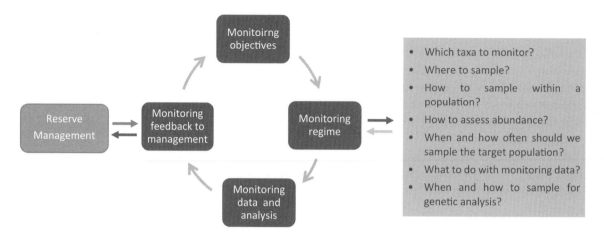

Figure 8.5 The genetic reserve monitoring regime.
(Adapted from Maxted *et al.*, 1997c)

Where to Sample?

Even when monitoring a single species within a single reserve, it will rarely be possible to census the population and record every individual present, so a decision must be made over where to sample the selected taxa. This decision has two components: (a) assuming the species is present in more than one location, which populations to sample; and (b) within each population, which individuals to sample? However, in practice, the answer to both questions is randomly so that statistical conclusions can be drawn concerning all populations or individuals. There are three basic approaches to random sampling:

Random sampling – This implies that all populations or individuals within the range of the species or within the reserve are equally likely to be sampled. The actual locations for sampling are found by superimposing a grid onto a map of the distribution or reserve, then using a random number generator to identify the precise sampling locations. This is the most statistically robust method of sampling and is the most widely applied (Elzinga *et al.*, 2001; Iriondo *et al.*, 2008).

Systematic sampling – This means samples are taken at regular intervals. Again a grid is superimposed on a map of the distribution or reserve and say every 5th,

20th or 100th grid square is sampled, with the first sampling point being selected randomly. As biological phenomena are often spatially auto-correlated, this method avoids over-sampling of 'uninteresting' areas at the expense of 'interesting' ones that might be selected at random. If the population or individuals are obviously correlated with a natural gradient, e.g. altitude or rainfall, then the systematic sampling should follow the direction of the gradient (Hayek and Buzas, 1997). However, applying systematic sampling may limit the statistical tests that can be applied, and care must be taken to avoid over-sampling ecological features that may be coincidentally correlated with the sampling interval applied. Systematic sampling is believed to be more efficient than simple random sampling, particularly if populations or individuals are spread over large areas, because of the decreased setup and travel time between sampling locations (Iriondo *et al.*, 2008).

Stratified random sampling – This involves dividing the taxon range or reserve into distinct homogeneous zones (= strata), then sampling each zone randomly in proportion to the size of the zone. Zones could, for example, be identified based on different aspects, vegetation or soil type, or even site management practices. This method is often applied if populations or individuals are spread over large areas and within those areas there are distinct habitats. This method ensures each habitat is sampled according to its size; therefore, larger habitat patches will have more associated random sampling points (Elzinga *et al.*, 2001). This type of sampling would be employed when collecting samples for genetic analysis, as it could be assumed that there might be a correlation between diverse habitat types and patterns of genetic diversity (Iriondo *et al.*, 2008). This is perhaps the most common sampling strategy applied.

How to Sample within a Population?

Once the sampling site has been located, the next question is how should individual plants within the target populations be identified? There are three sampling location methods:

Quadrat sampling – This involves creating a plot, usually square encompassed by wood, metal or tape and provides a standard area for sampling within which the target populations are recorded. Larger quadrats may be subdivided to make recordings and avoid missing individuals. Different sizes of quadrats are used to sample different sorts of vegetation, but having begun monitoring, the quadrat size should be kept constant to permit easy comparison of data from subsequent surveys.

Transect sampling – This involves the samples of vegetation being taken along an intercept, either a line or belt transect. A line transect involves the recording of species that touch the line and recording may be continuous along the line or at random or stratified intervals, whereas a belt transect involves the recording of species in quadrats spread out at random or stratified intervals along the transect. Normally the target populations are recorded randomly along the transect but may be stratified where there is a rapid change in management, vegetation or marked environmental gradient.

Plotless sampling – This is used when species are sparsely distributed as in forests, Mediterranean garrigue, or semi-desert and alpine communities. The nearest individual is the simplest method for plotless sampling where random sampling points are taken throughout an area and the measurement of the distance to the nearest set number of individuals is recorded. Successive distance measurements are taken, and the whole procedure is then repeated for a series of random points.

The most frequently used method to sample vegetation and individual species is the quadrat, but whatever method is applied, the recorded plants should be selected randomly, so conscious or unconscious selection of vegetation or habitat types that were positively associated with the target species is avoided or areas not containing the target population or large percentages of rock or bare soil are never included. If using a plot-based method, the plots may be selected afresh, randomly on each recording visit, or be fixed so the same plots are successively recorded, or be a mixture of fresh and permanent plots. Fresh plots are considered more statistically valid because it is assumed that each plot recorded is independent from all other recordings. However, statistical methods exist to analyze time-series data from fixed plots (e.g. based on fitting models separately to the data from each sampling visit and then comparing the parameters of the models from all the sampling visits), and using permanent plots may be easier and quicker if the survey area is large and the terrain is difficult. Use of permanent plots means they must be relocated each recording season and individual plots or plants can be photographed, GPS located or tagged with some form of marker to enable easy relocation over time.

The actual size of the plot will vary depending on the size and longevity of the target taxa: the physically smaller species require smaller plots to record (e.g. 0.5–1.0 m^2), bushes require medium size plots (e.g. 50–100 m^2) and larger trees require large plots (200–1000 m^2). There are three critical points: plot size, consistency of plot size and number of plots. The plot should be small enough to be searched easily and permit enough replicates in the time and with the resources available, but large enough to accommodate whole plants of the target species. Goldsmith (1991) proposed a rule of thumb, that the target taxa should have a frequency of 20–70% in the plots; if the taxa

are found in all plots then the plot is too large.
However, whether optimal quadrat size for vegetation
surveys will also be optimal for an assessment of the
abundance of individual target species is not clear.
Having established the plot size, it should remain
constant for successive recordings to facilitate
statistical analysis. There are statistics that will permit
changes in plots size, but this may limit the statistical
analysis. The larger the number of plots the more
reliable the result, but the extra information gained
from recording an additional plot will diminish as the
total number of plots increases (Goldsmith, 1991).
However, in practice the number of plots is a
compromise between the number that are required to
generate meaningful statistics and the resources
available for monitoring.

How to Assess Abundance?

Within the sampling location, whether a quadrat or
some form of transect is to be used, there is a need to
assess abundance of the target populations and collate
the data for time-series comparison. There are various
means of assessing abundance but the most
commonly used are density, frequency and cover.

Density – This is the number of individuals per unit
  area and is the preferred recording assessment for
  demographic studies of plant populations. It is easy
  to use when individuals exist as discrete plant units,
  such as orchids or most trees, but is more difficult,
  if not impossible, when plants grow in clumps or
  propagate vegetatively, such as grasses or clovers.
  In the context of genetic conservation, it is
  desirable to identify distinct genetic entities, which
  may be challenging and destructive if dealing with
  clonally propagating, clumped species and
  distinguishing individuals requires uprooting the
  plants.
Frequency – This records the percentage of plots
  occupied by the target species within a sampled
  area, so that if a grid is laid over the sampling
  location, the percentage of plots occupied by the
  species is the frequency. The method records simply
  species presence (+) or absence (–) across the plots
  within the sampling location; it is the simplest and
  quickest form of data collection. It is sometimes

combined with the frequency symbols: Dominant,
Abundant, Frequent, Occasional or Rare (DAFOR) to
allow an estimation of relative abundance.
Frequency is often used for monitoring annual
CWR, whose density fluctuates substantially from
year to year and for rhizomatous or clonal species
where presence only is recorded so there is no
necessity to record individual numbers (Elzinga
et al., 2001). Another advantage of this approach is
that frequency can be measured with minimal
training; the recorder needs only to be able to
recognize the target taxa and record their presence
in the plot (Iriondo *et al.*, 2008).
Cover – This is defined as the vertical projection
  within a quadrat that is occupied by the above-
  ground parts of a species; the actual percentages
  could be more than 100% as multiple species may
  overlay each other. Cover can be estimated visually
  as a percentage or converted to a value on a cover
  scale. Two commonly used cover scales are shown
  in Table 8.3. This method of estimating abundance
  is subjective as the estimation is done by eye and is
  prone to error, particularly if different recorders are
  used. Cover can also be estimated using a cover pin
  frame. This method measures cover objectively, as
  opposed to the above subjective method. It is most
  commonly used for studying grassland species as
  individuals do not need to be recorded; however,
  the cover estimates may change as the season
  progresses and therefore cover should be estimated
  at the same stage in the season in successive
  assessments.

However, it is assessed population characteristics
are a manifestation of demographic processes, such as
the births and deaths occurring in a population.
Therefore, detailed demographic monitoring will also
involve the recording of rates of growth, reproduction
and survival. Useful data to collect on individuals
include size, age and reproductive stage. This will
allow the age structure of a population, that is, the
proportion of seedlings, juveniles and mature adults,
to be investigated. An absence or low number of
plants in an age class, particularly juveniles, may
indicate that the population is in decline. Only by
studying the dynamics of key populations over

Table 8.3 **The Braun–Blanquet and Domin cover scales**

Braun–Blanquet cover scale		Domin cover scale			
0	<1% cover	+	1 individual	6	25–33%
1	1–5% cover	1	1–2 individuals	7	34–50%
2	6–25% cover	2	<1%	8	51–75%
3	26–50% cover	3	1–4%	9	76–90%
4	51–75% cover	4	4–10%	10	91–100%
5	76–100% cover	5	11–25%		

several generations can the effects of any management initiative be assessed and modified appropriately. An experiment where comparisons between a control population, where no management is in place, and one or more populations that are subject to different management regimes could be designed.

When and How Often to Sample?

It is important that successive monitoring events occur at the same seasonal time to be able to record comparable results; earlier in the season there may be large numbers of seedlings that later in the season will have suffered extensive mortality. Iriondo *et al.* (2008) suggest target taxa should be monitored when they are easiest to locate, and for most species this is when they are flowering. The frequency of sampling is often dictated by available knowledge of the target taxa and the strength and nature of the threats the population faces. The less available the knowledge and greater the threat, the greater the frequency of monitoring. Monitoring on a 5-year cycle is enough for most herbaceous plant species, though for a long-lived perennial species the monitoring may be less frequent with a 20-year cycle. Though if the reserve is newly established, monitoring may be more frequent to build up a baseline of census data. The most appropriate management prescription is unlikely to be known for a newly established reserve and changes or adjustments to the management prescription will be more frequent. As the management regime is established, so the frequency of monitoring can be extended. Also, if the species is rare and/or very threatened, monitoring may occur monthly through several growing seasons until the appropriate management is identified and enacted (Harper, 1977).

How to Analyze Monitoring Data?

To enable comparison of monitoring data and trend identification, data recording must be consistent between populations, within populations and over successive samplings. The type of analysis will depend on the type of data collected. Routine statistical analysis of the data sets will indicate whether there has been significant change in population size or density over time and whether any trend is becoming apparent (see Elzinga *et al.*, 2001 for detailed discussion of different forms of analysis). However, as noted above, care must be taken to distinguish between the natural ranges for population characteristics and those induced by management or other intervention.

The analysis of the time-series data from the successive monitoring events should reveal trends in the target taxa. If the populations are roughly stable with any deviation from the norm being explained by natural stochastic events, then the current management regime is appropriate, but if the trend indicates a deterioration of the conservation status of the target species, then the management plan is inappropriate, and the prescription requires revision. Thus, monitoring data analysis generates information that feeds back to the management regime stimulating necessary changes. This feedback process can be illustrated by the classic example described by Shands (1991) concerning maize relative, *Zea diploperennis* Iltis *et al*. The perennial teosinte was discovered in the 1970s as an endemic to the Cerro de San Miguel, Sierra de Manantlan, Jalisco, Mexico; it was diploid ($2n = 20$) and had agronomic significance as a gene donor to maize (Iltis *et al.*, 1979). Given the new species significance, *Z. diploperennis in situ* conservation was incorporated into the management plan of the tropical forest protected area of Sierra de Manantlan, Mexico and the plants were protected by a fence from grazing.

Routine monitoring of population sizes within the reserve indicated that the halting of all grazing had allowed succession to occur, and if the original prescription was not amended, the other forest plants would out-compete the wild maize. The fence was removed, and controlled grazing was allowed, and the target taxon thrived. This clearly demonstrates why monitoring conserved populations is so important; without monitoring and trend identification the threat would have passed unrecognized and the conserved population could have been lost. The routine monitoring process identified the threat, acting as a feedback mechanism triggering changes in the management of the reserve and ensuring that genetic resources are safely conserved.

8.5.2 Practical Genetic Monitoring

The previous section focused on maintaining a sufficient population of the target taxon to maintain its demographic and by inference its genetic health; however, there is a fundamental assumption that demographic health is equivalent to genetic health, which is often but not always true. It is possible that number of individuals remains constant but unseen genetic diversity is significantly changed or lost. Therefore, there is a need for genetic as well as demographic monitoring. However, genetic analysis should begin before the reserve is established.

Ideally, before a genetic reserve site and population are selected the ecogeographic and gap analysis will have identified a shortlist of potential locations where the target taxon exists, or group of taxa coincident in large numbers, and then the genetic diversity at the competing sites will have been quantified and compared. The designation of genetic reserves should be evidence based; sites and populations are deliberately selected for inclusion in a genetic reserve network because they are known foci of genetic diversity for the priority taxa. This will involve an estimation of relative fitness of the competing populations, the ability of an individual to contribute its genetic diversity to the next generation through the production of male and female gametes, fertile seeds, and the number of reproducing plants that in turn produced seeds (Frankham *et al.*, 2014). Low population fitness is associated with inbreeding depression, loss of diversity or heterozygosity and ultimately population extinction, although taxa fitness can be bolstered by gene flow from other populations, estimated by a populations *F* statistic (coefficient of inbreeding) in the genetic analysis. The site selection analysis will also determine the level of out- or in-breeding. For example, where the target taxon is predominantly out-breeding, there will be much greater variation partitioned within populations and less between populations, in which case given the choice, one single large reserve would be preferable to several remotely located smaller reserves. However, in terms of management, predominantly in-breeding populations may be managed with less need to promote inter-population gene flow. Two key issues remain:

Sampling for genetic analysis – About 50 plant or seed samples are taken randomly from the population, though von Bothmer and Seberg (1995) suggest 20–30 individuals would be enough if necessary and the population is very rare. If the population has obvious stratification, then the samples would be selected to ensure samples from each distinct sub-population were included. The samples collected in the field are stored in a sealed container (e.g. plastic specimen tube) with a drying agent and clearly labelled ready for subsequent analysis.

Frequency of sampling – Although the cost of genetic analysis is reducing annually, it can still be significant, and there is a need not to over-sample and analyze the populations more than is necessary (Iriondo *et al.*, 2008). There is, as stressed above, a need to sample and analyze for genetic diversity as part of the site/population evidence-based selection process. This will establish the genetic baseline for the reserve that subsequent analyses can be measured against. If in establishing the genetic reserve there has been a need for significant intervention or the population has suffered a trauma (e.g. flooding, pest attach, over-harvesting, loss of associated keystone taxa), then it would be wise to sample and analyze the population to check the impact of those interventions on the target population's genetic diversity. There have been so few existing genetic reserves established providing a recommendation on the necessity or periodicity of routine genetic monitoring. Iriondo *et al.* (2008) argue 'it is almost certainly unnecessary to undertake routine monitoring', but given the extended time frame of *in situ* genetic conservation and the value of the conserved resource, it would seem wise to periodically re-sample and analyze the genetic diversity of the target populations and compare them to the initially established genetic baseline, but how often? Well not as often as for routine demographic monitoring (5-year cycle for herbaceous species and 20 years for long-lived perennial species), so once every 10 generations. However, it should be noted that there is no evidence base to support such a periodicity for genetic population monitoring, and research of this issue is advisable to avoid wastage of limited conservation resources. Further, as with demographic monitoring, fluctuations in natural background levels of genetic diversity are to be expected and the genetic analysis will need to distinguish between natural cycles of variation and significant genetic drift or erosions, again a subject that requires further research to aid practical *in situ* conservation. As noted above, in the future it is likely that germplasm users will access germplasm direct from the *in situ* source and, therefore, genetic monitoring might henceforth also be linked to population characterization and utilization promotion.

The actual technique for assessing relative genetic diversity and fitness change as techniques develop, so no specific techniques for genetic analysis are recommended here, but Iriondo *et al.* (2008) list the characteristics of the technique to be applied – it should be polymorphic, show co-dominant

inheritance to permit discrimination of homozygotes and heterozygotes, be representative of all parts of the genome, be focused on expressed genes if a narrow question requires an answer, be easy, fast and cheap to detect, and be reproducible (see Chapter 5). Box 8.7 illustrates how knowledge of the pattern of genetic diversity within species can help in their *in situ* maintenance.

8.6 Utilization and Complementarity

8.6.1 Genetic Reserve User Communities

Four groups of *in situ* genetic reserve users were recognized by Hawkes *et al.* (1997):

Traditional users – No reserve is established in an anthropogenic vacuum; local communities will have previously managed and often sustainably exploited the site prior to reserve designation. Local farmers, wild plant gatherers, firewood collectors and other members of the local community will have used the site, possibly for millennia previously. Therefore, to avoid conflict and promote local community support for the conservation action continued use by the local community should be encouraged that is compatible with the site/population's conservation objectives, particularly in the buffer or transition zones (Lewis, 1996; De Pourcq *et al.*, 2016; Mazaika, 2016). To ensure the conservation objectives are realized local community access and any harvesting, hunting or other uses may need to be regulated, even if it is necessary to reach a compromise between traditional utilization and conservation objectives, so facilitating the success of the reserve. Local people should be encouraged to become involved with the planning, establishment, management and monitoring of the reserve. This could be achieved through volunteer warden schemes and where possible reserve staff should be employed from within the local community (Borrini-Feyerabend *et al.*, 2004).

Reserve visitors – Increasingly all protected areas are under pressure from national funding authorities, not only to conserve but also to demonstrate the

Box 8.7 | *In Situ* Conservation of Rare Military Orchids Based on Genetic Diversity Information

The British are famous for their love of plants, and in the case of the military orchid (*Orchis militaris* L. Orchidaceae) it was almost loved to extinction. It is one of those species likely to have been collected almost to extinction in the mid-19th century and thought extinct by 1914, but was rediscovered in 1947 (Marren, 1999; Foley and Clarke, 2005). It grows currently in Europe from southern Sweden and Estonia south to central Spain, central Italy, Bulgaria, southern Russia and Anatolian Turkey, and despite its wide distribution is locally rare throughout its range (Rankou, 2011). However, today in the UK, only three populations survive in open and shaded habitats including woodland or scrub overlaying chalk, in Suffolk, Buckinghamshire and Oxfordshire. The latter two populations are only 9 km apart, and the Oxfordshire population now only contains six plants. The two main populations each contain 200+ individuals, though the precise numbers vary annually. The Oxfordshire population was thought to be a new colony established from the relatively geographically close Buckinghamshire population, and so the Oxfordshire population was considered of minor importance in maintaining species diversity (Fay, 2003). However, AFLP and microsatellite analysis (Qamarus-Zaman *et al.*, 2000, 2002) showed the three populations are equally distinct and possibly result from independent colonization events from the continental European populations, and that the Oxfordshire population is not derived from the Buckinghamshire population. Further the Suffolk population has limited inter-individual diversity, certainly compared to the Oxfordshire population where all plants are easily distinguishable. Therefore, the management recommendation based on the genetic analysis is that increased effort should be focused on sustaining suitable habitat for the individuals of the Oxfordshire population with their significant individual plant diversity and hand pollination stimulates seed set. Suitable habitat management that maintains the semi-open habitat is suggested for the Buckinghamshire population and the less genetically diverse Suffolk population. Fay *et al.* (2004) argue the Suffolk population resulted from a very small founder population, even possibly a single plant, by long-distance seed dispersal from the continent. All populations should have seed collected and placed in the Millennium Seed Bank, Royal Botanic Gardens, Kew as the national wild species gene bank, but that seed, if possible, should be collected from each of the six individuals in the Oxfordshire population.

value to society of the conserved resource. One indicator of relevance is to record the numbers of visitors visiting the protected area over a reporting period. These may include ecotourists. The IUCN (Ceballos-Lascuráin, 1996) defines ecotourism as: 'environmentally responsible travel and visitation to relatively undisturbed natural areas, in order to enjoy and appreciate nature (and any accompanying cultural features – both past and present) that promotes conservation, has low visitor impact, and provides for beneficially active socio-economic involvement of local populations'.

Stolton *et al.* (2010) reviews the associated benefits to local communities and cites the generation of additional income for local people as the hospitality (restaurants, gift shops and accommodation) required by the ecotourists is labour intensive, providing proportionately higher opportunities for women and has a high multiplier ratio for benefit to the local community. However, if the reserve is to benefit from ecotourism, it should take account of the needs of visitors and interested members of the public, and perhaps provide natural trails and guided walks within the buffer and transitional zone, educational boards, reserve information packs and even visitor centres with a lecture hall for guest lectures and an ecotourist lodge. If the reserve is specifically associated with agro-biodiversity, then an associated restaurant that utilizes the resource being conserved would be ideal.

General users – The broader population, whether local, national or international, should be encouraged to feel involved with reserve conservation (Dudley *et al.*, 2010); their support may be essential to the long-term political and financial viability of the reserve. Therefore, ethical and aesthetic justification for species conservation is of increasing importance to conservationists. As much reserve-based conservation occurs on state-owned land and ultimately it is financed by taxpayers, so the reserve should be open to visitors for recreation and education, even if this may turn in part the focus toward the conservation of flagship species, such as orchids, rare species or appealing 'woodland glades'. Funding for conservation projects is always limited but would be even scarcer if the public saw reserve-based conservation as being of purely academic interest. With the public on the conservationist's side helping lobby the authorities for adequate funding, continued reserve support may be underwritten. Therefore, education of the public in the importance of nature conservation and the link to beneficial ecosystem services, specifically food security, is essential to any project that wishes to receive long-term public support.

Professional users – Professional agro-biodiversity utilization from the genetic reserve should be comparable to professional utilization of germplasm conserved *ex situ*. However, one perceived disadvantage of *in situ*, as opposed to *ex situ*, agro-biodiversity conservation is that it is more difficult for the user, notably plant breeders or farmers, to gain access to the *in situ* conserved germplasm. This problem can be reduced by ensuring that germplasm within the reserve is characterized, evaluated and publicized. The quantity and level of documentation has a direct relationship with the potential of the germplasm for exploitation (FAO, 2011a). Reserve managers, just as gene bank managers, should promote the utilization of the material in their care, as well as the reserve site itself. Gene banks and botanic gardens often publish catalogues of their collections, so potentially the reserve could publish a catalogue and description of the germplasm held to inform potential users. The level of documentation of passport, characterization and evaluation data recorded should be just as extensive for germplasm conserved *in situ* as for *ex situ*.

In many cases, the work of professional users, the general public and local people can be linked through partnership within non-governmental organizations, especially those involved in sustainable rural development as conservation volunteers, or in the use of resources in accordance with traditional cultural practices. All partners will therefore share the goals of sustainable use of biological resources, while acknowledging local social, economic, environmental and scientific factors that form a cornerstone to the nation's proposals to implement the CBD (2010b) Strategic Plan for Biodiversity 2011–2020 and the UN Sustainable Development Goals (United Nations, 2015).

8.6.2 Linking *In Situ* and *Ex Situ* Conservation with Utilization

The increased demand by professional users for greater breadth of genetic diversity to sustain cultivar production within changing cultivation ecosystems (McCouch *et al.*, 2013; IPCC, 2014b) means germplasm user's demand for diversity can only be met from the broader range of diversity found in nature and *in situ* conservation, as well as *ex situ* conservation, and that all three sources of diversity have improved links facilitating resource access to utilization. Although Aguirre-Gutiérrez *et al.* (2017) following SDM climate modelling concluded the current range of protected areas would not guarantee long-term CWR conservation in Europe, Maxted *et al.* (2016a) also concluded the required range of diversity could not be met by the current sample of diversity held *ex situ* alone. Both *in situ* and *ex situ* genetic conservation need to 'up their games' to meet this growing challenge to supply the diversity users require. In the context of *in situ* genetic conservation, simply locating genetic reserves or on-farm activities using a climate smart approach that models a site's relative predicted climate change impact and using this evidence to locate sites to maximize population sustainability will significantly improve effectiveness.

However, there has been increasing debate recently concerning use of plant resources held *in situ* in genetic reserves (Valdani Vicari & Associati *et al.*, 2015, 2016; Maxted *et al.*, 2015, 2017). The conclusion was that without the utilization link there would be a lower priority to conserve the wild plant resource and any *in situ* conservation would be less sustainable; conserved resource utilization is therefore key to systematic and sustainable conservation. But specifically, concern has been raised over the potential additional and significant financial burden that would be placed on gene banks if they were required to incorporate *in situ* back-up samples into their *ex situ* collection and make them available to users (Valdani Vicari & Associati *et al.*, 2016). Maxted and Palmé (2016) suggested a potential model for how *in situ* and *ex situ* CWR conservation and utilization might be better integrated, and this original model has been enhanced by further discussion with stakeholders (Figure 8.6). The model proposes a distinction between standard long-term *ex situ* sampling of CWR/LR diversity and populations sampled specifically as *in situ* back-up, and for *in situ* back-up there are two distinct options: backbox safety back-up and *in situ* back-up. Backbox safety back-up would involve sampling *in situ*/on-farm back-up populations in virtual 'black box' samples within the gene bank, where the samples would only be available to the original donor. These black box samples would not undergo routinely gene bank monitored procedures, be regenerated or be made available to the user community. Such an approach would significantly reduce the potential cost of *in situ* back-up where resources were limiting, but would not assist in making the *in situ* resource available to the user community. Initially it was thought that the *in situ* or on-farm resource would be made available by the *in situ* resource maintainers (protected area manager or farmer), but realistically these communities have very limited experience of supplying germplasm samples along with the appropriate SMTA to ensure benefit sharing. Therefore, a practical third option is suggested in Figure 8.6, i.e. *in situ* back-up. Here a germplasm sample would be transferred to the gene bank and pass through the normal registration and documentation, cleaning and drying, germination testing and then packed and banked, but the sample would not be regenerated (so reducing maintenance costs) but would be made available. As the *in situ* back-up would be distributed to users, so further samples could be supplied by the *in situ* maintainer. The *in situ* back-up recorded in the gene bank's documentation system would be flagged to the user community, and those wishing to obtain an *in situ* sample could then contact the gene bank to supply a sample. In this way the function of the gene bank would be enhanced to cover access to both germplasm conserved *ex situ* and *in situ*, and they might better be termed plant genetic resource centres rather than gene banks containing only *ex situ* samples. The regular re-sampling of the *in situ* population would obviate the need for germination monitoring or regeneration, and

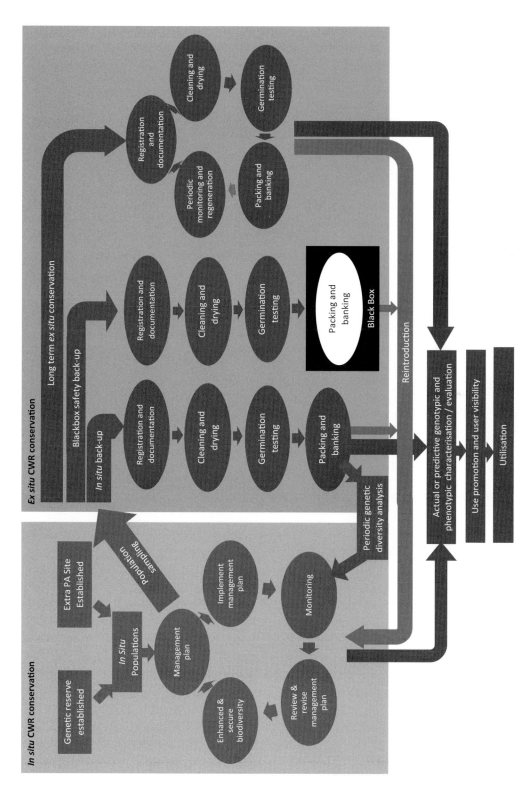

Figure 8.6 Integration of *in situ* and *ex situ* CWR conservation with utilization. PA, protected area. (A black and white version of this figure will appear in some formats. For the colour version, refer to the plate section.)
(From Maxted, 2020)

the lack of the latter would significantly reduce the financial burden of *in situ* germplasm supply on the plant genetic resource centre, and it would facilitate access to the *in situ* conserved resource and avoid direct contact with the *in situ* site manager or farmers.

However, it should be noted that the *in situ* site manager's or farmer's interest would be preserved by the Access and benefit sharing (ABS) agreement signed between the resource manager and the plant genetic resource centre.

9 On-Farm Conservation

9.1 Introduction

The continuous cultivation and management of a diverse set of populations by farmers in the agro-ecosystems where a crop has evolved.

Bellon *et al.* (1997)

In situ conservation involves the maintenance of plant genetic diversity in habitats where the plant species has evolved or occurs. *In situ* conservation can involve maintenance in either natural ecosystems or on-farm, depending on the plant genetic material under consideration. When this type of conservation involves crop genetic diversity in agricultural production systems, it is referred to as on-farm conservation. *In situ* and on-farm conservation are highly dynamic, involving the genetic material being exposed to human and natural selection pressure, as opposed to the semi-static nature of *ex situ* conservation. This allows the continued evolution of plants and crops over time, and their adaptation to the environments where they are grown. For some forms of biodiversity, *in situ* conservation is the only option. One of the main reasons given for choosing *in situ* conservation over *ex situ* is the need to maintain the evolutionary potential of species and populations. There have been many definitions (Box 9.1) of *in situ* and on-farm conservation. Sthapit *et al.* (2012) describe 'on-farm conservation as a highly dynamic form of plant genetic resources management, which allows the processes of both natural and human selection to continue to act in the production system. Farmer's ability to search for new diversity, selection of new traits and exchange of selected materials with friends and relatives is the processes that allows genetic material to evolve and change over time'.

On-farm conservation considers entire agro-ecosystems and how farmers manage them and includes not only locally useful species such as cultivated crops, forages and agroforestry species but also their wild and weedy relatives that grow in and around farms (Jarvis *et al.*, 2000). Farmers who maintain high levels of traditional varieties are often those who farm in more marginal environments. They are also often farmers who are relatively resource-poor, with small areas of cultivation (Box 9.2).

Several abiotic and biotic factors influence the extent and distribution of crop diversity in agro-ecosystems including: temperature, rainfall, light intensity and soil properties, topography, altitude and slope, incidence and severity of pest and disease including weeds and invasive species, natural enemies, pollinators and below ground biodiversity (Jarvis *et al.*, 2016). The on-farm conservation of a diverse range of traditional varieties and the necessary more extensive farming system also contributes directly to enhancing ecosystem functions including regulation of pest and diseases, promotion of below ground biodiversity and pollinator diversity. In a similar vein, it is suggested that the presence of crop genetic diversity on-farm increases the long-term stability of the ecosystem which in turn promotes continuous biomass maintenance in the ecosystem reducing soil erosion, and enhancing regulating and supporting services such as CO_2 sequestration (Jarvis *et al.*, 2016).

There are many possible objectives or benefits to be considered for projects and programmes aiming to strengthen on-farm conservation (Box 9.3). On-farm conservation includes both private and public benefits and goods. The maintenance of high levels of well-adapted traditional varieties in agricultural production systems is a global public good in terms of coping with changing climate, while at the same time providing farmers with options for private benefits through targeting markets with varieties that might have particular niche opportunities, such as those with geographic indication or other such labelling or even certification.

Box 9.1 | Definitions of *In situ* Conservation and On-Farm

(Extracted from Jarvis *et al.*, 2000)

- "*In situ* conservation of agricultural biodiversity is the maintenance of the diversity present in and among populations of the many species used directly in agriculture, or used as sources of genes, in the habitats where such diversity arose and continues to grow." (Brown, 2000)
- "*In situ* conservation specifically refers to the maintenance of variable populations in their natural or farming environment, within the community of which they form a part, allowing the natural processes of evolution to take place." (Qualset *et al.*, 1997).
- "*In situ* conservation refers to the maintenance of genetic resources in natural settings. For crop resources, this means the continued cultivation of crop genetic resources in the farming systems where they have evolved, primarily in Vavilov centres of crop origin and diversity." (Brush, 1991).
- "*In situ* conservation means preserving, in their original agroecosystem, varieties cultivated by farmers using their own selection methods and criteria." (FAO 1989; Bommer 1991; Keystone Centre, 1991, in Louette and Smale 1996).
- "On-farm conservation is the sustainable management of genetic diversity of locally developed traditional crop varieties, with associated wild and weedy species or forms, by farmers within traditional agricultural, horticultural or agri-silvicultural cultivation systems." (Maxted *et al.*, 1997b).

The CBD, the ITPGRFA and the Second Global Plan of Action for PGRFA, prepared under the aegis of the Commission on Genetic Resources for Food and Agriculture, all identify *in situ* and on-farm conservation as a priority activity. The Second Global Plan of Action, which responds to the needs and priorities identified in the *Second Report on the State of the World's PGRFA*, reported some improvement in the number of on-farm conservation initiatives compared to the first global report published a decade earlier (Box 9.4).

On-farm crop diversity provides a portfolio of assets and options which play an important role in the livelihood strategies and well-being of farmers and communities. Traditional varieties, among their many benefits, adapt over time to better suit the marginal and agro-ecological heterogeneity environments in which they are cultivated (Figure 9.1), and as such

they act as insurance against environmental risk, help meet changing market demands and opportunities, provide protection and management against the ravages of pests and disease, help meet cultural and religious needs, and may be kept for their dietary or nutritional value, taste or for the price premiums (Jarvis *et al.*, 2011; Bellon *et al.*, 2015a, 2015b).

However, there is no guarantee farmers will always maintain a diversity of traditional varieties on-farm. Doing so may entail some private costs to the farmer. The availability of improved varieties and associated external chemical inputs can foster greater specialization on-farm towards a few varieties, exposure to more efficient marketing channels can lead to the disappearance of market niches, while changes in food preferences and dietary transitions and the greater availability of new food products can all reduce the demand for diverse local varieties

| Box 9.2 | **Small Farmers Maintain World's Crop Diversity** |

(Modified from EurekaAlert!, www.eurekalert.org/pub_releases/2015-02/ps-wcd021115.php).

As much as 75% of global seed diversity in staple food crops is held and actively used by a wide range of smallholders – farm workers of less than 3–7 acres – with the rest in gene banks. Examining new census data from 11 countries in Africa, Asia and Latin America shows that it is small farmers, in many cases women, who are the ones preserving landraces of food crops. A landrace is a locally adapted, traditional variety of a domesticated species. Depending on the crop, farmers may plant anywhere from 1 to 15 different landraces. While the livelihoods of small land users are often precarious, these landraces provide vital farm and food resources. For maize, farmers plant one to three varieties because this crop readily outcrosses to form new varieties, producing too many new hybrids for the farmers to evaluate. Nevertheless, the next farm over would probably plant different landraces, so some more diversity is available. Not all small farmers produce the high-agro-biodiversity landraces, but those that do are often connected through networks of seed and knowledge exchanges. They are creating many new or so-called emergent agro-biodiversity systems rather than strictly relict or vestigial carryovers of heirloom crops as is often depicted from the outside. For crops like potatoes that outcross rarely, as many as 25–30 landraces (sometimes referred to as morphotypes) could be planted in a farm plot. Rice, which outcrosses infrequently, shows a pattern like maize with few varieties in each field, but significant diversity across neighbouring fields.

The importance of location types for the small farms goes beyond environmental factors with socio-economic factors also playing an increasing role. Peri-urban locations, places where people's livelihoods depend significantly on urban space and activities, are becoming more important for preserving and actively using the diversity of food crops in all major regions of the world. These are hybrid spaces socially and agriculturally, with influences from the modern city but where people might speak indigenous languages. They have good access to urban markets where restaurants and customers prefer local foodstuffs, like maize and quinoa varieties as in Peru. On the other end of the spectrum from peri-urban areas are the marginal locations in remote rural areas. These areas often have declining populations so there are not enough growers and during drought or other disasters, there may not be enough seeds to replenish the base of diversity, which is usually very high. In between peri-urban and marginal areas are a range of environments where small farmers grow crops and preserve diversity. Knowledge of potential problems in these areas and plans for responses to potential disruptions of agriculture are important to preserve diversity and improve food security.

Having stressed the global predominance of on-farm conservation in marginal agro-environments in developing countries, it should be noted that on-farm conservation is also practised in developed countries. In these situations, on-farm conservation is retained in marginal agro-environments, or where there is a distinct niche market that cannot be met by 'intensive' production. Further, increasingly a form of 'novel' management, as opposed to strict conservation, is being introduced by farmers addressing the organic market or where there is a market for produce from a more genetically diverse production system.

Box 9.3 | **The Multiple Benefits of On-Farm Conservation**

(Modified from Jarvis *et al.*, 2000)

Conserves the Processes of Evolution and Adaptation

On-farm conservation of crop diversity, as well as other aspects of agro-biodiversity, helps maintain the ongoing processes of evolution and adaptation of crops to their niches and environments within farming systems. This idea of dynamic conservation extends to all aspects of the farming system, including the wild and weedy plant species that may interact with their cultivated relatives to produce unique genetic diversity.

Conserves Diversity at all Levels

On-farm conservation applies the principle of conservation to all three levels of biodiversity: ecosystem, species and genetic (intra-specific) diversity. In conserving the structure of the agroecosystem, with its different niches and the interactions among them, the evolutionary processes and environmental pressures that affect genetic diversity are maintained, as are the diverse interactions of crop populations.

Integrates Farmers into the National Plant Genetic Resources Conservation System

Farmers know and understand the nature and extent of local crop resources better than anyone through their daily interactions with the diversity in their fields. They are experts with considerable local knowledge built up over many years. Given their expertise, incorporation of farmers into the national PGR system helps to create more productive and effective partnerships for all involved such as including farmers as partners: in the maintenance of selected germplasm; in national PGR dialogues and exchange of information; in linking with national gene banks and making genetic material more easily accessible to farmers, from gene bank to field, but also ensuring that threatened and unique genetic diversity on-farm is conserved in gene banks.

Conserves Ecosystem Services

On-farm conservation may be an important way to maintain local crop management systems for agroecosystem sustainability by ensuring improved soil formation processes, enhanced pollinator populations and reducing pest and disease damage, thereby reducing the need for external chemical inputs with their concomitant environmental impacts.

Improves the Livelihoods of Resource-Poor Farmers

On-farm conservation programmes can make multiple contributions to farmer and farming communities' livelihoods and well-being. On-farm conservation can be combined with local infra-structure development or the increased access for farmers to useful germplasm held in national gene banks, which can be the basis for initiatives to increase crop production or secure new marketing opportunities.

Box 9.3 | (cont.)

Strengthens the Control and Access of Farmers over Genetic Resources

On-farm conservation also serves to empower farmers to control the genetic resources in their fields. On-farm conservation recognizes farmers and communities as the curators of local genetic diversity and the indigenous knowledge to which it is linked. In turn, farmers are more likely to reap any benefits that arise from the genetic material they have conserved.

Public and Private Benefits (Socio-Economic, Ecological and Genetic)

The importance of conservation of agro-biodiversity for the future of global food security lies in its potential to supply other farmers, crop breeders' and other users' future needs for germplasm. In addition to these 'public' genetic benefits, on-farm conservation can provide other benefits to society and to the farmers who maintain crop diversity. Society can benefit from the agroecosystem stability and sustainability.

Box 9.4 | On-Farm Conservation, the State of Play

(Modified from FAO, 2010a)

The on-farm management and conservation of traditional crop varieties in production systems has gained much ground since the publication of the first State of the World Report. Many new national and international programmes have been set up around the world to promote on-farm management and the published literature over the last ten years has resulted in a clearer understanding of the factors that influence it. New tools have been developed that enable this diversity and the processes by which it is maintained, to be more accurately assessed and understood.

Efforts to measure genetic diversity within production systems have ranged from the evaluation of plant phenotypes using morphological characters, to the use of new tools of molecular biology. Considerable variation exists among production systems and many country reports pointed out that the highest levels of crop genetic diversity occurred most commonly in areas where production is particularly difficult, such as in desert margins or at high altitudes, where the environment is extremely variable and access to resources and markets is restricted.

Little information was available from country reports regarding actual numbers of traditional varieties maintained in farmers' fields. The Georgia country report mentioned that 525 indigenous grape varieties are still being grown in the mountainous countryside and isolated villages, while in the Western Carpathians of Romania, more than 200 local traditional varieties of crops have been identified.

In contrast to the country reports, published scientific literature since the first State of the World Report contains a considerable amount of information on numbers of traditional varieties grown on-farm. A major conclusion from these publications is that a significant amount of crop genetic diversity in the form of traditional varieties continues to be maintained on-farm even through years of extreme stress. In a study in

Box 9.4 | (cont.)

Nepal and Vietnam of whether traditional rice varieties are grown by many households or only a few, and over large or small areas, it was found that more than 50% of traditional varieties are grown by only a few households in relatively small areas.

Farmers' variety names can provide a basis for estimating the actual numbers of traditional varieties occurring in a given area and, more generally, as a guide to the total amount of genetic diversity. However, different communities and cultures approach the naming, management and distinguishing of local varieties in different ways and no simple, direct relationship exists between varietal identity and genetic diversity.

Figure 9.1 The Huehuetenango region, in the Cuchumatanes highlands of Western Guatemala, a marginal and heterogeneous environment which is an important centre of diversification for maize.
(Photo courtesy of Bioversity International/G. Galluzzi)

(Bellon *et al.*, 2015a, 2015b). Increased levels of migration to cities and the need to pursue off-farm labour opportunities can impact on-farm conservation because it is relatively labour intensive, all contributing to considerable technical, economic, social and cultural change, which in many scenarios can reduce the value of maintaining crop diversity on-farm.

Yet, on-farm conservation of traditional varieties also provides a unique public good value to society in that it contributes to the maintenance of crop diversity that allows our agriculture and food systems to adapt to changing conditions (Bellon *et al.*, 2015a, 2015b). However, we cannot simply expect resource-poor farmers to forego opportunities for economic development and improved well-being just to maintain such public good values for wider society. Therefore, if we are to value the public values that on-farm conservation of crop diversity can provide to society through the provision of a more resilient agriculture and healthier food system, then there is a need for outside intervention to support and promote on-farm conservation by farmers and farming communities (Bellon *et al.*, 2015a). Guidance on providing such support for on-farm maintenance of genetic diversity is a large focus of the following sections of this chapter.

9.2 Establishing Support for On-Farm Conservation

The purpose of an on-farm conservation project or programme is to provide support in various forms to farmers and farming communities that will help them to maintain the cultivation of traditional varieties in their cultivation practices and systems, thereby conserving and perpetuating plant genetic diversity. It has only been in the last 30 years or so that on-farm conservation has attracted scientific attention, though it still struggles to draw enough funds and support for implementation in national plant genetic resources programmes compared to *ex situ* conservation programmes. Most on-farm conservation projects remain donor-funded and time-limited or externally driven, and the extent of this is paltry despite knowledge of the multiple benefits that can accrue.

The International Plant Genetic Resources Institute (IPGRI, now Bioversity International) in collaboration with the national Plant Genetic Resource Programmes of eight countries was among the first global projects which specifically set out to study and analyze on-farm conservation with the aim to strengthen its scientific rational and basis as well as enhance the capacity of national partners to continue to research and implement on-farm conservation activities (Box 9.5).

The United Nations University (UNU) GEF-funded project, People, Land Management and Environmental Change (PLEC), was another major initiative that began to look more closely at understanding the scientific basis for on-farm conservation and which resulted in a number of landmark publications (Brookfield *et al.*, 2001, 2002, 2003; Kaihura and Stocking, 2003). PLEC included a network of locally based clusters in areas of high biological diversity, thus having Brazil, China, Ghana, Guinea, Jamaica, Kenya, Mexico, Papua New Guinea, Peru, Tanzania, Thailand and Uganda as participating

countries. Around the globe, PLEC had some 30 on-farm project sites, where farmers could collaborate with scientists, other professionals and policy-makers to show how and why 'agrodiversity' (which is how PLEC referred to a wider interpretation of agricultural biodiversity and its management) was worth supporting. The PLEC approach was to work with the most skilled or 'expert' farmers in devising ways of using natural resources that combined superior production along with enhancement of biological diversity at the farm and community level.

NGOs have played a key role over the same period in supporting and improving our understanding of on-farm conservation activities especially through farmer and farming community institutions. The Community Biodiversity Development Conservation (CBDC) programme involved a number of countries in Africa, Latin America and Asia, and was spearheaded by several local and international NGOs who brought together governmental institutions and NGOs at the global, regional and national level. Like the above projects, it had a major focus on on-farm conservation and sustainable use of the traditional varieties maintained by farmers. The CBDC programme was largely focused on civil society partnerships to support the ongoing work of farmers and farming communities in on-farm conservation and improved livelihoods. Among the CBDC founding partners were: Community Technology Development Association, Zimbabwe; CPRO-DLO Centre for Genetic Resources, The Netherlands; Genetic Resources Action International (GRAIN), Spain; Centro de Educación y Technología (CET), Chile (on behalf of the Latin American Consortium for Agroecology and Development (CLADES)); Norwegian Centre for International Agricultural Development, Norway; Plant Genetic Resources Centre, Ethiopia; Rural Advancement Foundation International (RAFI), Canada; and, South-East Asian Regional Institute for Community Education (SEARICE), The Philippines.

Since the development of these global programmes, there have been a few examples, largely country-based projects, of other on-farm conservation projects or case studies, some of which have been described or reviewed by Sthapit *et al.* (2012) and De Boef *et al.* (2013b). The

Box 9.5 | **Strengthening Scientific Basis of *In situ* Conservation of Agricultural Biodiversity On-Farm**

(Modified from Jarvis *et al.*, 2008b; Sthapit *et al.*, 2012)

The first global on-farm conservation project involved Burkina Faso, Ethiopia, Hungary, Mexico, Morocco, Nepal, Peru and Vietnam collaborating with Bioversity International. The project investigated the extent and distribution of genetic diversity in 27 crops and explored with farmers and rural communities the management practices used to maintain traditional varieties. The research products included: tools to assess the amount and distribution of crop genetic diversity in production systems; an increased understanding of when, where and how this diversity is maintained by farmers; identification of practices, communities and institutions that support maintenance and evolution of crop genetic diversity in production systems; and provided possible mechanisms for ensuring that the custodians of these systems and genetic materials benefit from their actions. Participating countries worked together to collate datasets from biologically and culturally diverse sites into a small number of globally applicable diversity indices that could be used for comparison across farmer households and communities. Data were collected on varieties representing 27 crop species from five continents. This data was analyzed to determine overall trends in crop varietal diversity on-farm. Measurements of richness, evenness and divergence demonstrated that considerable crop genetic diversity, in the form of traditional crop varieties, was maintained on-farm. It was found that major staples had higher richness and evenness than non-staple crops. Variety richness for clonal species was much higher than that of other breeding systems. The research suggested that crop diversity was maintained as an insurance to meet future environmental changes or social and economic needs and underscored the importance of many small farms adopting distinctly diverse varietal strategies as a major force that maintains crop genetic diversity on-farm.

first comprehensive training guide for planning and implementing an on-farm conservation initiative was published by Jarvis *et al.* (2000) and to map out how to establish an on-farm conservation project by Maxted *et al.* (2002). *Genes in the Field* (Brush, 2000), *Managing Biodiversity in Agricultural Ecosystems* (Jarvis *et al.*, 2007), *European Landraces: On-Farm Conservation, Management and Use* (Veteläinen *et al.*, 2009a) and *Crop Genetic Diversity in the Field and on the Farm* (Jarvis *et al.*, 2016) represent some additional landmark publications dealing with the topic. Bellon *et al.* (2017) have briefly reviewed the practices and evidence for on-farm conservation.

9.3 Supporting On-Farm Conservation

While there are many ways in which farmers can benefit from a greater use of local crops and varieties, in many instances, farmers will need support if they are to find ways of making their traditional landraces compete with modern and bred cultivars of major crops. The selection of options to promote the conservation of genetic diversity on-farm are many and can only be formed after consultation with farmers and the local community, and after an assessment of the socio-economic and political factors in the target region. The goal is to promote an environment where the farmer wishes to maintain a diverse range of traditional varieties on his/her farm but not to dissuade the farmer from also adopting new crop varieties that may also increase food availability and income and general well-being. The second State of the World's Report on PGRFA identifies several potential interventions that can be implemented to help strengthen and promote on-farm conservation and increase competitiveness of traditional varieties. These include (FAO, 2010a):

- Adding value through improved characterization of traditional varieties;
- Improving traditional varieties through plant breeding and seed processing;
- Improving market incentives and public awareness to increase consumer demand;
- Improving access to information and materials; and
- Enhancing enabling environments through more supportive policies, legislation and incentives.

Adding Value through Improved Characterization of Traditional Varieties

While work has been carried out in several countries on characterization of local materials, traditional varieties are often inadequately characterized, especially under on-farm conditions. There is some indication from the country reports which contributed to the State of the World's Report on PGRFA (FAO, 2010a) that greater efforts have been made to characterize traditional and local varieties over the past decade since the first report. An example of how this has occurred was the development of the commercial cultigen Grindstad for Timothy grass in Norway, which was originally a local landrace (Marum and Daugstad, 2009). One topical, yet a much neglected, subject in relation to this is the huge difference between crop species and varieties in terms of their nutritional compositional value. Although some work has been carried out in this area, the level of nutritional analysis research, especially at the level of traditional varieties, is negligible. However, the topic is one of much current interest in relation to diverse and healthy diets and improved nutrition and health and could add considerable value and interest to the use of traditional varieties (Hunter *et al.*, 2015; Kennedy *et al.*, 2017).

Improving Traditional Varieties through Breeding and Seed Processing

Improvement of local materials can be achieved through plant breeding and/or through the production of better-quality seed or planting material, as illustrated by the case of 'Muchamiel' and 'De la Pere' tomatoes in Spain (Ruiz and Garcia-Martinez, 2009). The improvement of traditional varieties (in terms of yield, quality and disease resistance) through local plant breeding programmes, such as participatory plant varietal selection and breeding, is a focus of Chapter 17, and the approaches, methods and tools described therein are very applicable in the context of on-farm conservation. Effective participatory plant varietal selection and breeding can make the cultivation of traditional varieties more competitive compared with modern varieties and exotic crops. It can also enhance the diversity of traditional varieties maintained on-farm. The collaboration of farmers with plant breeding institutes and gene banks would also improve the farmer's access to new sources of genetic variation by repatriating lost traditional varieties or introducing ones from another region.

Improving Market Incentives and Public Awareness to Increase Consumer Demand

Raising public awareness of local crops and varieties can help build a broader base of support. This can be achieved in many ways, for example, through personal contacts, group exchanges, diversity markets and fairs, poetry, music and drama festivals and the use of local and international media. Some of these approaches were used to develop new markets and enhance the supply chain of the landrace 'Bere' barley in the Scottish islands (Martin *et al.*, 2009). Other ways of income generation through market incentives include promoting agritourism and ecotourism, and branding products with internationally accepted certificates of origin or similar for niche markets. The creation and development of specialized niche markets for traditional varieties would also encourage farmers to continue growing traditional varieties. This may require active promotion or marketing of the traditional varieties to stimulate consumer demand for diverse food crops. Facilities that would reduce the drudgery of cultivating certain traditional varieties including their processing and preparation are important, for example, the recent work on this in relation to minor millets in India (Bergamini *et al.*, 2013), as is storage and transportation of traditional varieties to market that would increase the net profits for the farmer.

Improving Access to Information and Materials

The importance of maintaining and managing information and knowledge about diversity at the community or farmer level is recognized in many country reports of the second State of the World Report (FAO, 2010a). A number of initiatives have been developed through the NGO community, aiming to strengthen indigenous knowledge systems, for example 'Community Biodiversity Registers' in Nepal, that record information on varieties grown by local farmers. Diversity fairs also allow farmers to see the extent of diversity available in a region and to exchange materials and have proven to be a popular and successful way of strengthening local knowledge and seed supply systems. To facilitate the access and exchange of genetic diversity in the form of new traditional varieties from other regions or to replace old varieties that have been lost, Community Seed Banks (Vernooy *et al.*, 2015, 2017) and farmer networks could be established or improved.

This form of farmer-based conservation, or what is commonly referred to as community biodiversity management (De Boef *et al.*, 2013; De Boef and Subedi, 2017), is at a local community level where the farmers have control over their own genetic resources. The emphasis is that farmers are rewarded for displaying the widest range of diversity, not for the biggest or best specimens. Community-based management and diversity fairs should focus on native crops and provide an opportunity for farmers to share knowledge of farming practices, and exchange and purchase seed that they may have lost or wish to experiment with on a trial basis. Public recognition of those farmers who cultivate a wide range of traditional varieties may act as an incentive to other farmers to recover or adopt local varieties. Such 'custodians' and the ways and means of supporting them have been

described (Sthapit *et al.*, 2016, 2017). Chapter 10 contains a detailed description of the various practices of CBM that can be used to strengthen and support on-farm conservation.

Within the UK, the Scottish islands are an area with a particularly high concentration of traditional landrace cultivation in on-farm systems, but the relative old age of the maintainers, the difficulty of maintaining the market and a series of poor harvests led to a significant loss of genetic diversity in the 1990s/2000s. To help sustain long-term production the Scottish Landrace Protection Scheme was set up at Science and Advice for Scottish Agriculture (SASA) in 2006 (Green *et al.*, 2009) with the aim to store seed of all landraces currently grown in Scotland; to collect passport information about donor and landrace; to seek growers' consent for general distribution of the seed; to characterize the accessions; and to provide the donor access to stored seed for continued use of landraces, i.e. to form a safety net for the landrace. Growers who participate in the scheme can donate some of their seed crop each year for storage in the SASA gene bank and have access to some of this seed in the event of harvest failure. In addition, growers are advised on the germination of seed donated. This provides a valuable illustration of the formal and informal genetic conservation sectors working together to promote long-term maintenance of a unique genetic resource.

Enhancing Enabling Environments through More Supportive Policies, Legislation and Incentives

Agricultural policies that include perverse incentives which discourage the conservation of agro-biodiversity, for example, providing credit for the cultivation of modern cultivars only, will need reform if on-farm conservation is to be successful. However, governments may for various political and economic reasons be reluctant to remove perverse incentives. The pressure of public opinion can be used to help persuade the government, which underlines the importance of on-farm conservation projects in raising public awareness and education of conservation issues. Positive incentives that encourage the maintenance or sustainable use of traditional varieties may have to be introduced or recommended, as has recently been brought in to establish quality label or protected designation of origin (Veteläinen *et al.*, 2009b), or attempted in India to promote minor millets in India (Notaro *et al.*, 2017) or for underutilized native biodiversity in Brazil (Kennedy *et al.*, 2017). Policies could include paying direct subsidies to farmers or improving the amenities and local facilities in a village.

Traditional varieties are also generally dynamic and evolving entities (FAO, 2010a), characteristics that need to be recognized in policies designed to support their maintenance. Recent years have seen several countries enact new legislation to support the use of traditional varieties as reported by the second State of the World Report. In Cyprus, for example, the Rural Development Plan 2007–2013 is the main policy instrument covering the on-farm management of PGRFA. It contains a range of different measures to promote the conservation and use of diversity in agricultural and forestlands within protected areas. In Hungary, the National Agri-Environment Programme (NAEP) has adopted a system of Environmentally Sensitive Areas (ESA) through which areas of low agricultural productivity that have, however, high environmental value are designated for special conservation attention. While in the European Union specific legislation on 'conservation varieties' (Commission Directive 2008/62/EC 20 June 2008) has attempted to promote landrace maintenance across the EU.

On-farm conservation of traditional varieties by farmers can only be sustainable if the contribution that farmers have made to crop diversity and its maintenance is recognized and incentivized by the international community. Commercial varieties are often the product of applying breeder's technologies to an individual farmer's variety. The innovative work expended in producing a new variety by a plant breeder in this way is only too often rewarded and protected by enforcing Plant Breeders' Rights. The property protection over commercial varieties provided by Plant Breeder's Rights provides financial returns to the breeder, but no reward for the farmer from whom the original genetic material was collected and used. The concept of Farmers' Rights, already

described in Chapter 2, was developed in recognition of this inequality. Farmers' Rights are now internationally recognized by the FAO; however, the system of compensating farmers for their efforts in maintaining crop diversity has yet to be effectively implemented.

In a detailed study assessing the impact of efforts to promote on-farm conservation, Bellon *et al.* (2015b) documented 79 interventions across five projects in Ecuador, Peru (2) and Bolivia (2) with the highest number of specific interventions provided by any one project being 19 in Ecuador. These interventions were grouped across 13 thematic areas including:

- providing new knowledge about the native crop diversity held beyond the household and community;
- providing access to additional diversity of target crops;
- providing new knowledge, skills and practices for the agronomic management of target crops;
- providing new knowledge, skills, practices and technologies for managing important pests of target crops;
- providing new harvesting knowledge, practices and technologies for target crops;
- providing new knowledge, skills, practices and technologies for storing or processing target crops;
- providing new knowledge, skills and practices for preparing and consuming target crops;
- providing new knowledge, skills, practices and organization on marketing target crops;
- providing new knowledge, skills, practices and organization for participating in agro-tourism;
- training local farmers to provide advice to others on agricultural matters;
- disseminating information to other farmers on agricultural matters;
- providing new knowledge and skills to support farmer organization; and
- providing new knowledge and skills on agro-forestry.

Jarvis *et al.* (2011) identify an equally large number of possible interventions, about 60 in total, that projects can use to promote and support on-farm conservation. Based on this they have proposed a heuristic framework (Figure 9.2) which they believe

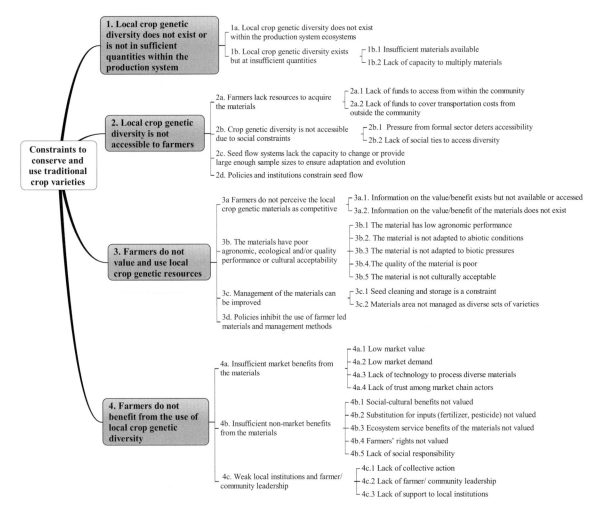

Figure 9.2 Heuristic framework for identifying actions to support on-farm conservation and use of traditional crop varieties. (From Jarvis *et al.*, 2011)

conservation and development practitioners working with farmers and farming communities can use to determine which options and actions will be the most relevant in different contexts.

The framework is based on grouping into four main categories the issues and challenges that make it difficult for farmers to benefit from on-farm conservation, while identifying a myriad of potential actions that can be explored to overcome these constraints. The four main groups are:

- the lack of enough diversity of traditional crop varieties within the production system;

- the lack of access by farmers to available diversity;
- the limitations in information on and the performance of varieties available in key aspects; and
- the inability of farmers and communities to realize the true value of the materials they manage and use.

Jarvis *et al.* (2011) also highlight an intervention category or theme not yet discussed in this context. The option of Payment for Ecosystem Services or the payment for agro-biodiversity conservation services provided by on-farm conservation (Box 9.6).

Box 9.6 | Payment for Agro-biodiversity Conservation Services

(Modified from Bioversity International: www.bioversityinternational.org/pacs/)

The benefits of agricultural biodiversity are not fully reflected by the market. This creates a bias in favour of commercially-driven, often highly specialized monocultural, agricultural systems designed to maximize output. The resulting displacement leads to many other important plant (and animal) genetic diversity becoming increasingly threatened as conserving biodiversity is not considered commercially attractive. Payment for Agro-biodiversity Conservation Services (PACS), the application of Payment for Ecosystem Services specifically for agricultural biodiversity conservation, is a novel idea currently being developed and tested in several countries including Peru, Bolivia, Ecuador, India and Nepal, to test the potential of competitive tenders in creating cost-efficient strategies to conserve priority endangered species and also improve indigenous farmer livelihoods.

Agricultural biodiversity's unrecognized values in the commercial market-place mean that conservation is often carried out on-farm by poor smallholder farmers, who maintain certain crop species/varieties at their own cost for reasons other than just high output. Smallholder farmers, especially those on marginal lands, are often much more interested in minimizing risk than in maximizing productivity. They need to feed themselves and their families. A surplus for sale is good, but not the key to a sustainable livelihood. For example, maintaining a variety of different crops can reduce the risk of complete loss in the event of harvest failure. A particular species/variety may also be socially and culturally valuable, used as part of a traditional cuisine or ceremony such as a wedding. Yet maintaining crop diversity at the 'on-farm' level generates benefits at the local, national or even global level.

The costs of maintaining diversity for local, national and global benefit is currently borne by the smallholder farmer. This cannot continue if we want to secure socially desirable levels of conservation for the greater public good and protect the priority crops that are at the most risk of extinction. Recognition of the value of farmers' work in maintaining such agricultural biodiversity, and the provision of positive incentives that adequately compensate them for doing so (as called for by the Convention on Biological Diversity), is urgently needed – a form of Payment for Ecosystem Services.

9.4 Implementing an On-Farm Conservation Project

The development of a single methodology for on-farm conservation that can be adapted to any crop group is challenging, if not impossible. The scenarios, contexts and conditions for on-farm conservation are so diverse and unique, each with their own peculiar challenges and

opportunities, it makes any blueprint or prescriptive approach doomed to failure. The social, institutional, economic, cultural and political environment influencing farming systems makes this endeavour so complex, and it is the farmer and farming community, through the cultivation of traditional varieties who undertakes the conservation of agro-biodiversity, not the scientists or other development practitioners. Much of the decision-making and conservation actions are beyond the direct control of researchers, and conservation and development practitioners. The focus of any on-farm conservation project must be the farmer and community and must involve a participatory and consultative approach which aims to empower farmers and communities, and strengthen livelihoods and well-being. Only by doing this will there be any chance of sustaining on-farm actions.

It is also stressed that after 30 years of research and practice there is now a much better information and knowledge base to assist and guide the implementation of on-farm conservation projects. Anyone involved in or attempting to undertake an on-farm conservation project is strongly advised to make themselves familiar with this literature, especially the landmark publications highlighted in Section 9.2. Of particular value in this context is *A Training Guide for In Situ Conservation On-farm* (Box 9.7, Jarvis *et al.*, 2000), which was written for key actors and stakeholders involved in national programmes interested in promoting and supporting the conservation of agricultural biodiversity on-farm. The guide attempts to cover the basic skills and tools required to build institutional capacity and partnerships to implement an on-farm conservation programme. It addresses the kind of information required which can form the baseline situation, as well as the practical steps that can be taken for implementation.

The on-farm conservation approach, compared to other *in situ* and *ex situ* approaches, is unique, in that the conservationist's role is restricted, and it is the farmer and farming communities that take the lead in maintaining the PGRFA resource. Although, as noted above, it is not possible to propose a single methodology for on-farm conservation, there are certain options and steps that will always take place in order to better support an on-farm conservation project or programme.

9.4.1 On-Farm Project Planning and Establishment

Identification of Project Sites

Ecogeographic surveys and survey missions for target crop species as well as consultations with farmers and farming communities help identify potential agro-ecological regions of high genetic diversity where on-farm conservation projects could potentially be initiated (see Chapter 1). Potential sites are likely to be situated in primary and secondary centres of crop diversity. Several factors have been identified that may indicate that an area is suitable for the establishment of an on-farm conservation project. These include:

- Fragmentation of land holdings
- Marginal agricultural conditions
- Upland areas
- Heterogeneous abiotic conditions (e.g. soils, rock types)
- Economic or physical isolation
- Distinct ethnic groups
- Distinct cultural values
- Preference for diversity

Each of these factors may be associated with the evolution of localized genetically distinct alleles and so should be considered when identifying on-farm project sites.

Selecting sites will involve assessment of the level of genetic diversity at competing potential sites for the target taxon (i.e. crop or crops) as the conservation goal is to maximize conserved genetic diversity. This may either be achieved directly by genomic comparison, but where resources are limited, ecogeographic, including environmental heterogeneity, and morphometric techniques can be used as a proxy for comparison of the actual genetic diversity found in villages and individual farms (Figure 9.3). A useful tool to aid selection would be ethnobotanical studies using formal and informal interviews and discussions with farmers. Such ethnobotanical surveys should only be conducted after permission from the chief, village elder or elected official has been granted. Questions incorporated in the interviews may include:

- Which crop species are grown on the farm?
- How many varieties per species are cultivated on the farm?
- What are the local names used for the traditional varieties?
- What are the specific uses of the traditional varieties?

Box 9.7 | A Training Guide for *In Situ* Conservation On-Farm

(Modified from Jarvis *et al.*, 2000)

This Guide covers disciplines ranging from genetics to ecology to anthropology, and topics include sampling, data analysis and participatory methods. Science, project management and development are all included in the following framework:

- *Introduction* – brief overview of *in situ* conservation on-farm, detailing why it is important and how it differs from *ex situ* conservation strategies.
- *Social, cultural and economic factors and crop genetic diversity* – discusses the 'human' side of crop genetic resources management and how this influences farmers' decision-making.
- *Agroecosystem factors: natural and farmer-managed* – covers agroecological factors and their role in shaping crop genetic diversity.
- *Agro-morphological character, farmer selection and maintenance* – highlights the importance of farmer selection of agro-morphological characteristics in the cultivation of intra-species crop diversity and the measurement of the characters through field and lab trials.
- *Crop population genetics and breeding systems* – covers their role and influence in on-farm conservation.
- *Seed systems* – including supply and storage and importance for effective on-farm conservation.
- *Building an on-farm conservation initiative* – discusses the national institutional and disciplinary frameworks necessary for the creation of an on-farm project, based on partnerships between diverse personnel and institutions.
- *Getting started: preparation, site selection and participatory approaches* – discusses the process of implementing research and conservation by diverse disciplines and the range of criteria involved in selecting sites and target crops, as well as carrying out diagnostic surveys.
- *Sampling, structuring, documenting and presenting information for action plans* – discusses the process of documenting the results for use by managers and policy-makers with a particular focus on the importance of returning information to the community
- *Enhancing the benefits for farmers from local crop diversity* – discusses potential strategies to support farmers involved in on-farm conservation activities and which are critical for sustaining the maintenance of crop diversity.

- Are certain traditional varieties sent to market and others retained for home consumption, and if so, why?
- Which modern varieties are cultivated?
- What proportion of land is used to grow traditional varieties compared with modern cultivars?
- Which cultivation techniques are used?
- Which traditional varieties, if any, are currently on trial in home gardens, etc.?

- Who are the suppliers for seed purchase, replenishment or exchange?

The factors that will impact site/farm/variety inclusion in the on-farm conservation project include:

a. *Farm diversity*: The various farmers within one village may grow different, even unique, traditional varieties; it is therefore recommended that as broad a range of farms (ethnic, wealth,

Figure 9.3 Some local maize varieties from the Huehuetenango region, in the Cuchumatanes highlands of Western Guatemala. (A black and white version of this figure will appear in some formats. For the colour version, refer to the plate section.)
(Photo courtesy of Bioversity International/G. Galluzzi)

cultural, etc.) are surveyed for possible inclusion in the on-farm project.

b. *Varietal number*: The number of traditional varieties maintained by each farmer is seen as an estimate of genetic diversity; therefore, farmers who grow a larger number of traditional varieties are assumed to maintain higher levels of genetic diversity.

c. *Morphometric diversity*: Relative diversity can also be assessed by morphometric examination, which involves direct observation of traditional varieties in the fields and discussions with the farmer on how each traditional variety is identified and differentiated. Traditional varieties are usually identified by a combination of morphological, agronomic and use characteristics. How the 'folk' taxonomy used by farmers corresponds to the systematic nomenclature used by scientists will also require assessment.

d. *Genetic diversity*: Where resources are available, 'true' genetic diversity can be assessed using biochemical and molecular markers. The amount of genetic diversity and its distribution amongst and within populations of traditional varieties in and between villages and regions will be dependent on many factors including the breeding system of the crop (clonal, self-fertilizing or outcrossing) and crop management. The measurement of genetic diversity and its partitioning over the target crop's distribution should indicate which sites and the ideal number of sites to represent most genetic variation found and therefore where the on-farm project should be sited.

e. *Ecogeographic diversity*: Where resources are unavailable, ecogeographic heterogeneity can be used as a proxy for 'true' genetic diversity. This will involve selecting sites for establishing on-farm conservation projects based on spatial or temporal heterogeneity (e.g. traditional varieties with different flowering times), which should be given priority over homogeneous areas. The wider the range of ecogeographic diversity in a potential site the better, because genetic diversity is commonly associated with ecotypic adaptation and therefore allelic distinction. For example, do the fields cultivated by the farmer span a range of elevations? Are certain soil types associated with different fields? Direct observation of the fields and field characteristics will be necessary.

f. *Relative cost of implementation*: Pragmatically, the relative cost of inclusion of one site as opposed to another in the on-farm conservation project will also affect site selection. If two sites contain roughly equivalent levels of genetic diversity, then the site with lower establishment and implementation costs should be chosen. To determine the relative establishment and implementation costs it is advisable to undertake some form of cost–benefit analysis before the actual on-farm conservation project site is finally selected.

g. *Site genetic uniqueness*: To facilitate long-term conservation, it is advisable that more than one location for on-farm conservation is established for any crop. Thus, if genetic diversity is lost at any one location, it will have a lesser effect on the conserved gene pool as a whole. Locations that complement each other in terms of traditional variety composition and local ecogeographic conditions should be selected where possible. This will permit the conservation of diverse ecotypes representing the entire gene pool that it may not be possible to conserve at any one location.

h. *Inter-species crop diversity*: The main objective of setting up an on-farm conservation project will be to maximize the conservation of the genetic diversity of the target crop(s). As well as having a high level of intra-species diversity of the selected crop, the participating farmer may also grow a wide range of other crops, vegetables or herbs (inter-species diversity) for subsistence or sale. The level of inter-species diversity at each site should also be considered when selecting sites. Obviously selecting such farms will enhance broader plant genetic resource conservation.

i. *Integration with other conservation and rural development initiatives*: Where possible, the on-farm conservation project should be integrated into existing local community development or conservation projects.

Project Sustainability

Sustainability is a fundamental concept for any on-farm conservation project just as it is for genetic reserve and other conservation activities. Compared to *ex situ* conservation, *in situ* techniques are not an inexpensive option. Once the on-farm conservation project has been established, there is an ongoing high level of commitment in terms of financial and personnel resources to sustain the selected site and its farming system. Monitoring of crop populations, agricultural practices and socio-economic changes will be necessary as well as measures to promote sustainable traditional variety cultivation.

Socio-economic and political factors will place constraints on the biologically ideal location and sustainability of any on-farm project. The level and effects of these factors on agro-biodiversity and its maintenance by farmers at included sites will have to be assessed in the planning stage of the project, so that any negative influences that may adversely affect levels of genetic diversity can be ameliorated or any conflicts minimized. The assessment will involve research in the form of interviews and surveys at the farmer level, community level and government level. The approach should be inter-disciplinary, involving aspects of ethno-botany, economy and sociology.

The threat factors that will impact site/farm/variety maintenance within the on-farm conservation project include:

a. *Loss of agro-biodiversity:* An assessment of the factors that may threaten the maintenance of agro-biodiversity and lead to the abandonment of traditional varieties will be required. The sorts of factors that threaten agro-biodiversity sustainability are a subset of those associated with genetic erosion and will include:
    o Introduction of government subsidies for improved varieties or exotic crops
    o Extension services that promote improved varieties or exotic crops
    o Changes in land use
    o Socio-economic change or upheavals
    o Changes in agricultural practice

    However, these factors may cause genetic erosion but not total extinction of localized genetic diversity. Modern cultivars of potato were released in Peru in the 1950s and today can be found in most villages in the highlands. Although Andean farmers have adopted modern cultivars and modern technological inputs, they continue to cultivate them alongside traditional varieties.

b. *Rural development plans:* The on-farm conservation project should be established where there are no imminent threats from rural development projects, such as the construction of dams for hydroelectricity, drastic changes in land use or plans to subsidize modern cultivar production. An assessment of future government proposals for regional development should be a routine prerequisite of on-farm site selection.

c. *Multiple projects and* ex situ *conservation:* The establishment of multiple sites for the on-farm conservation project will increase the sustainability of the project. The loss of agro-biodiversity at one site will have less impact on the overall on-farm project. Even in the most well-researched location, genetic erosion may become a problem. Therefore, it is important to ensure a sample of germplasm from each traditional variety is conserved *ex situ* using an appropriate technique.

d. *Socio-economic factors:* There are many socio-economic factors that contribute to the genetic erosion or maintenance of agro-biodiversity. Modernization of agriculture and associated technological improvements has led to the production of high-yielding crop cultivars and increased mechanization of farming processes. This process has been necessary to keep up with the increasing demands imposed by a growing human population. In many rural areas, traditional varieties grown in traditional agro-ecosystems have been replaced by improved high-yielding cultivars. These modern cultivars, which are generally produced for wide-scale adoption, are suited to uniform agronomic conditions in favourable environments where high agricultural inputs (fertilizers, pesticides, etc.) are necessary. Resource-poor farmers living in heterogeneous and marginal environments, however, have tended to continue to rely on traditional varieties for their subsistence. By growing a range of diverse traditional varieties and crops, the farmer reduces the risk of large-scale crop failure and increases household food security, an essential feature of subsistence agriculture. The reasons for the adoption of modern cultivars and the abandonment of traditional varieties are complex.

There are many examples of farmers growing both traditional varieties and modern cultivars side by side to suit their diverse needs. Traditional varieties are often grown for home consumption and are still favoured because of their agronomic qualities, culinary preferences and storage properties. Modern cultivars are commonly grown for commercial sale in local or remote markets where product uniformity and value are prime concerns.

e. *Political factors:* Primary and secondary centres of crop diversity are often located in developing countries. On-farm conservation projects will therefore be primarily focused in these countries. Government food and agricultural policy in these countries will almost invariably be directed towards self-sufficiency and hence increased food production. There may therefore be a conflict between the goals of on-farm conservation of traditional varieties and government policy, where incentives may be given to farmers to grow modern, high-yielding cultivars or exotic (cash) crops. An assessment of the current agricultural policy of the government and its effect on agricultural biodiversity will be required as a prerequisite to the establishment of the on-farm project. The government policy on national commodity subsidies will need to be reviewed and its effect on the on-farm project assessed. Other areas of agricultural policy that will need investigation include the role of agricultural extension services and the availability of agricultural packages, whereby farmers must adopt modern cultivars to obtain fertilizers or pesticides free or at reduced costs.

f. *Flexibility of management plans:* Unforeseen threats and natural catastrophes, such as periods of drought or famine, cannot be planned for in advance. It is fundamental therefore that monitoring is rigorous, and the management plan is flexible.

Identification of On-Farm Conservation Project Partners

The on-farm conservation project partners are likely to include:

a. *Farmers*: These are likely to be small landholders who already grow traditional varieties of the target crop species. They should value agro-diversity for cultural or agronomic reasons, understand the importance of conservation and be willing to be involved in the project.

b. *Agricultural extension workers*: Their role may be critical in identifying potential farmers to be involved in the project, helping implement the project, as well as bridging the gap between the farmer and the conservationist in the project.

c. *Non-governmental organizations*: There are many NGOs working in rural communities on developmental and agricultural projects. Their experience could be invaluable to the success of the on-farm project.

d. *Community leaders*: Most rural communities have some form of organized local leadership. The success of the on-farm project will involve these local authorities, who are often essential 'gate keepers' facilitating access to the local community.

e. *Researchers*: Their role will be project management and implementation, providing scientific and technical backstopping including assessment, measurement and monitoring of genetic diversity, training and capacity building of farmers, community groups, national agricultural research and extension staff and others.

Each individual project is likely to include a unique mix of the various partners involved. Having identified that a particular village or district has a wealth of genetic diversity in the target crop, it may not be necessary to instigate an on-farm project for the whole village or area. It is more likely that specific farmers will be selected to be involved in the project. The farmers selected could be chosen because they:

a. *Maintain high levels of agro-biodiversity*: Farmers who already have high levels of diversity and an appreciation of this diversity will be an obvious choice for inclusion in the project.

b. *Are older and more experienced*: There may be a correlation between the length of time a farmer has been farming and the percentage of traditional varieties grown.

c. *Are younger*: If the project is to be sustainable, then younger farmers will also need to be selected. This will allow the cultivation skills and knowledge

from more experienced, older farmers to be passed down to younger generations.

d. *Are relatively wealthy*: This group of farmers may wish to continue growing traditional varieties as a hobby or for more sentimental reasons.

e. *Are relatively poor*: This group of farmers may not have the economic resources to change from traditional varieties to high-yielding bred cultivars, particularly if it is associated with an increased level of inputs (fertilizers, herbicides, etc.)

f. *Are from different ethnic groups*: Certain traditional varieties may be associated with ethnic groups with different cultures and traditions.

g. *Are chiefs or village elders*: They may wish to continue growing traditional varieties because they are the major custodians of traditional values within the village. There may always be pragmatic reasons for selecting the chiefs or headmen among those included in the project because of their key role within the village; they act as a model or leader showing the importance of traditional variety conservation for the other farmers.

Thus, the farmers selected will be a true cross-section of their local community, though the precise mix will vary from project to project. The farmers and communities selected for the on-farm project should all show an interest in conservation and be willing to collaborate with ethno-botanists and scientists.

9.4.2 On-Farm Project Management and Monitoring

Formulation of Project Activities

The design of the on-farm project should promote the sustainable conservation of the target crop species at the selected site. Before formulating an on-farm project, the conservationist must have a basic understanding not only of the biology of the target crop species but also the role of the farmer in conserving agro-biodiversity. Only when we have an understanding of the reasons why the farmers continue to grow traditional varieties, their farming system and the factors which are likely to promote traditional varieties cultivation or abandonment can we begin to develop strategies and incentives to encourage and facilitate the conservation of agro-biodiversity.

The information required for formulating the on-farm project should be an extension of the data already collated during ecogeographic surveys and survey missions. On-farm conservation projects require more detailed input from ethnobotanical and social research than other conservation techniques. This information may be obtained from farmer questionnaires and surveys. Research will also be based on direct observation of the farmers and their cultivation techniques.

Basic information requirements for the on-farm project include:

- *Farmer description:* The type of information to be included under this heading is a description of the farmer's household, the smallholding, his/her social status and his/her role in the village community. Although the man may be the head of the family, he is often not responsible for cultivating crops or managing the home garden. In many countries, it is women who are responsible for providing food for the household, hence the farmers are women. This may affect the mode of data collection, especially in some cultures, where it may be unacceptable for male ethno-biologists to collate cultural information from female farmers.
- *Abiotic/environmental features of the site:* The basic abiotic features of the site, for example, the geology, type of soils, climate and drainage, should be known. The elevation over which the farmer's fields extend is also important, as is the proportion of lowland and upland fields.
- *Details of agro-biodiversity:* The traditional varieties should be examined for diversity in morphological characters. This will involve sampling germplasm from traditional varieties identified by farmers, growing out the accessions in field plots under controlled conditions at a research institute, followed by characterization and analysis. Any characterization of morphological characters should use the standard Bioversity International descriptor lists for the appropriate crop.
- *Agricultural practice:* This will involve gathering information on methods of ploughing, sowing, pest

management, selection of seed for the following season, threshing and storage. Types of questions that could be included are:
  - Is the traditional variety intercropped with other crop species?
  - Does the farmer use crop rotation?
  - If the farmer grows a mixture of modern bred cultivars and traditional varieties, does he/she use different methods of cultivation?
  - Does the farmer apply fertilizer, herbicides or pesticides to the crop and if yes, what dosage?
  - Does the farmer employ extra manual labour at peak times?
- *Reasons for growing particular traditional varieties*: There are many reasons why farmers may continue to grow traditional varieties even when adopting modern bred cultivars. The reasons may be agronomic, cultural, socio-economic, due to government policies or aesthetic. Some examples are given below:
  - The traditional variety may perform better in marginal environments.
  - It may be favoured for a particular culinary feature, for example, superiority in flavour or baking.
  - It may fetch a higher price in the market than modern bred cultivars.
  - It may be associated with cultural traditions, as gifts, payment for work or use in rituals.
  - It may store better for human consumption.
  - It may have a multi-purpose use, for example, for human consumption and animal fodder
- *Agro-ecosystem dynamics*: For effective conservation of wild species and natural habitats, we must have knowledge of ecosystem dynamics and interactions. This is also true for crop species within agro-ecosystems. The habitat of crop species is the farming unit itself or in some cases the local community. The fundamental difference is that the farmer has a central role in engineering the environment of the farm. The farmers decide which modern cultivars or traditional variety to plant out each season; they are responsible for planting, harvesting and selecting seed for the following year.

- *Crop usage*: The percentage of the traditional variety or bred cultivars used for consumption, sale or saved for seed is important in understanding the household economics of the farmer.
- *Threats to agro-biodiversity*: There are many threats to agro-biodiversity associated with the loss of genetic diversity, or genetic erosion. One major threat is the replacement of traditional varieties by modern cultivars due to agricultural development and modernization. Other general factors that promote the loss of genetic diversity were discussed in Chapter 1.
- *Conservationists as observers*: When formulating the on-farm project, it is imperative to remember that the conservationist's role is largely a passive one. The farmer is responsible for the maintenance of agro-biodiversity on their own farm. The conservationist's role is to promote and facilitate the conditions under which the farmer can conserve the genetic diversity on his/ her farm, but not to intervene against the wishes of the farmer. The management plan should not be formulated without extensive consultation with the farmers to identify which factors contribute to, promote or constrain the cultivation of traditional varieties.
- *Monitoring activities*: These will not only focus on monitoring the genetic diversity of the target crop but also the socio-economic, cultural and political environment in the region that will in turn affect the crop itself. The frequency of sampling germplasm for genetic analysis, the type of sampling technique and the protocols for measuring diversity should be included in this section.
- *Research activities*: There will inevitably be areas within the on-farm project that will require further research, whether of a scientific or a social nature. These should be incorporated within the management plan with details on any collaborating institution, NGO or agency. Field trials could be set up where one or two aspects of farmer management are applied to an experimental crop population and compared with a control population.
- *Information management:* The data that are collected and collated over the years will have to be managed so that access and reference to the results of surveys and monitoring is used efficiently. It is critical and ethical that such information is made available to farmers and communities. The information can be organized into:
  - *A basic library* – with essential textbooks, manuals and journals.
  - *Project files* – including complete reports on research and monitoring.
  - *Project database* – which could include details and maps of traditional variety cultivation and agricultural practice.
  - *Appropriate information tools* – that are appropriate and accessible to farmers and farming communities.
- *Education and public awareness*: This is an essential component of the on-farm project. Education in the benefits of conserving agricultural biodiversity, both inter-specific and intra-specific, should be aimed at the local farmer and community level, the public and government officials (the policy-makers). Publicity should be produced that highlights the benefits to local food security and commodity independence, environmental protection and nutritional balance. Education at a local level could include visits to schools and community halls, as well as having stalls at local agricultural fairs and markets, or even organizing specific seed fairs to display the diversity of traditional varieties to farmers.
- *Personnel and resource management*: The personnel and resources that will be required to monitor and promote the conservation of agro-biodiversity in the on-farm conservation project will have to be costed and budgeted for.
- *Collaboration with NGOs and extension services*: There may already be NGOs and agencies in the community where the on-farm project is located, which are working on sustainable agriculture and rural development. Collaboration is beneficial as resources and knowledge can often be shared. The role of these organizations in the on-farm project should be included in the management plan.

Project Management and Monitoring

Like the initial implementation of the management plan of a genetic reserve, the initiation of the on-farm management plan will be experimental and will, at least initially, require regular review. Thus, the initiation of the management plan will require careful introduction, combined with evaluation, revision and refinement in the light of its practical application (Figure 9.4). Therefore, the initial level of management will be high, with intensive and extensive monitoring procedures, and thus the plan will need to be flexible.

The purpose of on-farm monitoring is to determine how changes in agricultural practice and management by the farmer affect the genetic diversity of crop populations. Factors such as the size of the crop population planted out each year, the source of the seed supply and the selection criteria used by the farmer will all affect genetic diversity. These farmer-based decisions are essentially dynamic and will be influenced by current environmental, socio-economic, political and cultural factors. On-farm monitoring will therefore involve monitoring:

- Environmental, socio-economic, political and cultural factors
- Agricultural practice and management
- Genetic diversity

These three components are closely interrelated, and emergent patterns should be linked to feedback for the on-farm conservation plan.

Environmental, Socio-economic, Political and Cultural Factors

The management prescriptions that outline and promote the sustainable conservation of traditional varieties are unlikely to have immediate effects on farmer decision-making as they are based on more long-term aims. Key elements that should be monitored include:

- Local and regional market economies, e.g. crop prices
- Consumer demand between and within crops
- Patterns in rural emigration

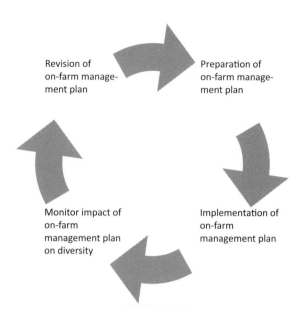

Figure 9.4 The on-farm management implementation cycle.

- National and international environmental and agricultural policy
- Household and community economics

Agricultural Practice and Management

It is useful to apply population genetic theory to farmer's management of agro-biodiversity on-farm. Changes in gene frequencies, and hence genetic diversity in a polymorphic population such as a traditional variety, result from the combination of environmental and human selection, mutation, migration and genetic drift (see Chapter 4). The on-farm agro-ecosystem is dynamic. Monitoring of agricultural practice will be based on ethnobotanical surveys and questionnaires coupled with direct observation in the field. Monitoring should coincide with the key stages of the agricultural calendar, for example, sowing, harvesting selection, and processing (threshing). It may even be necessary to stay for several months of the year in the village where the on-farm project is sited. It is important that monitoring is not too intrusive, that it is gender sensitive, and that the culture and wishes of the farmer are fully respected.

Farmers will continue to change the varieties they plant on an annual basis to meet their needs for subsistence and sale. Observations and detailed notes should be made on which variety the farmer chooses to adopt and which they abandon with the reasons for abandonment or adoption. The nature of the introduced variety, whether it is a high-yielding cultivar or a traditional variety, will affect the genetic diversity of the agro-biodiversity on-farm.

Monitoring Genetic Diversity

In order to monitor temporal changes in the genetic diversity of crops, populations will have to be sampled on an annual basis and analyzed using biochemical or molecular marker techniques. Sampling populations before and after each selection cycle can assess the effects of farmer selection on genetic diversity. By combining and analyzing data patterns between farmer decision-making, farmer management and genetic diversity, any patterns should become apparent.

- *Seed source*: The introduction of seed from a new source (migration) would have the effect of introducing new genetic variation into the on-farm conservation project, which may in turn lead to extinction of locally endemic variation. Therefore, the introduction of novel germplasm into the on-farm project should be thoroughly considered and the effects on native genetic diversity monitored.
- *Selection*: Environmental and human selection will cause the loss of less adapted or undesirable genotypes, which in turn will lead to changes in gene frequencies. The farmer decides where to plant a particular traditional variety each season. Each field on the farm will have a different microhabitat; for example, differences in drainage, nutrient status, temperature and light. Changes in the environmental conditions that a crop population is subjected to over time may affect the genetic diversity of the crop population. So, genetic diversity and gene frequencies within the on-farm project will 'naturally' change over time, and this must be borne in mind when reviewing the results of the monitoring.

Observations should be made on the selection criteria and agro-morphological characters used by the farmer. The number of plants sampled from the population to serve as seed for the next season and the estimated number of seeds collected are important parameters for estimating the effective population size. Farmers will also unconsciously be selecting certain genotypes through their planting, harvesting and processing, and seed storage methods used.

- *Population size*: The size of the population of a particular traditional variety that the farmer maintains will affect the genetic diversity. Small populations are more likely to be subject to the effects of genetic drift, inbreeding and stochastic events, resulting in the loss of alleles and hence genetic variation. Crop population sizes may be reduced for several reasons including:
  - The farmer's decision to reduce the acreage allocated for the traditional variety because of changing personal requirements or markets
  - Losses due to pest or pathogen damage
  - Poor storage conditions
  - Seed designated for planting may be consumed in times of severe drought or civil wars
  - Estimations of population size of seed that the farmer saves each season for the following year should be made, as well as the population size of the actual seed planted out.
- *Gene flows*: If the target taxon is outcrossing, such as maize or sorghum, then there is a high potential for gene flow between different varieties. The rate of exchange of genes will be greater if plots of crop are near each other and if the flowering times of different varieties coincide. Some gene flow between selfing species may also occur but will be lower. Introgression between inter-breeding complexes of wild relatives of the crop species, weedy types and traditional varieties may be an important mechanism for the generation of genetic variation in traditional varieties.

9.4.3 On-Farm Diversity Utilization

The establishment and management of the on-farm project is not an end in itself. There is an explicit link between genetic conservation and utilization: genetic conservation must facilitate utilization, either now or in the future. Utilization of the material conserved on-farm may be divided among traditional, general and professional users.

Traditional Users

The most direct users of the germplasm conserved on-farm are the farmers who traditionally plant, grow and harvest the traditional varieties. The on-farm project must be sensitive to the needs of the local communities in which they are working. The project managers should not attempt to pressurize the farmers into growing traditional varieties that may yield a lower gross income for their families. The success of the on-farm project will certainly depend on the cooperation and support of the farmer and the local community. The on-farm project should aim to improve the livelihoods of the farmers and rural communities involved; therefore, a compromise between conservation ideals and the needs and desires of farmers and local communities may have to be reached. The support of local communities may be raised by offering to employ local people to assist in the on-farm project, for example, in monitoring traditional variety diversity or by organizing local seed shows, thus the community will further benefit from the project. Other farmers within the region will also benefit from the project by having access to a range of traditional varieties facilitated by the on-farm project.

General Users

The population at large, whether local, national or international, may provide essential support that will aid the long-term political and financial viability of the project. The ethical and aesthetic justification for conservation is of increasing importance to the public. The on-farm project should be used to raise general public awareness of the need for conservation in general, but also specifically the need to conserve crop genetic resources. Members of the public may wish to visit the project, and this should be encouraged as an educational exercise. If such agri-ecotourist visits are correctly managed, the local people may receive

further direct financial benefits for their community in terms of merchandising, as well as providing refreshment and accommodation. The on-farm management plan should consider the needs of visitors to the project and the establishment of visitor centres, nature trails, lectures and the provision of various media information packs.

Professional Users

Professional utilization of germplasm conserved in the on-farm project will be similar to the professional utilization of germplasm conserved *ex situ*, which is largely for plant breeding. One of the main disadvantages of *in situ* as opposed to *ex situ* conservation is that it is more difficult for the plant breeder to gain access to germplasm. To ameliorate this problem the on-farm project should ensure that traditional varieties are characterized and evaluated, and the conserved material could be publicized in catalogues. The level of documentation of passport, characterization and evaluation data recorded should be just as extensive for germplasm conserved *in situ* as for *ex situ*.

In many cases, the work of professional users, the public and local people can be linked through partnership within non-governmental organizations, especially those involved in sustainable rural development, conservation volunteers or use of resources in accordance with traditional cultural practices. All partners will therefore share the goals of sustainable use of biological resources taking into account social, economic, environmental and scientific factors, which form a cornerstone to the nations' proposals to implement the objectives of the CBD, GPA and the Sustainable Development agenda.

Linkage to Ex Situ *Conservation, Research, Duplication and Education*

To provide a backup to the on-farm conservation efforts, the germplasm should always be sampled and deposited in appropriate *ex situ* collections. Although both *ex situ* and *in situ* techniques have their advantages and disadvantages, the point is re-emphasized here that the two strategies are not alternatives or in opposition to one another, they are complementary. A good on-farm project manager should ensure that the germplasm present in the on-farm communities is duplicated in *ex situ* collections to provide long-term security. Multiple on-farm projects should be established, where possible, to effectively duplicate the conservation of the material *in situ*.

The reader who is interested in more information for on-farm conservation should also refer to the other relevant chapters of this book. In particular, Chapters 10 and 17 provide the rationale for employing community- and participatory-based approaches with farming communities and farmers, respectively. They also describe in considerable detail the approaches, tools and methods that make up the basket of options made available through community biodiversity management (CBM), including participatory varietal selection (PVS) and participatory plant breeding (PPB), which can bring many benefits to farmers involved in on-farm conservation. Chapters 7–10 of the Jarvis *et al.* (2000) training guide (Box 9.7) focus on the practical aspects of designing and implementing an on-farm conservation project and contain useful information on better understanding the national and institutional context for on-farm conservation as well as a range of techniques that can be used to carry out participatory research. Friis-Hansen and Sthapit (2000) provide additional guidance for participatory approaches for conservation and use of plant genetic resources.

9.5 The Impact of On-Farm Conservation

To date there has been limited evaluation of the impact of on-farm conservation projects that make a rigorous assessment of project interventions over and above what farmers would normally conserve on their own. Bellon *et al.* (2015a) presents a conceptual framework for analyzing on-farm conservation projects. Bellon *et al.* (2015b) apply this framework in one of the few studies to assess the effectiveness of on-farm conservation projects by analyzing five projects that were implemented in the High Andes of South America, where many on-farm conservation

projects have been implemented by NGOs, universities and national research organizations supported by different donors, from national governments to foundations and international agencies. The five projects analyzed in the Bellon *et al.* (2015b) study were implemented in Ecuador, Peru and Bolivia, and represent a range of implementing institutions, donors and partners. The projects involved six native crops, quinoa (*Chenopodium quinoa* Willd., Figure 9.5), cañihua (*Chenopodium pallidicaule* (Allen)), potato (*Solanum tuberosum* L.), oca (*Oxalis tuberosa* Mol.), ulluco (*Ullucus tuberosus* Caldas) and mashwa (*Tropaeolum tuberosum* R.&t P.), and in one project site up to 137 plant species were involved. This included a large range of varieties for each crop. Across all five projects there were a total of 79 interventions, grouped across 13 themes, implemented covering a range of options from implementation of fairs for seed exchange to establishment of a community museum.

Although there were shortcomings in all five projects' design, for example, none of the projects had predetermined control groups and neither baseline nor end-line data were available, the study was able to employ a methodological approach to deal with these and other limitations (Bellon *et al.*, 2015b). Data from household surveys demonstrated that the number of farmers who applied intervention options provided by projects was much higher than anticipated from *a priori* information used to draw the sample of participants from project records, clearly indicating spill-over effects and that project interventions were addressing real needs. The study found that on average those participating in the on-farm conservation projects had implemented between 20% and 40% of the total myriad of options provided by a project. Furthermore, the fact that farmers were able to describe specific examples of how they put actual options into practice was taken as an indication that they were not simply ticking yes or no boxes but were able to articulate how the implementation of options had led to specific behavioural changes.

The findings of this study not only show that there is robust evidence demonstrating that the options provided through on-farm conservation projects can

Figure 9.5 Traditional cultivars of quinoa from Peru. (A black and white version of this figure will appear in some formats. For the colour version, refer to the plate section.)
(Photo courtesy of Bioversity International/A. Drucker)

lead to significant levels of application, but they also have important implications which must be considered in any discussions or efforts to scale up on-farm conservation projects (Bellon *et al.*, 2015b). All five projects implemented an array of different types of interventions – 19 being the highest in one project in Ecuador and 12 the lowest in one project in Peru. These interventions, addressing multiple aspects of production, consumption and marketing of native crops, provided a basket of options approach designed in such a way to the specific social and agro-ecological conditions of project sites, as opposed to seeking out one or two limited options.

Like the approaches of participatory plant breeding and varietal selection, the value of a basket of interventions is to provide diverse choices, some of which may be more important to certain farmers as opposed to others, depending very much on their contexts and circumstances. All of this has important implications for scaling up and which assumes such efforts cannot be successfully pursued simply by trying to apply the same interventions and options

over large areas or groups of farmers. Rather it depends for success on a process of 'systematic contextualization' whereby diverse options drawn from different types of interventions are assembled and targeted to fit different contexts, letting farmers and farming communities choose which fit their needs and circumstances best (Bellon *et al.*, 2015b).

9.6 Landrace Conservation

As argued throughout this text, crop LR are a critical resource for crop improvement because of the breadth of diversity they contain and the fact that, unlike CWR, there is no breeding barrier between the crop LR and the elite lines the breeder may use in producing a new cultivar. Therefore, the value of on-farm conservation is not only in maintaining a diversity-based farming system, but also in maintaining the allelic diversity found in individual LR as a genetic resource for future utilization. While it is recognized that on-farm conservation is a dynamic conservation system and

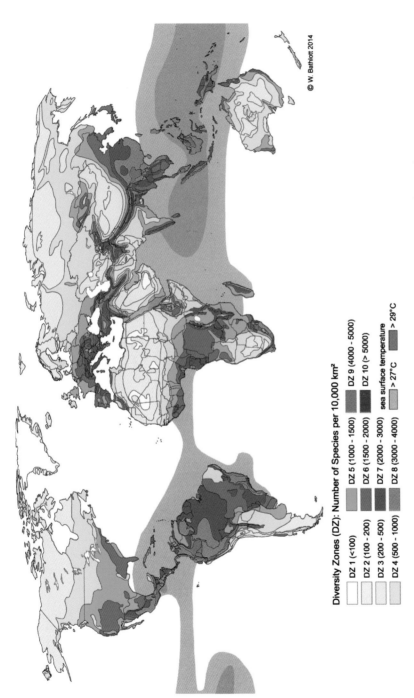

Diversity Zones (DZ): Number of Species per 10,000 km²

DZ 1 (<100)
DZ 2 (100 - 200)
DZ 3 (200 - 500)
DZ 4 (500 - 1000)

DZ 5 (1000 - 1500)
DZ 6 (1500 - 2000)
DZ 7 (2000 - 3000)
DZ 8 (3000 - 4000)

DZ 9 (4000 - 5000)
DZ 10 (> 5000)

sea surface temperature
> 27°C
> 29°C

Figure 1.5 Species diversity globally of vascular plants. (A black and white version of this figure will appear in some formats.) (Reproduced from Barthlott et al., 2014.)

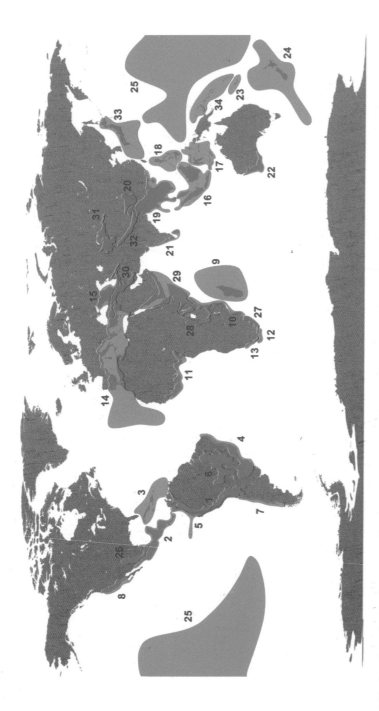

Figure 1.6 The location of areas of exceptionally high biodiversity richness – biodiversity hotspots. (A black and white version of this figure will appear in some formats.) (Reproduced from Mittermeier et al., 1999.)

Verisk Maplecroft Food Security Index 2019-Q1

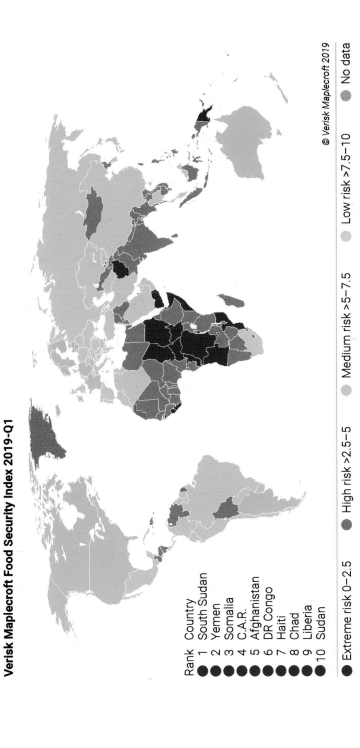

Rank	Country
1	South Sudan
2	Yemen
3	Somalia
4	C.A.R.
5	Afghanistan
6	DR Congo
7	Haiti
8	Chad
9	Liberia
10	Sudan

© Verisk Maplecroft 2019

● Extreme risk 0–2.5 ● High risk >2.5–5 ● Medium risk >5–7.5 ● Low risk >7.5–10 ● No data

Figure 1.8 Food security risk index 2013. (A black and white version of this figure will appear in some formats.) (From Maplecroft, 2013.)

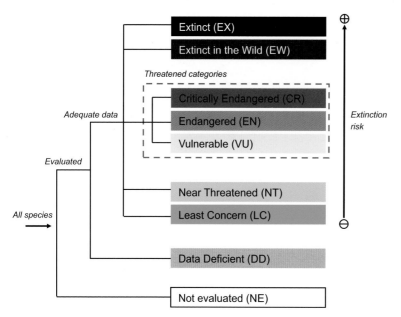

Figure 1.9 Structure of IUCN Red List Categories. (A black and white version of this figure will appear in some formats.)
(Reproduced from IUCN, 2001.)

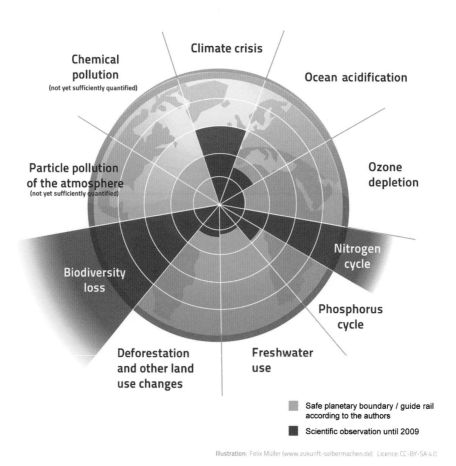

Illustration: Felix Müller (www.zukunft-selbermachen.de) Licence: CC-BY-SA 4.0

Figure 2.1 The planetary boundaries concept where the red areas represent human activities that have exceeded safe margins. (A black and white version of this figure will appear in some formats.)
(From Steffen *et al.*, 2015, after Johan Rockström, Stockholm Resilience Centre et al. 2009)

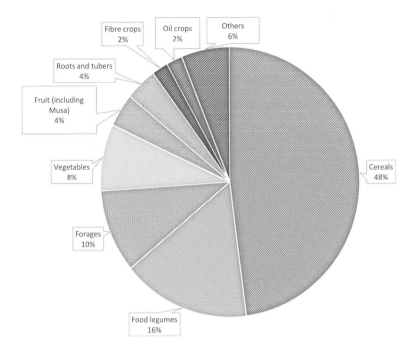

Fibre crops
2%

Oil crops
2%

Others
6%

Roots and tubers
4%

Fruit (including
Musa)
4%

Vegetables
8%

Forages
10%

Food legumes
16%

Cereals
48%

Figure 2.4 Contribution of major crop groups to total *ex situ* collections. (A black and white version of this figure will appear in some formats.) (Reproduced from FAO, 1998.)

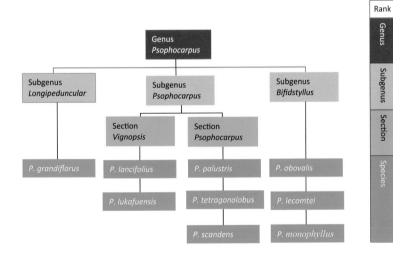

Rank

Genus

Subgenus

Section

Species

Genus
Psophocarpus

Subgenus
Longipeduncular

Subgenus
Psophocarpus

Subgenus
Bifidstyllus

Section
Vignopsis

Section
Psophocarpus

P. grandiflorus

P. lancifolius

P. palustris

P. obovalis

P. lukafuensis

P. tetragonolobus

P. lecomtei

P. scandens

P. monophyllus

Figure 3.2 Organizational view of *Psophocarpus* classification. (A black and white version of this figure will appear in some formats.) (From Fatihah et al., 2012.)

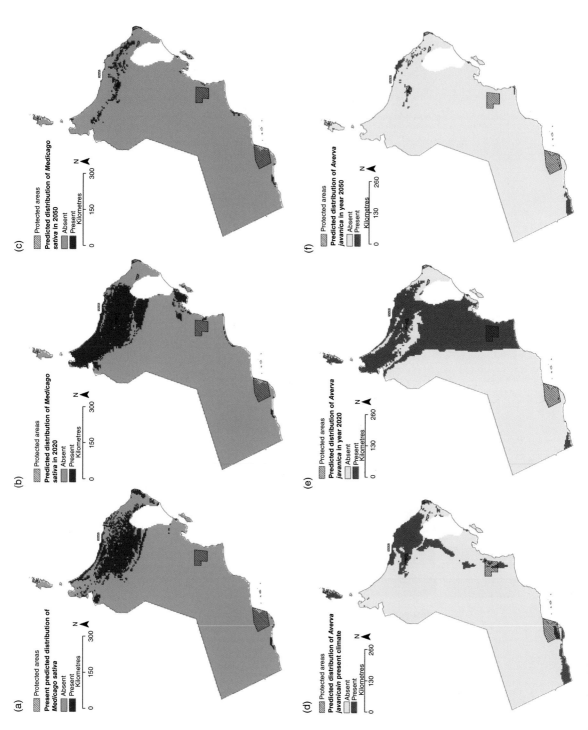

Figure 6.5 Predicted distribution of *Medicago sativa* (a, b and c) and *Aerva javanica* (d, e and f) at present and two future times (red area indicates potential distribution). (A black and white version of this figure will appear in some formats.)

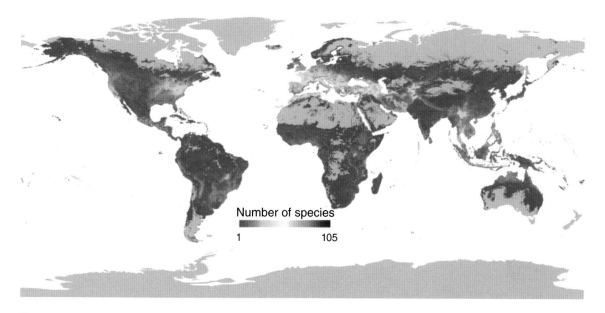

Figure 6.7 Priorities for global *ex situ* CWR conservation. Hotter colours indicate greater CWR concentration. (A black and white version of this figure will appear in some formats.)
(From Castañeda-Álvarez et al., 2016a.)

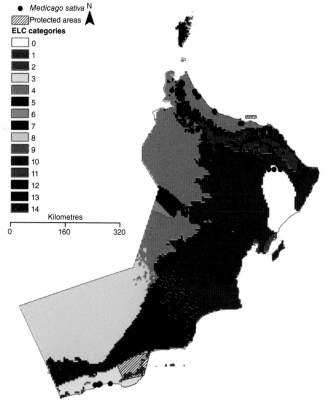

Figure 6.10 Specific ELC map for *M. sativa* locations in Oman. (A black and white version of this figure will appear in some formats.)

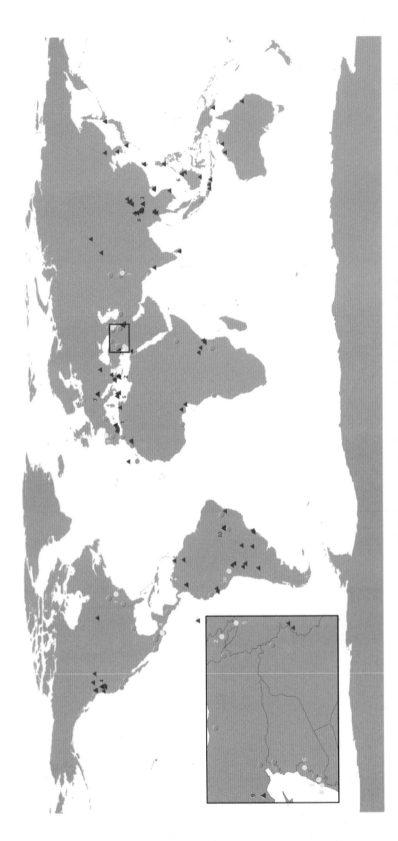

Figure 6.8 Priorities for global *in situ* CWR conservation, with the inset map showing the priority sites in the Fertile Crescent and Caucasus. The top 10 sites within existing protected areas are shown as magenta triangles, and the remaining 90 priority sites within protected areas are represented by blue triangles; the top 10 sites outside of existing protected areas are shown as yellow circles, with the remaining priority 40 sites outside of protected areas represented by turquoise circles. (A black and white version of this figure will appear in some formats.) (From Vincent et al., 2019.)

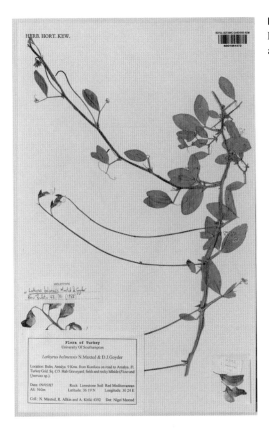

Figure 6.12 An example of a typical herbarium specimen (Royal Botanic Gardens, Kew). (A black and white version of this figure will appear in some formats.)

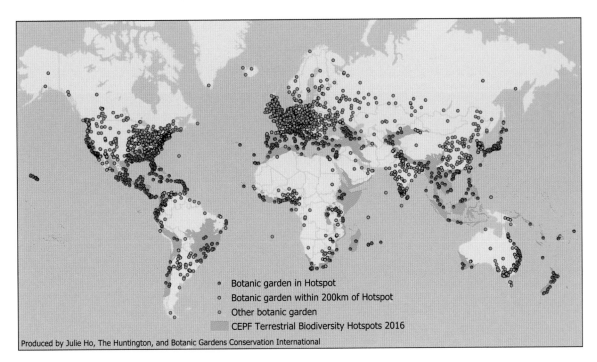

Figure 7.3 Botanic gardens (dots) and biodiversity hotspots (orange areas). (A black and white version of this figure will appear in some formats.)

(Image prepared by Julie Ho, The Huntington and BGCI).

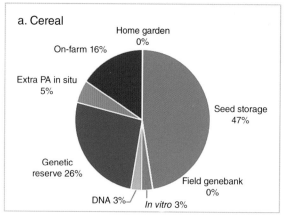

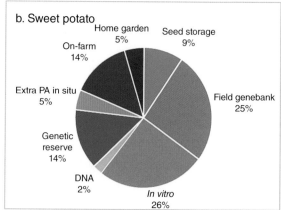

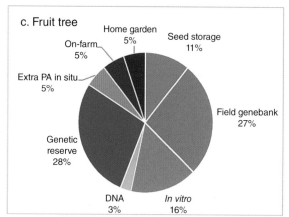

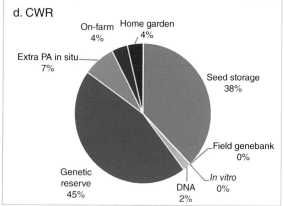

Figure 7.4 Hypothetical representation of the proportions of the gene pool conserved using eight *in situ* and *ex situ* conservation techniques for different plant groups: (a) cereal crop, (b) sweet potato, (c) fruit tree and (d) CWR. (A black and white version of this figure will appear in some formats.)
(Adapted from Maxted et al., 1997b.)

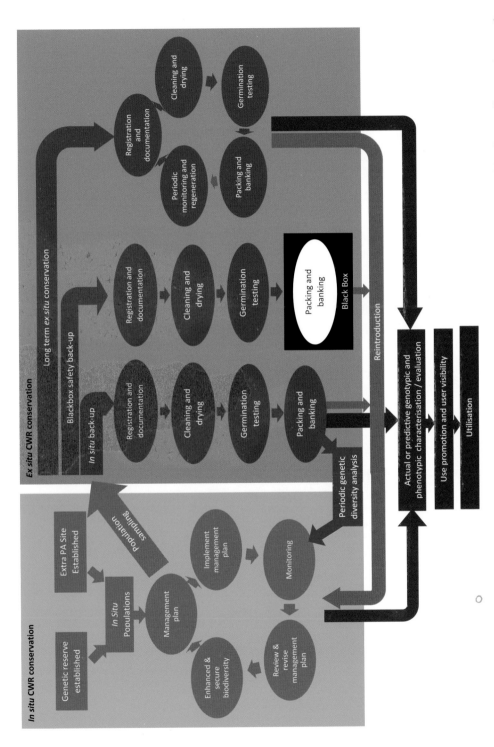

Figure 8.6 Integration of *in situ* and *ex situ* CWR conservation with utilization. PA, protected area. (A black and white version of this figure will appear in some formats.) (From Maxted 2020.)

Figure 9.3 Some local maize varieties from the Huehuetenango region, in the Cuchumatanes highlands of Western Guatemala. (A black and white version of this figure will appear in some formats.)
(Photo courtesy of Bioversity International/G. Galluzzi).

Figure 9.5 Traditional cultivars of quinoa from Peru. (A black and white version of this figure will appear in some formats.)
(Photo courtesy of Bioversity International/A. Drucker.)

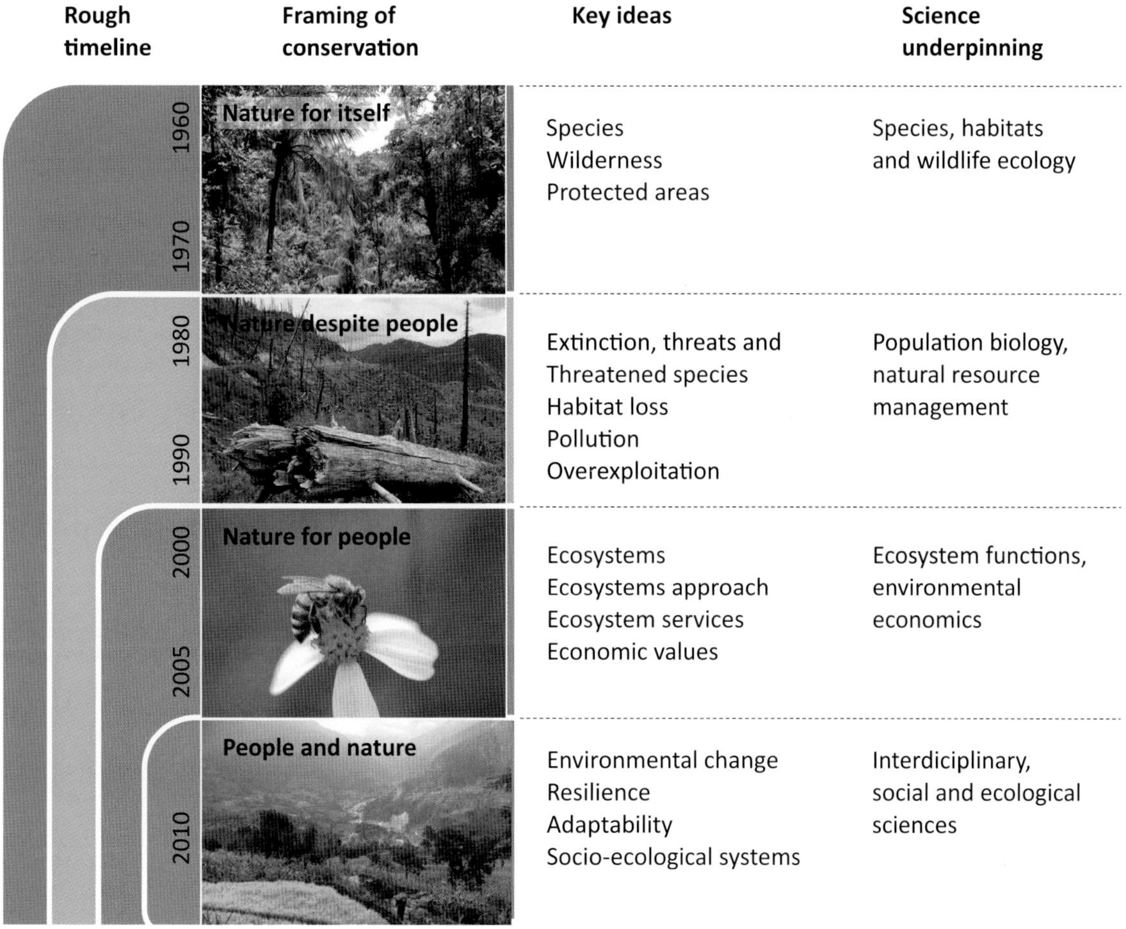

Rough timeline	Framing of conservation	Key ideas	Science underpinning
1960 – 1970	**Nature for itself**	Species Wilderness Protected areas	Species, habitats and wildlife ecology
1980 – 1990	**Nature despite people**	Extinction, threats and Threatened species Habitat loss Pollution Overexploitation	Population biology, natural resource management
2000 – 2005	**Nature for people**	Ecosystems Ecosystems approach Ecosystem services Economic values	Ecosystem functions, environmental economics
2010	**People and nature**	Environmental change Resilience Adaptability Socio-ecological systems	Interdiciplinary, social and ecological sciences

Figure 10.1 The multiple framings of conservation over the past 50 years. (A black and white version of this figure will appear in some formats.)
(From Mace, 2014.)

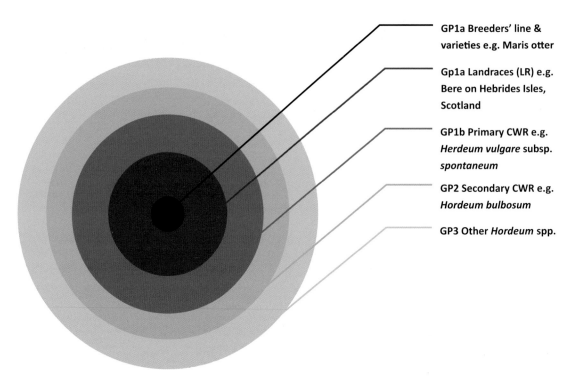

GP1a Breeders' line &
varieties e.g. Maris otter

Gp1a Landraces (LR) e.g.
Bere on Hebrides Isles,
Scotland

GP1b Primary CWR e.g.
Herdeum vulgare subsp.
spontaneum

GP2 Secondary CWR e.g.
Hordeum bulbosum

GP3 Other *Hordeum* spp.

Figure 11.2 Estimated relative genetic diversity held in each component of the barley gene pool with bulk of diversity found in landraces, primary and secondary CWR. (A black and white version of this figure will appear in some formats.)

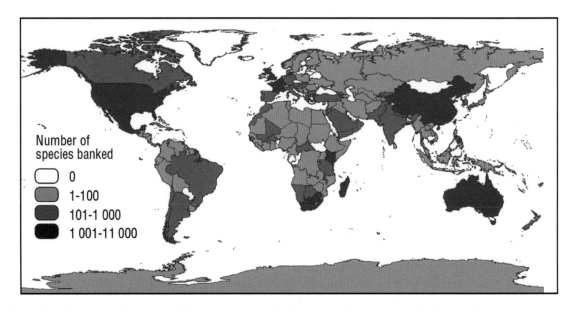

Number of
species banked

0

1-100

101-1 000

1 001-11 000

Figure 13.2 Number of wild taxa seed banked per country. (A black and white version of this figure will appear in some formats.) (O'Donnell and Sharrock, 2015.)

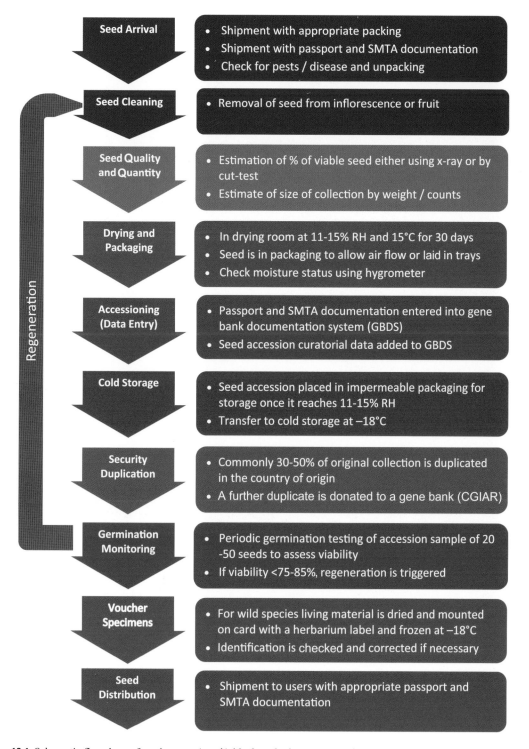

Figure 12.1 Schematic flowchart of seed processing. (A black and white version of this figure will appear in some formats.) (Adapted from Terry *et al.*, 2003.)

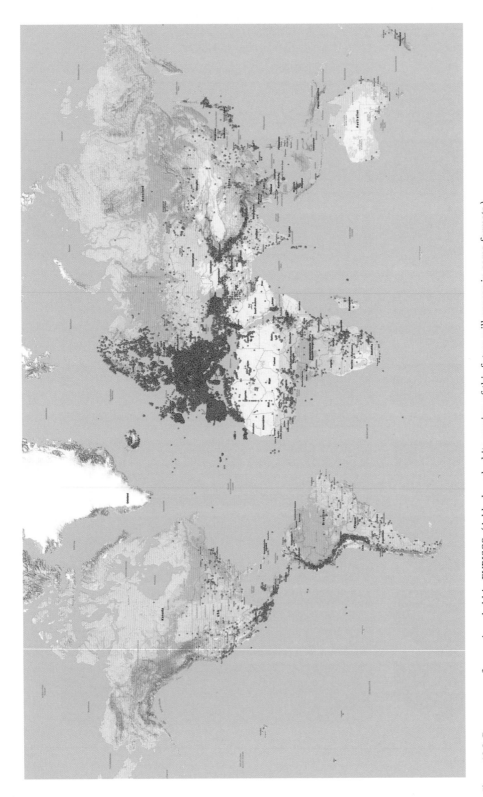

Figure 18.2 Provenance of accessions held in EURISCO. (A black and white version of this figure will appear in some formats.)

patterns of genetic diversity will change over time, it is desirable that alleles which are unique to any particular LR are maintained within that LR population. Although the only way to guarantee the latter would be to take a sample and place it as an *in situ* backup sample in an *ex situ* gene bank – reinforcing the need for complementary *in situ* and *ex situ* techniques to provide systematic conservation.

It has been argued that LR is the most severely threatened element of all biodiversity (Maxted, 2006). Perhaps at first sight compared to many IUCN Critically Endangered assessed taxa, a rather rash statement, but Maxted justifies the statement as follows:

- We have no idea globally or nationally how many LR exist.
- Landrace maintainers are almost always older, and their number is dwindling each year (the average age of LR maintainers on Scottish islands was 65; Scholten *et al.*, 2008).
- Farmers who maintain LR are ultimately commercial growers; they cultivate to sustain them and their family therefore economic viability must be their prime goal, they are *not* conservationists whose prime goal can be conservation of the LR irrespective of income.
- Seed companies and government agencies actively promote the replacement of LR with modern cultivars that may be able to generate increased economic margins for the farmer.
- In most countries no species conservation agency has direct responsibility for LR maintenance.
- No country has a comprehensive inventory of extant LR. The most comprehensive national LR inventory is for Italy, but even that is not fully comprehensive (Negri *et al.*, 2013).

Maxted concludes that unless action is taken immediately, LR loss will continue and complete extinction of much LR diversity is the only possible conclusion.

It is a truism to state that to conserve any wild animal or plant first you must have some concept of the group to be conserved. This usually takes the form of a taxonomic classification and so you at least have a list of accepted taxa (Davis and Heywood, 1973), but for LR there is almost invariably no concept of what

LR exist or little knowledge of where they are concentrated. Therefore, the first step to conserve LR is to produce an inventory – we need to know what exists before we efficiently plan conservation (Maxted *et al.*, 2007). The 'checklist' approach was first developed by Hammer and colleagues (Hammer, 1990, 2001; Hammer *et al.*, 1999) and by Negri (2003), who focused primarily on interviewing farmers and gardeners to record the LR diversity found in their gardens or fields. Subsequently Maxted *et al.* (2007, 2009) reviewed the approaches that might be taken to creating an LR inventory, and Veteläinen *et al.* (2009a) contains discussion of methodologies used to create national LR inventories and progress that has been made thus far.

An initial question to address is what constitutes a landrace for the purposes of the inventory? This is more than just which LR definition is to be used, and includes how distinct LR are recognized, the scope of the inventory and the scale of the cultivation required. The definition of LR is complex and still a matter of debate with the plant genetic resource community (see Chapter 7), but broadly speaking it is a dynamic population of a cultivated plant species that often has: historical origin, distinct identity, not been formally bred, intrinsic genetic diversity, locally adapted to a geographic location, associated with traditional cultivation systems and has cultural associations. It is also true that there are very few LR where all seven criteria are present, even though experts would agree they were representatives of an LR, so each LR itself may have 4 out of 5 of the criteria and still be regarded as an LR. It is also worth noting that lack of agreement over what constitutes an LR has slowed LR conservation; therefore it is recommended that a pragmatic approach is taken to the definition of LR to be used in preparing the inventory but the definition that is used is stated so inventory users are aware of the concept applied.

The next question is what constitutes landrace diversity? Is it going to have a nomenclatural or genomic basis? Is LR going to be recognized on the basis of nomenclatural or genomic identity? The distinction is practical but does have implications: a nomenclatural-based inventory would include all LR with different names, whereas a genomic-based inventory would include all genetically identified different entities. In terms of genetic conservation, our goal is to maximize the genetic diversity conserved so we would favour the approach based on assessment of genetic diversity, but to take such an approach would require that the genetic integrity of all LR are tested for uniqueness, which at this time is still unviable in terms of the volume of genomic analysis required. Therefore, pragmatically we employ the proxy of nomenclatural identity, meaning if the LR has a unique name then we assume it has a unique genetic identity.

Inventories are likely to vary in their taxonomic and geographic scope depending on the specific requirements of the commissioning agency. In terms of taxonomy, the scope may vary over the breadth of crops included, for example all crops groups recognized by FAO (2010b) or a subset of say cereals, or grain legumes and cereals or forages, and may even include less obvious LR related to medicinal species and even semi-domesticated wild-harvested species. The geographic scope may practically be limited by the resources available for inventory creation. A notable example is the iterative work of the Negri research group at the University of Perugia which began with Italian provincial inventories working up to an inventory for Italy as a whole (Porfiri *et al.*, 2009; Panella *et al.*, 2012; Pacicco *et al.*, 2013).

Given that the goal of genetic conservation is to maximize the genetic diversity conserved and level of LR cultivation recordable, Maxted *et al.* (2007) discuss whether inventory inclusion should be restricted to LR that are commercially grown for sale, or grown by a single farmer or single home garden maintainer. Their conclusion is that no matter the scale of production the fact that even an LR grown by a single home garden maintainer could contain unique alleles means that ideally all LR should be included. However, resource availability may force a more restrictive view, as ensuring every LR grown by a single home garden maintainer will be time consuming.

Having defined the scope of the inventory, the researcher can collate the available information of the extant LR within the chosen geographic range of the inventory. Finding where LR are present may involve:

- *Surveying* ex situ *collections* – By collating LR passport information from gene banks, community seedbanks, field gene banks and botanic garden databases you can at least identify where LR existed historically, though current existence at these locations should be confirmed because the LR population may post-collection have been lost. Historically obtaining this information would have involved visiting individual institutions, but today much information can be easily collated from seed search on-line databases like EURISCO or GENESYS.

- *Expert advice* – Staff associated with gene banks, national testing centres, research institutes, agricultural extension divisions, farmers' NGOs, agricultural statisticians and other professionals can often provide localities for LR cultivation or suggest where surveying might usefully occur.

- *Commercial companies* – Large, medium and small-scale companies involved in seed production, brewing and milling may themselves hold historic LR material or have knowledge of where it could be found.

- *Scientific literature* – Reviews of historical literature, research reports, crop-based studies and papers may contain information on historic LR presence or regions where LR have traditionally been cultivated.

- *'Grey literature' archival materials* – Many gene banks, research institutes and seed companies have libraries or data stores of unpublished collecting reports or other survey data that may include localities where LR are reported to have been grown.

- *Internet searches* – These may turn up further reports of LR cultivation from less formally published sources such as crop-based or farmer or grower NGO newsletters, or other information bulletins.

- *Official documents* – Today when some countries choose to subsidize the production of LR, official agricultural statistics will contain information on who and for what people are being subsidized; the EU Common Catalogues for vegetable and agricultural varieties and National Lists may also contain some LR material.

- *Farmer interviews* – Farmers/growers may also be surveyed either indirectly through advertisements, articles in farmers'/growers' magazines and local newspapers, and directly via personal contacts. Questionnaires may be sent by mail or email and interviews conducted by phone or in person. The approach for recording information from farmers is quite specialized to ensure the farmer/grower provides impartial data rather than what they think the reviewer wishes to hear, and includes participatory survey techniques such as focus group discussions, ranking and sorting exercises (ranking: preference ranking of variety, rating: provided by people and sorting: Organizes items into groupings), mapping and modelling, four-cell analysis, and establishing a memory bank or community biodiversity register (see Cunningham, 2001; Tuxill and Nabhan, 2001).

The records of LR presence can be collated into a web-enabled database (see an example for Italy, Negri *et al.*, 2013; http://vnr.unipg.it/PGRSecure/start .html). Maxted *et al.* (2009) reviewed the types of data to be recorded during production of an inventory:

- Scientific name
- Name of landrace
- Maintainer details (e.g. name, contact details, age, gender, family structure, education, main source of income, owned or rented land, size of farm, organic status, arable or mixed farming system)
- Geographic location (e.g. province, nearest settlement, latitude, longitude, altitude)
- Landrace characteristics (e.g. characterization and evaluation details, maintainer perceived value, length of seed saving, relationship to other landraces)
- Cultivation details (e.g. area currently sown, history of area sown, time sown, time harvested, cultural practices, cultivation inputs, method of seed saved selection, method of seed storage, maintainer exchange frequency, other and non-landraces material grown, maintainers comparison to modern varieties, local or national maintainer incentives)
- Relative uniqueness of landrace (e.g. grown on single farm or more widespread, genetic distinction)
- Usage (e.g. description of main usage, secondary usage, home consumption or marketed, marketing, current and past values, member of grower or marketing cooperative)
- Threats (e.g. perverse incentives, lack of sustainability of farming system, lack of market)

These basic descriptors were revised and extended by members of the ECPGR On-Farm Working Group (Negri *et al.,* 2012). The latter descriptors are extensive, so it is unlikely that all descriptors will be recorded for each landrace within the national inventory. Once the inventory is completed, it should be published; examples are available on the ECPGR On-Farm Working Group website (www.ecpgr.cgiar .org/working-groups/on-farm-conservation/). The production of the inventory is not an end in itself. It should then be used to help plan LR conservation using standard techniques – gap analysis of current active *in situ/ex situ* activities against the inventory representation of total LR presence and gap-filling can commence to ensure systematic conservation and promote LR utilization.

10 Community-Based Conservation

10.1 Introduction

The last 50 years or so have seen a very dynamic period for biodiversity conservation, agriculture and rural development. Each has dealt with a major paradigm shift that has resulted in greater consultation and working with local communities. For example, in conservation this has seen shifts in thinking about the role of local communities in conservation from that of non-consultation, even exclusion, to approaches that are now more inclusive or even community-led. Likewise, in agriculture and rural development major changes have also occurred among researchers and extension practitioners from innovation and technology development processes that involved little consultation with the end-users to much more participatory processes that involve farmers and local communities much more in decision-making processes influencing innovation and technology development – a much more farmer-led process. During this period, we have also seen the rise of social movements and alliances of rural communities demanding more control over their environment and farms, and the food production process including greater control over genetic resources and eschewing corporate control of both. These social movements have drawn attention to farmers' rights to seed, land, water and basic rights to food and good nutrition, all of which are captured under the concept of food sovereignty. These parallel developments have all contributed to transformative change in how professional conservationists or other scientists interact with small-holder farmers and indigenous and local communities. In many ways, there has been much cross-fertilization in terms of ideas, approaches and methodologies.

In a relatively short period, many conservation and rural development activities have shifted from researchers 'know best' to 'let the locals lead' approaches. However, many conservationists and scientists remain unconvinced about the benefits of community-based conservation or farmer-led approaches, or the handing over of control for land and natural resource management to local communities (Pearce, 2014). The reasons for this divide in beliefs for and against community management and control of natural resources is ideologically driven and can be traced to arguments immersed in Hardin's '*tragedy of the commons*' (Lloyd, 1833; Hardin, 1968) versus those of Elinor Ostrom, who highlighted that communities, if given control over their own resources, can work out practical rules and institutions for sharing and sustainably managing natural resources (Ostrom, 1990). Yet there remains considerable resistance to community control or community-based conservation despite a growing body of evidence that highlights how positive outcomes can be for forest and wildlife conservation.

In the case of conservation, these arguments and different perspectives have led to a pluralistic framing of conservation approaches which continue to evolve and develop. But there has undoubtedly been increasing involvement of local communities in conservation activities and specifically in the conservation of agro-biodiversity in agro-ecosystems. These changes are most apparent in relation to on-farm conservation, community biodiversity management and other approaches such as participatory plant breeding. It is this shift in thinking and practice which may be generically referred to as community-based conservation.

The definition of community-based conservation, provided by Western and Wright (1994) 'includes natural resources or biodiversity protection by, for, and with the local community'. In addition to conservation or protection objectives community-based conservation also integrates development objectives by improving the access to and sharing of benefits that might arise from conservation or the management and sustainable use of biodiversity or natural resources, and increasingly addresses issues of social justice and equity.

Box 10.1 | **What Is Community?**

Community is a term that is widely used but rarely defined. It may comprise any number of types of individuals, from traditional residents to immigrants and, in some cases, even visitors. It is a term that is employed in both developing and developed communities, in rural and urban settings. It is important to realize that individuals in a community may have completely different attitudes, perspectives and approaches to their local environment and its resources, among many other things. These attitudes may stem from historical, cultural and socio-economic differences affecting the needs of the people and the degree to which the local environment can satisfy those needs. Another term often used in conservation and development practice is 'locals'. This group has been defined as people who have in common regular face-to-face contact and interpersonal relationships and experience of a locality, including family or kin groups, communities (villages) and localities (sets of communities). The 'community' has been variously defined on the following characteristics:

- ethnicity and indigeneity;
- traditions;
- length of time that a group has lived in the area;
- sense of common purpose;
- geographic context alone.

In the latter case, the community would have to include immigrants, cultures in transition, and those with no ancestral ties to the land or to each other. However, the community is defined, you should realize that (as development professionals have discovered) even traditional communities are rife with internal conflicts and divergent interests and often are split along economic, gender and social lines.

It is difficult, even misguided, to generalize about what constitutes a community (Laird and Noejovich, 2002). A community might be considered as rural, small (in number) and marginalized (Box 10.1). Alternatively, a community may be in peri-urban or urban areas with many of its members living or working in cities. Communities are often represented as homogeneous entities, but they are stratified into different groups with different levels of power, wealth and control. They represent a diversity of perspectives and understandings, views and agendas, as there are different types of groups and individuals – old and young, male and female, able and disabled, sick and healthy, wealthy and poor, the doctor, the teacher, the

farmer; all commonly will have different perspectives, motivations and agendas. Employing a generalized approach to working with communities is therefore inherently dangerous.

As this chapter highlights, it is important to appreciate that many communities and community leaders have put their lives on the line in protecting their environment and resources against powerful outside influences and forces, and that the stakes involved are often high and deadly. 'On Dangerous Ground' (Global Witness, 2016) highlights the tragic news that 2015 was the most dangerous year on record for murders of land and environmental defenders.

This chapter briefly reviews historical development of approaches to conservation and agricultural and rural development and how changes in attitudes to community involvement has meant transformative change to how we as conservationists and scientists interact with small-holder farmers, and indigenous and local communities, and how this demands that we change our attitudes and behaviours in an effort to develop a much-needed 'new professionalism' (de Boef *et al.*, 2013b). The chapter also highlights the multiple benefits that can accrue to greater community involvement in conservation and natural resource management as well as many of the incentives to facilitate this shift. Finally, the chapter concludes with some key guidance for planning community-based conservation and a brief outline of some common approaches, tools and methods that can help strengthen the role of local communities in conservation.

10.2 Conservation Perspectives: From Exclusion to People-First

Community-based conservation is ancient: all that is new is the recognition by the formal conservation sector of its validity and potential. Local and indigenous communities in biodiversity-rich countries have been closely linked to their natural environments for millennia and have intimate knowledge of habitats and their wild plant and animal species – a relationship that has often been disrupted by conventional conservation approaches (United Nations, 2009). Professional conservationists have often failed to appreciate the role that local communities have successfully played in conserving genetic diversity of crops, CWR or other native plant species within their own environment (Díaz *et al.*, 2019). Professional conservationists have also seldom seen the practical benefit that might arise from involving local communities in PGRFA conservation activities. Although local communities and professional conservationists may have completely different reasons for conservation, they both manage biodiversity to maintain its diversity. The formal conservation sector has too often arrived at a location

for conservation actions 'armed' with a 'top-down' solution that is then enforced without consultation on the local community. The local community network is often shunned and underutilized, oftentimes even banished from their customary lands with loss of access rights and the traditional service the habitat provided. Some professional conservationists still mistakenly believe that they can achieve their conservation goals without local community participation and, furthermore, they may even have considered the involvement of local people as a distraction or an impediment to the achievement of their scientific conservation goals.

The need for the professional conservation sector to respect, preserve and maintain the practices of conservation and use employed by indigenous and local communities is incorporated into Article 8(j) of the Convention on Biological Diversity (CBD, 1992), as follows:

Each contracting party shall … respect, preserve and maintain knowledge, innovations and practices of indigenous and local communities embodying traditional lifestyles relevant for the conservation and sustainable use of biological diversity.

Therefore, it is no longer desirable, or even considered ethical, to exclude or marginalize local communities, but essential that professionals work together with local communities, if biodiversity is to be effectively conserved for future generations.

Here community-based conservation can be seen as both an independent and complementary paradigm to that enacted by the formal conservation sector; it involves the indigenous people or community of a particular region working either independently or in partnership with the formal sector to conserve biodiversity. The important point to note is that community-based conservation need not compromise the goals of formal conservation; community-based conservation can ensure the maintenance of taxonomic and genetic diversity. It is important to note however that community-based conservation seeks to meet the dual goals of biodiversity and livelihood sustainability.

Recent recognition of the role of communities in conservation by the formal sector has fostered attempts by the formal sector to value, promote and exploit traditional practices for the overall conservation goal. Various reasons, objectives and motives have been proposed to justify increased efforts to stimulate community involvement of biodiversity and genetic resources conservation. These have included promotion of:

- traditional use;
- genetic resources (e.g. protection of relatives of domesticated breeds and cultivars);
- ecosystem services and integrity (e.g. to buffer protected areas from ecological impoverishment);
- tourism resources (to provide extra income for local communities);
- hunting, gathering and subsistence use;
- medicinal use.

Much of the current formal conservation sector's interest in community participation in conservation projects has resulted from the failure of formal conservation projects that do not have the commitment or involvement of the local community. The realization that conservationists cannot dictate to local communities over their environment and natural resources has forced conservationists to work more closely with these communities, and this has in turn resulted in a realization of the traditional role of local communities in conserving their own environment. Nearly all local communities have themselves initiated and undertaken effective conservation of their particular natural resources for centuries, because their environment has socio-economic use or other value to them. For example, local communities throughout the world conserve tree species in sacred groves, but in many cases the trees are also harvested annually for their fruits (see Box 10.2, which provides a further illustration).

Box 10.2 | **Local Communities and Biodiversity**

(Source: Hawkes *et al.*, 2000)

It is now generally agreed that we would not have the wealth of domesticated and non-domesticated biodiversity that remains extant today were it not for the conscious effort of local communities over millennia to conserve biodiversity in all its forms. For example, indigenous farmers from the Andes maintain a gene pool of more than 3000 local cultivars of potatoes representing eight cultigen pools. While in Papua New Guinea, approximately 5000 local cultivars of sweet potatoes are grown, with one farmer growing up to 20 cultivars in a single garden. Community-based conservation has been particularly important during the last hundred years when both the threat of and actual genetic erosion have increased exponentially. Many traditional farmers, for example, still permit ancestral crop LR and their weedy relatives to coexist within their fields. This facilitates introgression within the species gene pools, thus generating new diversity.

However, modern intensive agriculture has taught the majority of farmers of the economic benefits of monoculture farming and therefore the coexistence of multiple taxa from the same gene pools in farmers' fields is becoming increasingly rare. In the light of the increasing threat of global plant genetic erosion and the realization that professional conservationists alone are unable to guarantee the safety of the world's biodiversity, the formal sector are realizing, perhaps rather late in the day, that local communities have an important role to play in PGR conservation.

Box 10.3 | Failures of Exclusionary Conservation Approaches

(Source: Kothari *et al.*, 2013)

Strict exclusion of local communities and resource users can have the following detrimental impacts on conservation and development goals:

- Alienating local communities from conservation efforts
- Removing any incentive to cooperate with protected area managers and regulation
- Losing the conservation and management benefits of traditional knowledge and resource management practices
- Losing ability of communities to make and enforce the rules that govern resource use
- Encouraging and intensifying illegal resource use
- Shifting resource use to other areas, with intensified impacts
- Increasing illegal use by 'outsiders' through removing the rights and presence of traditional custodians
- Upsetting complex food webs with unintended consequences on target conservation species
- Removing options for much-needed sustainable financing of protected areas

In other cases, outside intervention and incentives must be established to sustain community involvement. However, while many community conservation projects have been and still are started and managed by outsiders, the fact that most resource destruction is also imposed from outside (whether through direct or indirect pressures) must not be forgotten.

10.2.1 Excluding Local Communities and Users from Conservation Efforts

During the first half of the 20th century, biodiversity conservation, particular in protected areas, became increasingly exclusionary, discouraging local communities and in many instances forcefully removing them from their traditional lands. Far too often conservation policies and practices were centrally planned and 'top-down', with far too little consultation and planning around the needs or aspirations of local communities. Sadly, these 'command-and-control' approaches have often perpetuated poverty and inequality and limited the achievement of both conservation and development goals (Hunter and Heywood, 2011, see Box 10.3).

The early days of formal biodiversity conservation and the establishment of a string of protected areas that followed the establishment of Yellowstone National Park in 1872 set the tone for what was to follow: conservation models that were largely protectionist and exclusionary and which viewed human use as not conducive to conservation objectives (Kothari *et al.*, 2013). Increasingly government-managed protected areas were established without consultation and often with the expulsion of residents or user communities from what are their ancestral lands, e.g. the Maasai from the Serengeti in Tanzania, and the Karen from reserves in Thailand. The result of this was an increasing sense of dispossession and disempowerment among many local communities. Although such models and approaches have been increasingly challenged, they remain the norm in many parts of the world and often efforts made to reinstate customary rights to such disenfranchised communities are too often resisted by the conservation community.

10.2.2 A Re-examination of Conservation Approaches

As a consequence of the many lessons learned from exclusionary and top-down models, the last part of the 20th century has witnessed a re-examination of approaches to conservation, thereby leading to a major shift within conservation practice that has increasingly seen community participation being viewed as fundamental to achievement of not only conservation goals but also socio-economic and political equity (Kothari et al., 2013). This has resulted in a major paradigm shift that O'Riordan and Stoll-Kleeman (2002) refer to as moving from 'ecology first' to 'people-first' perspectives. With the result that it is increasingly considered neither politically feasible nor ethically justifiable to deny local communities the use of natural resources without providing them with alternative means of livelihood, or to manage protected areas without their empowerment and support.

Though by no means universal, the ability of local communities to sustainably manage natural resources and ecosystems and the need for greater attention to be given to this in conventional conservation and practice has been demonstrated by the growth of common property scholarship since the late 1980s (Kothari et al., 2013). These findings have increasingly built legitimacy for more community-based forms of conservation and protected area governance and management. However, this must be qualified by pointing out that not all local communities are in all situations conservation oriented. For some communities, there are many external and internal pulls and pressures affecting their traditional or customary ways which may force them to adopt unsustainable lifestyles. However, when contexts are in their favour, it appears that participatory, rights-based approaches need to be increasingly adapted for effective conservation.

As a result of these transformative shifts, conservation policy and practice is increasingly incorporating considerations of community participation and inclusion with a view to improving human well-being and conservation areas as drivers and providers of social and economic benefits. There

is increasing recognition of the rights and claims of indigenous peoples and local communities to their traditionally held lands and resources in conservation priority areas (Kothari et al., 2013). Recent surveys have highlighted that conservation professionals and managers now view community involvement and participation as one of the important factors for successful and effective management (Stoll-Kleeman and Welp, 2008), although whether this ultimately translates into economic benefits for local people is still questionable (Galvin and Haller, 2008).

Community-based natural resource management (CBNRM) is one of the more recognized inclusive conservation models to emerge during this period of transformation, representing a shift to more devolved and inclusive approaches. CBNRM captures a broad range of approaches and practices but at their core are local collective institutions or groups of people, organized formally or informally, managing and utilizing their lands, resources and common property. A recent assessment of the outcomes of CBNRM approaches in Africa has highlighted some notable ecological, economic and institutional achievements (Roe et al., 2009), such as significant increases in communal land management for sustainable wildlife production and the generation of $20 million in revenue for local communities and district administrations in just over a decade by the CAMPFIRE initiative in Zimbabwe.

Mace (2014) highlights four main phases in developing world nature conservation – nature for itself, nature despite people, nature for people and finally people and nature – in the framing of how conservation perspectives relate to the relationship between nature and people, which have occurred over a relatively short period, and that represent a diversity of views, ideas and roles for science now underpinning conservation, which all coexist today (Figure 10.1).

Danielsen et al. (2009) highlight the potential benefits of local community participation in conservation monitoring of species or habitats in developing countries where state agencies have small budgets, there are fewer skilled professionals or amateurs, and socio-economic conditions prevent development of a culture of volunteerism unlike what

Rough timeline	Framing of conservation	Key ideas	Science underpinning

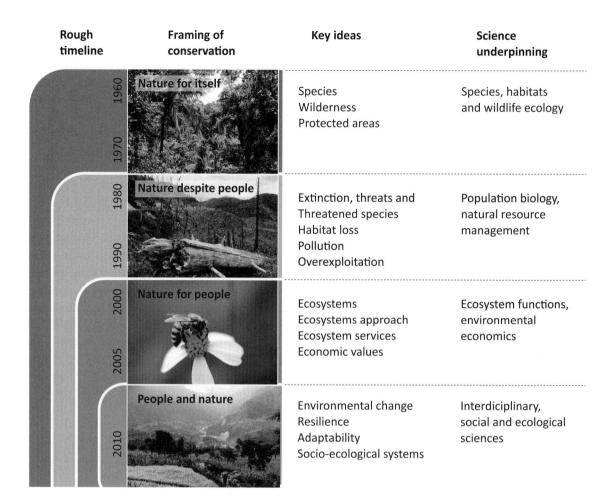

	Nature for itself (1960–1970)	Species Wilderness Protected areas	Species, habitats and wildlife ecology
	Nature despite people (1980–1990)	Extinction, threats and Threatened species Habitat loss Pollution Overexploitation	Population biology, natural resource management
	Nature for people (2000–2005)	Ecosystems Ecosystems approach Ecosystem services Economic values	Ecosystem functions, environmental economics
	People and nature (2010)	Environmental change Resilience Adaptability Socio-ecological systems	Interdiciplinary, social and ecological sciences

Figure 10.1 The multiple framings of conservation over the past 50 years. (A black and white version of this figure will appear in some formats. For the colour version, refer to the plate section.)
(From Mace, 2014)

is possible and often routine in developed countries. To address this constraint, they suggest a typology of five monitoring categories, defined by their degree of local community participation which ranges from no local community involvement with monitoring largely under the control of professional researchers to an entirely local effort with monitoring undertaken by local people. They proposed that local community-led monitoring was particularly relevant in developing countries where it could lead to prompt decisions about threats and empower local communities to better manage their resources and improve local livelihoods, though further study and protocols were required.

10.3 Development Perspectives: From Technology Packages to Farmer First

It is probably no coincidence, given the socio-political climate, that similar paradigm shifts in theory and practice were occurring in agricultural and rural development in the late 20th century around the same time as those in conservation. In some ways, the reasons for these transformations were similar but the context was different. Agricultural development had come on in leaps and bounds in the 1960s as a result of the Green Revolution (as highlighted in Chapter 2). The availability of improved cultivars meant that many millions of farmers were able to increase productivity allowing many to escape poverty and millions of people to avoid famine and hunger. However, as we have seen in Chapter 2, the Green Revolution did have its downside. The promotion of improved cultivars often leads to the displacement of traditional local varieties and landraces and increased genetic erosion. This prompted global action leading to increasing collection of genetic diversity and its conservation in gene banks. For the first time *ex situ* conservation was taking precedence over conservation in farmers' fields.

This was a period that also coincided with the increased exploitation of the world's genetic resources for commercial plant breeding and the emergence of contentious issues around ownership of genetic resources and the corporate control of seed systems. The issue of the global North's exploitation of the genetic resources of the global south played a major role in the development of the CBD, which in addition to conservation also had pillars recognizing sustainable use, and access and equitable benefit sharing. The CBD also made it clear that it favoured *in situ* conservation as opposed to *ex situ* conservation, which it viewed more as a back-up safety net for *in situ* conservation. This provided added support for the strengthening of on-farm conservation. These developments had major significance for the conservation of PGRFA/CWR, and it is from here that we can begin to trace the development of the discipline of on-farm conservation. This period saw for the first time the development of programmes and projects to explore and establish the scientific basis for on-farm conservation (Chapter 9). It also witnessed the rise of a growing social movement of civil society including peasants and indigenous peoples, smallholders and family farmers, pastoralists and forest dwellers, who wanted to retain control of genetic resources and challenged the increasing centralization of genetic resources in a few global gene banks and the

industrial and corporate control of genetic resources (see Section 10.4).

10.3.1 Farmer-Led Development

The Green Revolution also gave rise to the development of what became known as the transfer-of-technology (ToT) model, which was based on the premise that scientists with their expertise, knowledge and resources could develop the best technologies that farmers needed to increase productivity, and that there was little need for farmer involvement in the process. Rather like the exclusion of local communities from conservation approaches discussed above, farmers and rural communities were excluded from the innovation and technology development in farming systems. The researchers were working on the best resourced research stations, with access to the most fertile trial sites and an endless availability of inputs such as water, fertilizers and pesticides. Under such homogeneous and uniform conditions this allowed researchers to quite easily identify those new cultivars that responded best and yielded the most and which could then be transferred to farmers for evaluation and dissemination and presumably widespread adoption.

For a while this approach worked well. The researchers had a network of farmers, usually reasonably well resourced in terms of good land, labour, capital and inputs. The problem was that when it came to the vast majority of small farmers, often scraping out an existence on a plot having less than one or two hectares, with few resources cultivating in marginal environments and risk adverse, the new cultivars or technologies did not perform as well as the local cultivars. It was the complex, diverse and marginal environments where these farmers operated and the general lack of adoption of new technologies that gave rise to the next major paradigm shift in agricultural and rural development, which saw the emergence of participatory plant breeding (PPB; see Chapter 17). PPB provided the paradigm that allowed scientific breeders to work directly with local communities to produce crop material that suited the complex, diverse and marginal environments but still took advantage of breeder's elite lines with improved adaptive traits.

These limitations in research and development and limited acceptance of research outputs and new technologies led to new thinking by many on how agricultural research should be undertaken. This led to development of new approaches and methods that facilitated the greater involvement of farmers in the innovation and technology process. This resulted in the emergence of new research and extension models variously referred to as Farmer Participatory Research and Participatory Technology Development. The developments in this area, the trends and achievements that launched a new movement to encourage greater farmer participation in agricultural research and development, and the various models and approaches employed, have been captured in milestone books including *Farmer First* (Chambers *et al.*, 1989), *Beyond Farmer First* (Scoones and Thompson, 1994) and *Farmer First Revisited* (Scoones and Thompson, 2009). Each aimed at putting farmers first, at the core of agricultural development, and have had a major influence on how researchers work with and engage farmers and local communities, and have heralded major methodological, institutional and policy transformations.

One of the earliest applications of this *Farmer First* movement was in the areas of plant breeding, ultimately involving greater consultation and participation of farmers in breeding programmes (see Chapter 17). Participatory plant breeding soon became an alternative approach to conventional plant breeding, which allowed farmers to be involved in the planning and evaluation of materials in the complex, diverse and risk-prone environments where they farmed. This too represented a major change to breeding. Instead of trying to find that one size fits all cultivar that performed consistently across different environments, the search was now on to find the many different cultivars that performed well in many different environments.

Pretty (1995) provides a historical overview of the modernization of agriculture as leading to three distinct phases, the industrialized, the Green Revolution, and the agriculture of the vast majority

of smallholders which is complex, diverse and risk-prone, pointing out that modernization has contributed successfully to the first two yet not so much to the latter, and in some cases has possibly made smallholders and some farming households poorer. He highlights that the main problem is that many resource-poor farming households were not in a position to adopt modern agricultural practices, and that in many instances complete technological packages are not appropriate to the complexities of most rural communities, their farming systems or socio-economic contexts. Where they have tried to adopt such packages, it has led to dependence on external inputs and rising environmental and health problems related to chemical inputs, as well as financial problems and debt. He concludes that addressing the problems and constraints of most of the world's small farmers requires the adoption of an entirely different approach to agricultural and rural development, one that recognizes the skills and ingenuity of local people and communities.

10.3.2 The Rise of Participatory Approaches and Methods

The movement towards greater inclusion of farmers in agricultural research-for-development (R4D) was facilitated by critical reflection and analysis of the R4D process and exploration of methods and approaches which would promote farmer-based participatory research. The introduction in the late 1970s of a new research approach known as *Rapid Rural Appraisal* (RRA) quickly became popular with decision-makers in development agencies and NGOs. This approach was gradually refined and evolved into what is known as *participatory rural appraisal* (PRA). One reason for this was the view that the RRA approach could be extractive and did not necessarily empower communities. The role of local communities was limited to providing information, while the power of decision-making about the use of this information and subsequent follow-up actions and interventions was still decided by outsiders. This represented a significant change in philosophy and purpose of participatory approaches, with an ongoing commitment to follow-up action by those involved based on community concerns and interests and to build a process of inclusion that would lead to sustained actions and local capacities to intervene and address such concerns. In many ways, PRA empowered communities to better analyze and understand their situation and to decide on changes and action. Today, these shifts in practice are now reflected in *participatory learning and action* (PLA) (Chambers, 2007). The emergence and development of participatory methods have had a major influence beyond agriculture and rural development including on conservation practice.

Many factors can influence the degree to which participation is evident in conservation and rural development projects and programmes. The conditions under which participation flourishes will differ from place to place, from country to country and may depend on: traditions, cultural norms of social behaviour; the political environment and how this shapes participation and who can participate; local power structures and previous contact and interaction with development and conservation agencies. Typologies or ladders of participation have been developed by various practitioners that allow us to think analytically about the 'quality' of participation in conservation and rural development processes; Pretty's (1995) Typology (Table 10.1) and Arnstein's (1969) Ladder (Figure 10.2) are two of the most commonly used tools.

The application of participatory approaches in plant genetic resources conservation and use has been dealt with in detail by Friis-Hansen and Sthapit (2000) and Hunter and Heywood (2011). Importantly, much of the theory and practice which emerged during this dynamic period has also had an impact on the broader biodiversity and conservation field including in protected area management in responding to and addressing the challenges of 'exclusion' of local communities highlighted earlier (see Box 10.4). Nemarundwe and Richards (2002) provide a comprehensive review of the use of participatory approaches, methods and tools in the context of plant genetic resources from forests and trees.

10.3.3 The Need for a New Professionalism

One of the greatest challenges in supporting such paradigm shifts that engender citizen power is in changing the mind sets of professionals working at the nexus of conservation, agriculture and rural development. Much of what is required challenges power relations and hierarchies and often involves handing over control to non-professional communities. What is required is a 'new professionalism' that embeds the necessary changes in attitudes and behaviours (Box 10.5) as well as equipping professionals with the necessary array of tools, approaches and methods to facilitate real change and empower communities to undertake community-based conservation.

Capacity building is key to supporting effective community-based conservation and the effective levels of participation required to achieve this goal. Yet, the mainstreaming of such approaches into relevant programmes is still limited. Universities and colleges that offer formal courses, as well as other institutions offering non-formal programmes, are well placed to meet this need for quality support for community-based conservation and participatory development. Universities and colleges through teaching, training and research can play a pivotal role in the social, political and economic change necessary for community-based conservation and sustainable development, but only if they are responsive to the needs of the wider community. While several universities have developed courses that teach community-based conservation approaches and effective participatory teaching programmes, this has not happened as quickly as desired and consequently, opportunities for formal training and research are still limited. Many higher and adult education systems still lack the systemic capacity and innovative forms of learning and teaching that are required to address the learning needs of professionals and practitioners of community-based conservation and social change that ultimately promote the emergence of civil societies underpinned by good governance and human rights (Taylor and Hunter, 2008).

Almost three decades after the publication of *Farmer First* and its clarion call for a new

Table 10.1 **Typology of participation**

Passive participation	People participate by being told what is going to happen or has already happened. It is a unilateral announcement by an administration or project management without any listening to people's responses.
Participation in information giving	The information being shared belongs only to external professionals. People participate by answering questions posed by extractive researchers using questionnaire surveys or such similar approaches. People do not have the opportunity to influence proceedings, as the findings of the research are neither shared nor checked for accuracy.
Participation by consultation	People participate by being consulted, and external agents listen to views. These external agents define both problems and solutions and may modify these in the light of people's responses. Such a consultative process does not concede any share in decision-making, and professionals are under no obligation to take on board people's views.
Participation for material benefits	People participate by providing resources such as labour, in return for food, cash or other material incentives. Much on-farm research falls in this category, as farmers provide the fields but are not involved in experimentation or the process of learning. It is very common to see this called participation, yet people have no stake in prolonging activities when incentives end.
Functional participation	People participate by forming groups to meet predetermined objectives related to the project, which can involve the development or promotion of externally initiated social organization. Such involvement tends not to be at early stages of project cycles or planning, but rather after major decisions have already been made. These institutions tend to be dependent on external initiators and facilitators but may become self-dependent.
Interactive participation	People participate in joint analysis, which leads to action plans and the formation of new local institutions or the strengthening of existing ones. It tends to involve interdisciplinary methodologies that seek multiple objectives and make use of systematic and structured learning processes. These groups take control/ownership over local decisions, and so people have a stake in maintaining structures or practices.
Self-mobilization	People participate by taking initiatives independent of external institutions to change systems. Such self-initiated mobilization and collective action may or may not challenge existing inequitable distributions of wealth and power.

From Pretty (1995).

professionalism necessary to underpin that transformation, 'normal' professionalism is still the dominant practice in our universities and in downstream conservation and agricultural development practice (de Boef *et al.*, 2013b). Universities, colleges and other capacity building institutions that address conservation and sustainable use of plant genetic resources and biodiversity in their curricula still to a large extent depend on the existing body of knowledge which in the PGRFA context emphasizes *ex situ* conservation (Thijssen *et al.*, 2013). Rare are the cases where innovative approaches that address conservation and sustainable use within an empowerment, livelihood, resilience and sustainable development setting are the norm.

Box 10.4 | **Participatory Natural Resource Management and Agricultural R4D**

(Source: Chambers, 2007)

- Biodiversity, conservation and protected area management
- Agriculture, crops and animal husbandry
- Conservation and use of plant genetic resources
- Forestry, especially joint forest management, agro-forestry and community forestry
- Participatory irrigation management
- Participatory watershed management and soil and water conservation
- Conservation and use of plant genetic resources
- Integrated Pest Management

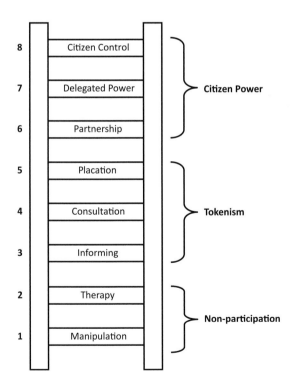

Figure 10.2 Ladder of Citizen Participation.
(From Arnstein, 1969)

10.4 Community Perspectives: The Rise of Social Movements

As highlighted in Chapter 2, it was events post-1492 and the Columbian Exchange that established the expansion of European imperialism and set in motion waves of explorers who were destined to search the New World and distant margins of the Old World for exploitable plant resources. It was this imperial expansion that also set in motion a series of devastating impacts that are still being felt by small holders, indigenous peoples and rural communities in many parts of the developing world today. All the colonial powers of the time saw it as their God-given right to scour 'their' territories across the globe looking for potential exploitable resources, while issues of ownership or compensation of local people for use of the resources exploited from their locality were completely ignored. In Chapter 2 we briefly touched upon some of these early negative impacts such as health of indigenous populations exposed to new exotic diseases to which they had very little, if any, resistance. There were also catastrophic impacts on the social and economic systems of indigenous peoples and rural communities including on their traditional farming systems, hunting and foraging grounds and other land management practices which

Box 10.5 | **Professionals with (New) Attitude!**

Worldwide, PRA practitioners and trainers have been finding that personal behaviour and attitudes are fundamental for true participation. Behaviour and attitudes matter more than methods, powerful though PRA methods have proved. At the personal level, practitioners and trainers have found that the major problem in development is not 'them' – local people, the poor and marginalized, but 'us' – the outsider professionals. Again and again, we have rushed and dominated, imposing our reality, and denying that of the weak and vulnerable. For the poor to be empowered requires us to change, to interact in new ways, to become not controllers, teachers and transferors of technology, but convenors, facilitators and supporters, enabling those who are weak and marginalized to express and analyse their realities, to plan and to act. For this we must behave differently; it is our attitudes that must change.

(Chambers, 2009, in Kumar, 1996)

were often encoded with observed rules and norms to support sustainable management and the fair and equitable sharing and utilization of resources. In many instances, these local systems of governance were misunderstood or ignored by colonial invaders or seen as inappropriate to the colonial powers' vision for ownership, wealth accumulation and intensive production, which was to have widespread ecological impacts. The impact on many indigenous peoples and local communities are still felt to this day.

Some of the most dramatic impacts have been on traditional farming systems. For example, the chinampa system as practiced by Aztec and Toltec societies prior to European arrival in what is now the Mexico City Basin were artificially created biodiversity-friendly islands in Lake Texcoco and were vitally important for food production, and which supported large populations of many millions of people (Perfecto *et al.*, 2009). Unfortunately, the Spanish colonialists did not understand or appreciate this productive and diverse farming system and by the early part of the 17th century had deliberately drained most of it. In recent years there have been moves to rehabilitate the chinampas of Mexico.

In a similar vein, swidden or shifting cultivation which represents one of the oldest forms of subsistence agriculture has often been dismissed out of hand, often demonized by colonial and post-

colonial governments, even criminalized yet it remains poorly understood and researched. However, it is still practised by millions of poor farmers in the tropics, and despite its supposed unsustainability it may in fact represent a highly adaptive land management practice (Cairns, 2015) and even play an important role in diets and nutrition (Ickowitz *et al.*, 2016).

Chapter 2 also outlined how subsequent events and technological innovations laid the foundations for the agricultural and industrial revolutions and concomitant economic growth. This was a period which also saw many innovations in agricultural technology being increasingly protected by patents, a situation somewhat mirrored in the 20th century by the increased corporate control of PGR and seed systems. This period culminated in the emergence of the industrial agribusiness and agri-food model, highly reliant on improved seeds, pesticides and fertilizer, which today still dominates food production and which places huge economic, political, social and ecological challenges for most of the world's small holders and indigenous people and local communities (Altieri, 2009). The latter despite such external pressures continue to maintain the bulk of the world's genetic resources (Box 10.6).

In addition to the problems wrought by colonial powers including through exotic disease and the

---

Box 10.6 | Small Farms Represent a Sanctuary of Agro-biodiversity

(Source: Altieri, 2009)

Traditional small-scale farmers tend to grow a wide variety of cultivars. Many of these plants are landraces, more genetically heterogeneous than formal modern varieties, and grown from seed passed down from generation to generation. These landraces offer greater defences against vulnerability and enhance harvest security during diseases, pests, droughts and other stresses. In a worldwide survey of crop varietal diversity on-farms involving 27 crops, scientists found that considerable crop genetic diversity continues to be maintained on-farms in the form of traditional crop varieties, especially of major staple crops. In most cases, farmers maintain diversity as insurance to meet future environmental change or social and economic needs. Many researchers have concluded that variety richness enhances productivity and reduces yield variability.

---

dismantling of traditional farming and land management systems, many smallholders and indigenous communities were disenfranchised of their communal lands, dislocated and disconnected from their natural environments and traditional food productions systems, excluded from protected and other areas, deprived of the basis of their livelihoods, and were subjected to the loss of their most fundamental civil and human rights. Not only were their lands appropriated and exploited but so too were their vast genetic resources plundered and utilized in alien ways of ownership and privatized goods and markets. Many smallholders were also to become locked into a vicious cycle of increasing fertilizer and pesticide use, linked to ill-health and in some cases crippling debt. Such disempowerment continues to this day with farmers who daily practice agroecology approaches driven off their lands and denied the right to apply a time-tested biodiversity-friendly approach to land management.

Perfecto *et al.* (2009) stress that conservationists and other development practitioners must deal with the often ignored long history of how small holders and indigenous groups have interacted with their landscapes, and the more recent exploitative treatment of their native lands and natural resources

by colonial powers and subsequent political, economic and cultural dominance:

The point here is not to take sides or make retroactive moral judgements about all the implications of these events. Rather, it is to insist that conservationists recognize that their work is ultimately governed by massive shifts in power and interests that they cannot control but must attempt to understand if their work is to endure. Conservationism without political and historical consciousness cannot hope to be well designed or well positioned. . . . Conservationists who want to do more than draw boundaries on maps and throw up fences patrolled by armed guards have to try to understand the complex intellectual and political terrain in which they move, just as they must understand the physical and biological landscape.

Due to these multiple historic impacts and ongoing domination of industrial agriculture and global food systems by a small number of corporations as well as modern-day 'land grabs', a number of rural social movements have arisen to meet these challenges and provide a voice and space for the thousands of landless, marginalized and disenfranchised smallholders and indigenous groups and local communities (Perfecto *et al.*, 2009). In Chapter 2, we have already highlighted one of the most

Box 10.7 | Declaration of the 2015 International Forum for Agroecology (Nyéléni, Mali)

We want local food producers to be at the heart of a participatory, inclusive decision-making process. We must defend collective rights; change laws and discriminatory policies and develop new legal frameworks that respect and protect Farmers' Rights to use, save and exchange and sell seeds and livestock breeds, putting the control of biodiversity and knowledge back in the hands of peasants. Policies need to value local knowledge and give us the opportunity to share our knowledge.

(IPC, 2015)

representative and powerful of these social movements, *Via Campesina*, a global coalition of over 100 smallholder and peasant organizations. In Brazil, the *Movimento dos Trabalhadores Rurais Sem Terra* (MST, The Movement of Landless Rural Workers) is the largest rural social movement in the world. In Mali, *Convergence of Rural Women for Food Sovereignty* (COFERSA) brings together 36 cooperatives of rural women in six provinces and supports biodiversity and agro-ecological production. While the main focus of these social movements is that of food sovereignty recognizing the rights of rural communities to decide what and how to produce including the basic right of farmers to land, water, seeds and food, there is also a strong recognition that the nature of small-holder agro-ecological food production is based on biodiversity-rich practices and that by supporting such rural movements it can help preserve biodiversity and habitats (Perfecto *et al.*, 2009).

In 2015, delegates representing diverse organizations and international movements of small-scale food producers and consumers, including peasants, indigenous peoples, communities, hunters and gatherers, family farmers, rural workers, herders and pastoralists, fisherfolk and urban people, gathered at the Nyéléni Center in Mali to pursue a common understanding of agro-ecology as a key element in the construction of food sovereignty, and to develop joint strategies to promote agro-ecology known as the Declaration of the International Forum for

Agroecology (see Box 10.7). Biodiversity is central to peasant subsistence livelihood strategies, and they have cultivated, developed and sustained millions of varieties and breeds as well as diverse aquatic biodiversity over many generations. Their production systems have nurtured populations of pollinators and pest and disease predators as well as below ground biodiversity (IPC, 2015).

Perfecto *et al.* (2009) in their landmark publication *Nature's Matrix* propose a radical new approach to conservation pointing out the striking similarities between 'natural systems' agriculture and the way that smallholders in the tropics manage and plan their farms. Based on this they argue that it provides a biodiversity-friendly model, '*Natures matrix*', as to how agriculture, conservation and food sovereignty are interrelated and how the goals of each might be more equitably and effectively achieved in the future. By supporting the multitude of smallholder farmers and indigenous and local communities who continue to locally produce about 70% of the world's food and who still practice 'natural systems' agriculture or what many farmers and social movements refer to as agro-ecology.

Some conservation organizations are slowly beginning to recognize the need to work with small holders and their social movements, for example the Institute of Ecological Research in the Pontal do Paranapanema fragments of the Atlantic coast rainforest in Brazil (Perfecto *et al.*, 2009). While challenging (see Box 10.8), working with the millions

Box 10.8 | Sources of Bias in Conservation Practice

(Source: Perfecto *et al.*, 2009)

Perfecto *et al.* (2009) summarize the sources of bias in conservation practice among conservationists, policy-makers, agricultural scientists and extension officers and ecologists when it comes to working with small holders and indigenous and local communities. They identify such bias at eight key levels:

1. The strong desire to maintain high levels of biodiversity and pristine habitats has led many, including those with considerable scientific expertise, to regard all types of agriculture as detrimental to conservation efforts and find it difficult to distinguish between significantly different types of agriculture in terms of their potential to contribute to and support conservation efforts.
2. The large number of scientifically trained agricultural scientists who still view the types of agriculture practiced by forest dwellers, indigenous communities and smallholder farmers as inefficient and unproductive and not deserving of their attention or research.
3. Too often conservationists are found to be associated with the formally educated and social elites of a region as opposed to those less powerful and marginalized. It is often these local elites who are most biased against the interests of the rural poor and very often transmit these biases to formal conservationists while vigorously protecting their own interests and mandates. Consequently, deep-seated prejudices can sometimes be found within conservation organizations.
4. Poor language skills and cultural knowledge on the part of agricultural development workers and conservationists often creates substantial gaps and barriers with those peasants, small holders and indigenous groups engaged in agricultural activities in biodiversity-rich landscapes.
5. Conservationists and policy-makers often fail to appreciate and understand the dynamic nature of small-holder management of agricultural landscapes and are all too often eager to quickly solve problems without proper historical and social analysis. Analysis of the present context is all too often preferred over efforts to understand the considerable historical change that many communities have been exposed to thus leading to poorly designed agricultural and conservation policies and programmes.
6. In a similar vein, analysis of local circumstances may be privileged over a better understanding of wider contexts at the regional, national and global context with inappropriate outcomes, policies and programmes.
7. Conservationists and scientists may lack the motivation and skills to engage in the necessary teamwork across scientific and social science disciplines again limiting the potential to design relevant programmes and projects and compromising long-term goals.
8. Conservationists are too often willing to apply simple Malthusian analysis to environmental problems. It is all too easy and often misleading to see biodiversity conservation simply as a function of a population numbers game, with growing populations 'relentlessly' destroying habitats and species.

of small holders worldwide and the many indigenous and local communities around the world and their growing number of representative social movements offers a multitude of opportunities to reverse biodiversity loss and rehabilitate biodiversity-rich habitats while maintaining healthy and sustainable food systems.

Finding mechanisms to support the continuity of dynamic agro-ecological management practices is one of the most important contributions that peasants, fisherfolk, indigenous peoples and local communities can make to enhancing conservation and development in the biodiverse-rich environments where they live. Commitments by peasants and smallholders are increasingly taking root in many countries worldwide and are being actively promoted by grassroots organization networks and other social movements that support local biodiverse and ecological food provision, such as The Latin American Agroecology Movement (MAELA), The Permaculture Network, the Participatory Ecological Land Use Management Association (PELUM) in Africa, The Farmer-Scientist Partnership for Agricultural Development (MASIPAG) in Asia, among others (IPC, 2015). This commitment by peasants, fisherfolk, indigenous peoples and local communities and their representative grassroots organizations and social movements represents a considerable opportunity for conservationists, agriculturalists and communities to work together to achieve more affective conservation.

In Brazil, the struggle to protect forestry territories includes not only indigenous people, but also forest people and other traditional communities. In the late seventies, Chico Mendes led a movement to protect the Brazilian Amazon. This movement gave visibility to the forest people or '*extrativistas*', those people that rely on harvesting forest products for their livelihood, especially non-wood forest products. While the struggle sadly took the life of Chico Mendes, it did lead to approval in year 2000 of a federal law establishing '*Reservas Extractivistas - Resex*' as one of the categories recognized by the Brazilian protected areas system. Now, *Extractivistas* living in the Brazilian Savanah are pushing for recognition and demarcation of their territories to stop land grabbing

by plantation and mining companies. After decades of struggle, in October 2014 local communities from north of Minas Gerais State obtained the recognition of their land rights through the creation of 'Reserva de Desenvolvimento Sustentável Nascentes Geraizeiras'. Now, local communities can continue their traditional farming systems that combine agriculture, cattle grazing and harvest of tree fruits while ensuring the maintenance of forest coverage that protects important Brazilian savannah biodiversity as well (IPC, 2015). However, these gains often arise because of appropriate enabling environments, national governance and leadership being in place and can dramatically change, or be reversed, with change of government.

In the last decades, indigenous peoples' and community conserved territories and areas (ICCAs) have become increasingly recognized as important areas and processes for the conservation of nature, especially in areas at risk from the wide range of economic and political forces already discussed (see Chapter 2). ICCAs are premised on the close association that exists between a specific indigenous people or local community and a specific geographic territory, area or body of natural resources, and upon which is established fair and effective local governance mechanisms that allow benefits that accrue to biodiversity conservation but also to the traditions, culture and livelihoods of the guardians of these areas and the custodians of its genetic resources. ICCAs are also a source of identity and culture, autonomy and freedom, health and nutrition. With such governance in place it allows for conservationists and researchers to work closely with indigenous people and local communities to revive or modify traditional practices, some of which are of ancient origin, as well as establishing new initiatives, such as restoration of ecosystems and management of traditional landraces and crop wild relatives. Hunter and Heywood (2011) have reviewed ICCAs and highlighted that the mosaics of natural and agricultural ecosystems which ICCAs encompass including globally significant biodiversity represent an exciting prospect for future community-based work on *in situ* conservation (Box 10.9).

Box 10.9 | The Potato Park, Peru

(Source: Hunter and Heywood, 2011)

Six Quechua communities in Peru have worked closely with the NGO Asociación ANDES for several years to establish a 'Parque de la Papa' (the Potato Park). The Potato Park is a centre of diversity for a range of important Andean crops in addition to the potato, including quinoa and oca. The park represents a community-based agro-biodiversity-focused conservation area – also described as indigenous community conserved areas (ICCAs) and indigenous biocultural heritage areas (IBCHAs) – and is home to a diversity of Andean crop landraces as well as CWR, along with many other species regularly harvested from the wild for food, medicine, and cultural and spiritual purposes. The park is also home to a host of endemic plant species. The park aims to ensure sustainable livelihoods of indigenous communities by relying on local resources to create alternative livelihoods while using customary laws and institutions to facilitate the effective management, conservation and sustainable utilization of biodiversity and ecosystems. The Potato Park, which is not an official protected area, and IBCH areas, in general, represent unique opportunities for CWR conservation practitioners to engage with communities and grassroots organizations to ensure that CWR issues and concerns are integrated into plans. Such work also presents opportunities for linkages between protected and agricultural landscapes to facilitate expected CWR species migration under climate change.

With the burgeoning interest in territorial and landscape approaches involving indigenous and local communities and smallholders as well as an array of supporting social movements and local NGOs and CBOs, the time seems ripe for plant conservationists to better mainstream agro-biodiversity conservation into ongoing activities and initiatives. In addition to ICCAs, initiatives such as indigenous Protected Areas, Globally Important Agricultural Heritage Areas (GIAHs), the Satoyama Initiative, Landcare, Sacred Sites Networks, Man and the Biosphere programme and Landscapes for People, Food and Nature (LPFN), to name a few, all seem to offer opportunities for agro-biodiversity practitioners to work more closely with social movements and local communities to realize the multiple benefits of biodiversity conservation and sustainable use.

10.5 Community-Based Conservation: Incentives and Benefits for Local Communities

In many parts of the world, the homelands of indigenous peoples and local communities are often the last remaining places of wilderness and biodiversity (Garnett et al., 2018). These communities often have significant and intimate knowledge concerning the habitats they have maintained for millennia, both cultivated and wild, including CWR. This is no coincidence. Many indigenous peoples have contributed to maintaining biodiversity and ecosystems in their territories in two important ways:

- They have lived in ways that have left the natural resource base intact – through the development of patterns of resource management and use that reflects intimate knowledge of the local environment, protecting particular areas as sacred, developing land use regulations, and adhering to customs that limit the impacts of subsistence resource use.

- In many cases indigenous peoples have helped maintain the ecological integrity of their homelands by fighting outsiders' efforts to lay claim to their territory or to economically exploit its natural resources.

Where indigenous people and local communities have left or been forced from their lands or have effectively been rendered powerless through changes in land ownership imposed by outside cultures and socio-economic systems, it has often been observed that large-scale environmental transformation follows to the detriment of biodiversity.

As a result of increasing outside pressure for economic, social and political change, the long-term integrity of the local/indigenous management system is far from guaranteed. There are numerous situations where local communities have used wildlife sustainably for long periods of their history, but the situation has become worse due to the system being put under additional pressure from outside. In the Luangwa Valley in Zambia, for example, the local perception is that the reduction in wildlife in the area is due to outsiders – which includes international criminal syndicates of ivory traders and safari hunters – although it is sustainable hunting by the indigenous people that has mostly been prohibited by the government in recent decades.

Some communities have shown remarkable resilience in the face of adverse pressures on their local resources: for example, peasant women in the lower Himalayas, who depend on the forest for wood for fuel and building, grass, vegetables, honey, medicinal herbs and fruits, developed the 'Chipko' Movement. Their most distinctive tactic has been to hug trees to prevent them being cut down by commercial loggers. The companies had been permitted to fell trees on the land used by these communities because the state did not recognize the rights to land and trees of these so-called *tribal* peoples.

And as we have already seen, the resistance of indigenous peoples and local communities by asserting their rights in defence of local natural resources is often met with violence by the state or those with a vested interest in exploiting those natural resources to discourage opposition, as with the death of Chico Mendes and other leaders of the National Council of Rubber Tappers in Brazil. Brazilian rubber tappers and native forest dwellers in the Amazon mobilized against repeated, large-scale incursions by cattle ranchers and other interests. They developed the concept of 'extractive reserve' as a means of protecting their livelihood (see Section 10.3.2). For many activists, such as Berta Cáceres Flores of Honduras killed in 2016, death is a constant companion.

10.5.1 Why Community-Based Conservation, Benefits and Incentives?

The incentives for community conservation activities vary widely but generally bestow cultural, ecological or economic rewards. At a minimum, however, any conservation activity should contribute to increase the awareness of local farmers, NGOs and government officials about the value of conserving/managing diverse plant resources. When conservation efforts include local communities and small holders whose agro-ecological practices help maintain nature and who have an empathy for diversity, these activities tend to directly or indirectly re-value local culture and values. This is particularly important and needs to be given a more central focus to reach the younger generation, the next generation of farmers. This could be characterized as a form of environmental education, and it is not only very valuable to the professional conservationist but also much appreciated by local communities and farmers.

Seed fairs (see Box 10.19), for example, include various incentives, from social recognition and the prestige associated with that, to prizes in the form of seeds and tools, plus the opportunity to access new plant materials from a wider geographic area. Winners of such fairs have often requested technical support to deal with pests and diseases of their crops, information about market alternatives, etc., and in some cases, these have been included as additional activities or add on initiatives, thus becoming an incentive for future participants.

Training about collection and maintenance of biodiversity is usually offered in tandem with conservation efforts and in many cases includes association with local educational institutions such as schools (e.g. the work of the Swaminathan Foundation in India). Farming technologies requiring little inputs have also been successfully implemented through conservation activities. Bolivian farmers working with the Fundación PROINPA clean seed program have learned to multiply virus free landrace seeds in protected seedbeds and using sprouts. Although the cost of cleaning each landrace is estimated to be about $US800, community-based organizations (CBOs) have readily met this expense as they have seen that cleaned landraces significantly increase yield. The access to seeds in cases of emergency, access to improved, low-input landraces and displaced crops are among other incentives that can accrue to community participation in on-farm conservation and seed bank schemes.

Many of the successful community-based conservation projects discussed above have the following features in common:

- Good planning to ensure the marriage of conservation objective with local community needs, desires and aspirations.
- Local communities must be involved in all aspects of the project (planning, implementation, routine management, project evaluation and feedback).
- The conservation actions should be flexibly implemented bearing in mind input from the local community, even if this involves compromising some conservation goals.
- The professional conservationist must be consistent in his or her approach; communities will lose interest in a project if they feel it is poorly managed.
- Attempt to tie conservation activities to alternative or additional means of meeting community development goals, e.g. new economic activities, ecotourism, cultivation of new crops and even sustainable wild harvesting, therefore providing

real economic advantages to the local community by being associated with the project.

- Pay attention to the more intangible benefits that can result from good community-based conservation such as community empowerment, empowerment of women, enhanced leadership and community organization, enhancing local institutions, business skills, etc.
- Fostering local project ownership, connection, commitment and responsibility for their biodiversity.
- Promotion of conservation education and raising public awareness of associated issues.

These points are not exhaustive but are likely to be the key feature of successful community-based conservation projects. Considering the *in situ* conservation of crop wild relatives, Hunter and Heywood (2011) highlight how community participation presents key opportunities and benefits for those involved (see Box 10.10).

10.5.2 Local Communities and Professional Conservation Collaboration

The key issues relating to collaboration between the local communities and the formal conservation sector are summarized in Box 10.11.

Jarvis *et al.* (2016) describe a range of strategies that support farmer communities in benefitting from the conservation and use of traditional varieties, which include improved processing technologies to deal with diversified, as opposed to uniform, materials and the creation and promotion of markets for diversity using a range of tools and approaches such as quality labels, geographic indications, and organic, fair trade and eco-labelling.

10.5.3 Other Types of Benefit and Incentives

Payments for Ecosystem Services (PES) and *Agro-biodiversity Conservation Services (PACS):* These are market-based incentives for the conservation of ecosystem services through charges, tradable permits, subsidies and market friction reductions permitting the capture of public conservation values at the farm

Box 10.10 | **Opportunities and Benefits of Community Participation: The Case of *In Situ* Conservation of Crop Wild Relatives**

(Source: Hunter and Heywood, 2011)

- Facilitate data gathering.
- Provide insights into CWR and indigenous knowledge such as ethnobotanical knowledge on uses, understanding of the distribution of CWR, patterns of the use of CWR and potential threats.
- Participatory approaches allow opportunities for local and indigenous communities to be involved in planning and partnerships.
- Scientists and organizations can work with communities to strengthen the management of habitats and CWR species both inside and outside protected areas.
- Capacity can be developed so that communities and grassroots organizations are involved in the implementation of national CWR action plans and protected area and species management plans.
- Communities can be involved in species and habitat monitoring.
- Working closely with communities presents opportunities to communicate knowledge on the importance of CWR and to raise awareness and build support for CWR conservation.

Box 10.11 | **Working Together: Local Communities and Conservation Professionals**

(Source: Hawkes *et al.*, 2000)

Professional conservationists have often failed to appreciate the role local communities have successfully played in conserving diversity of wild or domesticated species within their local environment. They also have not seen the practical benefit that might arise from involving local communities in conservation activities. Although local communities and professional conservationists may have completely different reasons for and modes of conservation, they both manage biodiversity for long-term maintenance and so should work together. Some professional conservationists still mistakenly believed they can achieve their conservation goals without local community participation and may even consider the involvement of local people a distraction or an impediment to the achievement of their scientific conservation goals.

The synergy that can result from professional conservationists and local communities working together can be shown both to facilitate and to enhance overall biodiversity conservation goals. Specifically, collaboration necessarily increases the efficiency of 'professional' conservation, because local communities have a broader local knowledge base

Box 10.11 | (cont.)

concerning biodiversity found in their area. They are therefore able to assist in the development of a more practical focused and hopefully efficient approach to locally based conservation. Importantly, employing a collaborative approach between the professional conservation sector and local communities empowers local people and can often engender increased awareness and appreciation of native biodiversity and its conservation and use. Rather than deferring responsibility for their environment to external science-based experts, they can retain a sense of environmental responsibility for and take greater pride and identity in their own environment if governance mechanisms allow for this.

Other than the various direct benefits of collaboration outlined above, there are general benefits that accrue to society from increased public awareness of conservation issues. It is easier to ignore conservation issues if they are enacted by unknown professionals and entirely centred on a remote biodiversity institute, whereas if your village neighbour is a community seed curator or rare pig breeder, then conservation and exploitation have more immediate meaning, thus ensuring greater sensitivity to conservation issues. Also, when these matters are debated publicly, the result must be that the public will clearly be better informed and more supportive of formal conservation sector activities.

As professionals, formal sector conservationists should be sensitive to the conservation and exploitation desires of traditional communities; respect their wishes and not 'patronize' or 'preach' to those they consider ill-informed or untrained. Local communities need to be real partners in conservation activities with their professional counterparts. This may mean on occasion it is necessary to compromise purely scientific expediency to ensure good participatory conservation management practice that involves and is in harmony with local community needs and desires. On balance, for instance, it may be preferable for the formal sector not to oppose the partial replacement of traditional, endemic diversity by farmer-selected exogenous bred varieties, if in return it ensures the maintenance of traditional farming systems and their diversity as well as ensuring the support of the local communities for local conservation activities.

Professional conservationists also have a role to play in lobbying governments on behalf of local community-based conservation: firstly, to halt perverse economic incentives that not only discourage but may even ban local communities from conserving genetic resources; and secondly, to promote conditions under which farmers are encouraged to cultivate diversity in the crops and wild species in their local environment, even establishing various incentives to maintain traditional techniques and cultures.

level (Jarvis et al., 2016). To date PES have been applied largely to natural habitats; however, the application of these approaches specifically for agricultural biodiversity conservation is a new idea. Work is currently underway in Peru, Bolivia, Ecuador, India and Nepal to test the potential of competitive tenders in creating cost-efficient strategies to conserve priority endangered species and also improve indigenous farmer livelihoods – this is referred to as PACs.

Legal incentives: The international legal impetus for promoting community involvement in conservation stresses the need for the professional conservation sector to respect, preserve and maintain the practices of conservation and use employed by indigenous and local communities is incorporated into Article 8(j) of the Convention on Biological Diversity (CBD, 1992).

Each contracting party shall … respect, preserve and maintain knowledge, innovations and practices of indigenous and local communities embodying traditional lifestyles relevant for the conservation and sustainable use of biological diversity.

Therefore, it is highly desirable and obligated that professionals work together with local communities, if biodiversity is to be effectively conserved for future generations. Further justification is provided by ongoing support and impetus in the global community since the Rio conference: For example, the *Global Biodiversity Forum* Reports stress the following:

Participation is vital because much of the action needed to conserve biological resources must be done on the ground by a diversity of players (such as NGOs, women, local communities, indigenous peoples…) which possess specialized and local knowledge …

Local Communities – it is critical that they perceive and receive both immediate and long-term benefits from conservation action …

… indigenous peoples (including indigenous peoples in the North) – are often excluded from or powerless within vital processes. Greater effort is needed to inform and involve them in an appropriate manner. In order to ensure respect for and participation of indigenous peoples, land tenure issues should be resolved so as to assure equitable resource sharing and recognition of customary rights.

Ethical and cultural incentives: With increasing attention to rights-based approaches there is growing uneasiness, both locally and internationally, with the notion that the local communities can be overruled, ignored and, frequently, forced to relocate to meet the needs of conservation – an ideal local community do not necessarily share. In practical terms, conservation initiatives that ignored local people have failed.

Economic incentives: Biological resources are often under threat because the responsibility for managing them has been removed from the people who live closest to them, and instead has been transferred to government agencies in distant capitals. But the costs of conservation still typically fall on the relatively few rural people who otherwise might have benefited most directly from exploiting these resources. It is interesting to note that in many cases around the globe, those rural people who live closest to the areas with greatest biological diversity are often among the most economically disadvantaged. Community-based conservation reverses the top-down, centre-driven conservation approach by focusing on the people who bear the costs of conservation and may not, in fact, reap any of the direct benefits. Much of the effort today at implementing and assisting community-based conservation is directed towards ensuring an equitable distribution of costs and benefits amongst those living in closest proximity to the biodiversity in question.

Community level incentives: Because of the multiple issues inherent in community social, economic and cultural structure and function, biodiversity conservation action can in many situations address multiple objectives simultaneously. The conservation objective itself may therefore be minor compared to the additional benefits, which the community derives directly from participating in conservation-friendly activities. These objectives include:

- To build the capacity of communities adjacent to protected areas to develop productive activities which do not deplete biological resources.
- To reduce agricultural pressure on marginal lands.
- To concentrate agricultural development on the most productive agricultural lands.
- To conserve traditional knowledge about the use of biological resources, and the cultural systems which hold such knowledge.
- To re-establish common property management institutions where these have been effective in the past.

Box 10.12 | **Some Typical Incentives for Community-Based Conservation**

(Source: McNeely, 1988)

- Disincentives (fines)
- Community organization initiatives
- Community development initiatives
- Limited access to biological resource permitted
- Land tenure granted to local groups
- Training offered
- Education
- Employment
- Direct fiscal benefits

- To compensate villagers for possible loss of income through restrictions on use of protected biological resources, or for damages suffered from the depredations of wild animals on crops or livestock.

McNeely (1988) provides a useful summary of economic incentives for conservation at the community level. These are categorized in terms of the incentives and disincentives they address (see Box 10.12).

Community-based conservation also helps countries implement many of the conservation actions necessary to meet their obligations and targets, as set out in international agreements and conventions such as the CBD and the ITPGRFA (Hunter and Heywood, 2011), and if implemented on a wide scale, could contribute significantly to the 2030 development agenda for sustainable development and many of the goals and targets of the SDGs (see Chapter 2).

However, it is important to bear in mind that community participation and involvement is not completely free of challenges and constraints. Some typical problems that may be encountered are:

- It is difficult to arrange the participation and cooperation of the large and diverse groups of individuals typically involved in environmental and conservation issues.
- It is difficult to make available a great deal of time for extensive rounds of consultations and discussion.
- It is difficult to make sure that participants fully understand the issues and the technical and scientific implications of options and proposals.
- It is difficult to be sure that those participating are representative of the people concerned and so prevent the hijacking of participatory processes in favour of special groups or classes.
- It is difficult to ensure that the diversity of perspectives, needs and expectations can be fairly and adequately captured in conservation plans.

Community involvement and participatory approaches will present challenges for scientists and conservation practitioners, particularly those schooled in 'normal' professionalism and who may be used to working with conventional, quantitative research approaches (Hunter and Heywood, 2011). For this reason, when planning for conservation actions that involve community participation, it is essential to formulate gender-balanced, cross-disciplinary teams that involve social and natural scientists. It is especially advisable that conservationists and researchers seek out those in their organization or elsewhere who have extensive skills and experience in using participatory methods and tools and facilitating participatory approaches with indigenous peoples and local communities.

The involvement of local communities and NGOs and social movements in PGR conservation may also be beneficial not least because NGOs have a greater freedom of advocacy and expression than the formal sector. The latter may be constrained on certain issues by government policy, whereas the informal sector is not restricted by financial expediency. Someone working in a formal biodiversity institute may be unwilling to speak out, for example, against the government policy of perverse incentives to persuade traditional farmers to grow modern cultivars, thus leading to the abandonment of traditional landraces or the promotion of genetically modified crops in all environments. Even if the formal sector worker believes these policies to be misguided for the longer term future of the country, he or she may feel incapable of speaking freely. NGOs are not reliant on governmental sources for funding and so have freedom of advocacy.

10.6 Planning for Community-Based Conservation

How exactly can local communities become involved in conservation? Broadly speaking they can benefit in two ways:

- *Intrinsic* – In this case the local community conserve biodiversity for their own reasons, the same reasons that they have been conserving biodiversity for millennia. Members of the local community set the priorities, and the benefits that accrue in the first instance at least are to them. For example, communities maintaining a mixture of LR that are locally adapted and always produce yield despite variation in the growing season.
- *Extrinsic* – In this case the local community work in collaboration with professionals to conserve biodiversity within their local region, but in this case the initiative for the collaboration is led from outside the community (the formal conservation sector). The priorities will be national, regional or global, and the benefits will accrue to society at large. If well thought through, the local community should benefit; however, their direct benefit is not the prime goal of the conservation project, for example, communities maintaining CWR populations despite them being of no direct use to that local community and possibly for crops not grown in that vicinity.

The role of local communities in plant conservation is two-fold. Not only to continue to conserve biodiversity using their own traditional practices for their own future direct and indirect benefit, but also to work in collaboration with the professional conservation sector to conserve broad-based biodiversity for their country and public good for all humankind's benefit.

As discussed above, the role of local communities in plant conservation is two-fold, not only to continue to conserve biodiversity using their own traditional practices for their own (intrinsic) future direct and indirect benefit, but also to work in collaboration with the professional conservation (extrinsic) sector to conserve broad-based biodiversity for their own community, country and humankind's benefit as a whole. The priorities for local community-based conservation will come from the local community, but their conservation priorities *per se* are likely to be secondary in comparison to wealth generation or development goals, whereas the priorities for the formal sector will be explicitly conservation based but are likely to have a broader national, regional or even global context, even if applied at the local level.

It is perhaps easy when discussing local community conservation activities for the professional conservationist to ignore the aspirations of local communities (see Box 10.13) and focus entirely on the science of conserving biodiversity. It is, however, always important to realize this approach is unlikely to gain the support of any local community. In fact, there are numerous examples of projects that have adopted this approach and failed.

Box 10.13 | Agro-ecology and *In Situ* Conservation of Native Crop Diversity

(Source: Altieri and Merrick, 1987).

Recommendations for *in situ* conservation of crop genetic germplasm have emphasized the development of a large system of village-level landrace custodians (a farmer–curator system) whose purpose would be to grow a limited sample of endangered landraces native to the region. To preserve crop plant diversity, Wilkes (1983) suggested that governments set aside carefully chosen 5-by-20 km strips at as few as 100 sites around the world where native agriculture is still practiced, areas where both indigenous crops and their close wild relatives may inter-breed periodically.

However, the idea of setting aside parks for crop relatives and landraces is obviously a luxury in countries where farmland is already at a premium, but to some this may be less costly than allowing native crop varieties to disappear. In many areas, the urgent short-term issue is survival, and it would therefore be totally inappropriate to divert the limited land available to peasants for conservation purposes *per se*, so that the germplasm could be used by industrialized nations, which some may see as an undesirable form of neo-colonialism. However, the LR and CWR conserved in such a global network of PGR diversity would be equally beneficial to international commercial breeders, national breeding institutes and even farmers. The fact is with climate change and the requirement for greater breadth of PGR diversity to sustain food production, the imperative to establish some form of public good global network of conservation sites is inevitable (FAO, 2013c).

Effective community-based conservation like most other conservation approaches will depend on putting together a team with the best blend of skills and experience as well as consideration of wider collaborators and partnerships. It will require developing and nurturing relationships and partnerships among a wide range of individuals and institutions, as well as mobilizing CBOs and NGOs and farmers' groups and organizations (see Box 10.14). These are fundamental steps in any approach to community-based conservation, be it on-farm conservation, community biodiversity management or wild species or crop wild relatives *in situ* conservation.

Managing the many different actors and stakeholders will be critical as is establishing credibility, trust and respect. As the focus is on community-based conservation in this instance, it is essential that any collaborations and partnerships ensure there are benefits to the communities involved and that the process is building local community capacity and empowerment. The identification and selection of the final partnership is critical and will depend on a range of factors that need to be considered and which will influence any outcome targeting community-based conservation (see Box 10.15).

In planning and implementation of a programme for on-farm conservation (Jarvis *et al.*, 2016) stress that it requires much more than resources and expertise. Working at nurturing partnerships involving a range of individuals and institutions is required, and the mobilization of CBOs is critical if on-farm conservation is to be successful. These authors identify six broad groups as being important for on-farm conservation: 1. Farmers and local

---

Box 10.14 | **Individuals and Organizations to Be Considered for Inclusion in Community-Based Conservation**

(Source: Hunter and Heywood, 2011)

- Community and indigenous leaders
- Community and indigenous groups
- CBOs, NGOs, farmer groups
- Community mobilizers
- Training specialists especially in community development
- Facilitators of participatory methods/approaches
- Political leaders and senior policy-makers
- Senior biodiversity, environment and agriculture decision-makers
- Heads of relevant organizations and institutes
- National and local policy planners
- Scientists and researchers
- Protected area managers
- Project management staff
- Field technicians
- University lecturers and postgraduate students
- Communications and public awareness specialists
- Extension and outreach specialists
- Information analysts and managers

Note it is critical to community-based conservation to ensure local communities are involved early in the planning process and equally contribute to planning conservation actions.

---

communities; 2. Ecologists or ecosystem health workers; 3. Conservationists and breeders; 4. National governments; 5. Private sector and 6. Consumers. The authors also point out that often many of these individuals and institutions are not used to working in multi-institutional collaborative frameworks but that it is critical to success that time and effort is made available to develop such collaborations even though this is oftentimes cumbersome and time-consuming. They also stress that farmers and local communities are given the capacity and space to participate on an equal footing with other partners to ensure an equitable collaboration.

De Boef et al. (2013a) discuss many of these issues concerning planning and implementation within the context of what is known as community biodiversity management (CBM), which aims to contribute to implementing in situ conservation and the on-farm management of agro-biodiversity. They describe CBM as a methodology with its own set of practices, which aims, through a participatory process, to strengthen community institutions and build their capabilities to effectively undertake the conservation and sustainable use of plant genetic resources (Box 10.16). Like the earlier authors describing the complexity of involvement when planning and establishing

Box 10.15 | **Issues to Consider When Planning Partnerships for Community-Based Conservation**

(Source: Hunter and Heywood, 2011)

- Common interests around community-based conservation
- Common goals for community-based conservation
- Reputation both nationally and internationally
- Level of expertise in community-based approaches
- Track record, including past achievements/problems, with local communities
- Clear objectives of what to achieve
- Power relations with other sectors and actors
- Experience and attitudes towards other partners
- Receptivity to public opinion
- What drives partners/limits them/enables them
- Their interests/revenues/rewards

Box 10.16 | **Practices Used as Part of the CBM Approach**

(Source: de Boef *et al.*, 2013a)

- Diversity/seed fair
- Diversity and culture
- Diversity block
- Diversity kit
- Reintroduction of gene bank accessions
- Local multiplication of gene bank accessions
- Agro-biodiversity and tourism
- Agro-biodiversity and education
- Traditional food fair/recipes
- Community seed bank
- Community biodiversity register
- Agro-biodiversity products
- Community-based seed production

partnerships for CWR and on-farm conservation, de Boef *et al.* (2013a) are quick to point out that those conservation and development organizations aiming to support CBOs in the implementation of CBM require knowledge, skills and expertise for building social institutions at community levels. CBM practitioners also call for a new professionalism for those working to empower CBOs.

Hamilton and Hamilton (2006) highlight important elements and steps of general *in situ* plant conservation projects that involve local communities, stressing that it is helpful that individuals and organizations involved pay particular attention to the benefits of long-term commitment as well as strengthening local capacity including that of local institutions or if these are not strong or there are gaps, making sure you pay attention to establishing relevant institutions. They also recommend avoiding 'rash promises' that might raise the expectations of local communities which you might not be able to meet. They point out how to assist individuals in local communities to enhance their knowledge and skills, as well as behaviours and attitudes, and that this can be provided through formal and non-formal education and training. Hunter and Heywood (2011) also highlight the importance of capacity-building strategies and plans in the context of *in situ* CWR conservation with a focus on key steps and approaches. Hamilton and Hamilton (2006) particularly stress the importance of targeting local institutions such as forest user groups, self-help groups, church groups, schools and local governments as these institutions are vital because of the roles they play in community social life and the need for strong and effective institutions to negotiate with outside agencies and stakeholders. Cunningham (2001) highlights how conservationists can gain a better understanding of the social, economic, ethical, religious and political factors so critical to the success of community-based approaches including methods which can be used to better understand tenure and resource use characteristics to establish common ground between local communities and conservationists.

Pretty (1995) stressed the critical role of local institutions in agricultural and rural development targeting resource conservation highlighting that without full participation, collective action of rural communities is unlikely to succeed. Local groups and institutions critical to success include: community organizations; natural resource management groups; farmer research groups; farmer-to-farmer extension groups; credit management groups; and consumer groups. Pretty (1995) described the rationale for collective action at the local level, the perils of ignoring local institutions, establishing self-reliant groups and approaches to scaling up from the local level.

A final word about the importance of clear communication. Partnerships convene a diversity of stakeholders, interests, motives and agendas which do not necessarily align with those of local communities. Promoting clear and open communication is important. Most problems that arise in conservationist–local community partnerships can be traced to poor communication. At the planning stage, it is useful for the partnership to consider developing a communications strategy which should also incorporate aspects of external communication and advocacy for the partnership in general, not just internal communication between partnership members. However, having stressed the need to work with local communities, the onus must be on the professional conservationists and scientists to ensure that misunderstandings do not arise with local communities. Almost invariably, local communities have innate and overwhelming pride in their local environment. Therefore, if they realize that their local environment contains a particularly rare or in some way special animal or plant, their natural pride is accentuated, and this can lead to misunderstandings. As is outlined in Box 10.17, drawing attention to a rare plant population in Syria had dire consequences for the plant population.

One of the biases or much criticized issues in conservation practice or any research involving local communities has been the lack of 'giving back' or returning the results and findings of projects and other activities to communities. In fact, this was one of the major critiques launched at early Rapid Rural Appraisal approaches. There are many reasons for this, including some researchers believing they have

---

Box 10.17 | **There Must Be Some Misunderstanding!**

(Source: Maxted, personal communication)

In 1986 a collecting team rediscovered the faba bean relative *Vicia hyaeniscyamus* widely considered extinct, as well as a new faba bean related species (*Vicia kalakhensis*), near Tel Kalakh, Western Syria. Due to the botanical interest of the site, the collecting team visited the site several times during the year on each occasion collecting fresh herbarium specimens and seed.

Obviously, the local community were interested in what the strangers were doing in and around their gardens, and wishing to explain and involve the local community, the conservationists took time to inform the local village community through an interpreter of the botanical treasure to be found within their community and the importance of this find to agriculture and botany. The realization that their community contained rare plants raised a lot excitement and the local people's pride in their environment. However, there was some difficulty in that the conservationists and local community did not speak each other's language; this hampered a complete understanding of the issues involved and thus resulted in two serious misunderstandings.

- Firstly, the local community offered to assist, but the professional collectors failed to explain clearly the principles involved, and the local people collected several thousand green immature pods, which were of no conservation use and greatly eroded the base populations.
- Secondly, in trying to explain why the plants were of importance, the professional collectors simplified the explanation by saying they were of use as animal fodder. The local community took this as being literally true and subsequently fed the two *Vicia* species found around their village to their animals.

These initial misunderstanding have led to a decline in the populations of these two species around Tel Kalakh over subsequent years. Further as *Vicia kalakhensis* has still only been found at this location, it resulted in the near extinction of the species. Here the fault clearly lies with the collecting team and their inadequate communication with the local community – a situation that should be a salutary warning to any conservationist.

---

done their job when their research has been peer-reviewed and published. Shanley and Laird (2002) point out that this is all too common in conservation projects and must be given close attention when designing projects so that returning research findings to local communities becomes routine and part of any on-going local capacity. If not already embedded, it must become an essential element of any equitable research relationship and an example of benefit

sharing with local communities. Many scientists might not necessarily have the necessary skills to do this, so they should seek out those who can and form strategic alliances so research data and findings in forms relevant to local communities can be given back. There are numerous forms of 'giving back', and these were reviewed by Hunter and Heywood (2011) who highlight the importance of rural drama and poetry, tools common to the CBM Toolkit.

Laird and Noejovich (2002) provide an overview of two elements which are important in establishing equitable research relationships with indigenous peoples and local communities. These are free, prior informed consent (FPIC) and research agreements. While Laird and Posey (2002) highlight how relevant professional organizations have developed approaches, standards and best practice that try to ensure ethical values are adhered to during research and practice such as the use of codes of ethics and research guidelines.

10.6.1 Do Not Patronize, Listen!

The point has already been made that local communities and professional conservationists working together to conserve plant diversity engender the ideal conservation milieu. The collaboration of local and professional communities can only be viewed as beneficial for biodiversity conservation. Other than the various direct benefits of collaboration outlined above, there are the general benefits that accrue to society from increased public awareness of conservation issues. It is easier to ignore conservation issues if they are enacted by unknown professionals and entirely centred on a remote biodiversity institute, whereas if one's village neighbour is a community seed curator or participant plant breeder, then plant conservation and exploitation should have more immediate meaning, thus ensuring greater sensitivity to plant genetic resource issues by all. This enhanced public awareness of biodiversity conservation issues among the general public will promote discussion and ensure that the public is better informed, and is able to hold politicians more accountable for environmental issues. It should also be acknowledged that interaction with local communities is almost exclusively life enhancing for the conservationist as a human being, conservationists get to travel off the 'beaten track', and it is a privilege and personally rewarding.

10.6.2 Compromise May Be Necessary

As professionals, the formal sector conservationists should be sensitive to the conservation and development needs and desires of traditional

communities, respect their wishes and not patronize or 'preach' to those they consider ill-informed or untrained. Local communities should feel that they are partners in conservation activities with the professional agencies. This may mean that often it is necessary to compromise purely scientific expediency to ensure participatory conservation management that involves and is in harmony with local communities. Or the conservationist can rethink the purely scientific approach to fulfil both scientific and local community benefits, as a supportive local community will always enhance conservation outcomes. On balance, for instance, it may be preferable for the formal sector not to oppose the introduction of some form of participatory breeding or even partial replacement of traditional, endemic diversity by farmer-selected exogenous, improved breeds or varieties, if in return it ensures the maintenance of traditional farming systems and their diversity as well as the support of the local communities for local conservation activities. Thus, conservation must also be seen to be central to human interests and aspirations for it to be effective.

Hunter and Heywood (2011) and the four CABI agro-biodiversity texts (Maxted *et al.*, 2008d, 2012b, 2016b; Iriondo *et al.*, 2008) provide a comprehensive overview of planning and partnership building as well as the application of participatory approaches for crop wild relative *in situ* conservation which includes checklists, tools and methods, good practices and lessons learned. For those readers interested in on-farm conservation of crop genetic diversity, Jarvis *et al.* (2016) provide guidance on strategies for collaboration, partnerships and implementation. De Boef *et al.* (2013a) provide a comprehensive overview of the CBM process, tools and methodology, and it is some of these processes and tools which we turn to in the next section.

10.7 Communities Working to Conserve Agro-biodiversity

There are numerous examples of local communities and farmers in the past and today working actively to conserve a broad range of biodiversity both for their own direct and indirect benefit. For example, farmers in Chiapas, Mexico, keep certain maize landraces as a form of luxury commodity; in Tunisia certain landraces are maintained as a family legacy; as a sacred gift from the gods or their ancestors for Indian religious festivals; to protect the farmer from natural or anthropogenic risk; to promote soil–water conservation; to produce the largest vegetable for village fairs in the UK or simply because they taste and cook better. Local communities continue to maintain biodiversity because it is in their interest to do so! But why is it in their interest? In an attempt to answer this question, the broad range of conservation activities in which communities engage in, or can engage in, is discussed in more detail below.

Experience over many years has taught subsistence and commercial farmers and local communities in general that maintaining diversity amongst and within domesticated species is essential to sustainable agriculture and environment, particularly in marginal lands. Diversity, farmers in marginal environments have found from experience, can ensure survival in periodically adverse conditions (e.g. droughts, floods, disease or pest out-breaks). One means of local communities encouraging their farmers to retain diverse agro-biodiversity and ensure seed security is to facilitate or establish local seed or diversity fairs.

The following examples offer a brief glimpse of the different kinds of approaches, tools and methods that can be used or strengthened to better support smallholder farmers and communities in the conservation and sustainable use of plant genetic resources.

10.7.1 Communities Taking the Lead

Seed Diversity Fairs

Seed fairs have been in existence for some time in various parts of the world, offering communities an opportunity to come together to sell, buy and exchange plant genetic resources (Jarvis *et al.*, 2016; Sthapit *et al.*, 2016). They also act as fora that facilitate interactions between rural communities, farmers, extension agents and others. In some instances, they can also permit the interaction between farmers and the private sector allowing the

formal seed sector to learn more about the preferences and needs of farmers as well as the opportunity for farmers to learn about what the formal seed sector has to offer them. Seed fairs are now commonly used by NGOs and CBOs to raise awareness about the importance of crop diversity, to address a range of issues including adaptation to climate change, enhancing food security and nutrition and PGRFA contributions to sustainable livelihoods.

The fairs focus on displays of native domesticated animals or crops and genetic diversity at the community level, and prizes are given to those exhibiting the widest range of diversity, not necessarily for the biggest or 'best' specimens. The fairs make farmers aware of the benefits of diversity, facilitate exchange and acquisition of desirable seeds and promote a more sustainable form of agriculture (see Box 10.18).

Seed fairs have been in existence in the Andes of South America for many centuries, where farmers congregate in set places on certain days of the year to exchange various goods, including germplasm from different valleys and regions (see Box 10.19). Weekly markets are held where general goods are exchanged, but specifically once a year, usually after harvest at a religious festival, the focus is on seed and farmer's knowledge exchange. In recent years, local NGOs have taken a more active role in organizing these fairs to encourage greater seed and crop diversity.

Seed fairs are not limited to biodiversity-rich countries; increasingly various forms of seed and diversity fairs are occurring in developed countries. In the UK, Garden Organic (www.gardenorganic.org.uk/) organize several 'Seedy Sundays', which aim to bring together seed savers, herb and home-grown vegetable enthusiasts, and local gardening, allotment and community groups to exchange locally grown, traditional variety vegetable seeds with other enthusiasts. Experts from the Garden Organic Heritage Seed Library are on-hand to provide advice to growers on choice of variety for the growers' conditions, discuss variety cultivation, to demonstrate the process of seed saving, and promote lobbying for agro-biodiversity and broader biodiversity conservation. From the genetic conservation perspective, it should be noted that less formal seed

Box 10.18 | Genetic Diversity in the Himalayas

(Source: Rijal *et al.*, 1998)

A specific example of the way in which local communities and farmer groups can establish seed fairs is illustrated in a project from Begnas, Nepal. The aim of the project was to:

- Recognize farmers who maintain large crop diversity and associated indigenous knowledge and who act as a source of information for others.
- Prepare an inventory of local crop genetic diversity and identify areas of maximum diversity.
- Identify and locate the most endangered landraces.
- Identify main sources of informal seed supply within communities.
- Understand the farmers' reasons for maintaining diverse genetic resources in terms of use, and economic, cultural, religious, breeding and ecological values.
- Empower local communities to take control of their own genetic resources and develop ownership using their community gene bank to link both the formal and informal seed supply systems.

They defined the following stages in preparing for the seed fair:

- Exchange of ideas during village meetings.
- Planning meetings at site level to decide on methods and norms for competition.
- Announcement of date and venue (1 month in advance).
- Information transfer through public awareness radio broadcasting and local papers.
- Formal invitations (along with the rules for competition) to the farmer groups.
- Orientation training on norms and procedures of the diversity fair for groups and site staff.
- Provision for packaging materials (by site staff).
- Field registration and verification of the registered materials.
- Evaluation and nomination of the award.
- Public display of materials in an open place (e.g. school or marketplace) with prize winner prominently labelled.
- Prize distribution ceremony at the end of the diversity fair and distribution of unique seed packets as a consolation prize to all participants in the competition/or close runner ups.
- Revisit custodians for future contact and collaboration.

The competition was judged by local and external experts. The criteria they used were:

- Number of local landraces displayed by the group or farmer (40%).
- Quality of information provided and its authenticity (30%).
- Style of presentation (15%).
- Degree of women's participation (10%).
- Participatory group dynamics (5%).

Box 10.18 | (cont.)

The winning group displayed 43 rice landraces, including upland and wild rice, 15 landraces of finger millet, 18 cultivars of sponge gourd and 7 cultivars of taro. The fair had two important results for the community in Nepal; firstly, it demonstrated the crop diversity found locally and made that diversity available for formal sector *ex situ* conservation by the Nepalese national programme, and secondly, it raised public awareness of plant genetic resources conservation and use issues at the grassroots level in the area.

Box 10.19 | Genetic Diversity in the Andes

(Source: Marleni Ramirez, personal communication)

In the Community of Quispillacta, Peru, seed fairs have increased the local diversity of crops grown in individual and communal farm plots. In fact, farmers from this community were more interested in acquiring new and or lost landraces than in the competition itself. This is noteworthy considering that farmers already maintain sizeable landrace collections. For example, the farmer with the most diverse collection at one of the diversity fairs showed 123 landraces representing 12 crops.

Seed fairs in the Andes are usually associated with a broader agenda of agro-ecological activities such as soil conservation, water management and the establishment of seed banks. Seed fairs at La Encanada (Cajamarca, Peru) watershed have been carried out since 1990, basically with the purpose of measuring agro-biodiversity in the watershed, to promote conservation activities and exchange of germplasm and knowledge. The persistence of these activities for 10 years is perhaps related to the fact that the idea for seed fairs emanated from the farmers' concerns with seed loss and the weakening of traditional seed exchange networks. Seed fairs have evolved over time from just counting farmer diversity to knowledge assessment and seed quality evaluation. The site of the fairs also rotated among the various communities within the watershed thus improving the participation of farmers from different agro-ecological zones.

Fairs at La Encanada follow the following steps:

- Two months before the fair a letter of invitation is sent to each village in the watershed and posters indicating the date and place are distributed around.
- At the fair, a registry of crops and landraces presented is developed for each farmer.
- Farmer judges are selected by fellow farmers and two external judges are also appointed.
- Evaluation is according to the diversity of species and landraces presented and includes the major subsistence crops, fruits and medicinal plants. The health of samples as well as the

Box 10.19 | (cont.)

knowledge about the various crops is evaluated. Although at La Encanada the male head of household is usually registered as the exhibitor, his wife and other family members contribute their knowledge about crops in the competition.

- All this information is entered in the collective memory bank of traditional landraces.
- Prizes are given out categorized by crop, groups of crops, by individual farmer and by groups of farmers.

The 10 years of records reveal that almost a quarter of the 1000 households in the watershed have exhibited in the Fairs, and about 15% of those exhibited more than once with a few exhibiting up to four times.

Elsewhere in Peru, in Aymara in the Central Andes, a biodiversity fair has been held annually since 1989, and although it originally highlighted the diversity of potato landraces, it now includes other crops. At Aymara, more families enter the competitions every year and the number of cultivars displayed has also increased. It is noteworthy that at a recent fair in nearby Colpar most of the participants were women. The increasing popularity and recognition of seed fairs appears to have enticed women to participate as exhibitors by themselves, which is good news since women are the most knowledgeable about crop landraces and their characteristics.

diversity initiatives often contain significant unique diversity that is not always fully duplicated in the more formal sector (Preston *et al.*, 2018).

The common feature of each of these fairs is that they engender the farmer's pride in retaining diverse crops and animal breeds, promote farmers' selection and facilitate inter-farmer exchange. The fairs also provide an opportunity for farmers to share knowledge of good farming practices, and to exchange and purchase animals or seed that they may have lost or wish to experiment with on a trial basis. Public recognition of those farmers who cultivate a wide range of diversity acts as an incentive to other farmers to recover or adopt local varieties or breeds and so conserve diversity.

Food Diversity Fairs

Increasingly food fairs are being employed by communities, CBOs, NGOs and other organizations to use food and food cultures as a way to highlight and promote local biodiversity. With the rapid growth in awareness about healthy foods and diets and the benefits of dietary diversity there has never been a better time to use such avenues as a way of strengthening CBM. Likewise, the rise in popularity of food tourism provides an opportunity to get involved in and benefit from such tourism linked to agro-biodiversity. Food fairs are also an important tool for increasing awareness about good nutrition through education as well as revalorizing traditional recipes and the exchange of these between different community groups.

Community Seed Banks

Community seed banks are local institutions that have been established in response to a range of factors, including genetic erosion and conflict, and aim to conserve and make available crop diversity to local communities (Jarvis *et al.*, 2016). They offer a more practical and accessible point of contact for PGRFA access than the more formal gene bank networks that less often supply the individual needs of farmers or growers. Though CSB offer an effective, low-tech alternative to formal gene banks, they are often supported by international or national organizations. They are centred in communities and perform multiple functions (Vernooy *et al.*, 2017), which might include:

- Awareness raising and education
- Documentation and dissemination of traditional knowledge
- Collection, multiplication, distribution and exchange of seeds
- Exchange of knowledge and experiences
- Promotion of agro-ecological approaches
- Supporting participatory plant breeding and farmer experimentation
- Income-generating activities for members
- Networking and policy advocacy
- Development of other community enterprises

In the only comprehensive review of the approach, Vernooy *et al.* (2017b) grouped the functions and activities of community seed banks into three core areas: conservation, access and availability, and seed and food sovereignty.

The Community Seed Bank at the M.S. Swaminathan Research Foundation is an *ex situ* facility for the medium-term conservation of seeds of landraces, wild crop relatives and medicinal plants collected from tribal and rural farm communities in southeast India. Farmers affiliated with NGOs or self-help groups participate in the collection, on-farm seed multiplication and screening trials of minor millets and traditional rice cultivars. Some of these seeds are then exchanged during traditional festivals. The Community Gene Bank, in cooperation with the National Bureau of Plant Genetic Resources for India, has also reintroduced some lost traditional rice paddy cultivars in Wayanad, which have been distributed to farmers for field testing. Simultaneous documentation of traditional knowledge and maintenance of voucher specimens at a Community Herbarium is meant to document the contributions of tribal and rural

communities to the diversity of plant genetic resources.

As with seed fairs, community seed banks are not limited to biodiversity-rich countries, and increasingly various forms of CSB are found in developed countries. The Heritage Seed Library of Garden Organic (www.gardenorganic.org.uk/) is a CSB meeting the needs of the UK gardeners that enjoy cultivating old, diverse vegetable varieties. It holds about 800 traditional varieties either never present or lost from commercial agriculture as a result of seed legislation in the 1960s and the establishment of national varietal lists (Preston *et al.*, 2018) (Box 10.20). Another CSB scheme in the UK is the Scottish Landrace Protection Scheme (Green *et al.*, 2009) established by the Scottish Agricultural Science Agency (SASA) in 2006 to provide a safety net for crofter (= small-scale farmer) production of crop landraces. The scheme involves SASA staff working with crofters to collect passport information concerning the donor and the landrace, gain permission to distribute seed samples (crofter may refuse, then the seed sample is maintained as a 'black box' sample), characterize the accession and maintain a duplicate of the donor accession that can be returned on request to the original donor if they lose their next season seed for planting. Thereby it permanently secures the landraces and its contribution to future crofting diversity.

Seed Saver Schemes

Many countries have established seed saver schemes that conserve traditional cultivars. These schemes promote the conservation and use of traditional landraces of crops. They are usually coordinated by a national agricultural conservation NGO, even working in contradiction of national legislation (see Box 10.20).

Probably the most comprehensive Seed Saver Scheme is that based in the United States, the Seed Savers Exchange (SSE) (www.seedsavers.org). The SSE is an American non-profit tax-exempt organization that is saving 'heirloom' (handed-down) garden seeds from extinction. SSE's 8000 members

Box 10.20 │ **Seed Saving of Vegetables in the United Kingdom**

(Source: Louise Daugherty, personal communication)

In the United Kingdom, such an operation is run by Garden Organic (www.gardenorganic.org .uk/hsl) and is referred to as the Heritage Seed Library (HSL). Its activities are primarily to focus on conserving old vegetable varieties. It is recognized that many old or traditional varieties are being replaced by modern cultivars, and European legislation has accentuated this process by making it illegal to trade in and offer for sale varieties not found on the 'National List' of approved cultivars. Registration for inclusion on this list is expensive so seed merchants have greatly reduced the number of varieties they offer for sale. The HSL aims to save old diverse or traditional varieties, and it currently has more than 800 vegetable varieties consisting of predominantly open-pollinated ones from throughout Europe. There are three main 'varietal categories': (a) heirloom varieties of old, traditional types; (b) local varieties that may be restricted to a specific location and (c) ex-commercial varieties.

The HSL collection is supported by a membership of 9321 amateur conservationists. This includes individuals, community groups and larger organizations. In 1999, based on accessions advertised in their catalogue, they distributed approximately 40 000 packets of seed. Members of the HSL each contribute £18 towards the running of the programme and in return receive a quarterly newsletter and can select up to six packets of free seed chosen from the yearly catalogue. The fact that a member does not actually pay for the seed means that the European legislation can be circumvented, and the vegetable varieties conserved. Of the 9321 members, 282 are classified as 'Seed Guardians'. Seed Guardians agree to raise seed of a chosen variety, ensuring that it remains pure and true to type. They can cultivate as many varieties as they wish to guard. The coordinators generally encourage Seed Guardians to specialize in preserving a few varieties, and first time Seed Guardians are offered varieties of species that are relatively easy to grow, e.g. tomatoes or peas. More experienced Guardians tend to guard more difficult crops. The Seed Guardians harvest and clean the seed and return it to the Seed Library for storage and further distribution to other members through the Seed Library Catalogue. In total, the seed returned from Seed Guardians plus that returned by other HSL members contribute up to 50% of the seed that is distributed in the following year's HSL catalogue. The staff of the HSL offers advice and training on seed saving and actively collect historical information associated with the varieties in the collection and characterize the collection.

Associated with the HSL are other conservation projects, such as the 'Adopt a Veg' appeal, which allows individuals to contribute financially to the longer term maintenance of named varieties and more generally support the activities of the Henry Doubleday Research Association (HDRA). The Adopt a Veg appeal also importantly generates public awareness about the need for vegetable landrace conservation and often leads to the adoptee becoming a member of HSL. The HSL also has 15 HSL displays in various gardens around the UK. The purpose of these HSL displays is to demonstrate the principles of seed saving and again to underline the need to conserve heritage vegetable varieties.

Source: www.gardenorganic.org.uk/hsl

grow and distribute heirloom varieties of vegetables, fruits and grains. SSE's focus is on heirloom cultivars that gardeners and farmers brought to North America when their families immigrated, and traditional cultivars grown by Native Americans, Mennonites and Amish. Since SSE was founded in 1975, members have distributed an estimated 750 000 samples of endangered seeds not available through catalogues and often on the verge of extinction.

The Heritage Farm is the SSE's 170-acre headquarters near Decorah, Iowa. It is a living museum of historic cultivars that is open to the public and houses the seed storage facilities and educational centre. Each summer an estimated 4000–5000 gardeners and fruit grower's tour Heritage Farm's large organic Preservation Gardens and Historic Orchard.

The Australian equivalent is the Australian Seed Savers' Network (www.seedsavers.net/), which aims to develop and promote:

- educational programmes for the preservation of open-pollinated (non-hybrid) seeds and the genetic diversity of plant varieties;
- non-profit seed exchange programmes;
- agricultural and horticultural programmes with particular emphasis on the propagation of open-pollinated plant varieties;
- preservation gardens for open-pollinated plant varieties;
- seed banks for non-hybrid plant varieties;
- scientific research relating to the above matters, either alone or in conjunction with a public university or other institution.

Indigenous Community Conserved Areas (ICCAs)

Indigenous and community conserved areas (ICCAs) are considered as natural and modified ecosystems containing significant biodiversity, ecological services and cultural values, and which are voluntarily conserved by indigenous and local communities, through customary laws or other effective means of governance or management. One well-known example of an ICCA with significance for plant genetic resources is the Potato Park (see box 10.9) in Peru declared in 2002 by the six Quechuan

agrarian communities, known as Chawaytiré, Sacaca, Kuyo Grande, Pampallaqta, Paru Paru and Amaru. The Potato Park focuses on protecting and preserving the critical role and interdependence of the indigenous biocultural heritage for the maintenance of local rights and livelihoods and the conservation and sustainable use of agro-biodiversity (Hunter and Heywood, 2011).

Agro-biodiversity and Tourism

A community's agro-biodiversity can also provide an entry point for linking to tourism. For example, in Nepal, the local rice Jethobudho is an aromatic variety considered excellent for its cooking qualities and is in high demand in the hotels and restaurants of the Pokhara valley (de Boef et al., 2013a). In Ecuador and other places ways have been found to link local agro-biodiversity, traditional food cultures and local knowledge to tourism and ecotourism activities. The example of the Potato Park in Peru provides an excellent example of how to link agro-biodiversity to tourism and community development. Many of the tools and approaches employed in CBM, such as rural drama, seed and diversity fairs, and food festivals are highly amenable to integration and linking to tourism (de Boef et al., 2013a).

Agro-biodiversity and Education

Communities where CBM is well accepted and developed, and where variety ownership is strong provide suitable locations for the mainstreaming of agro-biodiversity into local teaching and education. Many of the CBM tools such as community seed banks and diversity fairs can be excellent educational tools. Community biodiversity registers can also be maintained and updated as an educational tool for schools, but this is also a valuable element of the overall CBM approach. In Ecuador, an education course and guidebooks for rural teachers were prepared and used in local primary schools and used for student clubs also (de Boef et al., 2013a). In Samoa, university breeding clubs have been used to link participatory plant breeding to the teaching curricula of the local university (Iosefa et al., 2013).

10.7.2 Encouraging Diversity

Community Biodiversity Registers

A community biodiversity register represents an inventory or record of traditional crop varieties as well as the traditional knowledge associated with them (de Boef *et al.*, 2013a; Jarvis *et al.*, 2016). The CBR can be maintained by farmers, farmer or community groups including local schools. The CBR can be particularly important in raising awareness among communities of the extent of their biodiversity as well as helping them to monitor it. It can also be an important tool in preventing biopiracy. CBRs can contain information on the morphology and agronomic performance of varieties, how the variety is utilized, special features and traits of the variety, place of origin and those farmers considered as custodians of varieties.

Diversity and Culture

There are many cultural processes and traditions which are amenable to promoting the importance of conservation and sustainable use of plant genetic resources including diversity theatre and rural poetry (Hunter and Heywood, 2011; de Boef *et al.*, 2013b). In Nepal, plant genetic resources projects and programmes have employed local cultural groups and rural poets to raise awareness among local communities of the importance of PGR. Tools and approaches used have included biodiversity fairs, folksong (teej geet) competitions, rural poetry journeys and rural roadside drama. One effective example of this approach in Nepal is that of rural poetry journeys, a kind of participatory travelling seminar, in which selected teams of national and local poets visit diversity-rich areas and spend time with farmers, learning of the value of wild rice and reciting poems and songs in the evening in the village. The poets travel from community to community and at the end of journey, the poems were compiled and published as a book.

Diversity Block

A diversity block can have multiple functions at the community level including: the morphological characterization of selected varieties as well as their agronomic evaluation and organoleptic testing; cross-checking the consistency in naming farmer varieties; selection of prospective parent material for participatory plant breeding programmes; regeneration of seed from community seed banks; and creating community awareness about local biodiversity (Shrestha *et al.*, 2013b). A diversity block is usually established near public spaces, is accessible by roads and should have well-designed and visual signs. The block is a non-replicated experimental plot established and managed by farmers or farmer groups.

Diversity Kit

Diversity kits can be an effective tool in building community awareness and establishing social organization for CBM as well as enhancing the informal exchange of seed and information among farmers (Shrestha *et al.*, 2013b). Diversity kits consist of small amounts of seed of different crops and cultivars that are made available to farmers for the purpose of farmer research and evaluation. These kits give priority to that crop diversity considered underutilized and neglected, especially rare and unique crop varieties and species. Diversity kits therefore have an important role in enhancing the access of communities and farmers to crop diversity as well as raising awareness and understanding.

10.7.3 Communities and the Formal Conservation Sector

Reintroduction of Gene Bank Accessions

Reintroduction of previously endogenous LR gene bank accessions is a common approach in CBM whereby traditional cultivars that might have been previously found in a community or area but for various reasons are not found can be introduced back into the community from formal gene banks. Like several tools and approaches in the CBM Toolkit it can contribute significantly to a community regaining control once again over its genetic resources as well as enhancing food sovereignty as demonstrated by the repatriation of endogenous potato cultivars from the

CIP gene bank to the Potato Park. Farmers can also be encouraged or organized to visit formal gene banks to learn about crop diversity. For example, in Brazil, Embrapa has been working with indigenous groups to help them identify germplasm that is comparable to previous germplasm they have lost (de Boef *et al.*, 2013a). These projects are restorative in the sense that genetic diversity sampled from the region previously is returned to it following some form of disruptive event that led to the loss of the material in the original locality.

Following emergencies, whether natural (drought, flood, cyclone, earthquakes, etc.) or anthropogenic (civil unrest, economic and political failure, social upheaval, etc.), aid agencies often donate seed for subsistence planting. Unfortunately, this seed is often imported and not adapted to local conditions and thus may lead to further crop failures. Coordination between relief agencies and the conservation sector is highly desirable to restore lost endogenous agricultural diversity in war-ravaged areas. Therefore, it is important to stress that ideally, landrace restoration should be based on material originally collected from the area or region where it is distributed, restoring therefore the original range of genetic diversity. In Rwanda, for example, the objective of Seeds of Hope following the genocide in the early 1990s was to make seed aid more responsive to local communities' needs (Burucharu *et al.*, 2002). They not only provided some seed of beans, maize, potatoes, cassava and sorghum from ecogeographically similar neighbouring countries, but also assisted in the regeneration of previously grown varieties.

Participatory Community Conservation

Professional conservationists should take the initiative to always ensure that local communities are involved as participants not only in community-based conservation but also in more formal sector conservation projects. There is a range of ways in which these ideals can be put into practice. Professional conservationists can employ local people as guides ('gate keepers') to biodiversity knowledge, the local people acting to smooth the process of *ex situ*

or *in situ* conservation (Hawkes *et al.*, 2000). Often employment of local people can not only directly aid conservation objectives but can also help ensure the support of the whole community in a project that may be managed by 'outsiders'. This may be a problem with *ex situ* conservation projects, as the material once collected is transferred elsewhere for long-term storage, but local people can act as interpreters and guides or supplying specialist information on locations and local uses of plants or animals.

It is important for the professional conservationists to encourage traditional conservation practices, for example, the maintenance of home garden diversity or sacred groves and even certain taboos regarding the cutting or exploitation of certain species, as well as the traditional values that have maintained the diversity in the absence of professional conservationists until today.

Sacred Sites and Groves

Sacred sites also offer opportunities for conservation professionals to work closely with local communities to strengthen the conservation and sustainable use of plant genetic resources. Sacred sites represent an important traditional nature conservation approach, recognized as part of the religion-based conservation ethos of communities in many different parts of the world (Hunter and Heywood, 2011). Examples of sacred sites in Sri Lanka include the Yala National Park (Category Ia), which is highly significant to Buddhists and Hindus and requires high levels of protection for faith reasons, and the Peak Wilderness Park (Sri Pada-Adams Peak), a sacred natural site for Islam, Buddhism, Hinduism and Christianity, attracting many pilgrims of all these faiths annually.

Globally Important Agricultural Heritage Systems (GIAHs)

GIAHS represent a network of the world's important agri-cultural heritage systems that was started by the FAO in 2002 (www.fao.org/giahs/giahs-home/en/). GIAHS recognize and support agricultural systems and landscapes that have been created and nurtured by generations of farmers, herders and communities, which are local knowledge and biodiversity rich

GIAHS are important dynamic sites for the *in situ* and on-farm conservation and management of globally significant agricultural biodiversity and associated indigenous knowledge systems. They represent resilient agro-ecosystems. Among the multiple goals of GIAHS they are equally important for food security and well-being including sustaining the livelihood security for millions of poor and small farmers. The GIAHS initiative now has project interventions in Algeria, Azerbaijan, Bangladesh, Chile, China, Ethiopia, India, Indonesia, Iran (Islamic Republic), Japan, Kenya, Mexico, Morocco, Peru, Philippines, Sri Lanka, Tanzania, Tunisia and Turkey, and GIAHS sites in these countries offer good opportunities for conservationists to work with farmers and local communities to further the goals of conservation and sustainable use of plant genetic resources.

On-Farm Conservation

On-farm conservation (Chapter 9) has been defined as *the continuous cultivation and management of a diverse set of populations by farmers in the agro-ecosystems where a crop has evolved* (Bellon *et al.*, 1997) and concerns entire agro-ecosystems, including useful species (such as cultivated crops, forages and agro-forestry species), as well as their wild and weedy relatives that may be growing in nearby areas. On-farm conservation activities with farmers and local communities have multiple objectives including: allowing the processes of evolution and adaptation of crops to changing environments and needs to continue; the conservation of diversity at different levels – ecosystem, species, within species and within crops; facilitating the integration of farmers and local communities into the national plant genetic resources system; the conservation of ecosystem services; enhancing the livelihood and well-being of resource-poor farmers; and ensuring the maintenance or increase of farmers' control over and access to crop genetic resources as a contribution to food sovereignty (Jarvis *et al.*, 2000).

Participatory Plant Breeding

Participatory plant breeding involves farmers and local communities working with professional breeders

and researchers in the breeding process to ensure diversity enhancement and conservation (de Boef *et al.*, 2013a). Although traditional landraces may be resistant to pests and diseases or have other beneficial traits, some are relatively low yielding; consequently the trend has been towards the replacement of local landraces by externally produced higher yielding cultivars. This replacement has often been associated with a general decrease in locally adapted genetic diversity, including any desirable traits it had contained. One approach to maintain and conserve this diversity and its desirable traits is to conserve the entire traditional farming system in an on-farm conservation project. However, to avoid creating a village museum, one approach to make the local landraces more competitive with the higher yielding cultivars and thus further secure their sustainability is to introduce exogenous adaptive material (cultivars or breeder's lines) for use in local breeding programmes to improve the yield, or other characteristics of local types. PPB activities may be either:

- *Farmer-led* – Therefore under the control of the farmers themselves. The farmer makes deliberate or allows natural crosses between their LR and material with more desirable traits, such as disease and pest resistance.
- *Breeder-led* – There may be introduction of exotic germplasm by breeders to local communities for hybridization with the local material.

The important point is, however, that either strategy avoids the complete replacement of locally adapted material with more input-responsive cultivars, ill-adapted to low-input and stress conditions. Local communities use their own germplasm and their own selection skills with introduced professional breeder's material or landrace material from other regions to produce improved local landraces.

Two processes of farmer participatory varietal improvement are distinguished, participatory plant breeding (PPB) and participatory varietal selection (PVS) (de Boef *et al.*, 2013a).

- PPB is the selection of segregating materials at an early stage in the farmer's fields and/or bulk of heterogeneous materials to generate a new composite variety.

- PVS is the selection of finished products by farmers in their own fields.

Participatory plant breeding is likely to be more beneficial for conservation goals because it works with variable, segregating material that is derived from or similar to material already in the local farming system, while PVS is considered either unpredictable, as it could either increase or decrease diversity, or negative for conservation because it is based on the replacement of local populations with less variable ones from breeding programmes. As well as its role in genetic resource utilization, PPB can also be viewed as a form of community-based plant genetic resource conservation, as it ensures the conservation of at least some locally adapted genetic diversity that is otherwise likely to be displaced and lost. PPB is discussed in more detail in Chapter 17.

Germplasm Collecting

11.1 Introduction

Although the Convention on Biological Diversity Article 8 (CBD, 1992) recognizes *in situ* conservation as the primary approach to biodiversity conservation, Article 9 contextualizes *ex situ* conservation as 'predominantly for the purpose of complementing *in situ* measures'. However, as regards plant genetic resource conservation, this fails to appreciate the *status quo*; still today most active conservation is *ex situ* and use of conserved diversity remains almost exclusively from population samples held at *ex situ* locations. Although this situation is changing and *in situ* PGR is beginning to be systematically established, the predominance of *ex situ* applications is unlikely to change in the short term – farmer and breeders alike will continue to obtain their novel material from gene banks. However, the increasingly limited resources available to collect germplasm and maintain germplasm collections (FAO, 2010a) has led plant collectors and collection managers to adopt a more objective, scientific approach to plant genetic sampling – populations are deliberately chosen following analysis of user requirements and conservation planning.

The precise goal of the collecting expedition is to maximize the target diversity (i.e. taxa, populations or traits) sampled in the minimum number of samples with the minimum expenditure of resources. The minimum number of samples because the processing and incorporation of each sample has an associated cost, so there is a balance between collecting the required diversity while limiting the number of samples to be deposited. The choice of target diversity will impact whether seeds, rhizomes, bud-wood or tubers are sampled and moved *ex situ* to a remote location for conservation. There is a different emphasis when collecting plant genetic resources for food and agriculture (PGRFA) compared with a more general botanical collecting expedition. The former adopts a more selective approach to sampling the breadth of genetic variation found in the gene pool of the target taxon, rather than sampling the full range of plant taxonomic diversity found at any location. For any crop or related wild gene pool there is commonly too much genetic diversity to attempt to conserve all alleles *ex situ*; therefore, the conservationist is forced to sample from the range of diversity available. It would be impractical, if not impossible, to attempt to sample all the variation in every population, because this would involve collecting every individual and would thus be likely to eradicate the species you were trying to conserve. It is crucial that the conservationist identifies the most appropriate and effective pattern of sampling to ensure the genetic variability both between and within populations is conserved, so that:

- if the species becomes extinct or if the primitive landrace is replaced by modern cultivars, the original genetic diversity or a high proportion of it is conserved and remains available for utilization, and
- if a specific gene or gene complex exists in natural populations, even at low frequencies, and it is required for some form of utilization (e.g. breeding for host plant resistance to pathogens and pests), it is accessible and available.

But to know what and how to sample, the conservationist should know or have some idea of the amount and pattern of genetic variation within and between populations, population structure, breeding system (inbreeder or outbreeder), taxonomy and ecogeographic distribution of the target taxon in the target area. Local ecological or environmental factors will have a marked effect on patterns of local variation in traditional crops (from which seed is saved each generation for the next planting) and wild species. In these cases, species form ecotypes or agro-ecotypes, where local wild species populations or landraces, respectively, evolve genetic distinctions associated with soil or climatic factors (e.g. maximum and minimum average temperatures; precipitation; length of the growing season; light intensity; and day length) that enhance their suitability for growing in that environment. These factors may vary gradually over geographic distance causing gradual changes in genetic traits and such clinal variation is often noted in quantitative characters, e.g. the classic study of clinal pattern of height of *Pinus sylvestris* in Scotland, UK, which are impacted by photoperiod and temperature, become gradually shorter from south to north (Langlet, 1971). A patchwork or a mosaic pattern may also be superimposed on the cline, e.g. *Pisum sativum* subsp. *elatius* showing sharp differences in flowering times and plant height (Smýkal *et al.*, 2018). These patterns of genetic variation occur both in self- and cross-pollinating plants. If altitude or topographic factors are also considered, the cline often becomes more of a mosaic distribution, as was shown by another classic study of *Agrostis tenuis* and *A. stolonifera* distribution in relation to lead tolerance, which showed very sharp differences in relatively short distances of 50 to 100 m (Walley *et al.*, 1974).

Unfortunately, too often much of these genetic, morphological or environmental data are unavailable when planning conservation and field collection. The type of material being collected, seed or vegetative, will also affect the sampling strategy, handling techniques, quarantine and, ultimately, the method of storage. Historically, field collection has primarily focused on collecting seeds, being the natural storage organs that has evolved over millennia to facilitate germination only when conditions are appropriate for plant growth; because these are the natural storage organs, they are most suitable for collection and storage.

Further, historically populations as opposed to single plant samples have been most often collected, due to the simplicity of collecting the entire species population sample as one collection, as opposed to the additional collection effort, processing and storage cost in taking a collection from a single plant, which inevitably means a larger number of small samples are collected overall. Although single plant samples

are required if population genetics research is to be undertaken, as would be the case in screening for adaptive alleles conferring host plant resistance to pathogens and pests, or adaptation to extreme stressful environments.

Each of the factors listed above will impact the choice of collecting strategy recommended for the target taxon, but even today relatively little research has been undertaken on the most appropriate collection strategies for different groups, crops and wild taxa, even where the necessary data is available. Perhaps the most comprehensive statement of multi-crop/wild species collection strategies is provided by Guarino *et al.* (1995) and the 2011 update Guarino *et al.* (2012). This includes a generic chapter on sampling strategies (Crossa and Vencovsky, 2011), along with specific chapters on wild species collection (Bothmer and Seberg, 1995; Parra-Quijano *et al.*, 2011b), collecting seed in the field (Hay and Probert, 2011), collecting vegetatively propagated plants (Dansi, 2011), collecting vegetative material of forage grasses and legumes (Sackville Hamilton and Chorlton, 1995), collecting woody perennials (Schmidt, 2011), collecting indigenous knowledge along with germplasm (Guarino and Friis-Hansen, 1995; Quek and Friis-Hansen, 2011), collecting healthy plants (Macfarlane *et al.*, 2011) and the legal context for germplasm collection (Moore and Williams, 2011).

11.2 Prior to Collecting

The first collecting mission led by one of us (Nigel Maxted) had the aim of collecting sorghum and millet landraces and CWR in central and northern Uganda in 1985, when there was no conservation planning as we now know it today, and Nigel had never seen cultivated or wild sorghum and millet, had never been to Uganda previously and knew no local languages. Happily for PGRFA conservation today we are much better organized. The conservation is systematically planned (see Chapter 6), but if the collecting mission is to be efficient and effective, we also need to ensure the right experts are present, the timing is right for collection, local administrative requirements are met, the team has the necessary equipment, and where necessary preliminary survey missions have been undertaken.

11.2.1 Assembling the Collecting Team

Each member of the collecting team should ideally be selected to bring specific skills to the team, e.g. identification of the target taxon and the flora of the target area (taxonomist), background knowledge of the range of variation already existing in the target taxon (breeder), and expertise in local languages and customs (local extension worker or anthropologist). Often collecting teams comprise two elements: national or international experts and people with local knowledge, the former providing generic target taxon variation knowledge and the latter providing knowledge of the local flora, geography, languages (required to access indigenous knowledge from the local community) and customs. Hawkes (1980) reviews the characteristics of plant collectors and lists them as including:

- a good knowledge of target crop plants,
- a good knowledge of target crop plant's wild relatives and relationship to the crop,
- a good eye for variation in plant and environment,
- an understanding of gene pool concepts,
- knowledge of the need for random as well as non-random sampling, and
- an understanding of 'coarse-grid' and 'fine-grid' collecting strategy.

Ideally, the team should collectively be involved in the planning, and a good collecting mission should involve an element of training. The combination of experts in plant groups and other botanists provides an ideal opportunity for all the team to broaden their knowledge of less familiar plant groups. Practical experience using keys and descriptions with expert guidance is often the best way for a botanist to familiarize him or herself with fresh plant groups. Finally, when travelling in remote regions local guides with local political and cultural knowledge may prove invaluable in smoothing the collecting process.

11.2.2 When to Collect

If the objective, as is most often the case, is to collect seed, then the obvious time to collect is when the seed is at maturity, but prior to build up of pests, or fruit dehiscence or seed dispersal. In many cases, particularly for wild species, the so-called collecting window – the time period during which ripe seed can be collected – is very narrow, so it is crucial to predict the right time to collect for the success of an expedition, although allowances should be made for seasonal differences associated with broad geographic or climatic features (latitude, longitude, altitude, rainfall, aspect, etc.). In mountainous regions the collector may commence collecting at lower altitude and as time progresses move to higher altitudes along with the ripening season. Conservation planning will provide a general idea of ripening times, but precise timing will need to come from people with local access to the target taxa, though for larger shrubby and tree species remote sensing (e.g. routinely gathered remote sensing data are used for planning collection routes in Namibia) can also be used. Hawkes (1980) stresses that arriving at a site during the collecting window allows the team to:

- collect the maximum breadth of genetic diversity,
- collect from each distinct habitat, soil type, climate or altitude zone where the target taxon is found, and
- collect crop weedy forms and related wild species found in or around field margins.

If multiple taxa are being collected, it may be impossible to arrive at the ideal time for all species, because they are unlikely to all flower synchronously. In this case one should distinguish priorities within the target taxon and visit the appropriate area at the right time to collect the priority material.

11.2.3 Local Administrative Requirements

When planning and undertaking PGRFA collection, the collecting team should note and respect the regulations and customs of the countries in which they are working. The legal situation regarding PGRFA collection has changed drastically since the

Convention on Biological Diversity (CBD, 1992) and International Treaty on Plant Genetic Resources for Food and Agriculture (FAO, 2001), particularly access and benefit sharing via the Nagoya Protocol. The team will need to be granted Prior Informed Consent (PIC) from the relevant authorities before they commence collecting. This will stipulate the terms of access (i.e. the permission to transfer germplasm outside the country when the mission is completed). Access will normally be reliant on mutually agreed terms (MAT), which normally allows access on condition that any benefits that might arise from the commercialization and utilization of the germplasm is shared with the host country. Specifically, for crops and CWR listed in Annex I of the ITPGRFA, there is a Standard Material Transfer Agreement (SMTA) that needs to be concluded before any germplasm can be legally sent or received. An SMTA is only obligatory for ITPGRFA Annex 1 listed taxa, but most countries have MTA that require a signature for non-Annex 1 listed taxa, and many countries in Europe apply SMTAs to all PGRFA taxa to simplify the system. There are also distinct ethical principles involved in any collection of germplasm, particularly of LR diversity that has been maintained by farming communities for millennia, and the commercialization of that diversity without appropriate acknowledgement of the input of the original LR custodians and its appropriate reward (Engels *et al.*, 2010). The FAO Code of Conduct for Germplasm Collecting and Transfer (FAO, 1993; www .fao.org/nr/cgrfa/cgrfa-global/cgrfa-codes/en) also includes the equitable sharing of benefits arising from the use of collected germplasm. This Code was intended to help regulate the collecting and transfer of plant genetic resources and their associated information (including indigenous knowledge), with the aim of facilitating access to these resources and promoting their use and development on an equitable basis.

Commonly when undertaking an international expedition, the team will be composed of a mix of international and national experts, and the latter are likely to know the appropriate local authorities and how to obtain the PIC and MTA to prepare and implement the mission. In general, the local authorities should be fully aware of the objective of the mission, target taxa and mission itinerary. The entire team should have the required collecting permits and be aware of any restrictions on the export of collected materials and any local phytosanitary permits. The CBD clearing-house for the country should be able to provide specific advice on local regulations in specific countries, although much information is available on-line. It is almost always part of the PIC that a full duplicate of all the conserved material and copies of any publications resulting from the expedition will be deposited in appropriate institutes designated by the host country. Further, sound ethical principles suggest that the actual expedition should be seen as only one stage in the conservation, evaluation and utilization to use collaborative process that should be jointly undertaken by the international collection depositary, germplasm user and an appropriate host country authority.

11.2.4 Collecting Equipment

What collecting equipment is required will be dictated by the target taxon and area, e.g. whether collecting seed, tubers or vegetative material, whether the team will camp or stay in local accommodation, the mode of travel (four wheels or light plane or even helicopter). A detailed discussion of these points is provided in Hawkes (1980), including a checklist of basic collecting equipment. However, if collecting wild species Flora, revisions/monographs of the target taxon and a botanical glossary are essential. Which Flora to use in the target area can be identified using a country-based list of the world's Floras (Frodin, 2011). For areas where there is no adequate Flora or the Flora is written in an unfamiliar language, it may be possible to make use of that of a neighbouring region; for example, the Flora of Turkey lists many of the species found in Syria. Taxonomic or ecogeographic monographs and field guides may also provide useful collection information such as keys, descriptions and known distributional ranges to aid specimen identification. Historically, detailed maps (geographic, vegetation, soil and climatic) and more latterly global positioning systems would have been essential, but given the evolution of smart phones, many of the functions can now be met by simply making sure your phone is charged up and usable.

11.2.5 Preliminary Survey Missions

Despite the growing sophistication of conservation planning techniques, most is based on modelling and the collection expedition may be the first opportunity to ground truth the prediction and confirm whether or not the populations are found where they are predicted to be present. If there is a lack of ecogeographic data or previous survey data, it may be necessary to undertake preliminary survey missions to ground truth the predicted distribution for the target taxa. In which case a mixture of 'coarse' and 'fine' grid sampling can be used. Coarse-grid sampling involves travelling throughout the target region and sampling sites at wide intervals over the whole region. It also enables the conservationist to gather data from local agronomists, farmers and any other persons who possess some knowledge of the crop and its distribution, particularly relating to its adaptation to soils, altitude and farming processes. The actual size of the course grid interval is dependent on local environmental diversity but stopping to survey potential sites for target taxon presence every 30–50 km. The full-scale collecting expedition (fine-grid sampling) could then focus on the areas shown to contain the target taxa but stopping to collect samples more regularly at 1–15 km. Hawkes (1980) notes such an approach allows the sampling of specific, selected populations of the target taxon with desirable traits, as well as deliberately including in the sampling isolated or peripheral populations. Although the ecogeographic survey and gap analysis may highlight broad areas of interest to the conservationist, using the coarse and fine grid sampling approach will maximize the genetic diversity sampled.

11.3 Types of Collection Site

Plant germplasm (seed, fruit, vegetative organs or cuttings) can be collected from wherever it is found, but there are five basic types of location most commonly sampled. These are:

- *Farmer's'fields* – Most fields contain large-sized populations of genetically uniform crops grown by the farmer, but the collector is more likely to focus on sampling genetically heterozygous LR cultivated and seeds saved by the farmer and their ancestors for generations. These LR are often at risk of genetic erosion or extinction, because increasingly they are being replaced by modern high-yielding cultivars, although objective quantification of LR loss is largely absent (FAO, 1999). Given the relative ease of collecting from farmers' fields, the fact that LR are the main repository of crop diversity and that LR diversity is being rapidly lost means historically farmers' fields have been the focus of field collection. When collecting from the farmer's field, the collector must arrive at the right time of year (during the so-called collecting window) to find mature seed on the plants; too early and the seed will not be ripe, too late and the seed will have been harvested. Even in subsistence agriculture, farmer's fields are likely to contain large populations of a relatively small number of crops grown for commercial sale and home consumption, so providing the collector arrives in the collecting window, obtaining an adequate sample is straightforward.

- *Kitchen or orchard gardens* – These are the areas that farmers cultivate around their homes to produce food for their own consumption. They are more diverse than farmers' fields and are likely to contain larger crop diversity but of smaller population sizes. The LR grown are likely to be of minor crops, vegetables, horticultural crops, fruits, herbs, condiments and medical plants. As they are grown for personal consumption, there is less pressure to produce uniformity to meet specific market requirements and without the pressure to produce for sale they are less likely to suffer genetic erosion. Individual crop population sizes are likely to be small compared to fields sown for commercial sale; the collector may need to obtain samples from several neighbouring gardens to ensure an adequate collection sample size. In areas of severe genetic erosion, kitchen or orchard gardens may provide a refuge for local landrace germplasm, e.g. potato plots in Peru (Hawkes, 1978) or sorghum and bulrush millet from East Africa (Maxted *et al.*, 1986).

- *Farmer's store* – If the collector arrives too early or too late to collect a ripe sample from the field, the collector may still be able to obtain a sample from the farmer's store. This has the advantage that the timing of arrival is less critical, and it facilitates the sampling of other crops held in the store. However, it does not permit the collector to select a scientifically appropriate sample (see Section 11.4.4), as the sample can only be taken from the stored crop. It is therefore important to ask the farmer the following questions:
  - Is the material a single variety or LR or a mixture of the two or more varieties or LR?
  - If one variety or LR, was it harvested from one or several adjacent or remote fields?
  - What is the LR name?
  - What details are available of its original provenance?
- Samples obtained from the farmer's store are likely to be large if intended for commercial sale but if for home consumption, they will be smaller.
- *Markets* – A collector arriving at a location out of season may also be able to obtain a sample from a local market, although as with the sample taken from the farmer's store, this does not allow the selection of a scientifically appropriate population sample and a market holder may be selling and have mixed several LR from diverse locations. When sampling from a farmer's field or store, the collector can gain useful passport information and indigenous knowledge from interviewing the farmer, but this is less available from the market trader who is possibly remote from the original production region. However, the stallholder should be questioned to obtain the information they do possess. As with the sample taken from the farmer's store, it is important to try to ask the same questions about the materials being sold. Samples obtained from markets are likely to be large, but the coverage of crops may be limited, and certain varieties may not be sent to market and retained for private consumption.
- *Natural or semi-natural habitats* – Populations of wild species are often found in disjunct

populations, and it is rarely possible to encounter sufficiently large populations to permit the collection of large genetic samples. There is an obvious selective advantage to wild species in not being uniform, providing maximum likelihood of survival, but in terms of collecting this does mean that at any specific time only a limited proportion of the population will be ripe and large samples will be rare. In fact, the collection of too large a sample from any one wild population may itself erode the genetic basis of the target population. Therefore, the tendency when sampling wild species is to take smaller but more frequent samples, possibly returning to a population several times during the fruiting season. Many weedy relatives of crop plants grow in semi-natural habitats influenced by humans, and these must be sampled from the edges of fields and pathways, in the crops themselves or around dwelling places.

A summary of the characteristics of five basic types of collection locality is provided in Table 11.1.

11.4 Field Sampling

The personnel, time and finance available, as well as a consideration of the target taxa being sampled, will affect the practical sampling strategy employed during an expedition, but overall the aim is to sample the maximum quantity of genetic diversity within the minimum number of samples (Brown and Marshall, 1995). However, in addition five specific factors need to be considered when sampling in the field.

11.4.1 Distribution of Sites within the Target Area

Sampling sites selection is often governed by presence of the right habitat for the target taxa and local ecogeographic conditions. It is assumed that genetic variation is correlated with ecogeographic heterogeneity, so where the latter conditions vary greatly, more sites will need to be sampled to ensure any allelic diversity associated with distinct ecogeographic zone is captured. The collection team

Table 11.1 **Types of collection locality characteristics**

	Type of target taxa	Possible nos. of target taxa	Possible size of sample	Indigenous knowledge collection	Genetic purity of sample	Collecting window
Farmer's fields	Commercial crops	Small	Large	Good	Good	Narrow
Kitchen/ orchard gardens	Major & minor crops, herbs, condiments & medical plants	Medium	Small	Good	Good	Narrow
Farmer's store	Commercial crops	Small	Large	Good	Medium	None
Markets	Major & minor crops, herbs, condiments & medical plants	Medium	Large	Poor	Poor	None
Natural or semi-natural habitats	Wild species	Large	Small	Poor	Good	Broad

with finite time available needs to decide whether to spend more time at each site and collect larger samples or sample more sites but collect smaller samples at each site. Due to the common correlation between genetic and ecogeographic variation, covering a larger area and collecting at more sites is likely to increase the genetic variation sampled. Therefore, where there is no additional information on the distribution pattern of genetic variation, the collector should aim to sample as many distinct habitat types through the distribution range of the target taxon. Given this requirement, there are two approaches: using the course and fine-grid approach along a transect through the distribution range; or selecting clusters of sites, 4 or 5 sites per cluster, throughout the distribution range (Hawkes, 1980). Using transect sampling captures the maximum amount of variation associated with broad ecogeographic differences, while the cluster sampling procedure captures variation associated with micro-ecogeographic factors. The actual method may well depend on the amount of ecogeographic variation found in the target area. If a large area with diverse ecogeographic conditions is to be covered, then the transect method may be advisable, but if the area is smaller or there is little ecogeographic variation, then the cluster method may be more suitable. The transect method may involve taking samples approximately every 50-km or 200-m change in elevation, depending on the total distances or range of altitudes to be covered. The transect method is more advisable for major crops or annual species, where there may be significant mixing of the crop following harvest. The cluster method is favoured for wild or weedy species, or for those crop species where inter-varietal mixing is unlikely. In both latter cases gene flow will be less, allowing greater genetic adaptation to local habitats. Collecting at clustered sites is also thought to save overall travelling time between sites, force the collector to sample diverse micro-habitats and increase the value of the collection for population studies. However, as a rule the higher the level of between-site variation, the closer together the sites should be (Brown and Marshall, 1995).

However, the basic strategy should be modified if there is information on the distribution of genetic variance within the target taxon. For example, if there is little or no inter-population variation as in white clover (*Trifolium repens*) in the UK (Hargreaves *et al.*, 2010), then the collector would sample a large number of individuals from fewer populations, as opposed to sampling the related subterranean clover (*Trifolium subterraneum*) in Sicily (Piano *et al.*, 1993), i.e. if there are high levels of differentiation between sites, then the collector would sample a few individuals from a larger number of sites. The point should be made that this level of knowledge of the patterns of genetic variation in a target taxon is rare *a priori*. Where these patterns are not understood, it is prudent to sample as widely as possible over a range of space and habitat, possibly running a transect through the centre of diversity of the target species. Planning a diverse collecting route can be assisted by using temperature, rainfall, soil, and vegetation maps or overlays in GIS. Even the amount of within-site habitat variation may affect the selection of sites; for example, sites that are located on the interface between two different habitats or soil types are likely to contain higher levels of genetic variation.

In practical terms, however, the approach to selecting sites to sample must be flexible. All collecting expeditions will meet problems: transport breakdowns, bureaucratic hold-ups, illness, getting lost, roads shown on maps not actually existing or *vice versa*, areas covered by bandits or civil unrest, and numerous other extraneous factors that can influence the selection of sites sampled within the broad target area.

11.4.2 Number of Sites to Sample

If there is no information about the distribution of genetic variation in the target species in relation to geographic or ecological distribution, the collector must assume that each additional site provides the opportunity to sample additional alleles. Therefore, the optimum number of sites to sample is the maximum possible with the time and resources available. Even for an outbreeding species, this will hold true over distances greater than the pollen

migration and seed dispersal distance. Given unlimited resources, the addition of extra sites will start to add fewer novel alleles with each extra site added, and Brown and Marshall (1995) suggest that about 50 sites per species would capture 95% of all alleles in a species with a frequency greater than 0.05, which is all but very rare local alleles. In practice, however, the length of the collecting season, relative abundance of the target species and the resources available to the collecting team will practically restrict the number of sites sampled; Neel and Cummings (2003a) found that sampling from four populations captured 67 to 83% of all alleles in the species. As with the examples of white and subterranean clover discussed above, if there is information available concerning the distribution of genetic variance of the target taxon within the target area, then the number of sites sampled may need to be modified.

11.4.3 Delineation of the Site

It is relatively easy to define the edges of a site for a crop, because the farmer plants the crop in discrete stands, so the edges of the site would be delineated by the edge of the field or orchard. However, in the case of a minor crop grown in small populations by each farmer in a village, it may be necessary to bulk samples from more than one field or kitchen garden within the village system to obtain a sufficiently large seed sample (Hawkes *et al.*, 2000). For CWR found in and around cropped fields, the delineation of sites is more problematic. In this case the species are not planted in uniform stands but distributed naturally in more disjunct populations. CWR are also often found in a semi-obligate relationship with a crop species and may be harvested and planted with the crop, especially in traditional farming systems. For example, wild oat (*Avena fatua*) is a common weed of cereal crops and Narbon bean is often a weed of faba or broad bean fields in the Middle East (Bennett and Maxted, 1997). In this case, the edge of the site for a wild oat or Narbon bean population will be the edge of the cereal or faba bean field, so the collector is forced to make fairly subjective decisions about what constitutes a site, over what area the material of one taxon will be bulked and when the material would be better collected as separate samples. This decision is likely to be related to the size of the inter-breeding unit for the target taxon, which is in part a function of the number and density of plants, the mating system, and the level of pollen and seed dispersal.

The edges of the site may also be delineated by dominant habitat changes. For example, a site taken on the border of several distinct habitats (e.g. forest, open prairie, flood plain) would be more appropriately treated as a series of distinct sites, unless there is evidence of extensive gene flow between individuals in the different habitats. In practice, however, many of these factors are difficult to assess in the field, and collectors tend to stop collecting when they believe subjectively they have a good representation of the genetic diversity at the site.

11.4.4 Distribution of Plants Sampled at a Site

Within the site, there are two basic approaches to selecting which plant to sample from the population: random and selective. The random (or non-selective) approach involves the collector entering the field or site and taking a transect, walking back and forth across the site taking samples from the whole site at a predetermined number of paces, say every ten paces, depending on the size of the site and the abundance of the target species. If a population sample is being taken, then the samples are placed in the same collection bag. The starting point and direction of the transect and positions of the sampling points could be chosen using random procedures. Without knowledge of the genetic structure of the population being sampled, this method provides the best sampling of genetic variation (Brown and Marshall, 1995). Having walked the set number of paces, the collectors will take the sample from the nearest seed-bearing inflorescence to their right or left hand. If a random sample is being taken, the collector must not search around for the 'best' plants to sample. Conversely, selective (biased or non-random) sampling involves the conscious selection of phenological, physiological (e.g. drought adaptation, host plant resistance to pathogens and pests) or ecological types by the collector. The collector is actively looking for the most

valuable material to put into the population sample. The collector may also 'enrich' a random sample by specifically including rare phenotype variants that they wish to ensure are contained in the sample of the target population.

In practice, collecting plant genetic resources is often biased. Collectors, especially if they are breeders as well, look for the largest or disease-free fruits, or other 'good' characteristics, because of the explicit link between conservation and utilization. However, it should be remembered that it is only possible to select for visible characters in the field, while other characters, e.g. quality of pest resistance if the pest is absent, are impossible to select for in the field. Biased sampling has the advantage that it allows the collection of obvious variants, but it will be more time-consuming because of the need to search for off-types. If both random and biased samples are taken, then they should be kept separately, as the mixed sample will bias the sample and reduce its value for population research.

To increase sampling efficiency, whether randomly or selectively, the collector may wish to subdivide the site into distinct ecogeographically homogeneous areas depending on, *inter alia*, slope, soil, aspect or grazing. The collector will then take a stratified sample ensuring material is collected randomly or selectively from each stratified distinct area identified. This is particularly important if the plants are thought to be self-pollinated or have limited seed dispersal, because genetic mixing between plants at a site may be restricted. However, if the species is an obligate out-breeder, whether insect or wind pollinated, genes are likely to spread freely throughout the site, so it would not be advisable to collect heterogeneous areas as distinct samples. Whatever the situation, the collector should aim to thoroughly sample genetic diversity, ensuring the site's heterogeneity and any associated genetic diversity is reflected in the sample or samples collected.

11.4.5 Number of Plants per Site

Ideally, the collector should sample at least one sample of each different allele in the population. If alleles were distributed truly randomly among and within populations, then no collecting strategy would be needed; one would simply sample as many plants as possible. However, this is not the case and excessive sampling of one site will logically restrict the sampling at other sites. There is always a choice between collecting large population sample, which may take significant resources to collect and manage and restrict further population samples, and collecting several small population samples, which because they are too small result in random changes in allelic frequency that result in continuous fixation and loss of alleles, reducing the proportion of heterozygous individuals in the population (Crossa and Vencovsky, 2011). Therefore, in practical terms, Brown and Marshall (1995) propose the number of plants to sample is that which contain 95% of all the alleles at a random locus occurring in the target population with a frequency greater than 0.05. To achieve this goal, they consider that a random sample of 40 to 50 (but not more than 100) plants per population would be required, though for forest tree species Krusche and Geburek (1991) argue for a higher number. Brown and Hardner (2000) argue that a truly random sample of 59 gametes would achieve the collecting goal and 30 randomly mating plants for an outbreeding species and 60 for an inbreeding species. However, in practice, even numbers this low, especially for trees, may be difficult to achieve with wild species distributed in disjunct populations. When collecting wild species, this number should be treated as a goal, but if this number cannot be reached still collect, otherwise the collection of wild species may be too often restricted to common weed species found in large dense populations.

Brown and Marshall (1995) state that achieving the collecting strategy goal will ensure all but restricted and rare alleles are collected, which they argue are of less use to breeders anyway. However, breeders do use rare characteristics that would diminish the plant's competitive efficiency in the wild, e.g. dwarf stature, non-brittle rachis mutations in cereals, reduced tillering, open canopy and male sterility, each of which are likely to be restricted and rare alleles. Also, resistance genes seem often to be rare but are of great use to agriculture, e.g. *Barley mosaic virus* resistance has been found in only two collections of *Hordeum bulbosum* yet has proven extremely useful (Ruge-Wehling *et al.*, 2006). Subsequently many authors

have provided estimates of the most appropriate number of plants to sample from each population, but in each case, they assumed that the mating system and thus the amount of gene flow between and within populations of the target species was unknown. Therefore, the estimates were a worst-case scenario, where gene flow between individuals was very limited, as in the case of apomictic species, where the progeny are genetically identical to the parents, or among inbreeders where the population is homozygous at most loci. However, if the target taxon is self-incompatible or largely an outbreeding species, a smaller sample size could be safely collected.

Lawrence *et al.* (1995) go further and argue that a random sample of 172 seeds collected at random from throughout the gene pool of any species is enough to conserve most of the genetic variation in any species, irrespective of whether the species are self- or cross-pollinated. They also argue that if the collector visits multiple populations, the size of the sample drawn from each site need be no more than 172 divided by the number of populations visited. Potentially the implementation of this strategy could cause considerable resource savings in the collection and storage of materials in gene banks. But this number is too low for practical purposes; the point of collecting is to make the conserved resource available to users, so the sample needs to be sufficiently large to avoid immediate regeneration to permit the collection to be offered for distribution to users. Also collecting small samples will carry with it a greater burden of genetic drift when the material is necessarily regenerated. In addition, the number of seeds per seed parent, the number of seed parents per population and the number of populations need to be increased if (a) the sample is to be split between the collecting team or safety duplicated, (b) there is any suspicion that a proportion of the seed collected is unviable and (c) there is likely to be any loss of sample post-collecting from quarantine, viability testing, etc. It is also advisable not to collect more than 20% of the total mature seed available on the day of collection to avoid threatening regeneration of the population in subsequent years (Way, 2003). Given these provisos, there is increasing scientific consensus on the most appropriate general strategy for field collecting (see Box 11.1).

Box 11.1 | **A Generalized Collection Strategy**

(Hawkes *et al.*, 2000)

The following 10 points are recommended as an effective and practical sampling strategy:

- Ensure that the sites sampled represent the broadest range of environmental conditions under which the species is found.
- Sample between 5 and 50 sites per species (the greater the number of sites the more extensive the alleles captured), sites being relatively closer together if the local environment is very variable or the target taxon shows a high level of variation between sites.
- If a taxon is rare, then it is generally better to collect 10 plants from each of 100 populations rather than 100 plants from 10 populations.
- The size of a site will be defined by the level of habitat variation found at the site and the local level of gene flow for the target taxon; if that is known, each distinct habitat should be sampled as a distinct site.
- Collect individuals randomly throughout the site, collecting any morphological or other oddities as separate samples.
- Collect from 30 (cross-pollinators) to 50 (self-pollinators) individuals per site and collect 30–50 seeds from each plant (50 × 50 = 2500 seeds per sample) or up to 100 individuals if seed on individual plants is scarce.
- Sample enough seed or vegetative material per plant to ensure representation of each original plant in all duplicates.
- Take voucher specimens of the samples collected, especially if collecting wild species.
- Always talk to local people, especially farmers, to collate indigenous knowledge on the material collected, of both cultivated and wild taxa.
- Photograph the sites and ensure you take meticulous field notes on habitat, aspect, soil, presence of other species or utilization practice, etc.

11.5 Specialized Types of Collection

The general strategy recommended above has been largely developed for the collection of cereal and legume crops and will need to be adapted for the collection of other crop groups and wild species.

11.5.1 Collecting Fruit and Other Trees

Many temperate and tropical fruit trees (including the Rosaceous species such as apples, pears), coffee, tea, cocoa, rubber, many nut trees and most forest trees

possess 'recalcitrant' seeds, which cannot be stored long-term under normal gene bank conditions. In such cases, often cuttings are taken and planted in a field gene bank soon after collection to avoid loss of viability. In the wild, these species are not commonly found in dense stands, so the sampling strategy is amended. Random sampling would be inappropriate because of the low population size and therefore selective sampling occurs even though it may result in bias in the alleles collected. Foresters commonly use biased sampling, particularly when sampling for good timber characteristics (straightest bole), large or multiple fruits, resistance to disease, and adaptation to specific environments. This is of course practical for short-term utilization, but in the longer term can lead to a narrow genetic base and missing valuable disease and pest resistance which may not be apparent when the tree is sampled.

If it is impossible to collect seed from the species, then various types of vegetative cutting can be taken and later grafted onto rootstocks. These may be in the form of: (a) naturally layered shoots and rooted suckers, common for wild species; (b) leafy cuttings, with which care must be taken to prevent drying out; (c) hardwood cuttings, though it can be difficult to regenerate a healthy plant from these and (d) bud-wood cuttings, which are lignified but not mature (Sykes, 1975). For the latter, the most successful method is to wrap the sample in black plastic and keep it moist and cool in an icebox while on the collection expedition and get the sample back to base as soon as possible.

It is very easy to over-sample tree cuttings, then growing them out in the field in gene bank collections can cover a large area of land and be expensive to maintain. In many cases it may be more practical to conserve *in situ* conservation in their native habitats, unless these habitats are threatened, or the material is needed for immediate exploitation (Hawkes *et al.*, 2000). Another conservation option in this case would be to consider *in vitro* conservation (see Chapter 13). A general strategy for cultivated timber and fruit trees would be to sample seeds or fruits wherever possible. If this is not possible, then some form of cutting should be taken and kept moist and cool during the collection expedition. In this case, the collector should try to sample each distinct morphotype in a village, sampling 10–15 individuals per population with ideally 100–200 m between sampled trees, and sampling as many populations as possible. Further information on collecting woody species can be found in FAO (1995) and Schmidt (2011).

11.5.2 Collecting Forages

Forage crops are often selections rather than having been actively bred. If the market requires a species with higher cold tolerance, then the approach is often to use GIS and species distribution modelling to locate areas with slightly colder temperature, collect from the region, grow out the collection in the area where the greater cold tolerance is required and then selecting the most desirable line. Forage species often spread clonally, which makes it difficult to select individual plants, so the preferred method of collection is random sampling. Box 11.2 outlines a strategy for collecting forage legume germplasm. Further information on collecting forages can be found in Sackville Hamilton and Chorlton (1995) and Hanson (2011).

Box 11.2 | Collection of Forage Legume Germplasm

(Maxted and Bisby, 1989)

Sites are selected in the manner detailed in Box 11.1, and seed is collected randomly throughout the site. However, there may be difficulty in avoiding mixed seed collecting as different species or individuals form a dense entangled mat and at seed maturity all that remain are the crisp shattering plants and pods, which are neither easily distinguished nor disentangled. This problem can be avoided by visiting each site twice. The initial visit enables accurate identification of the material found at the site, selection of pure stands, the making of high-quality voucher specimens and the collection of rhizobia for legume species. Painted wooden or plastic stakes or even tagging plants can be used to mark target specimens or populations so they can be easily located later. During the second visit 5 or 6 weeks later, the seed can be collected from the populations identified previously, and other populations

Box 11.2 | (cont.)

missed during the first visit are subsequently easier to identify and collect. Visiting each site twice also enables a better estimate of the exact fruiting time of that population in that year, further facilitating the collection of larger samples. Personal experience has shown that it also helps collectors develop a 'search image' for the target taxon and the habitats they inhabit.

When collecting forage legumes, rhizobia cultures should also be sampled, as they may be necessary for future regeneration of healthy plants in remote locations. The nodules plus the associated root fragments should be placed in the culture vial with a drying agent, separated from it by a thin layer of cotton wool. The sample should then be placed in a dry, cool, dark place, such as a cool box during the expedition.

With certain forage species such as *Lolium* and *Trifolium*, vegetative material rather than seed is commonly collected to try to avoid mixed seed collections. This requires that collectors are very familiar with their material and have the appropriate vegetative identification keys. In the field, a clump of vegetation (= a divot) is sampled, and later it is separated into single vegetative units, grown on, and mixed components weeded out of the plot before anthesis. Material of these species is virtually impossible to identify if collected as mixed seed samples.

11.5.3 Collecting Vegetative Material

For vegetatively propagated crops, rhizomes, corms, bulbs, tubers and roots are normally collected rather than seed, the reason being the species may not reach sexual maturity or rarely produces seed. However, where possible, seed should also be collected along with the vegetative material; for example, potatoes do often produce small quantities of seed because it is easier to process and conserve. Due to the multiple factors resulting in a lack of seed, it is very difficult to recommend one sampling strategy for all vegetatively propagated crops.

Collecting vegetatively propagated material is complex because it is bulky, difficult to transport, and the propagules often appear underground so must be dug up and then kept cool or they lose viability and quickly rot, or conversely dry out completely. It is often difficult to collect truly at random or to assess the genetic variation present in a population. As such, there may be one or multiple genotypes present in a field of potatoes in Peru. If the different genotypes can be selected based on variation in morphological characters, such as in tubers of potato or sweet potato, they are referred to as morphotypes. However, even though these plants may appear phenotypically identical, they may be genetically distinct. The reverse may also be true where the same apparent morphotype may be called by different names by local people, but in fact may be genetically identical. In general, it is best to enter the village and collect a sample of each morphotype, whether defined by the local farmers or the collecting team.

The vegetative propagules may be gently dried in the sun, then excess soil can be cleaned off and a fungicidal dusting or dip used to prevent deterioration. The propagules should then be wrapped in a semi-permeable material and kept under cool conditions, for example, in insulated cool boxes. Expanded polystyrene containers covered with reflective mirror foil are suitable. The material of some species may be fragile and so should be packed loosely rather than packed high on top of each other. The collecting of large samples, such as yams that can grow stem tubers up to 1 metre, can present practical problems of transport and storage when collecting. In these cases, it may be possible to dissect out and removed the meristem, possibly with a sprout, but they are then prone to attack by disease or disease transmission. Field tissue culture techniques have been developed to alleviate these problems. However, as the precise method of collection varies so much between species, the collector is advised to consult specialist literature for that species to find out the most appropriate sampling, harvesting and storage conditions.

A special difficulty arises when root or tuber crops are collected, as the plants cannot be collected until the tubers are mature, by which time the above-ground parts will have withered or dried completely (Box 11.3). Identification is therefore impossible in the field and can only be made by growing plants in the conservation or research station and waiting until the mature plants can be studied. Alternatively, many root or tuber plants will provide true seed. If these are collected, they will provide a better sample of the diversity of the population from which they were derived. There are also many vegetatively (or clonally) propagated species that are not collected as roots or tubers, e.g. bananas and sugarcane; some of these are listed in Table 11.2. Further information on collecting vegetatively propagated species can be found in Huamán *et al.* (1995) and Dansi (2011).

11.5.4 Collecting Wild Species

Most *ex situ* conservation activities to date have focused on conserving crop species. Only 10.5% of germplasm conserved *ex situ* in gene banks is of wild species and most of that has been collected from

---

Box 11.3 | Collection of Vegetatively Propagated Species

(Hawkes *et al.,* 2000)

Year	Activity	
I	(1)	Make a bulk sample of 2–4 fully grown, healthy, vegetative propagules of each morphotype present in each market area or each area where potatoes are cultivated. If morphotypes cannot easily be distinguished, take 10–15 randomly chosen vegetative propagules. Always collect seed if possible.
II	(2)	Grow out all samples in an experimental field and identify possible 'identical clones' using morphological comparison.
III	(3)	Re-grow each set of 'identical clones' adjacent to check their identical status.
IV	(4)	Use molecular techniques, if available, to establish clonal identity.
V	(5)	Rationalize and discard duplicate material. Self-pollinate remaining collections, where possible, to convert the vegetative collections to true seed accessions, that can be held in the gene bank. If self-incompatibility is discovered in a collection, cross-pollinate between collections obtained from the same area. Store collections as true seed wherever possible; if impossible, it may be necessary to store the material in a field gene bank or in vitro collection.

Table 11.2 **Selected vegetatively propagated herbaceous crops indicating the plant parts suitable for germplasm collecting**

	Stools	Rhizomes	Bulbs	Offsets	Shoots	Suckers	Tubers	Stem cuttings	Bulbils	Stolons	Storage roots
Acorus calamus	✓	✓									
Allium sativum			✓						✓		
Ananas comosus					✓						
Agave spp.						✓			✓		
Arracacia xanthorrhiza				✓	✓						
Asparagus officinalis		✓									
Canna edulis		✓					✓				
Colocasia esculenta						✓	✓				
Colocasia gigantea						✓					
Curcuma longa		✓									
Dioscorea esculenta							✓				
Dioscorea spp.									✓		
Dioscorea bulbifera								✓			
Eleocharis dulcis										✓	
Fragaria vesca										✓	
Hedychium coronarium		✓						✓			
Ipomoea batatas											✓

Species							
Manihot esculenta			✓				
Maranta arundinacea				✓			
Mirabilis expansa	✓		✓	✓		✓	
Musa spp.			✓	✓		✓	
Oxalis tuberosa			✓		✓		
Pachyrrhizus ahipa			✓	✓	✓		
Polymnia sonchifolia			✓			✓	
Saccharum spp.			✓				
Sansevieria spp.		✓	✓				
Solanum spp. (tuber-bearing)			✓	✓		✓	
Tacca leonto-pedaloides		✓					
Tropaeolum tuberosum			✓	✓			
Ullucus tuberosus			✓	✓	✓		
Xanthosoma nigrum			✓		✓		
Vanilla planifolia			✓				
Zingeber officinale		✓	✓				

From Huamán *et al.* (1995).

anthropomorphic environments, i.e. roadside, urban areas or areas that are under threat of development and 72% of global priority CWR are a high priority for further collection (Castañeda-Álvarez *et al.*, 2016a). Increasingly, further collection of major crop material has and is only likely to duplicate alleles already present in existing collections (Hawkes *et al.*, 2000). Tanksley and McCouch (1997) pointed out that domestication involved a significant loss of genetic diversity (Figure 11.1) and that 95% of genetic diversity in the tomato gene pool was located, for example, in wild *Solanum* species related to tomato. Perhaps the situation in tomato is extreme because of

the large numbers of CWR in the gene pool. The situation for the barley gene pool is shown Figure 11.2, where the relative size of the concentric circles equates to level of genetic diversity held in each component of the barley gene pool. Wild species contain significant additional and novel alleles, which are proving increasingly important to plant breeding, especially with the advent of biotechnology bridging the gaps between species and facilitating the inclusion of exotic CWR alleles in crop species. This realization has been reflected in increased sampling and *ex situ* conservation of CWR diversity. Forty percent of germplasm collected between the first (FAO, 1998)

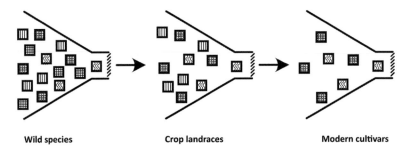

Wild species   Crop landraces   Modern cultivars

Figure 11.1 Genetic bottleneck imposed on crops during domestication and through modern plant breeding. Boxes represent allelic variation of genes originally found in the wild, but gradually lost through domestication and breeding (Tanksley and McCouch (1997))

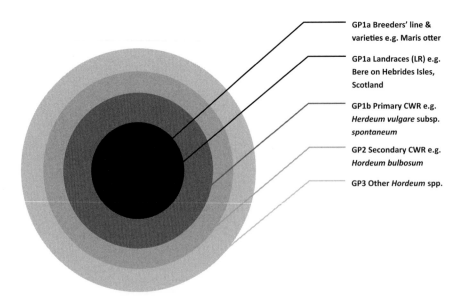

GP1a Breeders' line & varieties e.g. Maris otter

GP1a Landraces (LR) e.g. Bere on Hebrides Isles, Scotland

GP1b Primary CWR e.g. *Herdeum vulgare* subsp. *spontaneum*

GP2 Secondary CWR e.g. *Hordeum bulbosum*

GP3 Other *Hordeum* spp.

Figure 11.2 Estimated relative genetic diversity held in each component of the barley gene pool with bulk of diversity found in landraces, primary and secondary CWR. (A black and white version of this figure will appear in some formats. For the colour version, refer to the plate section.)

and second (FAO, 2010a) State of the World's PGRFA reports was due to CWR collection while the Crop Trust/Royal Botanic Gardens, Kew initiative to systematically collect global priority CWR diversity will have significantly improved the percentage of global CWR diversity actively conserved *ex situ* (Dempewolf *et al.*, 2013).

The generalized collection strategy must again be adopted when sampling wild species for *ex situ* conservation. Hawkes *et al.* (2000) summarized the differences between collecting wild and cultivated species as follows:

- Wild species generally have a broader genetic base than allied crops, due to the founder effect of domestication and subsequent disruptive selection, especially in those characters associated with domestication (e.g. heading and flowering dates, seed set, dispersal mechanisms).
- Natural populations show greater tolerance of geographic and ecological conditions, as they are not bred for uniformity in annual sowing or bulk harvesting.
- The population density of wild species is in general much less than that of cultivated species, and they are almost always found as part of mixed communities and not in dense mono-specific stands.
- Wild plants display a range of breeding systems, and there is a higher proportion of out-breeding species.
- Most crop plants are annuals, whereas most wild plants are either annual or perennial, and, as such, populations of wild species often show a complex age structure.
- Crop conservation is usually focused on one or two species, but wild species conservation tends to have a much broader focus on numerous species with disparate life strategies, or even habitats with whatever species are contained within; as a result, establishing conservation priorities is more complex.
- Cultivated material is highly mobile and can be spread rapidly by man, whereas wild material is unlikely to be spread so rapidly by man and so can develop highly localized patterns of gene distribution.

- Crops may be consciously spread by humans, but wild species are rarely spread consciously, although humans often unconsciously spread weedy species.

These differences mean it is difficult to recommend one overall strategy for collecting wild species, but the following points should be considered when formulating a strategy for any wild species or group of wild taxa:

- Resources are unlikely to be available to conserve all wild species, so priority should be given to those which are rare, endemic, threatened or those most closely related to economically important species and likely to be used in breeding.
- Because of their disparate geographic distribution, wild species collections should focus on the target taxa centres of diversity, but also include any significant outlying areas with known adaptive traits.
- Always sample the species from as diverse geographic and ecological sites as possible.
- Emphasis should be placed on collecting from more sites rather than larger numbers of plants per site.

Most wild species collection expeditions will commonly have multiple species as their target, and if the target taxon is taxonomically broad, the collecting team may be insufficiently experienced to identify all the species encountered. If the population is sufficiently dense, the collector should walk through the population taking random samples every few paces to make a bulk sample. As with forage species, wild species are likely to be highly adapted to the environments where they grow, so the collector should sample as many sites as possible and ensure the sites reflect the full breadth of the environmental range of the taxon.

11.5.5 Collecting Indigenous Knowledge

Increasingly germplasm users when choosing which germplasm to utilize will base their choice on the available additional information. Primarily this is passport data describing the provenance of the accession, but further additional information, particularly as regards biotic or abiotic stress resistance or cultivation practice, or potential for additional value generation, will sway the choice of germplasm requested and used. Therefore, field collectors, as well as collecting seeds, tubers, cuttings, etc., will also collate the conservation, cultivation and utilization knowledge held by local people on the plants found in their area. The knowledge that is held by traditional farmers or plant users is referred to as local or indigenous knowledge. Farmers give names to the different landraces they cultivate and know their properties and requirements; nomadic pastoralists know where and when the plants grow that their livestock like to eat; tropical rain forest dwellers know which nuts can be eaten and which bark can be pounded to make a poison. Customs, cults, rites, taboos, legends, myths and folklore all speak of the relationship between people and plants, a relationship that is based on long-term, intimate experience and is often crucial to survival. The availability of indigenous knowledge may enhance the conservation of the taxon itself by helping the plant collector decide what and where to collect and conserve.

The acquisition of indigenous knowledge is particularly important at present, as traditional knowledge about plants is just as severely threatened by erosion as the genetic diversity of the species themselves. The importance of this point is underscored in Article 8(j) of the Convention on Biological Diversity (CBD, 1992). Therefore, preserving and documenting the dwindling resource represented by indigenous knowledge is not an optional adjunct to the conservation of germplasm, but an integral and necessary part of the process. Guarino and Friis-Hansen (1995) state that indigenous knowledge aids germplasm collection by providing knowledge of:

- the vernacular names of landraces, wild plants and their pests;
- the local criteria for distinguishing amongst them, and their relationship to each other in any folk taxonomy;
- their appearance, properties, environmental preferences and uses;
- the places and habitats where they may be found, and the rules of access to them;

- the agricultural and management practices with which they are associated;
- the origin (history) of planting material, including any selection practices that may have been applied; and
- the character of any changes in farming practice, land management and natural habitats.

Knowledge of these factors in turn enhances the conservation and use value of the conserved genetic resources. The actual collection of indigenous knowledge ideally involves ethnographic, as well as botanical expertise, possibly using ethno-botany techniques such as Rapid Rural Appraisal (RRA) or participatory rural appraisal (PRA) techniques (Guarino and Friis-Hansen, 1995; Schultes and von Reis, 1995; Quek and Friis-Hansen, 2011; Martin, 2015).

Before leaving a discussion of the collation and use of indigenous knowledge, the ethical aspects of indigenous knowledge use by professionals must be addressed. Often rural people are, perhaps naively, eager to supply information on their traditional practices, but this process should not be exploitative in a unidirectional sense. Local people should benefit from the interaction with plant genetic resource professionals, and the wishes of local people should always be respected (Quek and Friis-Hansen, 2011). When genetic resources are so directly related to economic value, the professionals would be unable to exploit traditional diversity for breeding new cultivars or developing pharmaceutical drugs unless rural people continued to grow the full range of plant genetic resources and retain their knowledge of its diversity and use. Therefore, local people should participate in any exploitation, be aware of the results obtained and, where appropriate, benefit financially through profit-sharing agreements.

11.6 Identification of Collected Material

Collecting unidentified or mis-identified seed accessions or specimens is an expensive waste of resources, in that unidentified accessions are unlikely to be used until they are identified, and mis-identified accessions used will produce misleading results; therefore, the germplasm user requires accurately identified germplasm. There is a cost to keeping unidentified accessions in a gene bank. It costs about US$600 to collect a germplasm accession internationally, then US$275 to incorporate it into the gene bank, then a further US$5 per year to maintain the accession and once identified, US$15 to distribute the accession to an end-user (Smith and Linington, 1997). Material held in gene banks that is not identified is unlikely to be used, and therefore is a waste of space and funds. However, lack of identification is not normally a problem for cultivated species, since those collecting sorghum or wheat will rarely be unable to identify their target crop. Those collecting wild species, however, may be collecting a wide range of species, of which they are only fully familiar with a few and field identification will be necessary.

The primary approach to conserving accurately identified germplasm should be to identify material accurately in the field and avoid mixed species collections. In general, it is easier to avoid mixed seed collections than to undertake the difficult and costly process of distinguishing separate entities within a mixed collection during subsequent germplasm regeneration. If it proves impossible to identify the specimen in the field, then it is important to collect a good reference voucher specimen of the population as well as the seed sample; after all, it may prove to be a new taxon. This voucher specimen can then be identified later using some form of identification aid: an illustration, interactive computer key, but most commonly a printed dichotomous key contained in a local flora (see Chapter 3). Having stressed the need for field identification, any plant collector knows from experience that adhering to a desirable collection strategy is not always possible. It may not be possible to identify the specimen at maturity when the seed is ripe for collection, because there may be no flowers, or the fruit may have shattered, and these are required for positive identification.

Both mixed and unidentified collections are of little value unless steps are taken to separate or identify them. Resolving this problem involves two distinct tasks: (a) sorting the seed into the component species

and (b) the accurate identification of each component. Sorting out the component species in a mixture varies enormously in complexity. It can be trivially easy if the seeds fall into two or more distinct classes, especially if the classes are both distinct in colour, size, hilum or ornamentation, and each class is relatively uniform. Problems arise if the components are not visually distinct or if one or other of the components is very variable, in which case it may be necessary to grow out the material and identify the mature plants. Once the components are separated, identification can be undertaken using seed identification keys specifically written for identifying seeds, although this process is always time-consuming and using seed keys may not allow identification to species level, especially if the potential taxa are closely related.

If the seed cannot be identified using seed keys, it may need to be germinated, grown in plots and identified using relatively easily distinguishable flower or fruit characters. An additional bonus of growing out mixed collections is that a flowering or fruiting voucher specimen can be taken for each of the components of the mixture, since one voucher specimen cannot represent the entire mixed collection. However, growing out wild collections in plots away from their native range does have inherent problems. The ecogeographic conditions of the plots will place a selection pressure on the collections for those local conditions and thus lead to differential genetic erosion. In addition, if the species are outcrossing, close proximity of each plot may facilitate unwanted cross-pollination. This is not a problem for the self-pollinating annual species, but there would be little point in maximizing the genetic variability collected in the field if this variability is lost during the routine maintenance of the collection.

Finally, in practice the material the collecting team can identify in the field may practically establish the taxonomic breadth of a collecting mission. Broad-based collecting expeditions that collect taxonomically diverse taxa may be impractical because of the number of unidentified species remaining at the end of the expedition. The priorities established in the commission should always be kept in mind, but the collecting team should remain flexible.

11.7 Handling Collected Material

Commonly, the material gathered during a germplasm collecting expedition is of four types: seed, vegetative plants, voucher specimens and passport data.

11.7.1 Seed

Care should be taken to collect mature fruits, seed or seed heads from populations that have been identified. ENSCONET (2009) suggests if the seed can be easily dislodged from the fruit or the seed has recently changed colour, both indicate mature seed. Seed collected from dubiously identified or hybrid plants should be collected under different collection numbers from material where the identification is certain. The seed should be of the highest quality (seeds and filled and not infected with pests or diseases), as there is a direct correlation between initial seed quality and absolute seed longevity. Most species have seed that is borne in dry dehiscent or indehiscent fruits, which, provided the collector arrives at the appropriate stage of fruit maturity, are relatively simple to collect. The most problematic species to collect have fleshy fruits, and with these, the collector must decide whether to extract the seed from the fruit in the field or transport the seed in the fruit and risk possible fungal attack. There is no hard and fast rule, but extracting the seed may be preferable unless the collected fruit can be placed in an open cloth bag and returned rapidly to base for further processing.

In the field, each seed collection, whether of a population or single plant, should be placed in a separate bag (paper or cotton, depending on the size of the seed and collection). Normally seed is collected as a bulk sample; all the seed from different plants of a population is collected together into one sample bag. However, if genetic diversity research is to be undertaken, then seed from each plant should be collected separately, even though this will slow down collection speed and increase the number of bags, samples, etc. to be handled. If seed from different plants at a location is collected as one collection, then the seed should be thoroughly mixed before any division, to ensure each sample has the same genetic profile. Each seed lot is identified by a unique identifier (the collection number), which is assigned by the collector(s) sequentially. This should not be confused with the accession number, which is given by the collection manager to the germplasm when it enters the institute's collection (e.g. seed gene bank, botanic garden, field gene bank, *in vitro* collection). An alpha tag or piece of paper with the field identification and collection number should be placed inside the collecting bag, and these details plus the site number should also be written on the outside of the bag. It is useful to place all the collections taken at a site inside one larger site bag, with the site number prominently marked on the outside of the bag to help avoid mislabelling. However, if the seed is very moist during collection this practice is inadvisable as it will decrease natural ambient drying and may promote fungal growth.

When possible, samples should be collected from disease-free plants. During the expedition seed should either be maintained at ambient temperature and high moisture content, permitting any seed damage repair, or dried so that damage is reduced. Some non-dormant seeds will germinate if left in the fruit at ambient temperatures, so the seeds may have to be extracted from fleshy fruits. The seed of most species should be extracted from the fruit and dried to avoid loss of viability and moulds, etc., especially if the material is collected in wet conditions. However, placing the material in a cloth or paper bag, which allows air circulation, will facilitate drying; using artificial heat can enhance this, but the temperature should not exceed 20°C and 10–25% relative humidity, depending upon species (FAO, 2014b). The moister the conditions at harvest, the slower the drying process and lower the drying temperature. Seed samples should be routinely checked for fungal and pest attack, and if a problem is detected, the seed must be treated. Do not place seed in strong direct sunlight or future germination may be affected.

Seeds in fleshy berries (e.g. tomatoes, peppers, pumpkins, gourds, aubergines, potatoes) should be extracted, washed in a sieve and then spread on absorbent paper to dry. The seeds could be spread out in the vehicle, placed in the shade or placed in silica gel containers to assist drying, or the collector must make regular trips back to base and use appropriate drying equipment. The problems of recalcitrant seeds with 2- or 3-day viability can be difficult to overcome, but specific techniques are being developed for species groups, and laboratory-based techniques are being developed for the field. Unfortunately, these techniques are often taxon specific, so no general recommendations can be made for recalcitrant seeded species. If in doubt, collect the fruits and keep until the seeds can be extracted and sown in the gene bank garden. Some general recommendations as to how best to collect and handle seed collections in the field are summarized in Box 11.4 and in the gene bank in Box 11.5.

11.7.2 Collection of Live Plant Material

Vegetative material of certain forage species (e.g. *Lolium*, *Festuca* and *Trifolium*) is often collected, either because of the difficulty of obtaining seed or because the plants grow in such a dense mat that it would be difficult to ensure seed from only one plant was being collected. The collection of vegetative material requires the collectors to be very familiar with the species and have access to appropriate keys to the plants in their vegetative state. Different species require different sampling techniques, and information on the number of clones present at a site will obviously affect the number of vegetative units. Divots (vegetative samples) are usually taken, with

---

Box 11.4 | **General Recommendations for Collecting and Handling Seed Samples in the Field**

(Smith, 1995)

- For all species attempt to collect equal numbers of seeds from each plant sampled, and at the same stage of maturity, ideally when seed storage potential/desiccation tolerance is highest.
- Avoid damaged seeds (mechanical damage, introducing pest/disease attacks).
- If seeds must be cleaned during the trip, do so by hand to minimize the chance of mechanical damage.
- If it is possible to avoid quarantine seed treatments without breaking quarantine regulations, for example, through post-entry quarantine, do so.
- Ensure that seed arrives at the seed bank without undue delay.

For desiccation-intolerant seeds, which are more likely to be large seeds from dominant trees growing under wet conditions:

- Collect close to fruit fall. Do not collect from the ground unless you can be sure seeds are only recently dispersed.
- Keep seeds aerated and moist in inflated polythene bags, changing the air at least weekly by deflation and re-inflation.
- Do not allow such seeds collected in the tropics to cool below 20°C or heat up above ambient shade temperatures in the field or during transport.
- Plan your activities so that no more than 1 month elapses between collecting and reception by the seed bank.

For desiccation-tolerant seeds or their fruits, which are more likely to be small seeds from herbs growing under dry conditions:

Box 11.4 | (cont.)

- If meteorological data suggest more than 0.1 probit per month will be lost during the collecting trip, either modify your itinerary or prepare to dry actively with silica gel.
- For fleshy-fruited species, if logistically possible keep the seeds in the fruits and the fruits aerated and at ambient temperatures.
- If the above is not logistically possible, and for fruits that are dry dehiscent or indehiscent, air-dry the hand-cleaned seeds in a thin layer (to ensure aeration) under shade for 3 days or more (larger seeds need longer) to reduce the seed moisture content to equilibrium with ambient relative humidity before packing to use space more efficiently.

Box 11.5 | **Gene Bank Standards for Orthodox Seeds**

(FAO, 2014b)

1. *Standards for acquisition of germplasm*:
    1.1 All seed samples added to the gene bank collection have been acquired legally with relevant technical documentation (e.g. SMTA for ITPGRFA Annex 1 taxa or MTA for non-Annex 1 taxa).
    1.2 Seed collecting should be made as close as possible to the time of maturation and prior to natural seed dispersal, avoiding potential genetic contamination, to ensure maximum seed quality.
    1.3 To maximize seed quality, the period between seed collecting and transfer to a controlled drying environment should be within 3–5 days or as short a time as possible, bearing in mind that seeds should not be exposed to high temperatures (>30°C or 85% relative humidity) and intense light and that some species may have immature seeds that require time after harvest to achieve embryo maturation.
    1.4 All seed samples should be accompanied by at least a minimum of associated data as detailed in the FAO/Bioversity International multi-crop passport descriptors (Alercia *et al.*, 2012).
    1.5 The minimum number of plants from which seeds should be collected is between 30 and 60, depending on the breeding system of the target species.
2. *Standards for drying and storage*:
    2.1 All seed samples should be dried to equilibrium in a controlled environment of 5–20°C and 10–25% of relative humidity, depending upon species.
    2.2 After drying, all seed samples need to be sealed in a suitable airtight container for long-term storage; in some instances where collections that need frequent access to seeds or are likely to be depleted well before the predicted time for loss in viability, it is then possible to store seeds in non-airtight containers.

Box 11.5 | (cont.)

2.3 Most-original-samples and safety duplicate samples should be stored under long-term conditions (base collections) at a temperature of −18 ± 3°C and relative humidity of 15 ± 3%.

2.4 For medium-term conditions (active collection), samples should be stored under refrigeration at 5–10°C and relative humidity of 15 ± 3%.

3. *Standards for seed viability monitoring*:

3.1 The initial seed viability test should be conducted after cleaning and drying the accession or at the latest within about 12 months after receipt of the sample at the gene bank.

3.2 The initial germination value should exceed 85% for most seeds of cultivated crop species. For some specific accessions and wild and forest species that do not normally reach high levels of germination, a lower percentage could be accepted.

3.3 Viability monitoring test intervals should be set at one-third of the time predicted for viability to fall to 85%[1] of initial viability or lower depending on the species or specific accessions, but no longer than approximately 40 years. If this deterioration period cannot be estimated and accessions are being held in long-term storage at −18°C in hermetically closed containers, the interval should be 10 years for species expected to be long-lived and 5 years or less for species expected to be short-lived.

3.4 The viability threshold for regeneration or other management decision such as re-collection should be 85% or lower depending on the species or specific accessions of initial viability.

4. *Standards for regeneration*:

4.1 Regeneration should be carried out when the viability drops below 85% of the initial viability or when the remaining seed quantity is less than what is required for three sowings of a representative population of the accession. The most-original-sample should be used to regenerate those accessions.

4.2 The regeneration should be carried out in such a manner that the genetic integrity of a given accession is maintained. Species-specific regeneration measures should be taken to prevent admixtures or genetic contamination arising from pollen geneflow that originated from other accessions of the same species or from other species around the regeneration fields.

4.3 If possible, at least 50 seeds of the original and the subsequent most-original-samples should be archived in long-term storage for reference purposes.

5. *Standards for characterization*:

5.1 Around 60% of accessions should be characterized within 5–7 years of acquisition or during the first regeneration cycle.

5.2 Characterization should be based on standardized and calibrated measuring formats and characterization data follow internationally agreed descriptor lists (e.g. FAO/Bioversity International, UPOV and USDA) and are made publicly available.

---

[1] The time for seed viability to fall can be predicted for a range of crop species using an online application based on the Ellis/Roberts viability equations (see http://data.kew.org/sid/viability/).

Box 11.5 | **(cont.)**

6. *Standards for evaluation*:
    6.1 Evaluation data on gene bank accessions should be obtained for traits that are included in internationally agreed crop descriptor lists. They should conform to standardized and calibrated measuring formats.
    6.2 Evaluation data should be obtained for as many accessions as practically possible, through laboratory, greenhouse or field analysis as may be applicable.
    6.3 Evaluation trials should be carried out in at least three environmentally diverse locations and data collected over at least 3 years.
7. *Standards for documentation*:
    7.1 Passport data of 100% of the accessions should be documented using FAO/Bioversity International multi-crop passport descriptors (Alercia *et al.*, 2012).
    7.2 All data and information generated in the gene bank relating to all aspects of conservation and use of the material should be recorded in a suitably designed database.
8. *Standards for distribution and exchange*:
    8.1 Seeds should be distributed in compliance with national laws and relevant international treaties and conventions (i.e. ITPGRFA and CBD Nagoya Protocol).
    8.2 Seed samples should be provided with all relevant documents required by the recipient country.
    8.3 The time span between receipt of a request for seeds and the dispatch of the seeds should be kept to a minimum.
    8.4 For most species, a sample of a minimum of 30–50 viable seeds should be supplied for accessions with enough seeds in stock. For accessions with too little seed at the time of request and in the absence of a suitable alternative accession, samples should be supplied after regeneration/multiplication, based on a renewed request. For some species and some research uses, smaller numbers of seeds should be an acceptable distribution sample size.
9. *Standards for safety duplication*:
    9.1 A safety duplicate sample for every original accession should be stored in a geographically distant area, under the same or better conditions than those in the original gene bank.
    9.2 Each safety duplicate sample should be accompanied by relevant associated information.
10. *Standards for security and personnel*:
    10.1 A gene bank should have a risk management strategy in place that includes *inter alia* measures against power cut, fire, flooding and earthquakes.
    10.2 A gene bank should follow the local Occupational Safety and Health requirements and protocols where applicable.
    10.3 A gene bank should employ the requisite staff to fulfil all the routine responsibilities to ensure that the gene bank can acquire, conserve and distribute germplasm according to the standards.

between 25 and 50 dug up with small soil samples using a sharp knife or trowel from throughout the site (Sackville Hamilton and Chorlton, 1995). Each divot is placed in a polythene bag with the sample identification details, excess air is extracted, and the bag is sealed and placed in a dry, cool insulated box. Following completion of the expedition each divot must be separated into a single vegetative unit, grown up and mixed components weeded out before anthesis. Fruiting or non-flowering material may need to be grown out before the identification can be accurately made.

11.7.3 Voucher Specimens

When collecting wild species, time should be made available to collect dried herbarium voucher specimens as it facilitates identification and records features present in the population. Representative flowering and, if possible, fruiting specimens, representative of the population should be pressed. If at a later stage any query arises over the identification of a collection, the voucher specimen can be consulted, and a new identification made for the collection, if appropriate. For this reason, voucher specimens should be carefully preserved, i.e. specimens should be pressed firmly, pressing papers should be changed as required to aid drying and fungal growth avoided.

A herbarium voucher specimen should be a whole or part of a plant that when pressed flat fits onto a sheet of paper measuring 45 × 30 cm. Good specimens should be representative of the population the collector is sampling and should contain all parts of the plant, not just a flower or leaf of the specimen. Collect mature material, as there are often differences between juvenile and mature material. If the species is dioecious (male and female flowers on separate plants), you should take separate specimens of each sex. Dig the specimen out of the ground to ensure that at least part of the root system is included with the specimen, though note that roots of rare species should never be dug up. Extra plant parts (e.g. flowers, fruits or seeds) can be collected and placed in a packet on the mounted specimen.

The best herbarium specimens are obtained by drying them directly after plant collecting. Newspaper is almost as good for this purpose as proprietary drying paper and has the added advantage of being easily available and reasonably cheap to purchase. Thin flimsy sheets that remain around the specimen until it is mounted should still be used within the drying paper/newspaper to avoid damaging the specimen when changing drying papers, as the drying plant may stick to the drying sheet. Drying papers should be changed when damp, and more often if the plant is very fleshy or you are collecting in a humid zone. If you are away from your press in the field, you can place the specimens in a polythene bag, which will keep the material fresh for a few hours until you can press it. When arranging the specimen in the flimsy paper, ensure that it is laid out and that the upper and lower sides of leaves are shown, but pieces of plant do not stick out from the press. Use corrugates (cardboard or aluminium sheet) to envelop the pile of specimens to allow an even distribution of pressure over the specimens by the press and to ensure that the press itself does not damage the drying specimens. The press is most commonly composed of crossed wooden slats tightened by two cotton or nylon straps (see Figure 11.3).

Specimens should be labelled by attaching an alpha (jeweller's) tag, with the collecting number, provisional identification and possibly the site number written in pencil or permanent ink. Where a herbarium specimen and a germplasm collection are taken from the same population on the same occasion, then both should be given the same collection number. Both collection and site numbers are given sequentially to avoid confusion and allow distinction between specimens and sites. It is also useful to write the collection number on the flimsy paper, which encloses the specimen inside the drying papers. Check that no parts are sticking out of the bundle and that the material is evenly distributed; a dense specimen does not make a good specimen when mounted.

If the specimens are wet, and drying conditions are poor, you may need to place the press over a source of heat, for example, a charcoal or wood fire or a kerosene stove. Direct the heat through the press by inserting extra corrugates here and there in the press.

Figure 11.3 Pressing voucher herbarium specimens.

Material can also be preserved in alcohol. In this case, the fresh specimens are wrapped in newspaper; several sheets are then tied together and placed in a polythene bag. Alcohol is then poured into the bag so that the newspaper is completely soaked and then the polythene bag is sealed. The bag is then placed in a second bag and sealed. This will preserve the specimens for months or at least until the collector returns to the herbarium and the specimens can then be pressed normally.

Ideally, collect more than one specimen of each accession, and the extra specimens can be sent to specialists, international herbaria or offered for exchange. If collecting wild species, four duplicates are recommended, but there should be a minimum of two. However, if the population is very sparse or rare, the number of duplicates should be reduced to ensure the population is not threatened by over-collection at

that locality. Multiple specimens taken from a single population of a taxon are referred to as duplicates and are given the same collection number. The dried specimens should be deposited in several herbaria as an insurance policy against disasters. Alternatively, if you are collecting a well-known crop you may simply wish to take a single inflorescence to represent that collection.

For rare restricted endemic species, it may be desirable to collect a soil sample with the plant; this can be used at your institution to grow the plant or analyzed to show the species-specific soil requirements. It should be noted that several countries have restrictions on the transport of soil samples across national borders, which may restrict the possibility of taking soil samples.

The kind of data that should be included on a herbarium specimen label include:

- determination (= identification);
- location (country/state/region);
- latitude and longitude;
- altitude;
- habitat;
- colour of flowers;
- size of population;
- dominant plant associations;
- local names;
- local uses;
- collection name and number (unique identifier);
- date of collection;
- any field notes that could not be deduced from seeing the specimen, e.g. tree 25 m tall.

11.7.4 Passport Data

The potential usefulness of a collection will depend largely upon the accompanying passport data, which is the information on the identification, population characteristics and collection site description. This information is likely to include: identification, collector, country, geographic data, date and any pre-characterization or pre-evaluation details, as well as indigenous knowledge notes obtained by questioning local people. It has been estimated that there are more than 7.4 million accessions in germplasm collections

(FAO, 2010a), but that many of these are in terms of utilization useless because their provenance is so poorly documented, or they lack any identification. Germplasm can only be deemed adequately collected if the accompanying passport data are thoroughly recorded. Passport data are vital if populations are to be relocated or accessions with certain characteristics distinguishable. The recording of detailed passport data can provide useful pre-evaluation information via predictive characterization, which can then facilitate accession use.

To assist the field collector in collating uniform passport data for each collection, it is preferable to use a standard collection form, rather than some form of traditional collection notebook, as this encourages standard and comparative data recording, and facilitates data entry into the expedition database. Different authors have each advocated their collecting forms, but experience has shown that no single form will meet the requirements of all collecting expeditions for all plant species. Any general form will need to be adapted to meet the specific

requirements of the crop or wild species being collected, the region where the expedition is taking place, the expertise of the collecting team, and how the germplasm is to be used. However, to illustrate the general features of a collecting form, an example is presented in Figure 11.4. The collecting form should be easily duplicated and held in the field in loose-leaf binders or bound into booklets of, say, 100 pages. The sheets should be filled using a pencil as this can be easily erased when mistakes are made and does not smudge if collecting conditions are wet.

Essentially, the data recorded in the field will always be of two basic elements: (a) information concerning the site location and site ecogeography, and (b) the collections gathered at that site. All collections must be given a unique collection number and at least tentative field identification. Although the specific data recorded by each expedition will vary according to the target taxon or target area selected, some information will commonly be recorded:

Site Information:

Country .... Province .... Date ../..../20....

Site Number.................Nearest settlement...

Location ....

Altitude (m) .... Latitude .... Longitude .... Rainfall (cm) ....

Site Physical Description...

Site Vegetation ....

Coded Environmental Information

PR	pH	TS	AS	SL	%C	DS	WR	AP	GP	%R	RT	%T	TT	Photo Nos.

Taxon Collection Information:

Collector(s) .... Coll. Nos. ....

Genus .... Species .... Subsp. / Var. ....

Petal Colour Standard .... Wing .... Keel ....

Habitat .... Pop. Description ....

Herb. Spec. Y/N   Nos. Duplicates .... Herb. Date ...../...../20..... Rhizobia Y/N

Seed Coll. Y/N   Coll. Size .... Nos. Plants Sampled .... Seed date 1 ...../...../20..... 2 ../...../20...

Collector(s) .... Coll. Nos. ....

Genus .... Species .... Subsp. / Var. ....

Petal Colour Standard .... Wing .... Keel ....

Habitat .... Pop. Description ....

Herb. Spec. Y/N   Nos. Duplicates .... Herb. Date ...../...../20.... Rhizobia Y/N

Seed Coll. Y/N   Coll. Size .... Nos. Plants Sampled .... Seed date 1 ...../...../20.... 2 ../...../20...

Figure 11.4 An example of a collection form.
(From Maxted and Bisby, 1989)

Key to Environmental Data Codes

Parent Rock (PR)
A = Peat and coal
B = Conglomerate
C = Sandstone
D = Shales - mudstone
I = Siliceous
J = Limestone
K = Laterites
L = Granites
M = Dolerites
N = Phyolites
O = Basalts
P = Hornlels
Q = Slate
R = Schist
S = Quartzites
T = Alluvium
U = Dunes

Type Of Soil (TS)
A = Calcic brown
B = Terra rossa
C = Heavy black
D = Woodland brown
E = Alluvial
F = Sandy loam
G = Clay

Depth Soil (DS)
A = 0 -10cm
B = 10-20cm
C = 20-40cm
D = > 40cm

Slope (SL)
L = Level 0-3%
U = Undulating 3-8%
R = Rolling 8-16%
M = Moderate 16-30%
S = Steep > 30%

Water Relations (WR)
F = Free draining
R = Run-off
WM = Wet meadow
S = Swamp

Agricultural Practice (AP)
S = Pasture
A = Fallow
C = Crop
G = Grassland
F = Forest
W = Woodland
R = Roadside
P = Protected enclosure
D = Disturbed
PN = Photograph number

Grazing Pressure (GP)
A = Nil
B = Light
C = Moderate
D = Severe

Soil Texture (T)
G = Gravel
S = Sand
Y = Sandy loam
L = Loam
M = Clay loam
C = Clay

Rock Type (RT)
A = Flat
B = Rocks
C = Boulders
D = Large boulders

Type of Tree Or Shrub (TT)
1 = Shrubs
2 = Small trees
3 = Medium trees
4 = Large trees

pH = Estimate of soil acidity

AS = Aspect

%C = % site with ground cover

%R = % site covered by rocky outcrops   %T = % site covered by trees & shrubs

Figure 11.4 (*cont.*)

- curatorial data (expedition identifier, collector(s) name and number, date, type of material);
- collecting site location (country, province, precise location, latitude, longitude, altitude, ecological data, farmer's name);
- collecting site description and context (site disturbance, physiography, soil, biotic factors);
- sample identification (scientific name, vernacular name);
- sampling information (population estimate, sampling method); and
- population information (phenology, pests and diseases, uses).

The minimum requirement of associated data is detailed in the FAO/Bioversity International multi-crop passport descriptors (Alercia *et al.*, 2012).

The precise geographic location is often standardized into the format '15.2 km south-east of

Yerevan on road to Goris'. Even now, when collecting expeditions will have geographic positioning systems (GPS) available to establish the precise latitude and longitude of collecting sites, it is wise to record the location in this standard format in case the GPS is not working properly. For plants growing wild, you should include some comments on the dominant vegetation of the site and if a crop, you should record details of the rotation used by the farmer. The point is well worth emphasizing that the more complete the passport data collected in the field, the more useful the germplasm accession will prove to potential users.

As the information noted on the form is likely to be transferred to a database on completion of the expedition, it may be a good idea to record the data in the field in the coded format that will be typed into the database, using sets of standard codes for the environmental data. Increasingly the collection form will not be printed on paper but held on data loggers or laptop PCs. However the collection form is completed, it is wise to duplicate completed collecting forms periodically during the expedition to safeguard against loss of the passport data.

11.8 Conservation and Duplication

Once sampled, the collected materials require storage. The seed collections should be fumigated (if required), threshed, cleaned, duplicates divided (if required), and dried as soon as possible after collection to avoid any deterioration in quality of the sample. Each time the seed sample is transferred from one container to another there is potential for errors in labelling or mixing of samples, so it is advisable to make the minimum number of container changes from the original collecting bag to the gene bank seed container. Therefore, it is wise to avoid having bags from several collections open at any one time. Before placing the samples in the gene bank, it may be advisable to multiply any samples deemed to be too small; the quantity of seed deemed too small for banking will vary from gene bank to gene bank but is likely to be based on a level sufficient to allow long-term storage as well as routine distribution of samples. Multiplying the samples before deposition

has the added advantage of ensuring that the material conserved is of the highest quality, allowing some characterization/evaluation and enabling the field identification to be checked. It is recommended that seed of orthodox seeded species is stored under long-term conditions (base collections) at a temperature of $-18 \pm 3°C$ and relative humidity of $15 \pm 3\%$, while for medium-term conditions (active collection), samples should be stored under refrigeration at $5-10°C$ and relative humidity of $15 \pm 3\%$ (FAO, 2014b). The seed sample in long-term storage should be regenerated when the sample size falls below 1500 for an inbreeding species or 3000 for an outbreeding species to avoid genetic erosion of the accession or when the viability level has fallen below 85% (FAO, 2014b). With each regeneration cycle there is likely to be some genetic erosion, even if enough seed is sown, because the regeneration environment is unlikely to be similar to the original provenance location and so there will be selection for adaptation to the local environment during each regeneration cycle.

The dried voucher specimens should be re-identified on return to the laboratory to check the field identification, and the laboratory identification should be entered into the gene bank database. The voucher specimens can then be mounted on card and the herbarium label produced from the database automatically, saving time and avoiding errors in typing repetitive information. An example of a herbarium specimen label is provided in Figure 11.5. Duplicates of voucher specimens should be placed with the institute where the scientist who identified the material is based, the national herbarium of the host country and an important regional or international herbarium. If the specimen, on arriving at a herbarium, is re-identified, then a determination slip containing the new identification, name of person identifying the specimen and date should be added to the mounted specimen. The new collection identification should be relayed to the gene banks holding the germplasm and noted in the gene bank management database.

The data sheets containing the two basic kinds of information (location and taxon data) should be transferred to the gene bank management database. Once the passport data have been transferred

Flora of Turkey

University of Southampton

Vicia eristalioides Maxted

Province - Antalya   **Nearest Settlement**: Cavus   **Collection date**: 25/04/87
Directions: 67 kms from Antalya on road to Kumluca (road D400).

Flora Turkey Grid. Sq. C/3   **Altitude**: 550m   **Latitude**: 36º 21′ N   **Longitude**: 30º25′ E
Rock type: Limestone   **Soil**: Red Mediterranean   **Aspect**: North west
Habitat: Young coniferous plantation (*Pinus brutei*) and hillside scrub (*Rubus*, *Quercus* and *Ficus* sp.).

Collectors: N. Maxted, R. Allkin and A. Kitiki 4256   **Identification**: Nigel Maxted

Figure 11.5 An example of a herbarium specimen label.

accurately from the data sheets to the database collection, management and user access to the accession is facilitated. Labels for the gene bank seed containers, herbarium specimens and passport data reports to accompany seed requests can be generated directly from the gene bank management database.

An important element of any expedition is the distribution of duplicates of the collected material; the germplasm, voucher specimens and passport data should each be safety duplicated. Field exploration is costly, so once collected, the material must be safely conserved and made available for utilization by the national and international community. Normally seed duplicates are divided between the host country conservation institute(s), other participating institutes and an internationally accredited centre (such as the appropriate CGIAR centre). The voucher specimen duplicates, once mounted, labelled and determined, should be distributed between the appropriate host country herbaria and major international herbaria with an interest in the host country. Duplication of the passport data is relatively easy once the expedition database has been built, and it is important that the commissioning agency, the host country PGRFA authority, the other participating institutes and other interested parties should hold copies. The distribution

of duplicate sets of material acts as a safeguard against accidental loss (fire, economic, social unrest, etc.) of material from any one institute.

11.9 Publicizing Germplasm Collections

If use of the conserved material is to be promoted, then the appropriate user communities must be informed that the material has been collected and is available for utilization. The incorporation of the collecting information in the gene bank management database may first alert users to new accessions being available and the basic information may subsequently be enhanced by additional characterization and evaluation data. In many countries if there is a decentralized gene bank system, there will be a national germplasm documentation system that collates the data found in individual institutes or agencies. Further, in Europe and other regions there is another iteration of accession data collation and merging to provide ease of access to the putative germplasm users (see EURISCO: https://eurisco.ipk-gatersleben.de/) and possibly even a further iteration of accession data collation and merging into the global portal for accessing information on ex situ germplasm collections (see GENESYS: www .genesys-pgr.org).

The collector or gene bank is likely to have to provide some form of collection report to the collection sponsor, whether the expedition was funded externally or internally to the collector or gene bank's home institution. The precise format of this report is likely to be dictated by the commissioning agency, but if no specific format is recommended, then the 12 points detailed in Box 11.6 should be considered. The report may then be edited into an appropriate format for a published popular article or

---

Box 11.6 | Points to Include in a Germplasm Expedition Report

(Hawkes *et al.*, 2000)

1. Summary: objectives, timing, human groups involved, materials conserved, highlights, locations of duplicated material.
2. Introduction: discussion of the choice of target taxon and target area, threat of genetic erosion, why taxon and area have been given priority, background details of the groups involved.
3. Ecogeographic background to target area:
   ◦ physical environment: description and maps of the target area, sub-regions, relief, soils, etc.
   ◦ climate: description and maps of the target area, sub-regions, precipitation, humidity, etc.
   ◦ agriculture and vegetation: discussion of predominant agriculture and vegetation types, rotation practice employed, etc.
4. Collection timing: discussion of why the expedition visited areas at certain times.
5. Expedition personnel: complete list of the personnel, with institutional addresses and communication details, actively taking part in or involved in the expedition.
6. Chronological itinerary: daily account of the progress of the expedition, which areas were visited when, and who was involved.
7. Site selection and sampling strategy: discussion of how certain areas were selected for sampling and which population sampling strategy was adopted.
8. Material collected:
   ◦ Germplasm: details of how seed, fruit or vegetative materials were sampled and the collection numbering system used.
   ◦ Herbarium voucher specimens: details of how specimens were sampled, and the numbers of duplicates taken.
   ◦ Passport data: details of the passport data collected for each collection and how these data were recorded (if the data set is very large, the report may contain a summary rather than the complete data set).
9. Collection processing:
   ◦ Germplasm: discussion of the processing and distribution of duplicate material to base and active collections, as well as for safety duplication.

Box 11.6 | (cont.)

- ○ Herbarium voucher specimens: discussion of the processing and distribution of duplicate material.
  - ○ Passport data: discussion of the structure of the expedition database.
10. General discussion: assessment of the match between objectives and actual conservation activities undertaken during the expedition, a brief overview of the material collected (highlighting any particularly interesting material), comments on sample sizes, discussion of any agronomic or utilization potential contained in the material, discussion of expedition procedures employed, discussion of any important contacts made during the expedition, reassessment of genetic erosion of the target taxon in the target area, assessment of local *in situ* conservation activities.
11. Review of future conservation priorities: discussion of future ex situ and *in situ* conservation priorities, possible follow-up expeditions in the light of the results of the current expedition.
12. Details of collections made: list of taxa encountered, number of seed or vegetative samples collected.
13. References cited.

scientific paper in an appropriate journal, such as the *Genetic Resources Newsletter, Genetic Resources and Crop Evolution* or *Plant Genetic Resources Characterization and Utilization*. Throughout this text, the point has been repeatedly stressed that there is a necessary link between conservation and utilization. Therefore, there would be little point in placing efficiently collected germplasm in the gene bank if no attempt was subsequently made to draw the attention of potential user communities to it; therefore, publication of a collecting report summary should be an essential final stage in the germplasm collection process.

12 Seed Gene Bank Conservation

12.1 Introduction

There are about 1750 gene banks worldwide that store around 7 million accessions, of which at least 30% may be unique (FAO, 2010a). The storage of genetic diversity in true seed form dried and frozen is the most widely used and convenient approach for *ex situ* conservation because seeds are the natural storage organs of plants. However, not all species' seed can be stored in frozen conditions. Roberts (1973) distinguished between two basic types of storage behaviour for seeds: 'orthodox', which can tolerate desiccation and when dry freezing temperatures, and 'recalcitrant' or 'desiccation sensitive', which lose their viability after drying, e.g. seed from some shrubs and woody species of the tropics. Subsequently seeds of other species that can undergo drying to some extent but also cannot survive storage at low temperatures, thus having intermediate storage characteristics, e.g. avocado, cacao, citrus, coffee, mango or rubber, among other perennial crops were identified. Likewise, seeds of wild relatives do not always behave in the same way as their domesticated species, so knowledge of the optimal storage conditions must be determined for each (Smith *et al.*, 2003; FAO, 2014b). There is considerable variation among species in the length of time their seed can be stored without losing viability. Cereal seeds have longer periods of longevity, while legume seeds show intermediate longevity, and vegetable seeds are known for their short periods of longevity, tomato being an exception (Roos and Davidson, 1992).

Bioversity International, FAO and other gene banks' publications (e.g. Walters *et al.*, 2005; Lee *et al.*, 2013; van Treuren *et al.*, 2013) provide information regarding seed storage characteristics and behaviour, which led to a compendium of available data for about 7000 species (Hong *et al.*, 1996; Engels *et al.*, 2001; Rao *et al.*, 2006). The most important requirements for storing orthodox seed of annual and biennial crops are:

- Careful production of quality seeds to ensure maximum longevity;
- Adequate drying because very low moisture content is not optimal and requires extra care, i.e. 3% and 5% or more moisture content for 'oily' and 'floury' seeds, respectively;
- Appropriate storage temperature: $-18°C$ is recommended for long-term storage.

Hermetically sealed containers should be used to ensure that seeds remain dry. There are various products that absorb the humidity of the seed, e.g. the cheap silica gel, which changes colour as moisture absorption increases, or an aluminium silicate clay packaged in non-dusting, air-permeable bags.

Initially seed storage was focused on formal gene banks, often representing national or crop-specific collections or for wild species held in botanic gardens. Community seed banks emerged at the end of the 1980s for less formal and less expensive conservation of plant genetic resources (Vernooy *et al.*, 2015) (see Chapter 11). They evolved and today, alongside formal gene banks, provide access to seeds and serve as platforms for community development by ensuring food sovereignty through the sustainable use of agro-biodiversity. Other areas of interest in these community seed banks are plant health, pursuing high edible yields and marketing of agricultural diversity products. Community and other informal gene banks should be regarded as biorepositories that promote and propagate seed of traditional landraces and other local cultivars of various crops, including medicinal plants.

12.2 Storage of Seeds

A refrigerator is a good place to store dry seed, which can last for hundreds of years. Many gene banks therefore maintain seeds in a cold storage, which needs a reliable supply of electricity. The container keeping the seeds at very low temperatures should not be open until it matches the ambient temperature to prevent moisture from condensing. When it is not possible to control the temperature and humidity to the maximum, it is advisable to follow the practical rule that the sum of the storage temperature in degrees Fahrenheit and the humidity of the air in percentage must be equal to or less than 100. Likewise, seed should be kept below 0°C when it is impossible to control the atmospheric humidity (Tyagi and Agrawal, 2015).

> **The ideal environmental conditions for seed conservation** are low relative humidity (15 ± 3%), storage temperature of –18 ± 3°C, absence of light and solar radiation, and low oxygen content (FAO, 2014b).

An alternative procedure to store seed for long periods at room temperature should rely on drying seeds to a moisture content of 1% or about 3% for oily and floury seeds, respectively, and packing them tightly. Places with fresh and constant temperature are the best to preserve these seeds. If the ambient humidity is low (> 50%), the temperature could be that of the environment in a site with a mean daily temperature below 15°C (Rao *et al.*, 2006)

A drying room kept at 15°C, having 10–15% relative humidity and good air recirculation allows drying of seed to 5% moisture content. Such an environment can be achieved by using an air-dehumidifier with refrigeration to lower the temperature and remove heat generated by the former. Another very simple procedure for drying freshly harvested seeds is to place them in a sealed container with an equivalent amount by weight of silica gel. Both silica gel and seeds are kept in separate paper bags. After 7 or 8 days, the seeds will be at the required moisture level for prolonged storage. A small amount of the silica gel (5–10%) is left together with the seed and the container sealed again for storage. These containers are thereafter stored in a dry and well-ventilated room, in the shade, taking care that the temperature inside the room does not exceed 20°C. In hot locations or in the summer it will be necessary to ventilate the environment or use air conditioners during daylight (Rao *et al.*, 2006).

12.2.1 Storage Behaviour

Seed is the most convenient way of preserving most plant genetic resources (Ellis, 1988), those of vegetatively propagated plants being the main exception and which seldom produce viable seeds (e.g. banana), or species with short-lived and recalcitrant seeds. Some temperate recalcitrant seeds may be stored at or slightly below freezing, while tropical recalcitrant seeds are unable to survive at low temperatures. Seeds with 'intermediate' storage behaviour can undergo drying at low moisture levels but are sensitive to low temperatures used for storing orthodox seeds. Orthodox seed storage may be affected by the genetics at various levels: genus, species and cultivars. For example, the seed of many legume species are known for their hard, impermeable seed coats that allow their longevity, though the seeds of the oily groundnut show a short lifespan (Rao *et al.*, 2006).

12.2.2 Seed Condition before Storage

Seeds should reach maturity before placing them in storage (Roos, 1989). Seed maturity is the stage of development at which maximum dry weight has been attained. Many crops, however, produce mature seeds after some days or even weeks, thus making standardization difficult during collecting. There are pre- and post-harvest factors that affect seed lifespan, e.g. moisture or temperature extremes during maturation, harvesting and processing reduce seed viability. Weather during collecting, ripening and harvesting plays an important role in viability loss of cereal seed; seed collected in damp conditions means the seed may have to be artificially dried before it can be banked.

12.3 Factors Controlling Longevity during Storage

Seed age is not an indicator of its viability because seed viability depends mostly on the environment storing the seed (Ellis and Jackson, 1995). Moisture content and temperature together are the two most important factors maximizing seed longevity in gene banks. As noted above, seed lifespan lengthens by reducing both humidity and storage temperature. FAO (2014b) recommend drying orthodox seeds to equilibrium in a controlled environment of 5–20°C and 10–25% moisture content, and thereafter placed in sealed containers for further storing at –18°C. Likewise, anoxia may prolong the longevity of *ex situ* conserved seed (Groot *et al.*, 2015), because oxygen accelerates the ageing of dry seeds under storage.

Seed moisture content (SMC) is the weight of water contained in a seed as a percentage of the total weight of the seed before drying or the wet-weight (wb) or fresh-weight basis, thus:

$$SMC\ (\%wb) = 100\frac{(wet\ weight - dry\ weight)}{wet\ weight}$$

SMC can also be given as dry weight percentage (db) as follows:

$$SMC\ (\%db) = 100\frac{(wet\ weight - dry\ weight)}{dry\ weight}$$

12.4 Assessment of the Storage Behaviour of Seeds

A screening protocol allows the determination of whether seeds of unknown species are orthodox or recalcitrant (Rao *et al.*, 2006). This protocol measures water content for individual component parts of the seed, e.g. axis versus storage tissue or other parts. Orthodox seeds are those whose initial moisture content of all seed parts is 15% or below, while recalcitrant seeds are those with moisture content of 15–20% or above (Rao *et al.*, 2006). Furthermore, this protocol considers whether the seed is desiccated at ambient temperatures, in a drying oven or with silica gel at 15% relative humidity. Last but not least, it also

includes taking a sample at regular intervals of weight loss and carrying out standard germination testing.

12.5 Recommended Storage Conditions for Orthodox Seeds

The FAO in 1973 and the International Board for Plant Genetic Resources (IBPGR, now Bioversity International) in 1974 defined two broad types of conservation in gene banks: long-term base collections and medium-term active collections, both really relating to seeds. Tyagi and Agrawal (2015) provide an update regarding standards for gene bank management of seeds.

12.5.1 Base Collection

The seeds of the 'base collection' are seldom distributed to users because they are held in gene banks for posterity. Hence, their seed are not drawn upon except for viability testing and subsequent regeneration, or when urgently required of any accession not available from any other source. The distinct accessions therein are as close as possible to the original sample and are preserved long-term. The base collection for a cultigen pool for any species may be shared across several gene banks within a crop network.

12.5.2 Active Collection

Medium-term 'active or working collections' include seed samples for ease of distribution to users. Their seeds are used for multiplication, regeneration, characterization and evaluation. Seed storage for active collections is often less stringent for purely practical and economic reasons. Active collections store, therefore, seeds with a minimum standard requirement that ensures their viability remains above at least 65% for 10–20 years (Rao *et al.*, 2006).

Storage under optimal conditions results in maintaining seeds with high viability over a long period, though with a decreasing viability that differs between seed lots (Ellis, 1988). The rate of seed viability loss should be estimated to determine how long it may take for any seed sample to fall below a viability threshold, which would trigger the need for regeneration (Roberts, 1973); i.e. growing plants to get fresh seed for further storage.

12.6 Seed Storage Facilities

Seed should be stored in a controlled environment room, where the temperature can be maintained at at least –18°C (or below). Regulating the relative humidity for the whole of this room will be technically difficult and expensive, thus most seeds are stored in hermetically sealed containers in gene banks. These may be glass vials, metal cans or laminated aluminium foil packets. The seed samples, already dried to the required moisture content, are placed within these containers which are then sealed.

There are various designs for building large and complex seed storage facilities, but there are trade-offs between efficiency and costs that should be considered. Small-scale seed storage may be achieved using domestic chest freezers, and after pre-drying, seeds placed in a sealed container can effectively maintain viability. Any storage facility depends on a constant supply of electricity to ensure their functioning. However, if not available, orthodox seed of some species may be stored at up to 20°C provided their drying ensures ultra-low 2–3.7% moisture content (Rao et al., 2006). A useful precaution will be having an auxiliary power supply (standby generator or battery) for the storage facility that kicks in automatically when the normal power supply is lost.

The required cold storage space for a gene bank may be estimated considering the probable number of accessions to store. Although building a large cold room may be cheaper than several small cold rooms with the equivalent required volume, it will be worth dividing the store because it gives protection against any equipment failure. A cold room having a volume between 50 and 300 m^3 may store about 7000–70 000 accessions, respectively (Rao et al., 2006).

Box 12.1 | **CIMMYT's Wellhausen–Anderson Genetic Resources Center**

(Taba et al., 2004)

The International Maize and Wheat Improvement Center (CIMMYT) inaugurated the Wellhausen–Anderson Genetic Resources Center (GRC) in September 1996. This state-of-the-art complex houses CIMMYT's maize and wheat gene bank. It is a two-floor structure constructed using reinforced concrete walls and holds about 150 000 wheat accessions from more than 100 countries, i.e. the largest germplasm collection in the world for a single crop. There are about 28 000 maize accessions including the world's largest collection of landraces – cultivars bred by farmers – along with CWR such as wild Zea spp. teosinte and the related Tripsacum, and of improved cultivars. On the GRC's main floor is a chamber maintained at –3°C and 25–30% relative humidity that contains the active maize and wheat collections. Seed here has an average shelf life of approximately 30 years for maize and 30 to 50 years for wheat. On the lower level is an equivalent chamber maintained at –18 °C. Similar rows of movable shelving are used to store the base and black box collections. Seed stored in this chamber has a shelf life of approximately 60 years for maize and wheat.

Access to the storage chambers is through a multi-locked entry system. Entry into the hallway leading to the main germplasm bank entrance is through a glass door that is opened

Box 12.1 | (cont.)

via an electronic key card. Only authorized personnel have such a key card. Access to the gene bank anti-chamber is through a steel and aluminium door with a numeric coded lock. Again, only authorized personnel have the code necessary to enter the gene bank. Inside the anti-chamber are stairs and a freight elevator leading to the lower floor and storage room. Access into each storage chamber is via a sliding steel and aluminium, thermal insulated door, again with a numeric coded lock.

Temperature and relative humidity are monitored via remote sensing devices in several locations in both chambers. Gene bank staff monitor these daily for any fluctuations. Alarms are installed to indicate when either chamber deviates from the set point. A diesel generator provides 24/7 automatic dedicated back-up power to the gene bank lighting, air-conditioning and access locks during power outages. The GRC complex also houses areas for seed preparation, short-term storage, seed drying and germination testing. For the growth of plants for either observation or regeneration, the GRC has access to greenhouses, net-houses and field space in CIMMYT's headquarters at El Batan and research stations in Mexico: Ciudad Obregon and Toluca for wheat, and Agua Fria and Tlaltizapán for maize. Each of these sites provides appropriate climatic conditions for the regeneration of most maize and wheat accessions held in the gene bank. Certain landraces, especially of maize, require regeneration outside Mexico, and these are done in collaboration with national programmes in a country near to the site of the original collection.

12.7 Management of Seed Collections in Gene Banks

There are various tasks associated with management of seeds under storage (Box 12.1). They include in-country collecting, introducing genetic resources from elsewhere (which requires plant quarantine), seed extraction and cleaning, regeneration, registration, seed drying, packaging, storage, seed viability monitoring through germination testing, characterization, evaluation, documentation and distribution. They are briefly described below. The workflow of a gene bank begins with acquisition of seeds through collecting or arriving from elsewhere. Registration should occur after the seeds arrive, and thereafter they undergo cleaning, viability testing and preparation for storage in either the base or the active collections or sent to another gene bank for safety duplication. When the seeds are insufficient for storage, they should undergo multiplication. Seeds samples are also used for both characterization and evaluation tasks (Tyagi and Agrawal, 2015).

12.7.1 Germplasm Acquisition and Conservation

This activity involves the initial step for obtaining plant genetic resources for conservation in a gene bank. It requires agreements and permits showing that the acquisition (through collecting or other approach) of plant genetic resources was in accordance with all applicable laws, including consent from government and landowners, when appropriate. Passport data are the minimum information required for each accession to guarantee their proper management. A unique accession number is given to all incoming material, whose seed may be placed under drying conditions at 15% within 4 weeks after collecting. Any pathogen-infested seed and debris are removed prior to drying. For further details see Chapter 11.

12.7.2 Germplasm Exchange

The import and export of plant germplasm should comply with national laws, the International Treaty of Plant Genetic Resources for Food and Agriculture (IT-PGRFA), the Nagoya Protocol on Access to Genetic Resources and the Fair and
Equitable Sharing of Benefits Arising from their Utilization to the Convention on Biological Diversity, and agreed safety guidelines. It may require some documents such as, *inter alia*, a phytosanitary certificate, additional declarations, a certificate of donation, a certificate of no commercial value and import permit. The minimum sample for shipping should include 30 to 50 viable seeds. The list of the material and associated information (passport data as a minimum) should be provided to the recipient along with the material transfer agreement stating clearly access and benefit sharing.

12.7.3 Germplasm Characterization and Evaluation

A gene bank is not a PGR museum, thus accessions held therein should be characterized and the data web-enabled to facilitate a preliminary selection of germplasm by *bona fide* users. Characterization describes plant germplasm by determining the expression of highly heritable morphological, physiological or agronomical characters or by using DNA markers, while evaluation refers to the description of gene bank accessions for characters – most often multi-genic – that are important to plant breeding, e.g. stress tolerance, host plant resistance to pathogens and pests, or edible yield, among others. Gene banks should aim to characterize 60% of the accessions after 5 years of acquisition or during the first regeneration cycle. Bioversity International and the International Union for the Protection of New Varieties of Plants provide descriptor lists for

characterizing various crops' cultigens and their wild relatives (see www.bioversityinternational.org/e-library/publications/descriptors/).

12.7.4 Data Management

The gene bank curatorial database management system should be able to provide information about the origin, morpho-agronomic characterization and evaluation traits of germplasm accessions; the data should be web-enabled facilitating selection based on predetermined criteria by users. Information from this gene bank database must be freely available to *bona fide* users worldwide. An integrated information system should therefore link all gene bank operations related to the conservation and management of plant genetic resources, e.g. GRIN-Global. Thus, gene banks deploy databases and assist scientists to employ best, open-access data stewardship practices to maximize the impact of their research. They may also provide technical advice in the design and analysis of experiments for plant genetic resources research.

12.7.5 Seed Accession Processing

Seed Arrival and Cleaning
This activity improves seed lot by separating the seeds of weed and inert matter and eliminating poor-quality seed and off-types. Physical damage to the seeds and loss of good seeds is minimized during seed cleaning, especially when seeds are very dry (Rao *et al.*, 2006).

Seed Quality and Quantity
This measures the percentage of seeds that are alive and can be regenerated into normal plants after germination (Ellis, 1988). The National Seed Storage Laboratory at Fort Collins, Colorado, USA, has shown the loss of seed viability after storing for 50 or more years, e.g. tomato, wheat and sorghum seed lose 0.4%, 0.5% and 1.9% viability per year, respectively. Seed germination tests are therefore done regularly (and before seeds are packed and placed into storage) to ensure they remain viable and vigorous after cleaning

and drying, or at the latest within 12 months after receiving the sample. A simple test consists of placing two layers of absorbent paper in a dish, pouring water and removing any excess, and putting a known number of seeds (400 are often required) on the paper. Seeds are then covered with another wet paper towel and thereafter with another inverted dish or foil, before placing them in a warm place. Germination is checked daily and water is added to conserve moisture when necessary, because seeds should not be left to dry. Seeds for most species often germinate between 2 and 21 days. The initial seed germination threshold is above 85% for most crops and wild species, below 85% if plant establishment during regeneration was adequate, and 70% or less when viability of 85% in newly replenished seed is rarely achievable. It is also worth assessing seed vigour, which refers to the ability of the seed to germinate rapidly with a normal growth rate and resistance to pathogens. A seed loses vigour before it dies completely; i.e. losing its viability.

Seed Health Evaluation

Seed health refers to the status of a seed sample regarding the presence or absence of pathogen-causing organisms and pests (Mezzalama, 2012). Crops may get infected by seed-borne pathogens that are not always visible or easily recognized. Seed-borne inoculums are known to reduce storage longevity and cause poor seed germination or further field establishment. Seed-borne inoculums also promote disease in the field, reducing the value of crops. The exchange of infected seeds may allow spread of pathogens and pests into new regions. A gene bank should therefore ensure that all seeds imported to, or exported from any locations, as well as all seeds stored therein meet international phytosanitary standards to minimize the risk of spreading seed-borne pathogens and pests (Box 12.2). Materials that do not meet health requirements may undergo regeneration but taking the necessary steps to avoid disease recurrence. A Seed Health Unit must check all seed shipments by determining their status regarding any pathogen or pest affecting the cultigen or their wild relatives. A phytosanitary certificate issued by the plant health authority of the country should accompany these seed shipments. Hot water

and fumigation treatments may be used as prescribed by the plant health authority of the recipient country.

Drying and Packaging

Seeds require a final drying because they may take in moisture during cleaning. For some crops such as rice, seeds are placed in a drying room at 15% relative humidity and 15°C for 1 week. It is unnecessary to determine moisture levels of individual accessions following these predetermined conditions. Moisture levels should be determined according to the rules of the International Seed Testing Association (ISTA).

This is done to keep each accession separate and to prevent absorption of water from the surrounding atmosphere after drying. Seeds can be stored in moisture resistant, rustproof aluminium cans for the base collection, while heat re-sealable laminated aluminium foil bags are used for storing seeds of an active collection that will frequently be retrieved and sampled. Small seed packets of the same gene bank accession are often used for sending out to *bona fide* users upon their request.

Accessioning

All data related to seed sample processing should be kept after recording them in each step indicated above. The information includes *inter alia* gene bank accession number, determination methods, weight of stored seed, temperature for drying and storage, times and dates for processing each step, seed amount (number and net dry weight in grams), and moisture content.

Security Duplication

This activity is the duplication of a genetically identical sub-sample of the accession and its related information to mitigate the risk of any partial or total loss caused by natural or human-made catastrophes. Hence, every gene bank accession should be kept in a geographically distant area, under at least the same conditions as those of the original gene bank. These safety duplicates are genetically identical to the long-term base collection, referred to as the secondary most-original-sample and are stored following a

Box 12.2 | **Seed Viability and Germination Tests at CIMMYT and IRRI**

(Taba *et al.*, 2004; IRRI, 2017)

The CIMMYT gene bank maintains seed viability as high as possible during seed storage. The initial germination test of seed lots of new introductions or seed increased in the introduction and regeneration nurseries is conducted *after* the seed is dried to the optimal moisture content (6–8% for maize; 5–7% for wheat). The initial germination test of seed accessions must exceed 90% germination to be stored in the gene bank. All new introductions stored in the gene bank will become new storage units, with the same bank identification number that was given during sample registration. Regeneration may be repeated two to three times to produce sufficient quality seed. In the course of seed preservation for the active collection, if seed viability of the accessions drops below 85%, or the number of seeds falls below 1500, the accession is regenerated or multiplied. The first monitoring of seed viability is conducted after 10 years of storage in the active collection; then, after every 5 years as recommended by the standard gene bank operation guidelines. CIMMYT follows the rule from the International Seed Trade Association (ISTA) to count the seeds that have normal and abnormal germination after 4 days and 7 days to determine per cent of germination.

  The International Rice Research Institute (IRRI) holds in trust 126 782 (as of January 2015) types of rice that include modern and traditional cultivars, and wild *Oryza* species. It is the largest collection of rice genetic diversity in the world. Its International Rice Gene bank stores seed of each rice type in both long-term storage at −20°C (base collection) and an active collection at 2–4°C for distribution. Seed viability for rice declines quickly over months when storing as Asian rice farmers do, e.g. seed viability would be 1% after 5 months if rice seed samples with initial viability of 95% are kept at 30°C with 70° relative humidity and seed moisture content of 13.5%. Rice seeds have the potential to remain viable for many decades in the International Rice Gene bank that does viability testing every 5 years for seed stored in the active collection, or every 10 years for seed stored in the base collection. Seed viability of all accessions is determined prior to long-term storage at −20°C. If viability falls below 85%, a sample of the remaining seeds is planted under lowest pest pressure to produce high-quality fresh seeds for storage. Seeds are thereafter threshed, dried and then hand cleaned. Seeds equilibrate to around 6% moisture content in the seed drying room before being packed in large aluminium foil packets containing about 500 g of seed for storage in the active collection; 10-g samples are stored in aluminium foil packets for germplasm distribution. Seed samples (120 g) in two aluminium cans are stored for the base collection. Their actual seed viability determines the schedule and frequency of further monitoring. The IRRI continually assesses management procedures in the International Rice Gene bank to ensure that the best practices are employed to conserve this vital genetic resource for future generations.

'black box' approach (Rao *et al.*, 2006). Under this arrangement, the repository gene bank has no prerogative to use and distribute this germplasm, which is returned when the original collection is lost or destroyed only on request from the depositor, who is the only one that can withdraw the seeds and open the boxes. The depositor must also ensure that the material to be included in the safety duplication is of high quality, they monitor seed viability over time, and they use their own base collection to regenerate the collections as their seed begin to lose viability. Gene banks for safety duplication need ≥500 viable seeds for outcrossing species and 300 seeds for in-breeding accessions. Initial seed viability should be >85% and kept at 3–7% moisture. The packaging material may be 12 μm of polyester, 30 μm of aluminium foil and an inner layer of high-density polythene of 80 μm thickness. The storage temperatures for safety duplication is between –18 and –20°C.

The ultimate global black box system is that of the Svalbard Global Seed Vault (Box 12.3), which was established as a fail-safe seed storage that will stand the test of time as well as any challenges brought by natural or human-made disasters.

Regeneration

It is a costly operation that should be undertaken when seed viability drops below 85% or the remaining seed is below what is required for three sowings of a representative sample of the accession (Rao *et al.*, 2006). In practice, regeneration happens when the seeds are not enough for long-term storage, e.g. 1500 for a selfing species and 3000 for outcrossing species. Regeneration methods may vary considerably according to the species' reproductive system (selfing or outcrossing), costs (labour, space, time) and principles of population genetics aiming at maximizing the effective population size. For example, pair-wise crossings between a designated number of individuals from one accession will be one way of maximizing the effective population size and maintaining genetic variation for outcrossing species, but their costs are high *vis-à-vis* uncontrolled crossing. The sample size for regeneration should include a minimum number of plants able to capture at least 95% of alleles with a minimum frequency of 0.05, and should consider the breeding systems of the species and the degree of heterogeneity of the accessions. Appropriate

---

Box 12.3 | **Global Seed Vault**

(/www.croptrust.org/our-work/svalbard-global-seed-vault/)

The Svalbard Global Seed Vault in a remote island in the Svalbard archipelago, halfway between mainland Norway and the North Pole, is an example of a secure facility for 'black box' PGR safety duplication. This Vault is located far beyond the Arctic Circle and 130 m deep inside a frozen mountain, whose permafrost provides a suitable environment for long-term secure conservation, which will only be accessed in case of disaster or loss of the samples from the original gene bank. The Vault can hold 4.5 million accessions at storage temperatures of –18°C. Seeds are stored and sealed in custom-made three-ply foil packages, which are wrapped inside boxes and stored on shelves inside the Vault, whose low temperature and moisture levels ensure a low metabolic activity, thus keeping the seeds viable for long periods. Each accession may contain on average 500 seeds, so a maximum of 2.5 billion seeds will be stored in the Vault. There were 860 000 accessions (originating from every country of the world) deposited at the Svalvard Global Seed Vault towards the beginning of 2016.

Box 12.4 | **Potential Duplicates of Nordic Barley Accessions in Gene Banks**

(Lund *et al.*, 2003, 2013)

Redundant duplication among putative Nordic spring barley material held at 12 gene banks worldwide were assessed using passport data and microsatellites spread in the entire barley genome. The similarity of accession name was initially used to partition 174 repatriated accessions into 36 potential duplicate groups, and one group containing 36 apparently unique or unrelated accessions. This partitioning was effective at identifying even accessions with relatively small average genetic distances within potential duplicate groups. Genetic distances based on descriptors that detect genetic heterogeneity show reduced sensitivity compared to microsatellites. Furthermore, grouping ensuing from either may reflect distinct diversity patterns resulting from varying mutation rates and selection intensity, thus complementing each other.

---

isolation methods should be used during regeneration to avoid gene flow, and due care to avert admixtures and ensure genetic integrity. The most efficient and cost-effective way of maintaining genetic integrity will be keeping the frequency of regeneration to a minimum. It is also worth ensuring that the regeneration site mimics the collecting site for crop wild relatives to minimize any genetic drift.

Identifying Duplicated Accessions

Duplicate accessions may be common (when derived from a common original population with all alleles in common), partial (if accessions are from the same original population but only partially have alleles or genotypes in common) and compound (in which all the alleles are included in one of the accessions). Parental duplication also occurs when both parent and offspring are part of the germplasm collection. Putative duplicates may be identified on the basis of passport data, or by characterizing them with a descriptor list in the field or with DNA markers (e.g. see Box 12.4). Bulking duplicate accessions prevents the loss of alleles in case of partial duplication in a gene bank.

12.7.6 Distribution under the IT-PGFRA

The international exchange of plant germplasm of 64 globally important crops falls under the International Treaty on Plant Genetic Resources for Food and Agriculture[1] (IT-PGRFA), whose objectives are the conservation and sustainable use of all plant genetic resources for food and agriculture and the fair and equitable sharing of the benefits arising out of their use, in harmony with the Convention on Biological Diversity, for sustainable agriculture and food security. There were 140 contracting parties to the Treaty (139 states and the European Union) as of December 2016. On ratifying the IT-PGRFA, countries agreed to make available their genetic diversity and related information about the crops stored in their gene banks to all through the Multilateral System (MLS). The crops included in this MLS account together for 80% of all human consumption. Access to plant genetic resources of the MLS is through the collections available in the world's gene banks. A Standard Material Transfer Agreement (SMTA) is used by gene banks in order to send seed samples

---

[1] ftp://ftp.fao.org/ag/cgrfa/it/ITPGRe.pdf

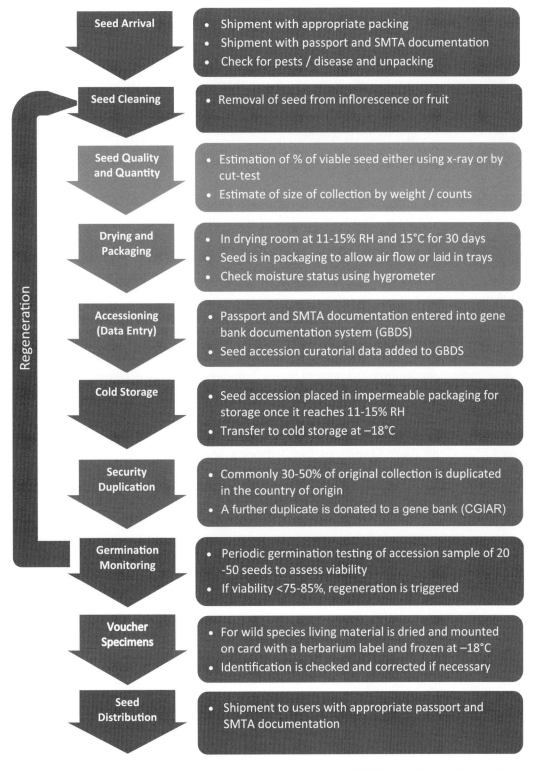

Seed Arrival
- Shipment with appropriate packing
- Shipment with passport and SMTA documentation
- Check for pests / disease and unpacking

Seed Cleaning
- Removal of seed from inflorescence or fruit

Seed Quality and Quantity
- Estimation of % of viable seed either using x-ray or by cut-test
- Estimate of size of collection by weight / counts

Drying and Packaging
- In drying room at 11-15% RH and 15°C for 30 days
- Seed is in packaging to allow air flow or laid in trays
- Check moisture status using hygrometer

Accessioning (Data Entry)
- Passport and SMTA documentation entered into gene bank documentation system (GBDS)
- Seed accession curatorial data added to GBDS

Cold Storage
- Seed accession placed in impermeable packaging for storage once it reaches 11-15% RH
- Transfer to cold storage at −18°C

Security Duplication
- Commonly 30-50% of original collection is duplicated in the country of origin
- A further duplicate is donated to a gene bank (CGIAR)

Germination Monitoring
- Periodic germination testing of accession sample of 20-50 seeds to assess viability
- If viability <75-85%, regeneration is triggered

Voucher Specimens
- For wild species living material is dried and mounted on card with a herbarium label and frozen at −18°C
- Identification is checked and corrected if necessary

Seed Distribution
- Shipment to users with appropriate passport and SMTA documentation

Regeneration

Figure 12.1 Schematic flowchart of seed processing. (A black and white version of this figure will appear in some formats. For the colour version, refer to the plate section.)
(Adapted from Terry *et al.*, 2003)

from desired MLS accessions to *bona fide* users, who must agree to certain terms and conditions for using such germplasm, e.g. accepting to share freely any new developments with others for further research, or paying to a common fund a percentage of any commercial benefits deriving from keeping the developments to themselves. The Governing Body of the IT-PGRFA established in 2008 this benefit sharing fund to support conservation of plant genetic resources and further development of agriculture in the developing world.

The activities involved in processing seed accessions on entry to the gene bank are summarized in Figure 12.1 (adapted from Terry *et al.*, 2003).

12.8 Conclusion

The best approach for preserving *ex situ* genetic resources of most plant species is through seed storage. Although *ex situ* conservation may be regarded as 'freezing' genetic diversity compared to that existing when their seeds were taken during

collecting, it allows an easy access to germplasm for further use by others. Orthodox seeds kept in gene banks tolerate a sizeable decrease in moisture content. However, recalcitrant seeds are unable to tolerate such loss because low internal moisture causes irreversible chemical processes that kill the seed, thus, they are not easily kept by gene banks. Monitoring seed viability is required to determine the need for regeneration of accessions under storage. In this regard, image analysis emerged recently as a promising approach for seed quality testing (Tanwar *et al.*, 2018). Moisture content and temperature affect short- and long-term storage of active and base collections, respectively. Both require drying of seeds that are kept in hermetically sealed and moisture-proof containers. The lifespan of seed under storage of many plant species increases when their moisture content decreases from 12 to 5%. Seeds under mid-term storage may be kept at 5% moisture content at 15°C or below. Orthodox seeds can be stored for up to 100 years at −18°C. The gene bank standards for orthodox seeds (FAO, 2014b) are highlighted in Box 12.5.

Box 12.5 | Gene Bank Standards for Orthodox Seeds

(FAO, 2014b)

1. *Standards for acquisition of germplasm*
   1.1 All seed samples added to the gene bank collection have been acquired legally with relevant technical documentation.
   1.2 Seed collecting should be made as close as possible to the time of maturation and prior to natural seed dispersal, avoiding potential genetic contamination, to ensure maximum seed quality.
   1.3 To maximize seed quality, the period between seed collecting and transfer to a controlled drying environment should be within 3–5 days or as short a time as possible, bearing in mind that seeds should not be exposed to high temperatures and intense light and that some species may have immature seeds that require time after harvest to achieve embryo maturation.
   1.4 All seed samples should be accompanied by at least a minimum of associated data as detailed in the FAO/Bioversity multi-crop passport descriptors: 4.1.5 The minimum number of plants from which seeds should be collected is between 30 and 60 plants, depending on the breeding system of the target species.

Box 12.5 | (cont.)

2. *Standards for drying and storage*

2.1 All seed samples should be dried to equilibrium in a controlled environment of 5–20°C and 10–25% of relative humidity, depending upon species.

2.2 After drying, all seed samples need to be sealed in a suitable airtight container for long-term storage; in some instances where collections that need frequent access to seeds or likely to be depleted well before the predicted time for loss in viability, it is then possible to store seeds in non-airtight containers.

2.3 Most-original-samples and safety duplicate samples should be stored under long-term conditions (base collections) at a temperature of –18 ± 3°C and relative humidity of 15 ± 3%.

2.4 For medium-term conditions (active collection), samples should be stored under refrigeration at 5–10°C and relative humidity of 15 ± 3%.

3. *Standards for seed viability monitoring*

3.1 The initial seed viability test should be conducted after cleaning and drying the accession or at the latest within 12 months after receipt of the sample at the gene bank.

3.2 The initial germination value should exceed 85% for most seeds of cultivated crop species. For some specific accessions and wild and forest species that do not normally reach high levels of germination, a lower percentage could be accepted.

3.3 Viability monitoring test intervals should be set at one-third of the time predicted for viability to fall to 85% of initial viability or lower depending on the species or specific accessions, but no longer than 40 years. If this deterioration period cannot be estimated and accessions are being held in long-term storage at –18°C in hermetically sealed containers, the interval should be 10 years for species expected to be long-lived and 5 years or less for species expected to be short-lived.

3.4 The viability threshold for regeneration or other management decisions such as re-collection should be 85% or lower depending on the species or specific accessions of initial viability.

4. *Standards for regeneration*

4.1 Regeneration should be carried when the viability drops below 85% of the initial viability or when the remaining seed quantity is less than what is required for three sowings of a representative population of the accession. The most-original-sample should be used to regenerate those accessions.

4.2 The regeneration should be carried out in such a manner that the genetic integrity of a given accession is maintained. Species-specific regeneration measures should be taken to prevent admixtures or genetic contamination arising from pollen geneflow that originated from other accessions of the same species or from other species around the regeneration fields.

Box 12.5 | (cont.)

4.3 If possible, at least 50 seeds of the original and the subsequent most-original-samples should be archived in long-term storage for reference purposes.

5. *Standards for characterization*

    5.1 Around 60% of accessions should be characterized within 5 to 7 years of acquisition or during the first regeneration cycle.

    5.2 Characterization should be based on standardized and calibrated measuring formats, and characterization data follow internationally agreed descriptor lists and are made publicly available.

6. *Standards for evaluation*

    6.1 Evaluation data on gene bank accessions should be obtained for traits that are included in internationally agreed crop descriptor lists. They should conform to standardized and calibrated measuring formats.

    6.2 Evaluation data should be obtained for as many accessions as practically possible, through laboratory, greenhouse and/or field analysis as may be applicable.

    6.3 Evaluation trials should be carried out in at least three environmentally diverse locations and data collected over at least 3 years.

7. *Standards for documentation*

    7.1 Passport data of 100% of the accessions should be documented using FAO/Bioversity multi-crop passport descriptors.

    7.2 All data and information generated in the gene bank relating to all aspects of conservation and use of the material should be recorded in a suitably designed database.

8. *Standards for distribution and exchange*

    8.1 Seeds should be distributed in compliance with national laws and relevant international treaties and conventions.

    8.2 Seed samples should be provided with all relevant documents required by recipient country.

    8.3 The time span between receipt of a request for seeds and the dispatch of the seeds should be kept to a minimum.

    8.4 For most species, a sample of a minimum of 30–50 viable seeds should be supplied for accessions with sufficient seeds in stock. For accessions with too little seed at the time of request and in the absence of a suitable alternative accession, samples should be supplied after regeneration/multiplication, based on a renewed request. For some species and some research uses, smaller numbers of seeds should be an acceptable distribution sample size.

9. *Standards for safety duplication*

    9.1 A safety duplicate sample for every original accession should be stored in a geographically distant area, under the same or better conditions than those in the original gene bank.

    9.2 Each safety duplicate sample should be accompanied by relevant associated information.

Box 12.5 | **(cont.)**

10. *Standards for security and personnel*

    10.1 A gene bank should have a risk management strategy in place that includes *inter alia* measures against power cut, fire, flooding and earthquakes.

    10.2 A gene bank should follow the local Occupational Safety and Health requirements and protocols where applicable.

    10.3 A gene bank should employ the requisites to fulfil all the routine responsibilities to ensure that the gene bank can acquire, conserve and distribute germplasm according to the standards.

13 Whole Plant, Plantlet and DNA Conservation

13.1 Introduction

Conservation in field gene banks, plantations, botanic gardens and arboreta are interrelated, because in each case the collected germplasm is maintained as a living collection, often geographically remote from the place of origin. However, there are at least two essential differences between conservation in field gene banks/plantations and botanic gardens/arboreta related to the numbers and focus of species conserved. In the case of field gene banks and plantations, the number of species conserved is restricted, possibly to a single species, and there will normally be many individuals per species; they also focus on conserving crop or forestry species. Conversely, in botanic gardens and arboreta the emphasis is placed on conserving large numbers of species from throughout the plant kingdom, but as a result often only one or a few individuals can be conserved for each species.

13.2 The Problems of Conserving Vegetatively Propagated Crops

The conservation of vegetatively propagated crops, such as potato, cassava, yams, sweet potato, sugar cane, taro and many temperate fruit trees amongst others, presents special problems for *ex situ* conservation (Huamán *et al.*, 1995). Although some of these crops are sexually fertile, it is often not convenient to propagate them commercially from seed because of high levels of heterozygosity, and breeders and horticulturalists commonly require uniform clones (Hawkes *et al.*, 2000). Many vegetatively propagated crops are, however, sexually sterile, or at the very least have reduced fertility, which precludes the possibility of seed storage.

The vegetative organs which are stored, namely tubers, rhizomes, corms, cuttings etc., are relatively short-lived and often deteriorate rapidly after harvest, unless ideal storage conditions are provided, but at best they will only last in storage from one growing season to another. Storage of vegetative organs of tropical and temperate crops is quite different. For instance, the tubers of yams and the tuberous roots of sweet potato store best under conditions of reduced temperature (12–16°C) and 85–90% relative humidity, but at temperatures lower than 12°C there is tissue decomposition (Devereau, 1994). High temperatures, while beneficial in the early stages of storage for hastening the physiological processes of curing during wound healing, also promote tissue deterioration, including increased sprouting. On the other hand, potato tubers can be stored adequately for 5–7 months at 4–5°C and 90% relative humidity (Potato Council, 2017).

In all these cases it is important to realize that all the organs for storage are metabolically active. Their life spans follow closely controlled cycles associated with the natural cycle of the species. Therefore, the storage methods indicated here only represent short-term strategies for the species. Annual regeneration is not without risk, in that germplasm is exposed to both the rigours of weather and disease at the regeneration site, which will result in some genetic erosion and differential loss of genotypes. For longer term storage other methods such as *in vitro* tissue culture or cryopreservation must be adopted.

It is extremely difficult to keep vegetatively propagated plants free from viruses, which represent a major phyto-pathological threat. Infection with virus diseases leads to degeneration of clonal stocks (Hawkes *et al.*, 2000). Only a few virus diseases are transmitted sexually through pollen and true seed, and most can be eliminated by propagation of a crop through true seed. Unfortunately, this is not always possible because of the lack of fertility, nor always desirable because of the need to preserve specific genotypes. One practical way developed to overcome these problems and permit the conservation of vegetatively propagated crops is the use of field gene banks.

13.3 Field Gene Banks and Plantations

The conservation of germplasm in field gene banks or plantations involves the collecting of material from one location and its transfer and cultivation in a second location. Both field gene banks and plantations commonly are species restrictive, each containing one or a limited number of species. Field gene banks have traditionally provided the answer for (a) recalcitrant seeded species (those species whose seeds cannot be dried and frozen without loss of viability), (b) sexually sterile species or (c) those species where it is preferable to store clonal material. Field gene banks are currently used to conserve species such as cocoa, rubber, coconut, mango, coffee, banana, cassava, sweet potato, taro and yam. Plantations are more commonly used for maintaining or conserving tree species by foresters but are not necessarily restricted to recalcitrant species. Field gene banks and plantations can also be established to maintain working collections of living plants for experimental and research purposes, including germplasm evaluation and characterization (see Box 13.2), in addition to basic conservation. There is also a need to link conservation to utilization and therefore field gene banks should ideally be located within the boundaries of the natural diversity of the crop and close to ongoing plant breeding activities to ensure the conserved resource is directly linked to use.

Examples of Field Gene Banks

- The International Cocoa Genebank, Trinidad (ICGT) was established in the 1980 by combining the

various collections held by the Faculty of Agriculture of the University of the West Indies (UWI) and now holds approximately 2400 cocoa accessions, the largest and most diverse in the world. The Cocoa Research Centre (CRC) at the St. Augustine campus of UWI is regarded as a centre of excellence for cocoa research and conservation (Figure 13.1),
- A very comprehensive field gene bank of pomegranate *(Punica granatum)* with a collection of over 760 unique varieties exists at an Agricultural Research Station near Yazd in Iran. The pomegranate plant, a native of the region, has great significance for Iranians as it is deeply embedded in their religious rituals, folklore, culture and pharmacopoeia.

The maintenance of large living collections in field gene banks requires large inputs of labour and land. Likewise, the conservation of fruit trees, usually in the form of orchards, is beset by many problems. These include space considerations associated with the need to conserve an adequate sample of genetic variation. As detailed above, the International Cocoa Genebank, Trinidad has 2400 cocoa accessions and for most accessions has 16 duplicates grown over 37 hectares. The field gene bank is remote from the CRC at the St. Augustine campus of UWI, and the CRC is struggling to maintain such an extensive field collection even though it is an invaluable and unique resource for their research. Even in a wealthy country like the UK, one of the major limitations associated with the UK National Fruit Collection (www.nationalfruitcollection.org.uk/) is the amount of land growing out the collection requires: for 3500 apple, pear, plum, cherry, bush fruit, vine and cob nut accessions 61 hectares are required even if each accession is only duplicated twice. Therefore, space is always a serious limitation when conserving plant genetic diversity in field gene banks. See Box 13.1 for a review of the underutilized crop breadfruit (*Artocarpus altilis* (Parkinson) Fosberg), and Figure 13.1 and Box 13.2 for guidelines on field gene bank establishment, management and germplasm distribution.

Box 13.1 | **The Breadfruit: An Underexploited Crop**

(Ragone, 1997)

The breadfruit (*Artocarpus altilis*) is a crop plant with limited commercial value but large potential and can be regarded as an underutilized or neglected crop. Despite this, it is of considerable importance as a staple crop in many parts of the Pacific and is beginning to be exported from the Caribbean. While its origin is in the Western Pacific, the greatest diversity is found in the eastern Pacific in Polynesia. It has many uses as a food, but primarily its importance is as a staple source of starch from the fresh fruit, which can be preserved before use. In addition, the bark has been traditionally used for making cloth, the leaves are used for wrapping, cooking and serving food, as well as cattle fodder, the flowers are pickled for eating and the sticky latex has many uses including medicinal ones. This is a versatile crop plant which has been subject to considerable genetic erosion, possible in part because most germplasm is held in small independent germplasm collections. The seeds of those cultivars which do produce seeds (some do not!) have no dormancy, germinate immediately and cannot be desiccated. Further details can be found at http://globalbreadfruit.com/our-team/dr-diane-ragone/ and https://ntbg.org/breadfruit/.

Figure 13.1 The International Cocoa Gene bank, Trinidad. (Photograph: N. Maxted)

13.4 *Ex Situ* Living Collections of Wild Plant Species

Most of the developments of *ex situ* conservation techniques has focused on conserving a relatively small number of plant species, primarily crops, of immediate utilization potential. The FAO State of the World Report estimates that there are 30 species which provide 95% of the world's calorie intake (and of these wheat and rice provide 49%), about 120 are important on a national scale and about 7000 in total are used in some way for food and agriculture (FAO, 1998). With limited resources available for conservation and the fact that even in the 21st century there remains widespread starvation in many areas of the world, it should not be surprising that funds and effort has been focused on conserving crops (and their close wild relatives), forage, pasture and some forest tree species.

Seed, *in vitro*, DNA, pollen and field gene bank storage are, however, all just as appropriate for wild species as they are for crop species, but the effort expended on conserving the breadth of plant diversity at the species level or developing specific

Box 13.2 | **Gene Bank Standards for Field Gene Banks**

(FAO, 2014b)

1. *Standards for choice of location of the field gene bank*:
   1.1 The agro-ecological conditions (climate, elevation, soil, drainage) of the field gene bank site should be as similar as possible to the environment where the collected plant materials were normally grown or collected.
   1.2 The site of the field gene bank should be located so as to minimize risks from natural and manmade disasters and hazards such as pests, diseases, animal damage, floods, droughts, fires, snow and freeze damage, volcanoes, hails, thefts or vandals.
   1.3 For those species that are used to produce seeds for distribution, the site of the field gene bank should be located so as to minimize risks of geneflow and contamination from crops or wild populations of the same species to maintain genetic integrity.
   1.4 The site of the field gene bank should have a secured land tenure and should be large enough to allow for future expansion of the collection.
   1.5 The site of the field gene bank should be easily accessible to staff and supplies deliveries and have easy access to water, and adequate facilities for propagation and quarantine.
2. *Standards for acquisition of germplasm*:
   2.1 All germplasm accessions added to the gene bank should be legally acquired, with relevant technical documentation.
   2.2 All material should be accompanied by at least a minimum of associated data as detailed in the FAO/Bioversity International multi-crop passport descriptors.
   2.3 Propagating material should be collected from healthy growing plants whenever possible, and at an adequate maturity stage to be suitable for propagation.
   2.4 The period between collecting, shipping and processing and then transferring to the field gene bank should be as short as possible to prevent loss and deterioration of the material.
   2.5 Samples acquired from other countries or regions within the country should pass through the relevant quarantine process and meet the associated requirements before being incorporated into the field collection.
3. *Standards for establishment of field collections*:
   3.1 A sufficient number of plants should be maintained to capture the genetic diversity within the accession and to ensure the safety of the accession.
   3.2 A field gene bank should have a clear map showing the exact location of each accession in the plot.
   3.3 The appropriate cultivation practices should be followed taking into account micro-environment, planting time, rootstock, watering regime, and pest, disease and weed control.
4. *Standards for field management*:
   4.1 Plants and soil should be regularly monitored for pests and diseases.
   4.2 Appropriate cultivation practices such as fertilization, irrigation, pruning, trellising, rootstock and weeding should be performed to ensure satisfactory plant growth.

Box 13.2 | (cont.)

    4.3 The genetic identity of each accession should be monitored by ensuring proper isolation of accessions wherever appropriate, avoiding inter-growth of accessions, proper labelling and field maps and periodic assessment of identity using morphological or molecular techniques.

5. *Standards for regeneration and propagation*:

    5.1 Each accession in the field collection should be regenerated when the vigour and/or plant numbers have declined to critical levels in order to bring them to original levels and ensure the diversity and genetic integrity is maintained.

    5.2 True-to-type healthy plant material should be used for propagation.

    5.3 Information regarding plant regeneration cycles and procedures including the date, authenticity of accessions, labels and location maps should be properly documented and included in the gene bank information system.

6. *Standards for characterization*:

    6.1 All accessions should be characterized.

    6.2 For each accession, a representative number of plants should be used for characterization.

    6.3 Accessions should be characterized morphologically using internationally used descriptor lists where available. Molecular tools are also important to confirm accession identity and trueness to type.

    6.4 Characterization is based on recording formats as provided in internationally used descriptors.

7. *Standards for evaluation*:

    7.1 Evaluation data on field gene bank accessions should be obtained for traits of interest and in accordance with internationally used descriptor lists where available.

    7.2 The methods/protocols, formats and measurements for evaluation should be properly documented with citations. Data storage standards should be used to guide data collection.

    7.3 Evaluation trials should be replicated (in time and location) as appropriate and based on a sound statistical design.

8. *Standards for documentation*:

    8.1 Passport data for all accessions should be documented using the FAO/Bioversity International multi-crop passport descriptors. In addition, accession information should also include inventory, map and plot location, regeneration, characterization, evaluation, orders, distribution data and user feedback.

    8.2 Field management processes and cultural practices should be recorded and documented.

    8.3 Data from points 8.1. and 8.2 should be stored and changes updated in an appropriate database system and international data standards adopted.

9. *Standards for distribution*:

    9.1 All germplasm should be distributed in compliance with national laws and relevant international treaties and conventions.

Box 13.2 | (cont.)

    9.2 All samples should be accompanied by all relevant documents required by the donor and the recipient country.

    9.3 Associated information should accompany any germplasm being distributed. The minimum information should include an itemized list, with accession identification, number and/or weights of samples, and key passport data.

10. *Standards for security and safety duplication*:

    10.1 A risk management strategy should be implemented and updated as required that addresses physical and biological risks identified in standards.

    10.2 A gene bank should follow the local Occupational Safety and Health (OSH) requirements and protocols.

    10.3 A gene bank should employ the requisite staff to fulfil all routine responsibilities to ensure that the gene bank can acquire, conserve and distribute germplasm according to the standards.

    10.4 Every field gene bank accession should be safety duplicated at least in one more site and/or backed up by an alternative conservation method/ strategy such as *in vitro* or cryopreservation $\pm$ where possible.

conservation protocols for wild species has been relatively low key. The reason for this appears to be almost entirely economic. *Ex situ* conservation has a *real financial cost* compared with the option of a minimal intervention *in situ* conservation in a reserve, and with a limited conservation budget, policy makers have tended to focus their budget on species that humankind can use now rather than in the future. Therefore, historically, the bulk of wild plant species have been conserved *in situ* in various forms of parks or reserves where the conservation cost per unit species is much lower than for *ex situ* forms of conservation.

There are, however, two *ex situ* conservation techniques that have been widely used to conserve the entire range of plant diversity, namely botanic gardens and arboreta. This point is illustrated by the analysis of botanical garden holdings, which show that all have living collections but approximately 7% also have seed banks for wild species, holding about 2000 CWR in Europe alone (FAO, 2010a).

13.5 Botanic Gardens and Arboreta

Botanic gardens and arboreta in the traditional sense are living collections of plants held for public display, educational benefit, economic exploitation, scientific botanical and taxonomic enquiry, and policy development. However, defining what constitutes a botanical garden today is increasingly difficult as many gardens are expanding their roles to include threat assessment, species distribution modelling, *ex situ* gene banking, micro-propagation units and supplying material for reintroduction and *in situ* conservation (Oldfield and Kapos, 2017). Here we discuss their more traditional role as these other facets of their current activities are dealt with in other chapters, so the focus here is on the maintenance and documented living plant collections, maintaining access for the public and provision of educational information about their collections, as well as undertaking botanical and horticultural research. The extended and enhanced role of current botanic gardens is reviewed by Maunder (2008), Donaldson (2009) and Blackmore and Oldfield (2017).

Early botanic gardens were often associated with physic gardens or displays of botanic curiosities. Their initial establishment in the 16th and 17th centuries grew out of the collections of earlier herbalists, who studied the diversity of plants for their medicinal properties. By 1545 there were already gardens in Padua, Florence and Pisa, though similar institutions are thought to have existed earlier in Eastern Asia though the actual dates of their establishment are not recorded. Colonial gardens in the tropics soon followed: the Dutch established a garden in Cape Town in 1694; the French in Mauritius in 1733 and through the 18th century the British established gardens in St. Vincent, Jamaica, Calcutta and Penang (Heywood, 1987). These gardens were often initially established as nurseries or introduction centres for screening of exotic or native species for use in local agriculture or forestry. Consequently, many of these early gardens were established for explicitly economic reasons, rather than education or academic study and certainly not conservation as is often cited as their focus today.

Today there are more than 2500 botanic gardens worldwide, containing more than 1.4 million registered living accessions of approximately 549 000 taxa. Of these 15 689 taxa have been IUCN Threat Assessed and 26 are Extinct in the Wild, 1527 are Critically Endangered, 2584 are Endangered and 3919 are Vulnerable, and of the threatened taxa in botanical gardens seed is available for 3201 accessions (BGCI, 2018a). Between the two FAO State of the World of PGRFA reports (FAO, 1998, 2010) the numbers of global botanic gardens increased by 1000. This in part is thought to reflect growing interest in floristic study in the centres of plant diversity and an acknowledgement of their economic potential for utilization. Although botanic gardens have always been concerned with medicinal, ornamental or agro-forestry species, many gardens report a growing focus on systematic conservation of native wild flora, though even here the focus may be on those taxa most associated with direct socio-economic value or cultural importance, explicitly including CWR taxa (FAO, 2010a).

One recognizable feature of botanic gardens is their living collection, which often displays broad coverage of the plant kingdom. They also often contain specimens of species of localized cultural or economic importance, which may not be conserved *ex situ* elsewhere. Living collections may also be biased towards those species that are ornamentally attractive or of special interest to the public (orchids and carnivorous plants are particularly well represented in many botanical gardens!). Equally, the actual species grown must be horticulturally amenable, in that they must be adaptable to cultivation in the conditions provided by the garden. Some gardens have specialist collections of economically important plants, their wild relatives, medicinal and forest species, and many gardens have sections devoted to the flora of phyto-geographic regions of the world. However, in these cases they may be presented primarily for their educational value for the public, rather than providing adequate genetic conservation of the flora or comprehensive monographic of a specific plant group.

In more recent years, with increased public awareness of environmental and conservation issues, there has been a movement towards increased conservation activity for wild species within almost all botanic gardens. However, with finite resources it is impossible to attempt to conserve all wild species, so botanic gardens have tended to focus on rare or endangered species of either exotic or native origin (Heywood, 1987). In Australia, for example, about a third of the rare native species are found as living collections grown in botanic gardens (Meredith and Richardson, 1991). Chen *et al.* (2017) argue for an increased role for botanic gardens in *in situ* plant conservation, outlining their potential contribution by providing (a) horticultural skills, such as assisted propagation to bulk-up rare populations and the development of sustainable land management practices facilitating buffering and connectivity of protected areas, (b) species distributional modelling skills to aid climate change proofing of existing protected area networks, identifying which sites should be added to networks and which given lower priority, (c) mapping, surveying and prioritizing marginal plant populations in anthropomorphic environments for developing management plans for active conservation and (d) high-quality genetically diverse plants for restoration projects or translocation in assisted migration to help plants avoid the adverse impact of climate change (Heywood, 2011). It is also the case that 15.3% of botanic gardens listed by the BGCI PlantSearch database have natural *in situ* conservation within their garden and many are also responsible for managing external natural habitats, so they are in a good place to contribute to both *ex situ* and *in situ* conservation (Chen *et al.*, 2017).

In terms of CWR conservation the role of botanic gardens should be highlighted. As noted in Chapter 7, Maxted *et al.* (2010) reviewed European *ex situ* seed collections in botanic gardens via the European ENSCONET portal and found CWR taxa accounted for 61.8% of total germplasm holdings in European seed banks, also that the 5756 CWR species included represent about a third of the 17 495 priority CWR species found in Europe. These figures are significantly higher than CWR holdings in agricultural gene banks, but as Maxted *et al.* (2010) concluded, the CWR found in European agricultural gene banks represent those most likely to be used by breeders, rather than attempting to provide breadth of taxonomic coverage as is the aim of botanic garden seed banks.

Having drawn attention to the expanding role of botanic gardens, they are not losing their fundamental role in maintaining living conservation collections (Dosmann, 2006; Crane *et al.*, 2009; Oldfield, 2009, 2010). Cibrian-Jaramillo *et al.* (2013) argue that botanic gardens are an underutilized conservation resource, and their role in supplying planting material for ecological restoration projects should be expanded. They state that the botanical living collection role is to maintain the greatest plant diversity at the least economic and logistic cost and propose a methodology that could be used to assess the efficiency of botanic gardens in this role. Their strategy combines three indicators of relative collection value: information on species threat, genetic representation and the operational costs associated with maintaining genetic representation. Although not yet widely used, such an approach would improve the conservation of botanic garden living collections.

There are elements of current living collection management that could be improved to increase their value. First, due to space limitations the number of accessions per species and number of samples per accession is small, and therefore cannot adequately represent species genetic diversity. Second, many botanic garden accessions are assembled without attention to their genetic diversity; samples are deliberately selected to complement each other say by choosing samples from different ecogeographic zones or even just different countries therefore are not an adequate genetic sample of the diversity found in the species. Third, there is strong practical selection pressure for the living collection to maintain species that grow well within the botanic gardens available, for example, semi-tropical species that will survive in temperate conditions or *vice versa*. Fourth, living collections are poorly managed in terms of avoiding crossing between outcrossing species and so unnecessarily threatens the conserved samples' genetic integrity. In this context it is interesting to note the divergent terminology used between agricultural gene banks and botanic garden seed banks when conserving seed *ex situ*. The different terminology is possibly associated with their relative focus on conservation of genetic diversity both between and within species and species conservation *per se*. As such it may seem at first harsh to criticize botanic gardens for not applying the principles of genetic conservation when that is not their primary goal. However, if their claim is to save plant species, it should be recognized that to conserve plant species requires conservation at the three biodiversity levels, plant habitat, plant species and plant genetic diversity (CBD, 1992).

Lack of genetic diversity in living collections limits the collection's value for genetic conservation, and subsequent reintroduction programmes subject introduced germplasm to a severe genetic 'bottleneck' that could threaten its chance of adequate establishment in the host site. It is a paradox that if botanical gardens are ambitious and aim to conserve the breadth of botanical diversity, they unfortunately limit their possible success in terms of genetic conservation. The number of species that can be conserved will limit the numbers of plants of each species that can be grown because of finite space: the greater the species count, the fewer the number of individuals per species. However, it should be noted that without conservation in living collections many species would now be extinct, both in the wild and in cultivation, so although conservation in living collections can be criticized in terms of efficiency of genetic diversity maintenance, they do provide some security for otherwise threatened species.

In recent years many botanic gardens have moved towards a more statutory role in providing policy advice and technical support to national governments, in part to help meet the national requirement for implementing international policy such as the CBD, ITPGRFA, Nagoya Protocol and UN SDGs. The growing need for an informed policy context and the growth of conservation activities within botanical gardens led in 1987, to the IUCN establishing the Botanical Garden Conservation Secretariat (BGCS) based on coordinating and promoting the role of botanic gardens in conservation. In 1994, BGCS became independent of IUCN as Botanical Garden Conservation International (BGCI) and currently has more than 800 botanic

garden members from more than 100 countries (BGCI, 2018b). BGCI, with its base at the Royal Botanic Gardens, Kew in London, is the world's largest plant conservation network and provides a global voice for all botanic gardens, championing and celebrating their work (www.bgci.org/).

One of the major achievements of BGCI, along with other global conservation agencies, has been the creation of the Global Strategy for Plant Conservation: 2011–2020 (GSPC) (Box 13.3), whose vision is 'of a positive, sustainable future where human activities support the diversity of plant life (including the endurance of plant genetic diversity, survival of plant species and communities and their associated habitats and ecological associations), and where in turn the diversity of plants support and improve our livelihoods and well-being' (Convention on Biological Diversity, 2012). Specifically, it includes 16 Targets of which Target 8 relates specifically to botanic gardens-based conservation (see Chapter 2). An interim assessment of progress in achieving Target 8 reported that thus far only 29% of plant species on the IUCN Red List of Threatened Species are in *ex situ* collections, so there is still some distance to go before

---

Box 13.3   GSPC Objectives and Targets

(Convention on Biological Diversity, 2012)
   The Strategy consists of five objectives:

Objective I: Plant diversity is well understood, documented and recognized
- **Target 1**: An online flora of all known plants.
- **Target 2**: An assessment of the conservation status of all known plant species, as far as possible, to guide conservation action.
- **Target 3**: Information, research and associated outputs, and methods necessary to implement the Strategy developed and shared.

Objective II: Plant diversity is urgently and effectively conserved
- **Target 4**: At least 15% of each ecological region or vegetation type secured through effective management and/or restoration.
- **Target 5**: At least 75% of the most important areas for plant diversity of each ecological region protected with effective management in place for conserving plants and their genetic diversity.

Box 13.3 | (cont.)

- **Target 6**: At least 75% of production lands in each sector managed sustainably, consistent with the conservation of plant diversity.
- **Target 7**: At least 75% of known threatened plant species conserved *in situ*.
- **Target 8**: At least 75% of threatened plant species in *ex situ* collections, preferably in the country of origin, and at least 20% available for recovery and restoration programmes.
- **Target 9**: 70% of the genetic diversity of crops including their wild relatives and other socio-economically valuable plant species conserved, while respecting, preserving and maintaining associated indigenous and local knowledge.
- **Target 10**: Effective management plans in place to prevent new biological invasions and to manage important areas for plant diversity that are invaded.

Objective III: Plant diversity is used in a sustainable and equitable manner
- **Target 11**: No species of wild flora endangered by international trade.
- **Target 12**: All wild-harvested plant-based products sourced sustainably.
- **Target 13**: Indigenous and local knowledge innovations and practices associated with plant resources maintained or increased, as appropriate, to support customary use, sustainable livelihoods, local food security and health care.

Objective IV: Education and awareness about plant diversity, its role in sustainable livelihoods and importance to all life on Earth is promoted
- **Target 14**: The importance of plant diversity and the need for its conservation incorporated into communication, education and public awareness programmes.

Objective V: The capacities and public engagement necessary to implement the Strategy have been developed
- **Target 15**: The number of trained people working with appropriate facilities sufficient according to national needs to achieve the targets of this Strategy.
- **Target 16**: Institutions, networks and partnerships for plant conservation established or strengthened at national, regional and international levels to achieve the targets of this Strategy.

the 75% target is achieved and given that <6% of the 400 000 plant species that exist have been IUCN threat assessed (Sharrock *et al.*, 2014). While O'Donnell and Sharrock (2015) surveyed global seed banks and reported 421 botanic garden or other non-agricultural institutions are involved in seed banking of wild plants in 97 countries, Central Africa, South America and South-East Asia were noted as regions of low wild species seed banking activity. Many are connected through the Millennium Seed Bank Partnership led by the Royal Botanic Gardens, Kew (www.kew.org/wakehurst/attractions/millennium-seed-bank).

O'Donnell and Sharrock (2015) also highlight the differential numbers of wild taxa banked per country (Figure 13.2). Given the obvious value of botanic gardens as centres for conservation, it is a concern that most wild seed is collected in countries remote from plant hotspots and further that botanic gardens are too often located in urban areas of Europe and North America (Figure 7.3). Regions of high floristic

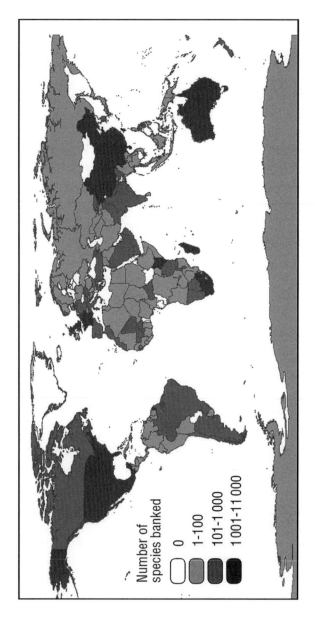

Figure 13.2 Number of wild taxa seed banked per country. (A black and white version of this figure will appear in some formats. For the colour version, refer to the plate section.) (O'Donnell and Sharrock, 2015)

diversity have significantly fewer gardens which are often poorly resourced and lack sufficiently trained staff. To attempt to overcome this problem, at least in part, larger temperate botanic gardens exhibit and attempt to conserve plants from the tropics in expensive to build and maintain glasshouses, where space for growing genetically representative samples of plants of a species is severely limited, when a better global geographic coverage of botanic gardens would provide a more efficient and egalitarian alternative. The important and unique role that botanical gardens can play in the conservation of rare or even otherwise extinct species is illustrated in Box 13.4 for *Sophora toromiro* from Easter Island.

13.6 Tissue Culture Techniques

For many recalcitrant species, whose seed cannot be stored in gene banks at $-18°C$, storing tissue samples *in vitro* offers an alternative to field gene banks. *In vitro* conservation is defined as the maintenance of explants in a sterile, pathogen-free environment and is widely used for vegetatively propagated, intermediate and recalcitrant seeded species (Hawkes *et al.*, 2000). The techniques of tissue culture can be summarized as a sequence of four distinct steps:

a. Mother plant selection and preparation.
b. Establishing an aseptic culture.

Box 13.4 | **The Conservation of *Sophora toromiro*: The Role of Botanic Gardens in Its Conservation**

(Maunder and Culham, 1997)

The Easter Island endemic tree species *Sophora toromiro* is now, using IUCN Red List Categories, 'Extinct in the Wild'. A taxon is regarded as Extinct in the Wild if exhaustive surveys in previous locations or likely habitats have failed to locate any individuals, but the species is still known from specimens held in *ex situ* collections. The last remaining *Sophora toromiro* specimen on Easter Island was chopped down for firewood in 1960. However, the decline of toromiro can be traced to ecological and social changes related to early Polynesian settlement and subsequent land clearance around 400 AD, and the final blow to the species probably came with the introduction of rabbits, sheep, pigs, horses and cattle in 1866.

Current surviving cultivated specimens were initially thought to be found in 18 botanical collections in eight countries, but on closer morphological and documentary study only nine specimens were correctly identified (Figure 13.3). Six of these collections are based on a single pod of the tree, which was collected by the explorer, Thor Heyerdahl, on his 1955–1956 expedition. This material was deposited in Gothenburg Botanical Garden and was subsequently distributed to other gardens. Living plants can also be found in Viña del Mar in Chile, Missouri Botanic Garden in Hawaii and the Royal Botanic Gardens in Melbourne, though the origin of these specimens is not so well documented.

It seems likely that the pod collected by Thor Heyerdahl was probably produced by self-fertilization of the last remaining specimen. Attempts to reintroduce the species have proved unsuccessful and therefore the Toromiro Management Group, a collaborative consortium of botanic gardens, geneticists, foresters and archaeologists, was established to coordinate conservation and reintroduction activities. Molecular screening of accessions using random amplified polymorphic DNA (RAPD) and microsatellites confirmed identifications and

Box 13.4 | (cont.)

Figure 13.3 *Sophora toromiro* (shared under CC BY-SA 4.0 licence).

assessed the genetic relationship between the remaining individuals. The future survival of the species requires breeding between the most genetically diverse individuals, a careful reintroduction programme and the establishment of viable populations. Without the collections held by botanic gardens around the world, this species would now be 'Extinct', and there would be no chance of the toromiro tree returning to its homeland and once again flowering and fruiting on Easter Island.

c. Multiplication or proliferation of the propagules.
d. Preparation and establishment of the propagules for an independent existence by hardening and acclimation.

Aseptic cultures are established within special cabinets called 'laminar flow cabinets', and each propagule or explant is initially surface-sterilized with a solution of bleach. Explants are then placed in a glass or plastic vessel on a special tissue culture medium, the basic components of which are: nutrients (carbohydrates such as sucrose), mineral salts (nitrate, magnesium sulphate, phosphate), vitamins (thiamine or B1), plant growth regulators (auxins and cytokinins) and matrix (aqueous medium or semi-solid agar). The most commonly used and commercially available medium is called Murashige

and Skoog medium. The process of *in vitro* propagation is presented in Figure 13.4. A photograph of an *in vitro* culture collection is shown in Figure 13.5 and a growing banana shoot tip in Figure 13.6. Although these are the generalized steps involved, the basic steps are often amended for the propagation of specific species, including callus medium, meristem and anther/pollen culture, as well as the maintenance of isolated cells in suspension and protoplasts (FAO, 2014b). Recommendations for the physical requirements for a plant tissue culture laboratory and its staffing and maintenance, as well as the establishment of a tissue culture system for a specific taxon are provided in detail by Reed *et al.* (2004).

The application of *in vitro* methods of plant genetic conservation has several supplementary advantages (Ashmore, 1997). During the culturing

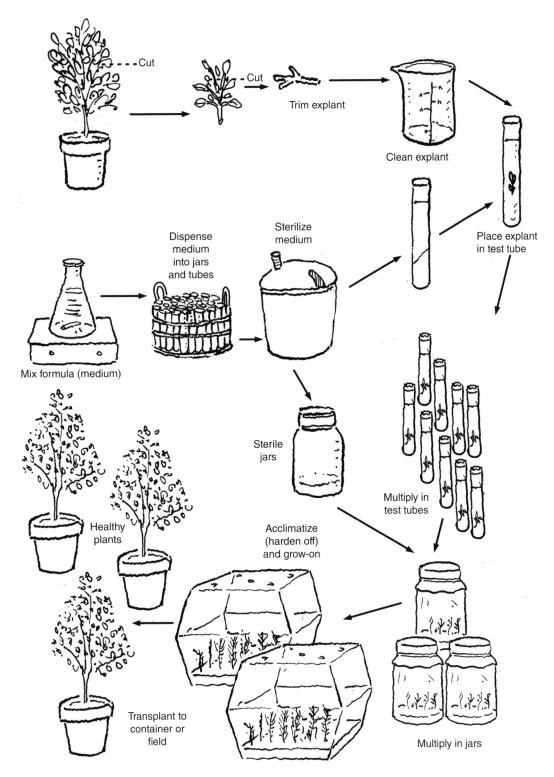

Figure 13.4 Sequence of shoot tip *in vitro* propagation.
(After Kyte and Kleyn, 1996)

Figure 13.5 *In vitro* gene bank.
(Bioversity International)

of vegetatively propagated species, many serious diseases, particularly viruses, can be eliminated and disease-free stocks maintained. This has important implications for germplasm exchange and international plant quarantine. Where the aim is to eliminate systemic pathogens, such as viruses, the choice of 'true' apical meristems (as opposed to larger shoot tips with more than two or three leaf primordia) as the explant is advisable. Such explants are frequently free of viruses even when these have been detected in the parent plants. The chances of obtaining pathogen-free cultures can be increased when the plants are subjected to thermo-therapeutic treatments, which involve growing plants for several weeks at elevated temperatures in the range of 36–40°C. Considerable success has been achieved with crops such as potatoes. *In vitro* cultures can also be extremely useful for exchanging germplasm on a local and international scale. For example, they are now generally considered to be the only acceptable method for the international transfer of banana and plantain germplasm (Heslop-Harrison and Schwarzacher, 2007).

13.6.1 Normal and Minimal Growth in Storage

Tissue cultures can be maintained under normal growth conditions virtually indefinitely provided that nutrients are supplied and accidental contamination is avoided (Kyte and Kleyn, 1996). Such systems where growth and proliferation of cells and tissues are at a high level require frequent sub-culturing to avoid the tissue sample out growing its test-tube or flask; therefore, is not ideal for long-term conservation. Also, in unorganized systems such as callus cultures, mutation rates are likely to be undesirably high because of the high rate of cell division. Minimal growth strategies, in which shoot tip cultures and plantlets from meristems grow at very slow rates, have the greatest application in genetic conservation. Clearly these methods of storage have considerable advantages in that the stored material is readily available for use, can be easily seen to be alive and cultures may be replenished when necessary. Methods used to induce minimal growth may involve the following:

- Alteration of the physical conditions of culture, most often by a reduction of the temperature.

Figure 13.6 Cultured shoot of banana (*Musa acuminata*) in tissue culture storage.
(Bioversity International)

Generally, tropical species do not tolerate as big a temperature reduction as temperate species can. For example, crops such as potatoes, apples, strawberries and grasses can be stored at 0–6°C, while crops such as bananas, cassava and sweet potato are stored within the range 13–20°C, which is still low enough to reduce their growth rates substantially.

- Alterations to the basic medium, by the reduction of some factor essential for normal growth (such as sucrose, the carbon source).
- Use of a growth retardant, such as abscisic acid, or compounds with osmotic effect such as the sugar alcohols, mannitol and sorbitol.

Success is often achieved using not just one, but also a combination of the above methods. These methods are applicable for short- or medium-term storage of germplasm, but for long-term storage the complete elimination of routine sub-culturing is desirable and can only be achieved by cryopreservation.

13.6.2 Cryopreservation

Cryopreservation, the freeze preservation of cultured animal cells (including spermatozoa and ovarian and embryonic tissues), as well as whole animal embryos, has had a successful history of several decades compared with the research on plant tissues. The potential for the long-term storage of plant germplasm using freeze preservation is vast (Hawkes *et al.*, 2000). The process of cryopreservation may be summarized into a series of steps: (i) selection, (ii) preculture, (iii) cryopreservation techniques, (iv) retrieval from storage and (v) seedling or plantlet establishment. These are summarized in Figure 13.7, and cryovats of liquid nitrogen for plant cryopreservation are shown in Figure 13.8.

During the freeze preservation of plant tissue (usually meristems, embryogenic tissue or anthers) there is a complete cessation of cell division; it is important that damage by ice crystal formation within the cells is prevented altogether or at the very least, minimized, as the temperature of the culture is reduced to that of liquid nitrogen (−196°C). In the past there have been two main methods developed to achieve this, namely the ultra-rapid freezing and the slow, stepwise freezing of cultures (Engelmann, 2000; Sakai, 2000). In the former, ice crystals form within the cells, but are extremely small and therefore cause no disruption. The slow freezing method depends for its success on extra cellular freezing for protection. In both systems, however, the tissue is often treated with a cryoprotectant such as DMSO (dimethyl sulphoxide) or glycerol, which reduces damage in several ways. Cryoprotectants can do various things like reduce the size and growth rate of ice crystals and lower the freezing point of intracellular contents and enable cells to be subjected to very low temperatures without disruption of the cell membrane or contents. Most cryo-damage is related to membrane structure and function; in contrast, nucleic acids are very resistant to freezing and thawing. Once frozen, the samples

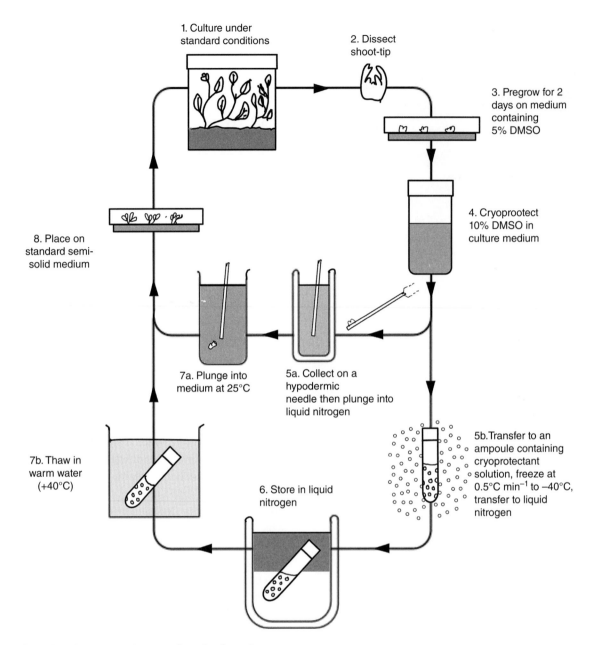

Figure 13.7 Cryopreservation procedures for shoot tips.
(Withers and Williams, Pers. Comm.)

may be transferred to a liquid nitrogen refrigerator for long-term storage.

To access the germplasm the cryopreserved tissue needs to be thawed, and there are a several methods that can be used. Most commonly the cryoprotectants are removed by washing following thawing, but this is now known to be deleterious in some cases

(Engelmann, 2000). The storage of imbibed seeds and zygotic embryos dissected from imbibed seeds is worthy of consideration, since it may have potential application to recalcitrant seeds. Such tissues have a high degree of actual and potential organization and contain both root and shoot meristem regions. Many difficulties have been encountered in the

Figure 13.8 Cryovats (containing liquid nitrogen) are used at National Board for Plant Genetic Resources (NBPGR) (New Delhi) to cryopreserve zygotic embryos. (Photo: B. Ford-Lloyd)

cryopreservation of embryos from plants such as maize and carrot, but this may be that the appropriate protocols have yet to be developed. Embryogenic tissue has been found to be very useful for cryopreserving germplasm of sweet potato (Jenderek and Reed, 2017). Embryogenic cells are formed on axillary buds, and these are encapsulated in alginate gel beads. They can then be 'precultured' on a high-sucrose medium and dehydrated before rapid freezing in liquid nitrogen. However, slow cooling improves tissue survival considerably.

13.6.3 Genetic Stability

In contrast to seed conservation, where the objective is to conserve genes and alleles, the aim of *in vitro* plant genetic conservation is the long-term preservation of genotypes (for instance, of cultivars of vegetatively propagated crop species). Therefore, conservation systems which are prone to genetic change are undesirable. The maintenance of a consistent 'germ-line' through several generations of tissue culture requires an ordered process of chromosome replication in the cells of the different layers of the apical and other meristem regions. However, callus cultures are prone to genetic instability, so-called somaclonal variation, because of their relatively unorganized differentiation. Somaclonal variation is defined as the variation that

arises *de novo* during the period of dedifferentiated cell proliferation that takes place between the culture of an explant and production of regenerants (IPGRI, 1991). Certain tissue culture techniques result in the production of somaclonal variation. Plants regenerated from protoplasts, leaf pieces, leaf petioles or callus may include varying levels of somaclonal variants. Those that are regenerated directly from meristems, shoot tips, axillary or apical buds, however, are normally genetically uniform. Genetic stability is likely to be maintained however where continued organized growth of the meristem in culture follows culture initiation. Somaclonal variation has been observed among regenerated tissue from many species, and its origins and causes have been widely investigated. The mutation types that are likely to be the cause of such phenotypic variation have been described following chromosome breakage and rearrangement during mitosis including changes in chromosome number, single base changes, changes in copy numbers of repeated sequences and alteration in DNA methylation patterns (Larkin and Scowcroft, 1981).

When deciding upon a conservation strategy employing tissue culture systems, the relative advantages of each should be evaluated (Hawkes *et al.*, 2000). A system that provides optimum storage conditions may not be entirely appropriate for plant regeneration after storage, and vice versa. The genetic stability and potential for both quantitative losses of plant material and qualitative changes (genetically) are all factors which will determine which tissue to utilize and which are the most appropriate storage conditions. Undoubtedly cryopreservation offers the best hope for long-term storage. Under such conditions there is only one possible source of genetic variation, namely mutation induced by background radiation, although at ultra-low temperatures, genetic repair processes will not occur, and therefore mutations could accumulate over time.

13.6.4 New and Promising Techniques

Recent research has provided indications of the ways in which germplasm might successfully be preserved on a large scale in the future (Andersson *et al.*, 2006;

Crane *et al.,* 2009; Volk, 2011; Sahoo *et al.,* 2012; Benson *et al.,* 2013; Jenderek and Reed, 2017). These techniques are only just beginning to be applied in gene banks. One such technique involves modification of the gaseous environment of the cultures to reinforce slow growth, by reducing the quantity of oxygen available to the cultures (Buddendorf-Joosten and Woltering, 1994). This can be achieved by covering the explants with paraffin, mineral oil or liquid medium. Another involves desiccation of cultures. Dehydration is followed by storage under reduced relative humidity; the dehydration prolongs the survival of the cultures in storage. If this dehydration is applied to somatic embryos, then they behave like 'real' seed embryos and can be stored for longer periods of time (Sahoo *et al.,* 2012). These are called 'synthetic seeds' if they are coated and protected within an alginate bead. Dehydration may also be used to improve storage of

meristems, which can also be encapsulated in alginate beads (Ray and Bhattacharya, 2008). Furthermore, desiccation can improve the prospects for subsequent cryopreservation.

Vitrification involves firstly the treatment of samples with cryoprotective agents, then dehydration with very concentrated solutions and finally rapid freezing. Following this, intracellular solutes form a glassy state (vitrify), avoiding the formation of intracellular ice crystals, which can reduce cell survival. Vitrification procedures have been developed for use with cell suspensions, somatic embryos and shoot apices (Pâques, 1991). This has been successful with more than 20 different species, including asparagus, citrus, sweet potato, apples, plums, rice and potato. The major drawback is that the cryoprotective mixtures are often toxic in the high concentrations used. See Box 13.5 for guidelines on in vitro culture and cryopreservation.

Box 13.5 | **Gene Bank Standards for *In Vitro* Culture and Cryopreservation**

(FAO, 2014b)

1. *Standards for acquisition of germplasm*:
   1.1 All germplasm accessions added to the gene bank should be legally acquired, with relevant technical documentation.
   1.2 All material should be accompanied by at least a minimum of associated data as detailed in the FAO/Bioversity multi-crop passport descriptors.
   1.3 Only material in good condition and of consistent maturity status should be collected, and the sample size should be large enough to make gene banking a viable proposition.
   1.4 The material should be transported to the gene bank in the shortest possible time and in the best possible conditions.
   1.5 All incoming material should be treated by a surface disinfectant agent to remove all adherent microorganisms and handled so that its physiological status is not altered, in a designated area for reception.
2. *Standards for testing for non-orthodox behaviour and assessment of water content, vigour and viability*:
   2.1 The storage category of the seed should be determined immediately by assessing its response to dehydration.
   2.2 The water content should be determined individually, on separate components of the propagule, and in a sufficient number of plants.

Box 13.5 | (cont.)

    2.3 The vigour and viability should be assessed by means of germination tests and in a sufficient number of individuals.

    2.4 During experimentation, cleaned seed samples should be stored under conditions that do not allow any dehydration or hydration.

3. *Standards for hydrated storage of recalcitrant seeds*:

    3.1 Hydrated storage should be carried out under saturated RH conditions, and seeds should be maintained in airtight containers, at the lowest temperature that they will tolerate without damage.

    3.2 All seeds should be disinfected prior to hydrated storage and infected material should be eliminated.

    3.3 Stored seeds must be inspected and sampled periodically to check if any fungal or bacterial contamination has occurred, and whether there has been any decline in water content and/or vigour and viability.

4. *Standards for* in vitro *culture and slow-growth storage*:

    4.1 Identification of optimal storage conditions for *in vitro* cultures must be determined according to the species.

    4.2 Material for *in vitro* conservation should be maintained as whole plantlets or shoots, or storage organs for species where these are naturally formed.

    4.3 A regular monitoring system for checking the quality of the *in vitro* culture in slow-growth storage, and possible contamination, should be in place.

5. *Standards for cryopreservation*:

    5.1 The explants selected for cryopreservation should be of highest possible quality and allow onward development after excision and cryopreservation.

    5.2 Each step in the cryo-protocol should be tested individually and optimized in terms of vigour and viability in retention of explants.

    5.3 Means should be developed to counteract damaging effects of reactive oxygen species (ROS) at excision and all subsequent manipulations.

    5.4 Following retrieval, explants should be disinfected using standard sterile procedures.

6. *Standards for documentation*:

    6.1 Passport data for all accessions should be documented using the FAO/Bioversity multi-crop passport descriptors. In addition, accession information should also include inventory, orders, distribution and data user feedback.

    6.2 Management data and information generated in the gene bank should be recorded in a suitable database, and characterization and evaluation data (C/E data) should be included when recorded.

    6.3 Data from points 6.1. and 6.2 should be stored and changes updated in an appropriate database system and international data standards adopted.

7. *Standards for distribution and exchange*:

    7.1 All germplasm should be distributed in compliance with national laws and relevant international conventions.

Box 13.5 | (cont.)

7.2 All samples should be accompanied by a complete set of relevant documents required by the donor and the recipient country.

7.3 The supplier and recipient should establish the conditions to transfer the material and should ensure adequate re-establishment of plants from *in vitro*/cryopreserved material.

8. *Standards for security and safety duplication*:

8.1 A risk management strategy should be implemented and updated as required that addresses physical and biological risks identified in standards including issues such as fire, floods and power failures.

8.2 A gene bank should follow the local Occupational Safety and Health requirements and protocols. The cryo-section of a gene bank should adhere to all safety precautions associated with using liquid nitrogen.

8.3 A gene bank employs the requisite staff to fulfil all routine responsibilities to ensure that the gene bank can acquire, conserve and distribute germplasm.

8.4 A safety duplicate sample of every accession should be stored in a geographically distant gene bank under best possible conditions.

8.5 The safety duplicate sample should be accompanied by relevant documentation.

13.7 DNA and Pollen Storage

Storing germplasm by way of DNA is now a reality (Adams, 1997), even though there are some limitations to this as a conservation technique. One limitation is that current technology will only allow for the recovery of single genes, and not whole genomes, genome segments, gene complexes or sets of genes which control quantitative traits (quantitative trait loci or QTL). Also, there is the further problem of the need to identify single genes of interest, for them to be subsequently cloned and transferred to a living plant by genetic engineering techniques. However, the counter rationale for DNA banking is: (a) species, particularly in tropical forests, are becoming extinct, even before being taxonomically described, and DNA extraction and storage is cheap and quick; (b) there is no limitation with recalcitrant seeded species; (c) DNA can be stored in good condition within tissues which are deep frozen, or can be extracted and then stored for long periods of time and (d) genes can currently be

isolated and cloned from extracted DNA, and transferred to a living plant, but the technology is developing so quickly it seems that the reconstruction of plant species from DNA is likely to be feasible. One of the largest global DNA banks is GenBank (Box 13.6).

DNA is often not deliberately sampled for conservation but is an artefact of genomic research, particularly in the age of high-throughput sequencing. Leftover DNA can be stored as a future resource once the research project has finished. It can be maintained at −20°C for short- and mid-term storage (i.e. up to 2 years), and at −70°C or in liquid nitrogen for longer periods (Andersson *et al.*, 2006). However, DNA held in a bank has a relatively short life compared to seed. A life span of 9 years has been seen for coffee, though the lifespan may be improved by storing the DNA on treated cellulose-based cards (Ebert *et al.*, 2006). DNA samples can be easily and relatively cheaply stored and exchanged without the problems of transmitting pests and diseases. They can act as a safety backup and duplicate to other

Box 13.6 | GenBank

(Benson *et al.*, 2013)

GenBank (www.ncbi.nlm.nih.gov) is a comprehensive database that contains publicly available nucleotide sequences for almost 260 000 animal and plant species, which provides retrieval and analysis services via the NCBI home page: www.ncbi.nlm.nih.gov. The included sequences are obtained primarily through submissions from individual laboratories and batch submissions from large-scale sequencing projects, including whole-genome shotgun (WGS) and environmental sampling projects. To help ensure global coverage, GenBank has daily data exchange with the European Nucleotide Archive (ENA) and the DNA Data Bank of Japan (DDBJ). GenBank is accessible through the NCBI Entrez retrieval system, which integrates data from the major DNA and protein sequence databases along with taxonomy, genome, mapping, protein structure and domain information, and the biomedical journal literature via PubMed. BLAST provides sequence similarity searches of GenBank and other sequence databases. Complete bimonthly releases and daily updates of the GenBank database are available at the website.

conservation techniques, but today are widely used as a reference source rather than an active means of germplasm maintenance (Ebert *et al.*, 2006). The problems over DNA storage longevity and the rise of gene editing techniques means, increasingly, it may be viewed that knowledge of the sequence data will be just as valuable as the DNA itself, and the material will in future be conserved *in silico*.

Similarly, pollen samples can be stored as an alternative or additional *ex situ* conservation method, and it is obviously suitable for recalcitrant species. The ease of pollen storage and shipment and the potential for its immediate use provide researchers with increased options when designing breeding programmes (Volk, 2011). However, a recipient female plant of the same species will always need to be available for fertilization with the pollen in the future to utilize the germplasm. In the case of tree field gene banks, it does obviate the need for growing male trees in breeding orchards so reducing the space requirement. It also facilitates wide hybridization across seasonal and geographic limitations, and reduces the coordination required to synchronize flowering and pollen availability for use in crosses

(Bajaj, 1987). Pollen grains are relatively small so can be stored efficiently within small sample sizes, and yet they still maintain genetic diversity. Pollen can also be shipped internationally, often without threat of disease transfer. The disadvantages of storing pollen are: some species produce little pollen and it is difficult to gather; there are few standardized processing or viability testing protocols available; and as with seed, pollen deteriorates with time, but unlike seed it cannot be regenerated, necessitating fresh pollen collection from mother plants (Volk, 2011).

Individual or population samples of pollen can be collected but collecting pollen from one tree captures only that genotype. Namkoong (1981) suggests collecting a population sample from a minimum of 68 trees. Pollen should be harvested soon after anthesis, usually in the morning, but it is often more practical to collect anthers, extract pollen in the laboratory, then process the pollen immediately to ensure maximum potential longevity. Pollen processing varies depending on the species, but for desiccation-tolerant species, pollen is dried to water contents of 0.05 g H_2O g^{-1} dry weight using silica gel at room temperature (Volk, 2011). The pollen of many species can then be stored between 4°C and –20°C for anywhere between a few days to a year. Longer term viability can be achieved by storage at –80°C or LN temperatures (–196°C). Pollen rehydration can be achieved by placing open vials of pollen in 100% humidity environments for 1–4 hours at room temperature (Hanna and Towill, 1995).

Pollen survived for between 10 and 100 times less than seed when stored dry at room temperature: there are plant species with pollen that has a storage life of a few days only. However, high-quality pollen dehydrated to an optimal moisture content and stored at LN temperatures has been documented to store for well over 10 years (Panella *et al.*, 2009). Pollen storage has not proved a widely used method of *in vitro* conservation.

Part IV

Plant Exploitation

14 Plant Uses

14.1 Introduction

If we look at J.C.T. Uphof's amazing *Dictionary of Economic Plants* and delve into just one page at random, say page 95, we find among the entries:

- *Calamus ovoideus*, a palm from Sri Lanka whose young leaves are edible, either raw or cooked;
- *Calmus rotang* from Bengal and other parts of India, whose stems are made into rattan furniture, ropes and baskets.
- *Calandrinia balonenesis* from Australia, which has fruit eaten by both indigenous and settler Australians; and
- *Calanthe mexicana*, an orchid from Central America and the West Indies whose ground petals are effective in stopping nosebleeds!

Uphof's book contains more than five hundred pages and describes another ten thousand unfamiliar plant species known to be useful to at least one human culture. The dictionary lays out a veritable treasure trove of foods, beverages, medicines, fibers, dyes and construction materials far greater than most of us can imagine.

Beattie and Ehrlich (2001)

Plants are used for so many purposes that we often forget how central they are to our everyday lives. These uses are summarized in Box 14.1 and Figure 14.1.

The intimate link between conservation and the use of plants has been stressed throughout this text. People worldwide depend on plants for their livelihoods, whether through direct harvesting of local resources for subsistence use, or through commercial exploitation by selling products locally or trading on international markets. Furthermore, plants through their contribution to biodiversity are equally important in the provision of many ecosystem services that we depend on. The value of plants in terms of plant breeding, and biotechnology, lies in their immediate and future use for improving crops and food security and nutrition. Plant genetic resources conserved *in situ* in genetic reserves, on-farm or in home gardens (see Box 14.2), and *ex situ* in gene banks are therefore critical for the survival of humankind, from the securing of livelihoods and well-being of indigenous and local communities to the diverse sources of genetic materials for plant breeders for the improvement of crops, especially in the face of growing uncertainty with changing climate.

14.2 Plants for Food

One of the most fundamental uses of plant diversity to humankind is in supplying the world's food. Plants, both wild and cultivated, provide food for humans in a wide range of forms, including fruit, vegetables, nuts, seeds, herbs and spices, colourings, beverages and preservatives. They are also important in feeding the animals we increasingly rely on for our own food, by providing forages and fodder the animals eat. It is surprising, however, that despite the vast range of species so far identified, so few are regularly used for food with recent research showing that national global food supplies and diets globally have become much more uniform in composition largely dependent on a few crops, including greater amounts of major oil crops and lesser amounts of regionally important staples (Khoury *et al.*, 2014).

For instance, although estimates vary greatly, it is speculated that: the true number of flowering plants may be more than 400 000 (Willis, 2017); of these, only about 7000 have been cultivated or collected by humans for food at some time in human history (FAO, 1998); of these, less than 200 species have been domesticated for food; and only 30 species are crops of major economic importance, accounting for more than 90% of the world's calorie intake (FAO, 1998).

Box 14.1 | **Uses of Plant Genetic Resources**

(Cook, 1995)

- Food: crop species, wild species, including beverages
- Food additives: including processing agents and other additives used in food and beverage preparation
- Medicines: human and veterinary
- Animal foods: the fodder or forage species eaten by vertebrate and invertebrate animals
- Materials: such as wood, fibres, cork, cane, tannins, latex, resins, gums, waxes, oils
- Poisons
- Environmental and ecological services: these include species that are ornamentals, recreational, hedges, shade plants, windbreaks, soil improvers, plant for regeneration, erosion control, soil fertility, water quality, indicator species (e.g. pollution, underground water)
- Gene donors: plants that contain desirable traits that can be transferred to other species to improve their use

Figure 14.1 Approximate number of plant species by use category.
(From Royal Botanic Gardens, Kew, 2016)

Box 14.2 | **Keeping It Close to Home (1): Home Gardens and Plant Use**

(Pushpakumara *et al.*, 2020)

Home gardens represent an important livelihood strategy for many rural dwellers worldwide. They also are an important element of any strategy to conserve plant diversity. Home gardens represent a dense, multi-storied arrangement of plant diversity at the species and intra-specific level, as well as other compatible biodiversity. Examination of all the different uses of plants described in this chapter highlights that this diversity of utility is well represented in home gardens, that they are spaces where people grow a wide variety of vegetables, fruits, roots and tubers for food as well as fodder and medicinal plants and other economically important plant species providing spices, herbs, gums and resins, canes and oils, beverages, flowers and building materials. In many ways, these are the plants that are important in the day to day lives of rural people (but also in urban and peri-urban locations), and for this very reason they like to keep such resources close to hand, and what better way than in your home garden. Home gardens are important locations for significant intra-specific plant diversity and are often home to important and unique traditional varieties and landraces. In fact, home gardens are often sites of many species, rare and unique varieties and landraces that are not commonly found in larger fields or the wider farming system.

Furthermore, we have come to depend on a dwindling number of commercial cultivars of cereals, roots and tubers, vegetables and fruits in the last hundred years or so. Thousands of heirloom varieties and farmer landraces have disappeared. It is hard to know exactly how many have been lost over the past century, but a study conducted in 1983 by the Rural Advancement Foundation International (RAFI) gave a clue to the scope of the problem. It compared USDA listings of selected vegetable and fruit seed varieties sold by commercial US seed houses in 1903 with those in the US National Seed Storage Laboratory in 1983. The survey, which included 66 crops, found that about 93% of the varieties had gone extinct (Figure 1.9). Not only is this critically important from a genetic diversity perspective, it is also important from a food and nutritional one. More and more research is demonstrating that nutrient composition among different varieties of the same food can differ dramatically (Hunter *et al.*, 2015; Kennedy *et al.*, 2017). For example, sweet potato varieties can vary in their carotenoid content by a factor of 200 or more;

protein content of rice varieties can range from 5% to 14% by weight; and provitamin-A carotenoid content of bananas can be less than 1 μg/100 g for some varieties and as high as 8500 μg/100 g for others. Meaning that the intake of one variety rather than another can make the difference between micronutrient deficiency and micronutrient adequacy among individuals (Burlingame *et al.*, 2009).

At the global level, cereals are the most important single class of commodity utilized by humans for food, providing around 50% of our daily calorific needs, and about 45% of protein. Of these, rice and wheat together provide around 40% of the world supply of both calories and protein. In contrast, meat provides around 15% of the protein, and fishery products only some 6%, or 15% of all animal protein (Groombridge and Jenkins 2002).

14.2.1 Major Food Crops

Although many plant products are still collected from the wild, e.g. gum Arabic (from *Acacia senegal*), which is used as a thickener for food, textiles and paints, or oregano (*Origanum vulgare*) and many other culinary herbs, most of the staple foods and many industrial products are produced on a large agricultural scale. These major crops include plants grown for human or animal food, such as cereals, root and tuber crops, vegetables, fruits and pulses, as well as other crops such as plantation crops, sugar, tea, banana and coffee (FAO, 1998).

Relatively few botanical families account for the world's major domesticated plants; of the 511 plant families currently recognized, only 173 have domesticated representatives (Hawksworth and Kalin-Arroya, 1995). Of these, the two most important families are Poaceae (wheat, barley, rice, maize, etc.) and Fabaceae (faba bean, chickpea, lentil, etc.). Table 14.1 lists the main plant families that account for a high percentage of the world's crops, showing the number of species utilized within each family. It is important to note that these figures vary depending on the source and should be taken as approximate numbers only.

However, when data on food energy supplies are analyzed at a sub-regional level, a greater number of crops emerge as significant. For instance, while cassava (*Manihot esculenta*) accounts for 1.6% of plant-derived energy at global level, it supplies more than 50% of plant-derived energy in Central Africa. Pulses (see Box 14.3) plantain and banana (*Musa* spp.), groundnut (*Arachis*), pigeon pea (*Cajanus*), cowpea (*Vigna*), yam (*Dioscorea*) and taro (*Colocasia esculenta*) are examples of other crops that comprise the staple diet of millions of the world's poorer people (FAO, 1998). Most of these crop species are included in Annex 1 of the Treaty (along with important animal forages) and are covered under the multi-lateral system (see Chapter 2).

14.2.2 Neglected and Underutilized Species

The term neglected and underutilized species, or NUS, includes both cultivated and wild species (as well as species with other important non-food properties). It is difficult, if not impossible, to define NUS to everyone's satisfaction. In addition

Table 14.1 **Number of cultivated species per family**

Plant family	Number of species
Poaceae	379
Fabaceae	337
Rosaceae	158
Solanaceae	115
Compositae	86
Cucurbitaceae	53
Labiatae	52
Rutaceae	44
Cruciferae	43
Umbelliferae	41
Chenopodiaceae	34
Zingiberaceae	31
Palmae	30

From Hawksworth and Kalin-Arroya (1995).

to being called neglected and underutilized, they are also referred to as 'orphan', 'minor', 'promising', 'niche', 'potential' and 'traditional' species to name a few options. Oftentimes they are also promoted as superfoods or wonder foods (see Box 14.4).

What they do seem to have in common is that they are locally important and are often well adapted to marginal and risk-prone environments so can grow well in environments which other crops, especially exotics, might find difficult, like many of the pulses highlighted earlier (Frison *et al.*, 2011; Lin, 2011). They are species that have often been selected because they have many traits which are important for traditional diets such as relating to taste, texture and appearance as well as local livelihoods and food security (Box 14.5). Importantly, a growing number of studies are demonstrating they are often nutritionally superior to their counterpart exotic species (Fanzo *et al.*, 2013; Hunter *et al.*, 2015).

Box 14.3 | **Finger on the Pulse**

(FAO website: www.fao.org/pulses-2016/about/en/)

Pulses are considered so important for food security and improving nutrition that the UN General Assembly declared 2016 the International Year of Pulses. They are the crops domesticated from the legume family for their large grains and include in the tropics *Arachis, Cajanus, Phaseolus, Vigna, Glycine* and *Psophocarpus* spp.; and in the temperate world *Lupinus, Cicer, Vicia, Lens, Lathyrus, Pisum* spp.

   The Food and Agriculture Organization of the United Nations (FAO) in collaboration with Governments was nominated to lead the charge on implementation of the Year and to improve understanding of the diversity of pulses which exist, especially awareness of their nutritional benefits as part of sustainable food production and healthy diets and the contribution they make to food security and combating malnutrition. Pulses really do pack a nutritional punch and are naturally rich in quality protein and key minerals such as iron and zinc. Pulses are also abundant in B vitamins and an excellent source of complex carbohydrates and fibre. Pulses such as lentils, beans, peas and chickpeas are a critical part of the general food basket in many countries providing a vital source of plant-based proteins and amino acids for people as well as animals.

   Pulses also provide many additional benefits. As leguminous plants, they have critically important nitrogen-fixing properties, which contribute to improving soil fertility and have a positive impact on the environment. Therefore, they are important as inter- and cover-crops and in agricultural crop rotations. As an example of the diversity which exists among pulses, taking common beans alone the International Center of Tropical Agriculture (CIAT) in Colombia holds more than 36 000 accessions on its database. Pulses have many other noble attributes. They require less water to grow than many other crops, and many are suited to marginal environments and may be increasingly important under changing climate. They also have a positive effect on soil biodiversity and can enhance soil microbial biomass.

There have been limited efforts at developing value chains specifically for neglected and underutilized crops even though many of these crops or species can fetch higher market prices than their exotic counterparts with evidence to indicate that a more equitable share of the profit can go to smallholders (Weinberger and Pichop, 2009). This can be an important incentive for farmers and communities to grow and conserve a rich biodiversity of locally important species and combined with the highlighted nutritional and health benefits of much NUS is often the stimulus necessary to create demand and production. Various organizations working with local partners in Kenya and in collaboration with Uchumi Supermarkets have been able to strengthen market linkages for communities and farmers who produce neglected indigenous African leafy vegetables. Results have been quite astonishing with a growth in sales of more than 1100% in just 2 years and networks of more than 300 growers linked to urban markets (Gotor and Irungu, 2010). A total of about 210 African leafy vegetable species have been recorded in Kenya, including *Cleome gynandra, Solanum villosum, Cucurbita moschata, Vigna unguiculata, Amaranthus*

Box 14.4 | SuperFoods, Wonder Foods? Just Eat Food, Mostly Plants

(Agricultural Biodiversity: http://agro.biodiver.se/2016/04/how-to-get-over-your-quinoa-guilt-trip-kinda/; Spinney, 2014)

All too often neglected and underutilized species have been marketed as the next superfood; recall quinoa during 2014 being the focus of the UN International Year, or other wonder foods; think breadfruit. Do such food fads, promotions and trends that reduce what we eat to a focus on single foods or even nutrients really do much for the promotion of plant diversity, or for the diversity and sustainability of our food systems and diets for that matter? Or do such approaches need serious debunking? Take the quinoa example. While there were initial fears that the boom in consumption and exports of quinoa would push up prices and impact on poor Latin American farmers and nutrition, it now seems that such claims that rising quinoa prices were hurting those who had traditionally produced and consumed were false. However, it is often other consequences that result from such booms and fads that might be more troublesome, such as sustainability issues. In the case of quinoa this has come to pass. Firstly, the boom in export markets focused on very few of the 3000 or so cultivars of quinoa. So, what happens to all those other neglected cultivars, which hold the future to further adaptation of quinoa as environmental conditions continue to change? Secondly, the sustainability of quinoa growing in the high Andes is in doubt because more intensive practices are resulting in more and more soil erosion and degradation. Maybe we should be concentrating our efforts on promoting the importance of a wider agro-biodiversity and plant diversity for our agriculture, food systems and diets and take a leaf out of Michael Pollan's book(s) and just eat food, mostly plants. Just add a dash of diversity and you are good to go. It does not seem like rocket science.

blitum, Corchorus olitorius, Solanum scabrum, Crotalaria ochroleuca, Crotalaria brevidens and *Brassica carinata*, but only a small proportion have been researched or exploited to date (Box 14.6).

A recent survey assessed the current context and potential for underutilized crops in Africa, the American continent, Asia and the Pacific and the Near East, and found that more than 250 crops were highlighted; fruits had the most potential in three of the regions surveyed, closely followed by vegetables. The survey reported on various initiatives underway for expanding market opportunities, including strengthening cooperation among producers, street fairs, organic farming, niche variety registration systems, school initiatives and product labelling initiatives (FAO, 2010a).

As part of its programme of work on agricultural biodiversity, the CBD has developed a Cross-cutting Initiative on Biodiversity for Food and Nutrition at COP8 in Brazil in 2006. This enables collaboration between the CBD, FAO and Bioversity International and other partners to better promote the nutritional value of locally important, but neglected food species. One global project associated with this initiative is the multi-country Global Environment Facility (GEF) funded project, *Biodiversity for Food and Nutrition* (www.b4fn.org) that supports sustainable biodiversity conservation and use for improved human nutrition and well-being by enabling planners and practitioners from agriculture, health and environment sectors to work together to mainstream locally important but neglected food crops and wild species into local,

---

Box 14.5 | What Are Neglected and Underutilized Species?

(Crops for the Future: www.cropsforthefuture.org/)

Neglected and Underutilized Species (NUS) are plant species that for some reason or other are not reaching their full utility potential yet have most or all of the following attributes:

- Unrealized potential for contributing to human welfare, in particular to:
  - income generation for the world's poor,
  - food security and nutrition,
  - reduction of 'hidden hunger' (caused by the micronutrient deficiencies resulting from uniform diets that rely on a limited number of food sources).
- Strongly linked to the cultural heritage of their places of origin, or of places to which they have been introduced in the distant past.
- Long history of mainly local production or wild species whose distribution, biology, cultivation and uses are poorly documented.
- Adaptation to specific agro-ecological niches and marginal land.
- Weak or no formal seed supply systems.
- Much intra-specific diversity (landraces).
- Traditional and diverse uses and processing that vary locally.
- Presence in traditional production systems with little or no external inputs, or collected from the wild.
- Receive little attention from research, extension services, farmers, policy and decision-makers, donors, technology providers and consumer.
- Nutritional, culinary, medicinal or other properties that are little-known or underappreciated.

A comprehensive list of species considered to be neglected and underutilized can be found on the Crops for the Future (CFF) website. The CFF website also contains downloadable key publications on NUS as a comprehensive set of online resources on useful plants.

---

national and global nutrition, food, and livelihood security strategies and programmes (Hunter *et al.*, 2016). As part of this project, the countries involved prioritized more than 150 plant species with nutritional potential with the intention of mainstreaming this biodiversity into various national policies and programmes such as public procurement and school feeding.

Global attention on the potential of neglected and underutilized crops seems to be on the rise. The recent IAASTD (2008) report acknowledges that greater investment in agricultural knowledge can result in enhanced sustainable productivity of subsistence foods including orphan and neglected crops upon which much of the rural poor depend (Heywood, 2013). Foley and colleagues (2011) in discussing solutions to the challenge facing global food production highlight that 'significant opportunities may also exist to improve yield and the resilience of cropping systems by improving *orphan crops* and preserving crop diversity, which have received little investment to date'. A comprehensive survey of key neglected and underutilized crops important for improving food and nutrition can be found in the

---

Box 14.6 | **Super Vegetables**

(Cernansky, 2015)

Long neglected in many parts of Africa it is not only the supermarket shelves where African leafy vegetables are starting to reappear. If you walk into a restaurant or hotel in Nairobi, increasingly you are likely to see a steaming plate of African nightshade leaves or a delicious tasty stew containing amaranth being passed back and forth. This return to the supermarket shelves and menus of cafes and restaurants is very welcome news for agricultural researchers and extension staff as well as others including nutritionists. Like the pulses mentioned before, African leafy vegetables have many similar attributes that make then valuable for consumption and combating malnutrition – they are generally rich in calcium and folate as well as vitamins A, C and E – but also, they are much hardier and better adapted to local environments or the risk-prone and marginal environments where they are cultivated. The demand for these vegetables means that researchers are increasingly focusing their efforts on them and seed companies are increasingly looking at ways to improve the quality of seed supply. But with more than 2000 plants that can be considered as leafy vegetables worldwide, prioritization is an issue given the limited budgets of national and international research organizations working on them.

---

recent book, *Diversifying Food and Diets: Using Agricultural Biodiversity to Improve Nutrition and Health* (Fanzo *et al.*, 2013).

14.2.3 Wild Food Species

Wild plant foods take a variety of forms and range from fruits, leafy vegetables, woody foliage, bulbs and tubers, cereals and grains, nuts and kernels, saps and gums, mushrooms and seaweeds. Often underestimated in terms of value they are important to households and communities in many ways and consumed widely by millions of people, in both developed and developing countries. Wild plants make an important contribution to food production and security in many agro-ecosystems worldwide and more than 10 millennia after the shift from foraging to settled agriculture, millions of rural smallholders in most geographic regions of the world continue to be reliant on wild products from foraging forests and wild lands for their subsistence and livelihoods (Wunder, 2014).

Wild plants add important diversity to peoples' diets and enhance nutrition and health. They are particularly important in providing the micronutrients that other staple foods may not. Thus, they contribute significantly to combating the scourge of 'hidden hunger' or micronutrient deficiency. Wild edible plant species are also important for income generation and livelihoods strategies and are frequently sold in both informal and formal markets. Wild plants also provide a key coping strategy for many rural households. They are particularly important during the 'hungry season' or 'lean season', when food in stores from the previous cropping season begin to decline, and the forthcoming season's crops have not yet matured (Hunter *et al.*, 2015; Kennedy *et al.*, 2017). Such periods can be crippling for individuals and households typified by hunger, declining calorific intake and a mundane and monotonous diet characterized by low diversity of foods in the diet. Wild food plants also help households and rural communities deal with shocks, such as crop failure due to pest and diseases or drought or other natural disasters such as cyclones or loss of regular cash income. In fact, they are known to be particularly important in households dealing with serious illness or death of the family breadwinner because of HIV/AIDS. Due to the dearth of information and data on wild plant distribution, harvesting and consumption, it is often difficult to put in place programmes that support sustainable management and utilization (Hunter *et al.*, 2015; Kennedy *et al.*, 2017).

The FAO estimates around 'one billion people use wild foods in their diet', and forests alone provide livelihoods and food for around 300 million people in the form of non-timber forest products (NTFP). Despite their immense value and contribution to the 'global food basket', such wild foods continue to be absent from official statistics on economic values of natural resources (Bharucha and Pretty, 2010). A recent survey by these authors summarizing information from 36 studies in 22 countries highlights that wild biodiversity still plays an important role in local contexts with around 90–100 wild species being used per place and community group, and in some instances individual country estimates of wild food utilization are reaching 300–800 species. The mean use of wild species was found to be 120 per community for indigenous communities in both industrialized and developing countries. The survey also found that a great many of these species are actively managed indicating that the distinction between cultivated and wild is not always clear-cut. Like much of plant diversity, the existence and availability of these sources of wild food is declining dramatically as their habitats come under increasing pressure from development activities including agricultural and industrial expansion, urbanization, over-harvesting and the exclusion of communities from conservation areas.

The Bharucha and Pretty (2010) study highlights the following: in India, at least 600 plant species are known to be important for food and many of these often have multiple uses in addition to providing food; in Nepal, 80% of 62 wild food plants were also found to have multiple uses; and in Tanzania, Batemi agro-pastoralists use 31 species as food, 6 as thirst quenchers, 7 for chewing, 2 as flavourings and 1 for honey beer. A further 35 wild edible plants were found

to be cultivated in Tanzania (Johns *et al.*, 1996). In terms of the nutritional value, wild species can be equally as important as their NUS counterparts described earlier in this chapter in providing much-needed nutrients in times and places of scarcity. To date, several studies have found that wild foods are important sources of micronutrients, although their energy-density is generally low (Bharucha and Pretty, 2010). In the Sahel, several edible desert plants are sources of essential fatty acids, iron, zinc and calcium. In the arid Ferlo region of Senegal, some 50% of all plants have edible parts, and those that are commonly consumed are critical suppliers of vitamins A, B2 and C, especially during seasonal lean periods. Among the plants used by the Fulani in Nigeria, those available during the dry season (and thus important for ensuring year-round nutritional security in the face of possible food shortages) were superior in energy and micronutrient content compared with those from the wet season (Bharucha and Pretty, 2010).

In a separate review study in Brazil, the indigenous wild leafy vegetables *caruru*, *mentruz*, *taioba*, *serralha* and *beldroega* were analyzed for carotenoid content and found to be a richer source than commercially produced leafy vegetables (Kobori and Rodriguez Amaya, 2008). Twenty percent of harvested food comes from the wild in Ghana. In Kenya, around 100 wild species are collected, some purposely managed, near homesteads. Native Americans in southwest USA have a variety of uses for about 350 wild plants (Pretty, 2007). Kuhnlein and Turner (1991) provide a comprehensive study of the nutritional properties of wild plants used by indigenous peoples in Canada. Turner and colleagues (2011) have also recently carried out a comprehensive review of the use of edible and tended wild plants. Wild mushrooms and seaweeds are traditionally used by various cultures, and a few species are produced commercially (Vaughan and Geissler, 2009). Globally it is estimated that more than 1000 species of wild fungi are consumed as important sources of protein and income though many are commercially produced. The wild field mushroom, *Agaricus campestris*, is commonly collected in Britain. Some seaweeds are often used as foods, as a garnish or seasoning or in salads and soups. In Japan about 50 species from

29 genera are used. Generally, they contain little protein or fat and no starch but can be high in iodine. They do contain carotene, vitamins E and C, and a trace of B vitamins. Dulse, *Palmaria palmate*, is commonly used as a snack food, while laver seaweed, *Porphyra umbilicalis*, is commonly used in the production of laver bread (Vaughan and Geissler, 2009).

It is only recently that there has been a decline in the collection and use (and associated knowledge) of wild plants in many developed countries, though countries like Italy and Turkey still maintain strong food cultures based on foraging for wild plants and foods. In order to understand the importance of wild foods in Europe, Schulp *et al.* (2014) recently analyzed the availability, utilization and benefits of wild animals, wild plants and mushrooms in Europe. They recorded 81 species of vascular plants and 27 species of mushrooms still collected and consumed throughout the EU. It has also been observed that during the recent economic downturn, which affected most European countries, there was growing interest in food and cooking in the media associated with foraging for wild plant foods (Box 14.7). Richard Mabey's landmark book, *Food for Free*, first published in 1972 remains among the definitive texts on the subject, describing more than 200 types of food that can be collected from the wild in Britain alone (Mabey, 1972).

Forests and their non-timber forest products (NTFP), either through direct or indirect provisioning for human nutrition, contribute substantially to food security, particularly in developing countries (Vinceti *et al.*, 2013; Hunter *et al.*, 2015). Wild foods from forests, including products from trees, herbs, mushrooms and animals, contribute in many ways to improving food security by providing ready accessibility to affordable and often highly nutritious food (Figure 14.2). These foods are often considered a component of the hidden harvest because their use and consumption is difficult to capture in statistics and databases yet are considered essential for combating malnutrition (see Box 14.8).

Many indigenous fruits are nutrient-rich with high vitamin and mineral contents (see Table 14.2). For example, a young child eating 40–100 g of *Grewia*

tenax (Forrsk.) Fiori berries receives almost 100% of their daily iron requirements. Tamarind's (*Tamarindus indica* L.) and baobab's (*Adansonia digitata* L.) high sugar content make them important sources of energy (Box 14.9). Further, the fruits of *Dacryodes edulis* (G. Don) H.J. Lam, and the seeds of *Irvingia gabonensis* (Aubrey-Lecomte ex O'Rorke) Baill., *Sclerocarya caffra* Sond. and *Ricinodendron rautanenii* Schinz are rich in fat, which is comparable or higher than that of peanuts.

While there are relatively very few rural communities in the world that currently rely on forest foods to provide their complete diet, they can help maintain household nutrition during the lean or hunger seasons and at times of low agricultural production, periods of climate-induced vulnerability and food gaps due to other cyclical events. For this purpose, researchers developed seasonal calendars (Figure 14.3) of indigenous fruit tree species which allow the continuous availability of ripe fruits from at least two species per month and which could be used to meet specific nutrient requirements (Vinceti *et al.*, 2013). The diversity, value and role of forests and their non-timber forest products has been comprehensively reviewed by Ingram *et al.* (2017) and Boshier *et al.* (2017).

The use of wild plants is still particularly prevalent in most indigenous communities and forms part of indigenous knowledge systems and practices that have been developed over many generations and which play an important part in decision-making in local agriculture, food production, human and animal health and management of natural resources (Slikkerveer, 1994). Indigenous peoples' food systems and cultures are good examples of the complexity and remarkable diversity of food availability and utilization demonstrating an intimate awareness and knowledge of wild plants and their multiple uses. Furthermore, many traditional communities often actively manage the wild resources that they use, and in many cases, customary methods of resource management may be fundamental to the sustainable conservation of habitats and species. For centuries, indigenous peoples have been the custodians of most of the planet's food and genetic resources, and stewards of the diverse ecosystems and cultures that

Box 14.7 | **Food for Free, but Not a Free Lunch**

Since the late 1980s, wild plants have also begun to make a perceptible impact on the commercial food business. In France, which has rather more materials at its disposal, there is a distinct school known as *cuisine sauvage*. And in Britain it is now not uncommon to find samphire, nettles, dandelions, bitter-cress, borage, wild strawberries, bilberries and ramsons (*Allium ursinum*) served in some form or other in both smart metropolitan restaurants and local pubs. Perhaps most encouraging is the fact that they are no longer regarded simply as rough peasant foods but are being used as ingredients for modern styles of cooking: wild herbs and fruits flavouring oils and vinegars; spring greens – garlic mustard, nettle, sorrel – stir-fried; flowers added to salads.

(Mabey, 1996)

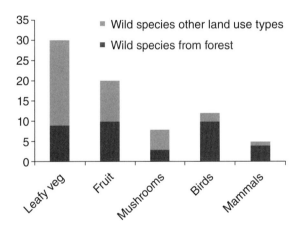

Figure 14.2 Total number of wild species from the forest and other land use types consumed by households surveyed in the East Usambara Mountains, Tanzania.
(From Powell et al., 2011)

have shaped these resources. Indigenous peoples' food systems, often rich in wild foods, have often provided for healthy and resilient diets and prior to colonization and development have ensured food security and nutrition (Hunter *et al.*, 2015).

Sadly, the connection of indigenous people to their land and traditional ways has been adversely affected by the forces of globalization and development to the extent that their food systems, diets and health have been decimated. Irrespective of geography, indigenous people are often the most disenfranchised,

marginalized and poorest members of wider society, and they are targeted by most governments for health improvement and development. Such development often leads to dietary change, including increased reliance on 'market foods', which are often highly processed and contribute to increased risk of chronic disease, including obesity and diabetes. This reduced reliance on traditional foods, especially wild-collected foods, has also led to an erosion of traditional food resources and associated indigenous knowledge. With obvious outcomes for food security, this has significantly affected the welfare, vulnerability and marginalization of indigenous communities (Hunter *et al.*, 2015).

Many of these devastating impacts could be moderated with increased attention to the principles of diet and health already contained within the culture, and with the recognition of the nutrient properties of traditional food resources, which includes a vast repository of wild plant resources, and how these foods can be used to best advantage for health promotion (Egeland and Harrison, 2013). To that end, many groups and organizations such as the indigenous Partnership for Agro-biodiversity and Food Sovereignty and the Centre for indigenous Peoples' Nutrition and Environment (CINE) at McGill University are working towards raising awareness about the cultural and nutritional value of indigenous peoples' foods. CINE has developed a procedure for

---

Box 14.8 | Hidden Harvest, Hidden Hunger

(Ickowitz *et al.*, 2016)

Hidden hunger, or micronutrient deficiencies of required vitamins and minerals, remains a serious problem in many parts of the world stifling physical and mental development and preventing many individuals from reaching their full potential. It is estimated that about 2 billion people worldwide suffer from such mineral and vitamin deficiencies. Micronutrient deficiencies and their impacts remain a serious problem in Indonesia with approximately 100 million people, or 40% of the population, suffering from one or more micronutrient deficiencies. It is thought that forests, trees and NTFP (often referred to as the Hidden Harvest) in rural areas with poor infrastructure and market access may be a critical source of nutritious foods. With the massive destruction and loss at unprecedented rates of forests and other tree-based systems in Indonesia, establishing such a relationship is both critical and urgent. As we have seen, forest foods including fruits and vegetables are generally rich in many micronutrients, low in fat and high in fibre, while other wild foods such as bushmeat, fish and insects are considered excellent sources of bioavailable micronutrients. Agriculture, in contrast, with a focus on the production of a handful of staple crops struggles to provide a balanced diet with enough micronutrients. Furthermore, forest foods have been documented to contribute to enhanced food security and nutrition in many regions and countries all over the world. Results from a recent study in Indonesia show that different tree-dominated land classes were indeed associated with the dietary quality of people living within them in the provinces where they were dominant. Areas of swidden/agroforestry, natural forest, timber and agricultural tree crop plantations were all found to be associated with more frequent consumption of food groups rich in micronutrients in the areas where these were important land classes. In particular, the swidden/agroforestry land class was the landscape which was found to be associated with more frequent consumption of the largest number of micronutrient-rich food groups.

---

documenting indigenous people's food systems, both wild and domesticated. Case studies covering 12 indigenous communities in different global regions including Ainu (Japan), Awajun (Peru), Baffin Inuit (Canada), Igbo (Nigeria), Ingano (Colombia), Karen (Thailand), Maasai (Kenya), Nuxalk (Canada) and Pohnpei (Federated States of Micronesia) clearly demonstrate the importance of wild plant foods. For example, the Ingano and Dalit food systems revealed diets highly reliant on wild plant foods (Kuhnlein *et al.*, 2009).

14.3 Plants for Medicine

Although a comprehensive list of all plant species used as medicines does not exist, we do know for sure that many thousands of plants have been used or are being used as medicines in various home remedies, ritual cures, poisons and tonics. In fact, the earliest texts were often 'Herbals', descriptions of plants and their medicinal uses (Brunsfels, 1530; Bock, 1539; Turner 1551; Gerard, 1597). Today these have been replaced by comprehensive pharmacopoeias that

Table 14.2 **Nutrient content of some African indigenous and exotic fruits per 100 g edible portion (high values are highlighted in bold)**

Species	Energy (Kcal)	Protein (g)	Vit C (mg)	Vit A (RE) (µg)	Iron (mg)	Calcium (mg)
Indigenous fruits						
Adansonia digitata L.	340	3.1	150–500	0.03–0.06	1.7	360
Grewia tenax (Forrsk.) Fiori	NA	3.6	NA	NA	**7.4–20.8**	**610**
Sclerocarya birrea Hochst.	225	0.5	68–200	**0.035**	0.1	6
Tamarindus indica L.	270	**4.8**	3–9	0.01–0.06	0.7	260
Ziziphus mauritiana Lam.	21	1.2	70–165	0.07	1.0	40
Exotic fruits						
Guava (*Psidium guajava* L.)	68	2.6	**228.3**	0.031	0.3	18
Mango (*Mangifera indica* L.)	65	0.5	27.7	0.038	0.1	10
Orange (*Citrus sinensis* L. Osbeck)	47	0.9	53.0	0.008	0.1	40
Pawpaw (*Carica papaya* L.)	39	0.6	62.0	**0.135**	0.1	24

Note: RE is retinol equivalents.
From Kehlenbeck *et al.* (2013)

provide details for identification and preparation of drugs derived from plant and non-plant sources. The NAPRALERT® database, launched in 1975, provides a compendium of published information about natural products covering more than 250 plant genera, including more than 15 000 species with known medicinal uses and potential applications (Dulloo *et al.*, 2014). However, because of ongoing efforts around identification of new plants and new chemical properties and constituents and medical applications, the number of medicinal plants globally is very much a moving target. In 2002, the global value of plant-based medicines was estimated at about US$30 billion with a value of US$941 million in 1997 alone for just one drug, Taxol, the anti-cancer drug obtained from the Pacific yew (Lewington, 2003).

A 4000-year-old Sumerian clay tablet, the earliest known medical document, is thought to depict the first use of a plant as a medical curative, and by the time of the rise of Egyptian civilization an extensive wealth of knowledge existed on the use of medicinal plants, as it did in ancient China also (Levetin and McMahon, 2012). The Pun-tsao, an ancient pharmacopoeia, contained thousands of botanical cures, many of which are believed to have existed over 4500 years ago. Likewise, in ancient India, medical plants and their curative powers are a rich tapestry of the sacred verses of the Hindu Rig-Veda. Hippocrates (460–377 BC), the father of Western medicine, actively used plant remedies as did Dioscorides, a Roman army medic, who compiled the *De Materia Medica* containing reference to more than

---

Box 14.9 | **Africa's Wooden Elephant**

(Gebauer *et al.*, 2016)

The African baobab (*Adansonia digitata*) is a tree that stands out majestically against the dramatic backdrop of the African savanna, scrubland and semi-desert. It is also a highly diverse and well-adapted and multi-purpose tree that has great potential to support communities in vulnerable dryland ecosystems facing the onslaught of climate variability. The baobab is sometimes referred to as one of the wooden 'Big Five' of Africa. Not only is there considerable diversity between trees, there is also significant diversity in terms of the high nutritional value of its fruits. Fruit pulp is the most important food from baobab, and it is rich in vitamins and minerals (Table 14.2). It has far higher levels of vitamin C, calcium and iron than more common tropical fruits such as mango and orange, but there is large variability in the levels of vitamin C in fruits of different trees. However, the nutritional value of the baobab does not stop at its fruit. The baobab also produces leaves that are used as nutritious vegetables and edible seeds, from which oil for consumption and cosmetic purposes are produced. This makes the baobab ripe for the picking when it comes to product development, market development and income generation. The potential of baobab for nutrition and income generation is a good example of a new product with high potential in the European market, given the acceptance of baobab as a novel food, often a bottleneck in promoting under-researched foods, in 2008. In fáct, a recent survey (2014) highlighted that more than 300 baobab products or products with baobab parts as an ingredient can already be found in Europe. The kinds of food products found included soft drinks, sandwich spreads, chocolates, sweets and muesli bars.

---

600 species of medicinal plant. Way before the first Europeans reached the New World, indigenous American Indians had compiled a formidable knowledge of useful medical plants, as did the indigenous communities of Australia and elsewhere. Scientists in the 19th century began to try to unlock the 'secret' ingredients that gave medicinal plants their curative properties, and Fredrich Sertuner became the first person to isolate morphine from the opium poppy in 1806. Although not as prominent as in earlier times, the interest in medicinal plants is still important, especially in developing countries, and represents an important coping strategy and a source of livelihood for some (Levetin and McMahon, 2012).

Today, at a local level, an extremely wide range of plant species are still used for medicinal purposes (Box 14.10). The World Health Organization (WHO) has listed more than 21 000 plant names (including synonyms) that have reported medical uses around the world (Dulloo *et al.*, 2014). Schippmann *et al.* (2006) suggest that somewhere between 50 000 and 70 000 plant species (representing about 20% of the global flora) are used in traditional or modern medicine systems, of which around 3000 are traded globally. Traditional medicine is still the basis of primary health care for approximately 80% of the population of developing countries (WHO/CBD 2015). Medicinal plant species are still to a large extent harvested from the wild and relatively few are cultivated as crop plants. The estimated annual value of wild-collected medicinal plants worldwide is in the region of US$10 billion. Plant breeding has only taken

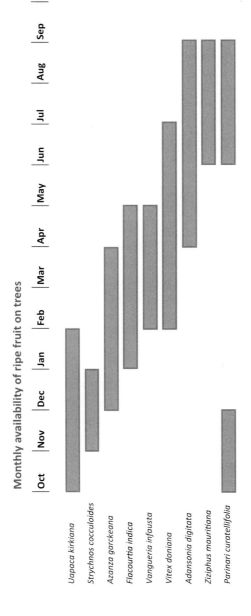

Figure 14.3 Harvest calendar of indigenous fruit tree species from Southern Africa. (Vinceti et al., 2013, modified from Jamnadass et al., 2011)

Box 14.10 | **Wild Medicine**

(Beattie and Ehrlich, 2001)

Rainforests are well-known reservoirs of plants with potential useful medical properties: in Peru, the quinine plant, which produces a treatment for malaria; in Madagascar, the rosy periwinkle and the powerful anti-cancer drugs which have arisen from its active ingredient; and in Australia, the impressive Moreton Bay chestnut has thrown up a chemical compound, castanospermine, which could potentially lead to better treatments for AIDS. A West African tree, *Millettia* spp., contains an active ingredient which may be helpful in fighting the battle against schistosomiasis, one of the world's biggest killers. We have all used an aspirin or two, which is based on salicylic acid originally obtained from the willow tree.

---

place with the commercially important plants such as *Papaver somniferum*, *Chamomilla recutita* and *Mentha piperita*.

As part of ongoing efforts to document and monitor medicinal plants, a global checklist is being compiled to support the work of the Medicinal Plant Specialist Group of the Species Survival Commission, IUCN (www.iucn.org/ssc-groups/plants-fungi/plants/medicinal-plant). Currently the checklist holds more than 27 000 accepted species names, distributed among five divisions, eight classes and more than 400 families in the plant kingdom. Furthermore, a Red List Index to track biodiversity used for food and medicine is included among the global indicators being used to measure progress towards the 2010 biodiversity target agreed in CBD decision VIII/15. Rather alarmingly, work on developing this indicator highlights that less than 3% of species included on the Global Checklist of Medicinal Plants have yet to be assessed for their threat status. Part of the problem is that information about most of the world's medicinal plants is scarce and fragmented (Dulloo *et al.*, 2014).

Traditional medicine, based on plant-based extracts is formally recognized in many countries today, including Sri Lanka, China, India, Vietnam and Thailand. Regions that are known to have important concentrations of major medicinal plants include Mexico and Central America, the west-central region of South America (Colombia, Ecuador, Peru), the Indian Subcontinent, China, west Asia and parts of north-eastern Africa. Boxes 14.11 and 14.12 illustrate how widely used wild plants are in traditional medicine in Ethiopia and China. In China alone it is estimated that more than 10 000 plant species are used in traditional medicine, with 500 species utilized in modern health practice (Lewington, 2003).

It is thought that around 80% of Africa's population utilize, largely because of convenience and low cost, traditional plants in their health care. In a study to document indigenous knowledge of medicinal plants for the treatment of reproductive health care among traditional healers and elders in the West Region in Cameroon, Central Africa, it was found that a total of 70 plant species from 37 families (mostly of the Asteraceae, Euphorbiaceae, Acanthaceae and Bignoniaceae) are used in the treatment of 27 reproductive ailments, with the highest number of species being used against venereal diseases and infertility problems of both sexes (21). Sixty per cent of the medicinal plant material were obtained from wild ecosystems (Tsobou *et al.*, 2016).

Approximately 25% of prescriptions used in the United States contain products derived from plants, while 119 pure chemical substances that have been extracted from 90 species of higher plants are used in medicines worldwide. For example, the rose periwinkle (*Catharanthus roseus*) of Madagascar has

Box 14.11 | **Integration of Traditional Phyto-Therapy into General Health Care: An Ethiopian Perspective**

(Dagne, 1998)

Ethiopia's diverse topography provides habitats for approximately 7000 higher plant species, among them a rich endemic flora with diverse therapeutic and other uses. Many of these plants are readily available at local markets, where medicines are customarily sold side by side with spices and other food items. At the principal market in Addis Ababa, for example, one finds leaves of the highly regarded stimulant *khat – Catha edulis* (Celastraceae), berries of the molluscicide *endod – Phytolacca dodecandra* (Phytolaccaceae), flowers of the well-known anthelmintic *koso – Hagenia abyssinica* (Rosaceae), stem bark of the anti-tumour agent *bissana – Croton macrostachyus* (Euphorbiaceae), roots of the analgesic *dingetegna – Taverniera abyssinica* (Leguminosae), rhizomes of the fumigant *kebericho – Echinops kebericho* (Compositae), aerial parts of the cough medicine *tossing – Thymus schimperi* (Labiatae), and leaves of the anti-dysenteric *attuch – Verbena officinalis* (Verbenaceae). These and other popular plant medicines are used extensively by indigenous populations who rely on both local healers and market vendors to prescribe the formulation and use of the crude drugs. An important aspect of the sustainable use of these botanical resources is to enhance our knowledge of their biological and pharmacologic effects and to characterize their constituents structurally. Such studies document the pharmacodynamics of Ethiopian medicinal plants and help to explain their continued use in traditional therapeutic contexts. They also contribute to the discovery of new natural products for medicinal and other applications and better equip traditional healers, modern health professionals and the public at large to confront the challenge of integrating plant medicines into the formal health sector.

yielded two important drugs that have been effective in the treatment of Hodgkin's disease, leukaemia and other blood cancers, and the bark of the Pacific yew, *Taxus brevifolia*, has been found to contain high concentrations of paclitaxel (marketed as Taxol), a chemical that has proved very effective in preliminary tests for the treatment of ovarian and breast cancer (Levetin and McMahon, 2012). It has been estimated that approximately 7000 plant-based medicinal products are known. However, the 120 or so regularly used around the world come from fewer than 100 species. There is thus a major effort by pharmaceutical companies to identify new sources of medicinally useful compounds. This is known as bioprospecting, or biopiracy when it breaches national and international law. Some examples of plants with important medicinal properties are shown in Table 14.4.

Plants are increasingly recognized as the source of natural pharmaceutical and medical products or, through the agencies of genetic transformation, they are used to produce plant-derived, novel pharmaceutical products not found in the original species. Pharmaceutical companies are becoming ever more aware of the existence of plant-based medicinal chemicals which could replace or complement existing drugs. Plants may also produce chemicals that inhibit plant pests or diseases, either by giving the pest deterrent signals or acting as natural toxins.

Box 14.12 | Use of Medicinal Plants in China

(Xiao and Peng, 1998)

China, a country with 22% of the world's population, has been making great efforts to utilize its rich ethno-pharmacologic experience and abundant medicinal plant resources for primary health care needs. The importance of traditional Chinese drugs, of which medicinal plants are the constituting 'backbone', is reflected in the 1995 Pharmacopoeia of the People's Republic of China. The first of two volumes is devoted exclusively to traditional Chinese drugs and their preparation. It contains a total of 920 items (crude drugs and prescriptions), which are estimated to account for about 40% of the total medicament consumption in China. Ethno-pharmacologic studies and research on medicinal plants seem to be the most important measure for promoting and mobilizing the utilization of traditional Chinese drugs.

Recent nationwide surveys in China reported 11 118 plant taxa used for medicinal purposes. Among these species, 467 belong to the thallophytes (excluding bacteria), 43 are bryophytes, 455 pteridophytes and 126 gymnosperms. But more than 90% of the total species are angiosperms, comprising 8598 dicotyledons and 1429 monocotyledons. Several nationwide surveys have collected more than 40 000 items (including plant parts used, dosage forms and traditional usages) of ethno-pharmacologic information on Chinese medicinal plants. These were subsequently analysed systematically and the number of genera and species within particular plant families used for medicinal purposes determined (Table 14.3).

Many important modern drugs are based on the uses and knowledge of indigenous people. For example, the rose periwinkle, *Catharanthus roseus*, was discovered through its traditional use to treat diabetes, while the Pacific yew, *Taxus brevifolia*, was used by indigenous people in North America for the treatment of some non-cancerous conditions. *Chondrodendron tomentosum* is used by indigenous peoples in South America in an arrow poison known as 'curare' (Levetin and McMahon, 2012). The plant yields the compound d-tubocurarine, which is used as a model for a series of similar synthetic neuromuscular-blocking agents, some of which are now routinely used as anaesthetics in surgical operations. *Rauwolfia serpentina* yields the anti-hypertensive alkaloid reserpine; known as the Indian snake root, it has a long history of traditional medical use, including the treatment of mental disorder, snake bites and as a tranquillizer. Reserpine revolutionized

Western medical treatment of hypertension in the 1950s, and caused massive over-harvesting of wild populations in India, thereby leading to its inclusion in CITES Appendix II in 1990 (Laird and ten Kate, 1999).

Biological communities are being continually surveyed for organisms that may be valuable in fighting human diseases, notably cancer and AIDS. Ethno-botanists play an important role here in consulting with local indigenous people about their uses of local plants. As each species is biochemically unique and thus may potentially provide a major scientific breakthrough, this provides a powerful argument for biodiversity conservation. Modern pharmaceutical companies have rapid automated systems for screening chemicals for biological activity. Currently, much of their screening of species is random, whereas concentrating on ethnobotanical information could be a more effective approach.

Table 14.3 **Number of genera and species within plant families used in Chinese medicine in China**

Family	Total genera (China)	Genera used (%)	Total species (China)	Species used (%)
Saxifragaceae	24	100	427	36
Compositae	227	68	2323	33
Ranunculaceae	41	83	737	57
Liliaceae	67	69	401	89
Rubiaceae	75	79	477	46

From Xiao and Peng (1998).

Table 14.4 **Examples of plants with important medicinal properties**

Medicinal property	Plant
Antibiotic activity	*Sphagnum* moss; garlic (*Allium*)
Decongestants for coughs and colds	*Eucalyptus*; ipecac; liquorice
Pain killers	Poppies (*Papaver*); willow (*Salix*)
Contraceptives	Yams (*Dioscorea*); cotton (*Gossypium*)
Heart drugs	Foxgloves (*Digitalis* spp.)
Laxatives	Senna (*Cassia* spp.)
Anti-tumour drugs	Rose periwinkle (*Catharanthus roseus*), mandrake
Anti-AIDS drugs	*Castanospermum australe*

The trade in medicinal plants is an important contributor to the income and livelihoods of millions of households in the developing world. However, over-exploitation of medicinal plants extracted from the wild is leading to problems of genetic erosion in some regions. Unsustainable harvesting is aggravated by other factors contributing to land degradation, as well as changing climate (Klein *et al.*, 2008). In India, for example, where 2500 plant species are used by traditional healers, species of *Aconitum*, *Dioscorea* and *Ephedra* are just some of the medicinal plants under threat in the wild as a result of over-harvesting.

Recognizing such threats, among others, practitioners and researchers have been active in promoting community-based approaches to the conservation and sustainable use of medicinal plants (Box 14.13).

Laird and ten Kate (1999) provide some indicators of the extent of harvesting and use of wild species in the botanical medicine trade in Europe:

- Of the 1200–1300 (botanical) species native to Europe in trade, at least 90% are still wild-collected; wild harvesting is particularly common in Albania, Turkey, Hungary and Spain.

Box 14.13 | **Keeping It Close to Home (2): Herbal Home Gardens**

(WHO/CBD, 2015)

The Foundation for Revitalization of Local Health Traditions (FRLHT) in India illustrates an example of good practice to validate and revitalize traditional medicinal practices and which has a focus on the promotion, conservation and sustainable use of medicinal plants. In a programme covering 6000 rural villages and comprising 150 000 household groups, herbal home gardens have been established and seedlings of a set of 20 prioritized medicinal plants from 12 to 15 species that are useful for common ailments grown and sold to rural households by Women Self-Help Groups in the FRLHT network. These groups were trained by FRLHT in raising, distributing and demonstrating the use of the plants for the relevant conditions. Through participatory rural appraisal, a list was developed of plants specific to each of the regions in which the project was conducted. The FRLHT approach has become a successful and replicable model for a self-reliant community-based health programme, and with its medicinal plant conservation focus it is promoting and linking sustainable use. In most rural communities, knowledgeable women take care of certain primary health needs of the family members, and the gardens become a handy resource for them. Some women, by taking on the role of suppliers of seedlings for the programme, also earn supplementary income.

- Of the 1560 plant species traded in Germany, 50–70% by volume are wild-crafted, representing 70–90% of species; only about 135 species originate from cultivated sources, with just 5–20% of these purchased from German farmers.
- By volume, 30–50% of medicinal and aromatic plant material in trade in Hungary is wild-collected; 75–80% in Bulgaria; and almost 100% in Albania and Turkey.
- The overall volume of wild-collected material in Europe is estimated to be 20 000–30 000 t annually.

Bioprospecting is another spectre that has cast a long shadow over the largely developed world's relentless search for active ingredients with commercial potential, among the plants of forests and other ecosystems of the biodiversity-rich countries largely in the developing world. In the past, this has led to blatant acts of biopiracy (see Chapter 2) with large profits from patented drugs going to developed countries and the pharmaceutical companies based

there. It was the issue of bio-prospecting and unfair exploitation of poor countries' biodiversity in such a manner which contributed to the need for a CBD. Although equitable sharing of benefits arising from products based on biodiversity is fundamental to the convention, implementation is no easy task. However, there is cause for a certain amount of optimism as several agreements between corporations, national governments and local people have begun. For example, this has led to the development of the anti-viral compound prostatin, which was isolated from the Samoan tree *Homalanthus nutans*. This agreement stipulates that 20% of the profits go back to Samoa (Lewington, 2003). For a comprehensive review of medicinal plants and traditional medicine see WHO/CBD (2015).

14.4 Plants for Recreation and Amenity

As well as providing food and medicines, plants are bred and cultivated extensively for a wide variety of uses for recreation and amenity, such as for horticulture (cut flowers, house plants, interior landscaping, bedding plants, ornamental trees and shrubs, and large-scale landscaping). The cultivation of ornamental plants has a long history. For example, lilies have been cultivated in China for decorative purposes for around 2000 years. In Roman times, roses, lilies, violets, anemones, narcissi and lavender were grown as garden plants in Europe. Today, the diversity of decorative plants established in cultivation around the world far surpasses the number of plants grown for food.

There is a major worldwide demand for decorative plants for homes, gardens and public places both inside and out. New and exotic forms are constantly entering the market, and the market value of the material is high both for cut flowers and for potted material. Not only are breeders on the lookout for new species to grow and sell, but they are also keen to improve the quality of the plant in terms of its market appeal. These include appearance (colour, shape), scent, novelty value, shelf and active life, and ease of maintenance.

Many of these plants can be vegetatively propagated, but others must be maintained sexually through seeds. As with crop plants, selection for quality and abiotic stress resistance is important, although the stresses are more likely to be poor maintenance, air-conditioning or central heating in homes and public buildings. Pests and diseases are a problem, but because the plants are not grown for food, there is less reluctance to use chemical control. Nonetheless, natural resistance is always preferred as it is cheaper and longer lasting and is a good selling point.

The demand for good-quality ornamental plants has led to the existence of a lucrative market, and while the ornamental horticulture industry represents a modest market as compared to crop plants and vegetables, it is nonetheless significant in monetary terms. It has been estimated that the annual global market for ornamental horticultural products lies between US$16 to 19 billion. The ornamental horticulture industry consists of five main areas: herbaceous plants, woody plants, cut flowers, foliage plants and bulbs. However, like much of the trade in

medicinal plants, a significant part of the trade involving ornamentals including cycads, orchids, cacti and succulents and bulbs is largely uncontrolled, even illegal (Box 14.14) (Heywood 2013).

With increasing urbanization worldwide, there is always a necessary psychological health requirement for open spaces within our cities where people can relax, enjoy nature and breath fresh air (WHO/CBD, 2015). The parks we use for this purpose are a mixture of natural trees and shrubs and various exotic species introduced for their colour and shape. Such species need to be able to cope with a range of problems that they may not have encountered in their natural environment, such as pollution, disturbance from humans and their pets, and vandalism. Robust species of shrubs, which are both decorative but protected with thorns or prickly leaves, are therefore more likely to survive in our city parks. Amenity grasses need to be resistant to trampling by walkers or the special problems associated with surviving on sports pitches or the public's pets. These problems include not only the direct physical damage but also the indirect effects on the compaction of the soil, lack of nutrients and regular mowing.

A wide range of plants is now being used as screening to hide unsightly objects and to create a greener and arboreal feel to a harsh urban or industrial landscape, even green walls where plants are grown vertically on the walls of buildings. New highway verges and central reservations are often planted with species for soil stabilization, screening or simply to make it more aesthetically pleasing for the highway users. Such roadside environments can provide valuable *in situ* reserves for native species, as they are relatively free from human activity, thus providing sites with the potential for whole plant communities to become established. They do however have to cope with problems such as pollution and high levels of salt.

Research by the Green Exercise Research Team at the University of Essex has highlighted the significant mental and physical health benefits which can arise as a result of exposure to 'green spaces' (Pretty, 2007). Green spaces can be natural or man-made environments where a diversity of plants are found, such as in forests, mountain landscapes, rural

---

Box 14.14 | Hidden Harvest, Hidden Trade

(BBC: www.bbc.com/news/science-environment-35699297; Phelps and Webb, 2015)

Unfortunately, the number of cases of illegal collection and trade in exotic plants continues to grow to keep in step with demand worldwide, as highlighted by a recent BBC news story highlighting the extent of the illegal trade in seeds of exotic plants collected from the Himalayas and being sold in the UK. All too often collectors fail to seek the permission of the relevant national authorities. Not only do such unregulated practices affect the conservation status of the species in question, it can also have negative effects on the plant's habitat. Many of the specimens are collected from protected areas. Such practices also prevent benefits being shared with or returned to local communities and contravenes national regulations and global conventions and agreements such as the CBD's Nagoya Protocol. In the BBC news story, collection of plant specimens is strictly prohibited from wildlife protected areas under the Wildlife Protection Act of 1972, and even from the other reserved forest areas. The CBD Nagoya Protocol, an international treaty that came into being in 2014, prohibits the collection of plant materials without an agreement with host countries on the sharing of benefits arising from such resources. Unfortunately, data on the global extent of these illegal activities is very thin on the ground. However, the first in-depth study of the trade of wild-collected ornamental plants in continental Southeast Asia (focusing on the four largest wildlife markets in Thailand and involving trade in Lao PDR and Myanmar) has uncovered a huge commercial trade in wild, protected ornamental plants. Focusing primarily on orchids, the findings have highlighted that not only does the trade threaten hundreds of plant species but also other charismatic animal biodiversity. Surveys identified 347 orchid species in 93 genera, including many listed as threatened. The implication is despite three decades of broad restrictions on the international trade of all wild orchids, these results highlight a major conservation challenge that has been almost completely overlooked.

---

countryside, parks, household gardens, community gardens, farms, urban agriculture, allotments and so forth. The evidence demonstrates that reconnecting with nature through exercise or other forms of therapy involving vulnerable groups such as hospital patients or prison inmates can bring manifold benefits including improved mental health, physical well-being and social cohesion, and a sense of belonging, which can have significant impacts by reducing loneliness, aggression, crime, physical illness and disease, especially non-communicable diseases, as a result of lifestyle factors. Furthermore, the costs of such approaches are much less than conventional treatment by drugs or surgery. Other researchers have highlighted the benefits on the moods and comfort of workers by simply having plants in offices or the value of interacting with plants through gardening or on farms to mental health, which has given rise to specialized forms of treatment such as horticultural therapy and green care.

Without plants our range of recreational and leisure pursuits would no doubt be diminished. From the paper we rely on to read the next novel, or the vegetable oils for printing inks and artist's paint, to the woods and other plant-based materials for sporting bats and balls and musical instruments, for example the willow tree (*Salix alba*) for cricket bats or the African blackwood (*Dalbergia melanoxyla*), the wood of choice for clarinets and oboes (Lewington, 2003). The importance of plants in this regard is highlighted by the recent crisis facing the traditional sport of hurling in Ireland, which uses a stick made from ash (*Fraxinus excelsior*), and the impact of Chalara ash dieback disease caused by the fungus *Chalara fraxinea*. The recent outbreak of the disease in Europe and implementation of trade bans has seriously threatened the supply of the hundreds of thousands of hurling sticks needed each year.

14.5 Plants for Poisons

While Section 14.3 highlights the many curative properties of medicinal plants, there are also many plants which can be highly detrimental to the health and well-being of humans and other animals. Socrates, the philosopher of Ancient Greece, met his fate as a result of a poisonous plant. At that time, the usual form of capital punishment was to drink from a cup of hemlock (*Conium maculatum*), which contains the extremely toxic alkaloid conine (Mabey, 1996). Strychnine, another alkaloid, from the plant *Strychnos nux-vomica*, is a very powerful nerve toxin, which is still commercially used as a rodenticide and which ironically also has some medical curative properties. Other poisonous plants which can adversely affect the health of humans are the common castor oil plant (*Ricinus communis*) which produces ricin, a deadly poison (Levetin and McMahon, 2012). Some of our food plants can even be poisonous including cassava (*Manihot esculenta*), which contains hydrogen cyanide that blocks respiration (Levetin and McMahon, 2012), and grasspea (*Lathyrus sativus*), which contains the neurotoxin β-oxalyl-L-α,β-diaminopropionic acid that causes limb paralysis, so indigenous people have developed processing methods to eliminate the toxins.

Other plants are highly poisonous to animals, wild and domesticated (Levetin and McMahon, 2012). Several weeds can be problematic in livestock husbandry, Paterson's curse (*Echium plantagineum*) and St John's wort (*Hypericum perforatum*) are particularly problematic noxious weeds for livestock farmers in Australia. Curare, obtained from tropical *Chondrodendron* species, is still used commonly to smear the tips of arrows for hunting animal prey by indigenous tribes in South America. Of course, many plants evolved to produce chemical compounds which would provide them with defences against attack from predators, and this has been exploited by many cultures over the course of human history to protect food crops from pests and diseases. These plants and the poisons they provide are sometimes collectively referred to as botanical pesticides and include neem (*Azadirachta indica*), pyrethroids (*Chrysanthemum cinearifolium*) and rotenone from various tropical legumes (Levetin and McMahon, 2012).

14.6 Plants for Materials

There are a wide range of plant-based materials used for construction and industrial purposes (Lewington, 2003; Levetin and McMahon, 2012). These include wood and other building materials for timber, fibres, dyes, resins, gums, rubber, oil and waxes. Wood in particular forms a vital commodity for many communities dependent on fuel for heating and cooking. Oils and waxes may be used for lubricants, chemical feedstocks and other specialized uses. Some substances from individual species have unique properties that confer value. For example, scientists have now discovered that the jojoba plant produces oil with similar qualities to sperm whale oil. As this plant can be easily cultivated, this could reduce pressure on sperm whale populations. Many plants also provide the fibres essential for our cloth, textiles and other materials such as cotton (*Gossypium hirsutum*), coir rope and other coconut products (*Cocos nucifera*) and the traditional tapa cloth extensively used in the South Pacific and made from the paper mulberry (*Broussontia papyrifera*; Box 14.15). Many plants are critically important for transport with wood and wood products important for building boats and carts, rubber for tyres and plants such as sugar cane providing biofuels.

As we have seen earlier, many of these materials are often collectively grouped as non-timber forest

---

Box 14.15 | Unpacking Tapa: A Neglected Material

(RBG, Kew: www.kew.org/read-and-watch/unpacking-tapa)

Until about 150 years ago, barkcloth, made from the inner bark of trees such as wild fig, was the textile of choice in most of the tropics, a strong and enduring material with much cultural significance. Despite its physical and cultural strength, it was not strong enough to withstand the arrival of missionaries who were keen to promote more 'modest' styles of dress and the cotton fabrics imported from Europe, which led to a sharp reduction of barkcloth production in South America, West Africa and parts of the Pacific where a very small number of countries still continue the tradition. In the Pacific, tapa cloth is made by beating the inner bark of trees such as paper mulberry (*Broussonetia papyrifera*) with wooden clubs. However, the decline in its production has meant there has been a huge amount of lost knowledge about the kinds of tree that can be used, dyes, and the range of oils and resins used to prepare surfaces. But all is not lost. Scientists at the Royal Botanical Gardens at Kew, which has a rich history of research into useful plants, are applying their wisdom and skills to better understand this knowledge. Kew currently maintains around 100 pieces of barkcloth dating from the 1820s to 1930s, which include about 60 from the Pacific, where barkcloth is commonly known as tapa. Here the picture of current day use gets complex. Tapa production ended in islands such as Tahiti and Hawaii in the late 19th century, but has recently been revived, while production continues to flourish in Tonga, Fiji and Papua New Guinea. For Pacific peoples, tapa collections such as that at Kew are reservoirs of traditional knowledge, often containing materials or styles that are no longer made. It is intended that the results of this project be highly relevant to contemporary makers and users of tapa in the Pacific and help keep the wonderful tradition of tapa making alive and kicking for younger generations to come.

---

products (NTFP) and have largely been overlooked in mainstream forestry and conservation policies even though they contribute immensely to the economic, nutritional, health, social and cultural well-being of countless millions of households around the world (Box 14.16 and Table 14.5). To date, there have been few studies that have looked at the importance and value of NTFP. One such attempt was 'The Hidden Harvest' project of the International Institute for Environment and Development (IIED) in the UK which commenced in the early 1990s (IIED, 1997) and more recently from the Rainforest Alliance's NTFP Marketing and Management Project (Shanley *et al.*,

2002). Both initiatives have been important in developing approaches to the valuation and sustainable management of NTFP and bringing attention and better understanding of their importance and value to forest managers, policy-makers, donors and conservation organizations. Today, People and Plants International is one of the leading organizations working with local communities to improve sustainable management, livelihoods and governance of NTFP.

Many of the plants addressed in this section such as wood and NTFP are also subject to overexploitation and uncontrolled or often illegal trading.

Box 14.16 | **The Hidden Harvest: Valuing Woodland and Forest Products in Traditional Cultural Systems**

(Campbell and Luckert, 2002)

A 1996 report of a study of the value of woodlands in Namibia showed that the total value of woodland resources is around N$105.8 million per year. The four most significant uses of woodland resources are construction poles, fences, tourism, firewood and charcoal. Such valuation exercises are important because they demonstrate the importance of trees and woodlands to the national economy to a wider audience. However, the way in which wild resources contribute to people's livelihoods is often hidden to outsiders, difficult to quantify and receives little recognition from development practitioners. Many external factors need to be considered to fully understand the role that the sale of forest products plays in rural livelihoods, such as international prices, macro-economic conditions affecting tourism, unemployment levels and AIDS. Rural households face many obstacles in managing local resources and dealing with external markets. These include low availability of capital, lack of formal education, risks such as extreme weather conditions, family illness and changes in the external economic climate. Recently, there has been greater focus on finding appropriate valuation methodologies for the goods and services derived from wild plants and animals. Campbell and Luckert (2002) outlined a number of potential uses for value estimates of indigenous resources:

- Land use questions, e.g. what is the value of the indigenous resources lost through centrally planned attempts to improve livelihoods through replacement of indigenous species with agricultural species?
- Cost–benefit analysis is required to get as broad a measure of social welfare as possible for development projects involving indigenous species.
- Justification for a change in focus towards greater consideration of indigenous resources in rural development.
- Insights into people's resource management options and livelihood strategies.

14.7 Plants as Gene Donors

Farmers have carried out plant breeding for centuries. These years of breeding and selection have led to the production of many of the foods that we eat today and have been fundamental to the development of human society. Modern methods in breeding and selection for new varieties have now largely taken over from traditional methods in most regions of the world, and market pressure has forced many farmers to farm modern cultivars, rather than traditional varieties or landraces. This has led to the loss of many old landraces that were bred to suit local conditions, and there is now greater awareness of and emphasis on reinstating and conserving the surviving landraces, which contain important genetic diversity that is useful for the future development of crops. This characterizes the Conservation–Development Paradox (Hawkes et al., 2000). Modern cultivars are almost always homozygous being bred for uniformity while

Table 14.5 **Examples of the socio-economic importance of wild plant species use**

Wild resource	Estimated annual economic value	Estimated social significance
Fuel-wood and wood-based products	US$418 billion worldwide, or nearly 2% of the world's gross domestic product (GDP)	3 billion people depend on wood for household energy worldwide
Wild-collected medicinal plants	US$10 billion worldwide	Traditional medicine is the basis of primary health care for ca. 80% of the population of developing countries, or some 3 billion people
Non-timber forest products in Indonesia	More than US$1 billion	In West Bengal, 155 species in sal (*Shorea robusta*) forests are used by local communities, accounting for 55% of the total income of forest fringe dwellers

From Freese (1998).

landraces are generally genetically variable, thus uniformity is replacing diversity. However, plant breeders are dependent upon the availability of a wide pool of diverse genetic material as a basis for their breeding activities. Therefore, the success of their work, through landrace replacement, is unwittingly causing the genetic erosion of plant diversity that they themselves will need in the future – therein lies the paradox.

Breeders' requirement for breadth of genetic diversity is now limiting crop improvement, particularly in the context of breeding for climate change resilience (McCouch *et al.*, 2013). The production of new varieties is crucial to farmers in both developing and developed countries. Extreme conditions such as drought or susceptibility to pests and disease can cause the loss of whole crops, economic ruin and famine. As new strains of disease evolve or agro-environmental conditions change, so new varieties are bred to resist changing pests and disease and harsh environmental conditions, and so sustain crop production. As highlighted by Godfray *et al.* (2010), unexploited genetic material sourced from landraces and crop wild relatives which contain the majority of PGRFA genetic diversity will be important in helping plant breeders respond to the challenge of feeding the 9 billion people expected to be inhabiting the planet by the middle of the century.

Crop wild relatives are an important element of PGRFA particularly valuable for plant breeding especially in the last few decades (Hunter and Heywood, 2011) as these are the richest PGR sources of novel genetic diversity (Tanksley and McCouch, 1997). In a recent review of their use, Maxted and Kell (2009) cited 91 articles that reported the identification and transfer of useful traits from 185 CWR taxa into 29 crop species. Indeed, modern varieties of most crops now contain some genes that are derived from a wild relative. For example, genes from several wild species of *Aegilops*, which is closely related to *Triticum*, have been transferred to cultivated wheat, including those that confer resistance to leaf rust, stem rust, powdery mildew and nematodes. Breeders' use of CWR diversity in improving food production has been estimated at an annual value of $115–120 billion worldwide (Pimentel *et al.*, 1997) and subsequently the same value for 26 priority crop gene pools alone (PwC, 2013). An idea of the scale of benefits can be gained from examining selected crops. For example, the desirable traits of wild sunflowers (*Helianthus* spp.) are worth up to US$384 million annually to the sunflower industry in the United States alone; one wild tomato accession has contributed to a 2.4% increase in solids content worth US$250 million; and three wild groundnuts (or peanuts) have provided resistance to the root-knot nematode, which cost groundnut growers around the world US$100 million each year (Hunter and Heywood, 2011). The principal traits (or characteristics) to which plant breeding is applied are: yield, quality and resistance to abiotic and biotic stresses.

14.7.1 Breeding for Yield

Breeders look for a very wide range of potential attributes from germplasm, although not necessarily all at once. The main aim is to improve yield, but this is a complex, multi-factorial trait, that is, it is often controlled by many genes interacting with the environment, so is difficult to select for *per se*.

Yield is often thought of in terms of product per unit area grown, but it is not always that simple. The harvested crop has a value expressed in currency at the relevant marketplace, but it also has a cost to the farmer or grower. This cost is measured in terms of various factors such as ground rent, labour charges to grow and harvest the crop, fertilizers, chemicals to control pests and diseases, and transport. The currency value to the grower varies with the season and the state of the market; it is more valuable if it arrives at the market before that of competitors. For these reasons of cost and benefit, it may be difficult to be precise about the value of an extra tonne of yield per unit area, because it depends on how much extra expense was required to generate it and what the current profit margin is. Thus, although it might be more logical to talk of yield per unit of currency (e.g. yield per dollar spent), it is normally more practical to consider yield in terms of product per unit area in a given time. It is difficult to imagine that there are many financial situations where lower yield is better in cash terms. Many different traits affect yield. Therefore, breeders can improve grain yield by selecting for yield components, such as the size of the grain, the number of grains per plant, seed retention on the parent plant, resistance to lodging or changing the growing season.

14.7.2 Breeding for Produce Quality

One aspect of yield that is becoming increasingly important, particularly as societies become more affluent, is the quality of the product. Firstly, the appearance of the product at market, particularly if it is food, is important in its appeal to customers, e.g. fruit should look attractive, be free of blemishes, have good flavour and be of uniform size. It should be free of chemical residues from pest and disease control measures and safe to eat. Foodstuffs should have good storage qualities, travel well without bruising or discolouring, and be ripe on arrival at the market. This is becoming ever more important as fresh food is moved around the world to satisfy consumer demand for consistent supplies irrespective of season, as well as to cater for more exotic tastes. Temperate crops, such as apples and soft fruits, are regularly transported between the northern and southern hemispheres to supply 'out of season' demand.

The chemical constitution of the product should be acceptable for its purpose, e.g. the gluten level in wheat or the nitrogen content of malting barley. Many foodstuffs are now preferred because of their positive health-giving qualities, e.g. garlic or many brassicas (such as *calabrese*) contain antioxidants that have beneficial effects in preventing heart attacks or certain cancers. There is growing interest in the nutritional value of our crop foods so much so that many people are calling for a shift in emphasis from conventional measures of productivity and yield such as quantity per unit area to something that captures the 'nutritional yield', which is a measure of the nutritional quality of that production.

14.7.3 Breeding for Tolerance to Abiotic Stress

Adaptation or tolerance to abiotic stress is becoming an important factor in crop production. The stresses commonly perceived as being a problem include high salt, drought, water-logging, extreme temperatures, pH and the presence in the soil of heavy metals such as zinc or copper. However, there are many specific situations that would benefit from specially bred plants, such as wind damage, salt spray and soil erosion (as in sandy soils). Breeders are keen to identify genotypes that have inherent tolerance or resistance to these stresses. These stresses can create difficult factors to breed for, because they are difficult to reproduce reliably in the breeding station and, as is the case with drought or water-logging, are likely to affect the crop at different times of development in different seasons.

14.7.4 Breeding for Host Plant Resistance to Biotic Stresses

Biotic stresses are another important category of breeding objective – these are principally pests and pathogens. Pests range from large herbivores to small root-invading nematodes, but the principal category is probably insects that browse, burrow and remove nutrients from the plants. They can destroy whole plantings or create small blemishes in fruits or timbers. Plant pathogens cause many diseases, and they are mainly fungal, bacterial or viral, the latter being largely spread by insect vectors.

It is possible to control many of these biotic stresses by suitable topical applications of chemicals, but such approaches have two disadvantages: (a) in response to chemicals, pests and pathogens evolve stronger resistance so that correspondingly higher levels of chemicals have to be applied and (b) chemicals often have toxic effects on humans and other animals and plants, either during their application or as residues in the food or crop product. For these reasons, it is becoming more and more important to identify host plant resistance to biotic stresses, because such resistances will not be dangerous to the farmer and consumer but, in some instances, may create a more complex and difficult obstacle for the pathogen or pest to overcome. Tables 14.6, 14.7 and 14.8 illustrate in what material useful genes have been found within the rice (*Oryza sativa*) collection at IRRI, what pest and disease resistance genes have been identified, and how some of those have been transferred into the rice crop.

14.7.5 Other Traits

There are many other traits that plant breeders are searching for, and the nature of this list varies depending on the crop. A few, however, are common to most crops. Uniformity of product is one such trait. The market increasingly requires standard

Table 14.6 'Hot spots' of useful resistance to diseases and pests in *Oryza sativa* germplasm

Disease or pest	Country
Bacterial blight	Bangladesh, Philippines
Blast	Vietnam, Laos, Myanmar, Thailand
Sheath blight	Malaysia, Sri Lanka, Vietnam
Rice tungro disease	Bangladesh
Brown planthopper (Biotypes 1, 2 and 3)	Sri Lanka
White-backed planthopper	Laos
Green leafhopper	Bangladesh

From Jackson (1994).

products for which it can be assured of a buyer. It may be necessary to have a size of fruit or tuber that fits into a can or which meets the processing needs of the cannery. On the other hand, in many marginal areas, diversity may be important in order to capitalize on a variable or heterogeneous environment and ensure at least some yield despite the annual growing conditions. Earliness of maturity is another trait that determines the time of arrival of the crop at the market. Early maturity may also allow further crops to be sown within a growing season or make it possible to grow the crop in areas with a shorter season. You will learn about the techniques for germplasm evaluation and plant breeding in Chapters 15, 16 and 17.

14.8 Other Uses of Plants

Many plants have social and cultural significance, and have influenced the philosophy, language, art, religion and social structure of many societies. For example, the Cedar of Lebanon, *Cedrus libani*, is a national symbol and appears on the Lebanese flag;

Table 14.7 **Resistance to 13 disease and insect pests in *Oryza sativa* germplasm evaluated at IRRI**

Disease or pest	*Oryza sativa* accessions	
	Number	% Resistant
Bacterial blight	48 203	11.2
Blast	36 305	26.2
Sheath blight	22 754	9.2
Rice tungro disease	15 795	3.5
Rice ragged stunt virus	13 759	4.7
Brown planthopper biotype 1	47 268	1.6
Brown planthopper biotype 2	13 652	1.5
Brown planthopper biotype 3	16 643	1.9
White-backed planthopper	56 237	1.6
Green leafhopper	57 437	2.7
Rice whorl maggot	22 598	3.0
Zigzag leafhopper	2732	10.1
Rice leaf folder	8005	0.6

From Jackson (1994).

Table 14.8 **Transfer of useful characteristics from wild species into *Oryza sativa* using embryo rescue**

Species	Disease or pest resistance
O. australiensis	Brown planthopper (BPH)
O. brachyantha	Yellow stemborer
O. latifolia	BPH
O. longisteminata	Bacterial leaf blight
O. minuta	BPH, blast, bacterial leaf blight
O. nivara	Bacterial leaf blight
O. officinalis	BPH, tungro tolerance
O. ridleyi	Yellow stemborer
O. rufipogon	Tungro tolerance, acid sulphate tolerance

From Jackson (1994).

the country has recognized this endemic species as culturally important, and although the species distribution is now severely limited and populations depleted, the Lebanese government is taking action to reintroduce the tree into the wild. Flowers and plants with cultural significance are also frequently used in churches and religious ceremonies, and for adornment on traditional costumes. Kava produced by pounding or grinding the roots of *Piper methysticum* is a mild sedative drink consumed in many parts of the Pacific and which has strong cultural and political significance (Levetin and McMahon, 2012). There are also many different varieties of the plant that have varying degrees of alkaloids which contribute to the sedative effect.

With increasing population and prosperity, the world has witnessed a huge shift in demand for non-food crops, a demand often met to the detriment of biodiversity. Large areas of land have been converted to produce 'luxury' foods (e.g. coffee, tea, cacao), fibres (e.g. cotton), biofuels and oil (e.g. palm oil, rapeseed; Box 14.17). The problem is that cultivation and production of many of these crops occur by large-scale clearance of forests, in areas that largely occur in biodiversity hotspots. Consequently, the expansion of these crops has come at great expense to local biodiversity (WHO/CBD, 2015).

The utilization of plant diversity by humankind for its aesthetic values is another important use. This may be as simple as enjoying a walk in the country or nature reserve, bird-watching, botanizing or going on a photographic safari. This basic requirement for human recreation is an important objective and attraction in establishing national parks and reserves, and in recent years, this concept has been exploited in the ecotourism industry. Humankind's need for recreational spaces can be a powerful tool in biodiversity conservation.

Box 14.17 | **The Global Rise of Oil Palm**

(WHO/CBD, 2015)

A native of West Africa, oil palm *(Elaeis guineensis)* is cultivated on more than 13.5 million ha of tropical, high-rainfall, low-lying areas in zones naturally occupied by moist tropical forest and which constitute the most biologically diverse terrestrial ecosystems on Earth. Some plantations exceed 20 000 ha. The oil is used to make cooking oil, margarine, soap and cosmetics and has industrial applications as well. The huge area of production has meant extensive clearing and burning of carbon-rich forests and peat lands, which has contributed substantially to biodiversity loss, poor air quality often causing respiratory health problems particularly in Southeast Asia, and adding CO_2 to the atmosphere. When compared to shaded coffee, pasture and natural forest palm oil cultivation supports extremely low levels of birds, lizards, beetles and ant communities. Surprisingly there have been few studies of the impact of oil palm on plant biodiversity.

Plant genetic resources are also used in the biotechnology industries, which embrace many sectors, including pharmaceuticals, botanical medicines, agriculture, crop protection products, horticulture, cosmetics, perfumes, detergents and paper-making. Biotechnology companies apply enzymes and use biologically active compounds derived from genetic resources as an integral part of processes and products in almost every industry sector.

14.9 Exploitation and Sustainable Management of Wild Species

Many wild resources, including food plants, timber, medicinal plants, ornamentals and other NTFP, have been and continue to be over-collected from the wild and as such are the subject of uncontrolled and often illegal trade. Illegal logging on indigenous Indian lands in parts of Brazil of wild mahogany, *Swietenia macrophylla*, has led to its serious decline. Sometimes referred to as 'green gold', one cubic metre of this tree's timber can attract as much as US$1600. Widespread global demand for mahogany has fuelled the destruction of the Brazilian Amazon, and while an

indigenous Indian may be forced to accept around US$30 for a tree on his or her land, by the time that same tree has been made into 15 dining tables and sold in the fashionable stores of London or New York the tree is worth a staggering US$128 250 (Lewington, 2003).

The Fiji Sago palm (*Metroxylon vitiense*) and wild yams (various *Dioscorea* spp.) are examples of wild plants that are used for food but harvested unsustainably from the wild. The Fiji Sago palm is harvested for its edible palm heart and is also traded informally in the country, largely uncontrolled, and the current rate of harvesting poses a serious threat to current populations (Morrison *et al.*, 2012).

Ornamental orchids, bulbs and cacti are amongst the most significant plant groups in the international horticultural trade. Although most of the material traded is from artificially propagated stock, the demand for wild-collected plants still exists. The effect of the industry on wild populations of orchids and cacti has led to the inclusion of the entire plant families in the CITES Appendices (Box 14.18). Some genera have been the focus of obsessive attention for ornamental use, including the popular slipper orchids of the genus *Paphiopedilum*, where the demented collector may trash the entire population once they

---

Box 14.18 | **Convention on International Trade in Endangered Species of Wild Fauna and Flora**

(CITES: www.cites.org/)

The international wildlife trade is a highly lucrative business and involves a wide variety of animal and plant species, which are traded both as live specimens and as bi-products. CITES is a tool for regulating and monitoring trade in species of wild fauna and flora. This Convention entered into force on 1 July 1975, and 152 countries are party to it (as of November 2000). Parties act by banning commercial international trade in an agreed list of endangered species, and by regulating and monitoring trade in others that could become endangered in the future.

CITES Article II: Fundamental Principles

Appendix I shall include all species threatened with extinction that are or may be affected by trade. Trade in specimens of these species must be subject to particularly strict regulation in order not to endanger further their survival and must only be authorized in exceptional circumstances.

Appendix II shall include:

(a) all species which, although not necessarily now threatened with extinction, may become so unless trade in specimens of such species is subject to strict regulation in order to avoid utilization incompatible with their survival and

(b) other species which must be subject to regulation in order that trade in specimens of certain species referred to in sub-paragraph (a) of this paragraph may be brought under effective control.

Appendix III shall include all species which any Party identifies as being subject to regulation within its jurisdiction for the purpose of preventing or restricting exploitation, and as needing the cooperation of other Parties in the control of trade.

The Parties shall not allow trade in specimens of species included in Appendices I, II and III except in accordance with the provisions of the present Convention.

---

have their samples to avoid others making collections. *Paphiopedilum delenatii* is an endemic of southern central Vietnam. Averyanov (1996) recorded approximately 6 tonnes of *P. delenatii* as living plants were exported from Vietnam in the mid-1990s with dealers paying between US$1 and US$3 per kilogram of plants, which were subsequently sold-on in Taiwan at approximately US$10 per kilogram. The trade in *P. delenatii* has decreased in recent years, possibly because the species is Extinct in the Wild. Many species have been over-collected from Southeast Asia, and populations have been extirpated even from protected areas such as national parks and nature reserves. Most *Paphiopedilum* spp. are naturally rare, having restricted geographic distributions and narrow habitat preferences, and it is estimated that 25 of the 60 recognized species of *Paphiopedilum* are seriously endangered in the wild (Salazar, 1996).

Many medicinal plants are cultivated, but some are difficult to cultivate, or are more valuable in their wild form. Some species in the botanical medicines trade are common and widespread 'weedy' species found in many regions, and some of these can therefore withstand wild collection. However, the collection of wild plants for the botanical medicine industry can threaten species with limited distributions, and those with small populations. In extreme cases, whole populations are removed from the wild. Harvesting of species such as pygeum (*Prunus africana*) and yohimbe (*Pausinystalia yohimbe*) is currently destructive and unsustainable (Laird and ten Kate, 1999). Medicinal plants are frequently traded internationally, and this trade can be extremely lucrative. Some products are so valuable that trade continues even when the species becomes rare in the wild, and in many cases represents a serious threat to the species' survival. Some of these species are subject to international regulations such as CITES (Box 14.18), but the illegal trade continues.

Several agencies and conventions aim to promote the sustainable use and management of natural products including wild species. The Convention on International Trade in Endangered Species of Flora and Fauna (CITES) has promoted sustainable utilization of certain plant species through the monitoring of trade in plants and plant products. Around 200 medicinal plant species are listed on the CITES Appendices.

Sustainable use of biological resources is one of the pillars of the CBD (see Chapter 2) and is inherent throughout the text of the Convention. Article 10 of the CBD has a specific focus on sustainable use. The Global Strategy for Plant Conservation (GSPC, see Chapters 1 and 2), developed and implemented within the framework of the CBD, is one of the main international policy efforts aimed specifically at the sustainable use of plants. The GSPC includes Target 11, *No species of wild flora endangered by international trade* and Target 12, *All wild-harvested plant-based products sourced sustainably.* The FAO has developed a range of policies covering the sustainable use of NTFP. Certification schemes of relevance to the sustainable management of wild plants falling within the NTFP's category include the Forest Stewardship Council (FSC), the International Federation of Organic and Agriculture Movements (IFOAM) and Fairtrade Labelling Organizations International (FLO) (Shanley *et al.*, 2002). Harvested wild plants specifically fall under the IFOAM section entitled *Collection of Non-Cultivated Material of Plant Origin including Honey*, which states:

wild-harvested products shall only be certified organic if derived from a stable and sustainable growing environment. Harvesting or gathering the product shall not exceed the sustainable yield of the ecosystem, or threaten the existence of plant or animal species.

Shanley *et al.* (2002) provide valuable guidance for management and certification of a range of wild plant resources. While Ingram *et al.* (2017) also review measures to better govern wild plant resources.

The Chiang Mai conference of 1988 with the rallying cry '*Saving Plants, Saving Lives*' brought global policy attention to the need for sustainable management and conservation of medicinal plants. This was followed in 1993 with the joint publication by the WHO/IUCN/WWF of guidelines for the conservation of medicinal plants, which were most recently revised by WHO/IUCN/WWF/TRAFFIC. There has been much recent global attention to the problem of overexploitation of medicinal plants which has seen the involvement of UNDP, GEF, IUCN, BGCI, WWF, TRAFFIC, FAO, UNCTAD, UNIDO, UNESCO, Danida and GIZ (Gesellschaft für Internationale Zusammenarbeit), among others. This has contributed to much of the effort being fragmented, unfocused and uncoordinated, and despite such global political focus, there has been too little on community-based strategies, which needs to be reversed as more and more countries are in the process of recognizing customary or traditional resource rights of communities (Unnikrishnan and Suneetha, 2012). Article 10 of the CBD recognizes the need to protect and encourage customary use of biological resources in accordance with traditional cultural practices that are compatible with conservation or sustainable use requirements, and to support local populations to develop and implement remedial action in degraded areas where biological diversity has been reduced.

Conservationists are now increasingly looking at ways of using wild resources sustainably; however, no single set of sustainable use criteria can be applied universally. Working towards sustainable use of wild resources is a complex process involving many political, social, economic and biological factors, which vary according to region and culture.

A particular area of emphasis today is on initiation of community conservation projects (a topic we turn to in Chapter 10), where in simplistic terms local people become the owners and stewards of the resources, and therefore reap the benefits of utilization. Such 'bottom-up' initiatives have resulted from the recognition that sustainable utilization can only be achieved with local community support, and through collaboration between farmers, local interest groups, NGOs and government (Boxes 14.19 and 14.20).

Jenkins and Roberts (2000) highlight several important points regarding the sustainable use of wild species:

- **Problems with existing conservation legislation applying to species of wild fauna and flora.** In most cases, legislation has been designed primarily to conserve species and has tended to disregard the needs of people, placing the emphasis on strict protection as the principal strategy. Such an approach is unlikely to be effective or sustainable in most developing countries as local communities will be unsupportive. Laws and government policies will not be effective in conserving biodiversity if the local communities who live with the wildlife and use natural habitats do not see that conserving biodiversity is to their advantage and embrace conservation action.
- **Market demand directly influences the management regime and sustainable use of wild species** through excessive harvesting of wild populations, exhausting the capacity of wild populations to sustain off-take, and unregulated captive production systems, which have the potential to saturate markets and drive down the unit value of the resource. This in turn can have the negative effect of removing the economic incentive for local land-owners to manage and conserve the species and its ecosystem.

- **Permit and licence systems, which are commonly used by governments to control the activities of individuals involved in harvesting and using wild species, are ineffective in many developing countries.** This approach has tended to foster large bureaucracies to administer the management programme, and is compromised by limited administrative, technical and financial capacities in many developing countries. Further, the bureaucracies have no remit to ensure the development of local communities, so are negatively viewed as external agencies enforcing change irrespective of its impact on local communities.
- **Two options that have recently been explored as alternative and more effective strategies are:**
  - Partnership arrangements between governments and local communities and/or private companies; under these partnerships, many of the functions that were formerly the sole responsibility of government agencies are being addressed by institutions established by users (e.g. CAMPFIRE programme in Zimbabwe; Frost and Bond, 2008).
  - Government devolvement of effective 'ownership' of, or management responsibility for, wild species to private stake-holders or local people; thus, benefits can be focused locally, with greater potential to create social and economic conditions that favour sustainable use. See Chapter 10 for further examples of successful community engagement with formal sector conservation.

For centuries, many local communities have practised customary use of biodiversity in accordance with traditional cultural systems. Many of these systems have involved sustainable harvest regimes in order to conserve the natural resources on which they rely for their survival. However, although many traditional cultural systems incorporate sustainable management practices, and much can be learned from these systems, there are also cases where use of resources by local communities is unsustainable. The reasons for this may be complex and vary on a case-by-case basis. For example, historically, the Bedouin

Box 14.19 | **Community-Based Medicinal Plant Conservation in Laos**

(Unnikrishnan and Suneetha, 2012)

The Cham people of Ninh Thuan province of Lao PDR have a long history of medicinal plant use based on around 300 species from about 100 plant families. These plants are also traded with neighbouring countries. In recent times the populations of many of the medicinal plants involved have started to decline because of over-harvesting to meet growing demand, inefficient production methods and changing weather conditions. Using resources from a UNDP small grants project, a project was initiated that links to the traditional knowledge of Cham communities to sustain medicinal plant production as a livelihood activity. Activities include strengthening the knowledge and capacity of Cham communities on conservation strategies, developing production systems providing self-regulations for use of medicines, improving the commercial viability of medicines by developing brands for producing villages and exploring livelihood options through formation of craft villages.

Box 14.20 | **Investment in People: A Key to Enhance Sustainability: Lessons from Northern Pakistan**

(Ahmed and Khan, 1998)

Environmental degradation can be prevented only if people are able to protect and defend their local environment. This, however, requires empowerment of people and their skill enhancement to manage their own natural resources on a sustainable use basis. In northern Pakistan, the Aga Khan Rural Support Programme (AKRSP) established a network of more than 2000 Village Organizations (VO), institutions since 1982, with a view to organize people for collective action. The programme has helped VO enhance their skills and generate their own capital for improvement of their economic well-being. The investment in people is now paying dividends and the VO have become basic building blocks of supra help programmes in the field of primary health, education, conservation of natural resources, marketing, etc. The sustainable use of natural resources through community participation warrants investment in people on the pattern of AKRSP.

shepherds of the Middle East practise a form of rotation grazing centred on the Jebel Druse, a plateau on the Syrian/Jordanian border surrounded by desert. They started grazing their flocks at lower altitudes in the Spring and gradually moved with the growing season to higher altitudes, so flocks constantly arrive at fresh pastures as they were needed and were ready to sustain grazing. However, the so-called practice of

'Heif' was abandoned in the 1960s as increases in flock sizes made rotational grazing less practical, and individual countries started to dissuade shepherds from following their traditional seasonal migration routes which involved crossing international borders. As a result, the seasonal pastures have been seriously degraded from local over-grazing throughout the migration range, and there has been a significant loss of plant diversity as a result. This is particularly poignant as the biologically isolated plateau had numerous endemic species. The recent civil unrest in Syria has resulted in significant flock reduction and an unexpected boost for local biodiversity, but when the political situation is normalized, careful environmental management by local communities will be required to ensure the biodiversity recovery is sustained (N. Maxted, pers. comm.).

The socio-economic problems described above are not easily overcome, particularly regarding over-populated lands, people affected by the ravages of political unrest and war, and stochastic events such as drought, floods and earthquakes. However, it is generally agreed that environmental degradation can only be prevented if people are able to protect and defend their local environment. Empowerment of local communities to manage the natural resources that they use and upon which they rely is recognized as the way forward for effective conservation and sustainable use of natural resources in many areas. Again, this is reflected in the provisions of Article 10 of the CBD.

Empowerment of local communities involves an open democratic process which also provides opportunities for minority interest groups to be heard and have a say in policy-making, development of interest groups and lobbies, a change of attitude on the part of government to ensure that the interests and rights of communities are not ignored or overruled, capacity building of stakeholder organizations, strengthening the financial and legal status of stakeholder organizations, and support for new, spontaneous initiatives by stakeholders.

Empowerment may also be facilitated by decentralization, where the power and rights over the environment and natural resources is returned to local communities and groups. While the political ramifications of empowering local communities are often complex and difficult to resolve, many case studies illustrate the success of this approach. Box 14.20 is a summary of a successful programme empowering local communities in northern Pakistan.

Tenure and access rights have also been identified as a common factor across regions in promoting increased sustainability of uses of wild renewable resources. Tenure can have several meanings depending on the region: ownership of land, which may or may not include ownership of the resources associated with that land; access rights to selected resources, such as fishing rights; indigenous people rights. In some regions, tenure is an integral part of national law and policies, and in others, the concept is highly controversial in the context of local political institutions.

15 Germplasm Evaluation

15.1 Introduction

It has been stressed several times throughout this text, but it is worth repeating here, the end point of genetic conservation is not conservation *per se* but utilization. Therefore, it is critical that germplasm users are facilitated in their access to conserved plant genetic resources whether held *ex situ* or *in situ*, but what is unknown or uncharacterized cannot be used. Hence, the extent of phenotypic variation and genetic diversity of germplasm available in gene banks should be fully determined in order to maximize its utilization. This knowledge further allows the proper organization of a gene bank and facilitates the use of accessions therein in plant breeding (for population improvement of cultivar development), genetic resources research or other related purposes.

Gene banks accessions are numerous (depending on the species) and diffuse (located in geographically diverse locations), which has resulted in significant accession duplication and non-effective management and non-rational utilization of genetic resources. While some level of *ex situ* duplication between gene banks is desirable to ensure safety back-up, excessive duplication is wasteful of resources. Actions taken to reduce redundancy among accessions include: (a) retaining those from regions with unlikely recurring collecting, (b) reducing sample size within each accession by splitting them apart, (c) forming bulks of known redundant accessions and (d) randomly selecting redundant accessions to retain while discarding others. The second and fourth options result in losing alleles *vis-à-vis* the first option. Hence, gene banks should define a minimum set of accessions that keep most of the diversity available in the cultigen pool. This subset of accessions serves as an entry point to the whole collection and improves germplasm access for further use in plant breeding and research.

15.2 Identification of Subset(s) of Gene Bank Accessions for Evaluation

The probability (P) of including all the alleles during sampling taking into consideration the percentage of accessions containing the allele of interest is estimated as

$$P = [1 - (1 - q)^n]^L$$

where q is the frequency of the rarest allele in the collection, n is the number of accessions in the collection and L is the number of loci under investigation (Sedcole, 1977; Ryder, 1988). This equation determines the sample size to retain a P percentage of the alleles at L loci having a known q frequency or can be used to establish the chance of including all alleles with known q frequency at L loci by sampling n accessions from the gene bank.

The assessment of genetic diversity assists in sampling the gene pool available in a gene bank. This sampling should target selecting accessions that represent the genetic variability of the whole gene pool to assist in managing and using plant genetic resources held by a gene bank, commonly referred to as a core collection (Box 15.1). The subset must be assembled by clustering accessions and sampling thereafter within these groups (Box 15.2).

A core collection should include minimum redundancy of the genetic diversity of a crop species and its wild relatives (Brown, 1989). It often consists of 10% of the gene bank accessions (3000 being the maximum), which should include 70% of the alleles. The upper limit of 3000 accessions allows keeping alleles with a frequency of 10^{-4} in the species according to the theory of neutral alleles. This sampling should include rare, widespread alleles, while common widespread and localized alleles are very likely in the core subset, especially after successful clustering. Rare localized alleles may not be in the core subset because of the impracticability of

---

Box 15.1 | **Defining a Quinoa Core Collection with Passport Data and Descriptors**

(Ortiz *et al.*, 1998)

Quinoa is an Andean crop whose new cultivars require plant characteristics that may be available in genetic resources held in gene banks. A core subset of the whole quinoa gene bank (1029 accessions) of the Universidad Nacional del Altiplano (UNAP, Puno, Peru) was defined using available passport data (location and altitude) as well as both qualitative and quantitative descriptors. The core subset includes 103 accessions that are ecotypes or landraces capturing most of the genetic variability available in this Peruvian quinoa germplasm. These accessions were selected based on a geographically stratified non-overlapping sampling procedure. The number of accessions that were allocated to the core subset was determined using a proportional method adjusted by the relative importance of the quinoa crop in each geographic cluster as determined by its acreage. The sampling method also considered the morphological diversity within four geographic clusters of at least 100 accessions. The multivariate pattern of morphological variation was defined within each of these clusters by independent principal component analyses. A comparison of phenotypic diversity confirmed the proper sampling strategy for this core subset of Peruvian quinoa germplasm. The most important phenotypic correlations between quantitative descriptors – likely under the control of co-adapted gene complexes – were also preserved by the core subset, which was defined as an entry to the proper exploitation of quinoa genetic resources available in this gene bank.

---

conserving everything. Sampling for the development of a core subset considers the hierarchical structure of the gene pool; i.e. stratification into groups sharing common characteristics such as taxonomy, geographic or ecological origin, and neutral or non-neutral descriptors.

The core subset ensues from sampling using stratified non-overlapping clustering; i.e. taking a sample from each distinct region or taxon (Hamon *et al.*, 1995). Such a sampling is likely to include an equal number of accessions from each group (*K* or constant strategy), a fixed fraction of accessions for each group (*P* or proportional strategy) and a proportional number of accessions to the log size of the group (*L* or logarithmic strategy). The results from each strategy differ: *K* shows a bias towards the smallest group, while *P* could favour most numerous groups when the same proportion was taken for each

group, and *L*, which seems to be the most appropriate choice, is based on the relationship between sample size (η) and number of alleles at a locus (κ) as follows

$$\kappa \approx \mu \log \delta\eta$$

where μ and δ are constants depending on mutation rate and the population size, respectively.

Genetic marker data could provide additional information to maximize allele richness in the core subset during sampling (**Box 15.3**). For example, the *H* strategy depends on Nei's diversity index.

$$1 - \sum q_{ij}^2$$

where q_{ij} is the allele frequency in the ith locus of the jth group. This approach leads to a diverse and representative core subset when unequal diversity and differentiation among accessions exists.

Box 15.2 Selecting a Core Subset of Hexaploid Sweet Potato Based on Identifying Duplicates and Using Morphological Descriptors plus Eco-geographic Data

(Huamán *et al.*, 1999)

Sweet potato ranks among the seven most important food crops of the world. The Centro Internacional de la Papa (CIP) holds one of the largest hexaploid sweet potato gene banks with more than 5000 accessions, which are clonally maintained because these farmer-selected varieties have been asexually propagated for many years. Because of this, there are duplicate accessions of the same varieties. Almost 30% of the sweet potato accessions assembled in this gene bank are from Peru, thus the first step to select a sweet potato core subset was to identify duplicates therein using morphological characters and electrophoretic banding patterns of total proteins and esterases. Such an approach reduced the number of Peruvian accessions from 1939 to 673. The number of duplicates of the same cultivar ranged from 1 to 99 accessions. A total of 21 morphological descriptors were scored in all the distinct Peruvian accessions. The unweighted pair-group method using an arithmetic average (UPGMA) determined the pairwise distance for members of distinct clusters based on these morphological descriptors. A core subset was selected using the square root of the number of accessions for each Peruvian department and respective cluster, as defined by UPGMA. The core subset includes 85 accessions (12.6%) from all Peruvian departments except that of Madre de Dios (lacking collection), and from all agro-ecological zones except Paramo, which has only 0.5% of the gene bank accessions. The sampling for this core subset was suitable as determined by comparisons of means and frequency distributions for all morphological descriptors. This sampling was further validated by the partial assessment of this sweet potato germplasm for resistance to pathogens and pests, salt tolerance, storage root dry matter content and vegetative period. This multi-step approach defined a core subset after giving priority by region and clustering germplasm according to specific discriminating characters.

15.2.1 How Many Accessions to Include in a Trial?

The characterization of the germplasm is done on a sample that is sometimes very small, and the statistics from this sample are used for the entire population. A few plants grown in a single site may represent the sample characterizing an accession if the population is an accession. Available funding and site characteristics determine how many plants (N) can be grown in a trial. Very often, accession subsets for evaluation are based on experience or practical knowledge. The total number of accessions will be ½ N when having at least two replications per accession in the trial. Hence, it will be very important to choose a suitable number of accessions (e.g. a mini-core subset, **Box 15.4**) that provides a valid estimate of the range of variation available in the collection. Known accessions such as popular cultivars are also included as checks or controls to have benchmarks for comparing results.

15.2.2 Pre-screening Accessions

It may be possible to pre-screen accessions using an approach that allows identifying genetic diversity or

---

Box 15.3 | **Sampling Tetraploid Andean Potato Using Characterization, Evaluation and Isozyme Data**

(Huamán *et al.*, 2000, 2001; Chandra *et al.*, 2002)

The Centro Internacional de la Papa (CIP) maintains the largest gene bank of tetraploid Andean potato cultivars. After identifying duplicate accessions using morphological characters and electrophoretic banding patterns of total proteins and esterases, the number of accessions was reduced from 10 722 to 2379. The number of accessions of the same cultivar in the original collection ranged from 1 to 276. Morphological, geographical and evaluation data were used to construct a phenogram using a simple matching coefficient and the unweighted pair-group method using arithmetic averages. The accessions included in the core subset were chosen to represent the widest morphological diversity and to maximize geographic representation of the clusters distributed on the main branches of the morphological phenogram. The proportional sampling included approximately the square root of the number of accessions from each first geographic division of countries according to the collecting data for this germplasm. The representative accession of each cluster was chosen considering data on their resistance to pathogens and pests, dry matter content, and number of duplicate accessions therein. There are 306 accessions (12.86%) from eight countries from Mexico to Argentina in this core subset, whose sampling strategy was further validated by investigating the genetic structure in both the entire gene bank accessions and the core subset with genetically characterized isozyme markers. Only two rare allozymes with a gene frequency of 0.0002 were not included in the core subset, while the most frequent in the entire gene bank accessions had the highest frequencies in the core subset. The allozyme frequency distributions were homogeneous for all loci except two. A simulation approach using these isozyme data was also used to test five sampling strategies: constant (C), proportional (P), logarithmic (L), square root (S) and random (R). A core subset of 600 accessions selected using either the P or the R sampling strategy was adequate to represent the diversity available in a tetraploid potato gene bank of 1910 genetically distinct groups, as measured by allele frequency and heterozygosity level.

---

variability. The more similar two accessions are morphologically, the closer they may be genetically and *vice versa*. Hence, it is possible to choose a sample of accessions for evaluation to minimize the probability of wasting time and resources by unknowingly duplicating material. For example, a computer-based stratified core selection (**Box 15.5**), which has its origin in the core subset concept, facilitates users focusing on selecting germplasm with a broad diversity for the trait of interest.

Most algorithms and methods for forming diverse core subsets depend on either allele representativeness or richness. 'Core Hunter' is an algorithm based on genetic data for sampling plant genetic resources having high average distance between accessions, or rich diversity, or a combination of both (Thachuk *et al.*, 2009). It also optimizes any number of genetic measures simultaneously, based on the preference of the user (http://corehunter.org). Core Hunter may select more

---

Box 15.4 | Establishing a Mini-core of Chickpea Genetic Resources

(Upadhyaya and Ortiz, 2001).

In many crops the number of accessions contained in the gene bank runs to several thousands, thus a core subset consisting of 10% of total accessions would be an unwieldy proposition. A mini-core subset consisting of only about 1% of the entire gene bank may still represent the diversity of the entire varietal diversity available therein. A two-stage strategy was used to select a chickpea mini-core subset from 16 991 accessions held at the gene bank of the International Crop Research Institute for the Semi-Arid Tropics (ICRISAT). The first stage involved developing a representative core subset (about 10%) from the entire collection using all the available information on origin, geographic distribution, and characterization and evaluation data of accessions. The second stage involves evaluation of the core subset for various morphological, agronomic and quality traits, and selecting a further subset of about 10% accessions therein. At both stages, a standard clustering procedure was used to separate groups of similar accessions. A mini-core subset consisting of 211 accessions from 1956 core subset accessions, using data on 22 morphological and agronomic traits, was selected. Newman-Keuls' test for means, Levene's test for variances, the χ^2 test and Wilcoxon's rank-sum non-parametric test for frequency distribution analysis for different traits indicated that the variation available in the core subset has been preserved in the mini-core, which due to its drastically reduced size, will allow proper field testing of these genetic resources.

---

small core subsets that retain all unique alleles from a gene bank than other algorithms.

Predictive characterization is a group of related methodologies that can be used to pre-select accessions for utilization (Thormann et al., 2014). These techniques are predictive in the sense that they assign an increased likelihood of trait presence to uncharacterized germplasm (either ex situ or in situ). There are three basic techniques each matches: (a) known biotic and abiotic characteristics of a collecting site (= FIGS method) (b) ecogeographic information associated with a collecting site (= ecogeographic filtering method) and (c) previously recorded trait occurrence with a set of locations different from those where the germplasm being examined has been collected (= calibration method) (Figure 15.1). For each method a predictor is used to build a hypothesis that germplasm from a location will contain the desired adaptive trait. While use of

predictive characterization does not avoid the use of resource-expensive field trials, it significantly reduces the number of potential accessions the breeder needs to screen to find the desired adaptive trait. The first systematic application of finding a predictive link between a resistance trait and a set of environmental parameters was named the Focused Identification of Germplasm Strategy (FIGS) (Mackay and Street, 2004; Street et al., 2008) that used biotic and abiotic matching techniques. Further examples of predictive characterization applications are shown in Box 15.6.

15.3 Characterization and Evaluation

Table 15.1 summarizes the main features of germplasm characterization and evaluation. The former begins with collecting or introducing plant genetic resources, and it ends with the publication and

---

Box 15.5 | Facilitating Access to a Gene Bank through Core Selection

(Mahalakshmi *et al.*, 2003)

Static core subsets, which are *a priori* selected by a gene bank curator, are often of limited use to users who are interested in a specific character or domain, which is any set of germplasm with a specific character or purpose, e.g. host plant resistance to a pathogen or seed size, or combination of traits such as time to flowering and host plant resistance to a pest. Information technology makes it possible for any user to make such selections themselves through the web. A stratified selection methodology allows to choose accessions within the domain of interest. The division of a domain in genetically distinct groups depends on the nature of the descriptor for which the domain was defined. The user can choose from the list of descriptors available in the database. They are either qualitative (colour, texture, country of origin) or quantitative (days to flowering, stem height, leaf size). The stepwise division of groups of a qualitative nature can easily be accomplished, as they are distinct. Quantitative descriptors can be divided into groups based on the distribution and range value for the trait among the accessions. The choice of the number of groups in such cases is left to the user from a minimum of 5 to a maximum of 20 classes. The size of the stratified selection can be chosen to be between 1 and 10% of the total collection size. The user can choose the selection algorithms based on either the proportional (*P*) or logarithmic (*L*) sampling strategy. The system selects a minimum of one entry per group to ensure the representation of small groups. This approach provides users with more focused selection of the germplasm with the diversity of the character of interest than a core subset.

---

sharing of the information related to the gene bank accession(s). Characterize is a synonym of distinguish, which means to mark as separate or different, or to separate into kinds, classes or categories. Characterization is largely based on recording simply inherited descriptors that are often highly heritable and are expressed in all environments. The main objective of characterization is to describe and determine the value of plant germplasm, while specific objectives include the true taxonomic clustering, suitable morphological description and phenotypic variability.

Evaluation depends on recording characters showing more complex inheritance and is often influenced by the environment and relationships between characters. Its main objective is the accurate and precise assessment of the agronomic value of gene bank accessions. This activity may include the

following stages: identifying core (or mini-core) subset from the cultigen pool available in the gene bank, characterizing the core (or mini-core) subset, selecting target characters to be evaluated, designing suitable and efficient trials for assessing the variation in the core (or mini-core) subset, data analysis using proper statistics, and reporting results to potential users.

15.4 Descriptors: Characters for Measuring Variation and for Evaluation

The descriptors give a code or value to descriptor state present in the accessions. The state of the descriptor is each of the variables of a character. The descriptors can be double-state (e.g. simple or compound leaves) or multi-state (e.g. red, mauve, blue flower colour).

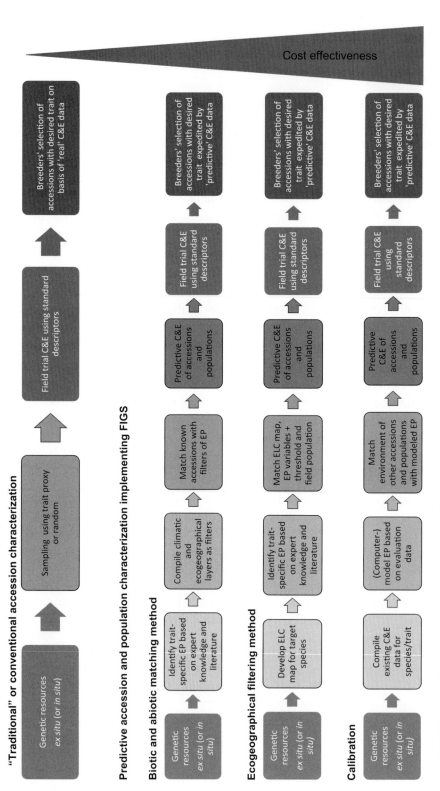

Figure 15.1 Different predictive characterization techniques. C&E = Characterization and Evaluation; EP = Environmental Profile; ELC = Ecogeographic Land Characterization; FIGS = Focused Identification of Germplasm Strategy. (From Thormann *et al.*, 2014)

Table 15.1 **Plant germplasm characterization and evaluation in gene banks**

Item	Characterization	Evaluation
Scope	Recording highly heritable and environmentally stable characteristics	Agronomic description of traits affected by the environment
Traits	Often qualitative and easy to measure and used for classification	Most often quantitative and less easy to measure and use for performance testing
Descriptors	Colour, pubescence and shape of plant parts; biochemical and DNA markers	Height, yield and components, time to flowering, maturity, protein or starch content, tolerance to stresses

---

Box 15.6 | **Examples of Predictive Association Studies and Identification of Pest- and Pathogen-Resistant and Drought-Tolerant Accessions**

- Powdery mildew resistance in wheat (Kaur *et al.*, 2008; Bhullar *et al.*, 2009).
- Sources of wheat resistance to Sunn pest, *Eurygaster integriceps* (El Bouhssini *et al.*, 2009).
- Predictive association between trait data and ecogeographic data for Nordic barley landraces (Endresen, 2010).
- Sources of resistance in bread wheat to Russian wheat aphid, *Diuraphis noxia* (El Bouhssini *et al.*, 2011).
- Predictive association between biotic stress traits and ecogeographic data for wheat and barley (Endresen *et al.*, 2011).
- Wheat stem rust resistance linked to environmental variables (Bari *et al.*, 2012).
- Resistance to stem rust (Ug99) in bread wheat and durum wheat (Endresen *et al.*, 2012).
- Traits identified related to drought adaptation in *Vicia faba* genetic resources (Khazaei *et al.*, 2013).
- Wheat yellow stripe rust resistance linked to environmental variables (Bari *et al.*, 2014).

Quantitative multi-state descriptors may be discontinuous as in leaf length. Descriptor standardization is required for characterization worldwide. Those dedicated to the characterization of germplasm should agree on the name, form and units to measure descriptors.

Bioversity International (2007) makes available almost 100 crop descriptor lists (www .bioversityinternational.org/e-library/publications/descriptors/) prepared by experts. These give the basic depiction of the characters of a crop and the different states that are expressed (characterization) or how to measure the range of their variation (evaluation), and include passport, characterization, evaluation and management descriptors. Passport data provide information on the sample identification and geographic location where the gene bank accession was collected (Table 15.2). Characterization descriptors (Table 15.3) are highly heritable, can be seen with the naked eye and are expressed without major variation across environments. Preliminary evaluation data (Table 15.3) refer to characters that

Table 15.2 **Most important gene bank passport data (GENESYS)**

Type	Description
Accession number	A unique identifier given to a gene bank accession and made up of a prefix, a sequence number and sometimes a suffix
Acquisition	Date accession was brought to the gene bank
Other identifiers	*Collected material*: accession name, collecting date and site plus collector *Breeder's material*: Pedigree (or selection history), names and identifiers used in breeding programme *Acquisition from other gene bank*: accession passport data as documented in the source
Taxonomy	Accession genus, species, species author, sub-taxon and sub-taxon authority
Storage and maintenance	*Method*: seed (short-, medium- and long-term), field, *in vitro*, cryopreserved, DNA

Table 15.3 **Main features of descriptor types**

Type	Scale	Measurement	Typical descriptor
Quantitative	Interval	Direct measure of a trait	Height, yield, days to flowering, maturity and harvest, protein content, yield and components
	Ratio	Combining two trait measures into a single descriptor	Harvest index, content percentage, length/width index
Qualitative	Ordinal	Relative value assigned within a standard scale	Host plant resistance (from 1: highly tolerant to 9: highly susceptible), shapes, overall quality
	Nominal	Qualitative state assigned into arbitrarily number-scale	Colour, seed pattern, growth habit
	Dichotomy	Trait presence or absence	Pubescence, colour spots
Other	Gene symbols	Known gene names assigned	*Italic fonts* given to known alleles

are required by germplasm users. Advanced evaluation descriptors are additional characters that are useful for plant breeding or are important for defining relationships between related species or amongst taxa of the same species. Management data include information that is essential for conserving gene bank accessions and for their regeneration.

The types of measurement data that may use intervals to quantify an attribute are ratios of two direct measurements or an index resulting from an inference, ordinal standard as a relative value from a scale, and nominal patterns assigning quantitative states into arbitrary classes. Accessions are evaluated and given a numerical score using the metric system if the character is quantitative. A numerical scale (1–9) is used for qualitative characters that vary continuously, e.g. host plant resistance to pathogens and pests or continuous colour variation. The value

---

Box 15.7 | **Choosing Quantitative Descriptors for Evaluation of Peruvian Highland Germplasm**

(Ortiz and Sevilla, 1997)

The main goal of this research was to determine the environmental (year and season) and measurement effects, plus the genotype–environment interaction (GEI), in quantitative variation of vegetative, tassel, ear and kernel descriptors used in maize racial classification and characterization. The most useful descriptors were those having both high broad-sense heritability (H^2) and repeatability (R), and low coefficients of variation (CV). Low R indicated large environmental effects and a significant GEI in the phenotype, whereas a high H^2 was noticed in descriptors having significant differences between accessions or low GEI. Likewise, low CV suggested that the measurement was carried out with minimum error or random variation was almost nil. The best descriptors were reproductive traits such as ear length, number of rows of kernels, cob diameter and kernel width.

---

0 should only be used when the descriptor has been measured, or it must remain blank in the records when the descriptor has not been evaluated.

Broad-sense heritability (H^2), or the ratio between accession and phenotypic variances, assists selecting descriptors for grouping germplasm (Box 15.7). H^2 has, however, little value in germplasm characterization because its calculation does not include the environmental variance. Repeatability, or the ratio of variance components due to differences between accessions and the sum of the corresponding components due to differences among environments and accession-by-environment interactions, allows choosing quantitative descriptors with wide quantitative trait variation, e.g. the best descriptors for germplasm catalogues are those easy to score and showing a similar phenotypic expression across environments; i.e. high repeatability due to low or nil environmental influence. Hence, descriptors for germplasm characterization should not be biased by the environment. Meanwhile, descriptors with high heritability due to low or nil genotype–environment interaction are more important for agronomic evaluation or selection, and for classification, although they may be affected by the environment. Descriptors should also be accurate or unbiased, and

precise or with minimum error measurements; i.e. showing low coefficient of variation.

Quantitative descriptors with continuous variation are used in classification, even though the environment or the genotype–environment interaction may significantly affect their phenotypic expression. The effects of the environment and the genotype-by-environment interaction may be reduced by assessing germplasm across environments and using the mean values for each accession, evaluating the germplasm across environments and defining similar phenotypic responses for accessions in each specific environment, and comparing accessions only with descriptors that are not influenced by the environment or the genotype–environment interaction.

15.5 Trials for Assessing Phenotypic Variation

The assessment and genetic characterization of diversity and variability for further use includes assembling potentially useful germplasm, using standard descriptors for characterization and evaluation, analyzing the number and types of useful polymorphisms among the descriptors, evaluating germplasm diversity using specific but variable descriptors, analyzing data with suitable statistical tools, and selecting the genetic material according to defined breeding goals. The resources and time spent on germplasm characterization and evaluation depend on the program objectives, financial support, available data on genetic diversity (including pedigree to estimate inbreeding coefficient), and knowledge plus importance of characters under research.

15.5.1 Design and Experimental Units

The design should be kept simple because adding other factors may reduce the power of focusing on the main objective of evaluation; i.e. testing if the accessions differ and, if so, by how much for a specific (quantitative) descriptor. The basic experimental unit may be a single plant, or it could be a plot having several plants. The efficiency of any experimental design raises as the number of its units increases while the size of the unit decreases (Gauch, 2006). An optimal design would therefore involve many plots having one plant (sometimes randomly taken) to be individually scored. The evaluation of quantitative descriptors such as edible yield uses, however, r plots of k plants. Although the larger the plot, the closer to the true farming situation, the design becomes less efficient statistically because using large plots means few replications.

15.5.2 Field Plot Techniques

The objective of an experimental design for estimating accession means aims to minimize the error variance for a given cost (Box 15.8). Estimates become more accurate as the number of times the measurement is taken increases; i.e. precision rises as the number of plants measured within a plot increases and the number of replications increases. The accuracy rises when the variance of averages ($\frac{\sigma^2}{n}$) is lower and this variation is proportional to

$$\frac{\sigma_E^2}{r} = \frac{\sigma_S^2}{rk}$$

where σ_E^2 and σ_S^2 are the size of the experimental error (or the variance of plots of the same accession) and

the variance between plants of a plot (within accession), respectively; while r and k are number of replications and number of plants of a plot, respectively. If it is not possible to increase the number of replications, it is necessary to increase the number of plants within the plot. In very homogeneous populations, the number of replications must be increased and the number of plants within the plot must be reduced, while the number of plants within the plot should increase and the number of replications should decrease in heterogeneous populations. The size of the sample to be used for evaluation depends, therefore, on the population variance, the level of heterozygosity and the allelic frequencies of polymorphic loci.

Optimum plot sizes for yield trials may be adjusted by adding fixed and variable costs of research and soil heterogeneity (b_i) of the experimental site. The coefficient b_i is estimated by solving the empirical relationship between the variance on a per experimental plot basis (σ_x^2) and the ratio between the variance among single plant plots (σ_j^2) and the number of plants per plot (k_i), thus

$$b_i = \frac{\log \sigma_j^2 - \log \sigma_x^2}{\log k_i}$$

The estimates of the variance components can be used to help improve the design of future field trials (Box 15.9). The phenotypic variance (σ_P^2) in several environments (e) and with several replicates is

$$\sigma_P^2 = \sigma_G^2 + \frac{\sigma_{GE}^2}{e} + \frac{\sigma_E^2}{er} + \frac{\sigma_E^2}{erk}$$

where σ_G^2 and σ_{GE}^2 are the genetic variance and the variance of the genotype–environment interaction, respectively. Hence, σ_{GE}^2, σ_E^2 and σ_E^2 decrease as e, r and k increase until reaching a minimum in which $\sigma_P^2 = \sigma_G^2$, i.e. $H^2 = 1$, thus enhancing the precision for the evaluation.

15.5.3 Replication, Randomization and Blocking

The cost and logistics of phenotyping impose limits on sample size. Allocation of resources needs, as noted above, should consider the number of testing

---

Box 15.8 | Optimum Plot Size for Banana Trials

(Ortiz, 1995; Nokoe and Ortiz, 1998)

Testing plant germplasm requires accuracy and precision. Uniformity trials of the banana cultivar Valery grown under two different management practices, namely alley cropping of banana with multispecies hedgerows and mono-cropping of banana in cleared land, and at two locations in the humid forest zone of West Africa were used to determine optimum plot size (i.e. number of plants per experimental plot) and the number of replications to detect significant true mean differences in bunch weight between accessions. The methods of maximum curvature and comparison of variances determined optimum plot size, while the number of replications was estimated by Hatheway's method. This statistical procedure considered an expected relative magnitude of differences between treatment means at a specific probability level and the relationship between the coefficient of variation and soil heterogeneity. The experimental plot should have a single row of 5 plants for banana in an alley cropping system, whereas single rows of 20 plants per plot should be used in a mono-cropping system. This finding indicates that alley cropping of banana with multispecies hedgerows may be an adequate system not only to maintain diversity but also for great efficiency in testing *Musa* germplasm. A table was developed, indicating the expected detectable true mean differences for bunch weight per plant for different number of treatments tested (5–30), plants per plot (5–40), replications (2–6) and resource management practices, to select minimum trial sizes. For example, a true bunch weight mean difference of 15% may be found to be significant when 30 accessions are included in a randomized complete block design with two replications of 10 plants per plot at alley cropping fields. Further research using segmented models, in which each plant from the data set was regarded as the basic experimental unit, suggested that 9–30 plants plot could be optimal for sole crop experiments. Plot size depended on target character and production cycle, e.g. about 13 ± 1 plants suffice to assess host plant response to black leaf streak, while 16 ± 3 plants are enough to evaluate growth characters and yield potential. Lower coefficients of variation were noted in the plant crop than in the ratoon, thus optimum plot size should be 13 ± 3 plants for the plant crop and 15 ± 2 plants for the ratoon.

---

environments and replications therein. Furthermore, appropriate replication number and suitable randomization determine to a large extent the efficiency of evaluation. Replication use often leads to accurate and precise estimates of the accession mean, or its average value, for a character. Precision increases linearly with the square root of the number of replications ($\sqrt{r}$). Hence, efficiency gains will be noted by increasing it.

It is essential to randomize accessions in the trial area to obtain an unbiased estimate of the error and accession variation. Accessions should therefore be allocated to a particular location in the trial field completely at random to avoid introducing any bias into their data recording or scoring. This should be organized before sowing, and the allocated position of each unit carefully recorded.

Box 15.9 | Minimum Resources for Phenotyping Quantitative Characters in Maize

(Ortiz *et al.*, 2008b)

Peruvian highland maize accessions were grown at two consecutive planting seasons in 2 years at one inter-Andean site in northern Peru. The trial data provided a means for calculating the variance components using the restricted maximum likelihood method for vegetative and reproductive maize characters. The variance components were assumed to be stable while the number of environments and replications varied to simulate phenotypic variation for each character. The least number of environments and replications, which does not affect the precision of phenotyping, was selected for assessing each character. Tabulated data provided the number of environments and replications that could be used to assess quantitative variation in plant and reproductive characters. The results indicated that fewer environments and less replications are necessary for reproductive characters than for plant characters because the former show higher heritability than the latter.

The trial area increases by adding replications, thus exposing plants of each accession to field heterogeneity. It is therefore advisable to introduce blocking to the randomization process by dividing the experiment into several equal units in space (or time). Any differences due to blocks are removed when doing data analysis after defining the appropriate experimental design. Spatial variability was accounted, among others, for by inter-block variability, which is a reality when using either complete or incomplete block designs in field trials. Unaccounted spatial variability may lead to erroneous conclusions (Singh *et al.*, 2003). Spatial methods of analysis for modelling local trends using nearest-neighbour residuals are therefore used to address this issue, thus increasing trial efficiency.

15.5.4 Provenance Effects

Seed quality, maturity and age should be the same between accessions included in a trial to estimate the true magnitude of differences between them. It is therefore inappropriate to raise plants from seed taken straight from a gene bank in a trial, because they will have been housed within the gene bank for varying time periods. Seed (or other propagules) from different

provenances often differ due to various factors, *inter alia*, quality of environment of seed parent, time of year the seed was set, or age. Trial data may differ from such factors instead of being due to genetic differences in provenance. Thus, it is important that seed of all accessions to be used in the trial is produced from contemporary plants raised together after taking their seed from the gene bank. Seeds are taken from these plants to be used in the trial because they are relatively free from any provenance bias.

15.5.5 Choice of Trial Sites

Initial evaluations are often undertaken in small nursery plots, while more detailed field trials are used to assess useful traits that show quantitative variation. Suitable screening techniques are required to assist in identifying sources of host plant resistance to main pathogens and pests. Advanced testing should be undertaken in the target population of environments (i.e. within the geographic and climatic span of the crop plant); under well managed farming; and with fairly homogeneous soil fertility, depth and moisture across the site; as well as sunshine and other factors (e.g. plant density, watering). Trial sites should therefore be representative of where cultivars or landraces will be grown, and reasonably homogeneous, but keeping some of the heterogeneity found in practice to ensure cultivars to be further released are suitable to the farming environment in which they are likely to be grown.

15.6 Statistical Analysis and Quantification of Variability

There are various textbooks (see *References*) and other publications devoted to the statistical analyses that may be used for field data. Sophisticated statistical analysis software (e.g. *Genstat, R* or *SAS*) are also available to provide a wide range of output information including summary tables and graphs. The analysis of variance (ANOVA) assists on hypothesis testing by identifying sources of variation based on a mathematical model of the response(s), which will also be useful for estimating confidence intervals. Linear models use the method of least squares to obtain estimates of their parameters as defined by the ANOVA. Such models include random error variables that account for all the sources of variability not explicitly defined in the model. Statistics should tell if the differences noted among accessions are significant for a character under study, thus allowing the best accessions for it to be identified. Probability values determine the significance of the results, but their interpretation may be influenced by selection bias after testing multiple hypotheses, fitting various models or choosing only interesting results after data recording (Ziliak, 2017). Data analysis should therefore refer to the reliability of the estimate regarding the performance of each accession for characters under study and may further tell any associations among them. It is worthy noting that significant results are not always biologically meaningful, particularly when using large sample sizes or having small variability.

15.7 Genomic Characterization

DNA markers are often unlimited descriptors that offer highly reproducible results due to the complete absence of environmental influence on their expression. Chapter 5 of this book provides details on DNA markers and polymorphisms and their analysis. It also gives examples on how they describe polymorphisms in crop gene pools including bred cultigens. Such polymorphisms, which depend on the DNA marker type, are used for measuring genetic diversity among and between defined gene pools. DNA markers are used solely or along with morphology-based plant identification with other descriptors for characterizing plant germplasm. Their use allows understanding the genetic differences among cultigens and between them and crop wild relatives.

The conservation and use of plant genetic resources need an accurate plant identification. DNA markers assist on determining genetic variation within a population, thus being a useful tool for identifying distinct gene bank accessions with maximum genetic diversity. In this regard, a DNA 'fingerprint' defines a unique profile for a gene bank accession. Genomic characterization of plant genetic resources through association genetics or DNA re-sequencing may further facilitate the finding of genes or identifying novel alleles therein.

15.8 Reporting Evaluation Results

Results must be shared with others through publications in journals or reports along with their databases, which should be easy to access through the web or as hard copy upon request. Such an approach facilitates the use of gene bank accessions. Hence, efficient germplasm conservation and use requires collecting, storing, characterizing and evaluating accessions, and thereafter disseminating any related results to identify accessions for further research or use for improving breeding populations, lines, clones or cultivars.

15.9 Data Management

The search for potentially promising gene bank accessions may be facilitated through free exchange of ideas and information among potential users. Accessing the most complete and accurate characterization and evaluation data on plant genetic resources is therefore a prerequisite to the selection of appropriate material for use, whether in research, breeding or direct use by growers. Characterization and evaluation data are of little use if they are not adequately documented and incorporated into an information system that can facilitate access to data. Linking information systems from various gene banks is essential to understand the extent of variation in plant genetic resources and to define what resources to target for collecting, characterization and management. For example, GENESYS, which was established in 2008, is a free online global portal to get information on plant genetic resources for food and agriculture. Such a single-entry gateway provides means for easily retrieving information on gene bank accessions. It manages 79 datasets that include about 3.9 million accessions from 473 gene banks, which share their data with users who are able to search using various criteria. The multi-crop European Search Catalogue for Plant Genetic Resources (EURISCO) – kept at the Leibniz Institute of Plant Genetics and Crop Plant Research (IPK, Gatersleben, Germany) – gives information on both passport and phenotypic data for about 1.9 million gene bank accessions preserved by almost 400 institutes. The web server Germplasm Resources Information Network (GRIN), which is managed by the National Germplasm Laboratory (Beltsville, Maryland) of the US Department of Agriculture's Agricultural Research Service, also provides passport information and characterization or evaluation data about plant germplasm.

15.10 Conclusions

Knowledge on the genetic variability the users want to tap is key to successful utilization of gene bank accessions. Potential users can thereafter search for the target character(s) and incorporate their variability into a more usable form for further use in plant breeding. Accessions of a gene bank are chosen for evaluation because all together are not easy to assess. Core subsets, which include with minimum redundancy the genetic diversity of a crop and its wild relatives, should therefore be defined to facilitate both the management and utilization of a germplasm held by a gene bank. Core subset accessions must be randomized and replicated in evaluation trials that should mimic the farming in a locality where the crop is often grown and in which all characters are assessed using informative scales. The results of such evaluations need to be shared with potential users promptly, otherwise the plant germplasm held in a gene bank may remain underexploited. Hence, data and other related information regarding characterization and evaluation of gene bank accessions should become available through a journal article or a catalogue, which may be easy to access and read through a web portal.

16 Plant Breeding

16.1 Introduction

Plant breeding is the science, art and business of changing and improving plants through the application of genetics. It is a science because its basis is genetics, while it may be regarded as art since plant breeders have the ability, owing to their experience, to identify the best methods and criteria to select genotypes for further use. Likewise, plant breeding is a business enterprise because any cultivars to be released should bring higher economic yields than traditional landraces. It was practiced for the first time when humans originally selected the best plants from wild species for seed for planting in the following year. As knowledge about how plants grew, humans made selections with increasing efficiency. After discovering sexuality, they were able to use hybridization as a breeding technique. Heredity was properly understood after the Augustinian friar Gregor Mendel's research on garden peas. Modern plant breeding, which began in the early years of the 20th century after the rediscovery of Mendel's inheritance laws, uses the principles of genetics – along with knowledge on plant pathology and pests as well as factors affecting plant adaptation to stressful environments or controlling quality characters – to increase the production and quality of crops per unit area, in the shortest time, with the minimum effort and at the lowest possible cost. Hence, plant breeding involves deliberate germplasm selection, crossing, testing and selection of offspring bearing desired target character(s). This main goal will be achieved by developing new cultivars producing more edible yield, feedstock or fibre in the smallest area of land, and that are suitable for target end-users. Plant breeding has contributed significantly to the increase of agricultural production and productivity since the very beginning of agriculture (Ortiz, 2015).

Plant breeding is in effect human-made evolution of plants, whose aim is to produce populations or cultivars with superior agronomic or economic characters. Why? New cultivars that produce high and good-quality yield and are pest resistant are necessary to feed the ever-increasing human population worldwide. How long is a cultivar life? Continuous cultivar turnover is key for plant breeding because genetic gains are delivered through varietal replacement in farmers' fields, but cultivar lifespans vary, e.g. every 3 years for a maize hybrid in the US 'Corn Belt' to 17 years for a maize cultivar in Kenya or 28 years for a rice cultivar in eastern India (Gary Atlin, Bill & Melinda Gates Foundation, Seattle, WA, pers. comm.). Hence, plant breeding enterprises should pursue a rapid cultivar substitution through product profiling (i.e. identify target traits), generating high-quality data to show convincingly new cultivar advantages and a strong seed system for release and dissemination. How much does it cost to produce a cultivar? Breeding systems (inbreeding or outcrossing), propagation (seed or clone), type (open-pollinated, hybrid, transgenic), selection methods and trait heritability determine the expenditures and time (often > 10 years) for developing a new cultivar.

A plant breeding strategy should be regarded as a plan that ensures genetic gains close to the optimum, in both the short and long term. This strategy should consider the goal(s) of crop improvement, the degree and pattern of genetic variation, the breeding system of the crop, the breeding method(s) and whether the target character may be easily improved by an alternative technology. Genetic variation in the target character and the effectiveness of selecting suitable parents to be used in crossbreeding are requirements for plant breeding to succeed. Breeding methods are based on crop breeding system (Box 16.1), germplasm introduction, gene recombination or mutation, and plant selection (after testing), which nowadays may be facilitated by precise phenotyping and high-throughput genotyping tracking within genome variation (Ortiz, 2015). These methods vary but share

---

Box 16.1 | **Breeding System Plays an Important Role in Choosing the Breeding Strategy**

The breeding system of the species also plays an important role in determining the procedure to follow to transfer the required trait. For instance, it would be easier to use selfing (pollinate the plant by its own pollen) when breeding self-pollinating, autogamous (e.g. wheat, barley or rice among others) species because they self-pollinate naturally. Therefore, repeated crossing will not only be difficult, time-consuming and costly in such cases, but will contribute little towards hastening the production of the new or improved cultivars because such cultivars will either be pure breeding lines or mixtures of closely related near-isogenic lines (inbred lines which differ for one or two genes only). Similarly, it will be futile to attempt selfing in outcrossing species (e.g. Brussels sprouts and other *Brassica oleracea* forms), which have in-built barriers against self-pollination.

---

features because their aim is to select the best genotypes within a population or breed new genotypes with previously defined characters. All methods (Box 16.2) are therefore designed to control flowering and pollination mechanisms, generate bred-seed whose offspring reproduce the desired genotype, make maximum use of the genetic variation available for target character(s) in the breeding population(s), and augment genetic variation for target character(s) through hybridization and gene recombination with the aim of obtaining new genotypes. These genotypes are then evaluated as segregating offspring and selected thereafter according to pre-defined breeding goals, and in which the effects of the environment are minimized, along with the genotype-by-environment (G×E) interaction and the experimental error, to improve the heritability of desired target character(s).

In more recent times plant biotechnology has offered tools to access genetic variation. For example, genetic engineering is a non-crossbreeding method of direct gene transfer within and between species, the products being genetically modified organisms. Its main uses in plant breeding, so far, are related to incorporating *Bt* genes from *Bacillus thuringiensis* that encode proteins with an insecticidal effect on some insects, or resistance genes to glyphosate herbicide in soybean, thus facilitating conservation agriculture based on direct seeding and zero tillage, crop rotation and residue management (Ortiz, 2015).

16.2 Germplasm Introductions and Multi-environment Testing

Introducing crop germplasm is an old plant breeding method, perhaps as old as agriculture itself! Germplasm is the raw material of the breeder, being defined as 'the genetic material which forms the physical basis of heredity and which is transmitted from one generation to the next by means of the germ cells' (IBPGR, 1991). Practically the breeder will most often access germplasm in the following forms from their own collections or from *ex situ* gene banks:

- *Current or obsolete cultivars* – produced by breeders in the recent past for commercial production by farmers but whose commercial exploitation is now declining or non-existent.
- *Breeding lines and genetic stocks* – produced by breeders and which have a known genetic composition.
- *Crop landraces (LR)* – produced by farmers over centuries via repeated cycles of planting,

Box 16.2 | Summary of Basic Breeding Methods

- *Single plant selection* (in- and out-breeders) – identification of superior plants with desirable traits then repeated cycles of selection and selfing to generate superior homozygous plants in inbreeding crops or in outbreeding crops retain heterozygosity but with high frequency of desired adaptive traits.
- *Mass selection* (in- and out-breeders) – several genetically diverse populations with many desired traits are sown in a mixture and allowed to cross randomly, off-types are destroyed and seed from selected individuals is composited following harvest and used to grow the next generation. Any more rigorous selection is delayed until after several successive randomly selected generations.
- *Pedigree breeding* (in- and out-breeders) – involves specific crosses between plants of two (or more) complementary accessions and selection through successive generations, until genetic purity is reached. Selection is based on desirable traits, performance and ancestry.
- *Mass-pedigree breeding* (in-breeders) – initially following mass breeding approach until after several generations, promising lines are identified and pedigree selection replaces the random mating.
- *Half-sib breeding* (out-breeders) – an open-pollinated variety using half-sib mating and various forms of progeny testing. The parents are selected from the source population. They can be either crossed to a common tester (e.g. top cross) or pollinated by a selected composite (e.g. an ear-to-row). The selected families are top crossed again, or simply selected and composite crossed. The families in either case are all related through one common parent (usually maternal). There are many variants in the method, depending on the crop species and the genetic traits being manipulated.
- *Line breeding* (out-breeders) – where lines are composited to create a variety, the lines having been tested by progeny testing.
- *Recurrent selection* (out-breeders) – where superior plants are selected from a heterozygous population and selfed. The selfed progeny are then intercrossed in all combinations to provide material for further cycles of selection and crossing.
- *Backcross breeding* (in- or out-breeders) – used for superior plant that is lacking in a specific trait; the superior plant is crossed with a plant possessing the desired trait then backcrossed to the superior plant, accompanied by specific selection for the specific trait.
- *Reciprocal recurrent selection* (in- or out-breeders) – selection involves two populations such that selections from each are both selfed and crossed as males to the other population selections. The progenies are then test-crossed to the source population and the best ones are selected as parents. Selfed seeds and selected parents may be grown in isolation or as lines, and intercrossed twice to make new source populations. It is a means of improving two populations simultaneously, in order to create parents for hybrids.

Box 16.2 | (cont.)

- *Family selection* (in- and out-breeders) – involves whole families selection based on the mean performance of their progeny.
- *Mutation breeding* (in- and out-breeders) – uses mutagenic genetics to create variability in a species and alter characteristics. Some of the altered characteristics may be agronomically useful and can be selected by the plant breeder.

cultivation, harvesting, selection and seed saving and more recently collected and made available via gene banks.

- *Crop wild relatives (CWR)* – wild species relatively closely related to cultivated species and crop progenitors which contain significant genetic diversity that can be introgressed into the crop.

Introduction of germplasm may occur by domesticating new crops from wild plants, developing new crops from old crops or adapting existing crops to new agro-ecosystems. The latter is often used in plant breeding for spreading new cultivars. Introducing crop germplasm may involve growing crop germplasm in remote but homologous agro-ecosystems elsewhere and providing thereafter cultivars or breeding lines from them to target populations of environments, where they are included in multi-site trials along with local cultivars or landraces used as checks. Testing may begin in research stations, but promising foreign cultivars are included thereafter in on-farm testing with growers participating in the evaluations of them with the aim of further cultivar release. The extensive assessment in farmer-managed trials should demonstrate the significant advantages of foreign germplasm *vis-à-vis* local cultivars (Box 16.3). This method is therefore very valuable in the early stages of developing a plant breeding programme, particularly when local breeding capacity lacks crossbreeding ability (Ortiz, 2015).

The G×E interaction often occurs because multi-genic complex quantitative characters, such as, *inter alia*, edible yield, biomass, plant height or vigour, are affected by the environment, and their final expression is due to the interaction of the genotype (G) and the environment (E) where the cultigen is grown. The G×E interaction contributes therefore to the phenotype (P) of any individual in each environment, because it influences the genetic expression of the genotypes. Thus, P = G + E + G×E. The G×E effects are only noted when different genotypes are grown across different environments, since the phenotypic expression of each genotype usually reveals itself in a different way in each environment. The variation arising from G×E seems to be very important for continuous quantitative characters, which must be considered in any agricultural research-for-development undertaking for the benefit of humankind.

In agriculture, multi-environment trials (METs) are essential for selecting the most productive and stable cultivars to be used at different sites. Stability analyses of genotypes across sites are used to select stable, high-yielding breeding lines or cultivars for a target population of environments (Box 16.4). Unstable genotypes may show high yields in some optimum environments, whereas lower yielding genotypes may be more stable across all environments. Several statistical methods can be used to study G×E. Quantifying the environmental factors and climatic variables that affect quantitative trait variation and G×E remains of paramount importance for understanding the stability of genotypes across different environments. Statistical models that incorporate numerous external co-variables into the MET analysis can be further employed for studying and explaining G×E. Factorial regression and partial least squares regression are useful tools for dissecting

---

Box 16.3 | **Stages in Evaluating and Releasing a New Species as a Commercial Crop**

Stage 1 Acclimatization and purification, evaluation of performance, quality and disease resistance, etc., and studies on the agronomic inputs (2–3 seasons in the breeder's field).
Stage 2 Small plot evaluation in breeder's trials and multiplication of seed for large trials (2–3 seasons).
Stage 3 National trials for DUS and agronomic recommendations (3–4 seasons).
Stage 4 Certification and release of seed for multiplication and cultivation (2–3 seasons).

Box 16.4 | **Multi-site Testing for Introducing Tomato Bred-Germplasm in Latin America**

(Ortiz, 1991; Ortiz and Izquierdo, 1992, 1994; Ortiz et al., 2007a)

Tomato is an important vegetable crop in Latin America and the Caribbean (LAC), where it originated, and shows great variation. LAC environments significantly affect the performance of tomato cultivars. Stability analysis was used to identify stable and high-yielding breeding lines or cultivars included in multi-environment trials (METs) across 18 LAC sites, which were found, following the analysis of variance, as high yielding (significantly higher than MET mean), average (equal to MET mean), low yielding (significantly lower than MET mean at $P = 0.05$) and very low yielding (significantly lower than MET mean at $P = 0.01$) for marketable fruit yield. Environmental factors such as days to harvest, soil pH, mean temperature, potassium available in the soil and phosphorus fertilizer accounted significantly for G×E interaction noted for marketable fruit yield in tomato, whereas trimming, irrigation, soil organic matter, and nitrogen and phosphorus fertilizers were important environmental co-variables for explaining G×E interaction for average fruit weight in tomato. Locations with relatively high minimum and mean temperatures favoured the marketable fruit yield of open-pollinated (OP) heat-tolerant breeding lines. The G×E interaction was significant in all the environment groups. The testing of cultivars varied significantly for marketable fruit yield in high- and low-yielding environments but not in very-low-yielding environments. An open-pollinated cultivar and an F_1 hybrid were the most stable cultivars for marketable fruit yield, which demonstrated that neither heterogeneous composition of an OP cultivar nor hybrid's heterozygosity *per se* account for yield stability across LAC sites. Hence, alleles providing broad adaptation may be required to achieve yield stability in tomato. Cultivars or breeding lines whose phenotypic stability for marketable fruit yield was high in high-yielding environments had high and stable marketable fruit yield in average and low-yielding

Box 16.4 | (cont.)

environments, which indicated the feasibility of identifying stable and high-yielding cultivars or breeding lines in sites where they reach their maximum yield potential. Heritability estimates were simulated using components of variance to determine the number of locations and replications to test new tomato introductions along with local check cultivar (s) in average to high-yielding sites. Tomato marketable fruit yield should be evaluated in two or three sites with at least three replications.

G×E, when environments are defined by climatic and soil factors rather than by their production means.

16.3 Germplasm Use in Improving Self-Pollinating Crops

Often farmers have cultivated landrace germplasm for a long time within their locality, where such germplasm shows a great variability and is often adapted to the local growing environment. Breeder-based landrace enhancement requires describing the genetic variability by testing them across sites and recording characters of interest (Dwivedi *et al.*, 2015; Box 16.5). Landraces may be further improved through mass selection, which consists of repeated growing out of a large population, then rogueing out off-types or plants with undesired characters and then bulking seeds of desired plants for further planting and testing with the aim in time of identifying and releasing the best performer(s) as new cultivars.

16.4 Methods of Producing Pure-Line Cultivars

Breeding self-pollinating crops depends on recombining genes to enlarge variation to thereafter select recombinants according to breeding aims, and then fixing that variation in breeding lines through repeated inbreeding. Mass selection, pure-line selection, pedigree selection, and bulk population and single seed descent (Box 16.6) are used for breeding self-pollinating crops with the aim of releasing inbred cultivars that are both homogeneous and homozygous, which may be achieved by selfing and selecting plants with desired characters (Ortiz, 2015). The proportion (%) of homozygotes considering one pair of heterozygous genes (n) with m generations of self-fertilization is

$$\frac{\dfrac{2^m - 1}{2^m}}{n} \, 100$$

i.e. 96.875% for one segregating locus, 48.4375% for two segregating loci and so on. Hence, as the number of loci increases, more selfing generations are required to reach homozygosity.

The Swedish botanist Hjalmar Nilsson established in 1890, while working at the Svalof Seed Association, that the whole plant is the right basis for selection and not just a spike or a single seed, thus practising pedigree selection in landraces and for breeding cultivars had clear advantages (Nilsson, 1909). Pedigree selection is still a widely used method of breeding self-pollinated species in which crossing among selected parents is used to generate variability in the base population, and thereafter by handling the segregating population that began by visually selecting best F_2 offspring (Ortiz, 2015) (Box 16.7). Superior plants are thereafter selected using spaced rows in the F_3 and their seed used for planting the F_4. After identifying superior rows in the F_4, some plants (3–5) are taken therein to establish F_5 family progeny rows, and further advanced to F_6 and F_7 by selecting

Box 16.5 | Searching and Using Landrace Diversity for White Lupin Breeding

(Christiansen *et al.*, 1999, 2000; Raza *et al.*, 2000)

White lupin originated in the western Mediterranean where it exhibits great diversity, although old landraces are low yielding with a medium high alkaloid content. It seems that tolerance to high pH or calcium content may be available in white lupin as noted while collecting germplasm in Egypt, thus showing its potential for growing in light soil and new reclaimed desert areas. White lupin landraces were further evaluated along with locally bred cultivars and selected foreign germplasm for grain yield and other characters at five sites with different climates and soil types in Egypt. Results support the use of local germplasm for further breeding of locally adapted lupins and suggest that local landrace germplasm may be an important source of alleles for shortening the vegetative period and reducing plant height and stem length, as well as for improving some yield components such as number of pods and seeds per plant. The genotype-by-site interaction was significant, but mass selection has the potential for enhancing yield, especially in white lupin germplasm adapted to newly reclaimed desert areas. A screening of host response to *Fusarium* rot was performed in a field and under controlled conditions in a greenhouse pot experiment. It found that these white lupin genetic resources from Egypt also possess partial resistance to the pathogen causing this disease, thus being another genetic source for incorporating such a character into the breeding pools of this crop.

Box 16.6 | The 'Miracle' Wheat of the Green Revolution

(Ortiz *et al.*, 2007b; Trethowan *et al.*, 2007)

The Green Revolution in wheat production, owing to the development of short-stature, high-yielding wheat cultivars in the 1960s and 1970s, was led by Nobel Peace Prize Laureate Norman E. Borlaug – a US forest pathologist. It ensued from a strategy involving shuttle breeding at two contrasting locations in Mexico, wide adaptation, durable rust and *Septoria* resistances, international multisite testing and the appropriate use of genetic variation to enhance yield gains of subsequently produced lines. Its genetic basis was very simple and involved the introduction into wheat of a few genes with major effects on plant height and the elimination of photoperiod response that along with improved host plant resistance to pathogens, greatly enhanced grain yield potential and its stability in the low-latitude wheat production environments of the developing world. The CIMMYT, Mexico, used pedigree selection from 1944 to 1985, modified pedigree/bulk from the mid-1980s to the mid-1990s,

Box 16.6 | (cont.)

and since the late 1990s, the selected bulk method. Pedigree selection of individual plants starts in F_2 followed by bulk selections until F_5, and pedigree selection in F_6. In the selected bulk method, the spikes of selected F_2 plants within a cross are harvested in bulk and threshed together, resulting in one F_3 selected seed lot per cross. This selected bulk process continues until F_5 and pedigree selection starts at F_6.

Box 16.7 | **An Example of Developing a Cultivar of Wheat Using Pedigree Breeding**

1963 4000 F_2 plants were raised under spaced planting to grow larger than normal plants, inoculated with yellow rust and selected for resistance to rust, short straw and high grain number. Small samples of seed from the 500 heads selected in this manner were subjected to a small-scale milling test and only 300 heads (= 7.5% of the F_2 population) retained for further breeding.

1964 300 F_3 families raised in single row plots and evaluated for short straw and high grain number. 10 heads were selected from the F_3. Assuming 40 plants per F_3 family and a head means a plant, 10 out of 12 000 plants is equal to 0.08% selection.

1965 10 ears to row (=F_4) families raised and best row selected to continue the pedigree. Seed from remaining rows bulked to provide enough seed for a yield trial at agricultural density.

1966 Selection continued among F_5 ear rows and the yield of bulked F_4 seed evaluated in breeders' small plot trial.

1967 Further selection among F_6 plants of the elite line and the bulked F_5 seed evaluated in breeders' small plot trial.

1968 Selection continued in the elite line for uniformity and the bulked F_7 seed evaluated in breeders' small plot trial.

1969 Reject seed from 1968 bulked to provide enough seed for larger trial to be carried out by the National Institute of Agricultural Botany (NIAB) for comparison with other promising lines. Purification of the elite line continued.

1970 Purification of elite line continued to F_9 and the rejected F_8 seed evaluated in NIAB trial.

1971 Purification continued to F_{10} and enough seed produced for a NIAB main trial over different sites and for seed multiplication by the National Seed Development Organization (NSDO). By this stage it had become clear that the line was good enough to be put on the market as a new variety.

1972/3 Seed multiplication and agronomic trials were continued.

1974 Limited quantities of seed of 'Maris Freeman' released for cultivation.

best plants in each generation. Preliminary yield trials may begin in the F_7 using a check cultivar as benchmark, while advanced trials – based on advancing only superior experimental material to the next generation – with replications and across sites and years may be undertaken in the F_8 or F_9 to identify breeding lines that are better than released cultivars used as trial checks.

Wheat research in the 1900s by the Swedish geneticist Herman Nilsson-Ehle led to developing bulk breeding to deal with the huge crossing numbers, F_n generations and plants therein (Åkerberg, 1986). He previously demonstrated that important (quantitative) characters in crops are inherited following Mendel's law and may be recombined through artificial crossing (Nilsson-Ehle, 1909). Nilsson-Ehle also found that quantitative characters are often multi-genic, thus providing the basis for quantitative genetics. In bulk breeding, natural selection is used in early generations (F_{3-4}), thus delaying the stringent artificial selection of superior plants until a late generation (often F_5). Seed of superior F_5 plants are used for planting F_6 progeny rows, which after testing, led to selecting progeny rows showing desired characters that are used further for planting preliminary yield trials in the F_7. Multi-environment testing across sites and years began with the F_8 and along with check cultivars. Thereafter, the superior breeding line(s) are included in the cultivar release process.

The Danish biologist Wilhem L. Johannsen proposed the concept of pure lines in 1900s while breeding beans and observing variation in seed size (Johannsen, 1903). There are repeated selfing cycles after the initial selection in a mixture of homozygous lines, while practising pure-line selection (Harris, 1911). He indicated that the variation in seed size was heritable and owing to a mixture of pure lines in the original seed lot, and that selection within pure lines was ineffective because variation among them was due to the environment. Johannsen also defined the concepts of gene, genotype and phenotype (Johannsen, 1905, 1911).

The single seed descent (SSD) method consists in advancing segregating F_n generations to a satisfactory level of homozygosity, by taking 1–3 seeds from each individual plant of one generation (beginning in the F_2) to establish the next generation (Ortiz, 2015). Such a procedure is repeated in subsequent generations until the desired homozygous level is achieved. The main characteristic of SSD is the reduction of the time required to obtain homozygous lines. Likewise, SSD keeps genetic drift at a low level due to female gametic control (vis-à-vis bulk breeding) and offers better protection against random loss of alleles during selfing, thus it retains most of the genetic variation of the original population in the advanced generations since an equal number of seeds are taken from each seed parent, while seeds are taken randomly in the bulk method.

Di-haploidy is a quick method for producing homozygous double haploid (DH) recombinants ensuing from heterozygous parents in a single generation, thereby allowing yield testing of candidate cultivars in a short timeframe (Dwivedi et al., 2015). After crossing, the resulting F_1 is grown in the field. The anthers of these F_1 plants are grown in vitro, and the number of chromosomes of haploid plants are doubled with colchicine to produce DH plants, which are thereafter evaluated in the field. Selected DH lines are further evaluated in trials to identify inbred lines for further cultivar release(s) or use as parents in the breeding programme. DH populations lack heterozygous genotypes and allow finding of rare genotypes, especially for recessive characters that are no longer masked by dominant alleles. These DH populations show greater additive genetic variation and absence of genetic variance due to dominance than segregating F_n populations. The success of using DH in plant breeding depends on the genetic variation available in the parental sources. Likewise, larger DH populations may be required to increase the probability of selecting desired recombinants, particularly if large numbers of genes control the character of interest.

Recurrent backcrossing is used if there is a cultivar lacking a character that must be bred and available from another cultivar or breeding line; i.e. the donor parent (Ortiz, 2015). This breeding method uses repeated crossing between the hybrid offspring and the cultivar to be improved (recurrent parent), during which the character for which improvement is sought is maintained by selection (Figure 16.1).

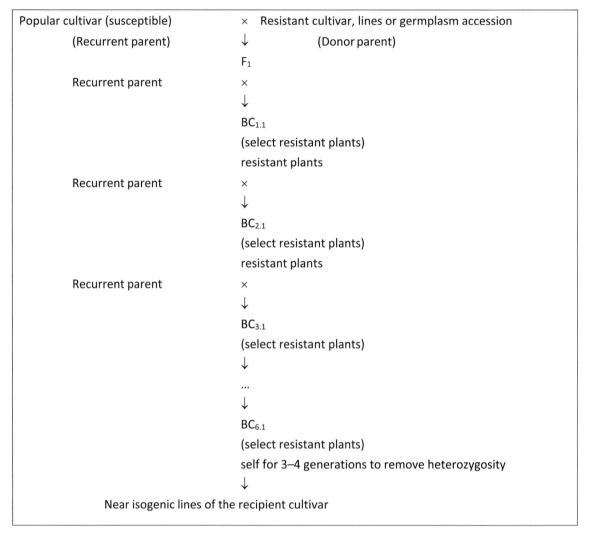

Figure 16.1 Recurrent backcrossing (BC_n) method.

The rate of genotype aa increases by ½ in each backcross (*b*); i.e. for one locus,

$$\frac{2^b - 1}{2^b}$$

or for *k* locus

$$\left[\frac{2^{b-1}}{2^b} \right]^k$$

Hence, the homozygous rate increases proportionally to the selfing rate.

Backcrossing has been widely used to obtain pest-resistant cultivars, or to modify morphological characters, colour and quantitative characters depending on few genes, such as early flowering or maturing, plant height and seed size and shape. Recessive genes are masked by their dominant counterparts available in the recurrent parent's genome. Hence selfing is used before to identify and select individuals carrying them, for further backcrossing. Each backcrossing cycle for recessive genes takes twice as long because selections can only be made after selfing the backcross-derived individuals.

Multilinear cultivars result after mixing isogenic lines produced by independent backcrosses from highly diffused cultivars that differ only by a single pest-resistance gene. The isogenic lines of the multilinear cultivars should have sufficient genetic diversity for host plant resistance, and be sufficiently uniform for their agronomic characters, thus making them compatible in the 'blend'.

The US geneticist Leslie J. Stadler proposed gametic selection for improving multi-genic characters. In this breeding approach, a series of gametes from a heterogeneous population with a superior inbred line or 'elite' germplasm are included in the crossing block (Stadler, 1945a, 1945b). If the frequency of upper zygotes is q^2, the frequency of gametes giving rise to zygotes will be q. Since $q > q^2$, then it would be easy to identify superior gametes than better zygotes. The crossing of the elite line by the source of gametes is done using the breeding line as female and the source as male. A random sample of pollen from the source is used, and since the line is inbred, the differences between F_1 plants obtained will be solely due to the differences between the gametic content of the pollen. The F_1 plants self-fertilize (hoping to recover the outstanding gamete) and cross simultaneously with a suitable tester. The plants whose crossbreeding tests give the best results are selected because they are undoubtedly those fertilized by superior pollen, thus being named gametic selection.

16.5 Germplasm Exploitation in Open-Pollinated and Hybrid Crops

Outcrossing plant populations are imperfect panmictic populations because they are not infinite and their fertilization is not entirely random. The percentages of selfing and outcrossing vary with the species, genotype within the species and with the environment. The breeding methods for outcrossing species, which consider above, are based on the choice of the individuals for producing the next generation and the way of crossing selected individuals to form the offspring that will constitute the next generation (Box 16.8). The different breeding methods depend, therefore, on selection and reproduction.

Selection in outcrossing species begins with a pre-existing cultivar or with a breeding population (Ortiz,

2015). Mass selection or recurrent selection are used if either includes different genotypes with a high degree of heterozygosity. Mass selection by the phenotype is effective if target characters are easy to observe or measure. Progeny tests with the offspring of selected plants are undertaken for those characters that cannot be evaluated by the individual phenotype of the plants. These offspring are derived from open pollination (i.e. without control of male gametes) or can be done by controlling reproduction.

Recurrent selection methods include cycles of selection and crossing with the aim of raising the frequency of favourable genes in the breeding population after crossing the best plants with each other (Ortiz, 2015). Recurrent selection methods share the same genetic basis; i.e. all plants in an outcrossing population showing a certain intensity of expression of a favourable character, regulated by major genes or multi-genes, do not necessarily have the same genotype. Selecting plants showing a phenotypic expression of the favourable character superior to a given level and then crossing them together leads to a new breeding population whose favourable gene frequency is superior to the initial population, and in which recombination has produced superior genotypes. A higher favourable gene frequency is further achieved after performing a new cycle of selection and recombination in this population. In theory, the limit will be when all plants of the population show the most favourable gene combination(s) *vis-à-vis* that at the beginning of the selection process. Intra-population recurrent selection may be based on phenotype, general or specific combining ability, while reciprocal recurrent selection allows improving simultaneously two breeding populations.

Recurrent selection may be based on either additive effects (or general combining ability, GCA), or non-additive effects such as dominance and epistasis (or specific combining ability, SCA), while reciprocal recurrent selection uses both GCA and SCA. The genetic gain after selection (Δ_G) is determined by

$$\Delta_G = \frac{c\ k\ \sigma_G^2}{y\ \sigma_P}$$

where c, k and y are the parental control, a function of the selection intensity, and number of years to

> ### Box 16.8 | Broad-Base Maize Populations
>
> (Ortiz *et al.*, 2010b)
>
> The CIMMYT, Mexico, bred a range of broad-based maize gene pools for tropical highlands and lowlands, as well as subtropical regions during the mid-1970s and throughout the 1980s with the goal of developing maize germplasm combining wide adaptation with plant types suitable for enhanced grain. These gene pools integrated genetic diversity from many useful germplasm accessions representing a broad genetic base for selection and recombination of various important agronomic characters. Superior families were integrated into breeding populations to broaden their genetic base. Full-sib offspring of the advanced breeding populations were then tested in diverse locations by national breeding programmes. Experimental cultivars were bred for enhanced performance within and across target environments. Population improvement schemes based on special purpose gene pools were used for enhancing host plant resistance to specific leaf pathogens and insect pests, as well as adaptation to drought and low nitrogen in the soil. Line and hybrid breeding were initiated at the same time using the same improved gene pools and populations. Diverse sources of donor germplasm were very important for enabling trait-based enhancement of these broad-base gene pools and to maintain overall genetic diversity.

complete one recurrent selection cycle, respectively, while σ^2_g and σ_P are the genetic variance among offspring or families, and the square root of the phenotypic variance, respectively. c is 1 if the evaluated materials are the selected offspring and further inter-mated.

16.5.1 Hybrid Cultivars

Hybrid cultivars are those whose F_1 is used as seed owing to heterosis (Ortiz, 2015). Such a hybrid ensues after crossing two pure lines while a double hybrid is obtained after crossing two F_1 hybrids. Double hybrid seed is cheaper than simple hybrid seed, since it is obtained on the plants of simple hybrids with high yield and very vigorous. A three-way hybrid results after crossing a simple hybrid as a female parent and a consanguineous line as a male. It has the advantage of the lower cost of the seed. Male sterility is used for crossing without emasculating (or castrating a strain by removing its anthers) of the parent that acts as a female (Box 16.9).

Male sterility is often under the control of a recessive gene *ms*. Crossing male sterile (*ms/ms*) and male fertile (*Ms/Ms*) inbred lines is the first step for producing hybrid seed. Some of the hybrid seed (A) is kept and the other used for planting, thus obtaining the following phenotypic proportion at harvest (B): ¼ *Ms/Ms* : ½ *Ms/ms* : ¼ *ms/ms*. This (B) seed is used for planting in a row, while the hybrid (A) seed kept is in the next row, and so on. Fertile plants (*Ms/Ms* or *Ms/ms*) are removed before the anthesis from the B rows, thus leaving only the *ms/ms* plants that are thereafter pollinated by plants from A row (*Ms/Ms*). Hence, the resulting hybrids will be either *Ms/ms* or *ms/ms*.

When using cytoplasmic male sterility (CMS), two kinds of females are included: the seed-producing, male sterile A-line with sterile cytoplasm (*s*) and non-restorer fertility genes (*rt/rt*), and its isogenic, maintainer, male fertile B-line with fertile (*N*) and non-restorer fertility genes (*rt/rt*). The male sterile A-line will be kept by crossing it with its isogenic B-line, which are maintained through selfing, sibling or open pollination in isolated crossing blocks. The pollen

Box 16.9 | Sorghum Hybrids through Genetic–Cytoplasmic Male Sterility

(Reddy *et al.*, 2004, 2005)

Sorghum breeding has been carried out at the International Crops Research Institute for the Semi-Arid Tropics (ICRISAT, India) since its inception in 1992. Its bred sorghum open-pollinated and hybrid cultivars have been released elsewhere. Several sources of male sterility were found in sorghum genetic resources. Recessive alleles (ms_i) in homozygous condition contributes to male sterility. ICRISAT maintains ms_3 and ms_7 in different bulks, and along with previously found cytoplasmic male sterility (cms) uses them for producing low-cost hybrid seed. Male sterility ensues from the interaction of *milo* cytoplasm (A_1 from Sudan and belonging to the Durra race) with sterility genes mostly found in the Kafir race (ms_1 and ms_2) and also in some cultivars of other sorghum races. Parents used as pollinators in hybrid breeding programmes restore fertility in the produced hybrids after crossing with the male sterile lines. There are other cms types, A_1, A_2, A_3 and A_4 (which consists of at least three variants) being the most common. The A_1 system is better than A_2, A_3 and A_4 (Maldandi), A_3 is better than A_2 and A_4 (Maldandi) and A_2 is better than A_4 (Maldandi) for maintaining the stability of male sterility across environments. The frequency of recovery of restorer plants was less on A_3 than on A_2, A_4 and A_1, thereby indicating that more genes were involved in controlling fertility restoration on A_3 than in the other systems. ICRISAT bred several high yielding male sterile lines using A_1 cytoplasmic genetic male sterility. These B-lines were further included in crossing blocks with resistant restorers and several male sterile lines showing host plant resistant to various pests, pathogens and parasitic weed *Striga*, and lines having special attributes were also bred. Likewise, some high yielding maintainer lines were used to breed male sterility using other cytoplasm.

parent for producing male fertile F_1 hybrid seed after crossing with the male sterile A-line is the fertility restorer or R-line whose genotype is *Rf/Rf*.

A synthetic cultivar is an advanced generation from seed obtained after free pollination among several genotypes of a crop. Such a genotype may be inbred lines, clones or populations selected by different breeding methods (including combining ability testing). The synthetic cultivar accumulates genes for different favourable characters, such as, *inter alia*, productivity, early maturing, host plant resistance to pathogens or pests, or produce quality. A synthetic cultivar may be used for its multiplication and delivery to the farmers, or as a reserve store of genes.

The yield of a synthetic cultivar may vary according to the number of genotypes entering its formation, the average yield of each of them in crossbreeding with the others, their combining ability, and the breeding system (i.e. selfing *versus* outcrossing). The expected yield in F_2 ($\overline{F_2}$) of a synthetic cultivar formed by any number of inbred lines may be

$$\overline{F_2} = \frac{(\overline{F_1} - \bar{P})}{n}$$

where $\overline{F_1}$ and $\bar{P}$ are the average yield of all F_1 hybrids in all combinations of inbred lines, and the average yield of inbred lines, respectively, while n is number of inbred lines.

Box 16.10 | **Saving the 'Poor' Cassava for Transforming African Agriculture**

(Nassar and Ortiz, 2007)

Cassava is a tropical root crop used as food by more than 800 million people in the tropics of South and Central America, Asia and Africa. There are 98 *Manihot* species, which are important sources of alleles for tuberous root yield, protein, essential amino acids, micronutrients, host plant resistance and stress adaptation. High-yielding cassava breeding clones may result from interspecific hybridization with some wild species. For example, *M. glaziovii* was used to produce cassava bred-clones resistant to cassava mosaic disease (CMD), which spreads by an insect vector and may be further distributed by infected plant cuttings, thus leading to low tuberous root yield from the use of such unhealthy planting materials. Such hybrids saved the whole of East Africa from CMD, which arrived there in the 1920s. In the 1970s, CMD threatened other West African locations in Nigeria and Democratic Republic of Congo. Clones such as Gold Coast Hybrid 7 (GCH-7 bred in Ghana) and 5318/34 (bred at Amani, Tanzania) were brought to the Moor Plantation (Ibadan, Nigeria) in the 1940s and 1950s as source material for further cassava breeding. One of the clones selected in Nigeria was 58308 – an important source of new hybrids bred in the 1970s at the International Institute of Tropical Agriculture (IITA), e.g. TMS 30572 and TMS 4(2)142, which are still widely grown in various African locations. In the 1990s African programmes incorporated IITA-bred materials into 80% of their released cassava cultivars, which led to 50% gains in cassava yields on average. They are an important contribution to Africa's food security, especially among the poor, because such cultivars raised per capita output by 10% continent-wide, benefitting 14 million farmers. Likewise, total benefits from a cassava partnership project between the National Agriculture Research Organization (NARO, Uganda) and IITA to fight a CMD pandemic in Ugandan districts was estimated to be ca. US$36 million over 4 years for an initial investment of US$0.8 million.

16.6 Improving Vegetatively Propagated Crops

Crossbreeding methods for asexual crops depend on sexual hybridization to get seed after crossing selected parents with the main goal of breeding clones that are phenotypically uniform but often highly heterozygous, particularly if non-additive gene action controls the character(s) of interest (Ortiz, 2004). Non-additive gene action may arise from intra- or inter-allelic (epistasis) interactions. The breeding of clones consists of selecting appropriate parents for crossing schemes, early or late selection in clonal generations,

which will be determined by the heritability of the targeted character(s), and multi-environmental testing of advanced breeding clones with the aim of selecting new cultivars (Box 16.10). The steps in a breeding scheme for vegetatively propagated crops by the International Institute of Tropical Agriculture are given in Figure 16.2.

Accelerated breeding schemes (ABS) allow the development of new cultivars in a short period (Box 16.11). ABS first selects suitable parents for producing F_1 seed, which are grown in a seedling nursery in the first year. Single plots of all genotypes from the seedling nursery are grown at several sites in

1. Source population (5 000 – 100 000 seedlings) after crossing selected parents
   ↓ *Defect elimination or mild-selection for specific attributes*
2. Single plots of (100 – 3 000 selected clones) for clonal evaluation
   ↓ *Screening for specific attributes as per breeding plan*
3. Preliminary yield trial (25 – 100 clones) with 2 replications
   ↓ *Screening to confirm attributes and early edible yield assessment*
4. Advanced yield trial (10 – 25 clones) with 3 to 4 replications in at least 3 sites
   ↓ *Further edible yield assessment*
5. Uniform yield trial (5 – 15 best clones) with 4 replications in many sites
   ↓ *Yield assessment and testing stability across location range*
6. On-farm participatory testing of elite materials (2 – 5 clones)
   ↓ *Farmer (and sometimes end-users') testing*
7. Multiplication of selected clone(s) and cultivar release

Figure 16.2 Breeding vegetatively propagated crops (propagule numbers are crop-dependent).

Box 16.11 | **Breeding Sweet Potato for Adaptation to Drought and Various End-Users**

(Andrade *et al.*, 2016a, 2016b, 2017)

Sweet potato provides household food security and is an important source of energy due to its ability to grow throughout the year. It takes 7–8 years to breed a suitable sweet potato cultivar. Drought – owing to uneven rainfall – causes significant storage root yield loss. The genotype × environment interaction (G×E) significantly affect sweet potato vegetative growth and storage root yield, thus making difficult selection of clones with broad adaptation, which delays cultivar release(s). Harvest index stability and the geometric mean are key to identify clones with high storage root yield and stability under full irrigation and without irrigation at the middle of the root initiation growth stage. The International Potato Center (CIP, Peru) and Instituto de Investigação Agraria de Mozambique (IIAM) used an accelerated breeding scheme to develop well-adapted orange-fleshed sweet potato (OFSP) cultivars combining high storage root yield with high β-carotene and dry matter contents. More than 198 500 seeds were germinated and rapidly multiplied for single plot trials at four breeding sites in the following year. Breeding clones with storage root yields above $10\,t\,ha^{-1}$ were advanced to preliminary and advanced trials across sites and for various years. Storage root yield and dry matter content for 15 selected OFSP breeding clones ranged from 14.9 to $27.1\,t\,ha^{-1}$ and from 24.8 to 32.8%, respectively; β-carotene content, iron and zinc $(mg\,100\,g^{-1})$ ranged from 5.9 to 38.4, 1.6 to 2.1 and 1.1 to 1.5, respectively. These OFSP breeding clones, thereafter, released as new cultivars in Mozambique, met the taste required by local consumers, thus demonstrating the feasibility of ABS. CIP and IIAM also used a bi-parental crossing block and an open-pollinated poly-cross to get seed for other sweet potato breeding populations (Andrade *et al.*, 2016a). After a series of trials across sites and on-farm testing, 3 other OFSP, 3 purple-fleshed sweet potato, and 3 food and fodder dual-purpose breeding clones – derived from such germplasm – were released as new cultivars.

the second year. The selected genotypes according to target character(s) are advanced to preliminary trials and subsequently to advanced trials. Uniform multi-site trials along with on-farm testing may be undertaken the following year. All data required for a cultivar release may become available by the end of year 4, which significantly reduced the time for developing bred-germplasm. Hence, temporal variation of testing environments equals to spatial variation of testing environments in early stages of an ABS for vegetatively propagated crops.

16.7 Using Crop Wild Relatives for Improving Crops

The term gene pool indicates the capability for gene exchange between the cultigen and its wild relatives (see Chapter 6). There are at least three gene pools (Harlan and de Wet, 1971). The primary gene pool includes all populations of the same species, e.g. landraces, bred cultivars and breeding lines or populations (Figure 6.1). There is both free gene exchange and absence of any barriers for intra-specific crosses within this gene pool. Close crop wild relatives belong to this gene pool as happens for most tuber-bearing *Solanum* species and potato. Likewise, the primary gene pool of some polyploid crops very often include their diploid ancestors. Hence, the primary gene pool often refers to the biological species of a crop, which are commonly split into a crop subspecies (e.g. *Hordeum vulgare* subsp. *vulgare*) and the crop progenitor subspecies (e.g. barley *Hordeum vulgare* subsp. *spontaneum*), and they are often completely inter-fertile. The secondary gene pool comprises populations that can exchange genes with the primary pool but after difficult interspecific hybridization through introgression as often occurs among tomato and related *Solanum* wild species. These species may cross with the crop to provide a source of gene transfer, after isolation barriers – mostly arising from hybrid sterility – are overcome. The tertiary gene pool could cross with the primary gene pool through special techniques such as 'bridge' species to facilitate interspecific hybridization and embryo rescue as noted between some non-tuber-

bearing *Solanum* species and potato. The hybrids resulting from crop and tertiary crosses are often anomalous, lethal or completely sterile, but may facilitate important trait transfer to the crop, as in the case of the barley tertiary CWR *H. chilense*, which has unique leaf rust resistance (Patto *et al.*, 2001; Martín and Cabrera, 2005). Gene pool cytogenetics can assist by providing insights regarding the gene pools through analyzing chromosome pairing to determine homology between genomes, which should always be viewed with caution because there are many simply inherited meiotic mutants.

16.7.1 Gene Transfer between Species

Germplasm enhancement or pre-breeding are often used to designate the phase between identifying a useful character, 'capturing' its genetic diversity, and putting those genes into 'usable' forms. Using traits from CWR is often resource-intensive, in part because of linkage drag, the transfer of maladapted along with beneficial traits when crossing crops with their wild relatives (Ellstrand, 2003). Therefore, application may be divided into a two-stage process, (a) pre-breeders (often state or CGIAR-funded) undertake the initial crosses between breeders' lines and CWR, and then (b) the breeders' lines containing the CWR-derived adaptive traits are used in their breeding programmes by commercial breeders. Many breeders' end products of germplasm enhancement may be deficient in certain characters but are still attractive to breeding programmes because they are improved *vis-à-vis* their original source of variation; i.e. a CWR or exotic unadapted germplasm. Knowledge on species variability and systematic relationships, maintenance of CWR, species cytology and on target characters are also usually required to develop the breeding strategy for a successful introgression of wild genes into the cultigen pool.

Some barriers must be overcome for succeeding when engaging in interspecific hybridization (Dwivedi *et al.*, 2008). For example, 'mentor pollination' with a mixture of dead-non-viable but compatible pollen and intended pollen may lead to effective pollination when there is incompatibility in the stigma for pollen germination. Likewise, identifying an optimum temperature for crossing or eliminating the stigma and pollinating directly the ovary or ovule assists on this endeavour. Moreover, reciprocal crosses for unilateral barriers, intra-specific variation in interspecific crossability, 'mentor pollination', the use of plant growth regulators, or by-passing barriers in the style through bud pollination of immature styles or by amputating the style and pollinating thereafter or applying pollen directly to the ovule are used for overcoming style barriers affecting supply of nutrients for pollen growth or timing due to distances. Furthermore, the application of exogenous growth regulators such as gibberellic acid followed by embryo rescue of hybrid seed may deal with partial failure of fertilization and abortion, while the removal of competing sinks, reciprocal cross, manipulating ploidy levels and endosperm dosage relationships or embryo rescue facilitate dealing with barriers in the embryo and endosperm during seed development (Ehlenfeldt and Ortiz, 1995).

Ploidy manipulation is the scaling up and down of complete chromosome sets using parthenogenesis or *in vitro* culture for haploid production (Ortiz, 2015; Box 16.12). Haploids are sporophytes with the gametic chromosome number. The source of allelic diversity or desired gene in wild or landrace germplasm are often available at the low ploidy level. Haploid-species (HS) hybrids are obtained through crossing, while sexual polyploidization can be achieved by using an HS producing $2n$ gametes (i.e. gametes with the sporophytic chromosome number), or chromosome doubling (CD) of HS through colchicine or tissue culture (i.e. asexual polyploidization for further crossing). Unilateral ($2n \times n$ or $n \times 2n$) and bilateral sexual polyploidization ($2n \times 2n$) are used to obtain polyploid hybrids. Ploidy manipulations (Box 16.12) are also used for pursuing an evolutionary breeding approach (Box 16.13).

16.7.2 Alien Chromosome Transfer

Amphidiploids are obtained after doubling the F_1 resulting from interspecific hybridization with colchicine, e.g. triticale, a hybrid between wheat and rye or for re-synthesizing wheat through crossing its putative ancestors or related species. The key for obtaining amphidiploids with sexual reproduction is fertility owing to regular diploid-like chromosome pairing.

16.7.3 Domesticating Wild Plant Germplasm

Domestication of plants should be regarded as the process of adapting wild germplasm for human use, e.g. as, *inter alia*, food, feed, fibre, feedstock and fuel. The release of a new crop should consider both the displacement of existing crops and the production of something unique not available among today's crops. For example, perennial plants are known to store more carbon, keep better soil and water quality, and manage nutrients more conservatively than annual plants. Direct domestication and interspecific hybridization are being used for breeding perennial grain crops. Knowledge regarding domestication syndrome traits may facilitate the fast-tracking domestication of new plants or accelerate the process for semi-domesticated crops (Box 16.15). The use of DNA markers for studying genetic diversity and species relationships and genomics research provide means for gaining insights into the evolutionary changes from wild plants into domesticated crops. Furthermore, DNA marker-aided breeding may permit selection of offspring with desired characters at the seed or seedling stage, thus allowing more diversity to be assessed, more crossing to be made, and increasing offspring to undergo screening without needing field testing over many years and in each election cycle.

16.7.4 Apomictic Seed

Apomixis is asexual reproduction in which sexual organs or related structures are involved but the (apomictic) seeds ensue without the union of the gametes, thus being vegetative in their origin. It results from the deregulation of the timing of sexual events rather than being due to apomixis genes. Nonetheless, there are mutations in genes controlling meiosis I and meiosis II entry that trigger production of $2n$ or apomeiotic mega-gametes (Dwivedi *et al.*, 2010). Apomixis is obligate when plants can only be reproduced by apomixis, thus producing a very

Box 16.12 | **Ploidy Manipulations, Wild Species and Landraces for the Genetic Enhancement of Potato**

(Ortiz *et al.*, 2009)

Potatoes were already grown near Lake Titicaca (Peru–Bolivia) around 4000 BCE. This tuber crop is today the third most important source of the world's food after rice and wheat. There are 189 species in the *Solanum* section *Petota*, which includes the potato cultigen. The Campbell-Bascom Professor Stanley J. Peloquin and his students at the University of Wisconsin defined a strategy to introgress specific characteristics and to broaden the genetic base of potato, in which chromosome sets are manipulated with haploids plus $2n$ gametes, and through interspecific–interploidy crosses. In this breeding strategy, wild *Solanum* species

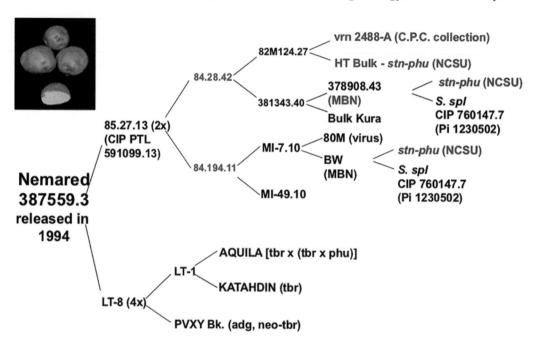

Figure 16.3 Highly heterozygous tetraploid ($2n = 2x = 4x$) potato cultivar 'Nemared' (CIP 387559.3) released in Burundi in 1994 due to its host plant resistance to root-knot nematode and bacterial wilt. 'Nemared' derived from unilateral sexual polyploidization following a $4x$–$2x$ cross between advanced tetraploid breeding clone 'LT-8' (bred by Humberto Mendoza *et al.* at CIP) and $2n$ pollen producing diploid '85.27.13' (bred by Masa Iwanaga and Rodomiro Ortiz at CIP), whose pedigrees are shown above and involve *Solanum tuberosum* cultivar groups Andigena (adg), Neo-tuberosum (neo-tbr), Stenotomum-Phureja (stn-tbr) and Tuberosum (tbr), and wild tuber-bearing *Solanum species S. sparsipilum (spl)* and *S. vernei (vrn)*.

(Diagram and photo: Merideth Bonerbiale, CIP, www.slideshare.net/cwr_use/cwr-cip-2012-finalpptx)

and diploid landraces are the source of genetic diversity, while haploids derived from adapted tetraploid cultivars capture this genetic diversity in crosses with the diploid germplasm. The haploid-species hybrids further transmit this genetic diversity to the adapted tetraploid breeding pool via $2n$ gametes. The efficiency of potato breeding using first division restitution (FDR) $2n$ gametes for multi-trait selection and progeny testing was shown after comparing

Box 16.12 | (cont.)

the 4*x* × 2*x* breeding scheme *vis-à-vis* the 4*x* × 4*x* method of breeding potatoes. Fewer replications and locations are required to evaluate tuber yield in the former than in the latter, while FDR diploid parents had better selection index scores than the tetraploid parents across testing sites. This ploidy manipulation approach has been used for breeding and transmission of host plant resistance to cyst nematodes, root-knot nematodes, bacterial wilt, early blight, late blight and potato tuber moth, as well as for producing high-yielding tetraploid genotypes with yield stability across environments. The tetraploid potato cultivar 'Nemared' (Figure 16.3), derived from a diploid breeding population from the International Potato Center (CIP, Peru), was released during the 1990s in Burundi due to its bacterial wilt and root-knot nematode resistances, plus desired agronomic traits.

Box 16.13 | **Evolutionary Plantain/Banana Breeding Led by Genetics Knowledge**

(Ortiz and Swennen, 2014)

The giant, perennial, herbaceous bananas (*Musa* spp. AAA), cooking bananas (*Musa* spp. AAB) and plantains (*Musa* spp. AAB) are native to the tropics of Asia and Oceania, but they are found today throughout the tropics and subtropics. West and Central Africa are the secondary centre of plantain diversification, whereas East and Central Africa are considered a secondary centre of diversity for highland matooke bananas (*Musa* spp. AAA). There are in excess of 1000 diverse *Musa* cultivars (mostly triploids or 2*n* = 33) derived from two ancestral diploid (2*n* = 22) species: *M. acuminata* (AA) and *M. balbisiana* (BB). Inter- and intra-specific hybridizations plus selection led to parthenocarpic diploid and triploid cultivars. The occurrence of 2*n* gametes (both 2*n* eggs and 2*n* pollen) in *Musa* suggests that the unilateral polyploidization (2*n* × *n*) can account for triploid cultivars. Further allele introgression from diploid to polyploids can occur through unilateral or bilateral polyploidization (2*n* × 2*n*). Triploid plantains provide an interesting example in which most of the variation observed in approximately 120 known West African cultivars ensued from mutations accumulated throughout the history of cultivation of this crop and farmer selection of a few strains.

Diploid banana species and triploid plantain producing 2*n* eggs were the tools for broadening plantain's genetic base. Tetraploid hybrids were obtained after hybridizing 2*n* eggs from plantains with *n* pollen from diploid banana wild accessions or cultivars. Plantain-derived diploid germplasm ensued also from such crossing schemes. The wild diploid banana 'Calcutta 4' – widely available in most *Musa* gene banks – has been extensively used in plantain and banana breeding programmes due to its resistance to black leaf streak, yellow Sigatoka, Panama disease, banana weevil and burrowing nematodes. For example, 'PITA 14' (triploid plantain 'Mbi Egome 1' × 'Calcutta 4') – bred by the International Institute of Tropical Agriculture (IITA) in south-eastern Nigeria in the 1990s – is one of the

Box 16.13 | (cont.)

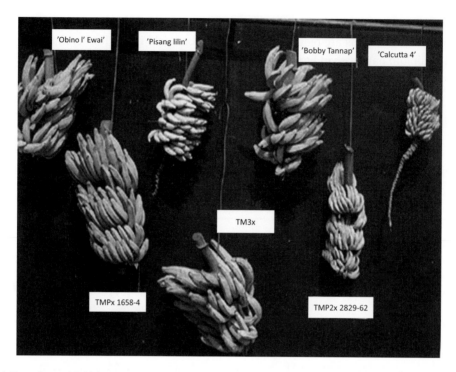

Figure 16.4 Secondary triploid ($2n = 3x = 33$ chromosomes) plantain–banana hybrid (bottom) ensuing after crossing primary tetraploid ($2n = 4x = 44$) hybrid (left, second row) with plantain-derived diploid ($2n = 2x = 22$) hybrid (right, second row). Both primary tetraploid and plantain-derived diploid hybrids result from crossing triploid ($2n = 3x = 33$) plantain cultivars (first and second bunch at the top) with either diploid ($2n = 2x = 22$) banana cultivars or wild species. (Photo: Abdou Tenkouano, IITA)

most promising plantain tetraploid hybrids because of its early fruiting, high bunch weight and fruit size. An impact study shows that each Nigerian farmer who grew it obtained about US$8.62 from 'PITA14' *versus* US$4.33 for their local plantain cultivar. The combination of host plant resistance to black leaf streak and increased bunch weight accounts for 'PITA-14' adoption potential. Advanced ploidy manipulations may lead to secondary triploid hybrids resulting from crossing selected tetraploid with elite diploid breeding stocks, both producing *n* gametes (Figure 16.4). In this regard, the National Agricultural Research Organization (NARO) of Uganda and IITA announced together in 2013, after several years of painstaking ploidy manipulations and field trials, the availability of 27 secondary triploid East African highland banana hybrids – known as NARITA ((triploid matooke bananas × 'Calcutta 4') × elite diploid breeding stocks) – due to their host plant resistance to pathogens and pests and high edible yield. Most NARITA bananas had an average bunch weight of 17.8 kg at trials in central Uganda, while the bunch weight of the widely grown local matooke was 11 kg. The edible yield of many NARITA bananas was significantly above their founder grandparent (i.e. heterosis ranging from 10 to 300%), which shows the significant breeding gains.

Box 16.14 | **The Re-synthesis of Wheat**

(Ortiz *et al.*, 2008a)

Hexaploid bread wheat (*Triticum aestivum*, $2n = 6x = 42$, genome constitution AABBDD) is known to have arisen as a natural hybrid between three ancestral diploid species ($2n = 14$), *T. urartu* (AA), *Aegilops speltoides* (SS = BB) and *Ae. squarrosa* (DD). Given the random nature of its origin, the limited genetic diversity could be overcome by the re-synthesis of wheat; i.e. deliberately selecting highly diverse material of the original three species to create a 'new' wheat species. Wheat re-synthesis ensued from combining novel and elite alleles from tetraploid *Triticum turgidum* and the wild diploid ancestor *Ae. tauschii* with the aim of broadening the hexaploid bread wheat cultigen pool. The CIMMYT, Mexico, has been pursuing the resynthesis of hexaploid wheat since the 1980s. This re-synthesis of wheat assists its genetic improvement by increasing useful variation for traits such as host plant resistance, spikes per plant, grain per spike, grain size, high grain yield potential, and tolerance to drought, heat, waterlogging, salinity and pre-harvest sprouting. Advanced breeding lines derived from resynthesized wheat lines improved adaptation worldwide, especially in drought-prone environments. For example, re-synthesized wheat-derived breeding lines performed well in many stressful environments. They out-yield commercial cultivars by 18–20% under rainfed. Hence, re-synthesized wheat provides useful genetic variability for stress adaptation and grain yield traits from the secondary gene pool.

Box 16.15 | **The Making of a Nordic Berry Crop**

(Hjalmarsson and Ortiz, 1998, 2001)

Lingonberry is a perennial, evergreen dwarf shrub that is indigenous to Scandinavia, where its pea-sized, bright-red berry is picked from wild stands. The first attempts for domesticating lingonberry included collecting wild germplasm for further testing. Promising ecotypes were propagated for further research and the first ever Nordic lingonberry cultivars – deriving from open-pollinated seed samples taken from the forest – were released in the 1970s in Germany and late 1980s in Scandinavia. At the beginning of this millennium there were still a dozen lingonberry cultivars selected from wild germplasm. Further research in a growth chamber using four populations originating from them revealed that spreading ability measured by rhizome number, growth, plant height and vegetative shoot number were under genetic control. Throughout the 1990s, promising breeding germplasm was selected owing to high yield, large berry size, uniform maturity and host plant resistance to 'little leaf disease' in Southern Scandinavia, where crossbreeding began in 1993 using a modification of a method previously used in blueberry that led to releasing the first lingonberry cultivars derived after crossing bred-germplasm.

Box 16.16 | **DNA Marker-Aided Search for Fertility Restorer Genes Facilitates Hybrid Rice Breeding**

(El-Namaky *et al.*, 2016)

Rice remains as the most important food crop worldwide. Hybrid seed technology has contributed significantly to improving rice grain yield. The cytoplasmic–genic male sterility system for producing hybrid rice seed includes cytoplasmic male sterility (CMS) lines, fertility restorers and maintainers. Various DNA markers enabled the determination of the chromosomal locations of fertility restoration (*Rf*) genes in a wild-abortive CMS system of rice. Microsatellites were used to detect the allelic status of *Rf* genes in rice cultivars and breeding lines, thus avoiding testcrossing. Such encouraging results show how DNA markers assist screening of breeding germplasm within a short period, thus saving time and resources.

---

uniform offspring, or facultative if plants can be reproduced by both apomixis and sexual reproduction, thereby producing variable progeny. The apomixis types are parthenogenesis, apogamy, apospory and diplospory. This type of reproduction and further propagation by seed should therefore be considered when selecting the breeding method to avoid confusion in the collection of seeds. Apomictic plants are often homogeneous and uniform populations if they are phenotypically and genotypically identical, thus being unable to practise any type of selection because any observed variation is due to the environment.

16.8 DNA Marker-Aided Breeding Using Plant Germplasm

DNA markers useful for plant breeding should show a high discriminatory power, lack of interaction with the environment, easy to reproduce results across research labs and consistency for estimating genetic distances (Dwivedi *et al.*, 2007). They are used for identifying and characterizing cultivars (Box 16.16), certifying purity of inbred lines and hybrids, allocating germplasm to heterotic groups, evaluating variability and diversity of breeding germplasm, developing genetic linkage maps, mapping genes and

quantitative trait loci for target characters (Box 16.17), analyzing quantitative character architecture (gene number, position, action, effect magnitude and interactions), aiding gene introgression through backcrossing, assisting selection at various stages and predicting breeding values.

Linkage mapping aims to identify DNA markers associated with certain genes of interest, such as host plant resistance to pathogens or pests, produce quality or edible yield, among other characters (Ortiz, 2015). For this purpose, parents bearing distinct gene(s) or showing different phenotypes are hybridized to obtain the F_1, which after selfing produces the segregating F_2 population. The segregating population (F_2 or above, e.g. recombinant inbred lines) along with the parents are included in field testing or greenhouse screening for phenotyping and their DNA used for genotyping with any DNA markers. Further analysis of both together leads to the association of certain DNA fragments (or markers) with target character. DNA markers surrounding the gene are the best aid for selection.

The use of DNA markers for gene introgression through backcrossing is among the best examples of marker-aided breeding. DNA markers linked to target genes facilitate the selection of desired genotypes (Dwivedi *et al.*, 2007). They also facilitate selection of

Box 16.17 | **Identifying Sources for Enhancing Adaptation to Salinity in Rice Landrace**

(Bimpong *et al.*, 2014)

Salinity is, after drought, the second most widespread abiotic factor affecting rice farming and productivity in rainfed and irrigated agro-ecosystems. Identifying diverse sources of adaptation to salinity in rice germplasm is a priority because of the need of breeding programmes to use selected tolerant germplasm. Finding quantitative trait loci (QTL) for adaptation to salinity will accelerate rice breeding for environments suffering such an abiotic stress. The salt tolerant Saudi landrace 'Hasawi' was used for crossing it with three African cultivars to generate three F_2 offspring, each of which included 500 individuals. F_2s and $F_{2:3}$s were evaluated for grain yield in saline fields. There were 75 QTL for different traits in the F_2s: 24 of them were shared while 31 were noted in two F_2s, and 17 in one F_2. 'Hasawi' contributed 49% alleles to these QTL. One marker with high LOD score and relatively large QTL effects was found to be suitable for further marker-aided breeding for improving rice adaptation to saline-prone sites.

Box 16.18 | **Marker-Aided Backcrossing for Breeding Kyrgyz Beans**

(Hegay *et al.*, 2013b, 2014)

Common beans are among the most important pulses in the world. Their host plant resistance to pathogens and pests enhances productivity elsewhere. The use of DNA marker-aided backcrossing speeds up host plant resistance breeding. For example, a dominant and a co-dominant DNA marker – tightly linked to target trait – were used along with inoculation-based selection to facilitate the introgression of the *Co-2* anthracnose-resistant gene in Kyrgyz beans. The co-dominant DNA marker was also useful for distinguishing among segregating materials and selecting those bearing the resistant gene. DNA markers were also used for pyramiding – through aided backcrossing together with detached leaf assays – genes that provide host plant resistance to bean common mosaic virus (BCMV), which is another seed-born disease affecting this crop by significantly reducing its grain yield in Kyrgyzstan. These examples show how DNA markers facilitate indirect selection through DNA marker-aided backcrossing.

genotypes similar to the recurrent parent, thus reducing the backcross generations necessary for its recovery and therefore accelerating the breeding of new cultivars (Box 16.18).

Linkage disequilibrium (LD) between loci may be used to find associations between characters of interest and DNA markers because it offers fine-scale mapping due to historical recombination (Ortiz, 2015).

Box 16.19 | Association Genetics and Genomic Prediction in Wheat Breeding

(Crossa *et al.*, 2007)

Association mapping was used in five historical wheat international multi-environment trials from the CIMMYT to find associations with host plant resistance to stem rust, leaf rust, yellow rust or powdery mildew, and grain yield. Marker–trait associations incorporating information on population structure and covariance between relatives were found after using linear mixed models. Several linkage disequilibrium clusters bearing multiple host plant resistance genes were revealed mostly in genomic regions having other previously noted genes or quantitative trait loci encoding the same traits. Many new chromosome regions for host plant resistance and grain yield in the wheat genome were also noticed. Furthermore, the modelling of the genotype × environment interaction facilitated the identification of markers contributing to both additive and additive × additive interaction effects of traits. Research shows the potential of whole-genome prediction to reshape wheat breeding because the estimated genetic gain per year using genomic estimated breeding values (GEBVs) seems to be several times that of crossbreeding. Practical guidelines are becoming available for implementing genomic prediction for selection after identifying the best strategy for such an undertaking in wheat breeding (Bassi *et al.*, 2016). Defined breeding schemes relying on GEBVs are F_2 recurrent mass genomic selection (GS), F_3 recurrent GS with phenotypic selection in F_2, and F_4 recurrent GS with phenotypic selection in F_2 and F_3. Each of them has advantages and disadvantages, and the one to be used should be that best suiting the needs.

This approach (Box 16.19) does not require, therefore, a specific mating design for developing experimental mapping populations, which are often time-consuming and expensive. It must be noted that covariance between DNA markers and target characters could ensue from population structure resulting from admixture, mating system and genetic drift, or due to artificial or natural selection throughout plant evolution, domestication and breeding. Hence, differentiating LD due to physical linkage from that arising from population structure is of the utmost importance for association genetics, which also requires phenotypic data for modelling genotype × environment interactions. Available data from nursery or advanced breeding trials and multi-environment testing are often used for association analysis defined by

$$T = C + (Q + K) + E$$

where T is phenotype (after measuring it in multi-environment testing), C refers to the polymorphisms revealed by high-density DNA markers, $(Q + K)$ measures the population structure using neutral DNA markers and coefficient of co-ancestry, and E is the residual.

Marker-aided breeding (MAB) only considers DNA markers tightly linked to target character(s), while association breeding uses LD mapping instead of testing and selection to identify and validate alleles. The success for breeding new cultivars depends often on minor genes with small effects, which may not be included when using MAB or association breeding. Genomic prediction became a suitable approach for estimating breeding values after becoming available both dense genome-wide markers, such as single nucleotide polymorphisms (SNP), as well as robust best linear unbiased prediction (BLUP) statistics that are of benefit for using various Bayesian methods, and which rely on more flexible assumptions for the distributions of SNP effects or variances. Whole-genome prediction methods allow defining genomic estimated breeding values (GEBV) for further use in selection (Abera Desta and Ortiz, 2014). DNA marker effects across the whole genome on a target population for desired character(s) are based on two but related groups, namely the training (TP) and breeding (BP) populations. TP undergoes both phenotyping and genotyping, while BP includes the descendants of a TP or related germplasm that are only used for genotyping. Those with highest GEBV in the BP are selected. Phenotyping becomes, therefore, the key informant for building up the accuracy of statistical models.

16.9 Genetic Engineering Issues

A transgenic plant has been genetically altered by genetic engineering to contain DNA from sources other than the host plant. Breeding transgenic-derived cultivars takes about 10 years before going to market. Their performance should be measured against environmental, economic and social impacts. Such cultivars must be integrated into suitable farming systems if both contributing to food security and reducing impacts of agriculture in ecosystems. Transgenic crops may further contribute to adapting agriculture to the changing climate and mitigating it by reducing emissions of greenhouse gases. Biosafety, risk and other safety assessments should be science-based and not used for delaying the delivery of a technology that is safe. Meta-analyses show that transgenic crops perform better than their conventional counterparts in terms of yield, production costs and gross margins, while reducing chemical pesticide use. Herbicide (glyphosate) tolerance, host plant resistance to insects (*Bt*) and viruses, crop composition and extended shelf life were bred through plant genetic engineering on the first generation of transgenic cultivars, while the pipeline includes traits related to host plant resistance to pathogens and insects, tolerance to other herbicides, better food and feed quality, enhanced adaptation to stressful environments, and improved input efficiency.

Genome editing appears to be a very promising new plant breeding technique for genetic modification. For example, research led to simultaneous editing of homoeoalleles in bread wheat that provided heritable host plant resistance to powdery mildew, or to improved soybean oil quality after targeted mutagenesis of the *fatty acid desaturase 2* gene family. Regulations may affect genome editing despite the fact that genetic changes can be made in the genome more precisely. Any regulatory systems should therefore be based on the characteristics of the bred-crops, irrespective of the method used to develop them.

16.10 Conclusion

The global food supply will have to increase significantly by 2050. It is estimated that 80% of this future growth will be on land currently in use as the potential for land expansion for agriculture remains limited (Hubert *et al.*, 2010). It is important to note that at the end of the 19th century a single farmer, on average, only fed 2.5 people, while at the end of the 20th century she/he fed 130 humans. To achieve and sustain this transformation, crops of the 21st century require resistance to pathogens and pests, have resistance to herbicides and adapt to abiotic stresses, possess better nutritional quality, and show a greater yield potential in order to feed the population of the future in roughly the same agricultural area. Over the past 10 000 years, humanity has used its knowledge about plants to improve food production. Plant breeding will assist on developing eco-efficient and resilient agro-ecosystems to meet end-user demands because it contributes *inter alia* to providing enough and safe food, enhancing human health through better nutrition, diminishing use of fossil fuels, or adapting crops to extreme weather and water stresses. The use of agro-biodiversity through plant breeding will further broaden the genetic base for farming, and for sustainably intensifying agriculture in agro-ecosystems (Ortiz, 2017). Today's plant breeding agenda should therefore focus on producing more food with less inputs, adapting agriculture to climate change, conserving agro-biodiversity through its use, improving the nutritional quality of the human diet, and adding value to the products produced.

Participatory Plant Breeding

17.1 Introduction

Institutional breeding appears to have failed to adequately meet the needs and requirements of 'difficult' environments. This stems largely from the fact that plant breeding is mainly directed at increasing yield in more favourable environments. While broad adaptability is a major objective, there are many more marginal environments in which improved varieties do not express their increased yield potential or do not satisfy other user requirements.

Hardon (1996)

The 1992 UNCED conference and the adoption of the Convention on Biological Diversity led to a fundamental shift in the traditional biodiversity conservation and use paradigm as outlined in Chapter 2, resulting in dramatic changes to the way plants, animals and their relationships to human society were viewed. Of course, many of these issues had been proposed and discussed much earlier by more far-sighted and visionary individuals, but UNCED and the CBD, and subsequent negotiations through the FAO leading to the adoption of the ITPGRFA, did raise these issues to the forefront of the international agenda. Among these issues was greater participation of local people in biodiversity conservation and use, which resulted in many of the shifts to more community-based conservation discussed in Chapters 2 and 10.

Community involvement in biodiversity conservation and use is not a new concept. It has been practised by subsistence cultures for millennia, but it has been increasingly recognized that as societies increased in economic wealth and developed and shifted away from their subsistence existence, there has also been a parallel transfer of local biodiversity conservation and use away from local communities to specialist conservationists and plant breeders. It is interesting to note what in pure economic

development terms would be considered the least advanced societies, that they still retain their fundamental link to nature. They continue, as they have always done, to manage, conserve and sustainably use biodiversity for their own benefit. For example, it has been shown that Amazonian Indians use in one way or another between 80–90% of the species in their locality (Prance, 1997).

Efforts to promote greater participation of local people in biodiversity conservation and use have resulted in significant changes in behaviours and attitudes among various professions working in conservation, agriculture and rural development. This has come with a proliferation of new participatory research approaches, methodologies and tools for those working with farmers and local communities (Pretty et al., 2003). One plant science discipline, which has seen as much change in this area as any other, is that of plant breeding. This has seen the emergence of participatory plant breeding (PPB) and the involvement of farmers and local communities working with professional breeders and researchers in the breeding process (Figure 17.1) (Ceccarelli et al., 2009). Although still a marginal activity in the larger scheme of conventional plant breeding, these changes supporting the greater role of farmers and farming communities in the management, conservation and use of plant genetic resources have been reflected in global policy fora including in both FAO Global Plans of Action for Plant Genetic Resources for Food and Agriculture to date (Box 17.1).

Such recommendations and aspirations represent a major change in attitude and mind-set around the development of agricultural technologies. While the First Report on the State of the World's Plant Genetic Resources for Food and Agriculture (FAO, 1998) highlighted limited involvement of farmers and farming communities in PPB and participatory research in general, the second Report (FAO, 2010a) more than a decade later did report some progress in

Figure 17.1 Farmers evaluate wheat varieties during participatory trials in Ethiopia (Bioversity International/C. Fadda)

Box 17.1 | Some Pointers toward the Participatory Approach

(FAO, 1991, 1996, 2011a)

Advanced technologies and local rural technologies are both important and complementary in the conservation and utilization of plant genetic resources.

FAO (1991)

Where appropriate, national research systems should consider strengthening local level capacity to participate in all stages of breeding, including onfarm selection and adaptation.

FAO (1996)

The Second Global Plan of Action for Plant Genetic Resources for Food and Agriculture (FAO, 2011a), compared to the first, is much more explicit and targeted in terms of its actions and recommendations on participation of farmers and communities, especially in the area of PPB, and recommends

Targeted and increased involvement of farmers and farming communities in national and local crop-improvement activities, including support for participatory research and plant breeding.

The Second GPA also develops a broad strategic framework comprising seven basic and interrelated aspects, one of which addresses strengthening of the role of farmers and farming communities in PPB:

Box 17.1 | (cont.)

Strengthening the efforts of, and partnerships between, public and private sector breeders to conserve and use PGRFA is essential. In addition, participatory breeding and selection, as well as participatory research in general, with farmers and farming communities, need to be strengthened and recognized more broadly as appropriate ways of achieving the sustainable and long-term conservation and use of PGRFA.

At a policy level, the Second GPA calls for:

Where appropriate, national policies should aim to strengthen the capacity of indigenous and local communities to participate in crop improvement efforts. Decentralized, participatory and gender-sensitive approaches to crop improvement need to be strengthened in order to produce varieties that are specifically adapted to socio-economically disadvantaged environments. This may require new policies and legislation – including appropriate protection, variety release and seed certification procedures for varieties bred through participatory plant breeding – in order to promote and strengthen their use and ensure that they are included in national agricultural development strategies.

The Second GPA also calls for:

Strengthened capacity building and training in participatory selection and breeding.

Improved rigorous, multidisciplinary scientific research in the area of crop improvement research, including participatory breeding, as a means of increasing crop yields and reliability without significant losses of local biodiversity.

The institutionalization of participatory, gender- and youth sensitive approaches to plant breeding as part of national PGRFA strategies in order to facilitate the adoption of new crop varieties.

Finally, the Second GPA recognizes the importance of PPB approaches to climate change adaptation by highlighting the importance of participatory evaluation, selection and improvement of farmers' varieties/landraces and early breeding lines as measures that could bring higher levels of diversity, adaptation and stability to crops.

the arena of PPB, though also highlighting these activities were still small-scale and localized and farmer involvement was limited (Box 17.2).

There are many ways and opportunities for farmers, plant breeders, researchers and others including agricultural extensionists to work together across the full scope of activities in a plant breeding programme. The level of involvement of farmers in PPB can take many different forms. From joint involvement in setting breeding or evaluation criteria to the evaluation and selection of varieties through to the actual selection of parents for crosses or breeding

lines or in some instances involvement in making parental crosses to produce segregating populations. A range of terminology is often used to describe PPB approaches. For example, Chapter 10 refers to farmer-led and breeder-led programmes. Others have proposed referring to the approach as 'consultative' or 'collaborative' or more generally as 'highly client-oriented breeding'. Chapter 10 also refers to participatory varietal selection (PVS) and participatory plant breeding (PPB), terms which you will frequently come across in this field of work, with the distinction depending on the degree and timing of

Box 17.2 | **PPB, the State of Play**

(FAO, 2010a)

According to the country reports, farmer's participation in plant breeding activities has increased in all regions over the past decade in line with Priority Activity Area 11 of the GPA. Several countries reported using PPB approaches as part of their PGRFA management strategies. As farmers are in the best position to understand a crop's limitations and potential within their own farming system, their involvement in the breeding process has obvious advantages. These have been noted in many of the country reports. Several developing countries, including Bolivia, Guatemala, Jordan, the Lao People's Democratic Republic, Mexico and Nepal, reported that for certain crops, participatory breeding approaches are the most suitable way to develop varieties adapted to farmers' needs. Several countries rely almost exclusively on participatory methods to develop improved varieties. In the Near East, 10 of the 27 countries that participated in the regional consultation indicated that they used participatory breeding approaches to improve different crops. In the Americas, the Latin America and the Caribbean regional consultation report stated: *Participatory plant breeding activities at the farm level are often mentioned as a priority, in order to add value to local materials and preserve genetic diversity.* Similar statements can be found in the reports of many countries in Asia, Africa and Europe.

However, 'In spite of the overall increase in PPB, farmer involvement has largely remained limited to priority setting and selection of finished crop cultivars. This a similar situation to that reported in the first State of the World Report in 1998'.

---

farmer involvement in the plant breeding process. When exploring the subject of PPB, it is best considered a continuum with PVS as a major element within PPB.

Even a cursory review of the literature on community participation shows that the role of local communities in biodiversity conservation and use is extremely diverse (see Chapter 10). In terms of PPB it may vary from simply providing local indigenous knowledge about a species to full-blown crossing of parental lines and landraces to generate segregating diversity for selection and production of improved breeders' lines. Local communities have repeatedly shown a desire to regain their direct link to their resource base and its development, not abrogate control to a remote multi-national commercial entity; they wish to play a central role in agro-biodiversity conservation and use. It is just as true that specialist conservationists and plant breeders have proved unable alone to halt the loss of ecosystems, species and genetic erosion. As professional conservationists

and plant breeders, there is a resource we are failing to tap – an opportunity to employ the enthusiasm and expertise of local communities in agro-biodiversity conservation and use. Although there have been great strides in this area in the last 30 years, there is much that still remains to change for these approaches to be properly institutionalized for scaling up. To do this will involve a dramatic change from the current conventional approach to a more inclusive, holistic participatory one, but with increasing interest in agroecology and healthy and sustainable food systems, and the impact of changing climate – one could say the time is right for some change.

The shift to a more participatory-based biodiversity conservation and use approach is unlikely to be straightforward. There are many ethical, policy and institutional factors that result in the professionals and local communities fearing and avoiding contact with each other. Value systems may differ and certainly the two approaches differ in their *modus operandi* and who controls the resource and who benefits from its exploitation. This remains one of the main challenges for PPB. How can it move from a breeding approach that has been largely project-driven, and which has had impact in a localized context, to a more mainstream approach fully supported by policy and legal frameworks and supportive institutions?

17.2 The Rationale for PPB

There are many motivations for establishing a PPB programme, some of which have already been highlighted. Primarily, most PPB programmes are aimed at developing new varieties more suited to the specific farming conditions of farmers and their varied other needs – varieties that farmers are more likely to adopt compared to cultivars released through conventional plant breeding. Other benefits and motivations can also include using PPB as a way of enhancing on-farm conservation and expanding genetic diversity (see Chapter 9), empowering farmers, especially women, and local communities (Box 17.3).

The main motivation for PPB has come largely from dissatisfaction over the slow pace of varietal change in many agricultural regions in developing countries served by conventional plant breeding. These all too long periods involved in the breeding and evaluation of new improved cultivars in research stations do not necessarily meet the diverse environments and needs of farmers (Box 17.4). Especially for those farmers cultivating in complex, diverse and risk-prone environments, the desire for a greater number or choice of cultivars or varieties goes beyond that which is available from research stations or national breeding programmes. In the case of a participatory selection programme on beans in Rwanda, the number of cultivars adopted over a 2-year experimental period (1988–1990) was 21, which happened to match exactly the total number of cultivars released by the national programme over the previous 25 years (Sperling, 1996). There is also evidence that demonstrates that when farmers and breeders carry out selection in the same environment, farmers' selections can be effective, validating that farmers have considerable knowledge, which has been largely ignored in conventional plant breeding

---

Box 17.3 | **The Multiple Benefits of PPB**

(Christinck *et al.*, 2005)

- The needs of specific user groups can be targeted more effectively.
- Yields in marginal environments where commercial cultivars are not adapted can be improved.
- Earlier adoption of newly developed cultivars.
- Level of biodiversity in farmers' fields can be increased through better access to germplasm and experimental lines, as well as links to local seed systems.
- If landraces are used in breeding parents, PPB can also contribute to on-farm management of traditional landraces which otherwise might be abandoned by farmers.
- Breeding programmes for marginal or underutilized crops can be developed through improved farmer–researcher collaboration based on local knowledge and experience.
- Further possible benefits include joint learning, capacity and skill-building, community empowerment, institutional development and improved development of policies and enabling environments.

Box 17.4 | **The Story of Taro PPB**

(Hunter *et al.*, 2001; Iosefa *et al.*, 2013)

An outbreak of the leaf blight disease in Samoa in 1993 resulted in the complete devastation of the country's staple crop, taro (*Colocasia esculenta*), and farmer's incomes from local and overseas markets dwindled almost overnight. This was a calamity for farmers and the livelihoods, diet and well-being of their families and communities. The preferred taro varieties at the time were all highly susceptible to the disease and attempts to solve the problem through fungicides and changed cultural practices had little impact. Efforts to evaluate imported exotic cultivars for disease resistance and to develop resistant taro through the formal national programme commenced in 1994. However, by the late 1990s the need for a more participatory approach to cultivar evaluation and plant breeding in Samoa was already being explored in discussions with farmers, who expressed dissatisfaction with the pace of release of resistant taro cultivars for their evaluation. Researchers at the local university were also concerned about the slow rate at which possible resistant taro cultivars were being identified and their release to farmers. The rigorous testing that took place over several years attempting to identify new cultivars proved of limited relevance to farmers' needs and environments.

These discussions and exploratory efforts were also prompted by developments occurring at the same time in PPB around the world highlighting much of the material that is rejected in conventional plant breeding has been found to have subsequent acceptance among farmers (Maurya *et al.*, 1988). It was also felt that the formal programme was doing little to increase the diversity of taro in the country by pinning its hopes on the identification of one or two candidate cultivars. A participatory approach to taro evaluation and breeding, the Taro Improvement Project (TIP), involving researchers, farmers and extension staff working together was officially put in place in 1999. The basis was:

1. Learning more about what farmers wanted from improved taro cultivars and involving them in the selection and evaluation process.
2. Involving many farmers under diverse environments, and providing them with a range of options so that they could select the best for their conditions, which would ensure that farmers gained quicker access to resistant taro cultivars.
3. Increasing the diversity of taro varieties grown by farmers in Samoa. This was an important perception in minimizing a repeat of the disease outbreak, the danger of relying heavily on one or a few genotypes.
4. Strengthening the linkages between researchers, extension staff and farmers.
5. Making more effective use of limited budgets, time and other resources of researchers, extension staff and farmers.

programmes (Ceccarelli *et al.*, 1996). In Colombia, following 15 years of low farmer adoption of cassava cultivars in low-input and low-rainfall areas, the implementation of a PPB programme resulted in the quick release of three farmer-evaluated cultivars (Sperling *et al.*, 2001).

17.3 The Development of PPB

The origins of PVS and PPB can be traced back to the dramatic changes and paradigm shifts associated with the early days of environmentalism and the sustainable use and conservation of genetic resources, which have already been described in considerable detail in Chapters 2 and 10. Those transitions that took place in development perspectives and approaches in the 1970s and 1980s are particularly important, which saw significant moves away from the widespread dissemination of pre-determined technology packages to more farming systems research and participatory and farmer-led approaches to technology development. More specifically, some working in the field of PPB point to the publication of the paper by Rhoades and Booth (1982) as being a pivotal moment in the beginnings of PPB.

However, the line between when PPB started or was officially recognized is difficult, if not impossible or even worthwhile, to pinpoint. Farmers having domesticated our major and minor food crops have always been involved in breeding and crop improvement and continue to be whether officially recognized by the formal sector or not. Furthermore, the documentation of explicit farmer participation in plant breeding or varietal development has been limited or ignored and was not at all well described in the 1970s and 1980s, or even earlier. Prior to official recognition of PPB, farmer involvement in breeding was probably more frequent than was being reported in the published literature, as highlighted by Walker (2006; Box 17.5).

While Raoul Robinson in his book *Return to Resistance* (Robinson, 1996) highlights the first known example of participatory varietal selection as that of Sang Ki Hahn, while working for the

International Institute of Tropical Agriculture (IITA) breeding cassava for West Africa in Nigeria in the 1970s and 1980s, it is believed that the terms PVS and PPB were first used at an IDRC workshop in 1995. The first joint use of PPB and PVS in the peer-reviewed literature was in the journal *Experimental Agriculture* the following year (Walker, 2006). Since this date, both the journals *Experimental Agriculture* and *Euphytica* have been very prominent in publishing on the theory, methods and practice of PVS and PPB. In 1996, the CGIAR established the Systems-Wide Initiative on Participatory Research and Gender Analysis (PRGA) in which PVS and PPB were prominent activities (Box 17.6). One of the most comprehensive reviews of farmers' participation in plant breeding, based on an assessment of 40 developing country 'cases', is that of Weltzien *et al.* (2003). Formal-led PPB is the focus of this review, whereas the role of NGO-led and informal-led PPB initiatives are the focus of publications by Vernooy (2003) and Almekinders and Hardon (2006). Jarvis *et al.* (2016) profile a number of 'PPB champions' who have been key to the development of theory, methods and practice with reference to their landmark publications. Capacity building and teaching materials suitable for plant breeders, development practitioners, biodiversity specialists, educators and students are now much more commonly available than they were 30 years ago and include a PPB training handbook (Christinck *et al.*, 2005) and two comprehensive PPB books (Ceccarelli *et al.*, 2009; Westengen and Winge, 2019).

17.4 The PPB Approach

17.4.1 Description of Approach

Modern plant breeding stands among the greatest scientific and human success stories of all time. However, in recent history, for much of the world, farmers have had little or no role in plant breeding. They were passive recipients of improved seed cultivars (the finished product) by a commercial seed company or government agency, and this genetically homozygous cultivar was bred for high performance

Box 17.5 | The Story of Canchan-INIAA in Peru

(Walker, 2006)

Farmers have been breeding and improving plants and crops since time immemorial. However, in the early days of PPB the nature of farmer participation was not that well described or articulated. However, such participation seems to have contributed to the successful identification of some cultivars. For example, farmers were involved in the development of what were to become widely adopted improved maize cultivars in a multi-decade CIMMYT-associated project in Ghana, but the extent of farmers' involvement was far from adequately described. In the mid-1980s, potato breeders in Peru evaluated advanced clonal material from a diverse late-blight-resistant population in farmers' fields in three hotspots for late blight disease. In return for their involvement, farmers were given half of the output of the trials. In the final evaluation, six of the most promising clones were identified with all participating farmers selecting seedling no. 380389.1 (from a cross made in 1979), which was subsequently released nationally as Canchan-INIAA in 1990. Among Canchan-INIAA's notable traits were earliness, late blight resistance, high yield potential and red flesh colour. Because the fields planted by the national potato programme had been badly affected by frosts, it was farmers' seed that was utilized for release. Worth indicating is that the pedigree of Canchan-INIAA includes hexaploid wild species *S. demissum* and the bitter diploid potato *S. ajanhuiri* plus the tetraploid cultigen *S. tuberosum* groups Andigena and Tuberosum. By the time of its national release many farmers were already growing Canchan-INIAA and a significant amount of seed had already been disseminated through informal seed networks. This was rather fortuitous given the national programme's increasingly tight operating budget for agricultural research. Despite a reported breakdown in its resistance to late blight and the lack of renewal of its seed, Canchan-INIAA remains a popular cultivar and is planted on 70 000 hectares accounting for about one-fourth of the potato area cultivated in Peru. While the Canchan-INIAA story highlights many of the elements of PPB, few, if any, of the plant breeders and researchers involved at the time are likely to have appreciated this. Essentially, what they were doing was a form of participatory varietal selection, and with additional foresight they could have involved farmers at an even earlier stage to effect a truly PPB programme. Who can really tell, but perhaps the earlier involvement of farmers in the setting of breeding criteria in the actual on-station research of the programme could have provided even greater returns.

in most environments, or broad adaptability. It was this paradigm of plant breeding and seed production that fuelled the Green Revolution of the 1960s and 1970s and saw, for example, the IRRI transform the production of Asian rice or CIMMYT semi-dwarf, photoperiod insensitive, rust-resistant, high-yielding cultivars changing the global wheat output elsewhere. However, the take up of high-yielding cultivars was not universal, and a significant proportion of farmers continued to grow locally adapted landraces.

Box 17.6 | **The CGIAR Participatory Research and Gender Analysis Programme**

(Prain *et al.*, 2000)

A CGIAR system-wide initiative on Participatory Research and Gender Analysis (PRGA) was put in place in 1996 in which PVS and PPB featured prominently since its beginning. PPB was one of three core components, the others being natural resource management and gender analysis. The PRGA was an attempt to bring together a wide range of actors and institutions from various CGIAR centres, national agricultural research systems (NARS) and NGOs in order to establish a credible institution of scientific excellence for furthering participatory research, and as an innovative institutional platform to build and strengthen effective partnerships. The PRGA helped pioneer the quality of science around the development of new participatory approaches and has had a significant output in terms of inventories, documentation and analysis of PPB programmes from around the world, including state-of-the art reviews of formal and farmer breeding systems, the role of biotechnology, guidelines for PPB and a strong suite of PPB small grants research activities. The PRGA was also successful in disseminating sociological and organizational methods which have been very useful for pioneer breeders involved in PPB. Among the landmark PRGA documents are:

- Plant Breeding Working Group/PRGA Program. *Guidelines for participatory plant breeding. version 3, April 2000* (PRGA Working Document No. 1).
- McGuire, S., Manicad, G and Sperling, L. *Technical and institutional issues in participatory plant breeding: done from the perspective of farmer plant breeding*, 1999 (PRGA Working Document No.2).
- Weltzien, E., M. Smith, L. Meitzner and L. Sperling. *Technical and institutional issues in participatory plant breeding from the perspective of formal plant breeding. A global analysis of issues, results and current experience*, 2003 (PRGA Working Document No. 3).

There are numerous reasons why a farmer may wish to continue to grow locally adapted landraces rather than switching to high-yielding cultivars, and these include:

- *Yield* – the high-yielding cultivar is unlikely to out-yield all local landraces, particularly when grown and compared in marginal environments where the landrace is locally adapted to, for example, drought or high altitude or may show saline tolerance.
- *Quality* – the local landraces may have some quality absent from the high-yielding cultivar, that may, for example, be associated with a method of

food preparation or fragrance associated with a specific religious ceremony.
- *Host Plant Resistance to Pathogens and Pests* – the high-yielding cultivars may not be resistant to all the pathogens and pests present in the target locality.
- *Requirement for high-input farming system* – the local farmer may not be sufficiently wealthy to invest in fertilizer or herbicides to get the maximum yield from the high-yielding cultivar.
- *Better crop insurance in marginal environments* – in marginal environments the genetic diversity

inherent in landraces ensures even in adverse years there is likely to be some production, but high-yielding cultivars under the same conditions are more likely to suffer complete crop loss.

- *Ease of harvest and storage* – the traditional landrace may be easier to harvest or store than bred cultivars within the traditional farming system.
- *Traditional use* – the local farmer may value a landrace because of its taste and cooking qualities, how fast a crop matures and the suitability of crop residues as livestock feed.
- *Tradition* per se – the farmer may have grown the landrace for generations and just be unwilling to change from a familiar reliable landrace to an unknown cultivar; alternatively, because the landrace is culturally important for sense of identity and place.

Whatever the reason, high-yielding cultivars have not always met the needs of some farmers, largely those in marginal environments. In more extreme environments, genotype × environment interactions play an increased role, and the novel cultivar must be amenable to both the local physical and socio-economic environment. It is unlikely that the conventional breeding paradigm would ever allow this level of specificity in breeding, thus a decentralized, participatory approach is required.

The traditional farming system is the exception to the conventional plant breeding and seed production paradigm (Figure 17.2A). The traditional farmer grows landrace material that has inherently high genetic diversity and each year keeps a proportion of 'best' seed to plant the following year. Over generations of sowing and harvesting these local landraces evolve to become adapted to the particular biotic and abiotic environment where they are grown. They develop particular ecotypic adaptation that is reflected in the genetic makeup of the landrace. Efforts to introduce improved cultivars by conventional plant breeding in this traditional system have been difficult.

There have been a significant number of farmers who have been exposed to high-yielding cultivars by conventional plant breeding and choose not to give up growing their traditional landraces because of the reasons outlined above. It was this realization that caused some professional plant breeders to re-examine their role in conventional plant breeding and move to encourage greater and earlier farmer involvement (Figure 17.2B, Figure 17.3 and Table 17.1). Rather than treating the farmer as a passive recipient, breeders in collaboration with the farmer have produced cultivars superior to their landrace material but retaining certain traits (e.g. taste, fragrance or host plant resistance to pathogens or pests) required by the farmer. The conventional and participatory approaches are contrasted in Table 17.1.

As highlighted, the distinction between PVS and PPB is often made. With PVS plant breeders retain more control of the breeding process, while with PPB the breeding process is decentralized from the breeding institute to the farmers' fields and enhances or provides added value to locally adapted landraces. There is added value in the sense of the generation of increased economic returns, as well as positive local environmental benefits of retaining sustainable agricultural systems. There can be several advantages accruing to the breeder and farmer alike in that best-performing landraces can be identified and evaluated, and the performance of modern recommended varieties can be compared to local germplasm. Additionally, more extensive diversity can be evaluated under the real conditions that farmer's experience.

Although the participatory approach has largely been focused on marginal agricultural environments, this approach is not limited to such environments or resource-poor farmers in the global south (see Section 17.6). Farmers in better-endowed environments may also benefit from more participation in the breeding process, for some of the same reasons as in marginal environments, especially from a choice between a wider range of varieties, and faster dissemination of products from breeding programmes (Box 17.7). Many of the environments that appear to have lost a lot of diversity may particularly benefit from more diversity (i.e. more varieties of diverse origin), for example for greater stability under pest and pathogen attack.

Farmers are often the passive recipient of 'finished' cultivars. However, collaboration between the breeder and farmer can occur at any stage of the breeding cycle. Although farmer participation is most often

(A)

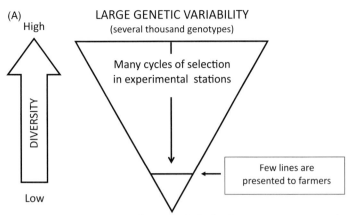

High

DIVERSITY

Low

LARGE GENETIC VARIABILITY
(several thousand genotypes)

Many cycles of selection
in experimental stations

Few lines are
presented to farmers

No or little genetic variability (composed of a few homozygous genotypes)

Figure 17.2 Representation of conventional (A) and participatory (B) breeding programmes.
(Adapted from Ceccarelli *et al.*, 1996)

(B)

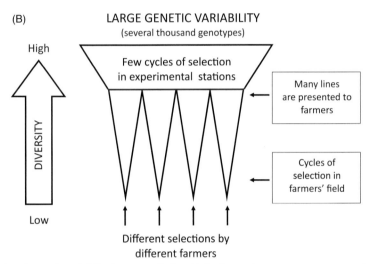

High

DIVERSITY

Low

LARGE GENETIC VARIABILITY
(several thousand genotypes)

Few cycles of selection
in experimental stations

Many lines
are presented to
farmers

Cycles of
selection in
farmers' field

Different selections by
different farmers

Reduced genetic variability (composed of multiple diverse, heterozygous genotypes)

associated with evaluation of new cultivars. The actual stage at which collaboration occurs will be determined by the breeding goals and the desired varietal outcome. The first stage is likely to involve joint priority setting between the breeder and local farmers and farm families. Women, who constitute the majority of the world's farmers, have proven essential to this initial discussion, as even in patriarchal societies women often play a key role in agricultural production and agricultural product exploitation. Once the goals have been set, plant breeders will probably play the major role in developing new breeding populations, involving both local materials and material from other sources. They will seek to widen the genetic diversity of the target crop through controlled crosses and to add new characteristics while maintaining a reasonable level of local adaptation. For an outcrossing species, the introduction of new genetic materials can also take place at the level of farmers through introgression between their landrace material and bred cultivars. Some early examples of farmer participation are detailed in Box 17.8, which illustrate the range of possible approaches. The chapter can only provide a brief overview of PPB, so readers who are interested in the details of specific projects, crops or geographies

Figure 17.3 Farmer evaluation of wheat field trial with 20 cultivars
(Bioversity International/A. Gupta)

are directed to the References section at the end of the book.

Farmers can have a key impact throughout the breeding cycle by:

- Formulating locally relevant breeding goals.
- Crossing their locally adapted landrace material with exotic germplasm and breeder's material.
- Crossing local germplasm of closely related wild species with exotic material.
- Allowing introgression between their landrace material and bred cultivars by growing the plants side by side and allowing or promoting gene flow between the two.
- Selecting segregating lines produced by crossing or introgression that are suited to local ecogeographic conditions.
- Establishing local field trails of segregating lines with desired traits, alongside the original local landrace material and conventional bred cultivars.
- Bulking up seed of the favoured lines for local adoption.
- Promotion of locally bred favoured lines in the local community.

Many of the applications of the participatory approach to breeding have been associated with traditional semi-subsistence farmers in marginal agricultural environments, but this need not always be the case. Ngoc De (2001) demonstrates using a case study from Vietnam that the participatory approach can also be beneficial in the high yield potential systems of the Mekong Delta, where the International Rice Research Institute have had such a dramatic effect in recent times. However, the participatory approach was not without its problems including:

- Low education level of farmers means they require more training and adoption of PPB is slow.
- Few farmers are interested in working with breeding and selection of segregating materials. Farmers are more willing to multiply the promising varieties than to select from segregating materials or make crosses.
- The number of collaborating farmers on PPB is limited, especially in the pedigree selection method and segregating material selection because the work is time consuming.
- Agriculture policy is more favourable to commercial production than to diversity.
- Owing to fast turnover of rice varieties by farmers (every 3–4 seasons), it is difficult to keep their interest and get their cooperation for the entire

Table 17.1 **General distinctions between conventional and participatory breeding**

		Conventional plant breeding	Participatory plant breeding
Methods	Plant breeding methods	Crossbreeding, hybridization and gene technology	Primarily mass selection
	Multiplication methods	Formal on-farm, cell and tissue culture	Farmer-based on-farm
	New variation (sources)	Globally, from induced mutations, potentially from unrelated species	Farmer exchange, introgression from diverse landraces
Results	Cultivars produced	Small number of genetically uniform cultivars selected	Large number of new varieties or landraces with inherent variation selected
	Adaptation	High adoption rates among intensive farmers with significant impacts on food security and national economies	Local adoption of new varieties or landraces with local impacts on food security and economy
	Yield potential	High in ideal agro-environments	Moderate in moderate or ideal agro-environments
	Production of varieties (time)	10 – 20 years	Continuous refinement with production

Adapted from FAO (1998).

selection process of segregating lines, which takes time to get results.

There were also several key lessons learned from the experience of PPB in Vietnam which included:

• Support from local authorities and organizations in terms of organization, management, additional funds and facilitation is very important.

• Cooperation with group/community on PPB/PVS gives better results than with individual farmers.

• Farmer Field School (FFS) and farmer field days on PPB/PVS are good ways to motivate the farmers' participation at community level.

• Farmers conserve and maintain plant genetic resources diversity to meet their needs related to home consumption, market economy and

---

Box 17.7 | Farmers As Researchers: The Rise of PPB

(Toomey, 1999)

Despite major advances in agricultural science, the rate of adoption of some new cultivars by hundreds of millions of small-scale farmers in developing countries had been low prior to the development of the System-wide Program on Participatory Research and Gender Analysis for Technology Development and Institutional Innovation of the CGIAR. Under this program, researchers in universities, national and international research institutes, NGOs, grassroots organizations, and other groups assessed and formulated new methods and organizational arrangements for participatory research in plant breeding and natural resource management. Jacqueline Ashby, coordinator of the programme states:

This has to do with bringing the farmer out of the field as a recipient of technology and into the screen houses, labs, and meeting rooms as a partner in research.

One problem with conventional breeding has been the tendency to focus heavily on 'broad adaptability' – the capacity of a plant to produce a high average yield over a range of growing environments and years. Unfortunately, candidate genetic material that produces very good yields in one growing zone, but poor yields in another, tends to be quickly eliminated from the breeder's gene pool. Yet, this may be exactly what small farmers need. Also 'improved' cultivars often require heavy doses of fertilizer and other agro-chemicals, which most poor farmers cannot afford.

Professional breeders, often working in relative isolation from farmers, have sometimes been unaware of the multitude of preferences – beyond yield, and resistance to pathogens and pests – of their target farmers. Ease of harvest and storage, taste and cooking qualities, how fast a crop matures, and the suitability of crop residues as livestock feed are just a few of the dozens of plant traits of interest to small-scale farmers, according to PPB specialists.

Despite this wealth of knowledge, in many cases farmers' participation in conventional breeding programmes has been limited to evaluating and commenting on a few advanced experimental cultivars just prior to their official release. Such token participation means that most farmers do not have any sense of ownership of the research and have not been able to contribute their technical expertise. Many of the cultivars reaching on-farm trials would have been eliminated from testing years earlier if farmers had been given the chance to critically assess them.

---

adaptation to local environments and farm resources.

- Biodiversity development should be considered on both a temporal and spatial basis at species, crop and agro-ecosystem levels. PPB/PVS increases plant genetic resources at a gene pool level and not at a specific varieties level.

- *In situ* and *ex situ* conservation and development are complementary.

Iosefa *et al.* (2013) illustrate the benefits of this last point in the context of a taro PPB programme in Samoa, which was linked to the wider Pacific region through a regional gene bank in Fiji and which ensured a two-way flow of genetic materials and a

---

Box 17.8 | Some Early Examples of Farmer Participation in Breeding Programmes

(FAO, 1998)

- In the International Crops Research Institute for the Semi-Arid Tropics (ICRISAT) pigeon pea breeding programme for resistance to insect pests, entomologists worked closely with women farmers in on-farm trials in drought-prone marginal environments in Andhra Pradesh (India). Participatory rural appraisal methods allowed farmers to compare improved pest-resistant pigeon pea (ICRISAT material) with their local varieties, using their own evaluation criteria. The traits women selected went well beyond the conventional yield and pest resistance used by most scientists and variety release schemes.
- Ethiopia's Biodiversity Institute has implemented a programme for restoring displaced farmer varieties and introducing appropriate landrace materials into various agro-ecological zones and farming communities. The programme incorporates the enhancement and improvement of farmers' varieties of barley, durum wheat and teff. The farmers are fully involved at all levels of varietal development. This strategy is in the process of being institutionalized through the National Plant Genetic Resources Policy.
- In the CIAT bean programme in Rwanda, expert farmers (i.e. superior seed selectors) evaluated selections with breeders under on-farm conditions. Results showed that farmers and breeders differed in their evaluations; farmers selected traits that would be suitable under specific conditions and in mixtures. As a result of the farmer's involvement, some of the selected genotypes were very quickly adopted, unlike on previous occasions.
- In a participatory rice breeding programme in Sierra Leone, part of the worldwide Community Biodiversity Development and Conservation Programme, local farmers and rice breeders from the Rokpur Rice Research Centre worked together to select superior rice varieties on-farm. This approach is expected to make improved cultivars available to farmers more quickly than via conventional approaches, and this programme can be used as a model for others.

---

much increased genetic diversity for the crop not only in Samoa and the Pacific region but also internationally (Figure 17.4). Farmers evaluate and select clones, which have been produced through crosses between cultivars from Samoa, Palau and the Federated States of Micronesia (Hunter *et al.*, 2001). Although selected clones enabled farmers to start growing taro again, there was concern that breeding from such cultivars with only a Pacific origin would not significantly broaden the required genetic base. Via the regional gene bank in Fiji farmers were able to obtain access to virus-indexed germplasm from Asia,

thus bringing together two distinct taro gene pools. The gene bank continues to play an active role by improving farmers' access to exotic cultivars and creating opportunities to distribute clones developed by the Samoa programme to other countries. In 2009, five new varieties were formally released and recommended in Samoa. Top selections from each breeding cycle in Samoa are tissue cultured and subsequently transferred to the regional gene bank and are available to farmers and breeders in countries that are contracting parties to the International Treaty on Plant Genetic Resources for Food and Agriculture. In 2009, the Pacific region

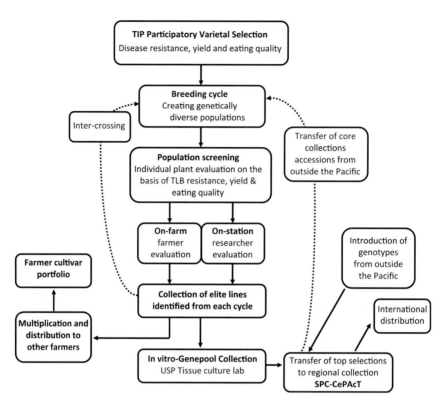

Figure 17.4 The Pacific Taro Improvement Project's strategy for participatory taro improvement, illustrating the links between farmer networks, *ex situ* conservation through the regional gene bank (SPC-CePAcT), and the formal regional and international plant genetic resources system.
(From Iosefa *et al.*, 2013)

agreed that the regional gene bank collections be placed in the Multilateral System.

17.4.2 Advantages and Disadvantages of the Participatory Approach

The participatory approach has several advantages and disadvantages. Among the advantages are:

- The fact that it facilitates the identification and enables the addressing of farmers' needs.
- It produces multiple heterogeneous varieties adapted to local conditions as opposed to the few uniform cultivars with broad adaptability produced by the conventional approach.
- The participatory approach is faster. The formal system needs 10–20 years from the initiation of a breeding programme until the result is available to the farmer. Some of this time is used for the

breeding itself, some for achieving uniformity and testing, as well as certain legal requirements before cultivar release. The participatory system and the nature of its products mean that some steps taken by the formal system can be dropped and seed can be made widely available more quickly.

However, there are also several disadvantages:

- Some breeders may legitimately or otherwise resist the added time and effort required to reach and interact with farmers.
- Breeders and farmers are not always located near one another.

Furthermore, while initial results may look promising, PPB does not yet have a proven long-term record that has been fully evaluated.

It should also be noted that early generation material from crosses is nearly always less productive than either the best of the starting (parent) material or the final selections from the breeding process, since they contain undesirable gene combinations. Thus, release of early generation (segregating) material could lower farmers' yields in the short term. The need to maintain such populations separate from the farmer's main crop imposes additional burdens of management on the farmer. Some farmers may not have the capacity or experience to carry out selection from highly segregating materials. For these and perhaps other reasons, in all the case studies analyzed so far, selections from early segregating materials have involved substantial breeder input and supervision; farmer-managed selections are carried out only on research stations, while on-farm trials are largely breeder-managed.

It should also be remembered that farmers like any other cross-section of society are not uniform; they are likely therefore to vary in their capacity, willingness and aptitude to collaborate in breeding and their goals (maximizing production versus ensuring stable minimum production, production for sale versus for subsistence, production for staples versus special culinary characteristics, etc.) are likely to vary greatly. Preferences are likely to vary according to wealth, social status, gender, age and experience.

In many communities, some farmers play a leading role in germplasm management. Depending upon the social structures, ensuring participation of these farmers in breeding programmes will not necessarily guarantee that all farmers in the community benefit, or even that the needs of other farmers will be identified. Often the poorest members of society may be the least likely to form groups or otherwise ensure their own representation. Therefore, the poorest and most vulnerable groups can still be marginalized by 'participatory' approaches, and it must be admitted that for the conventional breeder the participatory approach may prove 'challenging'. However, having made the final point, it should also be stressed that many conventional breeders might find working in collaboration with farmers equally rewarding.

The economics of the participatory approach are as yet unclear. There is a need for more methodological studies before this can be accurately estimated but

where breeding is required for specific, marginal environments, the decentralized, participatory approach has proven to be a highly feasible option. It has also been pointed out that decentralized and participatory approaches to plant breeding are likely to increase genetic diversity on-farm, in comparison with conventional approaches because:

(a) Different farming communities, working in different environments, will tend to adopt different 'solutions'; therefore, farm to farm diversity will be higher; and
(b) If farmers have access to, and can experiment with, more material, they are able to maintain more diversity within their farms, and they positively engage in genetic base broadening.

Both advantages can only be beneficial for plant conservation and exploitation. The participatory approach also benefits both farmers and breeders by encouraging genetic diversity conservation, as such it:

- Makes a wider range of genetic material available to farmers, directly as well as using a broader genetic base in formal breeding.
- It breaks the link between plant breeding/agricultural development and genetic erosion.
- It produces novel cultivars suitable for resource-poor farmers in marginal areas, which incorporate the genetic diversity of their own landraces and so engenders *in situ* on-farm conservation, rather than replacement, of local plant genetic resources.

17.4.3 Evolving Roles

A Continuing Role for Breeders
Breeders and farmers both have comparative advantages, which help define the role each assumes in PPB. Breeders have access to a wide range of genetic diversity as well as the scientific knowledge and methods to work scientifically on the development of improved germplasm. Farmers can be good at selecting material, which suits their environments and resource levels. Farmers also understand the qualities consumers will look for. To be able to work together, plant breeders and the

institutions for which they work will have to become more sensitive to the capabilities of farmers and the actual and potential contribution they can make to the breeding process. The need to rethink conventional breeding strategies and promote farmer participation, especially for the improvement of landraces, has often been identified.

Participatory plant breeding may increase the chances of breeding successes in the complex farming systems of these diverse and marginal environments. It may also contribute to the wider use of genetic diversity and increase the maintenance and development of locally adapted genetic resources. It may also speed up the process of making new varieties available. Finally, participatory approaches to plant breeding could help build a foundation for growth in productivity and prosperity, which could then provide a basis from which to attract investment from the private commercial sector

An Increasing Role for Farmers

Plant breeders lead most PPB projects, with farmers having a consultative or collaborative role; the researcher and the experiments designed by them almost always set the research agenda. However, a growing number of PPB programmes take a different approach and provide scientific and logistic support to the farmer's own plant breeding. These farmer-directed projects have been referred to as scientific-aided indigenous plant breeding or farmer-led PPB. These projects have two major characteristics:

(a) they seek to improve local indigenous plant breeding techniques and thus sensitize farmers to the value of plant genetic resources and

(b) they provide farmers with access to enhanced intra-species diversity (i.e. early generation plant populations derived from crosses of local varieties).

17.5 Impact of PPB

The obvious impacts from PPB are the considerable speed with which new cultivars are produced, and which meet the needs of resource-poor farmers and the complex, diverse and risk-prone environments they farm, and which in turn stand a considerable chance of adoption. The knowledge, skills and organizational capacities of farmers and farming communities have also been enhanced through PPB, as have the abilities and capacities of researchers and extensionists to effectively support farmers and communities. This all no doubt contributes to overall impacts in the efficiency of research and the technological innovation process. We have also seen in this chapter that PPB can improve the levels of plant genetic diversity on farms and contribute to the overall sustainability of agriculture.

Only a few studies have looked at the impact of PPB initiatives in a rigorous manner, and these have helped to determine what impacts have actually been achieved, and with which crops and geographic locations. Some of these studies were carried out under the umbrella of the CGIAR PRGA programme (Box 17.6 and 17.7). These studies have also helped shed light on where knowledge gaps on impact still exist and which areas of work would benefit from future study.

Ashby (2009), using a theory of change approach and using impact pathway frameworks for both PVS and PPB, and also considering the level and timing of farmer participation, summarizes the outputs, outcomes and impacts from a range of initiatives globally. Some of the findings and conclusions reported for the main impact pathways are:

- *PPB and PVS produce more desirable varieties leading to higher rates of adoption*: There are now numerous examples and studies which demonstrate that PVS and PPB have contributed to improving the acceptability of new varieties to resource-poor farmers living in marginal environments. In fact, the level of experience is so diverse and compelling across multiple crops, cultures and agroecological environments that it clearly demonstrates the benefits of participatory selection of varieties in such environments.

- *PPB leads to faster varietal release*: Introducing farmer participation at an early stage of the breeding process does lead to substantial reductions in the time required from initial parental crossing to farmers receiving materials for testing. Based on breeder experiences with different crops it was concluded that PPB reduced this time by 3–4 years.

- *PPB provides faster varietal release leading to earlier adoption and increases the stream of benefits to farmers*: An economic analysis of PPB barley breeding in Syria estimated benefits from conventional breeding in Syria of around US$22 million, whereas the estimates from three PPB approaches ranged from almost US$43 million to around US$114 million.
- *More desirable varieties and higher adoption rates improve research efficiency*: One of the concerns with PPB approaches is the debate around the possible increased research costs involved. It was concluded that there was a need for more analyses on the way PPB affects costs and at present it was not possible to conclude that PPB automatically represents a major increase in the costs for a breeding programme.
- *PPB fosters new skills, new knowledge and social capital that speed up innovation*: While farmers and breeders certainly acquire new skills and knowledge, the question whether this fosters new capacity to sustain this innovation remains poorly documented and is a key knowledge gap in assessments of the impacts of PPB to date.
- *PPB increases inclusion of the poor and disadvantaged, especially women, in R&D, leading to more equitable distribution of benefits*: This is another impact pathway which is poorly documented and with key knowledge gaps highlighting that unless participant selection targets a particular social group, a PPB approach does not automatically lead to benefits accruing to them. Depending on prevailing norms and customs, participation can lead to the exclusion of important target beneficiaries.

17.6 Perspectives and Prospects

Sperling *et al.* (2001) point out that to be effective PPB genetic base-broadening efforts will need to link to 'compatible' or 'like-minded' downstream research

and development field programmes if the diverse genetic products that PPB produces are to be taken to scale across thousands of site-specific contexts. To do so also means that many of the challenges and constraints highlighted in Section 17.7 will need to be fully addressed. While ongoing efforts to maintain the gains of PPB are crucial, so too are efforts to target new opportunities and initiatives that would benefit from, and are sympathetic to, PPB approaches which could take PPB to a larger scale including through global movements such as organics and agroecology. Both are growing rapidly and already embrace the principles and practice of PPB.

17.6.1 Agroecology and PPB

Market-driven agriculture with its use of high input selection environments has resulted substantially in modern plant breeding narrowing the genetic base of our crops along with trends towards uniformity and homogeneity. Uniformity and broad adaptation are also characteristics favourable to large-scale industrial agriculture and large-scale centralized seed production (Ceccarelli *et al.*, 1996). It has also meant that many of the crops we depend on for our food have been reliant on the use of pesticides and fertilizers as inputs, which have contributed substantially to numerous environmental and health problems. A considerable reduction in diversity available for diets and foods has negatively impacted nutrition. Many of these health and sustainability issues are at the heart of our agriculture and food system today.

Participatory plant breeding demonstrates clearly that it is possible to exploit genetic differences for specific adaptation to marginal environments under farmers' conditions and improve crop yields without additional chemical inputs (Ceccarelli *et al.*, 2009). This is not only better for the health of people and wildlife, but it also reduces many of the environmental problems associated with run-off of pesticides and fertilizers. Breeding for local adaptation under marginal conditions can also reduce the need for excessive irrigation. And as mentioned earlier, one of the important benefits of PPB is that it can contribute to increasing the levels of biodiversity in farmer's fields through a variety of ways including providing farmers with greater access to breeding lines and cultivars, as well as the importance it places on landraces and their on-farm conservation. Enhancing biodiversity in food systems in such a way contributes to better nutrition and health outcomes.

Participatory plant breeding is also about empowerment of communities, putting power back in the hands of farmers and farming communities so they control decisions over their natural resources and food systems. In all these ways, PPB is very much in line with the philosophy of agroecology. Agroecology works with millions of resource-poor smallholders, located in high-stress, heterogeneous environments, who cannot afford inputs and who have little power in terms of control over genetic resources and their food system (IPES-Food, 2016). By working with farmers and farming communities' own breeding and conservation processes and through knowledge and skill-building PPB can add much value to the agroecology movement to facilitate greater control over and accessibility to genetic resources, the breeding process and seed supply. Participatory plant breeding and genetic diversity is at the heart of agroecology.

17.6.2 Organic Agriculture and PPB

Participatory plant breeding and evolutionary plant breeding (EPB, see Section 17.6.4) have also been suggested as approaches to develop appropriate varieties specifically adapted to organic agriculture and other low-input agricultural systems in northern countries based on the ample experience and sharing to be gained from experiences with PPB in the global south (Desclaux et al., 2012). Production environments for organic agriculture are similar in several ways to the marginal environments which PPB has targeted in the global south. Production environments for organic agriculture are heterogeneous, and the needs of organic farmers are equally diverse. Organic farmers like their counterparts in marginal environments in the global south also do not rely on chemical inputs as a way of modifying or controlling these heterogeneous environments. In fact, organic status certification

precludes any such use. Like the marginal production environments of the global south discussed in this chapter, there is also a lack of appropriate organic plant varieties to address the diverse environments and needs of organic growers. For similar reasons, the breeding of specific varieties for organic agriculture is not viewed as a priority or profitable endeavour by the formal seed sector or companies.

In a recent survey of Wisconsin organic vegetable growers, finding high-quality cultivars in the form of organic seed was a major concern (Lyon et al., 2015). Insufficient cultivar availability was also reported as a challenge by larger scale growers. The need for more cultivar development focused on organic agriculture including increased organic seed production for existing preferred varieties was also highlighted. Addressing these needs and given the diversity and regional specificity of farmers' crop priorities and trait requirements and the limited formal research resources available, a decentralized, PPB approach was proposed as an effective way to serve the needs of organic growers. Participatory plant breeding for select organic crops is already underway elsewhere and has shown considerable potential, including for tomatoes in Italy (Campanelli et al., 2015).

17.6.3 Climate Change and PPB

Participatory plant breeding is also proposed as one strategy for adapting crops to climate change especially in the very same heterogeneous environments where it has to date been successful (Ceccarelli et al., 2010, 2013). Climate change is having the most profound impact in the kinds of marginal environments discussed, dramatically altering what can be grown and how. Yet, in these environments it is resource-poor farmers who find it most difficult to respond and adapt. PPB can help farmers in this context in many ways. Further, PPB may be the only viable option in such agro-environments. It is unlikely commercial plant breeders will invest in breeding suitable climate change resilient cultivars because they would be unlikely to gain suitable returns on their economic investment. If public plant breeders could help local communities in identifying new germplasm that can cope with the ravages of climate change and making this available rapidly to farmers for crossing with their locally adapted landraces, as well as building the capacity of communities and farmers, these communities could be better placed to withstand the challenges of climate change.

For the reasons already provided, it can be argued that conventional plant breeding producing uniform cultivars for broad adaptability and tied to an agricultural system where wealthier farms can modify their growing environment is not able to respond to the needs of small farmers in adapting to climate changes. However, what is essentially needed as part of any suite of approaches and technologies to deal with climate change adaptation are breeding strategies that allow a highly dynamic and efficient system of cultivar diffusion to farmers, that caters to their diverse needs and environments, and rapid adoption on-farm.

17.6.4 Evolutionary Plant Breeding

Many instances have been reported by farmers where mixtures of several different cultivars of a crop (and in some instances several crops) have provided good yield stability and a more desirable or higher quality product (Rahmanian *et al.*, 2016a, 2016b). Building on experiences of PPB, EPB has been developed as a dynamic and inexpensive strategy for farmers and farming communities – although ideas about EPB have been around for a long time (proposed by Sunenson, 1956) – to rapidly increase on-farm biodiversity and as a way to adapt their crops to climate change by planting and evaluating large mixtures of varieties they continue to cultivate as their main crop. Evolutionary plant breeding has also been suggested as a suitable breeding approach for organic agriculture.

Although there may be concerns about the quality of the final product because it is a mixture, the acceptability by farmers in Iran, and in France and Italy where the approach has been trialled, is good. These evolutionary populations have been referred to as a 'living gene bank' in farmers' fields (Rahmanian *et al.,* 2016a, 2016b), but in Iran the evolutionary populations introduced were composed of exotic mixed lines from Syria and they replaced local endogenous Iranian landraces, which were subsequently lost. It should therefore be noted that EPB should not itself cause genetic erosion. However, the continuously evolving populations and adapting material can be selected as new cultivars become increasingly better adapted to agronomic and climatic conditions, especially for vegetables where mixtures are less viable. Evolutionary populations still face the policy and seed law issues which are highlighted in the next section, though in places like Iran, if the mixtures are not being sold, it is acceptable to produce as farm-saved seeds and exchange. However, this is not the case everywhere and if a country like Iran was to sign up as a UPOV member, this situation would change. In the EU this scenario is likely to become less of a problem because the European Commission is now implementing a directive which makes it possible to market experimentally heterogeneous materials of different cereals (Dwivedi *et al.*, 2017).

Evolutionary participatory breeding has also been proposed as a useful tool for developing new quinoa genetic material in collaboration with farmers and that a global collaborative network could be the baseline for PPB programmes originating in developing or developed countries to meet the needs of farmers across a diversity of agro-ecosystems and environments (Murphy *et al.*, 2016).

17.7 Challenges and Issues

Despite progress, there are several areas which need to be addressed to strengthen PPB practice so that it can move beyond its research project base and to put it in a position where it is really seen as a legitimate alternative to conventional plant breeding.

17.7.1 Policy Issues

National policies and legislation obviously have a significant impact on the ways in which farmers and farming communities can benefit from their involvement in PPB programmes (FAO, 2011a). In many countries, cultivars can only be officially registered when they meet strict criteria, especially those of specific distinctness, uniformity and stability (DUS) standards. Seed laws for maintaining and multiplying registered seed also influence how farmers can participate in cultivar development. Nepal presents an example of how the national varietal release and registration committee of the national seed board supported the release and custodianship of a landrace.

17.7.2 Property Rights and Benefit-Sharing Issues

One unresolved major issue for PPB is the recognition of farmers' contributions and benefit sharing from the diffusion and use of PPB products. These are issues that have been largely avoided or 'simply skirted around' and are still not addressed by either the farmers' rights or formal breeders' rights debates (Sperling *et al.*, 2001). This poses a major obstacle to the long-term viability of the PPB. Suitable and workable benefit-sharing models do exist. Although legislation in China is not yet adequately formulated to fully support PPB, initiatives based on more than a decade of PPB work in southwest China has seen access and benefit-sharing issues being addressed through agreements and the labelling of products of particular geographic origin (Song *et al.*, 2016). The agreement recognizing farmers' contributions and including provision to fairly share commercial benefits from market exploitation are considered a first in China, and perhaps much more widely. There are also examples from developed countries. A long-term Dutch breeding model involving partnership between farmers and commercial breeding companies facilitates access to genetic materials and financial benefit sharing (Almekinders *et al.*, 2016).

17.7.3 Seed System Issues

Sperling *et al.* (2001) noted that it is often assumed that PPB breeding products will somehow find their way into formal seed system channels or local seed systems. However, in the case of the former this has happened very rarely to date with only a single example each for Nepal and India. As for the latter, the assertion that 'varieties move themselves', these

authors also highlight that this is not always backed up by the evidence either. In fact, they stress that we know little about the effectiveness of local seed systems in relation to PPB: how rapidly do they replace varieties, are the systems equitable, how many varieties are involved, what is the geographic coverage, and can this be scaled, or how many more new varieties can local systems handle? In deciding which seed system to focus on they suggest the following guiding questions:

- What are the numbers of varieties that would need to be disseminated?
- What is the geographic scale for diffusion, and to which target groups?
- What are the property rights of the seed system and are they acceptable to the partners involved?
- Is the final genetic material homogeneous or heterogeneous?

17.7.4 Institutional Issues and Scaling Up

Related to the last challenge, Bishaw and Turner (2008) in reviewing PPB and seed supply systems propose that practitioners need to give more attention to the multiplication and diffusion process of PPB products and that this cannot simply be left in the hands of resource-poor farmers or communities. They urge more attention be given to the institutionalization of PPB in national programmes which they feel would go a long way to resolving many of the constraints currently facing PPB, including the development of more appropriate national seed policies and regulations which recognize the role of PPB and the informal seed sector. Failure to address these issues they point out would mean failure to fully realize the potential wider technical and social benefits of PPB. These issues are also explored in Bishaw and van Gastel (2009).

The history of PPB to date is one largely externally driven by groups of donors, researchers and breeders. Efforts to bring PPB to scale have been limited and hampered by a myriad of problems and constraints, some of which have been touched upon here. Sperling *et al.* (2001) point out that those small number of cases where there has been some degree of taking PPB

Box 17.9 | **University Breeding Clubs**

(Robinson, 1996)

- Clubs would provide a new 'hands-on' approach to plant breeding in an effective group-learning context for students.
- Clubs could transfer plant breeding skills to many amateur breeders in communities working within a single agroecosystem involving a few thousand farmers.
- There would be a vast increase in breeding skills as graduates return to their villages and initiate local farmers' or amateur breeding clubs.
- Club members could disseminate new germplasm by taking it back to their communities for evaluation and further dissemination via informal seed networks and farmer-to-farmer exchange.
- Hundreds of plant breeding clubs worldwide could significantly improve crops by a huge increase in breeding activity.
- Clubs would re-establish links between researchers and farmers. High levels of farmer participation in plant breeding would result when farmers' children join university breeding clubs.

---

to scale have been working with farmers using varietal diversity, e.g. stabilized genetic materials, and that any involvement of farmers in further upstream breeding activities using segregating materials has been minimal. These authors also point out that effective decentralization of PPB efforts are usually beyond the resources of most public-sector research services, which have neither the budget or capacity. They suggest NGOs might have the edge, or the Farmer Field School (FFS) model linked to PPB initiatives could be a workable approach. Yet to date, efforts to scale up PPB approaches in any guise are limited, with the example of CIALs (local communities for agricultural research in Central and Latin America) as the main example. Using high schools and agricultural colleges (Sperling *et al.*, 2001) and university breeding clubs (Box 17.9) (Robinson, 1996, 1997, 2009) have all been suggested as possible organizational options to mainstream and scale up PPB, but while such options can be put in place (Hunter *et al.*, 2001), there is no evidence to suggest that they are viable long-term.

The fundamental question remains, while PPB has proven itself an effective model of participatory research that can clearly deliver novel products that farmers need for marginal environments, how can the approach be taken to scale to address the similar needs of millions of other farmers living in similar complex, diverse and risk-prone environments?

18 Conservation Data Management

18.1 Introduction

It is widely acknowledged within the PGRFA conservation and user community that one major factor hindering effective conservation and use of PGRFA diversity is the lack of easy access to data and obstacles to information exchange, due to the many different approaches and incompatible systems for managing data (FAO, 2010a). Still today, few countries have a fully operational documentation system with substantial accession data; most countries employ multiple bespoke systems developed over years for individual institutes. However, the situation is improving with time and effort from leading global stakeholders such as the International Treaty on Plant Genetic Resources for Food and Agriculture, FAO, Bioversity International, regional organizations including the European Cooperative Programme for Plant Genetic Resources (ECPGR) Documentation and Information Working Group (www.ecpgr.cgiar.org/working-groups/documentation-information/), and even national programmes like United States Department of Agriculture Genetic Resources Information Network and their GRIN-Global software system for gene bank management. The first State of the World of PGRFA report (FAO, 1998) stated that the majority of countries and collections lack detailed information on the accessions they conserved, and as high as 90% of conserved accessions lacked passport, characterization and evaluation data. But the second report (FAO 2010) found that there had been an overall improvement of accessibility to information, with dedicated information systems used in 60% of countries and 35% using generic database software; however, 16 countries still use paper-based documentation systems.

If we are to effectively conserve and use PGRFA, then consistent data collation and management is required. This process is likely to involve taxonomic, ecogeographic occurrences, and temporal distribution, threats and conservation status and genetic structure data, as well as the ability to track using time-series data and predicted demographic and genetic changes within a species in relation to land management and environmental factors (Maxted et al., 2013b). The data are too often inaccessible or not readily available and particularly for CWR are dispersed because of the broad taxonomic range of species and the fact that much data is held by those outside of the PGR community. Accessing such information is not only time-consuming but comparing datasets is often difficult due to the diversity of information management models used. If PGR is to be conserved and sustainably utilized, collating, analyzing and making information derived from the data accessible and in a standard format is required.

In the plant genetic diversity context, data are any meaningful information that relate to an organism: sets of numbers, names or other values that have both structure and context. If we have a dataset, we usually know where it came from, how it was collected and why. They are not arbitrary and have meaning for botanists and conservationists. Data are generally based on individual observations or measurements taken on the sampling unit (object of interest), for example, a plant, a leaf, a flower. When you locate a source of seed, record the species, estimate the altitude, you are making observations, recording fresh data that has conservation value. In conservation, it is customary to refer to the property measured by the individual observations as the character in a taxonomic context, or descriptor or variable in a gene banking or evaluation context. Data can be quantitative and dealing with numbers (e.g. 125 m for altitude) or they can be qualitative, that is, a description of the object being examined (e.g. yellow for petal colour or hair for leaf covering). For many descriptors there are a limited number of character or descriptor states. For instance, the character 'growth stage' might only have four descriptor states: seedling, vegetative plant, flowering plant and seed set. In other cases, the descriptor states are continuously variable, such as altitude. It is important to note that technically, information is not the same as data, although often mistakenly the terms are used interchangeably. Information has meaning, whereas data do not. For example, what does 'purple' or '94 g' mean? It might be that 'purple' is the fruit colour and '94 g' is the quantity of seed stored in the gene bank base collection.

Although the ultimate aim of plant genetic conservation is use, the users of conservation data fall essentially into two groups:

- *Internal users* – Much conservation information is used by the conservation community itself for setting conservation priorities and planning or managing activities. To organize activities, for example, the gene bank manager must annually regenerate a proportion of the germplasm in the gene bank; not all germplasm can be regenerated every year, yet there needs to be a constant number of gene bank accessions regenerated annually, so the manager decides which accessions to regenerate on the basis of seed stock levels, seed viabilities and how frequently particular accessions are distributed.
- *External users* – However, to promote use of the conserved resource information acquired and generated by a conservationist is of interest and value to the broader user community (scientific research, policy, farming and breeding). Collection managers will commonly distribute their accessions together with relevant information, such as passport information and information obtained from characterization and evaluation trials.

The conservationist needs to make these data available to both user communities and needs to ensure that the data are effectively managed to this end.

18.1.1 Data Sources

The sources of plant genetic diversity data may be any of the following:

1. *Historical, passport (provenance) and ecogeographic data.* These are traditionally generated as a result of collection and introduction activities. These data describe the location where the accession was collected, the population characteristics that were sampled, when it entered the country and the institution where it is maintained. In the *in situ* conservation context, data could be used to describe the location of the population that is being actively managed and monitored and its interaction with local biotic and abiotic factors. For example, a specimen label (see Figure 11.5) is composed of passport data, like a human passport, and tells us where the specimen was collected and some characteristics of the sampling location.

2. *Characterization or observational data.* Characterization consists of recording those characters that are highly heritable, can be seen easily by eye and are expressed in all environments (e.g. flower colour; number of fruits per inflorescence). These are used to describe the diversity within the genetic resources themselves; they are measured or recorded directly from observation. They are descriptive and may assist in the maintenance and use of the genetic material. For example, a specimen may have white flower petals, multiple leaflets per leaf or be perennial.

3. *Experimental or evaluation data.* Evaluation consists of recording those characters that are susceptible to environmental differences (e.g. fruit yield; drought susceptibility). Therefore, an accession can be evaluated in several different sites, giving similar or different results for several descriptors at each site. Evaluation characters also describe the diversity within the genetic resources themselves, but in this case the data are only observable by manipulation of the organism in the field or laboratory. They assist in the maintenance and particularly the use of the genetic material and are, together with passport data, the basis for the rationalization of the genetic resources held in *ex situ* collections. For example, a specimen may be highly susceptible to wheat yellow rust (*Puccinia recondita*), ripen early or have a large fruit size. Each of these data is not immediately obvious, and

the researcher would need to grow out the accession to ascertain the variable. The data recorded may have various forms:

- counts (discrete integer values, e.g. 2, 3, 4, etc.),
- measurements (continuous, e.g. 12.67, 39.4, 101.0, etc.),
- ordinal (categories or names with a logical sequence, e.g. cold, cool, warm, hot, etc.),
- nominal (categories or names with no sequence, e.g. blue, red, yellow, etc.),
- derived (e.g. ratios or indices calculated from other data, e.g. leaf length to width ratio).

Although characterization can be performed at the same time as regeneration or multiplication, evaluation cannot; for instance, you would not wish to assess pest or disease tolerance by deliberate infection of an accession when regenerating it or the seed produced may lack vigour.

4. *Management or curatorial data.* These are essential data that facilitate the maintenance of the accession or population. In the *ex situ* context these data are, among other things, used to monitor the condition (e.g. viability) and assist with the management (e.g. location of stock) of the genetic resources. While in the *in situ* context this would be closely associated with the management intervention affecting the target population being conserved and would necessarily be time-series data that would help the conservationist identify trends in population numbers.

18.1.2 Three Fundamentally Important Data

The Accession Number (or Unique Identifier)

Each conservation accession, whether it is in a protected area, gene bank, arboretum or tissue culture, will have its own, unique accession number, which distinguishes it from all other accessions. If an inappropriate or no numbering system is used, the consequences are serious: it will cause confusion, create much unnecessary work through mistakes being made, wastes resources and, ultimately, could lead to loss of diversity or even extinction. The rules should be as follows:

- Use a strictly numeric system that is sequential in operation. It should start at '1' and then increase as each new accession is received, that is, '2', '3', '4' and so on. So that you do not confuse the accession numbers with any other numbers, it is advisable to add your institutional acronym to the number. For example, if the acronym of your gene bank is 'EGRU', the accessions will be labelled EGRU 1, EGRU 2, EGRU 3 and so on.
- Do not incorporate additional information in the accession number. The accession number should be used for uniquely identifying the accession and nothing else. Do not try to incorporate additional information into the accession number, such as the year of collection or a crop or animal code, as it might cause confusion later on.
- Use only one numbering system. Operate a single numbering system, which is used for all accessions.
- Do not use the depositor's designation. If you only have a small gene bank, you might consider using the depositor's designation as the basis for your accession numbering system and therefore do not need to devise your own system. This approach can cause serious operational difficulties and should not be attempted. However, retain the depositor's designation within your system in case there is any confusion over the origin of the accession.
- Do not use the germplasm collector's number. If your gene bank is quite small but all your accessions are received from collecting missions, you might consider using the collector's numbers as accession numbers. As in the previous example, this approach should not be attempted, as it is likely to lead to confusion in the future, but also retain the depositor's designation to avoid any confusion over the origin of the accession.
- Do not use a different numbering system for each crop. Certain specialist gene banks that maintain very large collections of limited numbers of species operate separate numbering systems for each taxon. This approach is not recommended, as it is far simpler to operate a single, sequential and numeric accession numbering system.
- Do not use a 'reserved' numbering system. Some botanic gardens use a single accession numbering system but within the system they 'reserve'

numbers for particular plant species, for instance 1–100 for barley, 101–200 for wheat, 201–300 for oats and so on. This system can cause several operational difficulties in the long run and should not be attempted.

Remember that any accession numbering system should be operated carefully to prevent any errors from occurring. An important rule that should not be broken is that an accession number should never be re-used, even if an accession dies in storage. Even though the accession is dead, there will still be information about it in the documentation system and in printed reports, catalogues and scientific publications. So, keep it simple and give each new accession a new accession number.

A recent initiative by the International Treaty (www .fao.org/plant-treaty/areas-of-work/global-information-system/doi/en/) has involved the provision and use of Digital Object Identifiers (DOI), promoted by the DOI Foundation (2015) (www.doi .org). These are permanent unique identifiers that are used in the context of the Global Information System (GLIS) currently being developed by the International Treaty in relation to Article 17. These are unambiguously and permanently unique identifiers that are provided by the Secretariat of the International Treaty initially for *ex situ* PGRFA accessions held by gene banks, plant breeders, geneticists, other plant scientists, extension officers, seed companies, plant variety protection offices, gardeners, farmers, landowners and land managers. Given the broad range of accession holders, although GLIS has initially focused on *ex situ* PGRFA accessions, in the future it will also encompass CWR or LR populations maintained *in situ* or on-farm. GLIS itself is being constructed to (1) build on and facilitate linkage between existing systems and (2) allow for registration of DOIs applicable to all types of PGRFA, but (3) GLIS will not replace existing systems or duplicate their functionality but provide new services needed by the user community and missing from existing systems, (4) DOIs will be easy to implement, (5) GLIS will also accommodate DOIs created by other systems and (6) use of the system will be voluntary. Registration of DOIs for PGRFA will be voluntary,

and, except for a small number of essential metadata descriptors, most descriptors are voluntary (Alercia *et al.*, 2018).

The Scientific Name

All biodiversity is identified and categorized by its scientific name. To convey information about an accession in scientific papers, reports or newsletters or to exchange germplasm and information the scientific name is used. The scientific name is an internationally recognized name for the taxon and facilitates the collation of information on the taxon from many disciplines. The taxon to which the scientific name applies is likely to have numerous vernacular names in local languages, but these will inevitably vary from country to country. For instance, the crop rye, which has the single accepted Latin name of *Secale cereale*, has the following local names: rye, centeio, Roggen, centeno, Seigle, baraka and çavdar. There is only ever one accepted Latin binomial for each species, so the use of the accepted Latin name avoids confusion. This situation can be confused by synonyms, Latin binomials used for a species that are not accepted because they are not accepted binomial combinations according to the rules of the International Code of Botanical Nomenclature (see Chapter 3). The Plant List (www.theplantlist.org/) provides the accepted Latin name for most plant species with links to its most common synonyms. While it is perfectly acceptable to use the crop's common or local name for everyday use, it is important to use the accepted scientific name for all publications and communication with other conservationists and germplasm users.

The Date

The date may have several important links to the accession, possibly the date on which, for example, an accession was collected from the wild, entered the gene bank, was last regenerated, was last monitored in the *in situ* conserved population, or when the accession or population was characterized and evaluated. Once an accession enters the conservation system, there is the possibility that errors and confusion may occur; therefore, the addition of the date is important when any novel data are added to the system. If an error or inconsistency is suspected, then this may be traced via the date associated with the doubtful data.

18.1.3 Standardization

To avoid repetitive typing often data is codified when being entered in a conservation database, in which case it is important to adopt standard codes wherever possible to facilitate data transfer between databases. The use of standard, internationally accepted codes in a biodiversity database allows the following.

- *Multiple uses of the same dataset.* If data standards are applied, datasets can be easily exchanged between users and the same dataset can be viewed in more than one way. A database may be indexed on different kinds of items, or subsets of the database could be produced. It is possible to use Boolean searches ('and' – 'or' – 'not') to abstract subsets of data not conceived of by the original data collator.
- *Reduce any problem of text synonyms.* Text synonyms occur where two phrases have the same meaning but quite a different syntax. Compare, for example, 'leaves lanceolate' with 'lanceolate leaves' or even 'leaves long and lanceolate', or Spain, Espana, ESP and ES. In the former example, all three terms refer to a particular leaf shape and in the latter, all refer to the same geographic unit. Using an accepted standard would avoid this problem. It would also speed up any sub-string searches within the database and avoid synonymous text strings being missed when searching a database.
- *Greater consistency.* If one set of data standards were applied, it would be much simpler for everyone to understand and interpret the datasets. Also, if standards are enforced the data cannot be internally inconsistent, the same information will always be referred to in the same manner within the database.
- *Permits automatic checks for data integrity.* The centralized control implicit in the use of standards means that, prior to inclusion in the database, data must be standardized to standards being applied.

Therefore, the data are always uniform and standardized. This is of particular advantage if data are to be interchanged with other systems.
- *Permits lower transaction costs.* If data are available in standard formats, based on standard collection methodologies, the users can absorb them more easily into their work. Increasingly, if standards are not used, then data may be perceived as incompatible, inappropriately focused or otherwise unusable. The lower the transaction costs associated with accessing and using others' data, the greater the use made of the dataset.
- *Comparison of results.* Without agreement on data standards, organizations tend to employ their own methods of collecting and managing data, which inhibits interrogation. Even within an organization, methods may be applied inconsistently by different groups, or at different points in time. Data standards overcome this problem by enabling comparison of results in space and time, and between different sources.
- *Quicker data searching.* Standardized codes are by definition shorter than the original text they represent, and the shorter the text string that is searched for, the quicker the search time. However, with the speed of computers getting ever quicker this is of less advantage than it was 40 years ago.

Types of Data Standard

Standards may be applied to all aspects of data management, from data collection and storage, to quality assurance and distribution. They define accepted formats, structures, systems and procedures for managing data. Mostly, they define only the minimum requirements, but those using the standards would be encouraged to exceed the requirements as appropriate.

- *Collection.* Recording/measuring techniques for specific themes (e.g. biological records, human impacts, policy performance, sustainability); classification systems (e.g. soils, vegetation, climate, species names, agricultural practice); criteria for assessing threats to biological resources.
- *Storage.* These are the standards associated with the storing of the data in the database itself, such as the

core data model database structures; storage formats and media; methods of data retrieval; use of information technology; maintenance procedures.
- *Quality assurance.* Validation, maintenance and security procedures; documentation formats.
- *Distribution.* Product definitions (e.g. map keys, acknowledgements, symbols); reporting formats; data transfer formats (interchange standards); protocols for electronic communication of files. An example is the Darwin Core, which is a set of terms and definitions that facilitate the exchange of information about the geographic occurrence of organisms and the physical existence of biotic specimens in collections (see www.tdwg.org/standards/dwc/).
- *Description.* In the context of genetic conservation descriptive information is routinely used to help in describing, conserving and promoting use of the accession (see more detailed discussion in Section 18.2).

Setting Standards

It should be admitted that reaching agreement on data standards is a time-consuming, largely intellectual activity requiring concrete and determined action to succeed. However, the benefits outweigh the initial investment. Biodiversity Information Standards, formerly the International Working Group on Taxonomic Databases for Plant Sciences (TDWG), was established by the International Union of Biological Sciences in 1985 to facilitate data standardization and data exchange between botanical databases. Since establishment through a series of workshops involving the key international biodiversity database managers, they have produced sets of standard database codes for:

- authors of plant names
- economic botany data collection standard
- floristic regions of the world
- herbarium information standards and protocols for interchange of data
- Index Herbarium, Part 1: The herbaria of the world
- international transfer format for botanical garden plant records

- plant names in botanical databases
- plant occurrence and status scheme
- Taxonomic Literature edition 2 and its supplements
- users' guide to the DELTA system
- world geographic scheme for recording plant distributions
- XDF: a language for the definition and exchange of biological datasets

Information on the various sets of standards can be obtained from the TDWG website (www.tdwg.org/). This concept of easy data exchange has facilitated the development of large meta-databases, which contain little original data but rather link existing primary databases. Such an example is the Species 2000 database, which provides an easy means of linking biodiversity databases and avoiding unnecessary duplication. Regular exchange between databases will enforce the use of data standards to aid consistent data definitions.

Standards are also increasingly being used in designing biodiversity information architectures. The TDWG Technical Architecture Group describe three principles of biodiversity informatics when designing an informatics system, (a) the ontology, which describes a standardized descriptive vocabulary, (b) the data exchange protocols, which describe the collaborative processes and data publication and (c) the persistent identifier, which distinguishes each data item and enable linkages between data items. Perhaps the ones most familiar to plant genetic conservationists are ontologies, which are now routinely used to organize and categorize physical things such as specimens, gene bank accessions, genetic properties, alleles, institutions and people; or events such as the collecting of seed material, trait measurements and seed distribution (Endresen, 2017).

Applying database standards helps to ensure the quality of the data and the information that can be derived from the data. Data quality is a relative term, for which there is no precise definition; perhaps it is more helpful to think of the fitness of the data to serve the information needs of the conservation organization. It must be true, however, that the 'knowledge' (information) of the conservation organization is only as good as its data! However, it is not always feasible or even desirable for conservation organizations to adopt every existing data standard. They may have their own, highly effective ways of managing data, which could become compromised, possibly disrupted, by the blanket introduction of new and unfamiliar standards. Where increased efficiency is unlikely to follow the introduction of standards, they should not necessarily be pursued.

There is, however, one standard that will always bring efficiencies, at very little cost. These are Darwin Core interchange standards whose purpose is to streamline the transfer of data between databases. The introduction of interchange standards has very little impact on the way organizations collect and manage their own data but has had a strong impact on data mobility. Interchange standards focus mainly on the formats and media in which data are transferred.

18.1.4 Data Management and Analysis

Conservation, like many other branches of biology, is particularly data intensive: we generate a lot of data. However, it must also be reluctantly admitted that even today, much of current conservation data management is chaotically organized. Plants that are not identified in the field are unlikely to be collected or monitored; plants conserved without passport data are unlikely to be incorporated into an efficient gene bank and plants without characterization and evaluation data are unlikely to be utilized. Within plant genetic conservation, the management of data associated with collection, gene banking, *in situ* on-farm maintenance and characterization/evaluation of the samples is referred to as documentation, and documentation is defined by Painting *et al.*, (1993) as:

any process that involves the identification, acquisition, classification (ordering), storage, management and dissemination of information to users.

Effective conservation data management, particularly if combined with the application of a database management system, has provided a powerful tool to help resolve this problem. Data management can be viewed as a process that begins with the conception and design of the research project, continues through data capture and analysis, and culminates with

publication (that is, data archiving and data sharing with the broader scientific community). Efficient and effective data management provide added value to the conservation information held in the raw data. Effective data management at the institutional level can also facilitate research and increase productivity. The role of data management is continuing to change as research projects are becoming broader and more complex. There are a variety of reasons for implementing a data management system:

- It formalizes a process to acquire and maintain the products of a research project, including data, so that they may extend beyond the lifetime of the original investigator(s).
- It acilitates the resurrection of currently inaccessible historical data.
- It supports the preparation of datasets for peer review, publication in data journals, or submission to a data archive.
- It provides access to datasets that are commonly used by more than one investigator on a project.
- It provides access to datasets by the broader on-line scientific community.
- It reduces the time and effort spent by researchers in locating, accessing and analyzing data, thereby increasing the time available to synthesize results.
- It increases the scope of potential research projects by facilitating the investigation of broader scale questions that may require integration of multiple disparate datasets.
- It incorporates data from automated acquisition systems into ongoing analytical efforts such as ecological modelling.

It is undoubtedly true that many future advances in conservation science will hinge on the ability to integrate diverse digitized datasets, and clearly, carefully considered and applied data management practices are required. Although historically the documentation system would have been paper based, today computer-based systems are universal in an attempt to help avoid inconsistencies, facilitate data sharing, help enforce data standards, and allow repeated editing and remote access (FAO, 2010a). Many documentation systems are based on one of the proprietary database packages. The desirable features of such a documentation system are:

- *Data integrity* – Information retrieved from a documentation system must be accurate, reliable and up-to-date for it to be of value. It is necessary to think carefully about how the system can be designed and operated to facilitate the maintenance of accurate information.
- *Fast information retrieval* – If the system is well designed, retrieving information will be a simple and straightforward process. If it isn't well designed, you may spend hours, or worse still, not be able to find the information at all! If similar information is needed on a regular basis and it takes you several hours to locate the information each time, you are spending far too much of your time on information retrieval.
- *User-friendly operation* – Data do not appear in a documentation system as if by magic – the chances are that you or a colleague will have entered the data. In fact, you will be spending a fair amount of your time at the computer or writing up forms so anything that reduces this workload is helpful. It helps enormously if the documentation system is user-friendly. A user-friendly system requires the minimum of training. If the documentation system is user-friendly, you will find considerably fewer errors creeping in and it will be a lot more popular with the people who use the system.
- *Flexible operation* – The documentation system should not be rigid in its operation. It should be able to cope with different and evolving requests for information and accommodate change. You should try to anticipate information requests as far as possible.

When choosing a documentation system, it is best to avoid using a standalone bespoke PGR documentation management system. Hawkes *et al.* (2000) stress that conservationists are unlikely to have professional software development skills, as well as the time required to design and test an efficient and effective PGR documentation management system and inevitable a bespoke system would not allow the flexibility of a proprietary package. There are several proprietary packages available, so when selecting

which system to use, the following features should be taken into consideration (Hawkes *et al.*, 2000):

a. cost,
b. portability,
c. ease of use,
d. ease of use of programming language,
e. available applications,
f. multi-user access,
g. ease of data protection and security.

One such widely used conservation management system is described in Box 18.1.

18.2 Data Collation, Manipulation and Accessing

Conservation, perhaps more than any other branch of biology, requires a high investment in data capture, storage and retrieval (Hawkes *et al.*, 2000). In Chapter 1 we discussed a general methodology for plant genetic conservation (see Figure 1.11), and data management (collation, import, manipulation, generation, export) is essential for each step of the plant genetic conservation to use continuum. Ffour steps are particularly data intensive:

- Ecogeographic data (taxonomic, ecological, geographic and genetic: passport)
- Field population data (passport)
- Conservation management data (curatorial)
- Characterization and evaluation data (descriptive)

The integration of all four data types is necessary to develop an appropriate conservation strategy for a target taxon or taxa, describing what is known about the target taxon or taxa from the literature, where populations are in nature, how they are currently conserved and what traits might be included that will meet user's requirements for diversity. For each of these data types the data are recorded in some form of standardized descriptor. A descriptor may be defined as any attribute referring to a population, accession or taxon which the conservationist uses for the purpose of describing, conserving or promoting use of its diversity. As explained above, descriptors (e.g. flower colour, leaf length or yellow rust resistance) are

Box 18.1 | **GRIN-Global**

(www.grin-global.org/)

GRIN-Global is a freely available conservation management system designed for gene banks to store and manage information associated with plant genetic resources and to deliver that information globally developed by the Global Crop Diversity Trust and the United States Department of Agriculture (USDA).

For gene banks it provides a tool for managing germplasm collections and the respective data:

- A complete gene bank inventory management application.
- Enables flexibility in delegating user ownership and permission rights.
- Views, forms and wizards can be customized.
- Interface accepts data transfers from mobile applications.
- Can be displayed in any language (when translation is provided, English default).

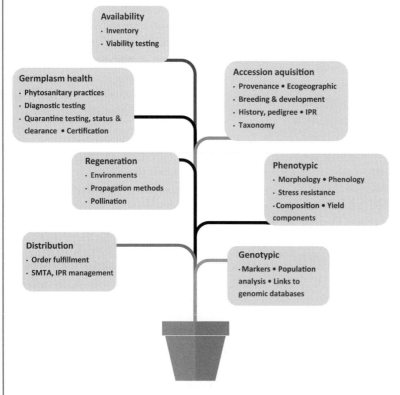

Figure 18.1 Data management capability of GRIN-Global.

While for researchers:

- Provides immediate access to PGR information.

Box 18.1 | (cont.)

- Incorporates easy-to-use interface for extracting and manipulating PGR information and requesting germplasm.
- Provides access to a worldwide plant taxonomy database.

Figure 18.1 indicates the kinds of data that can be stored and the data management tools provided by GRIN-Global.

GRIN-Global is an Open Source Project and there is an extensive helpdesk facility providing on-going support to users. It is being actively used or trialled by about 25 gene banks globally, but its recent successful implementation to provide the Portuguese PGR documentation system is reviewed by Barata *et al.* (2016).

abstract concepts; what is recorded and used are descriptor states (e.g. blue/yellow, 16–23 mm, partial/complete resistance) that conservationists record and utilize. Therefore, for each descriptor, the user records the descriptor state present in the taxon or population being assessed, usually in the form of numeric code. Originally the descriptor lists produced were for individual crops and for scientists working on those crops, but now they are used by farmers, curators, breeders, scientists and users, and facilitate the exchange and use of resources. They fall into six rough categories:

- *FAO/Bioversity International Multi-crop Passport Descriptors* – Originally published in 2001, the 'Multi-Crop Passport Descriptors' is widely used as the international standard to facilitate germplasm passport information exchange. Now expanded to include emerging documentation needs, this new version resulted from consultation with more than 300 scientists from 187 institutions in 87 countries.

   Alercia, A., Diulgheroff, S. and Mackay, M. (2015). *FAO/Bioversity Multi-Crop Passport Descriptors V.2.1 [MCPD V.2.1].* FAO/Bioversity International, Rome, Italy. Available at: www .bioversityinternational.org/e-library/publications/detail/descriptores-de-pasaporte-para-cultivos-multiples-faobioversity-v21-mcpd-v21/ (accessed 4 January 2019).

- *Descriptors for Genetic Marker Technologies* – A tool for researchers to generate and exchange genetic marker data that are standardized and replicable. They represent the minimum set of descriptors needed to describe a marker technology used with a particular plant species.

   de Vicente, M.C., Metz, T. and Alercia, A. (2004). *Descriptors for Genetic Marker Technologies.* FAO/Bioversity International, Rome, Italy. Available at: www.bioversityinternational .org/e-library/publications/detail/descriptors-for-genetic-markers-technologies/ (accessed 4 January 2019).

- *Key access and utilization descriptors for crop genetic resources* – Initial sets of characterization and evaluation descriptors for individual crops.

   Bioversity International and Agricultural Research Centre for International Development (CIRAD) (2009). *Key Access and Utilization Descriptors for Coconut Genetic Resources.* Bioversity International, Rome, Italy and Agricultural Research Centre for International Development, Montpelier, France. Available at: www.bioversityinternational.org/e-library/publications/detail/key-access-and-utilization-descriptors-for-coconut-genetic-resources/ (accessed 4 January 2019).

- *Descriptors for farmers' knowledge of plants* – The first attempt to capture and share information between farmers and scientists, integrating biology with traditional knowledge.

Bioversity and The Christensen Fund, 2009. *Descriptors for Farmers' Knowledge of Plants.*

Bioversity International, Rome, Italy and The Christensen Fund, Palo Alto, California, USA. Available at: www.bioversityinternational.org/e-library/publications/detail/descriptors-for-farmers-knowledge-of-plants/ (accessed 4 January 2019).

- Core descriptors for *in situ* conservation of crop wild relatives – Designed to facilitate the compilation and exchange of *in situ* conservation data, which are needed to develop and implement *in situ* conservation activities.

Bioversity International and University of Birmingham (2017). *Crop Wild Relative Checklist and Inventory Descriptors v.1.* Bioversity International, Rome, Italy. Available at: www .bioversityinternational.org/fileadmin/user_ upload/Crop_wild_relative_checklist_and_ inventory_descriptors_v.1_final.pdf (accessed 4 January 2019).

- Core descriptors for *in situ* on-farm conservation of crop landraces – Negri, V., Maxted, N., Torricelli, R., Heinonen, M. Veteläinen, M. and Dias, D. (2016). Descriptors for web-enabled national *in situ* landrace inventories. University of Perugia, Perugia, Italy. Available at: pgrsecure.bham.ac.uk/ sites/default/files/documents/helpdesk/ LRDESCRIPTORS_PGRSECURE.pdf (accessed 4 January 2019).

It is important to stress that standard lists of descriptors should be used when they are available. The use of well-defined, tested and rigorously implemented descriptor lists for scoring descriptors considerably simplifies all operations concerned with data recording, such as updating and modifying data, information retrieval, exchange, data analysis and transformation. The use of standard lists ensures uniformity, while reducing errors and problems associated with text synonyms. The latter is where different words are used to describe the same thing, e.g. 'ES, ESP, Espana and Spain' all represent the country on the Iberian Peninsula of Europe; or leaves lanceolate versus lanceolate leaves refer to narrow leaves with an acute apex. People can easily recognize text synonyms, but the computer must be explicitly warned, or confusions may arise. Conservationists using the same lists will be able to exchange data readily and interpret the data with few, if any, problems.

18.2.1 Ecogeographic Data

Data Description

Ecogeographic data analysis is the basis of conservation planning. It is the process of collating existing geographic, ecological, genetic and taxonomic data concerning a taxon or group of taxa and using that data to help establish current conservation actions (Castañeda-Álvarez *et al.*, 2011). Existing data is predictive, in that if a species has previously only been found growing over 2000 m in the Middle East on limestone scree, then if we wish to collect or *in situ* conserve that species now, then we will have to do so at a location over 2000 m in the Middle East where there is limestone scree for the population to grow on – we do not look for, in this case *Vavilovia formosa* (Stev.) Al. Fed. (a rare CWR of garden peas), at sea level on a sandy beach in South America. Further, as there are a limited number of mountains over 2000 m in the Middle East, any of which may have a *V. formosa* population that has never been sampled for *ex situ* conservation or which has not been designated as a managed protected area, there is a gap in the conservation coverage, and so filling this gap should be designated a priority for additional conservation action (see Chapter 6). In this way ecogeographic data is predictive of where conservation action should be targeted.

Data Collation

As noted by Maxted *et al.* (1995) and Castañeda-Álvarez *et al.* (2011), the data collated for ecogeographic analysis can be collated at two possible levels, taxon and population. Further individual population data are synthesized to produce taxon level data. For *V. formosa* as a species, we know the species always grows over 2000 m on limestone scree' because all the specimens of this species that have ever been observed have been observed from locations sharing these ecogeographic conditions; this is the species ecogeographic envelope. While an individual population has been observed from Kavussahap Daglaria, Van province in Turkey at 38 10 14N and 42 55 00E. Therefore, synthesized taxon-level-based data will largely be collected from literature sources, where the data from numerous known population locations are synthesized to produce the ecogeographic envelope of information for the taxon to which the populations belong, whereas population data can only be collated from individual populations or is associated with specimens collected from those populations. The sort of ecogeographic data that may be collated and used by the conservationist is summarized in Tables 18.1 and 18.2 for taxon and population level ecogeographic data, respectively.

Although for most plant taxa it would be possible to collate the data in Table 18.1 and for a population given time to observe all the data in Table 18.2, frankly it would be very unlikely to find this level of data completion available for any herbarium specimen or a gene bank accession. Table 18.3 shows the more likely level of data available for them.

Data Manipulation

Ecogeographic datasets for any taxon will vary in size and completeness, but it is undoubtedly true that the more comprehensive the dataset, the more sophisticated the data analysis and the validity of conservation actions proposed. Analysis may range from simple counts of populations found in different habitats to thorough ecological multivariant analysis, though increasingly ecogeographic analysis is GIS assisted, so being able to establish the precise provenance where the population is located, or the specimen/accession was collected, is essential for routine analysis. The kinds of ecogeographic analysis available are reviewed in Section 6.4.4 so will not be repeated here.

Data Accessing

The taxon-based taxonomic, genetic, geographic and ecological information is obtained from various sources, specialist publications (ecogeographic

Table 18.1 Taxon-level ecogeographic data

- Taxonomic/genetic data
  - accepted taxon name
  - classification
  - synonyms
  - locally used taxon name
  - taxon descriptions and illustrations
  - identification aids
  - genotypic and phenotypic variation
  - phenology
  - breeding system
  - taxonomic notes
- Geographic data
  - country distribution
  - provincial distribution
  - latitude and longitude of previously located populations
  - known geographic partitioning of genetic diversity
- Ecological data
  - habitat preference
  - topographic preference
  - altitude
  - soil type
  - geological preference
  - site slope and aspect
  - land use and/or agricultural practice
  - climatic and micro-climatic preference
  - biotic interactions (pests, pathogens, herbivores, competitive ability)
- Conservation and use data
  - threat status (using IUCN categories)
  - conservation status
  - plant uses
- Taxonomic/genetic data
  - accepted taxon name
  - classification
  - synonyms
  - locally used taxon name
  - taxon descriptions and illustrations
  - identification aids
  - genotypic and phenotypic variation
  - phenology
  - breeding system
  - taxonomic notes
- Geographic data
  - country distribution
  - provincial distribution
  - latitude and longitude of previously located populations

Table 18.1 (*cont.*)

- known geographic partitioning of genetic diversity
- Ecological data
  - habitat preference
  - topographic preference
  - altitude
  - soil type
  - geological preference
  - site slope and aspect
  - land use and/or agricultural practice
  - climatic and micro-climatic preference
  - biotic interactions (pests, pathogens, herbivores, competitive ability)
- Conservation and use data
  - threat status (using IUCN categories)
  - conservation status
  - plant uses

Tables 18.2. **Population level ecogeographic data**

- Herbarium, gene bank or botanical garden where specimen is deposited
- Collector's name and number
- Collection date (to derive flower and fruiting timing)
- Phenological data (does specimen have flower or fruit)
- Provenance, latitude and longitude
- Altitude
- Soil type
- Habitat vegetation type
- Site slope and aspect
- Land use and/or agricultural practice
- Phenotypic variation
- Biotic interactions
- Competitive ability
- Palatability
- Ability to withstand grazing
- Vernacular names
- Plant uses

surveys or studies; taxonomic revisions, monographs, checklists and Floras; conservation, phytogeographic and taxonomic journals) or specialist databases, educational materials (field guides, lectures, multimedia programmes) and taxonomic or ecogeographic experts. Increasingly, less traditional media such as microfiche, diskettes, multimedia CD-

ROMs and on-line abstracting services (PlantGeneCD, BIDS, BIOSIS, CABI, etc.) provide relatively easily accessible sources of data and information. Often multi-institutional database projects collate biodiversity information either for specific taxonomic groups or geographic regions. For example, ILDIS (International Legume Database and Information Service) established a botanical diversity database for the 17 000 legume species (Roskov *et al.*, 2019). The database can be queried using the Catalogue of Life at www.catalogueoflife.org/col. Similar projects have been established for several other taxonomic groups and regional Floras. The most comprehensive sources of biodiversity data are Global Biodiversity Information Facility (GBIF; www.gbif.org) and the Catalogue of Life. GBIF provides a free and open access to biodiversity data online. The data provided by many institutions from around the world is accessible and searchable through a single GBIF portal.

When planning conservation of any taxon, accessing the GBIF dataset for that taxon is an essential first step. The Catalogue of Life is the most comprehensive and authoritative global index of species currently available. It consists of a single integrated species checklist and taxonomic hierarchy. The Catalogue holds essential information on the names, relationships and distributions of more than 1.8 million species. But the bottom line is experts can be the most useful sources and can be identified by looking at the authorship of relevant scientific publications, by asking botanists at local herbaria, by consulting the Plant Specialist Index and/or through various internet resources, e.g. the TAXACOM discussion group (for a list, see the Internet Directory for Botany at www.botany.net/IDB). None of these on-line sources has all the information required to plan conservation completely, so there will be a need to synthesize additional information from specimens representing the target taxa. Such individual specimens are held in herbaria, gene banks, botanic gardens, field gene banks, arboreta, etc., one of which is introduced in Box 18.2. The sources of ecogeographic data and information are reviewed in Section 6.4.3 so will not be repeated here.

Table 18.3 **A typical ecogeographic database**

Genus	Species	COLLNAME	COLLNOS	HERBCODE	INFCHAR	LEGCHAR	COLLDATE
Vicia	*dionysciensis*	Maxted, Ehrman & Khattab	2498	SPN	+	–	15/04/2006
Vicia	*dionysciensis*	Maxted, Ehrman & Khattab	2507	SPN	+	–	15/04/2006
Vicia	*dionysciensis*	Maxted, Ehrman & Khattab	2560	SPN	+	–	16/04/2006
Vicia	*dionysciensis*	Mouterde	6937	G	+	–	23/04/1942
Vicia	*dionysciensis*	Maxted, Ehrman & Khattab	2582	SPN	+	+	16/04/2006

ISOCODE	PROVINCE	LOCALITY	LATITUDE	LONGTUDE	ALTITUDE	SOILTYPE	HABITAT
SYR	Suweida	Mimas	32 36N	36 43E	1380 m	Heavy black	Water meadow
SYR	Suweida	Salkhad	32 30N	36 42E	1320 m	Heavy black	Beside stream
SYR	Suweida	Sahouet El Khoder	32 33N	36 46E	1330 m	Heavy black	Water meadows
SYR	Suweida	Tell Ahmar	32 52N	36 29E	840 m	Heavy black	Hillside
SYR	Suweida	Mafailh	32 33N	36 46E	1330 m	Heavy black	Water meadows

Box 18.2 | EURISCO

(Weise *et al.*, 2017; https://eurisco.ipk-gatersleben.de/)

EURISCO is a web-based *ex situ* germplasm catalogue for European gene bank holdings hosted at and maintained by IPK Gatersleben on behalf of Bioversity International. It is

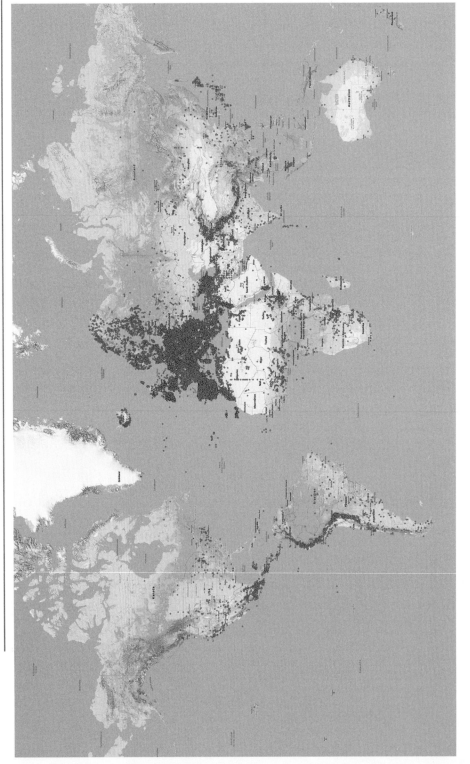

Figure 18.2 Provenance of accessions held in EURISCO. (A black and white version of this figure will appear in some formats. For the colour version, refer to the plate section.)
(From EURISCO, 2019)

Box 18.2 | (cont.)

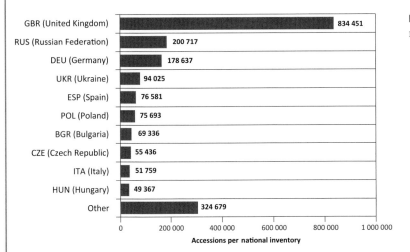

Figure 18.3 Accessions per national inventory.

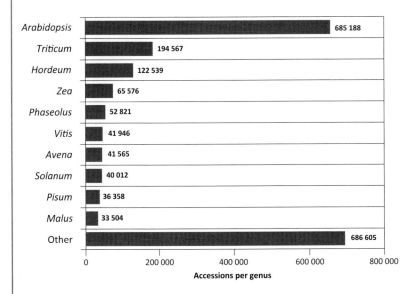

Figure 18.4 Accessions per genus.

dynamic in that it is regularly updated when the 43 member countries resubmit their National PGR Inventories via their National Focal Points (NFPs). The EURISCO Catalogue contains passport data for about 1.98 million samples of crop diversity representing more than 6392 genera and more than 43 445 species (genus–species combinations including synonyms and spelling variants) from 43 countries and 378 institutions.

These samples represent more than half the *ex situ* accessions maintained in Europe (EURISCO, 2019) and roughly 27% of total worldwide holdings. EURISCO's main function is

Box 18.2 | (cont.)

to provide an access point for user of germplasm to source the material they require, but is also a good source of ecogeographic and G × E information for European and Global food crops, forages, wild-and-weedy species, including cultivars, landraces, farmers' varieties, breeding lines, genetic stocks, research and CWR material. The international coverage of EURISCO is demonstrated by Figure 18.2, while the number of accessions per national inventory is shown Figure 18.3, the number of accessions per genus is shown in Figure 18.4 and the accessions per biological status in Table 18.4.

Table 18.4 **Accessions per biological status**

Biological status	Accessions	Percentage
400 (Breeding/research material)	501 535	25.29%
300 (Traditional variety/landrace)	289 419	14.59%
500 (Advanced or improved cultivar (conventional breeding methods))	256 629	12.94%
100 (Wild)	140 091	7.06%
110 (Natural)	96 640	4.87%
410 (Breeder's line)	58 043	2.93%
999 (Other)	24 936	1.26%
414 (Inbred line (parent of hybrid cultivar))	17 770	0.90%
412 (Hybrid)	9989	0.50%
120 (Semi-natural/wild)	8130	0.41%
420 (Genetic stock/mutant)	3997	0.20%
130 (Semi-natural/sown)	3607	0.18%
200 (Weedy)	3543	0.18%
421 (Mutant (e.g. induced/insertion mutants, tilling populations))	2673	0.13%
411 (Synthetic population)	2266	0.11%

18.2.2 Field Population Data

Field data form the basis of the information that describes the populations of the target taxon in the field and as such, relate to both *in situ* and *ex situ* conservation activities. They describe the site (Table 18.5) and the population of the target taxon found at that site (Table 18.6). The site and population data are linked in a relational manner by the site number.

Table 18.5 **Field population site data**

Site number	–	Consecutive number site designation
Country	–	Political country where the site is located
Province	–	Province within political country
Geographic location	–	Location in relation to nearest major town or city
Nearest settlement	–	Name of nearest human habitation
Latitude	–	Latitudinal location
Longitude	–	Longitudinal location
Altitude	–	Height above sea level (m)
Aspect	–	Compass direction the site is facing
Slope	–	Degree of slope of the site
Physical description	–	Physical description of the site
Habitat description	–	General habitat or field description of the site
Agricultural practice	–	Type of agricultural/silvicultural, etc. at the site
Grazing pressure	–	Degree of grazing at the site, if any
Dominant vegetation	–	Whether trees, shrubs, herbs, grass, etc.
Rainfall	–	Annual rainfall at site
Parent rock	–	Kind of parent rock at site
Rock size	–	Whether small rocks, boulders, sheet rock, etc. are seen
pH	–	pH of soil at site
Soil composition	–	Organocarbon, phosphate, potash, calcium and other mineral makeup
Soil texture	–	Whether clay, sand, loam, etc.
Depth of soil	–	Depth of soil at the site
Water relation	–	Whether free draining, boggy, etc.

Adapted from Hawkes *et al.* (2000).

Like the ecogeographic data used in conservation planning, there is a gap between the data that could theoretically be available for effective population management and the actual data that is routinely used in practical population monitoring, the latter being a subset of the former. As described in Chapter 8, field population data used for *in situ* or on-farm monitoring of a conserved population is analyzed to identify the effectiveness of current management practices (Iriondo *et al.*, 2008) and is often restricted to time-series demographic estimates of population size, density, frequency and cover. More exceptionally some form of periodic genetic diversity monitoring occurs. Globally there are so few active, long-term *in*

Table 18.6 **Field population target taxon site data**

Site number	–	Consecutive number site designation
Site managing authority	–	Name of authority and person responsible for site maintenance
Population number	–	Consecutive number of population within-site
Field identification	–	Taxon scientific (genus, species & subspecific name)
Vernacular name	–	Local name for the taxon
Habitat type	–	Particular habitat around population within the site
Population size	–	Size of population sampled (estimated numbers = demographic number)
Population description	–	Particular spatial characteristics of the population (clumped, dispersed, etc.)
Plants sampled	–	Whether representative population sample is held *ex situ*
Gene bank code	–	Link to gene bank and accession held
Organ sampled	–	Whether seed, corm, tuber, vegetative plant, etc.
Sample source	–	Whether the sample was obtained from a field, farmer's store, market, another conservation collection, etc.
Voucher specimen	–	Whether a voucher specimen of the sample was taken
Ethnographic data	–	information on the local farmers and cultivation practices

Adapted from Hawkes *et al.* (2000).

situ or on-farm PGRFA conservation activities that field PGRFA population data management has yet to be formalized, though consideration has been given to what such monitoring may involve and how such time-series data might be stored and manipulated (Iriondo *et al.*, 2008). Further development is a priority if national, regional or global *in situ* or on-farm networks are established (FAO, 2012c). Obviously once such a field population data management system is available, it will not only facilitate adaptive population management, but access to the data will also facilitate broader conservation planning.

18.2.3 Collection Data

Having stressed the lack of experience in field population data management, field data associated with collection and *ex situ* conservation does have an extensive literature. Collection data are generally recorded initially on paper, often using standardized forms, especially if the data are being recorded in the field. However, increasingly it is possible to use computerized data loggers, which avoids later transcription and the introduction of transcription errors. Data transcription is a common source of errors if it is not performed with great care. For this reason, the number of different manual transcriptions should be kept to a minimum. Whether using manual or computerized methods for recording raw data, it is essential that the data are recorded accurately and with the desired degree of precision. If data are recorded by hand using field notebooks or a standard form, it will require transcription into a collection database or gene bank management system. Still, many documentation procedures initially rely on the use of manual forms for recording raw data, as it may

University of Birmingham Collecting Expedition

Country ...... Armenia .... Province ...... Vayots Dzor ...... Date 30/05/18

Site Number 42 .. Nearest Village...... Yeghegnadzor ....

Location ...About 1km southeast of village toward Malishka

Altitude(m) 1208 Latitude 39°45'23" N Longitude 45°20'32" E Rainfall 300 ....mm

Site Physical ...... Terraced vineyard beside road ....

Site Vegetative ...... Weady areas in and around vineyard ....

Coded Environmental Information:

PR	pH	ST	DS	TS	AS	SL	WR	AP	GP	%R	RT	%C	%T	TT	Photo
J	8.2	Y	C	E	SE	U	F	C	A	S	2	70	—	—	016

Figure 18.5 Example of page layout for collecting data for one germplasm accession collected in Armenia.

not be possible to enter data directly into the database due to difficulty of the field condition and lack of a power source. It is important that any forms used, whether for a computerized or manual documentation system, are 'user-friendly' to ensure data integrity. Typically, you could use two sorts of design layout: page and column layout.

Page layout is commonly used where a large amount of data is recorded for a particular accession, as in the historic example of a forage legume collecting expedition to the Caucasus (Figure 18.5), where both textural and coded environmental information was recorded. Although many organizations have developed standard collecting forms, Hawkes *et al.* (2000) conclude that no one form is suitable for every plant species, whether crop, crop relative or wild species.

With a page layout you have considerably more freedom in where you can place descriptors on the page, and you can:

- use a mixture of columns or text boxes,
- use multiple choice questions to assist the user,
- add comments to assist the user,
- try to use pre-printed forms, otherwise use ruled paper,
- make each column or box large enough to accommodate data,
- aim for consistency in different layouts,
- include space for 'comments' and

- give your form a title and give indications for use.

Column layout takes the form of a table, for example Table 18.7 taken from the same historic forage legume collecting mission to the Caucasus, where columns represent characters and the rows represent individual collection sites. The following list suggests ideas on how to design a table layout with columns:

- Use pre-printed forms or ruled paper
- Arrange the columns to help data recording and retrieval
- Be consistent in the arrangement of columns
- Include a 'comments' column
- Give your form a title

As discussed above, manual data capture often forms the initial stage of documentation. The data would often be recorded in a predefined form on loose-leaf sheets, which are easy to generate once the page or column format has been defined. In the field loose-leaf sheets are easy to store and sort, especially if there is one accession per sheet and they can be held in a standard ring binder or wallet.

18.2.4 Accession Management Data

These data are associated with the germplasm collection once it reaches the plant genetic resource centre, botanic garden, field gene bank or tissue culture store and is given an accession number. Once

Table 18.7 **Example of column layout for collection site location data**

Site no.	Country	Nearest settlement	Location	Altitude (m)	Latitude	Longitude
42	Armenia	Yehegnadzor	North-eastern edge of village	1220	39 45	45 22
43	Armenia	Saravan	At 91/49 km sign on Azizbeklv to Sisian road	1540	39 43	45 39
44	Armenia	Shurnookh	At 29/33 km sign on Goris to Kafan road	1550	39 22	46 24
45	Armenia	Kadzharan	West edge of village destroyed by earthquake	1890	39 10	46 08
46	Armenia	Artsvanick	Northern edge of village	1115	39 15	46 28
47	Armenia	Foratan	At 44/19 km sign on Kafan to Goris road	1150	39 25	46 23

the population sample is actively conserved *ex situ*, it is referred to as an 'accession'. Each accession has a unique identifier that facilitates the accession's management within the collection. Management descriptors were defined by IPGRI (1997) as data associated with the 'management of accessions in the gene bank and assist with their multiplication and regeneration'. Broadly speaking, the following represent some generalized examples of procedures associated with the day-to-day activities of a seed gene bank that will generate data to be documented and managed:

- Accession registration (including date that the accession is registered)
- Seed cleaning
- Seed drying
- Seed viability testing (germination test)
- Seed packing and storage (including amount and location of stock)
- Seed distribution
- Seed monitoring
- Regeneration/multiplication
- Characterization/preliminary evaluation
- Accession distribution

The data management activities in a conservation collection will almost invariably be managed practically by some form of database management system, such as GRIN-Global described in Box 18.1.

18.2.5 Characterization and Evaluation Data

As noted throughout this text, to sustain conservation there is an explicit need to link conservation to use, so once an accession is actively conserved, it should be characterized and evaluated to facilitate utilization (see Chapter 15). For PGRFA the historic and current position is that only *ex situ* conserved accessions have been characterized and evaluated, but if *in situ* or on-farm conserved resources are to be sustainable, it has been argued that there is an equal need to promote characterization and evaluation of *in situ* or on-farm conserved resources (Maxted *et al.*, 2016a). Certainly, breeders considering which genetic resource to use will deliberately favour those with reliable descriptive information. The availability of characterization and evaluation data is used to assist in the maintenance and, together with passport data, is often used as a basis for the rationalization of genetic resources collections. Frankel and Brown (1984) originally advocated the development of a small, genetically representative sample of large germplasm collections for evaluation and use, and this gave birth to the concept of the 'core collection'. Core collection being selected based on either known genetic distinction/adaptive traits or ecogeographic diversity (based on the assumption that ecogeographic distinction is most likely to be coincident with genetic distinction; Hodgkin *et al.*, 1994). Characterization and evaluation data on a broader scale, combining data from many national germplasm collections for a crop, can provide the basis on which to plan international genetic resources activities to reduce duplication of effort in the areas of collecting, regeneration and multiplication. These data are at the heart of the European AEGIS initiative which aims to efficiently conserve and provide access to unique germplasm in Europe through the establishment of the European Collection (www.ecpgr.cgiar.org/aegis/).

Characterization and evaluation data describe the phenotypic and genotypic characteristics of an accession. They were both defined by IPGRI (1997):

Characterization descriptors: These enable an easy and quick discrimination between phenotypes. They are generally highly heritable, can be easily seen by the eye and are equally expressed in all environments. In addition, these may include a limited number of additional traits desirable by a consensus of users of the particular crop.

Evaluation descriptors: Many of these descriptors in this category are susceptible to environmental differences but are generally useful in crop improvement and others may involve complex biochemical or molecular characterization. They include yield, agronomic performance, stress susceptibilities and biochemical and cytological traits.

In practice, the distinction between characterization and evaluation descriptors is imprecise. IPGRI (1997) encourages gene bank curators to undertake characterization, but comments that evaluation will normally require growing accessions out in replicate sites and some form of time trial. Therefore, these activities will normally be better undertaken by those intending to utilize the material.

For many characterization and evaluation descriptors there are a limited number of descriptor states. For instance, the descriptor 'growth stage' might only have four descriptor states: seedling, vegetative plant, flowering plant, seed set. In other cases, the descriptor states are continuously variable such as plant height. An example of some descriptors for quinoa (Bioversity International *et al.*, 2013) is provided below:

Descriptor 7.3 – Growth habit (Figure 18.6):

1. Simple
2. Branched to bottom third
3. Branched to second third
4. Branched with main panicle undefined

Descriptor 7.4 – Plant height (cm):
Recorded at physiological maturity, from root collar to panicle apex. Average of 10 plants.

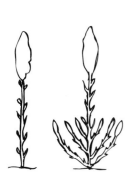

1   2   3   4

Figure 18.6 Types of growth habit. (From Bioversity International *et al.*, 2013)

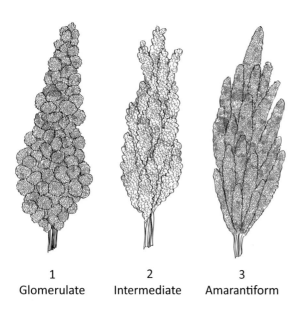

1
Glomerulate

2
Intermediate

3
Amarantiform

Figure 18.7 Types of panicle shape.
(From Bioversity International *et al.*, 2013)

Descriptor 7.8.4 – Panicle shape (Figure 18.7):

1. Glomerulate (glomerules are inserted in the primary axis showing a globose shape)
2. Intermediate (showing both shapes)
3. Amarantiform (glomerules are inserted directly in the secondary axis and have an elongated shape)

Recording Descriptors

It is important that the person recording the descriptor is familiar with the descriptors to ensure the data obtained are comparable. For example, another quinoa descriptor is Descriptor 7.8.7 Panicle density, which has the states: 1 Lax; 2 Intermediate; and 3 Compact (Bioversity International *et al.*, 2013); without an accompanying image, the recorder may not be able to consistently record the descriptor and the results are likely to be very different, non-comparable and misleading.

Arranging Data in a Useful Order

The collated descriptor data should be arranged in a useful order that aids retrieval and information provision. This order will not necessarily be by accession number or sequential order, and it may be necessary to sort the accessions in taxon order. In theory you could sort on the basis of any descriptor, but it would have to be both a useful and a practical order. Remember that an order that is useful for data retrieval may not always be practical for data recording and data modification: sometimes compromise is required.

Descriptor Lists

The use of well-defined, tested and rigorously implemented descriptor lists for scoring considerably simplifes all operations concerned with data recording, updating/modifying, information retrieval, exchange, data analysis and transformation. In other words, it is strongly advised that the existing and widely accepted descriptor lists are used if available. For many crops and some wild relative groups there are extensive existing descriptor lists available. Bioversity International lists 93 taxon-based descriptor lists as well as the generalized FAO/Bioversity Multi-Crop Passport Descriptors V.2.1 (Alercia *et al.*, 2015). Use of these pre-defined lists of descriptors and descriptor states can save data processing time, ensures uniformity and reduces error, as well as facilitating data exchange and easy multi-stakeholder data interpretation.

18.3 National Conservation Action

To more effectively promote and achieve the FAO Second Global Plan of Action for PRGFA priority activities (FAO, 2011a), the FAO has recently begun to campaign for the production of National Strategies for PGRFA (FAO, 2015b). The FAO defines a National Strategy as a 'blueprint for the management of a country's PGRFA as a continuum of interventions in order to achieve clearly defined timebound goals'. In other words, having a National Strategy means identifying a country's vision and goals, responsibilities and a timeframe and a process by which the National Strategy can be achieved. Preparing a National Strategy involves a series of steps (Box 18.3). Many steps involve stakeholder workshop discussion, but others can involve extensive

Box 18.3 | **Steps Involved in National Strategies for PGRFA Development**

(FAO, 2015b)

Stage A. Establish a coordinating mechanism at national level
      A1. Establish, maintain or strengthen the National Programme for PGRFA
      A2. Establish or confirm a National PGRFA Committee
      A3. Identify and involve PGRFA stakeholders
      A4. Promote linkages and partnerships
      A5. Convene ad hoc working groups and enlist expertise
Stage B. Establish the foundations for a National Strategy for PGRFA
      B1. Undertake or update a country assessment on the state of PGRFA
        B1.a. Coordination.
        B1.b. Preparation
        B1.c. Analysis
        B1.d. Review and validation
      B2. Define the scope of the National Strategy for PGRFA
        B2.a. Crops/taxa
        B2.b. Geographic coverage
        B2.c. Complementarity with national, regional and global strategies
        B2.d. Participatory approach
        B2.e. Content of the National Strategy for PGRFA
        B2.f. Complementarity within and between sectors in the PGRFA Continuum
        B2.g. Financial and human resources for implementation
      B3. Formulate a vision statement
      B4. Formulate goals and objectives
      B5. Formulate an action plan
      B6. Budget and mobilization of resources
      B7. Formulate a monitoring plan
      B8. Formulate a communication plan
Stage C. Finalizing and presenting the National Strategy for PGRFA
      C1. Prepare a draft National Strategy for PGRFA, forming the basis for stakeholder review
      C2. Coordinate the consultation process
      C3. Revision and final presentation of the National Strategy for PGRFA

data collation and analysis, if the data is not readily available such as B1; as will the final National Strategy implementation and monitoring.

Although FAO (2015b) PGRFA Strategy activities have focused at the national level, the efficiency of PGRFA conservation and use can be enhanced at multi-national and even global levels. One of the first regional approaches to PGRFA conservation and use was taken in the '*In situ* conservation and use of CWR in three ACP countries of the Southern African

---

Box 18.4 | Tools to Aid Wild Plant Conservation Planning

(Magos Brehm *et al.*, 2019)

- *'Interactive Toolkit for CWR Conservation Planning'* is an interactive toolkit that guides users through the steps involved in CWR conservation planning, shows examples of how a certain conservation planning step can be undertaken and provides many resources to help the user to plan CWR conservation (Magos Brehm *et al.*, 2017a). Available at www .cropwildrelatives.org/conservation-toolkit/
- *'Template for the Preparation of a NSAP for the Conservation and Sustainable Use of CWR'*, a *'Template for the Preparation of a Technical Background Document for a NSAP for the Conservation and Sustainable Use of CWR'*: the first document assists countries in preparing their National Strategic Action Plan for the Conservation and Utilization of CWR (NSAP) in a consistent and uniform manner; the second document helps countries to compile all the scientific and technical information that is behind the NSAP (Dulloo *et al.*, 2017; Magos Brehm *et al.*, 2017b).
- *'CWR Checklist and Inventory Data Template'* was developed to help users to establish a CWR checklist and inventory in a systematic manner and was based on the 'CWR checklist and inventory descriptors' (Thormann *et al.*, 2017; Bioversity International and University of Birmingham, 2017).
- *'Occurrence Data Collation Template'* was prepared to assist users to collate CWR occurrence data (from various sources such as herbarium vouchers, gene bank accessions, personal communications, field observations, etc.) in a systematic manner and standardize them to use in diversity and gap analyses, threat assessment and climate change analyses, as well as to prepare the data to undertake these analyses using the CAPFITOGEN tools (Magos Brehm *et al.*, 2017c).

Development Community (SADC) region' project, a 3-year project (2014–2016) co-funded by the European Union and implemented through ACP-EU Cooperation Programme in Science and Technology (S&T II) by the African, Caribbean and Pacific Group of States (www.cropwildrelatives.org/sadc-cwr-project/). The overall aim was to 'enhance the link between *in situ* conservation and use of CWR in the SADC region as a means of underpinning regional food security and mitigating the predicted adverse impact of climate change'. A key component was the production of National CWR Strategies for Mauritius, South Africa and Zambia, as well as the SADC region as a whole. The project produced several data management tools to aid conservation planning for CWR taxa, but as CWR are wild plant taxa that just happen to be related to crops, the use of these tools could be extended to all wild plant taxa (Box 18.4). Even more recently, another EU-funded project 'Joining forces for genetic resources and biodiversity management' (GenRes Bridge; www.genresbridge.eu) began in 2018 with the overall aim to 'strengthen conservation and sustainable use of genetic resources by accelerating collaborative efforts and widening capacities in plant, forest and animal domains'. Specifically, GenRes Bridge will build on the domain-specific knowledge and resources and unite them to develop an 'Integrated Strategy for genetic resources for food and agriculture', which will represent the backbone for a European-wide approach to the conservation and management of GenRes. This project will also be releasing data management tools that aid national and regional Strategy development and implementation.

18.4 Data Ownership

As a matter of principle, many would support the open, free availability of all data via the internet. Tim Berners-Lee (2005), the father of the internet, proposed a star system for best practice in open-access publishing, with: * indicating publishing the raw data on the net; ** indicating publishing the raw data on the net in a structured format; *** indicating publishing the raw data on the net in a non-propriety and open format structured format such as tab-delimited text (TDT), comma-separated text (CSV), extensible mark-up language (XML), or JavaScript Object Notation (JSON); **** if Uniform Resource identifiers (URI) are used; and ***** if the data adding a link between this data and other data available via the net. However, there are numerous occasions when data generated by one group is commercially exploited by another without recompense to the original data providers. Therefore, there has been a gradual raising of awareness of the issues involved in data ownership and access since the Convention on Biological Diversity (CBD, 1992), and particularly in Europe, more recently through General Data Protection Regulation (EU) 2016/679 (GDPR). This EU law regulation aims to give individuals control over their personal data and regulate its use in the business environment. Such legislation is now being discussed, prepared and enacted by many countries globally, and as final implementation varies nationally, it is beyond the scope of this text to provide detailed national guidance. However, the point has already been made above that PGRFA conservation and use is highly data intensive, and local national GDPR will impinge on PGRFA conservation and use activities. The principle is now well established that each individual should have control over their personal data, its use and exploitation.

Similarly, historical analysis of germplasm exploitation, whether in terms of crop improvement or pharmaceutical usage, has rarely resulted in direct economic benefit to the original country of provenance of the genetic diversity. With the development of gene editing the unscrupulous agent may no longer need access to the actual germplasm but just the data describing the adaptive trait, which could then be used to engineer the advantaged variety. It is basically the same problem over exploitation and benefit sharing that has been discussed by the PGRFA community for the last 50 years. Pre-CBD the international position over ownership of genetic resources was common heritage for plant genetic resources; all resources were owned by all humankind. This approach did not reward those farming communities or countries that had maintained diversity for millennia, so post-CBD the situation changed and now the primacy is for each country to have national sovereignty over its PGRFA and there is now further discussion over the descriptive data associated with those PGRFA. Some would argue this has limited PGRFA conservation, especially as the way is still open for agricultural or pharmaceutical companies to negotiate bi-lateral benefit-sharing agreements with nations or indigenous peoples for use of their indigenous knowledge and genetic resources. These issues are still a matter of active debate, and we look forward to a more mutually beneficial future where resource maintainers and resource users benefit appropriately from resource utilization.

ACRONYMS AND ABBREVIATIONS

AGIS	United States Agricultural Genome Information Server
AMOVA	Analysis of molecular variance
ANOVA	Analysis of variance
BC	Backcross
BGCI	Botanic Gardens Conservation International
Bioversity*	Bioversity International
CABI	Commonwealth Agricultural Bureau International (also abbreviated as CAB International)
CBD	Convention on Biological Diversity
CGIAR	Consultative Group on International Agricultural Research
CGRFA	Commission on Genetic Resources for Food and Agriculture (within FAO)
CIAT*	Centro Internacional de Agricultura Tropical
CIFOR*	Centre for International Forestry Research
CIMMYT*	Centro Internacional de Mejoramiento de Maíz y Trigo
CIP*	Centro Internacional de la Papa
CITES	Convention on International Trade in Endangered Species of Wild Fauna and Flora
CMS	Cytoplasmic male sterility
COP	Conference of the Parties (of the CBD)
CR	Critically Endangered (IUCN Red List Category of Threat)
CSIRO	Commonwealth Scientific and Industrial Research Organization, Australia
CUPGR	Conservation and Utilization of Plant Genetic Resources
CWR	Crop wild relative
DBMS	Database management system
DD	Data Deficient (IUCN Red List Category of Threat)
ECP/GR	European Cooperative Programme for Crop Genetic Resources Networks

ELC	Ecogeographic Land Characterizations
EN	Endangered (IUCN Red List Category of Threat)
EPS	Effective population size
ERIN	Environmental Resources Information Network
ETI	Expert-Centre for Taxonomic Identification
EUCARPIA	European Association for Research on Plant Breeding
EUFORGEN	European Forest Genetic Resources Programme
EW	Extinct in the Wild (IUCN Red List Category of Threat)
EX	Extinct (IUCN Red List Category of Threat)
FAO	Food and Agriculture Organization of the United Nations
FFI	Fauna and Flora International
GATT	General Agreement on Tariffs and Trade
GCDT	Global Crop Diversity Trust (also known as Crop Trust)
GEF	Global Environment Facility
GIAHS	Globally Important Agricultural Heritage Systems (within FAO)
GIS	Geographic information systems
GMO	Genetically modified organism
GP	Gene pool
GPA	Global Plan of Action for Conservation and Use of PGR (of FAO)
GPS	Geographic positioning systems
GRAIN	Genetic Resources Action International
GRID	Global Resources Information Database (of UNEP)
GRIN	United States Genetic Resources Information Network
GxE	Genotype by environmental interaction
HDRA	Henry Doubleday Research Association
IARC*	International Agricultural Research Centre

IBP	International Biological Programme		MLS	Multi-Lateral System of ITPGRFA
IBPGR*	International Board for Plant Genetic Resources (now Bioversity International)		MTA	Material Transfer Agreement
			MVP	Minimum viable population
			NBSAP	National Biodiversity Strategy and Action Plan
ICARDA*	International Centre for Agricultural Research in the Dry Areas		NE	Not Evaluated (IUCN Red List Category of Threat)
ICRAF*	International Centre for Research in Agroforestry		NGO	Non-governmental organization
ICRISAT*	International Crops Research Institute for the Semi-Arid Tropics		NIL	Near-isogenic lines
			NT	Near Threatened (IUCN Red List Category of Threat)
ICSU	International Council of Scientific Unions		NUS	Neglected and underutilized species
IITA*	International Institute for Tropical Agriculture		OTU	Operational taxonomic unit
			PA	Protected area
IK	Indigenous knowledge		PGR	Plant genetic resources
ILCA*	International Livestock Centre for Africa (now part of ILRI)		PGRFA	Plant genetic resources for food and agriculture
ILRAD*	International Laboratory for Research on Animal Diseases (now part of ILRI)		PGRP	United States National Plant Genome Research Programme
ILRI*	International Livestock Research Institute		PoW	Programme of Work
			PPB	Participatory plant breeding
INIBAP*	International Network for the Improvement of Banana and Plantain		PVA	Population viability assessment
			PVS	Participatory varietal selection
IPBES	Intergovernmental Science-Policy Platform on Biodiversity and Ecosystem Services		QTL	Quantitative trait locus
			RAFI	Rural Advancement Foundation International
IPCC	International Panel for Climate Change		RFLP	Restriction fragment length polymorphisms
IPGRI	International Plant Genetic Resources Institute (now Bioversity International)		RIL	Recombinant inbred line
IRRI*	International Rice Research Institute		RH	Relative humidity
ISTA	International Seed Testing Association		RP	Reciprocal parent
ITPGRFA	International Treaty for Plant Genetic Resources for Food and Agriculture (or International Treaty)		SBSTTA	Subsidiary Body on Scientific Technical and Technological Advice (of the CBD)
IUCN	International Union for the Conservation of Nature and Natural Resources		SCA	Specific combining ability
			SDG	UN Sustainable Development Goals
			SGRP	System-wide Genetic Resources Programme
KBA	Key biodiversity area			
LC	Least Concern (IUCN Red List Category of Threat)		SMTA	Standard Material Transfer Agreement
			SNP	Single nucleotide polymorphism
LD	Linkage disequilibrium		SoWPGRFA	State of the World's Plant Genetic Resources for Food and Agriculture report
LR	Crop landrace			
MAB	Man and Biosphere Programme of UNESCO		SSC	Species Survival Commission of IUCN
			SSR	Microsatellites or simple sequence repeats
MEA	Millennium Ecosystem Assessment			

TAC	Technical Advisory Committee (of the CGIAR)
TG	Taxon Group
TRIPS	Trade-Related Intellectual Property Rights
UKPGRG	United Kingdom Plant Genetic Resources Group
UNCED	United Nations Conference on the Environment and Development
UNDP	United Nations Development Programme
UNEP	United Nations Environment Programme

UPOV	Union for the Protection of New Varieties of Plant
VU	Vulnerable (IUCN Red List Category of Threat)
WARDA*	West Africa Rice Development Association
WCMC	World Conservation Monitoring Centre
WHS	World Heritage Sites
WRI	World Resources Institute
WWF	World Wide Fund for Nature
WWW	World Wide Web

* Indicates the institute is a CGIAR centre.

GLOSSARY

Accession number A unique identifier that is assigned by the curator when an accession is entered into a collection. This number should never be assigned to another accession.

Accessions Uniquely identified samples of seeds, plants or other germplasm materials that are maintained as an integral part of a germplasm collection.

Active collection A germplasm accession that is used for regeneration, multiplication, distribution, characterization and evaluation. Active collections are maintained in short- to medium-term storage and usually duplicated in a base collection maintained in medium- to long-term storage.

Adaptive management Involves regular monitoring of the impact of management interventions on target populations to test their outcomes and adopting the interventions most likely to ensure maximum conservation benefit.

Agro-biodiversity Elements of biodiversity – including plants, animals and micro-organisms – that benefit people; in the context of plants this includes all cultivated and wild plants associated with food and agriculture.

Allele Variant form of a given gene.

Allelic frequency Estimate of the relative number of alleles of each type in gametes.

Apogamy Embryo development from different haploid nuclei (antipodes, synergies, polar nuclei) to the ovules or from the fusion of two cells of the embryonic sac.

Apomorphy Derived condition of the character, synonymous with derived or advanced.

Apospory Reproduction in which the embryo sac is formed directly from a somatic cell without the reduction and formation of spores, i.e. without meiosis. Thus, the embryo develops directly from the diploid egg, without fertilization.

Artificial classification Classification based on few characters from one source of evidence.

Autapomorphy Character particular to one monophyletic terminal taxa; it is synonymous with diagnostic character to some extent.

Biodiversity Variability among living organisms from all sources including, *inter alia*, terrestrial, marine and other aquatic ecosystems, and the ecological complexes of which they are part; this includes diversity within species, between species and of ecosystems (CBD, 1992, Article 2)

Blocking Division of a very large experiment into smaller manageable subunits.

Botanic gardens Institutions holding documented collections of living plants for the purposes of scientific research, conservation, display and education.

Botanic garden conservation Collecting of seeds or living material from one location and its transfer, planting and cultivation in a second site as a living collection of large numbers of species for scientific, educational and ornamental purposes (for tree species it is an arboreta). Many contemporary botanic gardens are also involved in seed banking and other forms of conservation activity.

Character Any attribute (or descriptive phase) referring to form, structure or behaviour which the taxonomist separates from the whole organism for a particular purpose such as comparison or interpretation.

Characterization Recording of genetically controlled traits found in any environment where the species may occur, and it involves rating characters (also known as descriptors) that are often simply inherited and highly heritable, can easily be seen by eye and are expressed in all environments.

Characterization data Describes those characteristics of an accession or population that are highly heritable, can be seen easily by eye and are expressed in all environments (e.g. flower colour, number of fruits per inflorescence).

Classification Placement of taxa into groups on the basis of their relationships.

Climate change A change in climate that can be directly or indirectly attributed to human activity and that is in addition to natural climate variability over comparable time periods.

Climate smart conservation Conservation action that considers the implications of climate change, developing and implementing actions that address the additional threats, plan conservation to mitigate its impact and respond to increasing change and uncertainty.

Collection A group of germplasm accessions maintained for a specific purpose under defined conditions.

Combining ability (CA) Capability of a genotype or a population to give offspring characterized by the high expression of a character. CA is measured by the performance of the genotype or population after crossing.

Conservation Maintenance of the diversity of living organisms, their habitats and the interrelationships between organisms and their environment.

Conservation genetics It means characterizing the diversity, plus understanding the viability and further evolution of a population within a species.

Consultative Group on International Agricultural Research (CGIAR) A strategic alliance of countries, international and regional organizations and private foundations supporting 15 international agricultural research centres and their global partners through 12 well-defined CGIAR Research Programs (CRPs) and 3 supporting platforms.

Core subsets Representative accessions of the whole cultigen pool held in a gene bank.

Crop wild relatives (CWR) Wild species relatively closely related to crops, including crop progenitors.

Crossing Results from mating at least two parents and whose derived offspring ensues from the union of their gametes to form a zygote.

Cryopreservation Collection of embryo, pollen or tissue samples, their pre-treatment and storage at $-196°C$.

Cultivar Refers to a group of plants within a crop sharing common characters that distinguish them from other similar groups. A crop registration authority must certify a cultivar before its release for farming to guarantee it retains and transmits its distinct characters sexually or asexually.

Description Statement of an organism's morphological characteristics.

Descriptor An identifiable and measurable trait, characteristic or attribute observed in an accession that is used to facilitate data classification, storage, retrieval and use.

Descriptor list A collation of all the individual characters or descriptors used for a particular species or crop.

Descriptor state A clearly definable state which a descriptor can take (e.g. hair erect or flat; yellow or blue flower).

Descriptors Characters whose data define the features of a gene bank accession

Diagnosis Short morphological description covering only those characters (diagnostic characters) which are necessary to distinguish a taxon from related taxa.

Diplospory Having an embryo derived directly from the mother cell or megaspore. Pollination may be required to begin endosperm development.

DNA (deoxyribonucleic acid) Two-stranded molecule with a double helix-shape and containing the genetic code that provides the instructions used in the growth, development, functioning and reproduction of all known living organisms, including plants. Each DNA strand comprises sequences of the four bases (A, G, C, T), which are complementary to the opposite strand, thus forming the 'stairs' of the DNA 'ladder'. A pairs always with T, while C does with G, and both pairing through hydrogen bonds. Each DNA strand has a beginning (5′) and an end (3′), and the two strands run in opposite (or anti-parallel) direction to each other, i.e. one from 5′ to 3′ (sense strand) and other from 3′ to 5′ (anti-sense strand).

DNA storage Collecting of DNA and storage in appropriate, usually refrigerated, conditions.

Documentation system Any method of storing and maintaining data. A documentation system can use manual methods (such as handwritten records) or completely computerized methods for data storage and maintenance. The system is also designed for information retrieval.

Dominance Interaction between alleles of the same gene in a heterozygote. On the allele with this feature the dominant allele masks the expression in the phenotype of the recessive allele at the same locus.

Donor Organization or person responsible for providing a germplasm sample to a recipient.

Dormancy State in which certain live seeds do not germinate, even under normally suitable conditions.

DUS testing Means by which a newly bred cultivar differs from others already available (Distinctness), with characteristics or traits expressing unvaryingly (Uniformity) and not changing over generations (Stability).

Ecogeographic land characterization map Tool for quantifying and partitioning plant adaptation, which acts as a proxy for genetic diversity and facilitates conservation planning.

Ecogeography Science that studies the environmental effects on the distribution of living organisms. The definition given here is for ecogeographic survey.

Ecogeographic survey Process of gathering and synthesizing information on ecological, geographical, taxonomic and genetic diversity. The results are predictive and can be used to assist in the formulation of complementary *in situ* and *ex situ* conservation priorities. A study implies a more long-term research orientated investigation, while a survey is generally quick and collates kust the easily accessible information.

Ecosystem A dynamic complex of plant, animal and micro-organism communities and their non-living environment interactions as a functional unit.

Ecosystem diversity Refers to various ecosystems in a given (small or a large) area and includes a community of organisms and the environment where they live.

Ecosystem services Direct and indirect contributions of ecosystems to human well-being; categorized into four main types: provisioning services (e.g. Food, water, fuel); regulating services (e.g. Flood and disease control); supporting/habitat services (e.g. Nutrient cycling); and cultural services (e.g. Recreation).

Ecotypes A population or group of populations genetically adapted to a particular set of environmental conditions where they naturally occur.

Epistasis This occurs when the phenotypic expression of genes at one locus depends on the genotype at another locus, i.e. a non-allelic interaction.

Evaluation Refers to recording characters that may only be expressed by an individual in certain environments, and it consists of assessing characters that are often not simply inherited and which are susceptible to environmental differences.

Evaluation data Describes those characteristics of an accession or population that are susceptible to environmental differences (e.g. Fruit yield, drought or disease susceptibility).

Evolution Includes the biological and physical processes that lead to heritable changes of characters in populations over time. Genetic drift due to small sample size, gene flow (or migration), mutations and natural selection account for evolutionary changes.

Ex situ conservation Conservation of components of biological diversity outside their natural habitats.

Expected heterozygosity Refers to the number of expected heterozygotes assuming Hardy–Weinberg equilibrium, e.g. in a gene with two alleles would be 2pq.

Extinction Disappearance of a population due to its inability to evolve or adapt to a changing environment.

Extra PA *in situ* conservation Location, management and monitoring of genetic diversity in natural wild populations within defined areas designated for active, long-term conservation, but outside of more formal protected areas.

Field gene bank Collecting of living material for a restricted group of (recalcitrant) species and its transfer and cultivation in a second field site for conservation and utilization.

Fitness Can be calculated as the product of the relative survival time and average fecundity of a genotype. Natural selection favours 'survival of the fittest', thus facilitating the evolution of a population.

Flora Book or other work describing the flora of a given area, which usually contains means of identifying the taxa included.

Fragmentation Division of an ecosystem or habitat into distinct smaller parts, with a loss of connectivity between the parts. The parts may be large to retain the full complement of species without connection to other parts. Fragmentation can result from infrastructure development, such as roads and railways, or natural occurrences like forest fires.

Gap analysis A conservation planning technique that involves the comparison of the genetic and/or ecogeographic range of a taxon with the sample of that taxon which is actively conserved using *in situ* and *ex situ* techniques. The diversity that is not represented in the samples is the 'gap' and therefore the future conservation priority.

Gene Physical and functional unit of heredity (DNA sequence) found in the cell nucleus of living organisms.

Gene bank Facility where crop and wild plant species diversity is stored in the form of seeds, pollen, *in vitro* culture or DNA, or in the case of a field gene bank as plants growing in the field.

Gene flow Exchange of genetic information through pollen and/or seed dispersal in and between breeding populations of the crop, feral populations, weedy and wild relatives.

Gene pool Totality of distinct genes within an inter-breeding population.

General combining ability Relative capacity of an individual to transmit its genetic superiority to its offspring when crossing it with other individuals.

Genetic diversity Variety of genes within a species, i.e. individuals have their own particular genetic composition in the gene pool of a species. It also means the variation among individuals in a population due to changes in their DNA base pair sequences.

Genetic erosion Loss over time of genetic diversity caused by either natural or man-made processes.

Genetic reserve A site for the management and monitoring of genetic diversity of natural wild populations within defined areas designated for active, long-term conservation.

Genetic reserve conservation Location, management and monitoring of genetic diversity in natural wild populations within defined areas designated as protected areas for active, long-term conservation.

Genome Represents the entire complement of genetic material of an individual and includes the coding regions, the non-coding DNA, and the genomes of both the mitochondria and the chloroplast.

Genome sequence Refers to figuring out what is the order of nucleotides (A, G, C, T) in the DNA of an organism. A genome sequence provides the means for knowing where genes are, and how genes work together in an organism. A genome must be broken into small pieces for sequencing and thereafter reassembling them in the right order to get the whole genome.

Genotype Specific allelic composition of an individual for a particular gene.

Genotype frequency Measures the relative number of alleles of each type in adults.

Germplasm Genetic material that forms the physical basis of heredity and that is transmitted from one generation to the next by germ cells.

Heterozygous Genotype contains different alleles of a gene due to inheriting different alleles from its parents.

Hierarchy Group order into which taxa are placed.

Home gardens Sustainable management of genetic diversity of locally developed traditional crop varieties (landraces) by individuals in their home, back-yard or orchard gardens.

Homology Putative synapomorphy.

Homozygous Genotype indicates identical alleles are present in a particular gene because such an individual has received the same alleles from its parents.

Hybrid Offspring of a cross between parents differing in one or more genes.

Idealized population Described using a number of simplifying assumptions, e.g. Wright–Fisher population shows constant size and its members can mate and reproduce with any other member therein, or Moran model having overlapping generations, rather than the non-overlapping generations as the former. Hardy–Weinberg's idealized population assumes absence of selection, migration, mutation and random genetic drift to allow allele frequencies to remain constant over time, and having genotype frequencies related to allele frequencies according to a binomial square principle under random mating.

Identification Naming of an organism by reference to an already existent classification.

In situ conservation Conservation of ecosystems and natural habitats and the maintenance and recovery of viable populations of species in their natural surroundings and, in the case of domesticates or cultivated species, in the surroundings where they have developed their distinctive properties.

In vitro storage Collection and maintenance of plant tissue of usually recalcitrant species in a sterile, pathogen-free environment.

Inbreeding Results from mating between relatives (half- or full-sibs, cousins) or selfing, thus increasing homozygosity and the chances of resulting offspring being affected by recessive or deleterious traits.

Inbreeding coefficient (F) Measures the degree of consanguinity between two individuals and is calculated as a measure for the amount of pedigree collapse within an individual's genealogy. F estimates the expected homozygosity arising from a given system of breeding.

Inbreeding depression Refers to the reduction in fitness of offspring after selfing compared to the fitness of offspring ensuing from outcrossing within the same population.

Indicator A single measure of achievement or a description of the conditions that would show that a conservation action had been implemented successfully; good indicators are 'SMART', meaning they should be Specific, Measurable, Achievable, Realistic and Time-bound.

Information Distinct from data: it is the knowledge derived from the analysis, integration and interpretation of data, including 'expert' opinion. Unlike data, which may be applied to a range of purposes, information is produced for a specific purpose and has a short shelf life.

Landraces A dynamic population of a cultivated plant species that has historical origin, distinct identity and lacks formal crop improvement, as well as often being genetically diverse, locally adapted and associated with traditional farming systems, and often has cultural associations.

Linkage Refers to the association in transmission between genes on the same chromosome instead of displaying independent inheritance. Such genes are closely located on the same chromosome; thus, a chiasma does not often occur between their loci, which leads to a higher frequency of offspring carrying the parental combination of alleles than those containing a non-parental combination of alleles.

Linkage drag Where desirable and undesirable genes that are closely located are transferred from the donor to the recipient together, the undesirable genes must then be bred out.

Locus Refers to the specific place on a chromosome where a gene is located.

Long-term conservation Storage of germplasm for a long period, such as in base collections and duplicate collections. The period of storage before seeds need to be regenerated varies but is at least several decades and possibly a century or more. Long-term conservation takes place at sub-zero temperatures.

Management data Data that facilitate the maintenance of the accession or population in either the ex situ or in situ conservation context.

Medium-term conservation Storage of germplasm in the medium-term such as in active and working collections; it is generally assumed that little loss of viability will occur for approximately 10 years. Medium-term conservation takes place at temperatures between 0 and 10°C.

Microsatellites or simple sequence repeats (SSR) Short segments with di-, tri- or tetra-nucleotide tandem repeats in DNA sequences. The number of repeats varies in populations of DNA and within the alleles of an individual. SSR show a higher mutation rate than other areas of DNA, thus leading to high genetic diversity. These highly polymorphic genetic markers measure levels of relatedness between subspecies, groups and individuals.

Mitigation (of climate change) Climate change mitigation encompasses the actions being taken, and those that have been proposed, to limit the magnitude and/or rate of long-term human-induced climate change.

Moisture content (wet-weight basis) The weight of free moisture divided by the weight of water plus dry matter, expressed as a percentage.

Monophyletic group Group which contains all the descendants of a common ancestor: the group has a common ancestor unique to itself.

Multi-access key A key that can be entered at any point; the user is offered any choice or sequence of characters.

Multi-genes Many genes segregating simultaneously and affecting a quantitative character. Multi-genic inheritance arises when one characteristic is controlled by at least two genes.

Multiline cultivar Mixture of various pure lines showing alike phenotype but each having a distinct gene for target trait (often for host plant resistance).

Natural classification Classification based on numerous characters from many diverse sources of evidence, genomic, morphological, anatomical, palynological, etc.

Neutral theory of molecular evolution Suggests that most evolutionary changes at the molecular level and most of the variation within and between species are not caused by natural selection but by genetic drift of mutant alleles that are neutral.

Number of alleles per locus Means the number of alleles observed in the population.

On-farm conservation Sustainable management of genetic diversity of locally developed traditional animal breeds or crop varieties (landraces) by farmers within traditional agricultural, horticultural or agri-silvicultural cultivation systems, where the focus is the conservation of the genetic resource.

On-farm management Sustainable management of genetic diversity of locally developed traditional animal breeds or crop cultivars by farmers within traditional agricultural, horticultural or agri-silvicultural cultivation systems, where the focus is the conservation of the on-farm system itself.

Ontology A standardized representation, formal naming, and definition of the categories, properties and relations between the concepts, data and entities that substantiate one, many or all domains, enabling common understanding of data types, data attributes and data values.

Operational taxonomic unit (OTU) Lowest taxonomically ranked unit to be studied in a particular investigation, an element in a set that is to be classified.

Orthodox seeded species Seeds that can be dried to low moisture content and stored at low temperatures without damage to increase seed longevity.

Orthodox seeds Seeds that survive after drying (3–5% moisture content) and freezing for storage. Their lifespan will therefore be prolonged with low moisture content and chilly temperatures. Most arable and horticultural crop species have orthodox seeds.

Outbreeding depression Reduces the fitness of F_1 or F_2 individuals after crossing two species or populations due to genetic incompatibility or reduced adaptation to local environmental conditions.

Outcrossing Transfer of pollen from the anther of a flower to the stigma of another flower in a different plant of the same species by the action of wind, insects or other vectors, thereby enabling fertilization and reproduction, allowing for species diversity because it combines genetic information from different plants, and leading to heterozygosity if involving two genetically distinct plants.

Outgroup Group outside of the ingroup but closely related to it, which is used to polarize the tree in phylogenetic analysis.

Paraphyletic Group that includes the most recent common ancestor, but not all of its descendants.

Parthenogenesis Occurs when the embryo develops directly from an unfertilized egg. Resulting embryos and haploid plants are produced due to lack of chromosome duplication when the egg is unfertilized. Embryos and diploid plants may also arise if chromosomal duplication has been performed by some mechanism.

Participatory plant breeding (PPB) The farmer plays a central role. The production of novel varieties results from the collaborative effort of professional plant breeders and farmers. The degree of farmer involvement may vary from the farmer choosing from field trials the variety they wish to grow, to farmer's actually selecting material from breeder's segregating material, the choice being based upon their judgement as to what looks the best material in their fields.

Participatory varietal selection (PVS) Farmers are presented with a range of 'finished' bred cultivars and chooses the material from personal 'trials' on their farms.

Passport data Basic information about the origin of an accession, such as details recorded at the collecting site, pedigree or other relevant information that assists in the identification of an accession.

Phenetic classification Derived from the study of character combinations as they are now perceived.

Phenotype Describes the observable character of an organism determined by the interaction between its genotype and the environment.

Phylogenetic classification Derived from the study of characters that are thought to have been important in evolution; this classification incorporates the dimension of time and ancestry when relating organisms.

Phytosanitary certificate A certificate provided by government plant health personnel to verify that seed and other planting material is substantially free from pests and diseases.

Plant genetic resources (PGR) Taxonomic and genetic diversity of plants that is of value as a resource for the present and future generations of people.

Plant genetic resources for food and agriculture (PGRFA) Taxonomic and genetic diversity of plants that is of value to food and agricultural production as a resource for the present and future generations of people.

Plesiomorphy Primitive state of character, synonymous with ancestral.

Pollen storage Collecting of pollen and storage in appropriate, usually refrigerated conditions.

Polymorphism Occurs when the most common allele of a gene is at a frequency below a certain threshold (usually 95 or 99%).

Polyphyletic Group that does not include the common ancestor of all members of the taxon.

Population A group of individual plants or animals of the same taxon that share a geographic area or region and have common traits.

Primary data These are facts that result from measurements or observations about the world, referenced to stable, widely accepted standards. The latter include absolute measures, such as units of length, volume or density.

Protected area (PA) A clearly defined geographic space, recognized, dedicated and managed, through legal or other effective means, to achieve the long-term conservation of nature with associated ecosystem services and cultural values.

Randomization Non-systematic allocation of field positions to the replicates of various accessions in the experimental area.

Rank Applies to a taxon's levels of the taxonomic hierarchy.

Recalcitrant seeded species Seeds that are not desiccation-tolerant; they do not dry during the later stages of development and are shed at water contents in the range of $0.3–4.0$ g g^{-1}. The loss of water rapidly results in decreased vigour and viability, and seed death at relatively high water contents.

Recalcitrant seeds Do not survive drying (below 20–30% moisture content) and temperatures below $10°C$ because they lose their viability due to desiccation, thus being unable to be stored for long periods. Species producing recalcitrant seeds are mostly vegetatively propagated and their genetic resources often stored as growing plants rather than as seeds.

Relative humidity (RH) A measure of the amount of water present in the air compared to the greatest amount possible for the air to hold at a given temperature, expressed as a percentage. It differs from absolute humidity, which is the amount of water vapour present in a unit volume of air, usually expressed in kilograms per cubic metre.

Replication Should be regarded as the repetition of the set of all treatments included in an experiment. It does not mean repeated measurements of the same item.

Revision A novel analysis of the variation pattern within a particular taxon, considered in conjunction with information from the literature which results in the generation of primary and secondary revision products. The primary product is a novel classification of the taxon, which is complimented by a range of secondary products, such as descriptions, keys, synonymized lists, taxon illustrations, critical notes, etc.

Safety duplication A duplicate of a base collection stored under similar conditions for long-term conservation, but at a different location to insure against accidental loss of material from the base collection.

Sample A selected subset of a population taken randomly or not, thus being representative or non-representative, respectively.

Sampling variation Arises between samples of the population and measures the random error of the sampling technique used.

Secondary (or derived) data Data obtained from primary data by a process of classification or interpretation, either at the time of measurement or later. They may be referenced to absolute measures but more commonly relate to professionally agreed conventions and products that use an accepted structure and format, for example, latitude or longitude measurements.

Seed storage Collection of seed samples in one location and their transfer to a gene bank (seed bank) for storage; orthodox samples are usually dried to a suitably low moisture content of 5% relative humidity, then kept at sub-zero $(-18°C)$ temperature.

Seed viability Capacity of seeds to germinate under favourable conditions.

Selection Process that allows an increase in the proportion of certain alleles or genotypes in successive generations.

Selfing Occurs when pollen falls on to the stigma of the same flower, thus leading to homozygosity.

Single-access key A key that can only be entered by the first couplet, and then the user must progress sequentially through the subsequent couplet until an identification is made.

Single nucleotide polymorphism (SNP) Refers to the variation in a single base pair in a DNA sequence. Hence, an SNP represents a difference in a DNA nucleotide. A gene is described as having more than one allele when an SNP occurs within it, which may lead to variation in the amino acid sequence. SNP may also occur in non-coding regions of DNA, i.e.

they are not always associated with genes. SNP – which are mainly bi-allelic – can be used for measuring genetic diversity but they show lower polymorphism information content (PIC) than SSR. Thus a larger number of SNPs may be necessary to achieve the same level of discrimination among accessions as that provided by SSR.

Smart conservation Conservation planning or implementation that considers the likely impact of climate change on the conservation outcomes.

Species diversity Variety of species within a defined area hosting a range of species, i.e. all biotic organisms including plants.

Specific combining ability (SCA) Measures any deviation in the performance of a cross from the expected performance according to the GCA of its parents.

Storage life Number of years that a seed can be stored before seed death occurs.

Synapomorphy Shared apomorphy; this defines a monophyletic group.

Systematics Science of the diversity of living organisms and the study of how diversity arises.

Taxon Any group of organisms that is grouped together under one name; it can be applied to any rank, e.g. species, genus, tribe, family, etc. (plural taxa).

Taxonomy Scientific study, organization and understanding of biodiversity and how that diversity arose through time.

Varieties Cultivated plants that have been formally approved and registered.

Viability test A test on a sample of seeds from an accession that is designed to estimate the viability of the entire accession.

REFERENCES

Abera Desta, Z. and Ortiz, R. (2014). Genomic selection: genome-wide prediction in plant improvement. *Trends in Plant Science*, 19: 592–601.

Adams, R.P. (1997). Conservation of DNA: DNA banking. In: Callow, J.A., Ford-Lloyd, B.V. and Newbury, H.J. (eds.) *Biotechnology and Plant Genetic Resources*. CAB International, Oxford, pp. 163–174.

Aguirre-Gutiérrez, J., van Treuren, R., Hoekstra, R. and van Hintum, T.J.L. (2017). Crop wild relatives range shifts and conservation in Europe under climate change. *Diversity and Distributions*, 23: 739–750.

Ahmed, J. and Khan, S.S. (1998). Investment in people – a key to enhance sustainability: lessons from Northern Pakistan: enhancing sustainability – resources for our future. *SUI Technical Series* 1: 21–28.

Åkerberg, E. (1986). Nilsson-Ehle and the development of plant breeding at Svalof during the period 1900–1915. *Hereditas*, 105: 1–5.

Akimoto, M., Shimamoto, Y. and Morishima, H. (1999). The extinction of genetic resources of Asian wild rice, *Oryza rufipogon* Griff.: a case study in Thailand. *Genetic Resources and Crop Evolution*, 46: 419–425.

Al-Atawneh, N., Amri, A., Assi, R. and Maxted, N. (2008). Management plans for promoting *in situ* conservation of local agrobiodiversity in the West Asia centre of plant diversity. In: Maxted, N., Ford-Lloyd, B.V., Kell, S.K., Iriondo, J.M., Dulloo, M.E. and Turok, J. (eds.) *Crop Wild Relative Conservation and Use*. CABI Publishing, Wallingford, UK, pp. 338–361.

Al-Atawneh, N., Shehadeh, A., Amri, A. and Maxted, N. (2009). *Conservation Field Guide to Medics of the Mediterranean Basin*. ICARDA, Aleppo.

Alercia, A., Diulgheroff, S. and Mackay, M. (2012). *FAO/ Bioversity Multi-Crop Passport Descriptors (MCPD V.2)*. Food and Agriculture Organization of the United Nations, Rome, Italy and Bioversity International, Rome, Italy. Available at: www.bioversityinternational.org/uploads/tx_news/1526.pdf) (accessed 13 August 2018).

Alercia, A., López, F.M., Sackville Hamilton, N.R. and Marsella, M. (2018). *Digital Object Identifiers for Food Crops – Descriptors and Guidelines of the Global Information System*. Food and Agriculture Organization of the United Nations, Rome.

Al Lawati, A.H., Al Saady, N., Ghaloub, H.A., *et al.*, (2016). *Plant Agrobiodiversity Conservation Strategy for the Sultanate of Oman*. Oman Animal and Plant Genetic Resources Center, the Research Council of the Sultanate of Oman, Muscat, Sultanate of Oman.

Almekinders, C. and Hardon, J. (eds.) (2006). *Bringing Farmers Back into Breeding. Experiences with Participatory Plant Breeding and Challenges for Institutionalisation*. Agromisa Special 5. Agromisa, Wageningen.

Almekinders, C., Mertens, L., van Loon, J. and Lammerts van Bueren, E.T. (2016). Potato breeding in the Netherlands: successful collaboration between farmers and commercial breeders. *Farming Matters* (Special issue on Access and Benefit Sharing of Genetic Resources), May Issue, pp. 34–37.

Altieri, M.A. 2009. Agroecology, small farms, and food sovereignty. *Monthly Review*, 61(3): 102.

Altieri, M.A. and Merrick, L.C. (1987). *In situ* conservation of crop genetic resources through maintenance of traditional farming systems. *Economic Botany*, 41: 86–96.

American Museum of Natural History (1998). *National Survey Reveals Biodiversity Crisis – Scientific Experts Believe We Are in the Midst of the Fastest Mass Extinction in Earth's History*. Available at: www.mysterium.com/amnh.html (accessed 19 April 2013).

Amri, A., Ajlouni, M., Assi, R., *et al.* (2007). Major Achievements of the West Asia Dryland Agrobiodiversity Conservation Project. Proceedings of the International Conference on '*Promoting Community-Driven in situ Conservation of Dryland Agrobiodiversity*', ICARDA, Aleppo, Syria.

Andersson, M.S., Mesa Fuqen, E. and Carmen de Vicente, M. (2006). State of the art of DNA storage: results of a worldwide survey. In: Carmen de Vicente, M. and Andersson, M.S. (eds.) *DNA Banks – Providing Novel Options for Gene Banks?* International Plant Genetic Resources Institute, Rome, pp. 6–10.

Andrade, M.I., Alvaro, A., Menomussanga, J., *et al.* (2016a). 'Alisha', 'Ivone', 'Anamaria', 'Victoria', 'Lawrence', 'Bita', 'Caelan', 'Margarete' and 'Bie' sweetpotato. *HortScience*, 51: 597–600.

Andrade, M.I., Naico, A., Ricardo, J., *et al.* (2016b). Genotype × environment interaction and selection for drought adaptation of sweetpotato (*Ipomoea batatas* [L.] Lam.) in Mozambique. *Euphytica*, 209: 261–280.

Andrade, M.I., Ricardo, J., Naico, A., *et al.* (2017). Release of orange-fleshed sweetpotato (*Ipomoea batatas* [L.] Lam.) bred-cultivars in Mozambique through an accelerated breeding scheme. *Journal of Agricultural Sciences*, 155: 919–929.

Angiosperm Phylogeny Group IV. (2016). An update of the Angiosperm Phylogeny Group classification for the orders and families of flowering plants: APG IV. *Botanical Journal of the Linnean Society*, 181: 1–20.

Anikster, Y., Feldman, M. and Horovitz, A. (1997). The Ammiad experiment. In: Maxted, N., Ford-Lloyd, B.V. and Hawkes, J.G. (eds.) *Plant Genetic Conservation: The in situ Approach*. Chapman & Hall, London, UK, pp. 239–253.

Ansebo, L. (2015). *Ecosystem services: genetic resources and crop wild relatives*. Available at: www.nordgen.org/index .php/en/content/view/full/2934 (accessed 5 May 2017).

Antofie, M.M., Sand, M.P.C., Ciotea, G. and Iagrăru, P. (2010). Data sheet model for developing a red list regarding crop landraces in Romania. *Annals of Food Science and Technology*, 11: 45–49.

Arnstein, S.R. (1969). A ladder of citizen participation. *Journal of the American Planning Association*, 35: 216–224.

Ashby, J. (2009). The impact of participatory plant breeding. In: Ceccarelli, S., Guimara, E.P. and Weltzien, E. (eds.) *Plant Breeding and Farmer Participation*. Food and Agriculture Organization of the United Nations, Rome, pp. 649–671.

Ashmore, S.E. (1997). *Status Report on the Development and Application of in vitro Techniques for the Conservation and Use of Plant Genetic Resources*. IPGRI, Rome.

Avagyan, A. (2008). Crop wild relatives in Armenia: diversity, legislation and conservation issues. In: Maxted, N., Ford-Lloyd, B.V., Kell, S.K., *et al.* (eds.) *Crop Wild Relative Conservation and Use*. CABI Publishing, Wallingford, UK, pp. 58–68.

Averyanov, L.V. (1996). Endangered Vietnamese Paphiopedilums. Part 2. *Paphiopedilum delenatii. Orchids (Magazine of the American Orchid Society)*, pp. 1302–1308.

Azurdia, C., Williams, K., Williams, D., *et al.* (2011). *Atlas of Guatemalan Crop Wild Relatives*. Available at: www.ars .usda.gov/ba/atlascwrguatemala (accessed 5 May 2017).

Bachman, S., Moat, J., Hill, A.W., de la Torre, J. and Scott, B. (2011). Supporting Red List threat assessments with GeoCAT: geospatial conservation assessment tool. *ZooKeys*, 150: 117–126.

Bajaj, Y.P.S. (1987). Cryopreservation of pollen and pollen embryos, and the establishment of pollen banks. *International Review of Cytology*, 107: 397–420.

Balter, M. (2007). Seeking agriculture's ancient roots. *Science*, 316: 1830–1835.

Barata, A.M., Rocha, F., Oliveira, J., *et al.* (2016). Implementation of a PGR Global Documentation System in Portugal. In: Maxted, N., Dulloo, M.E. and Ford-Lloyd, B.V. (eds.) *Enhancing Crop Gene Pool Use: Capturing Wild Relative and Landrace Diversity for Crop Improvement*. CAB International, Wallingford, UK, pp. 441–452.

Bari, A., Amri, A., Street, K., *et al.* (2014). Predicting resistance to stripe (yellow) rust (*Puccinia striiformis*) in wheat genetic resources using focused identification of germplasm strategy. *Journal of Agricultural Science*, 152: 906–916.

Bari, A., Street, K., Mackay, M., *et al.* (2012). Focused identification of germplasm strategy (FIGS) detects wheat stem rust resistance linked to environmental variables. *Genetic Resources and Crop Evolution*, 59: 1465–1481.

Barthlott, W., Erdelen, W.R. and Rafiqpoor, D.M. (2014). Biodiversity and technical innovations: bionics. In: Lanzerath, D and Friele, M. (eds.) *Concepts and Values in Biodiversity*. Routledge, London/New York, pp. 300–315.

Bassi, F., Bentley, A., Charmet, G., Ortiz, R. and Crossa, J. (2016). Breeding schemes for the implementation of genomic selection in wheat (*Triticum* spp.). *Plant Science*, 242: 23–36.

Baudoin, J.P., Rocha, O.J., Degreef, J., *et al.* (2008). *In situ* conservation strategy for wild Lima bean (Phaseolus lunatus L.) populations in the Central Valley of Costa Rica: a case study of short-lived perennial plants with a mixed mating system. In: Maxted, N., Ford-Lloyd, B.V., Kell, S.K., *et al.* (eds.) *Crop Wild Relative Conservation and Use*. CABI Publishing, Wallingford, UK, pp. 364–379.

Baxter, G. (1974). *Fruits of the World in Danger*. Gotham Book Mart, New York.

Beattie, A. and Ehrlich, P.R. (2001). *Wild Solutions: How Biodiversity Is Money in the Bank*. Yale University Press, New Haven, CT.

Beentje, H.J. (2010). *The Kew Plant Glossary: An Illustrated Dictionary of Plant Identification Terms*. Royal Botanic Gardens, Kew, Richmond, UK.

Bellon, M., Gotor, E. and Caracciolo, F. (2015a). Conserving traditional varieties and improving livelihoods: how to assess the success of on-farm conservation projects. *International Journal of Agricultural Sustainability* 13: 167–182.

Bellon, M., Gotor, E. and Caracciolo, F. (2015b). Assessing the effectiveness of projects supporting on-farm conservation of native crops: evidence from the High Andes of South America. *World Development* 70: 162–176.

Bellon, M.R., Dulloo, E., Thormann, I, Sardos, J. and Burdon, J. (2017). In situ conservation – harnessing natural and human derived evolutionary forces to ensure future crop adaptation. *Evolutionary Applications*. https://doi.org/10 .1111/eva.12521.

Bellon, M.R., Pham, J.L. and Jackson, M.T. (1997). Genetic conservation: a role for rice farmers. In: Maxted, N., Ford-Lloyd, B.V. and Hawkes, J.G. (eds.) *Plant Genetic Conservation: The in Situ Approach*. Chapman & Hall, London, pp. 261–289.

Bennett, S. and Maxted, N. (1997). An ecogeographic analysis of the *Vicia narbonensis* complex. *Genetic Resources and Crop Evolution*, 44: 411–428.

Benson, D.A., Cavanaugh, M., Clark, K., *et al.* (2013). GenBank. *Nucleic Acids Research*. https://doi.org/10 .1093/nar/gks1195.

Bergamini, N., Padulosi, S., Ravi, S.B. and Yenagi, N. (2013). Minor millets in India: a neglected crop goes mainstream.

In: Fanzo, J., Hunter, D., Borelli, T., et al. (eds.) *Diversifying Food and Diets: Using Agricultural Biodiversity to Improve Nutrition and Health. Issues in Agricultural Biodiversity.* Earthscan from Routledge, London, pp. 313–325.

BGCI. (2018a). *PlantSearch database.* Botanic Garden Conservation International, Kew, UK. Available at: www.bgci.org/plant_search.php (accessed 6 August 2018).

BGCI. (2018b). *BGCI's Annual Member's Review 2017.* Botanical Garden Conservation International, Kew, UK.

Bharucha, Z. and Pretty, J. (2010). The roles and values of wild foods in agricultural systems. *Philosophical Transactions of the Royal Society B*, 365(1554): 2913–2926.

Bhullar, N.K., Street, K., Mackay, M., Yahiaoui, N. and Keller, B. (2009). Unlocking wheat genetic resources for the molecular identification of previously undescribed functional alleles at the Pm3 resistance locus. *Proceedings of the National Academy of Sciences of the United States of America*, 106:9519–9524.

Bilz, M., Kell, S.P., Maxted, N. and Lansdown, R.V. (2011). *European Red List of Vascular Plants.* Publications Office of the European Union, Luxembourg.

Bimpong, I.K., Manneh, B., Diop, B., *et al.* (2013). New quantitative trait loci for enhancing adaptation to salinity in rice from Hasawi, a Saudi landrace into three African cultivars at the reproductive stage. *Euphytica*, 200: 45–60.

Bimpong, I.K., Manneh, B., Diop, B., *et al.* (2014). New quantitative trait loci for enhancing adaptation to salinity in rice from Hasawi, a Saudi landrace, into three African cultivars at the reproductive stage. *Euphytica*, 200: 45–60.

Biodiversity Indicator Partnership. (2010). *Biodiversity indicators and the 2010 Biodiversity Target: Outputs, experiences and lessons learnt from the 2010 Biodiversity Indicators Partnership.* Available at: https://attachment.fbsbx.com/file_download.php?id=125191087651105&eid=ASvLJFDy-SfS_extnSRmO_foPRTc2X9gRJSbkvQHx9tIC75?OwO8m3Y_VeTxEBh8zjbE&inline=1&text=1383059566&hash=ASsHkIJbQqNBxkkt (accessed 20 October 2014).

Bioversity International. (2007). Guidelines for the development of crop descriptor lists. *Bioversity Technical Bulletin* 13. Bioversity International, Rome, Italy.

Bioversity International. (2016). *Safeguarding and using crop wild relatives for food security and climate change adaptation.* Available at: www.bioversityinternational.org/cwr/ (accessed 5 May 2016).

Bioversity International, FAO, PROINPA, INIAF and IFAD. (2013). *Descriptors for Quinoa (Chenopodium quinoa Willd.) and Wild Relatives.* Bioversity International, Rome, Italy; Fundación PROINPA, La Paz, Bolivia; Instituto Nacional de Innovación Agropecuaria y Forestal, La Paz, Bolivia; International Fund for Agricultural Development, Food and Agriculture Organization of the United Nations, Rome, Italy.

Bioversity International and University of Birmingham. (2017). *Crop wild relative checklist and inventory descriptors v.1.* Bioversity International, Rome, Italy. Available at: www.bioversityinternational.org/e-library/publications/detail/crop-wild-relative-checklist-and-inventory-descriptors-v1/ (accessed 4 January 2019).

Bishaw, Z. and Turner, M. (2008). Linking participatory plant breeding to the seed supply system. *Euphytica*, 163: 31–44.

Bishaw, Z. and van Gastel, A.J.G. (2009). Variety release and policy options. In: Ceccarelli, S., Guimara, E.P. and Weltzien, E. (eds.) *Plant Breeding and Farmer Participation.* Food and Agriculture Organization of the United Nations, Rome, Italy, pp. 565–587.

Blackmore, S. and Oldfield, S. (2017). *Plant Conservation Science and Practice: The Role of Botanic Gardens.* Cambridge University Press, Cambridge, UK.

Bock, H. (1539). *Kreuterbuch.* Strasburg.

Bommer, D.F.R. (1991). The historical development of international collaboration in plant genetic resources. In: van Hintum, T.J.L., Frese, L. and Perret, P.M. (eds.) *Searching for New Concepts for Collaborative Genetic Resources Management.* Papers of the EUCARPIA/IBPGR Symposium, Wageningen, The Netherlands. International Crop Network Series No. 4. IBPGR, Rome, pp. 3–12.

Borrini-Feyerabend, G., Kothari, A. and Oviedo, G. (2004). *Indigenous and Local Communities and Protected Areas: Towards Equity and Enhanced Conservation.* IUCN, Gland, Switzerland and Cambridge, UK.

Boshier, D., Loo, J. and Dawson, I. (2017). Forest and tree genetic resources. In: Hunter, D., Guarino, L., Spillane, C. and McKeown, P. (eds) *Handbook of Agriculture Biodiversity.* Routledge, London and New York, pp. 45–64.

Bowman, D.M.J.S. (2003). Australian landscape burning: a continental and evolutionary perspective. In: Abbott, I. and Burrows, N. (eds.) *Fire in Ecosystems of South-West Western Australia: Impacts and Management.* Backhuys Publishers, Leiden, The Netherlands, pp. 107–118.

Bradley, B.A., Blumenthal, D.M., Wilcove, D.S. and Ziska, L.H. (2010). Predicting plant invasions in an era of global change. *Trends in Ecology and Evolution*, 5: 310–318.

Breseghello, F. and Sorrells, M.E. (2006). Association analysis as a strategy for improvement of quantitative traits in plants. *Crop Science*, 46: 1323–1330.

Bronkhorst, S. (2014). *Adaptation must be conflict sensitive. conflict-sensitive adaptation: use human rights to build social and environmental resilience.* Brief 1. Indigenous Peoples of Africa Co-ordinating Committee and IUCN Commission on Environmental, Economic and Social Policy. Available at: www.iucn.org/downloads/tecs_csa_1_conflict_sensitive_adapation_bronkhorst.pdf (accessed 19 April 2017).

Brookfield, H.C. (2001). *Exploring Agrodiversity.* New York: Columbia University Press, New York.

Brookfield, H.C., Padoch, C., Parsons, H. and Stocking, M. (2002). *Cultivating Biodiversity: Understanding, Analysing and Using Agricultural Diversity*. ITDG Publications, Rugby.

Brookfield, H.C., Parsons, H. and Brookfield, M. (eds.) (2003). *Agrodiversity: Learning from Farmers across the World*. United Nations University Press, Tokyo.

Brooks, S. and Bubb, P. (2014). *Developing Indicators for National Targets As Part of NBSAP Updating: Examples of the Biodiversity Indicator Development Framework in Practice*. UNEP-WCMC, Cambridge, UK.

Brown, A.H.D. (1989). Core collections: a practical approach to genetic resources management. *Genome*, 31: 818–824.

Brown, A.H.D. (2000). The genetic structure of crop landraces and the challenge to conserve them *in situ* on farms. In: Brush, S.B. (ed.) *Genes in the Field: On-Farm Conservation of Crop Diversity*. Lewis Publishers, Boca Raton, FL, pp. 29–48.

Brown, A.H.D. and Briggs, J.D. (1991). Sampling strategies for genetic variation in *ex situ* collections of endangered plant species. In: Falk, D.A. and Holsinger, K.E. (eds.) *Genetics and Conservation of Rare Plants*. Oxford University Press, New York, pp. 99–119.

Brown, A.H.D. and Hardner, C.M. (2000). Sampling the gene pools of forest trees for *ex situ* conservation. In: Young, A., Boshier, D. and Boyle, T. (eds.) *Forest Conservation Genetics. Principles and Practice*. CSIRO and CABI, Canberra, Australia and Wallingford, UK, pp. 185–196.

Brown, A.H.D. and Marshall, D.R. (1995). A basic sampling strategy: theory and practice. In: Guarino, L., Ramanatha Rao, V. and Reid, R. (eds.) *Collecting Plant Genetic Diversity: Technical Guidelines*. CAB International, Wallingford, UK, pp. 75–91.

Brummitt, N. and Bachman, S. (2010). *Plants under Pressure: A Global Assessment*. The first report of the IUCN Sampled Red List Index for Plants. London: Natural History Museum.

Brunsfels, O. (1530). Herbarium vivae icons. *Argentorati*, 3 tomes.

Brush, S.B. (1991). A farmer-based approach to conserving crop germplasm. *Economic Botany*, 45: 153–165.

Brush, S.B. (2000). *Genes in the Field: On-Farm Conservation of Crop Diversity*. Lewis Publishers, Boca Raton, FL.

Buckler, E.S., Thornsberry, J.M. and Kresovich, S. (2001). Molecular diversity, structure and domestication of grasses. *Genetics Research*, 77(3): 213–218.

Buddendorf-Joosten, J.M.C. and Woltering, E.J. (1994). Components of the gaseous environment and their effects on plant growth and development *in vitro*. *Plant Growth Regulation*, 15: 1–16.

Burgman, M.A. and Neet, C.R. (1989). Analyse des risques d'extinction des populations naturelles. *Acta Oecologica*, 10: 233–243.

Burke, M.B., Lobell, D.B. and Guarino, L. (2009). Shifts in African crop climates by 2050, and the implications for crop improvement and genetic resources conservation. *Global Environmental Change*, 19: 317–325.

Burley, F.W. (1988). Monitoring biological diversity for setting priorities in conservation. In: Wilson, E.O. and Peter, F.M. (eds.) *Biodiversity*. National Academy Press, Washington, DC, pp. 227–230.

Burlingame, B., Charrondière, U.R. and Mouille, B. (2009). Food composition is fundamental to the cross-cutting initiative on biodiversity for food and nutrition. *Journal of Food Composition and Analysis*, 22: 361–365.

Burucharu, R.A., Sperling, L., Ewell, P. and Kirby, R. (2002). The role of research institutions in seed-related disaster relief: Seeds of Hope experiences in Rwanda. *Disaster*, 26: 288–301.

Cahill, A.E., Aiello-Lammens, M.E., Fisher-Reid, M.C., *et al.* (2012). How does climate change cause extinction? *Proceedings of the Royal Society B*, 280: 20121890. https://doi.org/10.1098/rspb.2012.1890.

Cairns, M. (ed.) (2015). *Shifting Cultivation and Environmental Change: Indigenous People, Agriculture and Forest Conservation*. Routledge, London and New York.

Camacho Villa, T.C., Maxted, N., Scholten, M.A. and Ford-Lloyd, B.V. (2005). Defining and identifying crop landraces. *Plant Genetic Resources: Characterization and Utilization*, 3: 373–384.

Campanelli, G., Acciarri, N., Campion, B., *et al.* (2015). Participatory tomato breeding for organic conditions in Italy. *Euphytica*, 204: 179–197.

Campbell, B.M. and Luckert, M.K. (eds.) (2002). *Uncovering the Hidden Harvest; Valuation Methods for Woodland and Forest Resources*. Earthscan, London.

Capistrano, G.C., Ries, D., Minoche, A., *et al.* (2014). Fine mapping of rhizomania resistance using *in situ* populations of the wild beet *Beta vulgaris* ssp. *maritima*. *Proceedings of the Plant & Animal Genome*, 22: 673.

Castañeda-Álvarez, N.P., de Haan, S., Juárez, H., *et al.* (2015). *Ex situ* conservation priorities for the wild relatives of potato (*Solanum* L. section *Petota*). *PLoS ONE*. https://doi.org/10.1371/journal.pone.0122599.

Castañeda-Álvarez, N.P., Khoury, C.K., Achicanoy, H.A., *et al.* (2016a). Global priorities for crop wild relative conservation for food security. *Nature Plants*, 2: 16022.

Castañeda-Álvarez, N.P., Khoury, C.K., Sosa, C.C., *et al.* (2016b). The distributions and *ex situ* conservation of crop wild relatives: a global approach. In: Maxted, N., Dulloo, E.M. and Ford-Lloyd, B.V. (eds.) *Enhancing Crop Gene Pool Use: Capturing Wild Relative and Landrace Diversity* for Crop Improvement. CAB International, Wallingford, pp. 149–160.

Castañeda-Álvarez, N.P., Vincent, H.A., Kell, S.P., Eastwood, R.J. and Maxted, N. (2011). Ecogeographic surveys. In: Guarino, L. Ramanatha Rao, V. and Goldberg, E. (eds.) Collecting Plant Genetic Diversity: Technical Guidelines. 2011 update. Bioversity International, Rome. Available at: http://cropgene bank.sgrp.cgiar.org/index .php?option=com_content&view=article&id=679 (accessed 6 March 2015).

CBD. (1992). Convention on Biological Diversity: Text and Annexes. Secretariat of the Convention on Biological Diversity, Montreal, pp. 1–34.

CBD. (2002). 2010 Biodiversity Target. Secretariat of the Convention on Biological Diversity, Montreal. Available at: www.biodiv.org/2010-arget/default.aspx (accessed 3 April 2007).

CBD. (2010a). Global Strategy for Plant Conservation. Secretariat of the Convention on Biological Diversity, Montreal.

CBD. (2010b). Strategic Plan for Biodiversity 2011–2020. Secretariat of the Convention on Biological Diversity, Montreal. Available at: www.cbd.int/undb/media/factsheets/ undb-factsheet-sp-en.pdf (accessed 6 March 2016).

CBD. (2010c). Identification, Monitoring, Indicators and Assessments. Secretariat of the Convention on Biological Diversity, Montreal. Available at: www.cbd.int/ indicators/intro.shtml (accessed 6 March 2016).

CBD. (2011). An Introduction to National Biodiversity Strategies and Action Plans. Secretariat of the Convention on BiologicalDiversity, Montreal. Available at: www.cbd .int/nbsap/guidance.shtml (accessed 6 March 2016).

CBD. (2014). Global Biodiversity Outlook 4. Secretariat of the Convention on Biological Diversity, Montreal.

CBOL Plant Working Group. (2009). A DNA barcode for land plants. Proceedings of the National Academy of Sciences, 106: 12794–12797.

Ceballos-Lascuráin, H. (1996). Tourism, Ecotourism, and Protected Areas: The State of Nature-Based Tourism around the World and Guidelines for Its Development. IUCN, Gland, Switzerland and Cambridge, UK.

Ceccarelli, S. (2014). GMO, organic agriculture and breeding for sustainability. Sustainability, 6: 4273–4286.

Ceccarelli, S., Galie, A. and Grando, S. (2013). Participatory breeding for climate change-related traits. In: Kole, C. (ed.) Genomics and Breeding for Climate-Resilient Crops, Vol. 1. Springer-Verlag, Berlin, Heidelberg, pp. 331–376.

Ceccarelli, S., Galié, A., Mustafa, Y., Grando, S. (2012). Syria: participatory barley breeding – farmers' input becomes everyone's gain. In: Ruiz, M. and Vernooy, R. (eds.) Custodians of Biodiversity: Sharing Access and Benefits to Genetic Resources. Earthscan, IDRC, London, pp. 53–66.

Ceccarelli, S., Grando, S. and Booth, R.H. (1996). Farmers and crop breeders as partners. In: Eyzaguirre, P. and Iwanaga, M. (eds.) Participatory Plant Breeding. International Plant Genetic Resources Institute, Rome, pp. 99–116.

Ceccarelli, S., Grando, S., Maatougui, M., et al. (2010). Plant breeding and climate changes. Journal of Agricultural Science, 148: 627–638.

Ceccarelli, S., Guimaraes, E.P. and Weltzien, E. (2009). Plant Breeding and Farmer Participation. Food and Agriculture Organization of the United Nations, Rome.

Cernansky, R. (2015). Super vegetables. Nature, 522: 146–148.

Chakraborty, S. and Newton, A.C. (2011). Climate change, plant diseases and food security, an overview. Plant Pathology, 60: 2–14.

Chambers, R. (2007). From PRA to PLA and Pluralism: Practice and Theory. Working Paper 286. Institute of Development Studies, London.

Chambers, R., Pacey, A. and Thrupp, L.A. (1989). Farmer First: Farmer Innovation and Agricultural Research. ITDG Publishing, London.

Chandra, S., Huaman, Z., Hari Krishna, S. and Ortiz, R. (2002). Optimal sampling strategy and core collection size of Andean tetraploid potato based on isozyme data – a simulation study. Theoretical Applied Genetics, 104: 1325–1334.

Chase, M.W. and Fay, M.F. (2009). Barcoding of plants and fungi. Science, 325: 682–683.

Chen, J., Corlett, R.T. and Cannon, C.H. (2017). The role of botanic gardens in situ conservation. In: Blackmore, S. and Oldfield, S. (eds.) Plant Conservation Science and Practice: The Role of Botanic Gardens. Cambridge University Press, Cambridge, pp. 73–101.

Chivian, E. and Bernstein, A. (eds.) (2008). Sustaining Life: How Human Health Depends on Biodiversity. Center for Health and the Global Environment. Oxford University Press, New York.

Christiansen, J.L., Raza, S., Jørnsgård, B., Mahmoud, S.A. and Ortiz, R. (2000). Potential of landrace germplasm for genetic enhancement of white lupin in Egypt. Genetic Resources and Crop Evolution, 47: 425–430.

Christiansen, J.L., Raza, S. and Ortiz, R. (1999). White lupin (Lupinus albus) germplasm collection and preliminary in situ diversity assessment in Egypt. Genetic Resources and Crop Evolution, 46: 169–174.

Christiansen, M.J., Andersen, S.B. and Ortiz, R. (2002). Diversity changes in an intensively bred wheat germplasm during the 20th century. Molecular Breeding, 9: 1–11.

Christinck, A., Weltzien, E. and Hoffman, V. (2005). Setting Breeding Objectives and Developing Seed Systems with Farmers. Margraf Publishers, Weikersheim.

Cibrian-Jaramillo, A., Hird Meyer, A., Oleas, N., et al. (2013). What is the conservation value of a plant in a botanic garden? Using indicators to improve management of *ex situ* collections. *The Botanical Review*, 79: 1–19.

Cochran, W.G. (1977). *Sampling Techniques*, 3rd ed. John Wiley & Sons, New York.

Convention on Biological Diversity. (2012). *Global Strategy for Plant Conservation: 2011–2020*. Botanic Gardens Conservation International, Richmond.

Cook, F.E.M. (1995). *Economic Botany Data Collection Standards*. Royal Botanic Gardens, Kew, Richmond, UK.

Cooper, D., Vellvé, R. and Hobbelink, H. (eds.) (1992). *Growing Diversity: Genetic Resources and Local Food Security*. Intermediate Technology Publications, London.

Cox, G.W. (1993). *Conservation Ecology*. W.C. Brown, Dubugue, IA.

Crane, P., Hopper, S.D., Raven, P.H. and Stevenson, D.W. (2009). Plant science research in botanic gardens. *Trends in Plant Science*, 14: 575–577.

Crossa, J. and Vencovsky, R. (2011). Basic sampling strategies: theory and practice. In: Guarino, L., Ramanatha Rao, V. and Goldberg, E. (eds.) *Collecting Plant Genetic Diversity: Technical Guidelines. 2011 Update*. Bioversity International, Rome. Available at: http://cropgene bank .sgrp.cgiar.org/index.php?option=com_content&view= article&tid=671 (accessed 13 August 2018).

Crossa, J., Burgueño, J., Dreisigacker, S., *et al.* (2007). Association analysis of historical bread wheat germplasm using additive genetic covariance of relatives and population structure. *Genetics*, 177: 1889–1913.

Crossa, J., Hernandez, C.M., Bretting, P., Eberhart, S.A and Taba, S. (1993). Statistical considerations for maintaining germplasm collections. *Theoretical and Applied Genetics*, 86, 673–678.

Crow, J.F. and Denniston, C. (1988). Inbreeding and variance effective population numbers. *Evolution*, 42: 482–495.

Cunningham, A. (2001). *Applied Ethnobotany: People, Wild Plant Use and Conservation*. Earthscan, London.

Dagne, E. (1998). Integration of traditional phytotherapy into general health care: an Ethiopian perspective. In: Prendergast, H.D.V., Etkin, N.L., Harris, D.R. and Houghton, P.J. (eds.) *Plants for Food and Medicine*. Royal Botanic Gardens, Kew, Richmond, UK, pp. 47–55.

Danielsen, F., Burgess, N.D., Balmford, A., *et al.* (2009). Local participation in natural resource monitoring: a characterization of approaches. *Conservation Biology*, 23: 31–42.

Dansi, A. (2011). Collecting vegetatively propagated crops (especially roots and tubers). In: Guarino, L., Ramanatha Rao, V., Goldberg, E. (eds.) *Collecting Plant Genetic Diversity: Technical Guidelines. 2011 update*. Bioversity International, Rome. Available at: http://cropgene bank .sgrp.cgiar.org/index.php?option=com_content&view= article&tid=666 (accessed 13 August 2018).

Davis, P.H. and Heywood, V.H. (1973). *Principles of Angiosperm Taxonomy*. Krieger, New York.

Davis, S., Heywood, V.H. and Hamilton, A.C. (1995). *Centres of Plant Diversity: A Guide and Strategy for Their Conservation*. WWF and International Union for Conservation of Nature and Natural Resources, Gland.

de Boef, W.S. and Subedi, A. (2017). Community biodiversity management. In: Hunter, D., Guarino, L., Spillane, C. and McKeown, P. (eds.) *Handbook of Agriculture Biodiversity*. Routledge, Abingdon, pp. 497–509.

de Boef, W.S., Subedi, A., Peroni, N., Thijssen, M. and O'Keeffe, E. (2013a). *Community Biodiversity Management: Promoting Resilience and the Conservation of Plant Genetic Resources*. Earthscan/Routledge, London.

de Boef, W.S., Thijssen, M. and Subedi, A. (2013b). New professionalism and governance in plant genetic resource management. In: De Boef, W.S., Peroni, N., Subedi, A. and Thijssen, M.H. (eds.) *Community Biodiversity Management: Promoting Resilience and the Conservation of Plant Genetic Resources*. Earthscan, London, pp. 353–364.

de Pourcq, K., Thomas, E., Arts, B., Vranckx, T. and van Damme, P. (2016). Understanding and resolving conflict between local communities and conservation authorities in Colombia. *World Development*, 93: 125–135.

de Vicente, M.C. and Andersson, M.S. (2006). *DNA Banks – Providing Novel Options for Gene banks?* Topical reviews in Agricultural Biodiversity. International Plant Genetic Resources Institute, Rome.

de Vicente, M.C., Metz, T. and Alercia, A. (2004). *Descriptors for Genetic Marker Technologies*. FAO/Bioversity International, Rome.

Dempewolf, H., Eastwood, R.J., Guarino, L., *et al.* (2013). Adapting agriculture to climate change: a global initiative to collect, conserve, and use crop wild relatives. *Agroecology and Sustainable Food Systems*, 38: 369–377.

Department of the Environment. (1996). *Towards a Methodology for Costing Biodiversity Targets in the UK*. Department of the Environment, London.

Deryng, D., Sacks, W.J., Barford, C.C. and Ramankutty, N. (2011). Simulating the effects of climate and agricultural management practices on global crop yield. *Global Biogeochemical Cycles*, 25: GB2006. https://doi.org/10 .1029/2009GB003765.

Desclaux, D, Ceccarelli, S., Navazio, J., *et al.* (2012). Centralized or decentralized breeding: the potentials of participatory approaches for low-input and organic agriculture. In: Mammerts van Bueren, E.T. and Myers, J.R. (eds.) *Organic Crop Breeding*. Wiley-Blackwell, Chichester, pp. 99–124.

Devereau, A.D. (1994). *Tropical Sweet Potato Storage: A Literature Review*. Natural Resources Institute, Chatham.

Díaz, S., Settele, J. and Brondízio, E. (2019). *Summary for policymakers of the global assessment report on* biodiversity and ecosystem services of the *Intergovernmental Science-Policy Platform on Biodiversity and Ecosystem Services*. IPBES Secretariat, Bonn. Available at: www.ipbes.net/sites/default/files/downloads/spm_unedited_advance_for_posting_htn.pdf (accessed 25 August 2019).

Dice, L.R. (1945). Measures of the amount of ecologic association between species. *Ecology*, 26: 297–302.

di Falco, S. and Chavas, J.P. (2006). Crop genetic diversity, farm productivity and the management of environmental risk in rainfed agriculture. *European Review of Agricultural Economics*, 33: 289–314.

DOI Foundation. (2015). *DOI Handbook, Version 5*. International DOI Foundation, London.

Donaldson, J.S. (2009). Botanic gardens science for conservation and global change. *Trends in Plant Science*, 14: 608–613.

Dosmann, M.S. (2006). Research in the garden: averting the collections crisis. *Botanical Review*, 72: 207–234.

Draper, D., Rosselló-Graell, A., García, C., Tauleigne Gomes, C. and Sérgio, C. (2003). Application of GIS in plant conservation programmes in Portugal. *Biological Conservation*, 113: 337–349.

Dudley, N. (ed.) (2008). *Guidelines for Applying Protected Area Management Categories*. IUCN, Gland.

Dudley, N., Ford-Lloyd, B., Kell, S.P., Maxted, N. and Stolton, S. (2006). *Food stores: using protected areas to secure crop genetic diversity*. A research report by WWF, Equilibrium and University of Birmingham.

Dudley, N., Hocking, M. and Stolton, S. (2010). Precious places: getting the arguments right. In: Stolton, S. and Dudley, N. (eds.) *Arguments for Protected Areas: Multiple Benefits for Conservation and Use*. Earthscan, London, pp. 251–264.

Dulloo, E., Hunter, D. and Leaman, D.L. (2014). Plant diversity in addressing food, nutrition and medicinal needs. In: Garib-Fakim, A. (ed.) *Novel Plant Bioresources: Application in Food, Medicine and Cosmetics*. Wiley-Blackwell, Chichester, pp. 1–21.

Dulloo, M.E., Guarino, L., Engelmann, F., *et al.* (1999). Complementary conservation strategies for the genus Coffea with special reference to the Mascarene Islands. *Genetic Resources and Crop Evolution*, 45: 565–579.

Dulloo, M.E., Labokas, J., Iriondo, J.M., *et al.* (2008). Genetic reserve location and design. In: Iriondo, J.M., Maxted, N. and Dulloo, E. (eds.) *Plant Genetic Population Management*. CAB International, Wallingford, pp. 23–64.

Dulloo, E., Magos Brehm, J., Kell, P.S, Thormann, I. and Maxted, N. (2017). *Template for the Preparation of a National Strategic Action Plan for the Conservation and Sustainable Use of Crop Wild Relatives*. https://doi.org/10.7910/DVN/QH9XWB, Harvard Dataverse, V1.

Dwivedi, S.L., Britt, A.B., Tripathi, L., *et al.* (2015). Haploids: constraints and opportunities in crop improvement. *Biotechnology Advances*, 33: 812–829.

Dwivedi, S.L., Ceccarelli, S., Grando, S., *et al.* (2017). Diversifying food systems in the pursuit of sustainable food production and healthy diets. *Trends in Plant Science* 22: 842–856.

Dwivedi, S.L., Crouch, J.H., Mackill, D., *et al.* (2007). Molecularization of public sector crop breeding: progress, problems and prospects. *Advances in Agronomy*, 95: 163–318.

Dwivedi, S.L., Perotti, E., Upadhyaya, H.D. and Ortiz, R. (2010). Sexual and asexual (apomixis) plant reproduction in the genomics era: exploring the mechanisms potentially useful in crop plants. *Sexual Plant Reproduction*, 23: 265–279.

Dwivedi, S.L., Stalker, H.T., Blair, M.W., *et al.* (2008). Enhancing crop gene pools with beneficial traits using wild relatives. *Plant Breeding Reviews*, 30: 179–230.

Ebert, A.E., Karihaloo, J.L. and Ferreira, M.E. (2006). Opportunities, limitations and needs for DNA banks. In: Carmen de Vicente, M. and Andersson, M.S. (eds.) *DNA Banks – Providing Novel Options for Gene Banks?* International Plant Genetic Resources Institute, Rome, pp. 61–68.

ECPGR. (2012). *Report of the 13th ECPGR Steering Committee Meeting* held at the Federal Ministry of Agriculture, Forestry, Environment and Water Management, Austria on 4–7 December 2012. Available at: www.ecpgr.cgiar.org/about-ecpgr/steering-committee/13th-sc-meeting/ (accessed 6 March 2015).

ECPGR. (2017). *ECPGR concept for on-farm conservation and management of plant genetic resources for food and agriculture*. European Cooperative Programme for Plant Genetic Resources, Rome.

Egeland, G. and Harrison, G. (2013). Health disparities: promoting Indigenous Peoples' health through traditional food systems and self-determination. In: Kuhnlein, H., Erasmus, B., Spigelski, D. and Burlingame, B. (eds.) *Indigenous Peoples' Food Systems & Well-Being Interventions and Policies for Healthy Communities*. FAO and CINE, Rome.

Ehlenfeldt, M.K. and Ortiz, R. (1995). On the origins of endosperm dosage requirements in *Solanum* and other angiosperma genera. *Sexual Plant Reproduction*, 8: 189–196.

El Bouhssini, M.E., Street, K., Amri, A., *et al.* (2011). Sources of resistance in bread wheat to Russian wheat aphid (*Diuraphis noxia*) in Syria identified using the focused identification of germplasm strategy (FIGS). *Plant Breeding*, 130: 96–97.

El Bouhssini, M.E., Street, K., Joubi, A., Ibrahim, Z. and Rihawi, F. (2009). Sources of wheat resistance to Sunn pest, *Eurygaster integriceps* Puton, in Syria. *Genetic Resources and Crop Evolution*, 56: 1065–1069.

Ellis, R.H. (1988). The viability equation, seed viability nomographs, and practical advice on seed storage. *Seed Science and Technology*, 16: 29–50.

Ellis, R.H, and Jackson, M.T. (1995). Accession regeneration in gene-banks: seed production environment and the potential integrity of seed accessions. *Plant Genetic Resources Newsletter*, 102: 26–28.

Ellis, R.H. and Roberts, E.H. (1980). Improved equations for the prediction of seed longevity. *Annals of Botany*, 45: 13–30.

Ellstrand, N.C. (2003). *Dangerous Liaisons? When Cultivated Plants Mate with Their Wild Relatives*. John Hopkins University Press, Baltimore, MD.

El-Namaky, R., Sedeek, S., Dea Moukooumbi, Y., Ortiz, R. and Manneh, A. (2016). Microsatellite-aided screening for fertility restoration genes (*Rf*) facilitates hybrid improvement. *Rice Science*, 23: 160–164.

Elzinga, C.L., Salzer, D.W., Willoughby, J.W. and Gibbs, J.P. (2001). *Monitoring Plant and Animal Populations*. Blackwell, Malden, MA.

Endresen, D.T.F. (2010). Predictive association between trait data and ecogeographical data for Nordic barley landraces. *Crop Science*, 50: 2418–2430.

Endresen, D.T.F. (2017). Information, knowledge and agricultural biodiversity. In: Hunter, D. Guarino, L., Spillane, C. and McKeown, P.C. (eds.) *Routledge Handbook of Agricultural Biodiversity*. Routledge, Abingdon, pp. 647–661.

Endresen, D.T.F., Street, K., Mackay,M., *et al.* (2012). Sources of resistance to stem rust (Ug99) in bread wheat and durum wheat identified using focused identification of germplasm strategy (FIGS). *Crop Science*, 52: 764–773.

Endresen, D.T.F., Street, K., Mackay, M., Bari, A. and de Pauw, E. (2011). Predictive association between biotic stress traits and ecogeographical data for wheat and barley landraces. *Crop Science*, 51: 2036–2055.

Engelmann, F. (2000). Importance of cryopreservation for the conservation of plant genetic resources. In: Engelmann, F. and Hiroko, T. (eds.) *Cryopreservation of Tropical Plant Germplasm*. Current Research Progress and Application. Japan International Research Centre for Agricultural Sciences, Tsukuba/International Plant Genetic Resources Institute, Rome, pp. 8–22.

Engels, J., Thormann, I. and Metz, T. (2001). A species compendium for plant genetic resources conservation. In: Knueppfer H, Ochsmann, J.(eds.) *Proceedings of Symposium Dedicated to the 100th Birthday of Rudolf Mansfeld*. Schriften zu Genetischen Ressourcen 18. ZADI/IBV, Bonn. www.researchgate.net/publication/270451042_A_species_compendium_for_plant_genetic_resources_conservation

Engels, J.M.M., Dempewolf, H. and Henson-Apollonio, V. (2010). Ethical considerations in agro-biodiversity research, collecting, and use. *Journal of Agriculture and Environmental Ethics*, 24: 107–126.

ENSCONET, (2009). *ENSCONET seed collecting manual for wild species*. Royal Botanic Gardens, Kew, Richmond, UK and Universidad Politéchnica de Madrid, Madrid. Available at: www.kew.org/sites/default/files/ENSCONET_Collecting_protocol_English.pdf (accessed 13 August 2018).

Esquinas-Alcázar, J.T. (1993). Plant genetic resources. In: Hayward, M.D., Bosemark, N.O. and Romagosa, I. (eds.) *Plant Breeding: Principles and Prospects*. Chapman & Hall, London, pp. 33–51.

Esquinas-Alcázar, J. T., Frison, C. and Lopez, F. (2011). Introduction: a treaty to fight hunger – past negotiation, present situation and future challenges. In: Frison, C., Lopez, F. and Esquias-Alcaza, J.T. (eds.) *Plant Genetic Resources and Food Security: Stakeholder Perspectives on the International Treaty on Plant Genetic Resources for Food and Agriculture. Issues in Agricultural Biodiversity*. Earthscan from Routledge, London and New York, pp. 1–23.

EURISCO. (2019). *European Search Catalogue for Plant Genetic Resources (EURISCO)*. Available at: https://eurisco.ipk-gatersleben.de/ (accessed 4 January 2019).

European Commission. (2010). *Monitoring the impact of EU Biodiversity Policy*. Available at: http://ec.europa.eu/environment/pubs/pdf/factsheets/biodiversity_fsh.pdf (accessed 25 October 2014).

European Environment Agency. (2012). *Streamlining European Biodiversity Indicators 2020: Building a Future on Lessons Learnt from the SEBI 2010 Process*. European Environment Agency, Copenhagen. Available at: www.eea.europa.eu/publications/streamlining-european-biodiversity-indicators-2020 (accessed 18 October 2014).

Excoffier, L., Laval, G. and Schneider, S. (2005). ARLEQUIN ver. 3.0: an integrated software package for population genetics data analysis. *Evolutionary Bioinformatics Online*, 1: 47–50.

Excoffier, L., Smouse, P. and Quattro, J. (1992). Analysis of molecular variance inferred from metric distances among DNA haplotypes: application to human mitochondrial DNA restriction data. *Genetics*, 131: 479–491.

Eyzaguirre, P.B and Linares, O.F. (2004). *Home Gardens and Agrobiodiversity*. Smithsonian Books, Washington, DC.

Falconer, D.S. and Mackay, T. (1996). *Introduction to Quantitative Genetics*. Longman Scientific Technical, New York.

Fanzo, J., Hunter, D., Borelli, T. and Mattei, F. (2013). *Diversifying Food and Diets: Using Agricultural Biodiversity to Improve Nutrition and Health. Issues in Agricultural Biodiversity*. Earthscan, London.

FAO. (1989). *Les Ressources Phytogenetiques: Leur Conservation in situ au service des besoins humains*. FAO, Rome.

FAO. (1991). *International Undertaking on Plant Genetic Resources. Annex III*. Food and Agriculture Organization of the United Nations, Rome.

FAO. (1993). *Code of Conduct for Germplasm Collecting and Transfer.* Food and Agriculture Organization of the United Nations, Rome, Italy. Available at: www.fao.org/nr/cgrfa/cgrfa-global/cgrfa-codes/en (accessed 6 March 2015).

FAO. (1995). *Dimensions of Need: An Atlas of Food and Agriculture.* Food and Agriculture Organization of the United Nations, Rome.

FAO. (1996). *Global Plan of Action.* Food and Agriculture Organization of the United Nations, Rome, pp. 1–510.

FAO. (1998). *State of the World's Plant Genetic Resources for Food and Agriculture.* Food and Agriculture Organization of the United Nations, Rome. Available at: www.fao.org/agriculture/crops/thematic-sitemap/theme/seeds-pgr/sow/en/ (accessed 6 March 2015).

FAO. (1999). *Technical Meeting on the Methodology of the World Information and Early Warning System on Plant Genetic Resources.* Food and Agriculture Organization of the United Nations, Rome. Available at: www.apps3.fao.org/views/prague/technical (accessed 13 August 2018).

FAO. (2001). *International Treaty on Plant Genetic Resources for Food and Agriculture.* Food and Agriculture Organization of the United Nations, Rome. Available at: www.fao.org/ag/cgrfa/itpgr.htm (accessed 6 March 2015).

FAO. (2008). *Climate Change and Biodiversity for Food and Agriculture.* Food and Agriculture Organization of the United Nations, Rome.

FAO. (2010a). *Second Report on the State of the World's Plant Genetic Resources for Food and Agriculture.* Food and Agriculture Organization of the United Nations, Rome. Available at: www.fao.org/agriculture/seed/sow2/en/ (accessed 25 July 2013).

FAO. (2010b). *World Programme for the Census of Agriculture.* Food and Agriculture Organization of the UN, Rome.

FAO. (2011b). *Introduction to the International Treaty on Plant Genetic Resources for Food and Agriculture.* Food and Agriculture Organization of the United Nations, Rome.

FAO. (2011a). *Second Global Plan of Action for Plant Genetic Resources for Food and Agriculture.* Food and Agriculture Organization of the United Nations, Rome, Italy. Available at: www.fao.org/docrep/015/i2624e/i2624e00.htm (accessed 6 March 2015).

FAO. (2011c). *Thirteenth Regular Session of the Commission on the Genetic Resources for Food and Agriculture, CGRFA-13/11/Report.* Food and Agriculture Organization of the United Nations, Rome. Available at: www.fao.org/docrep/meeting/024/mc192e.pdf (accessed 6 March 2015).

FAO. (2011d). *Satellite Technology Yields New Forest Loss Estimates.* Food and Agriculture Organisation of the UN, Rome. Available at: www.fao.org/news/story/en/item/95180/icode/ (accessed 1 August 2013).

FAO. (2012a). *Second Report on the Global Plan of Action for Plant Genetic Resources for Food and Agriculture.* Food and Agriculture Organization of the United Nations, Rome.

FAO. (2012b). *Food Security and Climate Change.* High Level Panel of Experts on Food Security and Nutrition Report. Food and Agriculture Organisation of the United Nations, Rome.

FAO. (2012c). *Towards the establishment of a global network for in situ conservation and on-farm management of plant genetic resources for food and agriculture.* Report from Technical Workshop. Available at: http://typo3.fao.org/fileadmin/templates/agphome/documents/PGR/Reports/Report-Technical_workshop_131112.pdf (accessed 11 March 2019).

FAO. (2013b). *Draft Standard Voluntary Reporting Format.* Food and Agriculture Organization of the United Nations, Rome. Available at: www.planttreaty.org/sites/default/files/gb5w18a1e_Reporting_format.pdf (accessed 6 June 2016).

FAO. (2013a). *Global Wheat Rust Monitoring System.* Food and Agriculture Organisation of the UN, Rome. Available at: www.fao.org/agriculture/crops/rust/stem/rust-report/stem-ug99racettksk/en/ (accessed 1 August 2013).

FAO. (2013c). *Towards the establishment of a global network for in situ conservation and on-farm management of PGRFA.* Report of Technical Workshop held in Rome, Italy, 13 November 2012. Food and Agriculture Organization of the United Nations, Rome. Available at: www.fao.org/agriculture/crops/core-themes/theme/seeds-pgr/itwg/6th/technical-workshop/en/ (accessed 5 April 2013).

FAO. (2014a). *Concept Note on Global Networking on in situ Conservation and On-Farm Management of Plant Genetic Resources for Food and Agriculture.* Information document to the 7th Session of the Intergovernmental Working Group on Plant Genetic Resources for Food and Agriculture (CGRFA/WG-PGR-7/14/Inf.3), Commission for Genetic Resources for Food and Agriculture, Rome, Italy. Available at: www.fao.org/3/a-ml477e.pdf (accessed 6 March 2016).

FAO. (2014b). *Gene Bank Standards for Plant Genetic Resources for Food and Agriculture.* Rev. ed. Food and Agriculture Organization of the United Nations, Rome. Available at: www.fao.org/3/a-i3704e.pdf (accessed 6 March 2018).

FAO. (2015a). *Voluntary Guidelines to Support the Integration of Genetic Diversity into National Climate Change Adaptation Planning.* Commission on Genetic Resources for Food and Agriculture and Food and Agriculture Organization of the United Nations, Rome, Italy. Available at: www.fao.org/3/a-ml4940e.pdf (accessed 9 June 2016).

FAO. (2015c). *Reporting Format for Monitoring the Implementation of the Second Global Plan of Action for*

Plant and Genetic Resources for Food and Agriculture. Food and Agriculture Organization of the United Nations, Rome. Available at: www.fao.org/3/a-ml478e.pdf (accessed 9 June 2015).

FAO. (2015b). *Guidelines for Developing a National Strategy for Plant Genetic Resources for Food and Agriculture.* Food and Agriculture Organization of the United Nations, Rome. Available at: www.fao.org/3/a-i4917e.pdf (accessed 13 April 2019).

FAO. (2016). *FAOSTAT.* Food and Agriculture Organization of the United Nations, Rome. Available at: http://faostat.fao.org/site/339/default.aspx (accessed 11 May 2016).

FAO. (2019). *First State of the World Report on Biodiversity for Food and Agriculture.* Food and Agriculture Organization of the United Nations, Rome.

Farming Matters. (2016). *Access and Benefit Sharing of Genetic Resources: Making it Work for Family Farmers.* ILEIA, The Netherlands.

Farooq, S. and El-Azam, F. (2004). Co-existence of salt and drought tolerance in Triticeae. *Hereditas*, 135: 205–210.

Fatihah, N.H.N., Maxted, N. and Rico Acre, L. (2012). Taxonomic study of *Psophocarpus* Neck. ex DC. (Leguminosae, Papilionoideae). *South African Journal of Botany*, 83: 78–88.

Fay, M. (2003). Using genetic data to help guide decisions about sampling. In: Smith, R.D., Dickie, J.B., Linington, S.H., Pritchard, H.W. and Probert, R.J. (eds.) *Seed Conservation: Turning Science into Practice.* Royal Botanic Gardens, Kew, Richmond, pp. 89–96.

Fay, M., Qamaruz-Zaman, F., Chase, M.W. and Samual, R. (2004). Military and Monkey Orchids – What Do We Have in England? *English Nature Research Reports, No. 607.* Proceedings of a Conservation Genetic Workshop held at the Royal Botanic Gardens, Kew, 27 November 2001. Natural England, Peterborough.

Felsenstein, F. (2016). *Theoretical Evolutionary Genetics.* University of Washington, Seattle, WA.

Ferguson, M.E., Ford-Lloyd, B.V., Robertson, L.D., Maxted, N. and Newbury, H.J. (1998a). Mapping the geographical distribution of genetic variation in the genus *Lens* for the enhanced conservation of plant genetic diversity. *Molecular Ecology*, 7: 1743–1755.

Ferguson, M.E., Robertson, L.D., Ford-Lloyd, B.V., Newbury, H.J. and Maxted, N. (1998b). Contrasting genetic variation amongst lentil landraces from different geographical origins. *Euphytica*, 102: 265–273.

Fisher, R.A. (1930). *The Genetical Theory of Natural Selection.* Claredon Press, Oxford.

Fitzgerald, H., Korpelainen, H. and Veteläinen, M. (2016). Developing a crop wild relative strategy for Finland. In: Maxted, N., Dulloo, E.M. and Ford-Lloyd, B.V. (eds.) *Enhancing Crop Gene Pool Use: Capturing Wild Relative and Landrace Diversity for Crop Improvement.* CAB International, Wallingford, pp. 206–216.

Foden, W.B., Butchart, S.H.M., Stuart, S.N., *et al.* (2013). Identifying the world's most climate change vulnerable species: a systematic trait-based assessment of all birds, amphibians and corals. *PLoS One* 8(6): e65427. https://doi.org/10.1371/journal.pone.0065427.

Foden, W.B., Mace, G., Vié, J.-C., *et al.* (2009). Species susceptibility to climate change impacts. In: Vié, J.-C., Hilton-Taylor, C. and Stuart, S.N. (eds.). *Wildlife in a Changing World – An Analysis of the 2008 IUCN Red List of Threatened Species.* IUCN, Gland, pp. 77–88.

Foley, M. and Clarke, S. (2005). *Orchids of the British Isles.* Griffin Press, Cheltenham, UK.

Foley, J.A., Ramankutty, N., Brauman, K.A., *et al.* (2011). Solutions for a cultivated planet. *Nature* 478: 337–342.

Ford-Lloyd, B. and Maxted, N. (1993). Preserving diversity. *Nature*, 361: 579.

Ford-Lloyd, B.V., Engels, J.M.M. and Jackson, M. (2014). Genetic resources and conservation challenges under the threat of climate change. In: Jackson, M., Ford-Lloyd, B.V. and Parry, M. (eds.) *Plant Genetic Resources and Climate Change – A 21st Century Perspective.* CAB International, Wallingford, pp. 16–37.

Ford-Lloyd, B.V., Kell, S.P. and Maxted, N. (2008). Establishing conservation priorities for crop wild relatives. In: Maxted, N., Ford-Lloyd, B.V., Kell, S.P., *et al.* (eds.) *Crop Wild Relative Conservation and Use.* CAB International, Wallingford, pp. 110–119.

Franco, J. and Crossa, J. (2005). The modified location model for classifying genetic resources. I: Association between categorical and continuous variables. *Crop Science*, 42: 1719–1726.

Franco, J., Crossa, J., Taba, S. and Shands, H. (2003). A multivariate method for classifying cultivars and studying group × environment × trait Interaction. *Crop Science*, 43: 1249–1258.

Franco, J., Crossa, J., Villaseñor, J., Taba, S. and Eberhart, S.A. (1998). Classifying genetic resources by categorical and continuous variables. *Crop Science*, 38: 1688–1696.

Franco, J., Crossa, J., Villaseñor, J., *et al.* (1999). A two stages, three-way method for classifying genetic resources in multiple environments. *Crop Science*, 39: 259–267.

Frankel, O.H. and Bennett, E. (1970). *Genetic Resources in Plants – Their Exploration and Conservation.* Blackwell, Oxford.

Frankel, O.H. and Brown, A.H.D. (1984). Current plant genetic resources – a critical appraisal. In: Bansal, A.H.C., Chopra, V.L., Joshi, B.C. and Sharma, R.P. (eds.) *Genetics: New Frontiers (vol. IV).* Oxford & IBH Publishing, New Delhi.

Frankel, O.H., Brown, A.H.D. and Burdon, J.J. (1995). *The Conservation of Plant Biodiversity.* Cambridge University Press, Cambridge.

Frankel, O.H. and Soulé, M.E. (1981). *Conservation and Evolution.* Cambridge University Press, Cambridge, UK.

Frankham, R., Ballou, J.D. and Briscoe, D.A. (2002). *Introduction to Conservation Genetics*. Cambridge University Press, Cambridge.

Frankham, R., Bradshaw, C.J.A. and Brook, B.W. (2014). Genetics in conservation management: revised recommendations for the 50/500 rules, Red List criteria and population viability analyses. *Biological Conservation*, 170: 56–63. https://doi.org/10.1016/j.biocon.2013.12.036.

Franklin, I.R. (1980). Evolutionary change in small populations. In: Soulé, M.E. and Wilcox, B.A. (eds.) *Conservation Biology: An Evolutionary-Ecological Perspective*. Sinauer Associates, Sunderland, MA, pp. 135–149.

Franzén, M., Schweiger, O. and Betzholtz, P.-E. (2012). Species-area relationships are controlled by species traits. *PLoS ONE*, 7(5): e37359. https://doi.org/10.1371/journal.pone.0037359.

Freese, C.H. (1998). *Wild Species As Commodities: Managing Markets and Ecosystems for Sustainability*. Island Press, Washington, DC.

Friis-Hansen, E. and Sthapit, B. (2000). *Participatory Approaches to the Conservation and Use of Plant Genetic Resources*. International Plant Genetic Resources Institute (IPGRI), Rome.

Frison, E., Cherfas, J. and Hodgkin, T. (2011). Agricultural biodiversity is essential for a sustainable improvement in food and nutrition security. *Sustainability* 3: 238–253.

Frodin, D.G. (2001). *Guide to the Standard Floras of the World*, 2nd ed. Cambridge University Press, Cambridge.

Frodin, D.G. (2011). *Guide to Standard Floras of the World: An Annotated, Geographically Arranged Systematic Bibliography of the Principal Floras, Enumerations, Checklists and Chorological Atlases of Different Areas*, 2nd ed. Cambridge University Press, Cambridge.

Frost, G.H. and Bond, I. (2008). The CAMPFIRE programme in Zimbabwe: Payments for wildlife services. *Ecological Economics*, 65: 776–787.

Gabrielian, E. and Zohary, D. (2004). Wild relatives of food crops native to Armenia and Nakhichevan. *Flora Mediterranea*, 14: 5–80.

Galvin, M. and Haller, T. (eds.) (2008). *People, Protected Areas and Global Change: Participatory Conservation in Latin America, Africa, Asia and Europe, Perspectives of the Swiss National Centre of Competence in Research (NCCR) North-South, vol 3*. University of Bern, Geographica Bernensia, Bern.

Gao, L., Chen, W., Jiang, W., *et al.* (2000). Genetic erosion in Northern marginal population of the common wild rice *Oryza rufipogon* Griff. and its conservation, revealed by the change of population genetic structure. *Hereditas*, 133: 47–53.

Garnett, S.T., Burgess, N.D., Fa, J.E., *et al.* (2018). A spatial overview of the global importance of indigenous lands for conservation. *Nature Sustainability*, 1: 369–374.

Gauch, H.G. (2006). Winning the accuracy game. *American Scientist*, 94: 133–141.

Gebauer, J., Adam, Y.O., Sanchez, A.C., *et al.* (2016). Africa's wooden elephant: the baobab tree (*Adonsonia digitata* L.) in Sudan and Kenya: a review. *Genetic Resources and Crop Evolution*, 63: 377–399.

Gerard, J. (1597). *The Herball or Generall Historie of Plantes*, 1st ed. John Norton, London.

Gillespie, J.H. (1998). *Population Genetics – A Concise Guide*. John Hopkins University Press, Baltimore, MD/London.

Gillman, M. (1997). Plant population ecology. In: Maxted, N., Ford-Lloyd, B.V. and Hawkes, J.G. (eds.) *Plant Genetic Conservation: The* in situ *Approach*. Chapman & Hall, London, pp. 181–185.

Global Witness. (2016). *On Dangerous Ground*. Global Witness, London.

Glowka, L., Burhenne-Guilmin, F., Synge, H., McNeely, J.A., Gündling, L. (1994). *A Guide to the Convention on Biological Diversity*. IUCN Environmental Law and Policy Paper No. 30.

Godfray, H.C.J., Beddington, J.R., Crute, I.R., *et al.* (2010). Food security: the challenge of feeding 9 billion people. *Science*, 327: 812–818.

Goldsmith, F.B. (1991). Vegetation monitoring. In: Goldsmith, F.B. (ed.) *Monitoring for Conservation and Ecology*. Chapman & Hall, London, pp. 77–86.

González-Orozco, C., Brown, A., Knerr, N., Miller, J. and Doyle, J. (2012). Hotspots of diversity of wild Australian soybean relatives and their conservation *in situ*. *Conservation Genetics*. https://doi.org/10.1007/s10592-012-0370-x.

Gotor, E. and Irungu, C. (2010). The impact of bioversity international's African leafy vegetables programme in Kenya. *Impact Assessment Project Appraisal* 28: 41–55.

Gottfried, M., Pauli, H., Futschik, A., *et al.* (2012). Continent-wide response of mountain vegetation to climate change. *Nature Climate Change*, 2: 111–115.

Gough, R. and Moore-Gough, C. (2011). *The Complete Guide to Saving Seeds*. Storey Publishing, North Adams, MA.

Graner, A., Andersson, M.S. and de Vicente, M.C. (2006). A model of DNA banking to enhance the management, distribution and use of *ex situ* stored PGR. In: de Vicente, M.C. and Andersson, M.S. (eds.) *DNA Banks – Providing Novel Options for Gene Banks?* Topical reviews in Agricultural Biodiversity. International Plant Genetic Resources Institute, Rome, pp. 69–75.

Green, N., Campbell, G., Tulloch, R. and Scholten, M. (2009). Scottish landrace protection scheme. In: Veteläinen, M., Negri, V. and Maxted, N. (eds.) *European Landraces: On-farm Conservation, Management and Use*. Bioversity Technical Bulletin 15. Bioversity International, Rome, pp. 233–243.

Gregory, P.J., Johnson, S.N., Newton, A.C. and Ingram, J.S.I. (2009). Integrating pests and pathogens into the climate change/food security debate. *Journal of Experimental Botany*, 60: 2827–2838.

Grobman, A., Salhuana, W., Sevilla, R. and Mangelsdorf, P.C. (1961). *Races of maize in Peru: their origins, evolution and classification*. National Academy of Sciences – National Research Council Publication 915, Washington, DC.

Groombridge, B. and Jenkins, M.D. (2000). *Global Biodiversity: Earth's Living Resources in the 21st Century*. Prepared for UNEP World Conservation Monitoring Centre. World Conservation Press, Cambridge.

Groombridge, B. and Jenkins, M. (2002) *World Atlas of Biodiversity*. University of California Press, Berkeley, CA.

Groot, S.P.C., de Groot, L., Kodde, J. and vanTreuren, R. (2015). Prolonging the longevity of *ex situ* conserved seeds by storage under anoxia. *Plant Genetic Resources Characterization and Utilization*, 13: 18–26.

Guarino, L. (1995). Mapping the eco-geographic distribution of biodiversity. In: Guarino, L., Rao, V.R. and Reid, R. (eds.) *Collecting Plant Genetic Diversity: Technical Guidelines*. CAB International, Wallingford, pp. 287–315.

Guarino, L. and Friis-Hansen, E. (1995). Collecting plant genetic resources and documenting associated indigenous knowledge in the field: a participatory approach. In: Guarino, L., Ramanatha Rao, V. and Reid, R. (eds.) *Collecting Plant Genetic Diversity: Technical Guidelines*. CAB International, Wallingford, pp. 345–366.

Guarino, L., Jarvis, A., Hijmans, R.J. and Maxted, N. (2002). Geographic Information Systems (GIS) and the Conservation and Use of Plant Genetic Resources. In: Engels, J.M.M., Ramanatha Rao, V., Brown, A.H.D. and Jackson, M.T. (eds.) *Managing Plant Genetic Diversity*. International Plant Genetic Resources Institute (IPGRI), Rome, pp. 387–404.

Guarino, L., Maxted, N. and Chiwona, E.A. (2006). *A Methodological Model for Ecogeographic Surveys of Crops*. IPGRI Technical Bulletin No. 9. IPGRI, Rome, pp. 1–58.

Guarino, L., Ramanatha Rao, V. and Goldberg, E. (eds.) (2012). *Collecting Plant Genetic Diversity: Technical Guidelines. 2011 Update*. Bioversity International, Rome. Available at: https://cropgene bank.sgrp.cgiar.org/index .php/procedures-mainmenu-242/collecting (accessed 13 August 2018).

Guarino, L., Ramanatha Rao, V. and Reid, R. (eds.) (1995). *Collecting Plant Genetic Diversity: Technical Guidelines*. CAB International, Wallingford.

Guisan, A. and Zimmermann, N.E. (2000). Predictive habitat distribution models in ecology. *Ecological Modelling*, 135: 147–186.

Guo, Q.F. (2014). Species invasions on islands: searching for general patterns and principles. *Landscape Ecology*, 29: 1123–1131.

Guralnick, R. (2007). Differential effects of past climate warming on mountain and flatland species distributions: a multispecies North American mammal assessment. *Global Ecological Biogeography*, 16: 14–23.

Haddad, N. (2000). Corridor length and patch colonization by a butterfly, *Junonia coenia*. *Conservation Biology*, 14: 738–745.

Hajjar, R. and Hodgkin, T. (2007). The use of wild relatives in crop improvement: a survey of developments over the last 20 years. *Euphytica*, 156: 1–13.

Haldane, J.B.S. (1932). *The Causes of Evolution*. Longmans, Green, & Co., London.

Hamilton, A. and Hamilton, P. (2006). *Plant Conservation: An Ecosystems Approach*. Earthscan, London.

Hammer, K. (1990). Botanical checklists prove useful in research programmes on cultivated plants. *Diversity*, 6(3–4): 31–34.

Hammer, K. (2001). Contributions of home gardens to our knowledge of cultivated plant species: the Mansfeld approach. In: Watson, J.W. and Eyzaguirre, P.B. (eds.) *Home Gardens and in situ Conservation of PGR in Farming Systems*. International Plant Genetic Resources Institute, Rome.

Hammer, K., Laghetti, G. and Perrino, P. (1999). A checklist of the cultivated plants of Ustica (Italy). *Genetic Resources and Crop Evolution* 46: 95–106.

Hamon, S., Dussert, S., Noirot, M., Anthony, F. and Hodgkin, T. (1995). Core collections: accomplishment and challenges. *Plant Breeding Abstracts*, 65: 1125–1133.

Hamrick, J.L. and Godt, M. (1996). Effects of life history traits on genetic diversity in plant species. *Philosophical Transactions of the Royal Society B: Biological Sciences*, 351: 1291–1298.

Hanna, W.W. and Towill, L.E. (1995). Long-term pollen storage. *Plant Breeding Reviews*, 13: 179–207.

Hannah, L. (2008). Protected areas and climate change. *Annals of the New York Academy of Sciences*, 1134: 201–212.

Hannan, R. and Hellier, B.C. (1999). Temperate legume conservation. In: Pavek, D.S., Lamboy, W.F. and Garvey, E.J. (eds.) *Ecogeographic study of Vitis species: Final Report for Caloosa and Sweet Mountain Grapes*. Unpublished Report, USDA Pullman.

Hanson, J. (2011). Forage grass genetic resources. In: Guarino, L., Ramanatha Rao, V., Goldberg, E. (eds.) *Collecting Plant Genetic Diversity: Technical Guidelines. 2011 Update*. Bioversity International, Rome. Available at: https://cropgenebank.sgrp.cgiar.org/index .php/crops-mainmenu-367/forage-grasses-mainmenu-27 (accessed 13 August 2018).

Hardin, G. (1968). The tragedy of the commons. *Science*, 162: 1243–1248.

Hardon, J. (1996). The global context: breeding and crop genetic diversity. In: Eyzaguirre, P. and Iwanaga, M. (eds.)

Participatory Plant Breeding. International Plant Genetic Resources Institute, Rome, pp. 1–2.

Hardy, G.H. (1908). Mendelian proportions in a mixed population. *Science*, 28: 49–50.

Hargreaves, S., Maxted, N., Hirano, R., *et al.* (2010). Islands as refugia of *Trifolium repens* genetic diversity. *Conservation Genetics*, 11: 1317–1326.

Harker, D., Libby, G., Harker, K., Evans, S., and Evans, M. (1999). *Landscape Restoration Handbook*. Lewis Publishers, Boca Raton, FL.

Harlan, J.R. and de Wet, J.M.J. (1971). Towards a rational classification of cultivated plants. *Taxon*, 20: 509–517.

Harper, J.L. (1977). *Population Biology of Plants*. Academic Press, London.

Harris, J.A. (1911). The biometric proof of the pure line theory. *American Naturalist*, 45: 346–363.

Harris, J.G. and Woolf Harris, M. (2001). *Plant Identification Terminology: An Illustrated Glossary*. Spring Lake Publishing, Payson, UT.

Hawkes, J.G. (1978). *Conservation and Agriculture*. Duckworth, London.

Hawkes, J.G. (1980). *Crop Genetic Resources Field Collection Manual*. IBPGR/EUCARPIA, Rome, pp. 1–37.

Hawkes, J.G. (1983). *The Diversity of Crop Plants*. Harvard University Press, Cambridge, MA.

Hawkes, J.G., Maxted, N., Ford-Lloyd, B.V. (2000) *The ex situ Conservation of Plant Genetic Resources*. Kluwer Academic Publishers, Dordrecht.

Hawkes, J.G., Maxted, N. and Zohary, D. (1997). Reserve design. In: Maxted, N., Ford-Lloyd, B.V. and Hawkes, J.G. (eds.) *Plant Genetic Conservation: The in situ Approach*. Chapman & Hall, London, pp. 210–230.

Hawksworth, D.L. and Kalin-Arroya, M.T. (1995). Magnitude and distribution of biodiversity. In: Heywood, V.H. (ed.) *Global Biodiversity Assessment*. Cambridge University Press, Cambridge, pp. 107–191.

Hawtin, G. and Fowler, L. (2011). The global crop diversity trust. In: Frison, C., López, F. and Esquinas-Alcázar, J.T. (eds.) *Plant Genetic Resources and Food Security*. FAO, Biodiversity International and Earthscan, Abingdon, pp. 209–221.

Hay, F.R. and Probert, R.J. (2011). Collecting and handling seeds in the field. In: Guarino, L., Ramanatha Rao, V. and Goldberg, E. (eds.) *Collecting Plant Genetic Diversity: Technical Guidelines. 2011 Update*. Bioversity International, Rome. Available at: http://cropgene bank .sgrp.cgiar.org/index.php?option=com_content&view= article&tid=655 (accessed 13 August 2018).

Hayek, L.C. and Buzas, M.A. (1997). *Surveying Natural Populations*. Columbia University Press, New York.

Hegay, S., Geleta, M., Bryngelsson, T., *et al.* (2014). Introducing host plant resistance to anthracnose in Kyrgyz beans through inoculation-based and marker-aided selection. *Plant Breeding*, 133: 86–91.

Hegay, S., Geleta, M., Bryngelsson, T., *et al.* (2013a). Comparing genetic diversity and population structure of common beans grown in Kyrgyzstan using microsatellites. *Scientific Journal of Crop Science*, 1: 63–75.

Hegay, S., Ortiz, R, Gustavsson, L., Persson, H. and Geleta, M. (2013b). Marker-aided breeding for resistance to bean common mosaic virus in Kyrgyz bean cultivars. *Euphytica*, 193: 67–78.

Heinonen, M. (2016). Landrace inventories and recommendations for *in situ* conservation in Finland. In: Maxted, N., Ehsan Dulloo, M. and Ford-Lloyd, B.V. (eds.) *Enhancing Crop Gene Pool Use: Capturing Wild Relative and Landrace Diversity for Crop Improvement*. CAB International, Wallingford, pp. 335–341.

Hellawell, J.M. (1991). Development of a rationale for monitoring. In: Goldsmith, F.B. (ed.) *Monitoring for Conservation and Ecology*. Chapman & Hall, London, pp. 1–14.

Hellier, B.C. (2000). *Genetic, morphologic, and habitat diversity of two species of* Allium *native to the Pacific Northwest, USA and their implications for* in situ *seed collection for the National Plant Germplasm System*. MSc thesis, Washington State University, Pullman, WA.

Hernandez, P., Graham, C.G., Master, L. and Albert, D. (2006). The effect of sample size and species characteristics on performance of different species distribution modelling models. *Ecography*, 29: 773–785.

Hertel, T.W., Burke, M.B. and Lobell, D.B. (2010). The poverty implications of climate-induced crop yield changes by 2030. *Global Environmental Change*, 20: 577–585.

Heslop-Harrison, J.S. and Schwarzacher, T. (2007). Domestication, genomics and the future for banana. *Annals of Botany*, 100: 1073–1084.

Heywood, V.H. (1987). The changing role of botanic gardens. In: Bramwell, D., Heywood, V.H. and Synge, H. (eds.) *Botanic Gardens and the World Conservation Strategy*. Academic Press, London, pp. 13–18.

Heywood, V.H. (1994). The measurement of biodiversity and the politics of implementation. In: Forey, P.L., Humphries, C.J. and Vane-Wright, R.I. (eds.) *Systematics and Conservation Evaluation*. Systematic Association Special Vol. 50. Oxford University Press, Oxford, pp. 15–22.

Heywood, V.H. (2011). The role of botanic gardens as resource and introduction centres in the face of global change. *Biodiversity and Conservation*, 20: 221–239.

Heywood, V.H. (2013). Overview of agricultural biodiversity and its contribution to nutrition and health. In: Fanzo, J., Hunter, D., Borelli, T., et al. (eds.) *Diversifying Food and Diets: Using Agricultural Biodiversity to Improve Nutrition and Health. Issues in Agricultural Biodiversity*. Earthscan, London, pp. 35–67.

Heywood, V.H. and Watson, R.T. (1995). *Global Biodiversity Assessment*. Cambridge University Press, Cambridge.

Hickey, M. and King, C. (2000). *The Cambridge Illustrated Glossary of Botanical Terms*. Cambridge University Press, Cambridge.

Hijmans, R.J., Cameron, S.E., Parra, J.L., Jones, P.G. and Jarvis, A. (2005). Very high resolution interpolated climate surfaces for global land areas. *International Journal of Climatology*, 25: 1965–1978.

Hijmans, R.J. and Graham, C.H. (2006). The ability of climate envelope models to predict the effect of climate change on species distributions. *Global Change Biology*, 12: 2272–2281.

Hijmans, R.J., Spooner, D., Salas, A., Guarino, L. and de La Cruz, J. (2002). *Atlas of Wild Potatoes*. Systematic and Ecogeographic Studies on Crop Gene pools 10. International Plant Genetic Resources Institute, Rome.

Hjalmarsson, I. and Ortiz, R. (1998). Effect of genotype and environment on vegetative and reproductive characteristics of lingonberry (*Vaccinium vitis-idaea* L.). *Acta Agriculturæ Scandinavica (Section B Soil and Plant Sciences)*, 48: 255–262.

Hjalmarsson, I. and Ortiz, R. (2001). Lingonberry: botany and horticulture. *Horticultural Reviews*, 27: 79–123.

Hodgkin, T., Brown, A.H.D., van Hintum, Th.J.L. and Morales, E.A.V. (eds.) (1994). *Core Collections of Plant Genetic Resources*. Wiley, Chichester.

Hoekstra, F.A. (1995). Collecting pollen for genetic resources conservation. In: Guarino, L., Ramanatha Rao, V. and Reid, R. (eds.) *Collecting Plant Genetic Diversity: Technical Guidelines*. CAB International, Wallingford. pp. 527–550.

Holmes, B. (2015). Quiet revolutions. *New Scientist*, 31 October, 31–35.

Holsinger, K.E. (2015). *Lecture Notes in Population Genetics*. University of Connecticut, Storrs, CT.

Hong, T.D., Linington, S.H. and Ellis, R.H. (1996). *Seed storage behaviour: a compendium. Handbooks for Gene banks 4*. International Board for Plant Genetic Resources, Rome.

Honnay, O., and van Nieuwenhuyse, A. (2018). Biodiversity and human health: mechanisms and evidence of the positive health effects of diversity in nature and green spaces. *British Medical Bulletin*, 127: 5–22.

Hopkins, J. and Maxted, N. (2010). *Crop Wild Relatives: Plant Genetic Conservation for Food Security*. Natural England, Peterborough.

Houde, A.L.S., Garner, S.R. and Neff, B.D. (2015). Restoring species through reintroductions: strategies for source population selection. *Restoration Ecology*, 23: 746– 753.

House of Lords, (2002). *What on Earth? The Threat to the Science Underpinning Conservation*. Select Committee appointed to consider Science and Technology, House of Lords, London. Available at: www.publications .parliament.uk/pa/ld200102/ldselect/ldsctech/118/11802 .htm (accessed 30 August 2011).

Huamán, Z., Aguilar, C. and Ortiz, R. (1999). Selecting a Peruvian sweetpotato core collection on the basis of morphological, eco-geographical, and disease and pest reaction data *Theoretical and Applied Genetics*, 98: 840–844.

Huamán, Z., de la Puente, F. and Arbizu, C. (1995). Collecting vegetatively propagated crops (especially roots and tubers). In: Guarino, L., Ramanatha Rao, V. and Reid, R. (eds.) *Collecting Plant Genetic Diversity: Technical Guidelines*. CAB International, Wallingford, pp. 457–466.

Huamán, Z., Ortiz, R. and Gómez. R. (2001). Selecting a *Solanum tuberosum* subsp. *andigena* core collection using morphological, geographical, disease and pest descriptors. *American Journal of Potato Research*, 77: 183–190.

Huamán, Z., Ortiz, R., Zhang, D. and Rodríguez, F. (2000). Isozyme analysis of entire and core collections of *Solanum tuberosum* subsp. *andigena* potato cultivars. *Crop Science*, 40: 273–276.

Hubert, B., Rosegrant, M., van Boekel, M.A.J.S. and Ortiz, R. (2010). The future of food: scenarios for 2050. *Crop Science*, 50: S33–S50.

Hughes, C.E. (1998). *Leucaena. A Genetic Resources Handbook*. Tropical Forestry Papers, 37. Oxford Forestry Institute, Department of Plant Sciences, University of Oxford, Oxford.

Hunter, D. and Heywood, V.H. (eds.) (2011). *Crop Wild Relatives: A Manual of in situ Conservation*. Issues in Agricultural Biodiversity. Earthscan, London.

Hunter, D., Burlingame, B. and Remans, R. (2015). Biodiversity and nutrition. In: Inís Communication (ed.) *Connecting Global Priorities: Biodiversity and Human Health, a State of Knowledge Review*. Convention on Biological Diversity/World Health Organization, Geneva, Switzerland.

Hunter, D., Guarino, L., Spillane, C. and McKeown, P. (eds.) (2017). *Routledge Handbook of Agricultural Biodiversity*. Routledge – Taylor and Francis Group, London and New York.

Hunter D., Iosefa, T., Delp, C.J. and Fonoti, P. (2001). *Beyond taro leaf blight: a participatory approach for plant breeding and selection for taro improvement in Samoa*. Proceedings of the International Symposium on Participatory Plant Breeding and Participatory Plant Genetic Resource Enhancement, Pokhara, Nepal, 1–5 May 2000. CGIAR Systemwide Program on Participatory Research and Gender Analysis for Technology Development and Institutional Innovation, Centro Internacional de Agricultura Tropical, Cali, pp. 219–227.

Hunter, D., Özkan, I., Beltrame D.M., *et al.* (2016). Enabled or disabled: is the environment right for using biodiversity to improve nutrition. *Frontiers in Nutrition* 3: 1–6

Hunter, M.L. (1990). *Wildlife, Forests and Forestry: Principles of Managing Forests for Biological Diversity*. Prentice Hall, Englewood Cliffs, NJ.

IAASTD. (2008). *Agriculture at a Crossroads: The Synthesis Report*. International Assessment of Agricultural

Knowledge, Science and Technology for Development, Washington, DC.

IBPGR. (1991). *Dictionary of Plant Genetic Resources.* Elsevier Science Publishing, New York.

Ickowitz, A., Rowland, D., Powell, B., Salim, M.A. and Sunderland, T. (2016). Forests, trees, and micronutrient-rich food consumption in Indonesia. *PLoS One* 11: e0154139. https://doi.org/10.1371/journal.pone.0154139.

IIED (1997). *Valuing the Hidden Harvest: Methodological Approaches for the Local-Level Economic Analysis of Wild Resources.* Sustainable Agriculture Programme Research Series, Volume 3, Number 4. Sustainable Agriculture Programme. IIED, London.

Iltis, H.H., Doebley, J.F., Guzmán, R.M. and Pazy, B. (1979). *Zea diploperennis* (Gramineae): a new teosinte from Mexico. *Science,* 203: 186–188.

Ingram, V., Vinceti, B. and van Vliet, N. (2017). Wild plant and animal genetic resources. In: Hunter, D., Guarino, L., Spillane, C. and McKeown, P. (eds.) *Handbook of Agriculture Biodiversity.* Routledge, Abingdon, pp. 65–85.

Iosefa, T., Taylor, M., Hunter, D. and Tuia, V. (2013). Supporting farmers' access to the global gene pool and participatory selection in taro in the Pacific. In: De Boef, W.S., Peroni, N., Subedi, A. and Thijssen, M.H. (eds.) *Community Biodiversity Management: Promoting Resilience and the Conservation of Plant Genetic Resources.* Earthscan, London, pp. 285–289.

IPBES. (2019). *Global Assessment Report on Biodiversity and Ecosystem Services.* Intergovernmental Science-Policy Platform on Biodiversity and Ecosystem Services. UNESCO, Paris.

IPC. (2015). *Biodiversity for Food and Agriculture: The Perspectives of Small-Scale Food Providers.* Thematic Study for FAO's Report 'State of the World's Biodiversity for Food and Agriculture'. International Planning Committee for Food Sovereignty, Agricultural Biodiversity Working Group.

IPCC (Intergovernmental Panel on Climate Change). (2007). Summary for policymakers. In: Parry, M.L., Canziani, O.F., Palutikof, J.P., van der Linden, P.J. and Hanson, C.E. (eds.) *Climate Change 2007: Impacts, Adaptation and Vulnerability.* Contribution of Working Group II to the Fourth Assessment Report of the Intergovernmental Panel on Climate Change. Cambridge University Press: Cambridge, pp. 7–22.

IPCC (Intergovernmental Panel on Climate Change). (2013). Summary for policymakers. In: Stocker, T.F., Qin, D., Plattner, G.-K., *et al.* (eds.) *Climate Change 2013: The Physical Science Basis.* Contribution of Working Group I to the Fifth Assessment Report of the Intergovernmental Panel on Climate Change. Cambridge University Press, Cambridge and New York.

IPCC (Intergovernmental Panel on Climate Change). (2014a). Summary for policymakers. In: Field, C.B., Barros, V.R.,

Dokken, D.J., *et al.* (eds.) *Climate Change 2014: Impacts, Adaptation, and Vulnerability. Part A: Global and Sectoral Aspects.* Contribution of Working Group II to the Fifth Assessment Report of the Intergovernmental Panel on Climate Change. Cambridge University Press, Cambridge, UK, pp. 1–32.

IPCC. (2014b). *Climate Change 2014: Synthesis Report: Longer Report.* Available at: www.ipcc.ch/pdf/assessment-report/ar5/syr/SYR_AR5_LONGERREPORT.pdf (accessed 3 November 2014).

IPES-Food. (2016). *From Uniformity to Diversity: A Paradigm Shift from Industrial Agriculture to Diversified Agroecological Systems.* International Panel of Experts on Sustainable Food Systems, Louvain-la-Neuve.

IPGRI. (1991). *Elsevier's Dictionary of Plant Genetic Resources.* Elsevier, Amsterdam.

IPGRI. (1993). *Diversity for Development.* International Plant Genetic Resources Institute, Rome.

IPGRI. (1997). *Annual Report.* International Plant Genetic Resources Institute, Rome.

Iriondo, J.I., Fielder, H., Fitzgerald, H., *et al.* (2016). National strategies for the conservation of crop wild relatives. In: Maxted, N., Dulloo, E.M. and Ford-Lloyd, B.V. (eds.) *Enhancing Crop Gene Pool Use: Capturing Wild Relative and Landrace Diversity for Crop Improvement.* CAB International, Wallingford, pp. 161–171.

Iriondo, J.M, Ford-Lloyd, B.V., De Hond, L., *et al.* (2008). Plant population monitoring methodologies for the *in situ* genetic conservation of CWR. In: Iriondo, J.M., Maxted, N. and Dulloo, E. (eds.) *Plant Genetic Population Management.* CAB International, Wallingford, pp. 88–123.

Iriondo, J.M., Maxted, N. and Dulloo, E. (eds.) (2008). *Conserving Plant Genetic Diversity in Protected Areas: Population Management of Crop Wild Relatives.* CAB International, Wallingford.

Iriondo, J.M., Maxted, N., Kell, S.P., *et al.* (2012). Quality standards for genetic reserve conservation of crop wild relatives. In: Maxted, N., Dulloo, M.E., Ford-Lloyd, B.V., *et al.* (eds.) *Agrobiodiversity Conservation: Securing the Diversity of Crop Wild Relatives and Landraces.* CAB International, Wallingford, pp. 72–77.

IRRI. (2017). The International Rice Gene Bank. International Rice Research Institute, Los Baños, Philippines. http://irri.org/our-work/research/genetic-diversity/international-rice-gene bank

IUCN. (1994). *The Convention on Biological Diversity: An Explanatory Guide.* Prepared by the IUCN Environmental Law Centre, Bonn.

IUCN. (2001). *IUCN Red List Categories and Criteria. Version 3.1.* IUCN Species Survival Commission. IUCN, Gland, Switzerland and Cambridge, UK. Available at: www.iucnredlist.org/documents/redlist_cats_crit_en.pdf (accessed February 2013).

IUCN. (2003). *Guidelines for Application of IUCN Red List Criteria at Regional Levels, Version 3.0.* IUCN Species Survival Commission. IUCN, Gland and Cambridge, UK. ii + 26 pp. Available from: www.iucnredlist.org/documents/reg_guidelines_en.pdf (accessed 1 August 2013).

IUCN. (2012). *Why Do We Need Biodiversity Indicators?* IUCN Species Survival Commission, Gland, Cambridge. Available at: www.iucn.org/about/work/programmes/species/our_work/biodiversity_indicators/ (accessed 1 June 2016).

IUCN. (2016). *IUCN Red List of Threatened Species.* IUCN Species Survival Commission. IUCN, Gland, Switzerland and Cambridge. Available at: www.iucnredlist.org/ (accessed July 2016).

IUCN. (2017). *Guidelines for Species Conservation Planning, Version 2.0.* IUCN, Gland, and Cambridge.

IUCN. (2018). *Red List of Threatened Species, Version 2018.* IUCN, Gland, Switzerland and Cambridge, UK. www.iucnredlist.org/resources/summary-statistics#Summary%20Tables (accessed 21 December 2018).

IUCN/SSC. (2008). *Strategic Planning for Species Conservation: A Handbook. Version 1.0.* IUCN Species Survival Commission, Gland and Cambridge.

IUCN/SSC. (2013). *Guidelines for Reintroductions and Other Conservation Translocations. Version 1.0.* IUCN Species Survival Commission, Gland.

Jaccard, P. (1901). Étude comparative de la distribution florale dans une portion des Alpes et des Jura. *Bulletin de la Société Vaudoise des Sciences Naturelles*, 37: 547–579.

Jackson, M.T. (1994). Care and use of rice biodiversity. In: *Food Security in Asia: Contributions of IRRI and British Science.* ODA, IRRI and BBSRC, London, pp. 7–10.

Jackson, M., Ford-Lloyd, B.V. and Parry, M. (eds.) (2013). *Plant Genetic Resources and Climate Change – a 21st Century Perspective.* CAB International, Wallingford.

Jain, S.K. (1975). Genetic reserves. In: Frankel, O.H. and Hawkes, J.G. (eds.) *Crop Genetic Resources for Today and Tomorrow.* Cambridge University Press, Cambridge, pp. 379–396.

Jamnadass, R.H., Dawson, I.K., Franzel, S., *et al.* (2011). Improving livelihoods and nutrition in sub-Saharan Africa through the promotion of indigenous and exotic fruit production in smallholders' agroforestry systems: a review. *International Forestry Review*, 13: 338–354.

Jarvis, A., Lane, A. and Hijmans, R.J. (2008a). The effect of climate change on crop wild relatives. *Agriculture, Ecosystems and Environment*, 126: 13–23.

Jarvis, A., Williams, K., Williams, D., *et al.* (2005). Use of GIS for optimizing a collecting mission of rare wild pepper (*Capsicum flexuosum* Sendtn.) in Paraguay. *Genetic Resources and Crop Evolution*, 52: 671–682.

Jarvis, D., Brown, A.H.D., Cuong, P.H., *et al.* (2008b). A global perspective of the richness and evenness of traditional crop-variety diversity maintained by farming communities. *Proceedings of the National Academy of Sciences USA* 105: 5326–5331.

Jarvis, D.I., Hodgkin, T., Brown, A.H.D., *et al.* (2016). *Crop Genetic Diversity in the Field and on the Farm: Principles and Applications in Research Practices.* Yale University Press, New Haven, CT.

Jarvis, D.I., Hodgkin, T., Sthapit, B.R., Fadda, C. and López-Noriega, I. (2011). An heuristic framework for identifying multiple ways of supporting the conservation and use of traditional crop varieties within the agricultural production system. *Critical Reviews in Plant Science*, 30: 125–176.

Jarvis, D.I., Myer, L., Klemick, H., *et al.* (2000). *A Training Guide for* in situ *Conservation On-farm.* Version 1. International Plant Genetic Resources Institute, Rome.

Jarvis, D.I., Padoch, C. and Cooper, H.D. (2007). *Managing Biodiversity in Agricultural Ecosystems.* Columbia University Press, New York.

Jarvis, S.G., Fielder, H., Hopkins, J., Maxted, N. and Smart, S. (2015). Distribution of crop wild relatives of conservation priority in the UK landscape. *Biological Conservation*, 191: 444–451.

Jenderek, M.M. and Reed, B.M. (2017). Cryopreserved storage of clonal germplasm in the USDA National Plant Germplasm System. *In Vitro Cellular and Developmental Biology – Plants.* https://doi.org/10.1007/s11627-017-9828-3.

Jenkins, R.W.G. and Roberts, S.R. (2000). *Sustainable Use of Wild Species – A Guide for Decision Makers.* IUCN, Cambridge and Gland, UK and Switzerland.

Johannsen, W. (1903). *Ueber Erlichkeit in Populationen und in reinen Linien: ein Beitrag zur Beleuchtung schweber Selektionsfragen.* Gustav Fischer, Jena.

Johannsen, W. (1905). *Arvelighedslærens Elementer.* Gyldendal, Copenhagen.

Johannsen, W. (1911). The genotype conception of heredity. *The American Naturalist*, 45: 129–159.

Johns, T., Mohoro, E.B. and Sanaya, P. (1996). Food plants and masticants of the Batemi of Ngorongoro District, Tanzania. *Economic Botany*, 50: 115–121.

Johnston, K.M.J., Reund, K.A.F. and Schmitz, O.J.S. (2012). Projected range shifting by montane mammals under climate change: implications for Cascadia's National Parks. *Ecosphere*, 3: 1–51.

Joint Nature Conservation Committee. (2014). *The Biodiversity Indicators.* Joint Nature Conservation Committee, Peterborough. Available at: http://jncc.defra.gov.uk/page-4233 (accessed 20 October 2014).

Joshi, B.K., Upadhyay, M.P., Gauchan, D., Sthapit, B.R. and Joshi, K.D. (2004). Red listing of agricultural crop species, varieties and landraces. *Nepal Agricultural Research Journal*, 5: 73–80.

Kaihura, F. and Stocking, M. (2003). *Agricultural Biodiversity in Smallholder Farms of East Africa.* United Nations University Press, Tokyo.

Karagöz, A. (1998). *In situ* conservation of plant genetic resources in the Ceyanpinar State Farm. In: Zencirci, N., Kaya, Z., Anikster, Y. and Adams, W.T. (eds.) *The Proceedings of International Symposium on* in situ *Conservation of Plant Diversity*. Central Research Institute for Field Crops, Ankara, pp. 87–91.

Kaur, N., Street, K., Mackay, M., Yahiaoui, N. and Keller, B. (2008). Allele mining and sequence diversity at the wheat powdery mildew resistance locus Pm3. In: Appels, R., Eastwood, R., Lagudah, E., *et al.* (eds). *11th International Wheat Genetics Symposium*. Sydney University Press, Brisbane.

Kehlenbeck, K., Asaah, E. and Jamnadass, R. (2013). Diversity of indigenous fruit trees and their contribution to nutrition and livelihoods in sub-Saharan Africa: examples from Kenya and Cameroon. In: Fanzo, J., Hunter, D., Borelli, T., *et al.* (eds.) *Diversifying Food and Diets: Using Agricultural Biodiversity to Improve Nutrition and Health Issues in Agricultural Biodiversity*. Earthscan, London, pp. 257–269.

Keiša, A., Maxted, N. and Ford-Lloyd, B.V. (2008). The assessment of biodiversity loss over time: wild legumes in Syria. *Genetic Resources and Crop Evolution*, 55: 603–612.

Kell, S.P., Ford-Lloyd, B.V. and Maxted, N. (2016). Europe's crop wild relative diversity: from conservation planning to conservation action. In: Maxted, N., Dulloo, E.M. and Ford-Lloyd, B.V. (eds.) *Enhancing Crop Gene Pool Use: Capturing Wild Relative and Landrace Diversity for Crop Improvement*. CAB International, Wallingford, pp. 125–136.

Kell, S.P., Laguna, E., Iriondo J.M. and Dulloo, M.E. (2008). Population and habitat recovery techniques for the *in situ* conservation of plant genetic diversity. In: Iriondo, J.M., Maxted, N. and Dulloo, M.E. (eds.) *Plant Genetic Population Management*. CAB International, Wallingford, pp. 124–168.

Kell, S.P., Maxted, N. and Bilz, M. (2012). European crop wild relative threat assessment: knowledge gained and lessons learnt. In: Maxted, N., Dulloo, M.E., Ford-Lloyd, B.V., *et al.* (eds.) *Agrobiodiversity Conservation: Securing the Diversity of Crop Wild Relatives and Landraces*. CAB International, Wallingford, pp. 218–242.

Kell, S.P., Qin, H., Chen, B., *et al.* (2015). China's crop wild relatives: diversity for agriculture and food security. *Agriculture, Ecosystems and Environment*, 209: 138–154.

Kelly, A.E. and Goulden, M.L. (2008). Rapid shifts in plant distribution with recent climate change. *Proceedings of the National Academy of Sciences of the USA*, 103: 11823–11826.

Kennedy, G., Stoian, D., Hunter, D., Kikulwe, E. and Termote, C., with contributions from Alders, R., Burlingame, B., Jamnadass, R., McMullin, S. and Thilsted, S. (2017). Food biodiversity for healthy, diverse diets. In: de Boef, W.,

Haga, M., Sibanda, L., Swaminathan, M.S. and Winters, P. (eds,) *Mainstreaming Agrobiodiversity in Sustainable Food Systems: Scientific Foundations for an Agrobiodiversity Index*. Bioversity International, Rome, pp. 23–52.

Khazaei, H., Street, K., Bari, A., Mackay, M. and Stoddard, F.L. (2013). The FIGS (focused identification of germplasm strategy) approach identifies traits related to drought adaptation in *Vicia faba* genetic resources. *PLoS ONE*, 8: e63107.

Khoury, C.K., Achicanoy, H.A., Bjorkman, A.D., *et al.* (2015a). *Estimation of countries' interdependence in plant genetic resources provisioning national food supplies and production systems*. Available at: www.planttreaty.org/content/research-paper-8 (accessed 11 May 2016).

Khoury, C.K., Amariles, D., Soto, J.S., *et al.* (2019). Comprehensiveness of conservation of useful wild plants: an operational indicator for biodiversity and sustainable development targets. *Ecological Indicators*, 98: 420–429.

Khoury, C.K., Bjorkman, A.D., Dempewolf, H., *et al.* (2014). Increasing homogeneity in global food supplies and the implications for food security. *Proceedings of the National Academy of Sciences of the USA*, 111: 4001–4006.

Khoury, C.K., Castañeda-Álvarez, N.P., Achicanoy, H.A., *et al.* (2015b). Crop wild relatives of pigeonpea (*Cajanus cajan* (L.) Millsp.): distributions, *ex situ* conservation status, and potential genetic resources for abiotic stress tolerance. *Biological Conservation*, 184: 259–270.

Kimura, M. (1968). Evolutionary rate at the molecular level. *Nature*, 217: 624–626.

Kimura, M. and Crow, J.F. (1963). The measurement of the effect of population size. *Evolution*, 17: 279–288.

King, R.C., Stansfield, W.D. and Mulligan, P.K. (2006). *A Dictionary of Genetics*, 7th ed. Oxford University Press, Oxford.

Klein, J.A., Harte, J. and Zhao, X.Q. (2008) Decline in medical and forage species with warming is mediated by plant traits on the Tibetan plateau. *Ecosystems* 11: 775–789.

Kobori, C.N. and Rodriguez Amaya, D.B. (2008). Uncultivated Brazilian green leaves are richer sources of carotenoids than are commercially produced leafy vegetables. *Food & Nutrition Bulletin*, 29: 320–328.

Koohafkhan, P. and Altieri, M.A. (2017). *Forgotten Agricultural Heritage: Reconnecting Food Systems and Sustainable Development*. Earthscan from Routledge, London.

Kothari, A., Camill, P. and Brown, J. (2013). Conservation as if people also mattered: policy and practice of community-based conservation. *Conservation and Society*, 11: 1–15.

Krebs, C.J. (2001). *Ecology: The Experimental Analysis of Distribution and Abundance*, 5th ed. Benjamin Cummings, San Francisco.

Kress, W.J., Wurdack, K.J., Zimmer, E.A., Weigt, L.A. and Janzen, D.H. (2005). Use of DNA barcodes to identify flowering plants. *Proceedings of the National Academy of Sciences*, 102: 8369–8374.

Krusche, D. and Geburek, Th. (1991). Conservation of forest gene resources as related to sample size. *Forest Ecology and Management*, 40: 145–150.

Küçük, S.A., Tan, A.Ş., Sabanci, C.O., *et al.* (1998). Ecogeographic and floristic differentiation of chestnut gene management zone in Kazdağ. In: Zencirci, N., Kaya, Z., Anikster, Y. and Adams, W.T. (eds.) *The Proceedings of International Symposium on in situ Conservation of Plant Diversity*. Central Research Institute for Field Crops, Ankara, pp. 135–148.

Kuhnlein, H.V., Erasmus, B. and Spigelski, D. 2009. *Indigenous Peoples' Food Systems: The Many Dimensions of Culture, Diversity and Environment for Nutrition and Health*. FAO, Rome.

Kuhnlein, H.V. and Turner, N.J. (1991). *Traditional Plant Foods of Canadian Indigenous Peoples – Nutrition, Botany and Use*. Gordon and Breach, Philadelphia.

Kumar, S. (1996). ABC of PRA: attitude and behaviour change. *PLA Notes*, 27: 70–73.

Kyte, L. and Kleyn, J. (1996). *Plants from Test Tubes: An Introduction to Micropropagation*, 3rd ed. Timber Press, Portland, OR.

Lacy, R.C. (2000). Structure of the VORTEX simulation model for population viability analysis. *Ecological Bulletin*, 48: 191–203.

Laird, S.A. and Noejovich, F. (2002). Building equitable research relationships with indigenous peoples and local communities: prior informed consent and research agreements. In: Laird, S.A. (ed.) *Biodiversity and Knowledge: Equitable Partnerships in Practice*. Earthscan, London, pp. 179–238.

Laird, S.A. and Posey, D.A. (2002). Professional society standards for biodiversity research: codes of ethics and research guidelines. In: Laird, S.A. (ed.) *Biodiversity and Knowledge: Equitable Partnerships in Practice*. Earthscan, London, pp. 16–38.

Laird, S.A. and ten Kate, K. 1999. *The Commercial Use of Biodiversity: Access to Genetic Resources and Benefit-Sharing*. Earthscan, London.

Langlet, O. (1971). Two hundred years of genecology. *Taxon*, 20: 653–722.

Larkin, P. and Scowcroft, W. (1981). Somaclonal variation: a novel source of variability from cell cultures for plant improvement. *Theoretical and Applied Genetics*, 60: 197–214.

Lawrence, M.J. (1996). Number of incompatibility alleles in clover and other species. *Heredity*, 76: 610–615.

Laurance, W.F., *et al.* (2012). Averting biodiversity collapse in tropical forest protected areas. *Nature*, 489: 290–294.

Lawrence, M.J. and Marshall, D.F. (1997). Plant population genetics. In: Maxted, N., Ford-Lloyd, B.V. and Hawkes, J.G. (eds.) *Plant Genetic Conservation: The in situ Approach*. Kluwer Academic Publishers, Dordrecht, Boston and London, pp. 99–113.

Lawrence, M.J., Marshall, D.F. and Davies, P. (1995). Genetics of genetic conservation. I. Sample size when collecting germplasm. *Euphytica*, 84: 89–99.

Lee, H.-S., Jeon, Y.-A., Lee, Y.-Y., Lee, S.-Y. and Kim, Y.-G. (2013). Comparison of seed viability among 42 species stored in a genebank. *Korean Journal of Crop Science*, 58: 432–438.

Lefèvre, F., Barsoum, N., Heinze, B., *et al.* (2001). *Technical bulletin: In situ Conservation of Populus nigra*. International Plant Genetic Resources Institute, Rome.

Lenoir, J., Gégout, J.C., Marquet, P.A., de Ruffray, P. and Brisse, H. (2008). A significant upward shift in plant species optimum elevation during the 20th century. *Science*, 320: 1768–1771.

Levetin, E. And McMahon, K. (2012). *Plants and Society*. McGraw-Hill, New York.

Lewington, A. (1990). *Plants for People*. The Natural History Museum, London.

Lewington, A. (2003). *Plants for People*. Eden Project Books

Lewis, C. (1996). *Managing Conflicts in Protected Areas*. IUCN, Gland.

Li, X., Takahashi, T., Suzuki, N. and Kaiser, H.M. (2011). The impact of climate change on maize yields in the United States and China. *Agricultural Systems*, 104: 348–353.

Lin, B. B. (2011). Resilience in agriculture through crop diversification: adaptive management for environmental change. *Bioscience* 61: 183–193.

Lindenmayer, D.B., Clark, T.W., Lacy, R.C. and Thomas, V.C. (1993). Population viability analysis as a tool in wildlife conservation policy: with reference to Australia. *Environmental management*, 17: 745–758.

Linnaeus, C. (1753). *Species Plantarum Vol 2*. Salvius, Stockholm.

Linnaeus, C. (1759). *Systema Natura*. Ed. 10. Salvius, Stockholm.

Lira, R. Tellez, O. and Davila, P. (2009). The effects of climate change on geographic distribution of Mexican wild relatives of domesticated cucurbitaceae. *Genetic Resources and Crop Evolution* 56; 691–703

Lloyd, W.F. (1833). *Two lectures on the checks to population*. Oxford University Press, Oxford.

Lomolino, M.V., Riddle, B.R. and Brown, J.H. (2006). *Biogeography, 3rd ed*. Sinauer Associates, Syracuse, USA.

Longin, C.F.H. and Reif, J.C. (2014). Redesigning the exploitation of wheat genetic resources. *Trends in Plant Science*, 19:631–636.

Loss, S.R., Terwilliger, L.A. and Peterson, A.C. (2011). Assisted colonization: Integrating conservation strategies in the face of climate change. *Biological Conservation*, 144: 92–100.

Louette, D. and Smale, M. 1996. *Genetic Diversity and Maize Seed Management in a Traditional Mexican Community: Implications for* in situ *Conservation of Maize*. NGR Paper 96-03. CIMMYT, Mexico, D.F.

Luck, J., Spackman, M., Freeman, A., *et al.* (2011). Climate change and diseases of food crops. *Plant Pathology*, 60: 113–121.

Lugo, A.E. (1988). Estimating reductions in the diversity of tropical forest species. In: *Biodiversity* (ed. Wison, E.O.). pp. 58–70. National Academy Press, Washington DC.

Lund, B., Ortiz, R., Skovgaard, I.M., Waugh, R. And Andersen, S.B. (2003). Analysis of potential duplicates in barley gene bank collections using re-sampling of microsatellite data. *Theoretical and Applied Genetics*, 106: 1129–1138.

Lund, B., Ortiz, R., von Bothmer, R. and Andersen, S.B. (2013). Detection of duplicates among repatriated Nordic spring barley (*Hordeum vulgare* L. s.l.) accessions using agronomic and morphological descriptors and microsatellite markers. *Genetic Resources and Crop Evolution* 60: 1–11.

Lyon, A., Silva, E., Zystro, J. and Bell, M. (2015). Seed and plant breeding for Wisconsin's organic vegetable sector: understanding farmers' needs. *Agroecology and Sustainable Food Systems*, 39: 601–624.

Mabberley, D.J. (2008). *Plant Book: A Portable Dictionary of Plants, their Classification and Uses, Third Edition*. Cambridge University Press, Cambridge.

Mabey, R. (1972). *Food for Free*. Collins

Mabey, R. (1996). *Flora Britannica*. Chatto and Windus

MacArthur, R.H. and Wilson, E.O. (1967). *The theory of island biogeography*. Princeton University Press, Princeton, New Jersey.

Mace, G.M. (2014). Whose conservation? *Science*, 345: 1558–1560.

Mace, M.G. and Ballie J.M. (2007). The 2010 Biodiversity Indicators: Challenges for Science and Policy. *Conservation Biology*, 21: 1406–1413.

Macfarlane, R., Jackson, G.V.H. and Frison, E.A. (2011). Plant health and germplasm collectors. In: Guarino L, Ramanatha Rao V, Goldberg E (eds.) *Collecting Plant Genetic Diversity: Technical Guidelines. 2011 update*. Bioversity International, Rome. Available online: http://cropgene bank.sgrp.cgiar.org/index.php?option=com_content&view=article&id=653 accessed 13.08.2018.

Mackay, M.C. and Street, K. (2004). Focused identification of germplasm strategy – FIGS. In: Black, C.K., Panozzo, J.F. and Rebetzke, G.J. (eds). *Cereals 2004. Proceedings of the 54th Australian Cereal Chemistry Conference and the 11th Wheat Breeders' Assembly*, 21–24 September 2004, Canberra, Australian Capital Territory (ACT). pp. 138–141.

Cereal Chemistry Division, Royal Australian Chemical Institute, Melbourne, Australia.

Magos Brehm, J., Kell, S., Thormann, I., *et al.* (2017a). *Interactive Toolkit for Crop Wild Relative Conservation Planning version 1.0*. University of Birmingham, Birmingham and Bioversity International, Rome. Available at: www.cropwildrelatives.org/conservation-toolkit/ (accessed 04.01.19).

Magos Brehm, J., Kell, S.P., Thormann, I., *et al.* (2017c). *Occurrence data collation template v.1*, doi:10.7910/DVN/5B9IV5, Harvard Dataverse, V1. Available here: https://dataverse.harvard.edu/dataset.xhtml?persistentId=doi:10.7910/DVN/5B9IV5 (accessed 04.01.19).

Magos Brehm, J., Kell, S.P., Thormann, I., Gaisberger, H., Dulloo, M.E. and Maxted, N., (2019). New tools for crop wild relative conservation planning. *Plant Genetic Resources*, 17: 208–212.

Magos Brehm, J., Kell, S.P., Thormann, I., Maxted, N. and Dulloo, E. (2017b). *Template for the Preparation of a Technical Background Document for a National Strategic Action Plan for the Conservation and Sustainable Use of Crop Wild Relatives*. doi:10.7910/DVN/VQVDFA, Harvard Dataverse, V1. Available here: https://dataverse.harvard.edu/dataset.xhtml?persistentId=doi:10.7910/DVN/VQVDFA (accessed 04.01.19).

Mahalakshmi, V., van Hintum, T.J.L. and Ortiz, R. (2003). Enhancing germplasm utilization to meet specific user needs through interactive stratified core selections. *Plant Genetic Resources Newsletter*, 136: 14–22.

Maplecroft, (2013). *Food Security Risk Index*. Available at http://maplecroft.com/about/news/food_security.html (Accessed 1 August 2013).

Marfil C.F., Hidalgo V. and. Masuelli, R.W. (2015). *In situ* conservation of wild potato germplasm in Argentina: Example and possibilities. *Global Ecology and Conservation* 3, 461–476.

Margules, C.R. (1989). Introduction to some Australian developments in conservation evaluation. *Biological Conservation*, 50: 1–11.

Margules, C.R. and Pressey, R.L. (2000). Systematic conservation planning. *Nature*, 405: 243–253.

Marren, P. (1999). *Britain's rare flowers*. T. & A.D. Poyser Natural History, London.

Marshall, C.R. and Brown, A.H.D. (1975). Optimum sampling strategies in genetic conservation. In: Frankel, O.H. and Hawkes, J.H. (eds.) *Crop Genetic Resources for Today and Tomorrow*. pp. 3–80. Cambridge University Press, Cambridge.

Martín, A, and Cabrera, A. (2005). Cytogenetics of *Hordeum chilense*: current status and considerations with reference to breeding. *Cytogenetic and Genome Research*, 109: 378–384.

Martin, G. (2015). *Ethnobotany: a methods manual*. Taylor and Francis, London.

Martin, P., Wishart, J., Cromoty, A. and Chang, X. (2009). New markets and supply chains for Scottish 'Bere' barley. In: Veteläinen, M., Negri, V. and Maxted, N. (eds.) (2009). *European Landraces: On-farm conservation, Management and Use.* Bioversity Technical Bulletin 15. pp. 251–263. Bioversity International, Rome.

Marum and Daugstad, (2009). Grindstad Timothy: the landrace that became a major commercial variety. In: Veteläinen, M., Negri, V. and Maxted, N. (eds.) (2009). *European Landraces: On-farm conservation, Management and Use.* Bioversity Technical Bulletin 15. Bioversity International, Rome, pp. 187–190.

Massawe, F., Mates, S. and Cheng, A. (2016). Crop diversity: an unexploited treasure trove for food security. *Trends in Plant Science,* 21, 365–368.

Maunder, M. (2008). Beyond the greenhouse. *Nature,* 455: 596–597.

Maunder, M. and Culham, A. (1997). Practical aspects of threatened species management in botanic garden collections. In: Tew, T.E., Crawford, T.J., Spencer, J.W., Stevens, D.P., Usher, M.B. and Warren, J. (eds.) *The role of genetics in conserving small populations.* pp. 122–130. Joint Nature Conservation Committee, Peterborough.

Maurya, D.M., Bottrall, A. and Farrington, J. (1988). Improved livelihoods, genetic diversity and farmer participation: A strategy for rice breeding in rainfed areas of India. *Experimental Agriculture,* 24: 311–320.

Maxted, N. (1990). A phenetic investigation of *Psophocarpus* Neck. ex. DC. (Leguminosae-Phaseoleae). *Botanical Journal of the Linnean Society,* 102: 103–-122.

Maxted, N. (2006). UK land-races – a hidden resource? *Plant Talk,* 44: 8.

Maxted, N. (2011). Aids to taxonomic identification. In Guarino L, Ramanatha Rao V, Goldberg E (editors). *Collecting Plant Genetic Diversity: Technical Guidelines. 2011 update.* Bioversity International, Rome. Available online: http://cropgene bank.sgrp.cgiar.org/index.php?option=com_content&view=article&id=652& Itemid=864&lang=english

Maxted, N. (2012). *Lathyrus belinensis*: a CWR discovered and almost lost. *Crop Wild Relative,* 8: 44.

Maxted, N., (2020). Another look at the *in situ / ex situ* CWR conservation linkage. *Crop Wild Relative,* 11: 22–25.

Maxted N., Amri. A., Castañeda-Álvarez, N.P., *et al.* (2016a). Joining up the dots: a systematic perspective of crop wild relative conservation and use. In: Maxted, N., Ehsan Dulloo, M. and Ford-Lloyd, B.V. (eds.) *Enhancing Crop Gene pool Use: Capturing Wild Relative and Landrace Diversity for Crop Improvement.* CAB International, Wallingford, pp. 87–124.

Maxted, N., Avagyan, A., Frese, L., *et al.* (2015). *Preserving diversity: a concept for* in situ *conservation of crop wild relatives in Europe Version 2.* Rome, Italy: *In Situ* and On-farm Conservation Network, European Cooperative Programme for Plant Genetic Resources, Rome. Available online: www.pgrsecure.org/documents/Concept_v2.pdf (accessed 11.05.17).

Maxted, N. and Bisby, F.A. (1989). Accurate identification of wild forage species. *Third ECP/GR Forage Working Group Meeting,* Montpellier. Appendix 5: 62–75. IBPGR, Rome.

Maxted, N., Dulloo, M.E. and Ford-Lloyd, B.V. (eds.) (2016b). *Enhancing Crop Gene pool Use: Capturing Wild Relative and Landrace Diversity for Crop Improvement.* CAB International, Wallingford.

Maxted, N., Dulloo, M.E., Ford-Lloyd, B.V., *et al.* (eds.) (2012b). *Agrobiodiversity Conservation: Securing the Diversity of Crop Wild Relatives and Landraces.* CAB International, Wallingford.

Maxted, N., Dulloo, M.E., Ford-Lloyd, B.V., Iriondo, J. and Jarvis, A. (2008a). Genetic gap analysis: A tool for more effective genetic conservation assessment. *Diversity and Distributions,* 14: 1018–1030.

Maxted, N., Esele, J.P. and Khizzah, B.W. (1986). Collection of sorghum and millets in Uganda. *Plant Genetic Resources Newsletter,* 64: 21–23.

Maxted, N., Ford-Lloyd, B.V. and Hawkes, J.G. (1997b). Complementary Conservation Strategies. In: *Plant Genetic Conservation: The* in situ *Approach* (eds. Maxted, N., Ford-Lloyd, B.V. and Hawkes, J.G.), pp. 20–55. Chapman & Hall, London.

Maxted, N., Ford-Lloyd, B.V., Jury, S., Kell, S.P. and Scholten, M.A. (2006). Towards a definition of a crop wild relative. *Biodiversity and Conservation,* 15(8): 2673–2685.

Maxted, N., Ford-Lloyd, B.V., Kell, S.P., *et al.* (eds.) (2008d). *Crop Wild Relative Conservation and Use.* CAB International, Wallingford.

Maxted, N., Guarino, L. and Dulloo, M.E. (1997c). Management and monitoring. In: Maxted, N., Ford-Lloyd, B.V. and Hawkes, J.G. (eds.) *Plant Genetic Conservation: The* in situ *Approach.* Chapman & Hall, London. pp. 231–258.

Maxted, N., Guarino, L., Myer, L. and Chiwona, E.A. (2002). Towards a methodology for on-farm conservation of plant genetic resources. *Genetic Resources and Crop Evolution,* 49: 31–46.

Maxted, N., Hargreaves, S., Kell, S.P., *et al.* (2012a). Temperate forage and pulse legume genetic gap analysis. *Bocconea,* 24: 5–36.

Maxted, N., Hawkes, J.G., Ford-Lloyd, B.V. and Williams, J.T. (1997b). A practical model for *in situ* genetic conservation. In: Maxted, N., Ford-Lloyd, B.V. and Hawkes, J.G. (eds.) *Plant Genetic Conservation: The* in situ *Approach.* pp. 545–592. Chapman & Hall, London.

Maxted, N., Hawkes, J.G., Guarino, L. and Sawkins, M. (1997a). The selection of taxa for plant genetic conservation. *Genetic Resources and Crop Evolution,* 44: 337–348.

Maxted, N., Iriondo, J., De Hond, L., *et al.* (2008c). Genetic Reserve Management. In: Iriondo, J.M., Maxted, N. and

Dulloo, E. (eds.) *Plant Genetic Population Management.* pp. 65–87. CAB International, Wallingford.

Maxted, N., Iriondo, J., Dulloo, E. and Lane, A. (2008b). Introduction: the integration of PGR conservation with protected area management. In: Iriondo, J.M., Maxted, N. and Dulloo, E. (eds.) *Plant Genetic Population Management.* pp. 1–22. CAB International, Wallingford.

Maxted, N. and Kell, S.P. (2009). *Establishment of a Network for the* In Situ *Conservation of Crop Wild Relatives: Status and Needs.* Commission on Genetic Resources for Food and Agriculture. Food and Agriculture Organization of the United Nations.

Maxted, N., Kell, S.P. and Magos Brehm, J. (2013a). Crop wild relatives and climate change. In: Jackson, M., Ford-Lloyd, B.V. and Parry M. (eds.) *Plant Genetic Resources and Climate Change - a 21st Century Perspective.* CAB International, Wallingford. pp. 114–136.

Maxted, N., Kell, S.P. and Magos Brehm, J. (2014). *Global Networking on* in situ *Conservation and on-farm Management of Plant Genetic Resources for Food and Agriculture.* Food and Agriculture Organization of the United Nations, Rome, 14 pp. Available online: www.fao.org/3/a-mm537e.pdf (accessed 11.05.17).

Maxted, N., Kell, S.P., Toledo, A., *et al.* (2010). A global approach to crop wild relative conservation: securing the gene pool for food and agriculture. *Kew Bulletin*: 65: 561–576.

Maxted, N., Labokas, J. and Palmé, A. (2017). *Crop wild relative conservation strategies. Planning and implementing national and regional conservation strategies.* Proceedings of a Joint Nordic/ECPGR Workshop, 19–22 September 2016, Vilnius, Lithuania. European Cooperative Programme for Plant Genetic Resources, Rome.

Maxted, N., Mabuza-Dlamin,i P., Moss, H., *et al.* (2004). An Ecogeographic Survey: African *Vigna. Systematic and Ecogeographic Studies of Crop Gene pools* 10. International Plant Genetic Resources Institute, Rome.

Maxted, N., Magos Brehm, J. and Kell, S.P. (2013b). *Resource book for preparation of national conservation plans for crop wild relatives and landraces.* Commission on Genetic Resources for Food and Agriculture. Food and Agriculture Organization of the United Nations, Rome. 457 pp. Available online: www.fao.org/agriculture/crops/thematic-sitemap/theme/seeds-pgr/resource-book/en/ (accessed 11.05.17).

Maxted, N. and Palmé, A. (2016). Combining *ex situ* and *in situ* conservation strategies for CWR to mitigate climate change. In: Valdani Vicari & Associati, Arcadia International, Wageningen UR: Centre for Genetic Resource, the Netherlands, Plant Research International and the socio-economics research institute, Fungal Biodiversity Centre of the Royal Academy of Arts and Science and Information and Coordination Centre for Biological Diversity of the German Federal Office for Agriculture and Food, (eds.) *The Impact of Climate Change on the Conservation and Utilisation of Crop Wild Relatives in Europe.* Directorate General for Agriculture and Rural Development, European Commission, Brussels, pp 6–7.

Maxted, N. and Scholten, M.A. (2007). Methodologies for the creation of National / European inventories. In: Del Greco, A., Negri V. and Maxted, N. (compilers) *Report of a Task Force on On-farm Conservation and Management*, Second Meeting, 19–20 June 2006, Stegelitz, Germany. pp. 11–19. Bioversity International, Rome.

Maxted, N., Scholten, M.A., Codd, R. and Ford-Lloyd, B.V. (2007). Creation and use of a national inventory of crop wild relatives. *Biological Conservation*, 140, 142–159.

Maxted, N., van Slageren, M.W. and Rihan, J. (1995). Ecogeographic surveys. In: Guarino, L., Ramanatha Rao, V. and Reid, R. (eds.) *Collecting Plant Genetic Diversity: Technical Guidelines.* CAB International, Wallingford, pp. 255–286.

Maxted, N., Veteläinen, M. and Negri, V. (2009). Landrace inventories: needs and methodologies. In: Veteläinen, M., Negri, V. and Maxted, N. (eds.) *European Landraces: On-farm Conservation, Management and Use.* Bioversity Technical Bulletin 15. Bioversity International, Rome, pp. 45–52.

Mazaika, K. (2016). *Assessing and addressing community conflict arising in conservation planning and management.* IUCN Social Science for Conservation Fellowship Programme Working Paper 6. Available at: www.iucn.org/sites/dev/files/v2_pdf_final_assessing_addressing_conflict_09_2016_0.pdf (accessed 14 August 2017).

McCouch, S., Baute, G.J., Bradeen, J., *et al.* (2013). Agriculture: feeding the future. *Nature*, 499: 23–24.

McDonald, T., Sokolow, J. and Hunter, D. (2018). Farmer and community-led approaches to climate change adaptation of agriculture using agricultural biodiversity and genetic resources. In: Yadav, S., Redden, R.J., Hatfield, J.L. *et al.* (eds.) *Climate Change and Food Security in the 21st Century.* Wiley-Blackwell International, Chichester.

McNeely, J. (1988). *Economics and Biological Diversity.* IUCN, Gland.

McNeill, J., Barrie, F.R., Buck, W.R., *et al.* (2012). *International Code of Nomenclature for algae, fungi, and plants (Melbourne Code) adopted by the Eighteenth International Botanical Congress Melbourne, Australia, July 2011.* Regnum Vegetabile 154. A.R.G. Gantner Verlag KG.

Meldrum, G. and Padulosi, S. (2017). Neglected no more: leveraging under-utilized crops to address global challenges. In: Hunter, D., Guarino, L., Spillane, C. and McKeown, P. (eds.) *Handbook of Agriculture Biodiversity.* Routledge, Abingdon, pp. 298–310.

Mendel, J.G. (1866). Versuche über Pflanzenhybriden. *Verhandlungen des naturforschenden Vereines in Brünn*, Bd. IV für das Jahr, *Abhandlungen*, 1865: 3–47.

Meredith, L.D. and Richardson, M.M. (1991). Towards an Australian botanic gardens conservation secretariat. In: Heywood, V.H. and Wyse Jackson, P.S. (eds.) *Tropical Botanic Gardens: Their Role in Conservation and Development*. Academic Press, London, pp. 35–44.

Mezzalama, M. (2012). *Seed Health: Fostering the Safe Distribution of Maize and Wheat Seed. General Guidelines*, 3rd ed. Centro Internacional de Mejoramiento de Maiz y Trigo, Mexico DF.

Michiels, F. (2015). *ABS Impact on the Plant Breeding Sector*. Essenscia seminar on Genetic Resources. Bayer Crop Science, Leverkusen.

Midgley, G.F., Hannah, L., Millar, D., Thuiller, W. and Booth, A. (2003). Developing regional and species-level assessments of climate change impacts on biodiversity in the Cape Floristic Region. *Biological Conservation*, 11: 87–97.

Mijatovic, D., Sakalian, M. and Hodgkin, T. (2018). *Mainstreaming Biodiversity in Production Landscapes*. United Nations Environment Programme. https://wedocs .unep.org/bitstream/handle/20.500.11822/26878/ biodivers_production.pdf?sequence=1&tisAllowed=y

Millennium Ecosystem Assessment (MEA). (2005). *Ecosystems and Human Well-Being: Biodiversity Synthesis*. World Resources Institute, Washington, DC.

Miller, S.E. (2007). DNA barcoding and the renaissance of taxonomy. *Proceedings of the National Academy of Sciences of the USA*, 104: 4775–4776.

Millstone, E. and Lang, T. (2008). *The Atlas of Food: Who Eats What, Where and Why?* Earthscan, London.

Mittermeier, R.A., Meyers, N., Robles, G.P. and Mittermeier, C.G. (1999). *Hotspots*. Garza Garcia N.L. CEMEX, Mexico, D.F.

Mittermeier, R.A., Robles, G.P., Hoffmann, M., *et al.* (2004). *Hotspots: Revisited*. Garza Garcia N.L. CEMEX, Mexico, D.F.

Moore G. and Williams, K. A. (2011). Legal issues in plant germplasm collecting. In: Guarino, L., Ramanatha Rao, V., Goldberg, E. (eds.) *Collecting Plant Genetic Diversity: Technical Guidelines. 2011 Update*. Bioversity International, Rome. Available at: https://cropgenebank .sgrp.cgiar.org/index.php/component/content/article/ 178-procedures/collecting/669-chapter-2-legal-issues- in-plant-germplasm-collecting (accessed 1 February 2020).

Morden, C.W., Doebley, J. and Schertz, K.F. (1990). Allozyme variation among the spontaneous species of *Sorghum* section *Sorghum* (Poaceae). *Theoretical and Applied Genetics*, 80: 296–304.

Morris, W.F. and Doak, D.F. (2003). *Quantitative Conservation Biology: Theory and Practice of Population Viability Analysis*. Sinauer Associates, Sunderland, MA.

Morrison, C., Rounds, I. and Wattling, D. (2012). Conservation and management of the endangered Fiji sago palm, *Metroxylon vitiense*, in Fiji. *Environmental Management*, 49: 929–941.

Motlhaodi, T., Geleta, M., Bryngelsson, T., *et al.* (2014). Genetic diversity in *ex situ* conserved sorghum accessions of Botswana as estimated by microsatellite markers. *Australian Journal of Crop Science*, 8: 35–43.

Murphy, D.J. (2007). *People, Plants and Genes*. Oxford University Press, Oxford.

Murphy, J.P. and Phillips, T.D. (1993). Isozyme variation in cultivated oat and its progenitor species, *Avena sterilis* L. *Crop Sciences*, 33: 1366–1372.

Murphy, K., Bazile, D., Kellogg, J. and Rahmaniam, M. (2016). Development of a worldwide consortium on evolutionary participatory breeding in quinoa. *Frontiers in Plant Science*, 7: 1–8.

Myers, N. (1988). Threatened biotas: hotspots in tropical forests. *The Environmentalist*, 8: 1–20.

Myers, N. (1990). The biodiversity challenge: expanded hot spots analysis. *The Environmentalist*, 10: 243–256.

Myers, N., Mittermeier, R.A., Mittermeier, C.G., da Fonseca, G.A.B. and Kent, J. (2000). Biodiversity hotspots for conservation priorities. *Nature*, 403: 853–858.

Myers, S.S., Zanobetti, A., Kloog, I., *et al.* (2014). Increasing CO2 threatens human nutrition. *Nature*, 510: 139–142.

Namkoong, G. (1981). Methods of pollen sampling for gene conservation. In: Franklin, E.C. (ed.) *Pollen Management Handbook*. USDA Agriculture Handbook No. 587. USDA, Washington, DC, pp. 74–76.

Nassar, N.M.A. and Ortiz, R. (2007). Cassava improvement: challenges and successes. *Journal of Agricultural Science*, 145: 163–171.

National Geographic. (2013). (Based on a study completed by Rural Advancement Foundation International, 1983). http://ngm.nationalgeographic.com/2011/07/ food-ark/food-variety-graphic (accessed 1 August 2013).

Neel, M.C. and Cummings, M.P. (2003a). Effectiveness of conservation targets in capturing genetic diversity. *Conservation Biology*, 17: 219–229.

Neel, M.C. and Cummings, M.P. (2003b). Genetic consequences of ecological reserve design guidelines: an empirical investigation. *Conservation Genetics*, 4: 427–439.

Negri, V. (2003). Landraces in central Italy: where and why they are conserved and perspectives for on-farm conservation. *Genetic Resources and Crop Evolution*, 50: 871–885.

Negri, V., Maxted, N. and Veteläinen, M. (2009). European LR conservation: an introduction. In: Veteläinen, M., Negri, V. and Maxted, N. (eds.) *European Landraces: On-Farm Conservation, Management and Use*. Bioversity Technical Bulletin No. 15. Bioversity International, Rome. pp. 1–22. Also available from: www .bioversityinternational.org/index.php?id=19&user_ bioversitypublications_pi1[showUid]=3252

Negri, V., Maxted, N., Torricelli, R., *et al.* (2012). *Descriptors for Web-Enabled National* in situ *Landrace Inventories.* University of Perugia, Perugia, Italy. Available at: https://pgrsecure.bham.ac.uk/publications (accessed 24 November 2018).

Negri, V., Pacicco, L., Bodesmo, M. and Torricelli, R. (2013). *The first Italian inventory of* in situ *maintained landraces.* Morlacchi Editrice, Perugia, Italy. Available at: http://vnr.unipg.it/PGRSecure/start.html (accessed 23 November 2018).

Nei, M. (1972). Genetic distance between populations. *American Naturalist*, 106: 283–291.

Nei, M. (1973). Analysis of gene diversity in subdivided populations. *Proceedings of the National Academy of Sciences of the* USA, 70: 3321–3323.

Nemarundwe, N. and Richards, M. (2002). Participatory methods for exploring livelihood values derived from forests: potential and limitations. In: Campbell, B. and Luckert, M. (eds.) *Uncovering the Hidden Harvest: Valuation Methods for Woodland and Forest Resources.* Earthscan, London, pp. 168–197.

Newton, A.C., Johnson, S.N., Lyon, G.D., Hopkins, D.W. and Gregory, P.J. (2008). Impacts of climate change on arable crops – adaptation challenges. In: *Proceedings of the Crop Protection in Northern Britain Conference.* The Association for Crop Protection in Northern Britain, Dundee, UK, pp. 11–16.

Ngoc De, N. (2001). Crop improvement at community level in Vietnam. In: Friis-Hansen, E. and Sthapit, B. (eds.) *Participatory Approaches to Conservation and Use of Plant Genetic Resources.* International Plant Genetic Resources Institute, Rome, pp. 103–110.

Nilsson, H.I. (1909). Aterblick på utsädesföreningens arbetsmetoder och de med dem vunna resultaten. *Sver. Utsadesjoren. Tidskr.* 18: 235–249.

Nilsson-Ehle, H. (1909). *Kreuzungsuntersuchungen an Hafer und Weizen.* Dissertation, Lund.

Nokoe, S. and Ortiz, R. (1998). Optimum plot size for banana trials. *HortScience*, 33: 130–132.

Notaro, V., Padulosi, S., Galluzzi, G. and King, I.O. (2017). A policy analysis to promote conservation and use of small millet underutilized species in India. *International Journal of Agricultural Sustainability* 15: 393–405.

O'Donnell, K. and Sharrock, S. (2015). Seed banking in botanic gardens: can botanic gardens achieve GSPC Target 8 by 2020? *BGjournal*, 12: 3–8.

Ocampo, J., d'Eeckenbrugge, C., Restrepo, M., *et al.* (2007). Diversity of Colombian Passifloraceae: biogeography and an updated list for conservation. *Biota Colombiana*, 8: 1–45.

Oka, H.I. (1988). *Origin of Cultivated Rice.* Elsevier Science Publishing Co., New York.

Oldfield, S. and Kapos, V. (2017). Botanic gardens band conservation impact options for evaluation. In: Blackmore, S. and Oldfield, S. (eds.) *Plant Conservation Science and Practice: The Role of Botanic Gardens.* Cambridge University Press, Cambridge, pp. 219–235.

Oldfield, S.F. (2009). Botanic gardens and the conservation of tree species. *Trends in Plant Science*, 14: 581–583.

Oldfield, S.F. (2010). *Botanic Gardens: Modern-Day Arks.* MIT Press, Cambridge.

O'Riordan,T. and Stoll-Kleemann, S. (2002). *Biodiversity, Sustainability and Human Communities: Protecting Beyond the Protected.* Cambridge University Press, Cambridge.

Ortiz, R. (1991). Una metodología de selección múltiple por productividad y estabilidad para cultivares de tomate. *Agro-Ciencia (Chile)*, 7: 135–142.

Ortiz, R. (1995). Plot techniques for assessment of bunch weight in banana trials under two systems of crop management. *Agronomy Journal*, 87: 63–69.

Ortiz, R. (2004). Breeding clones. In: Goodman, R.M. (ed.) *Encyclopedia of Plant & Crop Science.* Marcel Dekker, Inc., New York, pp. 174–178.

Ortiz, R. (2015). *Plant Breeding in the Omics Era.* Springer, New York.

Ortiz, R. (2017). Leveraging agricultural biodiversity for crop improvement. In: Hunter, D., Guarino, L., Spillane, C. and McKeown, P.C. (eds.) *Routledge Handbook of Agricultural Biodiversity.* Routledge/Taylor & Francis Group, Oxford, pp. 285–297.

Ortiz, R., Braun, H.-J., Crossa, J., *et al.* (2008a). Wheat genetic resources enhancement by the International Maize and Wheat Improvement Center (CIMMYT). *Genetic Resources and Crop Evolution*, 55: 1095–1140.

Ortiz, R., Crossa, J. and Sevilla, R. (2008b). Minimum resources for phenotyping morphological traits of maize (*Zea mays* L.) genetic resources. *Plant Genetic Resources: Characterization and Utilization*, 6: 195–200.

Ortiz, R., Crossa, J., Franco, J., Sevilla, R. and Burgueño, J. (2008c). Classification of Peruvian highland maize races using plant traits. *Genetic Resources and Crop Evolution*, 55: 151–162.

Ortiz, R., Crossa, J., Vargas, M. and Izquierdo, J. (2007a). Studying the effect of environmental variables on the genotype × environment interaction of tomato. *Euphytica*, 153: 119–134.

Ortiz, R., de la Flor, F., Alvarado, G. and Crossa, J. (2010a). Classifying vegetable genetic resources: a case study with domesticated *Capsicum* spp. *Scientia Horticulturae*, 126: 186–191.

Ortiz, R. and Izquierdo, J. (1992). Interacción genotipo por ambiente en el rendimiento comercial del tomate en América Latina y el Caribe. *Turrialba*, 42: 492–499.

Ortiz, R. and Izquierdo, J. (1994). Yield stability differences among tomato genotypes grown in Latin America and the Caribbean. *HortScience*, 29: 1175–1177.

Ortiz, R., Madsen, S. and Vuylsteke, D. (1998). Classification of African plantain landraces and banana cultivars using a phenotypic distance index of quantitative descriptors. *Theoretical and Applied Genetics*, 96: 904–911.

Ortiz, R., Ruiz-Tapia, E.N. and Mujica-Sanchez, A. (1998). Sampling strategy for a core collection of Peruvian quinoa germplasm. *Theoretical and Applied Genetics*, 96: 475–483.

Ortiz, R. and Sevilla, R. (1997). Quantitative descriptors for classification and characterization of highland Peruvian maize. *Plant Genetic Resources Newsletter*, 110: 49–52.

Ortiz, R., Sevilla, R., Alvarado, G. and Crossa, J. (2008d). Numerical classification of related Peruvian highland maize races using internal ear traits. *Genetic Resources and Crop Evolution*, 55: 1055–1064.

Ortiz, R., Simon, P., Jansky, S. and Stelly, D. (2009). Ploidy manipulation of the gametophyte, endosperm, and sporophyte in nature and for crop improvement – A tribute to Prof. Stanley J. Peloquin (1921–2008). *Annals of Botany*, 104: 795–807.

Ortiz, R., and Swennen, R. (2014). From crossbreeding to biotechnology-facilitated banana and plantain improvement. *Biotechnology Advances*, 32: 158–169.

Ortiz, R., Taba, S., Chávez Tovar, V.H., *et al.* (2010b). Conserving and enhancing maize genetic resources at global public goods – a perspective from CIMMYT. *Crop Science*, 50: 13–28.

Ortiz, R., Trethowan, R., Ortiz Ferrara, G., *et al.* (2007b). High yield potential, shuttle breeding and a new international wheat improvement strategy. *Euphytica*, 157: 365–384.

Ostrom, E. (1990). *Governing the Commons: The Evolution of Institutions for Collective Action*. Cambridge University Press, Cambridge.

Pacicco, C.L., Bodesmo, M., Torricelli, R. and Negri, V. (2013). Progress toward an Italian conservation strategy for extant LR: the first Italian official inventory of LR. *Landraces*, 2: 10. Available at: www.pgrsecure.bham.ac .uk/sites/default/files/documents/newsletters/Landraces_Issue_2.pdf

Pacifici, M., Foden, W.B., Visconti, P., *et al.* (2015). Assessing species vulnerability to climate change. *Nature Climate Change*, 5: 215–225.

Padulosi, S. and Dulloo, E. (2012). Towards a viable system for monitoring agrobiodiversity on-farm: a proposed new approach for Red Listing of cultivated plant species. In: Padulosi, S., Bergamini, N. and Lawrence, T. (eds.) *On-Farm Conservation of Neglected and Underutilized Species: Status, Trends and Novel Approaches to Cope with Climate Change*. Bioversity International, Rome, pp. 171–199.

Padulosi S., Heywood, V., Hunter, D. and Jarvis, A. (2015). Underutilized crops and climate change – current status and outlook. In: Redden, R., Yadav, S.S., Maxted, N., *et al.* (eds.) *Crop Wild Relatives and Climate Change*. John Wiley & Sons, Inc., Hoboken, NJ, pp. 507–521.

Painting, K.A., Perry, M.C., Denning, R.A. and Ayad, W.G. (1993). *Guidebook for Genetic Resources Documentation*. IBPGR, Rome.

Panella, L., Gigante D., Donnini, D., Venanzoni, R. and Negri, V. (2012). Progenitori selvatici e forme coltivate di Apiaceae, Chenopodiaceae, Poaceae e Rosaceae: primi risultati per il territorio dell'Umbria (Italia Centrale). *Quaderni Botanica Ambientale ed Applicata*, 23: 3–13.

Panella, L., Wheeler, L. and McClintock, M.E. (2009). Long-term survival of cryopreserved sugarbeet pollen. *Journal of Sugar Beet Research*, 46: 1–9.

Panis, B. (2019). 60 years of plant cryopreservation: from freezing hardy mulberry twigs to establishing reference crop collections for future generations. *Acta Horticulturae*. https://doi.org/10.17660/ActaHortic.2019 .1234.1.

Pâques, M. (1991). Vitrification and micropropagation: causes, remedies and prospects. *Acta Horticulturae* 289: 283–290.

Parmesan, C. and Yohe, G. (2003). A globally coherent fingerprint of climate change impacts across natural systems. *Nature*, 421: 37–42.

Parra-Quijano, M., Iriondo, J.M. and Torres, E. (2011a). Ecogeographical land characterization maps as a tool for assessing plant adaptation and their implications in agrobiodiversity studies. *Genetic Resources and Crop Evolution*, 59: 205–218.

Parra-Quijano, M., Iriondo, J.M. and Torres, E. (2012). Improving representativeness of gene bank collections through species distribution models, gap analysis and ecogeographical maps. *Biodiversity Conservation*, 21: 79–96.

Parra-Quijano, M., María Iriondo, J. and Torres Lamas, E. (2011b). Basic sampling strategies: theory and practice. In: Guarino, L., Ramanatha Rao, V. and Goldberg, E. (eds.) *Collecting Plant Genetic Diversity: Technical Guidelines. 2011 Update*. Bioversity International, Rome. Available at: http://cropgene bank.sgrp.cgiar.org/index .php?option=com_content&view=article&id=670 (accessed 13 August 2018).

Parra-Quijano, M., Torres Lamas, E., Iriondo Alegría, J.M. and López, F. (2014). *CAPFITOGEN Tools. Programme to Strengthen National Plant Genetic Resource Capacities in Latin America*. Food and Agriculture Organization of the UN, Rome. Available at: www.planttreaty.org/sites/ default/files/capfitogen_manualv1-2_es.pdf (accessed 23 May 2016).

Paton, A.J., Brummitt, N., Govaerts, R., *et al.* (2008). Towards Target 1 of the Global Strategy for Plant Conservation: a working list of all known plant species – progress and prospects. *Taxon*, 57: 602–611.

Patto, M.C.V., Aardse, A., Buntjer, J., *et al.* (2001). Morphology and AFLP markers suggest three *Hordeum chilense* ecotypes that differ in avoidance to rust fungi. *Canadian Journal of Botany*, 79: 204–213.

Pavek, D.S., Lamboy, W.F. and Garvey, E.J. (2003). Selecting *in situ* conservation sites for grape genetic resources in the USA. *Genetic Resources and Crop Evolution*, 50: 165–173.

Pearce, F. (2014). People power will save the world. *New Scientist*, 2 August, 14–15.

Pearce, T.R. and Bytebier, B. (2002). The role of an herbarium and its database in supporting plant conservation. In: Maunder, M., Clubbe, C., Hankamer, C. and Groves, M. (eds.) *Plant Conservation in the Tropics: Perspectives and Practice*. Royal Botanic Gardens, Kew, Richmond, pp. 49–65.

Pearson, R., Raxworthy, C., Nakamura, M. and Townsend Peterson, A. (2007). Predicting species distributions from small numbers of occurrence records: a test case using cryptic geckos in Madagascar. *Journal of Biogeography*, 34: 102–117.

Pearson, R.G., Thuiller, W., Araújo, M.B., *et al.* (2006). Model-based uncertainty in species' range prediction. *Journal of Biogeography*, 33: 1704–1711.

Peel, W. (2010). *Rainforest Restoration Manual for South-Eastern Australia*. CSIRO, Canberra.

Pence, V.C. and Engelmann, F. (2011). Collecting *in vitro* for genetic resources conservation. In: Guarino, L., Ramanatha Rao, V. and Goldberg, E. (eds.) *Collecting Plant Genetic Diversity: Technical Guidelines. 2011 Update*. Bioversity International, Rome. Available at: http://cropgene bank.sgrp.cgiar.org/images/file/procedures/collecting2011/Chapter24–2011.pdf

Perfecto, I., Vandermeer, J. and Wright, A. (2009). *Nature's Matrix: Linking Agriculture, Conservation and Food Sovereignty*. Earthscan, London.

Phelps, J. and Webb, E.L. (2015). 'Invisible' wildlife trades: Southeast Asia's undocumented illegal trade in wild ornamental plants. *Biological Conservation* 186: 296–305.

Phillips, J., Asdal, Å., Magos Brehm, J., Rasmussen, M. and Maxted, N. (2016). *In situ* and *ex situ* diversity analysis of priority crop wild relatives in Norway. *Diversity and Distributions* 22: 1112–1126. https://doi.org/10.1111/ddi .12470

Phillips, S.J., Anderson, R.P. and Schapire, R.E. (2006). Maximum entropy modelling of species geographic distributions. *Ecological Modelling*, 190: 231–259.

Phillips, S.J., Dudik, M. and Schapire, R.E. (2004). A maximum entropy approach to species distribution modelling. *Proceedings of the Twenty-First International Conference on Machine Learning*, 655–662.

Piano, E., Spanu, F. and Pecetti, L. (1993). Structure and variation of subterranean clover populations from Sicily, Italy. *Euphytica*, 68: 43–51.

Pimentel, D., Wilson, C., McCullum, C., *et al.* (1997). Economic and environmental benefits of biodiversity. *BioScience*, 47: 747–757.

Pingali, P.L. (2017). The Green Revolution and crop diversity. In: Hunter, D., Guarino, L., Spillane, C. and McKeown, P. (eds.) *Handbook of Agriculture Biodiversity*. Routledge, Abingdon, pp. 213–223.

Pinheiro de Carvalho, M.Â.A., Nóbrega, H., Freitas, G., Fontinha, S. and Frese, L. (2012). Towards the establishment of a genetic reserve for *Beta patula* Aiton. In: Maxted, N., Dulloo, M.E., Ford-Lloyd, *et al.* (eds.) *Agrobiodiversity Conservation: Securing the Diversity of Crop Wild Relatives and Landraces*. CAB International, Wallingford, pp. 36–44.

Pistorius, R. (2016). Access and benefit sharing of genetic resources for family farmers: theory and practice. *Farming Matters*, 6–13

Porfiri, O., Costanza, M.T. and Negri, V. (2009). Landrace inventories in Italy and the Lazio Region Case Study. In: Vetelainen, M., Negri, V. and Maxted, N. (eds.) *European Landraces: On-Farm Conservation, Management and Use*. Bioversity Technical Bulletin No. 15. Bioversity International, Rome, pp. 117–123. Available at: www .bioversityinternational.org/index.php?id=19&user_ bioversitypublications_pi1[showUid]=3252

Potato Council. (2017). *Humidification in Potato Stores*. Agriculture and Horticulture Development Board, Kenilworth.

Powell B., Hall, J. and Johns, T. (2011). Forest cover, use and dietary intake in the East Usambara Mountains, Tanzania. *International Forestry Review*, 13: 305–317.

Prain, G., Hambly, H., Jones, M., Leppan, W. and Navarro, L. (2000). *CGIAR Program on Participatory Research and Gender Analysis*. Internally Commissioned External Review. CGIAR, Washington DC.

Prance, G.T. (1997). The conservation of botanical diversity. In: Maxted, N., Ford-Lloyd, B.V. and Hawkes, J.G. (eds.) *Plant Genetic Conservation: The* in situ *Approach*. Chapman & Hall, London, pp. 3–14.

Preston, J.M., Ford-Lloyd, B.V., Smith, L.M., *et al.* (2018). Genetic analysis of a heritage variety collection. Plant *Genetic Resources: Characterization and Utilization*, 17, 232–244.

Pretty, J. (1995). *Regenerating Agriculture: An Alternative Strategy for Growth*. Earthscan, London.

Pretty, J. (2007). *The Earth Only Endures: On Reconnecting with Nature and Our Place in It*. Earthscan, London.

Pretty, J., Guijt, I., Thompson, J. and Scoones, I. (2003). *Participatory Learning and Action: A Trainers Guide*. IIED, London.

Pritchard, H.W. (2004). Classification of seed storage 'types' for *ex situ* conservation in relations to temperature and moisture. In: Guerrant, E.O, Havens, K. and Maunder, M. (eds.) Ex Situ *Plant Conservation: Supporting Species Survival in the Wild*. Island Press, Washington, DC, pp. 139–161.

Pritchard, H.W. and Dickie, J.B. (2003). Predicting seed longevity: the use and abuse of seed viability equations.

In: Smith, R.D., Dickie, J.B., Linington, S.H., Pritchard, H.W. and Probert, R.J. (eds.) *Seed Conservation: Turning Science into Practice.* Royal Botanic Gardens, Kew, Richmond, pp. 653–721.

Pullen, A. (2002). *Conservation Biology.* Cambridge University Press, Cambridge.

Pullin, A., and Knight, T. (2001). Effectiveness in conservation practice: pointers from medicine and public health. *Conservation Biology*, 15: 50–54.

Pullin, A.S. and Knight, T.M. (2003). Support for decision making in conservation practice: an evidence-based approach. *Journal for Nature Conservation*, 11: 83–90.

Pullin, A., and Stewart, G.B. (2006). Guidelines for systematic review in conservation and environmental management. *Conservation Biology*, 20(6): 1647–1656.

Pushpakumara, D.K.N.G., Sokolow, J., Stahpit, B., Sujarwo, W. and Hunter, D. (2020). Home gardens for biodiversity conservation. In: Dissanayake, H.G. and Maredia, K.M. (eds.) *Home Gardening for Enhanced Food Security and Livelihoods.* Earthscan, London.

PwC. (2013). *Crop wild relatives: a valuable resource for crop development.* Price Waterhouse Cooper. Available at: http://pwc.blogs.com/files/pwc-seed-bank-analysis-for-msb-0713.pdf

Qualset, C.O., Damania, A.B., Zanatta, A.C.A. and Brush, S.B. (1997). Locally-based crop plant conservation. In: Maxted, N., FordLloyd, B.V. and Hawkes, J.G. (eds.) *Plant Genetic Conservation: The* in situ *Approach.* Chapman & Hall, London.

Quek, P. and Friis-Hansen, E. (2011). Collecting plant genetic resources and documenting associated indigenous knowledge in the field: a participatory approach. In: Guarino, L., Ramanatha Rao, V. and Goldberg, E. (eds.) *Collecting Plant Genetic Diversity: Technical Guidelines. 2011 Update.* Bioversity International, Rome. Available at: http://cropgene bank.sgrp.cgiar.org/index .php?option=com_content&view=article&id=673 (accessed 13 August 2018).

RAFI. (1983). *Vegetable variety inventory: varieties from USDA 1903 list of American vegetables in storage at the national seed storage library* Unpublished report compiled by Chiosso, E., Rural Advancement Fund, Inc., Pittsboro, NC.

Ragone, D. (1997). *Breadfruit Artocarpus altilis (Parkinson) Fosberg. Promoting the Conservation and Use of Underutilised and Neglected Crops. 10.* Institute of Plant Genetics and Crop Plant Research, Gatersleben/IPGRI, Rome.

Rahmanian, M., Razavi, K., Haghparast, R., Salimi, M. and Ceccarelli, S. (2016a). Evolutionary plant breeding: a method for rapidly increasing on-farm biodiversity to support sustainable livelihoods in an era of climate change. In: Maxted, N., Ehsan Dulloo, M. and Ford-Lloyd, B.V. (eds.) *Enhancing Crop Gene Pool Use: Capturing Wild Relative and Landrace Diversity for Crop Improvement.* CABI, Wallingford, pp. 354–361.

Rahmanian, M., Salimi, M., Ravazi, K., *et al.* (2016b). Evolutionary populations: living gene banks in farmers' fields in Iran. *Farming Matters* (Special issue on Access and Benefit Sharing of Genetic Resources), May Issue, 24–29.

Ramirez-Villegas, J., Cuesta, F., Devenish, C., *et al.* (2014). Using species distribution models for designing conservation strategies of Tropical Andean biodiversity under climate change. *Journal for Nature Conservation*, 22: 391–404.

Ramírez-Villegas, J., Khoury, C., Jarvis, A., Debouck, D.G. and Guarino, L. (2010). A gap analysis methodology for collecting crop gene pools: a case study with *Phaseolus* beans. *PLoS One*, 5: 1–18.

Rankou, H. (2011). *Orchis militaris. The IUCN Red List of Threatened Species 2011.* IUCN, Cambridge and Gland.

Rao, N.K., Hanson, J., Dulloo et al. (2006). Manual of seed handling in genebanks. *Handbooks for Genebanks 8.* Bioversity International, Rome.

Ray, A. and Bhattacharya, S. (2008). Storage and plant regeneration from encapsulated shoot tips of *Rauvolfia serpentina* – an effective way of conservation and mass propagation. *South African Journal of Botany*, 74: 776–779.

Raza, S., Christiansen, J.L., Jørnsgård, B. and Ortiz, R. (2000). Partial resistance to a *Fusarium* root disease in Egyptian white lupin landraces. *Euphytica*, 112: 233–237.

RBG Kew. (2016). *State of the World's Plants Report – 2016.* Royal Botanic Gardens, Kew, Richmond.

RBG Kew. (2017). *State of the World's Plants Report - 2017.* Royal Botanic Gardens, Kew, Richmond.

Redden, R., Yadav, S.S., Maxted, N., *et al.* (eds.) (2015). *Crop Wild Relatives and Climate Change.* John Wiley & Sons, Inc., Hoboken, NJ.

Reddy, B.S.V., Ramesh, S. and Ortiz, R. (2005). Genetic and cytoplasmic-nuclear male sterility in sorghum. *Plant Breeding Reviews*, 25: 139–172.

Reddy, B.V.S., Rao, P., Deb, U.K., *et al.* (2004). Global sorghum genetic enhancement processes at ICRISAT. In: Bantilan, M.C.S., Deb, U.K., Gowda, C.L.L., *et al.* (eds.) *Sorghum Genetic Enhancement: Research Process, Dissemination and Impacts.* International Crops Research Institute for the Semi-Arid Tropics, Patancheru, pp. 65–102.

Reed, B.M., Engelmann, F., Dulloo, M.E. and Engels, J.M.M. (2004). *Technical Guidelines for the Management of Field and* in vitro *Germplasm Collections.* IPGRI Handbooks for Gene Banks No. 7. International Plant Genetic Resources Institute, Rome.

Reed, D.H., O'Grady, J.J., Brook, B.W., Ballou, J.D. and Frankham, R. (2003). Estimates of minimum viable population sizes for vertebrates and factors influencing those sizes. *Biological Conservation*, 113: 23–34.

Reid, W., Mcneely, J., Tunstall, D., Bryant, D. and Winograd, M. (1993). *Biodiversity Indicators for Policy-Makers*. World Resources Institute, Washington, DC.

Rhoades, R. and Booth, R.H. (1982). Farmer-Back-to-Farmer: A Model for Generating Acceptable Agricultural Technology. *Agricultural Administration*, 11: 127–137.

Rijal, D.K., Adhikari, N.P., Khatiwada, S.P., *et al.* (1998). *Strengthening the Scientific Basis for* in situ *Conservation of Agrobiodiversity: Findings of Site Selection in Bara, Nepal*. NP Working Paper No. 2/98. NARC/LI-BIRD, / IPGRI, Rome.

Roberts, E.H. (1973). Predicting the storage life of seeds. *Seed Science and Technology*, 1: 499–514.

Robinson, R.A. (1996). Return to Resistance. *AgAccess*, Davis, California.

Robinson, R.A. (1997). Host resistance to crop parasites. *Integrated Pest Management Reviews*, 2: 103–107.

Robinson, R.A. (2009). breeding for quanitative Variables. Part 2: breeding for durable resistance to crop pests and diseases. In: Ceccarelli, S., Guimaraes, E.P. and Weltzien, E. (eds.) *Plant Breeding and Farmer Participation*. Food and Agriculture Organization of the United Nations, Rome, pp. 367–390.

Roe, D., Nelson, F. and Sandbrook, C. (eds.) (2009). *Community Management of Natural Resources in Africa: Impacts, Experiences and Future Directions*. Natural Resource Issues No. 18. International Institute for Environment and Development, London.

Rogers, J.S. (1972). Measures of similarity and genetic distance. In: *Studies in Genetics* VII. University of Texas Publication, Austin, TX. 7213: 145–153.

Roos, E.E. (1989). Long term seed storage. *Plant Breeding Reviews*, 7: 129–158.

Roos, E.E. and Davidson, D.A. (1992). Record longevities of vegetable seeds in storage. *HortScience*, 27: 393–396.

Roskov, Y., Zarucchi, J., Novoselova, M. and Bisby, F. (eds.) (2019). *ILDIS world database of legumes* (version 12, May 2014). In: Roskov Y., Ower G., Orrell T., *et al.* (eds.) Species 2000 & ITIS Catalogue of Life, 24 December 2018. Digital resource at www.catalogueoflife.org/col. Species 2000: Naturalis, Leiden, The Netherlands.

Rosset, P.M. and Altieri, M.A. (2017). *Agroecology; Science and Politics*. Practical Action Publishing, Rugby, UK.

Rubio Teso, M.L., Kinoshita Kinoshita, K. and Iriondo Alegría, J.M. (2016). Optimized site selection for the *in situ* conservation of forage CWR: a combination of community and genetic level perspectives. In: Maxted, N., Dulloo, M.E. and Ford-Lloyd, B.V. (eds.) *Enhancing Crop Gene Pool Use: Capturing Wild Relative and Landrace Diversity for Crop Improvement*. CAB International, Wallingford, pp. 199–205.

Ruddiman, W.F., Ellis, E.C., Kaplan, J.O. and Fuller, D.Q. (2015). Defining the epoch we live in. *Science* 348: 38–39.

Ruge-Wehling, B., Linz, A., Habeku, A. and Wehling, P. (2006). Mapping of RYMl6Hb, the second soilborne virus resistance gene introgressed from *Hordeum bulbosum*. *Theoretical and Applied Genetics*, 113: 867–673.

Ruiz, J.J. and Garcia-Martinez, S. (2009). Tomato varieties 'Muchamiel' and 'De la Pera' from the Sout-east of Spain: genetic improvement to promote on-farm conservation. In: Veteläinen, M., Negri, V. and Maxted, N. (eds.) *European Landraces: On-Farm conservation, Management and Use*. Bioversity Technical Bulletin 15. Bioversity International, Rome, pp. 171–176.

Ryder, E.J. (1988). Efficient sampling from a collection. *HortScience*, 23: 82–84.

Sackville Hamilton, R. and Chorlton, K. (1995). Collecting vegetative material of forage grasses and legumes. In: Guarino, L., Ramanatha Rao, V. and Reid, R. (eds.) *Collecting Plant Genetic Diversity: Technical Guidelines*. CAB International, Wallingford, pp. 467–484.

Safriel, U.N., Anikster, Y. and Waldman, M. (1997). Management of nature reserves for conservation of wild relatives and the significance of marginal populations. *Bocconea*, 7: 233–239.

Sahoo, S.L., Rout, J.R. and Kanungo, S. (2012). Synthetic seeds. In: Sharma, H.P., Dogra, J.V.V. and Misra, A.N. (eds.) *Plant Tissue Culture: Totipotency to Transgenic*. Agrobios, India, pp. 101–114.

Saitou, N. and Nei, M. (1987). The neighbor-joining method: a new method for reconstructing phylogenetic trees. *Molecular Biology and Evolution*, 4: 406–425.

Sakai, A. (2000). Development of cryopreservation techniques. In: Engelmann, F. and Hiroko, T. (eds.) *Cryopreservation of Tropical Plant Germplasm. Current Research Progress and Application*. Japan International Research Centre for Agricultural Sciences, Tsukuba/International Plant Genetic Resources Institute, Rome, pp. 1–7.

Salafsky, N., Margoluis, R. and Redford, K. (2001). *Adaptive Management: A Tool for Conservation Practitioners*. BSP Publications, Washington, DC.

Salazar, G.A. (1996). Conservation threats. In: Hágsater, E. and Dumont, V. (eds.) *Orchids – Status Survey and Conservation Action Plan*. IUCN. Gland and Cambridge, pp. 6–10.

Sanchez-Velasquez, L.R. (1991). *Zea diploperennis*: Mejoramiento genetico del maiz, ecologia y la conservación de recursos naturales. *Tiempos de ciencia* 24 Qulio–septiembre: 1–8. University of Guadalajara, Jalisco.

Saunders, G. and Parfitt, A. (2005). *Opportunity Maps for Landscape-Scale Conservation of Biodiversity: A Good Practice Study*. English Nature Research Reports, Number 641. English Nature, Peterborough.

Sax, D.F., Gaines, S.D. and Brown, J.H. (2002). Species invasions cxcccd cxtinctions on islands worldwide: a comparative study of plants and birds. *American Naturalist*, 160: 766–783.

Scheldeman, X. and van Zonneveld, M. (2010). *Training Manual on Spatial Analysis of Plant Diversity and Distribution.* Bioversity International, Rome.

Schippmann, U., Leaman, D. and Cunningham, A.B. (2006). A comparison of cultivation and wild collection of medicinal and aromatic plants under sustainability aspects. In: Bogers, R.J. , Cracker, L.E. and Lange, D. (eds.) *Medicinal and Aromatic Plants.* Springer, Dordrecht, pp. 75–95.

Schmidt, L. (2011). Collecting woody perennials. In: Guarino, L., Ramanatha Rao, V. and Goldberg, E. (eds.) *Collecting Plant Genetic Diversity: Technical Guidelines. 2011 Update.* Bioversity International, Rome. Available at: http://cropgene bank.sgrp.cgiar.org/index .php?option=com_content&view=article&id=682 (accessed 13 August 2018).

Scholten, M., Maxted, N., Ford-Lloyd, B.V. and Green, N. (2008). Hebridean and Shetland oat (*Avena strigosa* Schreb.) and Shetland cabbage (*Brassica oleracea* L.) landraces: occurrence and conservation issues. *Plant Genetic Resources Newsletter*, 154: 1–5.

Schulp, C.J., Thuiller, W. and Verburg, P. (2014). Wild food in Europe: a synthesis of knowledge and data of terrestrial wild food as an ecosystem service. *Ecological Economics*, 105: 292–305.

Schultes, R.E. and von Reis, S. (1995). *Ethnobotany: Evolution of a Discipline.* Dioscorides Press, Portland, OR.

Scoones, I. and Thompson, J. (1994). *Beyond Farmer First: Rural People's Knowledge, Agricultural Research and Extension Practice.* ITDG Publishing.

Scoones, I. and Thompson, J. (2009). *Farmer First Revisited.* Innovation for Agricultural Research and Development. ITDG Publishing.

Sedcole, J.R. (1977). Number of plants necessary to recover a trait. *Crop Science*, 17: 667–668.

Seppä, H., Alenius, T., Bradshaw, R.H.W., *et al.* (2009). Invasion of Norway spruce (*Picea abies*) and the rise of the boreal ecosystem in Fennoscandia. *Journal of Ecology*, 97: 629–640.

Shafer, C.L. (1990). *Nature Reserves, Island Theory and Conservation Practice.* Smithsonian Press, Washington DC and London.

Shaffer, M.L. (1981). Minimum population sizes for species conservation. *BioScience American Institute of Biological Sciences*, 31: 131–134.

Shands, H.L. (1991). Complementarity of *in situ* and *ex situ* germplasm conservation from the standpoint of the future user. *Israel Journal of Botany*, 40: 521–528.

Shanley, P. and Laird, S.A. (2002). 'Giving back': making research results relevant to local groups and conservation. In: Laird, S.A. (ed.) *Biodiversity and Knowledge: Equitable Partnerships in Practice.* Earthscan, London, pp. 102–124.

Shanley, P., Pierce, A.R., Laird, S.A. and Guillen, A (2002). *Tapping the Green Market: Certification and Management of Non-Timber Forest Products.* Earthscan, London.

Shannon, C.E. (2001). A mathematical theory of communication. *ACM SIGMOBILE Mobile Computing and Communications Review*, 5: 3–55.

Sharrock, S., Oldfield, S. and Wilson, O. (2014). *Plant Conservation Report 2014: a review of progress in implementation of the Global Strategy for Plant Conservation 2011–2020.* Secretariat of the Convention on Biological Diversity, Montréal, Canada and Botanic Gardens Conservation International, Richmond, UK. Technical Series No. 81.

Shrestha, P., Gezu, G., Swain, S., *et al.* (2013a). The community seed bank: a common driver for community biodiversity management. In: De Boef, W.S., Peroni, N., Subedi, A., Thijssen, M.H. and O'Keeffe, E. (eds) *Community Biodiversity Management: Promoting Resilience and the Conservation of Plant Genetic Resources.* Routledge, Abingdon, pp. 109–117.

Shrestha, P., Subedi, A. and Sthapit, B. (2013b). Enhancing awareness of the value of local biodiversity in Nepal. In: De Boef, W.S., Peroni, N., Subedi, A. and Thijssen, M.H. (eds.) *Community Biodiversity Management: Promoting Resilience and the Conservation of Plant Genetic Resources.* Earthscan, London, pp. 72–76.

Singh, A.K., Varaprasad, K.S. and Venkateswaran, K. (2012). Conservation costs of plant genetic resources for food and agriculture: seed gene banks. *Agricultural Research*, 1: 223–239.

Singh, M., Malhotra, R.S., Ceccarelli, S., Sarker, A., Grando, S. and Erskine, W. (2003). Spatial variability models to improve dryland field trials. *Experimental Agriculture*, 39: 151–160.

Slikkerveer, L. (1994). Indigenous agricultural knowledge systems in developing countries: a bibliography. *Indigenous Knowledge Systems Research and Development Studies,* no. 1. Special Issue: INDAKS Project Report 1 in collaboration with the European Commission DG XII. Leiden Ethnosystems and Development Programme (LEAD), Leiden.

Smith, R.D. (1995). Collecting and handling seeds in the field. In: Guarino, L., Ramanatha Rao, V. and Reid, R. (eds.) *Collecting Plant Genetic Diversity: Technical Guidelines.* CAB International, Wallingford, pp. 419–456.

Smith, R.D., Dickie. J.B., Linington, S.H., Pritchard, H.W. and Probert, R.J. (2003). *Seed Conservation: Turning Science into Practice.* Royal Botanic Gardens, Kew, Richmond.

Smith, R.D. and Linington, S. (1997). The management of the Kew Seed Bank for the conservation of arid land and U.K. wild species. *Bocconea*, 273–280.

Smýkal, P., Trněný, O., Brus, J., *et al.* (2018). Genetic structure of wild pea (*Pisum sativum* subsp. *elatius*) populations in the northern part of the Fertile Crescent reflects moderate cross-pollination and strong effect of geographic but not environmental distance. *PLoS ONE*, 13: e0194056.

Sokal, R.R. and Michener, C.D. (1958). A statistical method for evaluating systematic relationships. *University of Kansas Science Bulletin*, 38: 1409–1438.

Song, Y., Yanyan, Z., Song, X. and Vernooy, R. (2016). Access and benefit sharing in participatory plant breeding in southwest China. *Farming Matters* (Special Issue on Access and Benefit Sharing of Genetic Resources), May Issue, 18–23.

Spellerberg, I.F. (1996). *Conservation Biology*. Longman Group Ltd, Harlow.

Sperling, L. (1996). Results, methods and institutional issues in participatory selection: the case of beans in Rwanda. In: Eyzaguirre, P. and Iwanaga, M. (eds.) *Participatory Plant Breeding*. International Plant Genetic Resources Institute, Rome, pp. 44–56.

Sperling, L., Ashby, J., Weltzien, E., Smith, M. and McGuire, S. (2001). Base-broadening for client-oriented impact: insights drawn from participatory plant breeding field experience. In: Cooper, H.D., Spillane, C. and Hodgkin, T. (eds.) *Broadening the Genetic Basis of Crop Production*. CABI, Wallingford, pp. 419–435.

Spinney, L. (2014). Wonder food. *New Scientist*, 28 June 2014, 40–43.

Stadler, L.J. (1945a). Gamete selection in corn breeding. *Journal of the American Society of Agronomy*, 36: 988–989.

Stadler, L.J. (1945b). Gamete selection in corn breeding. *Maize Genetics Cooperative Newsletter*, 19: 33–40.

Steadman, D.W. (2006). *Extinction and Biogeography of Tropical Pacific Birds*. University of Chicago Press, Chicago, IL.

Stearn, W. (1966). *Botanical Latin*. David & Charles, Newton Abbot and London.

Steffen, W., Richardson, K., Rockstrom, J., *et al.* (2015). Planetary boundaries: guiding human development on a changing planet. *Science*, 347: 6219.

Stein, B., Glick, P., Edelson, N. and Staudt, A. (2014). *Climate-Smart Conservation: Putting Adaptation Principles into Practice*. National Wildlife Federation, Washington, DC.

Sthapit, B., Lamers, H.A.H., Ramanatha Rao, V. and Bailey, A. (2016). *Tropical Fruit Tree Diversity: Good Practices for in Situ and On-farm Conservation*. Earthscan, London.

Sthapit, B., Ramanatha Rao, V., Lamers, H. and Sthapit, S. (2017). Uncovering the role of custodian farmers in the on-farm conservation of agricultural biodiversity. In: Hunter, D., Guarino, L., Spillane, C. and McKeown, P. (eds.) *Handbook of Agriculture Biodiversity*. Routledge, Abingdon, UK, pp. 549–562.

Sthapit, B., Subedi, A., Jarvis, D., *et al.* (2012). Community-based approach to on-farm conservation and sustainable use of agricultural biodiversity in Asia. *Indian Journal of Plant Genetic Research*, 25: 97–110.

Stoll-Kleemann, S. and Welp, M. (2008). Participatory and integrated management of biosphere reserves. *Gaia*, 17 (S1): 161–168.

Stolton, S., Dudley, N. and Kun, Z. (2010). Diverting places: linking travel, pleasure and protection. In: Stolton, S. and Dudley, N. (eds.) *Arguments for Protected Areas: Multiple Benefits for Conservation and Use*. Earthscan, London, UK, pp. 189–204.

Stolton, S., Maxted, N., Ford-Lloyd, B., Kell, S.P. and Dudley, N. (2006). *Food Stores: Using Protected Areas to Secure Crop Genetic Diversity*. WWF Arguments for Protection series. WWF, Gland, pp. 1–133.

Street, K., Mackay, M., Zuev, E., *et al.* (2008). *Diving into the gene pool – a rational system to access specific traits from large germplasm collections*. Available at: http://hdl .handle.net/2123/3390 (accessed 12 September 2014).

Suneson, C.A. (1956). An evolutionary plant breeding method. *Agronomy Journal*, 48: 188–191.

Sutherland, W.J. (2000). *The Conservation Handbook: Research, Management and Policy*. Blackwell Science, Oxford, pp. 1–278.

Swofford, D.L. (2002). *PAUP: Phylogenetic Analysis Using Parsimony, Version 4*. Sinauer Associates, Sunderland, MA.

Sykes, J.T. (1975). Tree crops. In: Frankel, O.H. and Hawkes, J.G. (eds.) *Crop Genetic Resources for Today and Tomorrow*. Cambridge University Press, Cambridge, pp. 123–137.

Taba, S., van Ginkel, M., Hoisington, D. and Poland, D. (2004). *Wellhausen-Anderson Plant Genetic Resources Center: Operations Manual, 2004*. Centro Internacional de Mejoramiento de Maíz y Trigo, El Batan.

Tadesse, W., Abdalla, O., Ogbonnaya, F., *et al.* (2012). Agronomic performance of elite stem rust resistant spring wheat genotypes and association among trial sites in the Central and West Asia and North Africa Region. *Crop Science*, 52: 1105–1114.

Tanksley, S.D. and McCouch, S.R. (1997). Seed banks and molecular maps: unlocking genetic potential from the wild. *Science*, 277: 1063–1066.

Tanwar, H., Sharma, S., Mor, V.S., Yadav, J. and Bhuker, A. (2018). Image analysis: a modern approach to seed quality testing. *Current Journal of Applied Science and Technology*, 27: 1–11.

Taylor, P. and Hunter, D. (2008). The Learning and Teaching for Transformation initiative; helping higher learning institutes to participate. *Policy and Practice* 3: 75–80.

ten Kate, K. and Laird, S.A. (1999). *The Commercial Use of Biodiversity: Access to Genetic Resources and Benefit Sharing*. Earthscan, London.

Terry, J., Probert, R.J. and Linington, S.H. (2003). Processing and maintenance of the Millennium Seed Bank collections. In: Smith, R.D., Dickie, J.B., Linington, S.H., Pritchard, H.W., Probert, R.J.(eds.) *Seed Conservation:*

Turning Science into Practice. Royal Botanic Gardens, Kew, Richmond, pp. 307–325.

Thachuk, C., Crossa, J., Franco, J., *et al.* (2009). Core Hunter: an algorithm for sampling genetic resources based on multiple genetic measures. *BMC Bioinformatics*, 10: 243. https://doi.org/10.1186/1471-2105-10-243.

Thijssen, M., de Boef, W.S., Subedi, A., Peroni, N. and O'Keeffe, E. (2013). General introduction. In: De Boef, W.S., Peroni, N., Subedi, A. and Thijssen, M.H. (eds.) *Community Biodiversity Management: Promoting Resilience and the Conservation of Plant Genetic Resources.* Earthscan, London, pp. 3–10.

Thomson, L., Graudal, L. and Kjaer, E. (2001). Selection and management of *in situ* gene conservation areas for target species. In: FAO, DFSC, IPGRI (eds.) *Forest Genetic Resources Conservation and Management, in Managed Natural Forest and Protected Areas, Vol. 2.* International Plant Genetic Resources Institute, Rome.

Thormann, I., Kell, S.P., Magos Brehm, J., Dulloo, M.E. and Maxted, N. (2017). *CWR Checklist and Inventory Data Template v.1.* doi:10.7910/DVN/B8YOQL, Harvard Dataverse.

Thormann, I., Parra-Quijano, M., Endresen, D.T.F., *et al.* (2014). *Predictive characterization of crop wild relatives and landraces. Technical guidelines version 1.* Bioversity International, Rome, Italy. Available at: www.bioversityinternational.org/index.php?id=244&ttx_news_pi1%5Bnews%5D=4967&tcHash=7cd3c6c2b8360927b83fa6ef7cc28d99 (accessed 17 July 2017).

Tittensor, D.P., Walpole, M., Hill, S.L., *et al.* (2014). A mid-term analysis of progress toward international biodiversity targets. *Science*, 346: 241–244.

Toomey, G. (1999). *Farmers as Researchers: The Rise of Participatory Plant Breeding.* International Development Research Centre, Ottawa.

Torricelli, R., Pacicco, L., Bodesmo, M., Raggi L. and Negri, V. (2016). Assessment of Italian landrace density and species richness: useful criteria for developing *in situ* conservation strategies. In: Maxted, N., Ehsan Dulloo, M. and Ford-Lloyd, B.V. (eds.) *Enhancing Crop Gene Pool Use: Capturing Wild Relative and Landrace Diversity for Crop Improvement.* CAB International, Wallingford, pp. 297–312.

Towill, L.E. (2004). Pollen storage as a conservation tool. In: E. Guerrent, K. Havens and M. Maunder (eds.) Ex situ *Plant Conservation: Supporting Species Survival in the Wild.* Island Press, Covela, CA, pp. 180–188.

Traill, L.W., Bradshaw, C.J.A. and Brook, B.W. (2007). Minimum viable population size: a meta-analysis of 30 years of published estimates. *Biological Conservation*, 139: 159–166.

Trauger, A. (ed.) (2015). *Food Sovereignty in International Context: Discourse, Politics and Practice of Place.* Routledge, Abingdon.

Trethowan, R.M., Reynolds, M.P., Ortiz-Monasterio, I. and Ortiz R. (2007). The genetic basis of the Green Revolution in wheat production. *Plant Breeding Reviews*, 28: 39–58.

Tsobou, R., Mapongmetsem, P.M. and Van Damme, P. (2016). Medicinal plants used for treating reproductive health care problems in Cameroon, Central Africa. *Economic Botany*: 70: 145–159.

Turner, N.J., Łuczaj, L.J., Migliorini, L.P., *et al.* (2011). Edible and tended wild plants, traditional ecological knowledge and agroecology. *Critical Reviews in Plant Science* 30(1-2): 198–225.

Turner, W. (1551). *A New Herball.* pt 1 Mierdman, London; pt 2 Barckman, Cologne.

Tuxill, H. and Nabhan, G.P. (2001). *People, Plants and Protected Areas.* Earthscan from Routledge, London.

Tyagi, R.K. and Agrawal, A. (2015). Revised gene bank standards for management of plant genetic resources. *Indian Journal of Agricultural Sciences*, 85: 157–165.

United Nations. (2009). *State of the World's Indigenous Peoples.* United Nations, New York.

United Nations. (2011). *World Population Prospects: The 2010 Revision.* United Nations, Department of Economic and Social Affairs, Population Division, New York.

Unnikrishnan, P.M. and Suneetha, M.S. (2012). *Biodiversity, Traditional Knowledge and Community Health: Strengthening Linkages.* UNU_IAS Policy Report.

Upadhyaya, H. and Ortiz, R. (2001). A mini core subset for capturing diversity and promoting utilization of chickpea genetic resources in crop improvement. *Theoretical and Applied Genetics*, 102: 1292–1298.

Valdani Vicari & Associati, Arcadia International, Wageningen UR: Centre for Genetic Resource, The Netherlands, Plant Research International and the Socio-Economics Research Institute, Fungal Biodiversity Centre of the Royal Academy of Arts and Science and Information and Coordination Centre for Biological Diversity of the German Federal Office for Agriculture and Food. (2015). *Better integration of ex situ and in situ approaches towards conservation and sustainable use of GR at national and EU level: from complementarity to synergy.* Workshop Report for Preparatory action on EU plant and animal genetic resources (AGRI-2013-EVAL-7). Directorate General for Agriculture and Rural Development, European Commission, Brussels.

Valdani Vicari & Associati, Arcadia International, Wageningen UR: Centre for Genetic Resource, the Netherlands, Plant Research International and the Socio-Economics Research Institute, Fungal Biodiversity Centre of the Royal Academy of Arts and Science and Information and Coordination Centre for Biological Diversity of the German Federal Office for Agriculture and Food. (2016). *The impact of climate change on the conservation and utilisation of crop wild relatives in Europe.* Workshop Report for Preparatory action on EU plant and animal genetic resources (AGRI-2013-EVAL-7). Directorate General for Agriculture and Rural Development, European Commission, Brussels.

Van Dyke, F. (2008). *Conservation biology: Foundations, Concepts, Applications.* McGraw-Hill, New York.

van Treuren, R., de Groot, E.C. and van Hintum, T.J.L. (2013). Preservation of seed viability during 25 years of storage under standard gene bank conditions. *Genetic Resources and Crop Evolution*, 60: 1407–1421.

van Zonneveld, M., Scheldeman, X., Escribano, P., *et al.* (2012). Mapping genetic diversity of cherimoya (*Annona cherimola* Mill.): application of spatial analysis for conservation and use of plant genetic resources. *PLoS One*, 7(1): e29845. https://doi.org/10.1371/journal.pone .0029845.

Vaughan, D.A. (1994). *The Wild Relatives of Rice, A Genetic Resource Handbook.* IRRI, Los Banos.

Vaughan, J.G. and Geissler, C.A. (2009). *The New Oxford Book of Food Plants.* Oxford University Press, Oxford.

Vavilov, N.I. (1917). 0 proiskhozhdenii kulturnoirzhi [On the origin of the cultivated rye]. *Bulletin of the Bureau of Applied Botany*, 10(7–10): 561–590 [Russian].

Vavilov, N.I. (1926). Tzentry proiskhozhdeniya kulturnykhrastenii [The centers of origin of cultivated plants]. *Works of Applied Botany and Plant Breeding*, 16 (2): 248 [Russian, English].

Vavilov, N.I. (1951). *The Origin, Variation, Immunity and Breeding of Cultivated Plants.* Transl. by K.S. Chester. Ronald Press, New York [English].

Vavilov, N.I. (1965). *Izbrannye trudy. Problemy proiskhozhdeniya, geografii, genetiki, selektzii astenii, rastenievodstva I agronomii.* [Selected works. The problems of origin, geography, genetics, plant breeding, plant industry and agronomy]. Vol. 5. USSR Academy of Science Press, M.-L. [Russian].

Veitch, C.R. and Clout, M.N. (2002). *Turning the Tide: The Eradication of Invasive Species.* Proceedings of the International Conference on Eradication of Island Invasives. IUCN, Gland.

Verdcourt, B. and Halliday, P. (1978). A revision of *Psophocarpus* (Leguminosae-Papilionoideae-Phaseolus). *Kew Bulletin*, 33: 191–227.

Vernooy, R. (2003). *Seeds That Give: Participatory Plant Breeding.* International Development Research Centre, Ottawa.

Vernooy, R. (2013). In the hands of many: a review of community gene/seed banks of the world. In: Shrestha, P., Vernooy, R. and Chaudhary, P. (eds) *Community Seed Banks in Nepal: Past, Present, Future. Proceedings of a National Workshop*, 14–15 June 2012, Pokhara, Nepal. Local Initiatives for Biodiversity, Research and Development, Pokhara, and Bioversity International, Rome, pp. 3–15. Available at: www .bioversityinternational.org/uploads/tx_news/

Community_seed_ banks_in_Nepal__past__present_and_ future_1642.pdf

Vernooy, R., Shrestha, P. and Sthapit, B. (2015). *Community Seed Banks: Origins, Evolution and Prospects.* Earthscan from Routledge, London.

Vernooy, R., Shrestha, P. and Sthapit, B. (2017a). Seeds to keep and seeds to share: the multiple roles of community seed banks. In: Hunter, D., Guarino, L., Spillane, C. and McKeown, P. (eds.) *Handbook of Agriculture Biodiversity.* Routledge, Abingdon, pp. 580–591.

Vernooy, R., Sthapit, B. and Bessette, G. (2017b). *Community Seed Banks: Concept and Practice (Facilitator Handbook).* Bioversity International, Rome.

Veteläinen, M., Negri, V. and Maxted, N. (eds.) (2009a). *European Landraces: On-farm conservation, Management and Use.* Bioversity Technical Bulletin 15. Bioversity International, Rome, pp. 1–359.

Veteläinen, M., Negri, V. and Maxted, N. (2009b). A European strategic approach to conserving crop landraces. In: Veteläinen, M., Negri, V. and Maxted, N. (eds.) *European Landraces: On-farm Conservation, Management and Use.* Bioversity Technical Bulletin 15. Bioversity International, Rome, pp. 305–325.

Vincent, H., Amri, A., Castañeda-Álvarez, N.P., *et al.* (2019). Modeling of crop wild relative species identifies areas globally for *in situ* conservation. *Communications Biology*, 2: 136. https://doi.org/10.1038/s42003-019- 0372-z.

Vincent, H., Castañeda-Álvarez, N.P. and Maxted, N. (2016). An approach for *in situ* gap analysis and conservation planning on a global scale. In: Maxted, N., Dulloo, E.M. and Ford-Lloyd, B.V. (eds.) *Enhancing Crop Gene Pool Use: Capturing Wild Relative and Landrace Diversity for Crop Improvement.* CAB International, Wallingford, pp. 137–148.

Vincent, H., von Bothmer, R., Knüpffer, H., *et al.* (2012). Genetic gap analysis of wild *Hordeum* taxa. *Plant Genetic Resources: Characterization and Utilization*, 10: 242–253.

Vincent, H., Wiersema, J., Kell, S.P., *et al.* (2013). A prioritised crop wild relative inventory as a first step to help underpin global food security. *Biological Conservation*, 167: 265–275.

Vinceti, B., Ickowitz, A., Powell, B., *et al.* (2013). Challenges and opportunities in strengthening the contribution of forests to sustainable diets. *Sustainability* 5: 4797–4824.

Vitt, P., Havens, K., Kramer, A.T., Sollenberger, D. and Yates, E. (2010). Assisted migration of plants: changes in latitudes, changes in attitudes. *Biological Conservation*, 143: 18–27.

Volis, S. and Blecher, M. (2010). Quasi *in situ*: a bridge between *ex situ* and *in situ* conservation of plants. *Biodiversity and Conservation*, 19: 2441–2454.

Volk. G.M. (2011). Collecting pollen for genetic resources conservation. In: Guarino, L., Ramanatha Rao, V. and

Goldberg, E. (eds.) *Collecting Plant Genetic Diversity: Technical Guidelines. 2011 Update*. Bioversity International, Rome. Available at: http://cropgene bank.sgrp.cgiar.org/index.php?option=com_content&view=article&id=654

von Bothmer, R. and Seberg, O. (1995). Strategies for the collecting of wild species. In: Guarino, L., Ramanatha Roa, V. and Reid, R. (eds.) *Collecting Plant Genetic Diversity. Technical Guidelines*. CAB International, Wallingford, UK and IPGRI, Rome, pp. 93–112.

Von Humboldt, A. and Bonpland, A. (1807). *Essai sur la géographie des plantes*. Paris, Chez Fr Schoelle Librairie, et Tubingue, chez J. G. Cotta, Librairie.

Wagner, S.C. (2010). Keystone species. *Nature Education Knowledge*, 3(10): 51.

Wainwright, W., Drucker, A.G., Maxted, N., *et al.* (2019). Estimating *in situ* conservation costs of Zambian crop wild relatives under alternative conservation goals. *Land Use Policy*, 81: 632–643.

Walker, T.S. (2006). *Participatory Varietal Selection, Participatory Plant Breeding, and Varietal Change*. Background paper for the World Development Report 2008. World Bank. Washington, DC.

Walley, K.A., Khan, M.S.I. and Bradshaw, A.D. (1974). The potential for evolution of heavy metal tolerance in plants. *Heredity*, 32: 309–319.

Walters, C., Wheeler, L.M. and Grotenhuis, J.M. (2005). Longevity of seeds stored in a gene bank: species characteristics. *Seed Science Research*, 15: 1–20.

Ward, J. (1963). Hierarchical grouping to optimize an objective function. *Journal of the American Statistical Association*, 58: 236–244.

Way, M.J. (2003). Collecting seed from non-domesticated plants for long-term conservation. In: Smith, R.D., Dickie, J.B., Linington, S.H., Pritchard, H.W. and Probert, R.J. (eds.) *Seed Conservation: Turning Science into Practice*. Royal Botanic Gardens, Kew, Richmond, UK.

Weinberg, W. (1908). Über den Nachweis der Vererbung beim Menschen. *Jahreshefte des Vereins für vaterländische Naturkunde in Württemberg*, 64: 368–382.

Weinberger, K. and Pichop, G.N. (2009). Marketing of African indigenous vegetables along urban and peri-urban supply chains in Sub-Saharan Africa. In: Shackleton, C.M., Pasquini, M.W. and Drescher, A.W. (eds.) *African Indigenous Vegetables in Urban Agriculture*. Earthscan, London, pp. 225–244.

Weise, S., Oppermann, M., Maggioni, L., van Hintum, T. and Knüpffer, H. (2017). EURISCO: The European Search Catalogue for Plant Genetic Resources. *Nucleic Acids Research*, 45(D1): D1003–1008.

Weltzien, E., Smith, M., Meitzner, L. and Sperling, L. (2003). *Technical and Institutional Issues in Participatory Plant Breeding from the Perspective of Formal Plant Breeding.*

A Global Analysis of Issues, Results and Current Experiences. PPB Monograph, No.1. PRGA, Cali.

Westengen, O. and Winge, T. (2019). *Farmers in Plant Breeding: Current Approaches and Perspectives*. Earthscan from Routledge, London.

Western, D. and Wright, R.M. (eds.) (1994). *Natural Connections*. Island Press, Washington, DC.

WHO/CBD. (2015). *Connecting Global Priorities: Biodiversity and Human Health, a State of Knowledge Review*. Secretariat of the Convention on Biological Diversity, Ottawa.

Wiens, J.A. (1989). *Processes and Variations. The Ecology of Bird Communities*, Vol. 2. Cambridge University Press, Cambridge.

Wilkes, G. (2007). Urgent notice to all maize researchers: disappearance and extinction of the last wild teosinte population is more than half completed. A modest proposal for teosinte evolution and conservation *in situ*. The Balsas, Guerrero, Mexico. *Maydica*, 52: 49–58.

Wilkes, H.G. (1983). Current status of crop plant germplasm. *CRC Critical Review of Plant Science*, 1: 133–181.

Williams, D.E. (2017). Agricultural biodiversity and the Columbian exchange. In: Hunter, D., Guarino, L., Spillane, C. and McKeown, P. (eds.) *Handbook of Agriculture Biodiversity*. Routledge, Abingdon, pp. 192–212.

Willis, J.C. (1922). *Age and Area; A Study in Geographic Distribution and Origin of Species*. Cambridge University Press, Cambridge.

Willis, K.J. (ed.) (2017). *State of the World's Plants 2017*. Report. Royal Botanic Gardens, Kew, Richmond.

Wilson, E.O. (1992). *The Diversity of Life*. Allan Lane, Penguin Press, London.

Wisz, M., Hijmans, R., Li, J., *et al.* and NCEAS Predicting Species Distributions Working Group. (2008). Effects of sample size on the performance of species distribution models. *Diversity and Distributions*, 14: 763–773.

Withers, L.A. (1995). Collecting *in vitro* for genetic resources conservation. In: Guarino, L., Ramanatha Rao, V. and Reid, R. (eds.) *Collecting Plant Genetic Diversity: Technical Guidelines*. CAB International, Wallingford, pp. 511–526.

Wittman, H., Desmarais, A.A. and Wiebe, N. (eds.) (2010). *Food Sovereignty: Reconnecting Food, Nature and Community*. Fernwood, Halifax.

Woodcock, P., Pullin, A.S. and Kaiser, M.J. (2014). Evaluating and improving the reliability of evidence syntheses in conservation and environmental science: a methodology. *Biological Conservation*, 176: 54–62.

Woodward, F.I. and Williams, B.G. (1987). Climate and plant distribution at global and local scales. *Vegetatio*, 69: 189–197.

Worboys, G.L., Lockwood, M., Kothari, A., Feary, S. and Pulsford, I. (eds.) (2015). *Protected Area Governance and Management*. ANU Press, Canberra.

Wright, S. (1922). Coefficient of inbreeding and relationship. *American Naturalist*, 56: 330–338.

Wright, S. (1931). Evolution in Mendelian populations. *Genetics*, 16: 97–159.

Wright, S. (1951). The genetical structure of populations. *Annals of Eugenics*, 15: 323–324.

Wright, S. (1965). The interpretation of population structure by F-statistics with special regard to the system of mating. *Evolution*, 19: 395–420.

Wunder, S. (2014). Forests, livelihoods, and conservation: broadening the empirical base. *World Development*, 64: 1–11.

Xiao, P.G and Peng, Y. (1998). Ethnopharmacology and research on medicinal plants in China. In: Prendergast, H.D.V, Etkin, N.L., Harris, D.R. and Houghton, P.J. (eds.) *Plants for Food and Medicine*. Royal Botanic Gardens, Kew, Richmond, pp. 31–39.

Zeven, A.C. (1998). Landraces: a review of definitions and classifications. *Euphytica*, 104: 127–139.

Ziliak, S. (2017). *P* values and the search for significance. *Nature Methods*, 14: 3–4.

Zimmerer, K.S. (2010). Biological diversity in agriculture and global change. *Annual Review of Environment and Resources*, 35: 137–166.

INDEX